Statistical Mechanics and Thermodynamics of Matter

Statistical Mechanics and Thermodynamics of Matter
An Introductory Survey

Joachim E. Lay
Michigan State University

1817
HARPER & ROW, PUBLISHERS, New York
Grand Rapids, Philadelphia, St. Louis, San Francisco,
London, Singapore, Sydney, Tokyo

Sponsoring Editor: Donald Childress
Project Editor: Ellen MacElree
Art Director/Cover Coordinator: Heather A. Ziegler
Cover Design: Wanda Lubelska Design
Production: Beth Maglione

Statistical Mechanics and Thermodynamics of Matter: An Introductory Survey

Library of Congress Cataloging-in-Publication Data

Lay, Joachim E.
Statistical mechanics and thermodynamics of matter : an introductory survey / Joachim E. Lay.
p. cm.
Includes bibliographical references.
ISBN 0-06-043884-3
1. Matter. 2. Statistical mechanics. 3. Thermodynamics.
I. Title.
TA403.L345 1990
620.11292—dc20 89-26976
CIP

90 91 92 93 9 8 7 6 5 4 3 2 1

To my wife, Marion Mae.

About the Author

Joachim E. Lay is professor of mechanical engineering at Michigan State University. Born in China, and educated in France, Switzerland, Belgium, and the United States, he received the Candidat Ingénieur degree from the University of Liège and a Ph.D. from the University of Michigan. He has been weight and stress engineer on the B-24 bomber at the Ford Motor Company (Willow Run), research engineer at Ford (Dearborn), assistant professor at the University of Detroit, associate professor at Michigan State, research associate at M.I.T., and consultant to the Office of Ordinance Research. He is the author of *Thermodynamics: A Macroscopic-Microscopic Treatment,* Merrill (1963, 1964), and co-editor of "Development in Mechanics," the proceedings of the Seventh Midwestern Conference on Solid Mechanics and Fluid Mechanics (1961). He is the inventor of a flying train, a flying crane, and a flying car; holds several patents on rotary engines; and is listed in *Time* magazine's special bicentennial issue (July 5, 1976) under "American Ingeniuity: Still Going Strong."

CONTENTS

PREFACE

This book is the outgrowth of teaching statistical mechanics and thermodynamics at the undergraduate and graduate levels for many years. It is for use in a statistical mechanics and thermodynamics course and for courses yet to appear in the engineering and science curricula. The explanation for this follows.

Modern teaching of thermodynamics has evolved along two distinct but equal branches of learning: the classical (macroscopic) and the statistical (microscopic). The former aproach is taught in the majority of schools, but the latter is, in many ways, more satisfying and provides a more accurate explanation of nature at the atomic level. At the author's own institution, statistical thermodynamics is offered as a separate course to supplement the classical approach provided by the mechanical engineering department or to fulfill the needs of those students in electrical engineering, chemical engineering, materials science engineering, physics, chemistry, Honors College, and other disciplines. In teaching such a course, a recurrent need is for a text that would be moderate in length and sophistication, contain enough worked-out examples, and motivate the student. Present and past books on statistical mechanics, statistical thermodynamics, kinetic theory of gases, solid state physics, and materials science are too lofty and comprehensive for undergraduate use. They are essentially treatises for the more erudite reader. On the other hand, the statistical material that may be found in a classical thermodynamics book (including the author's own previous book) is altogether inadequate and doesn't fit. Statistical mechanics and thermodynamics may be an unfamiliar and difficult subject for students, but it is a basic discipline that merits their full attention.

The present book consists of six chapters, each standing on its own. Chapter 1 deals with probability theory and distribution functions and generally provides the student with the necessary mathematical tools. Chapter 2 introduces statistical mechanics against a background of quantum mechanics. Chapter 3 covers classical systems, their usefulness and limitations. Chapter 4 deals with the basic division of nature's particles into bosons and fermions. Chapter 5 is the evaluation of thermodynamic properties in terms of the partition function. And, finally, Chapter 6 introduces solids, liquids, transport properties, band theory of metals, and atomic bonding and discusses applications of interest. The coverage of material takes the student to the threshold of materials science, solid state physics, and Boltzmann's transport equation in terms of the relaxation time. Throughout, many worked-out examples are included to give meaning to the theory, and a great number of problems is presented at the end of each chapter, with a solutions manual available upon request. For pedagogical reasons, many topics are repeated in varying degrees of rigor and sophistication from one chapter to another.

In his teaching and writing of the book, the author has consulted a vast number of texts on quantum mechanics, statistical mechanics, statistical physics, thermodynamics, physical chemistry, solid state physics, and materials science. It is not possible for him to single out the books he favors most, but those he has found to be helpful are listed in the bibliography and mentioned at appropriate places in the various chapters. In developing the subject of statistical mechanics, the author has elected to stay with Gibbs's method of ensembles. Information theory and Jaynes's principle of least prejudice are used, but not exclusively, and mostly to augment the ensemble method. Finally, the author feels that structuring the book into six chapters enables him to cover in a reasonable manner what he considers to be estimable and essential in science and engineering education. He would be happy to learn from readers if he has succeeded in this respect.

The author is grateful to the following people for their helpful and encouraging reviews of the book: Alan J. Brainard, University of Pittsburgh; Martin Crawford, University of Alabama; Charles M. Harman, Duke University; and Dr. James Pohl, Florida Atlantic University.

Joachim E. Lay

Chapter 1

Events and Probability Distribution

The purpose of statistical mechanics is to present a molecular approach to the study of macroscopic properties of matter from first principles. Macroscopic or bulk properties are those quantities obtained from measurements in the laboratory on samples of matter. They may be measured directly, as in the case of density, pressure, heat capacity, surface tension, vapor pressure, diffusivity, and viscosity, or they may be computed, as in the case of energy, entropy, and fugacity. Bulk properties are in this sense observable quantities.

Statistical mechanics provides the bridge between microscopic mechanics and macroscopic thermodynamics. It was developed during the latter part of the nineteenth century by Boltzmann, Maxwell, and Gibbs. For systems such as perfect gases, the calculations are not difficult, and values of thermodynamic properties obtained from spectroscopic data are often more accurate than the ones measured directly from calorimetry. For more complicated systems, especially those involving strong interaction between molecules, the theory is more involved and is the subject of on-going research. To understand and appreciate statistical mechanics, some knowledge of probability theory, classical and quantum mechanics, and mathematics is necessary. This chapter covers the probably theory and the mathematics. Chapter 2 covers the classical and quantum mechanics.

1.1 SIMPLE EVENTS

Probability theory is a well-established branch of mathematics; it is a self-consistent way of thinking about events and a logical method of examining the results of experimental measurements. The behavior of atomic systems, supplemented by the laws of probability, forms the basis for statistical mechanics.

It is a common knowledge that the outcome of an experiment is not always the same when the experiment is performed a number of times, even though the conditions of the experiment are kept as similar as possible. The reason is that some of the factors that contribute to the outcome of the experiment cannot be completely controlled. Examples of simple experiments are the rolling of dice, drawing of cards, tossing of coins, or any of the so-called games of chance. A typical feature of all experiments is that at the end of the operation, an event occurs that is of interest. The distinct (or mutually exclusive) results that are of interest are called *simple events*. The result of each experiment is always one, and only one, simple event. For simplicity these simple events are labeled with some index i. Thus, the two possible events when tossing a coin are heads or tails ($i = h, t$), whereas there are six possible events when a single die is rolled ($i = 1, 2, 3, \ldots, 6$), and so on. Simple events are represented by single points in a sample space (see Fig. 1.1-1).

Let a particular experiment be performed a number of times, say N times, and a particular event i found to occur n_i times. The ratio

$$F_i = \frac{n_i}{N} \tag{1.1-1}$$

is the fraction of the N experiments that result in the event i and is commonly known as the *frequency* of the event i. In general, F_i will be different for different groups of experiments. For example, if a coin is tossed 20 times ($N = 20$), the event "heads" may occur 8 times ($n_h = 8$), so that $F_h = 0.4$ for that sequence of tosses. If the coin is tossed 100 times, the coin might turn up heads 54 times, in which case $F_h = 0.54$ for that sequence of tosses. As N becomes larger and larger, the frequency F_h will approach the value 0.50. This limiting value is called the *probability* of heads (for the coin experiment). Thus, the formal definition for the probability P_i of an event i, is

$$P_i = \lim_{N\to\infty} F_i = \lim_{N\to\infty} \frac{n_i(N)}{N} \tag{1.1-2}$$

By this, it is meant that P_i equals the limiting value of F_i as N becomes arbitrarily large. This in turn equals the limiting value of n_i/N, where n_i also depends on the value of N.

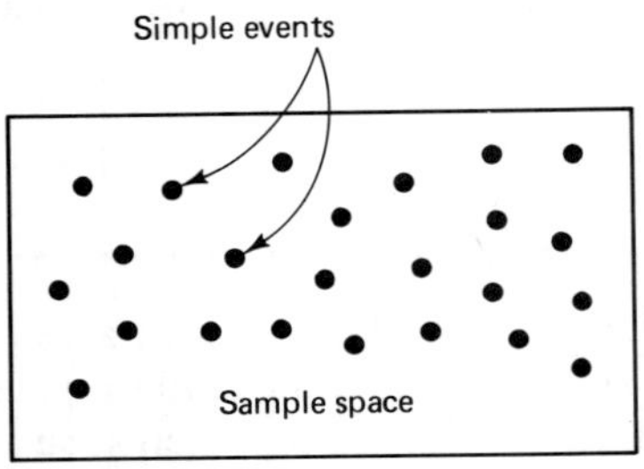

Figure 1.1-1 Simple events in sample space.

■ EXAMPLE 1.1-1

Discuss the sample space for the tossing of a coin, the rolling of a die, and the drawing of a card from a deck of 52 cards.

Solution. In developing a formal method for determining the probabilities associated with a random experiment, it is helpful to make a list of all the possible results. The technical name for this list is known as a *set*. This is just a collection of objects having some common characteristics, and the set members are surrounded by braces {. . .}. The set of all possible results is called the *probability space* or *sample* space. Examples of probability space for various experiments are given in the Table 1.1-1.

TABLE 1.1-1 EXAMPLES OF PROBABILITY SPACE

Experiment (event)	Probability space
Tossing a coin one time	{H, T}
Tossing a coin three times	{HHH, HHT, HTH, HTT, THH, THT, TTH, TTT}
Rolling one die	{1, 2, 3, 4, 5, 6}
Rolling two dice	{(1, 1), (1, 2) . . . (1, 6) (2, 1), (2, 2) . . . (2, 6)}
Drawing one card from a deck of 52 cards	{one hearts, one diamonds, . . . , two hearts, two diamonds, . . .} 52 elements in all

Capital letters are often used to stand for sets. For example, the set of all possible outcomes from rolling two dice such that the sum of the numbers on the two dice is 7 is

$$A = \{(1,6), (2,5), (3,4), (4,3), (5,2), (6,1)\}$$

Set A contains 6 outcomes. When the dice are rolled, there is an equal chance of getting any one of $6 \times 6 = 36$ outcomes. Since 6 of these outcomes give a total of 7 on the dice, the probability of getting a 7 is $6/36 = 1/6$. Set A is also called an *event*. In this case, the event consists of a group of outcomes. Set A is a subset of the probability space S. This means that every outcome in set A is contained in set S. ■

1.2 ERGODIC HYPOTHESIS

There are two ways of interpreting Eq. (1.1-2), both of which are used in statistical mechanics. First, a single physical system (say, a coin) can be chosen, and the same experiment performed on it over and over again (altogether, N times). The number n_i is, in this case, the number of times the event i occurs in this sequence of experiments. From this point of view the experiments are carried out at different times, one after the other. In practice, this is what is usually done.

There is, however, another way of looking at Eq. (1.1-2), one into which time does not enter. In this method, N identical systems are considered (for example, N identical coins, N decks of cards, or N particles of a gas). The systems are identical in the sense that they cannot be distinguished between them by any macroscopic means. Such a collection of many identical system is called an *ensemble*. Now, if the same experiment is performed on each of the N systems of the ensemble, then n_i becomes the number of these systems that yield the event i. Take, for example, an ensemble of 1000 identical coins. If all these coins are flipped at the same times, and it is found that 510 turn up heads, then $n_h = 510$, and so the frequency is $F_h = 0.51$. From this point of view, time is clearly unimportant in the experiment for determining F_h.

The fundamental assumption is that either method yields the same result. In statistical mechanics, this assumption is known as the *ergodic hypothesis*. Thus, frequency (and probability) can be thought of in terms of either a sequence of experiments on one system or one experiment on each member of an ensemble. Time plays no role in determining the probability. In other words, the time average of a dynamical quantity X (denoted by an overbar) is equal to the ensemble average or expectation value (denoted by brackets):

$$\overline{X} = \lim_{t\to\infty} X_i(t) = \langle X \rangle = \lim_{N\to\infty} X_i(N) \tag{1.2-1}$$

The foundation of classical statistical mechanics rests on the statement, that:

> The (sufficiently long) time average of a physical observable is equal to its ensemble average. (1.2-2)

Alternately, it will be shown in statistical mechanics, in conjunction with Liouville's theorem, that:

> For an isolated system, all accessible regions of phase space have equal a priori probability. (1.2-3)

These statements may be regarded either as deduced principles or as assumed postulates. The first follows from the fact that each system of an ensemble within the course of a sufficiently long period of time, assumes the complexion of every other member of the ensemble. The second follows from the fact that a representative point, of constant density, in accordance with Liouville's theorem; eventually passes through every point in phase space. It implies that for an ensemble of isolated systems, the systems are uniformly distributed throughout phase space.

1.3 COMPOUND EVENTS

It has been stated that the outcome of an experiment is known as an event, and that simple events can be represented as single points in a sample space. Each point represents a possible result of an experiment. A compound event, on the other hand, is a collection of points, as shown by the area A (see Fig. 1.3-1).

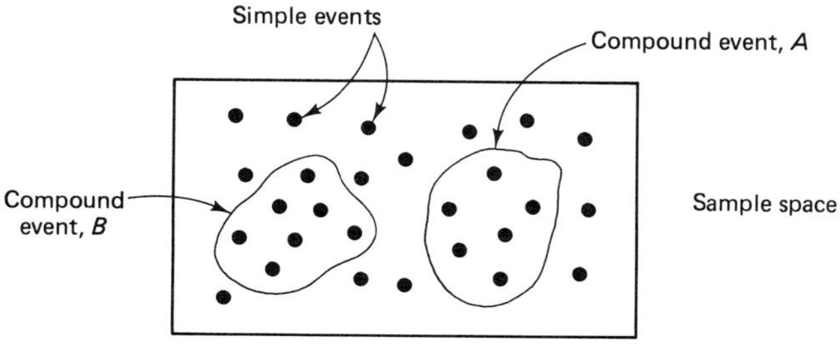

Figure 1.3-1 Simple and compound events.

To elaborate, consider the experiment of drawing a single card from a deck of cards. The possible simple events are 52 in number, and can be labeled $i = 1, 2, \ldots, 52$. The sample space in this case consists of 52 points, each representing one of the cards that may be drawn. Let three compound events A, B, and C be defined by the following characteristics:

A: all points representing hearts

B: all points representing a number three card

C: all points representing a one-eyed jack

Because there are 52 cards, and because $\Sigma P_i = 1$, the probability of each simple event equals 1/52. The respective probabilities of events A, B, and C may be determined as follows:

The probability of the compound event A is given by

$$P(A) = \sum_{i \subset A} P_i \tag{1.3-1}$$

where the sum is over all the points in the event (indicated by the symbol "$i \subset A$," which reads "all the i's contained in the event A"). Since there are 13 hearts, each with a probability 1/52, this gives

$$P(A) = \sum_{i \subset A} P_i = \frac{1}{52} + \frac{1}{52} + \cdots = 13 \times \left(\frac{1}{52}\right) = \frac{1}{4} \tag{1.3-2}$$

that is, the probability of drawing a heart is 1/4. In sample space, event A has 13 points, each with probability 1/4.

The probability of drawing a number three card is

$$P(B) = \sum_{i \subset B} P_i = 4 \times \left(\frac{1}{52}\right) = \frac{1}{13} \tag{1.3-3}$$

Event B has four points, each with probability 1/13.

The probability of drawing a one-eyed jack is

$$P(C) = \sum_{i \subset C} P_i = 2 \times \left(\frac{1}{52}\right) = \frac{1}{26} \tag{1.3-4}$$

Event C consists of two points, each with probability 1/26.

■ EXAMPLE 1.3-1

What is the probability of (a) getting one head when a coin is tossed three times; (b) showing a total of 5 when two dice are rolled; (c) drawing one card from a deck of cards to show an ace; (d) not getting a 7 when two dice are rolled?

Solution. The probability of an event A is the number of outcomes in A divided by the total number of outcomes in probability space S:

$$P(A) = \frac{N(A)}{S}$$

(a) When a coin is tossed three times, there are $2^3 = 8$ outcomes, so $S = 8$. The three outcomes that give one head are HTT, THT, TTH, so $N(A) = 3$. Thus, $P(A) = 3/8$.

(b) Here, there are 6×6 possible outcomes, so $S = 36$. The event B of getting a 5 upon rolling two dice contains 4 outcomes, that is,

$$B = \{(1,4), (2,3), (3,2), (4,1)\}$$

(c) The probability or sample space contains 52 possible outcomes. The event C of drawing an ace contains four outcomes:

$$C = \{\text{ace hearts, ace diamonds, ace clubs, ace spades}\}$$

Thus, $N(C) = 4$, and $P(C) = 4/52 = 1/13$.

(d) The event of not getting a 7 contains 30 outcomes as shown:

$$\begin{matrix} \{(1,1) & (1,2) & (1,3) & (1,4) & (1,5) & \\ (2,1) & (2,2) & (2,3) & (2,4) & & (2,6) \\ (3,1) & (3,2) & (3,3) & & (3,5) & (3,6) \\ (4,1) & (4,2) & & (4,4) & (4,5) & (4,6) \\ (5,1) & & (5,3) & (5,4) & (5,5) & (5,6) \\ & (6,2) & (6,3) & (6,4) & (6,5) & (6,6)\} \end{matrix}$$

Therefore, $N(D) = 30$, and since sample space is $S = 36$ when two dice are rolled, $P(C) = 30/36 = 5/6$. Note that, in general, if P is the probability that a particular event A will occur, then $1 - P$ is the probability that it will not occur.

A special name is given to the set that contains all of the A^c. (The little c stands for complement. The symbol A^c is read as "A complement."). Then,

$$P(A^c) = 1 - P(A)$$

This result is important for calculations, since sometimes it is easier to calculate the probability that an event will not occur. Here are some examples of complements:

If the experiment consists of tossing a coin, and A is the event of getting a head, then A^c is the event of getting a tail.

If the experiment consists of tossing four coins, and B is the event of getting four heads, then B^c is the event of getting at least one tail.

If the sample space is the set of all 52 cards in a deck of cards, and C is the event of drawing a red card, then C^c is the event of drawing a black card. ■

1.4 UNION AND INTERSECT

Two compound events may or may not have points in common. The events A, B, and C of the preceding are represented by their respective areas in Fig. (1.4-1), called a *Venn diagram*. This is a pictorial representation of the relationship between sets. Events A and B have a single point in common (namely, the point corresponding to the three of hearts), whereas events B and C have no points in common (there is no number three card that is also a one-eyed jack). If two events do not have any points in common, it is said that they are *disjoint*.

Consider first the disjoint events B and C. It is of interest to know what the probability is that the outcome of the experiment will be either event B or event C. What is the probability that the card drawn is either number three or a one-eyed jack? This is, of course, another compound event, which can be represented by $B \cup C$ (read: B union C). The union of B and C is simply the compound event consisting of all points that are either in B or in C (or in both B and C). The probability of this event $P(B \cup C)$ is just the sum of the probability of all points in that event. Since B and C have no points in common (disjoint), that is, are mutually exclusive, $P(B \cup C)$ equals the sum of $P(B)$ and $P(C)$, or

$$P(B \cup C) = P(B) + P(C) \tag{1.4-1}$$

More generally, if A and B are two events that are not disjoint, then $P(A \cup B)$ is still equal to the sum of the probabilities of all points in A and B. However, in order to express $P(A \cup B)$ in terms of $P(A)$ and $P(B)$, Eq. (1.4-1) must be amended. The events A and B intersect. (The intersection of two events is denoted by $A \cap B$ or AB. This is the set of points belonging to both A and B.) The reason Eq. (1.4-1) can no longer be used is that the sum of

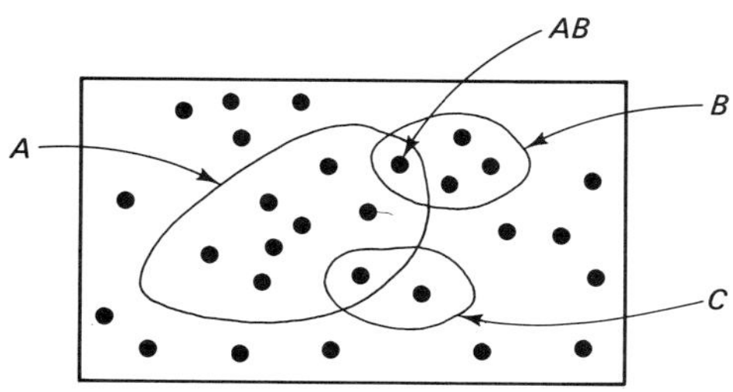

Figure 1.4-1 Intersect of compound events.

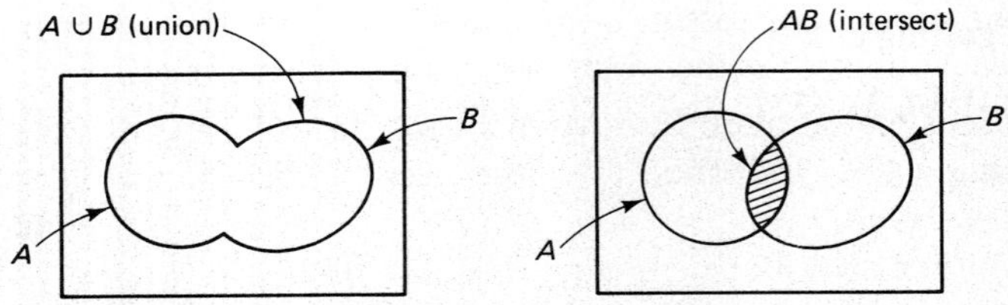

Figure 1.4-2 Union and intersect.

$P(A)$ and $P(B)$ adds the probabilities of the points that A and B have in common twice. To rectify this double counting, something needs to be subtracted. What must be subtracted is the probability $P(AB)$ of all the points that A and B have in common. This is illustrated in Fig. 1.4-2.

Thus, in the case of events A and B, the probability $P(A \cup B)$ is given by

$$P(A \cup B) = P(A) + P(B) - P(AB) \tag{1.4-2}$$

In the present example, $P(AB) = 1/52$, the probability of drawing a three of hearts. With $P(A) = 1/4$ and $P(B) = 1/4$ and $P(B) = 1/13$ from Eqs. (1.3-2) and (1.3-3), Eq. (1.4-2) gives

$$P(A \cup B) = 1/4 + 1/13 - 1/52 = 4/13 \tag{1.4-3}$$

On the other hand, the events B and C being mutually exclusive, the intersection $B \cap C$ or BC is an empty set (contains no points), $P(BC) = 0$, and Eq. (1.4-1) applies:

$$P(B \cup C) = 1/13 + 1/26 = 3/26 \tag{1.4-4}$$

that is, there are six cards in the deck that are either one-eyed jacks or numbered three. Summarizing, for any two compound events X and Y,

$$P(X \cup Y) = P(X) + P(Y) - P(XY) \tag{1.4-5}$$

which reduces to $P(X) + P(Y)$ if X and Y are disjoint, or mutually exclusive, $P(XY) = 0$.

■ EXAMPLE 1.4-1

A pair of dice is tossed. What is the probability of getting (a) either a total of 7 or 5; (b) either a total of 8 or doubles; (c) at least one 6?

Solution

(a) This is a case of the probability of a union. Let A be the event of getting a 7. Then A contains the outcomes

$$\{(1,6), (2,5), (3,4), (4,3), (5,2), (6,1)\}$$

If B is the event of getting a 5, then B contains the outcomes

$$\{(1,4), (2,3), (3,2), 4,1)\}$$

Let C be the event of getting either a 5 or a 7. Then C contains these outcomes:

$$\{(1,6), (2,5), (3,4), (4,3), (5,2), (6,1), (1,4), (2,3), (3,2), (4.1)\}$$

which is ten outcomes in all, so $P(C) = 10/36$.

This result will now be confirmed by rule. The set that contains all of the elements that are in either or both of two sets is called the *union* of the two sets. Let C be the union set of A and set B. This is indicated by a symbol that looks like a letter u: $\cup$. Therefore,

$$C = A \text{ union } B$$

or

$$C = A \cup B$$

The rule for a union is

$$P(A \text{ or } B) = P(A \cup B) = P(A) + P(B)$$

With $P(A) = 6/36$ and $P(B) = 4/36$, this gives

$$P(C) = \frac{6}{36} + \frac{4}{36} = \frac{10}{36}$$

as before. Note that the rule holds only when there is no possibility that event A and event B can occur at the same time (which is the case here).

(b) This is a case in which two events can happen together. It has to do with the probability of an intersection. Let A be the event of getting an 8. Then

$$A = \{(2,6), (3,5), (4,4), (5,3), (6,2)\}, \qquad P(A) = \frac{5}{36}$$

Let B be the event of getting a double. Then

$$B = \{(1,1), (2,2), (3,3), (4,4), (5,5), (6,6)\}, \qquad P(B) = \frac{6}{36}$$

The two events are not disjoint, since both an 8 and a double can be obtained, if the dice turn up $(4,4)$. The set that contains all of the elements that are in both of two sets is called the *intersection* (denoted by the upside-down union symbol: $\cap$). Thus, $C = A$ intersect B, or $C = A \cap B$. Now, when the outcomes $N(A)$ and $N(B)$ are added, the outcomes in $A \cap B$ are counted twice. So, in order to get the number of outcomes in $A \cup B$, the number $N(A \cap B)$ must be subtracted, like this:

$$N(A \cup B) = N(A) + N(B) - N(A \cap B)$$

or, in terms of probabilities,

$$P(A \cup B) = P(A) + P(B) - P(A \cap B)$$

This formula is good for any two events, whether or not they are disjoint. If A and B are disjoint, then $P(A \text{ and } B) = P(A \cap B) = 0$, and the same result obtained in case (a) is obtained here: $P(A \text{ or } B) = P(A \cup B) = P(A) + P(B)$. Returning to the present case, $P(A) = 5/36$, $P(B) = 6/36$, and $A \cap B$ is the event of getting (4, 4), which has a probability of 1/26. Thus, the probability of getting either an 8 or a double is

$$P(A \cup B) = \frac{5}{36} + \frac{6}{36} - \frac{1}{36} = \frac{20}{36}$$

(c) To obtain the probability of getting at least one 6 upon the roll of two dice, let A be the event of getting a 6 on the first die and B and the event of getting a 6 on the second die. Then the event of getting at least one 6 is $A \cap B$. Now, $P(A) = P(B) = 1/6$. Since $A \cap B$ is the event of getting a 6 on both dice, the probability of that happening is 1/36. Therefore, according to the rule,

$$P(A \cup B) = P(A) + P(B) = P(A \cap B) = \frac{1}{6} + \frac{1}{6} - \frac{1}{36} = \frac{11}{36} \quad \blacksquare$$

1.5 CONDITIONAL EVENTS

Two concepts are of importance in statistical mechanics: the notion of conditional probability and the notion of independent events. Conditional probability measures the effect (if any) on the occurrence of an event X when it is known that an event Y has occurred in the experiment. It is defined by

$$P(X \mid Y) = \frac{P(XY)}{P(Y)} \tag{1.5-1}$$

where $P(X \mid Y)$ is the probability that event X occurs given the fact that event Y occurred in the experiment. Clearly if events X and Y have no points in common (disjoint), then $P(XY) = 0$, so event X cannot occur if event Y occurred.

For example, the probability that an 8 will turn up when two dice are rolled is ordinarily 5/36. However, suppose that one die is rolled first, and a 5 is obtained. Then, to get a total of 8, all that is needed is to roll a 3 on the second die, and the probability of that happening is 1/6. Therefore, given the fact that the first die showed a 5, the chances of rolling an 8 have improved from 5/36 to 1/6.

On the other hand, suppose that the first die came up 1. Then there is no way to roll an 8, no matter what happens with the second die. Therefore, the probability of getting a total of 8 given that a 1 was rolled on the first die is zero.

The above situations are examples of *conditional probability.* A conditional probability says: What is the probability that one particular event will occur, if another specific event has occurred? In particular, if event B has occurred, what is the probability that event A will occur? This is written

$$P(A \mid B) \tag{1.5-2}$$

The vertical line means "given that."

In ordinary circumstances the probability that event A will occur is $N(A)/s$, where s is the total number of outcomes and $N(A)$ is the number of events in A. However, not all of these outcomes are possible. Since event B has occurred, only those outcomes in event B need to be considered. So the number of possibilities is $N(B)$. The next question is: How many of these remaining possibilities also have event A occurring? Ordinarily there are $N(A)$ ways for event A to occur, but not all of these are possible now. The outcomes that are in A but not in B cannot occur. Therefore, the number of possible outcomes in which event A can occur is equal to the number of outcomes that are in both event A and event B. But this is the event

$$A \text{ and } B = A \cap B \tag{1.5-3}$$

The event in which both A and B occur is A intersect B. The probability that event A will occur given that event B occurs is then

$$P(A \mid B) = \frac{N(A \cap B)}{N(B)} \tag{1.5-4}$$

This can be rewritten by dividing both the top and bottom by s:

$$P(A \mid B) = \frac{N(A \cap B)/s}{N(B)/s}$$

$$= \frac{P(A \cap B)}{P(B)} \tag{1.5-5}$$

In words, the probability that event A will occur, given that event B occurs, is equal to the probability that both A and B will occur, divided by the probability that B will occur. Note that this definition does not hold when $P(B) = 0$, because division by zero would result. Equation (1.5-5) holds for two events A and B which are independent.

■ EXAMPLE 1.5-1

Determine the type of event in the following situations:

(a) From a pack of shuffled cards, one card is dealt. Let event E_1 be "the card is a spade" and event E_2 be "the card is a heart." Evaluate the event $E_1 + E_2$.

(b) Two dice are rolled. Let event E_1 be "the sum is 7" and event E_2 be "at least one 3 turns up." Consider event $E_1 + E_2$, that is, the sum 7 and/or the number 3 on at least one die.

(c) Given that a roll of two dice has produced at least one 3, what is the probability that the sum is 7? Consider event E_1 to be "the sum is 7" and event E_2 to be "at least one 3 turns up."

(d) What is the probability of drawing two aces from a well-shuffled pack? Consider event E_1 to be "ace on the first card" and event E_2 to be "ace on the second card."

(e) A die is thrown. What is the probability that it will show 6 on the first throw and 2 on the second throw? Consider E_1 to be "6 on the first throw" and E_2 to be "2 on the second throw."

Solution

(a) These two events are mutually exclusive as one card cannot be both a heart and a spade. Mutually exclusive events have no common sample points, so the addition law for mutually exclusive events is

$$P(E_1 + E_2) = P(E_1) + P(E_2)$$

$$= \frac{13}{52} + \frac{13}{52} = \frac{1}{2}$$

(b) The events E_1 (sum is 7) and E_2 (at least one 3 turns up) are not mutually exclusive. The probability for the occurrence of at least one of the events is thus

$$P(E_1 + E_2) = P(E_1) + P(E_1E_2) - P(E_1E_2)$$

Inspection of the table in Example 1.3-1(d) shows that there are six instances where the sum is 7, eleven instances where at least a 3 shows up, and two instances where a 3 occurs in a sum that is 7. Thus, $P(E_1) = 6/36$, $P(E_2) = 11/36$, and $P(E_1E_2) = 2/36$. Thus,

$$P(E_1 + E_2) = P(E_1) + P(E_2) - P(E_1E_2)$$

$$= \frac{6}{36} + \frac{11}{36} - \frac{2}{36} = \frac{15}{36}$$

(c) This is the case of the occurrence of an event E_1 (the sum is 7) when it is known that event E_2 (a 3 has turned up) has occurred. The conditional probability of E_1, given that E_2 has occurred, is $P(E_1|E_2)$, read as $P(E_1$ given $E_2)$. From Eq. (1.5-1),

$$P(E_1|E_2) = \frac{P(E_1E_2)}{P(E_2)} = \frac{2/36}{11/36} = \frac{2}{11}$$

(d) The probability of E_1 (ace on the first card) is $P(E_1) = 4/52$. The probability $P(E_2|E_1)$ of ace on the second card is obtained by noting that if the first card shown is an ace, then there will be 51 cards left in the pack of which three are aces. Thus,

$$P(E_1E_2) = P(E_1)P(E_2|E_1)$$

$$= \frac{4}{52} \times \frac{3}{51} = 0.0045$$

(e) Here, $E_1 = 6$ on the first throw, $E_2 = 2$ on the second throw, but the occurrence or nonoccurrence of event E_1 has no bearing on whether or not E_2 occurs. The two events are independent, and

$$P(E_1E_2) = P(E_1)P(E_2)$$

$$= \frac{1}{6} \times \frac{1}{6} = \frac{1}{36}$$

■

■ EXAMPLE 1.5-2

Define the sample space and give the probability for the following: (a) tossing of a die; (b) tossing of two coins; (c) tossing of two coins with at least one head showing; (d) tossing of two dice.

Solution

(a) When a die is thrown, the sample space is 1, 2, 3, 4, 5, and 6. Each event has a probability of 1/6.

(b) If two coins are tossed, the sample space consists of four points or events: head–head, tail–tail, head–tail, and tail–head. Each event is equally likely to occur. Thus, denoting heads by H and tails by T,

Sample space	HH	HT	TH	TT
Probability	1/4	1/4	1/4	1/4

(c) If two coins are tossed, obtaining at least one head is an event, that consists of the three sample points (HH), (HT), and (TH). The probability of this event (denoted by E), is

$$P(E) = P(\text{HH}) + P(\text{HT}) + P(\text{TH})$$

$$= \frac{1}{4} + \frac{1}{4} + \frac{1}{4} = \frac{3}{4}$$

(d) If two dice are tossed, the sample space consists of the 36 combinations shown below.

1, 1	1, 2	1, 3	1, 4	1, 5	1, 6
2, 1	2, 2	2, 3	2, 4	2, 5	2, 6
3, 1	3, 2	3, 3	3, 4	3, 5	3, 6
4, 1	4, 2	4, 3	4, 4	4, 5	4, 6
5, 1	5, 2	5, 3	5, 4	5, 5	5, 6
6, 1	6, 2	6, 3	6, 4	6, 5	6, 6

Each of the 36 sample points is equally likely to occur and so each has probability 1/36. By inspection, it can be seen for example that

$$P(\text{sum of the two dice is 7}) = \frac{3}{36} = \frac{1}{6}$$

$$P(\text{sum is 2}) = \frac{1}{36}$$

$$P(\text{sum is 7 or 11}) = \frac{6}{36} + \frac{2}{36} = \frac{8}{36}$$

■

1.6 DISCRETE AND CONTINUOUS STOCHASTIC VARIABLES

To help understand the outcome of an experiment, the concept of a stochastic or random variable is useful. A quantity whose value is a number determined by the outcome of an experiment is called a stochastic variable. There are two kinds of stochastic variables—discrete and continuous—depending on whether they are represented by discrete or continuous values.

Consider first the case of discrete variables, that is, characteristics that are represented by a number and thus take on discrete values. For example, in rolling two dice, the event of interest is the sum s of the numbers on the two upturned faces. The sample space has eleven points, and the probability $P(s)$ of the sum s has six values as shown in tabular and graphic form in Fig. 1.6-1.

Each member of the ensemble has a definite value of the parameter. Let x_i be the value taken on by the ith member. The probability that characteristic x has the value x_i is then

$$P_i(x_i) = \frac{n(x_i)}{n} \tag{1.6-1}$$

Furthermore, since no member of the ensemble can have two different values of the parameter, the condition of mutual exclusiveness, Eq. (1.4-1) applies, and thus

$$\sum_{x_i} P_i(x_i) = \sum_{x_i} \frac{n(x_i)}{n} = \frac{n}{n} = 1 \tag{1.6-2}$$

That is, the sum of $P(x)$ over all possible values x_i of the characteristic variable yields unity. A probability $P(x)$ that satisfies Eq. (1.6-2) is said to be *normalized,* and Eq. (1.6-2) is often called the *normalization condition.* Because $P(x)$ determines how the probability is distributed over the events i, it is often called the *distribution function.*

Consider now the analogous one-dimensional distribution of points along

s	2	3	4	5	6	7	8	9	10	11	12
$P(s)$	$\frac{1}{36}$	$\frac{1}{18}$	$\frac{1}{12}$	$\frac{1}{9}$	$\frac{5}{36}$	$\frac{1}{6}$	$\frac{5}{36}$	$\frac{1}{9}$	$\frac{1}{12}$	$\frac{1}{18}$	$\frac{1}{36}$

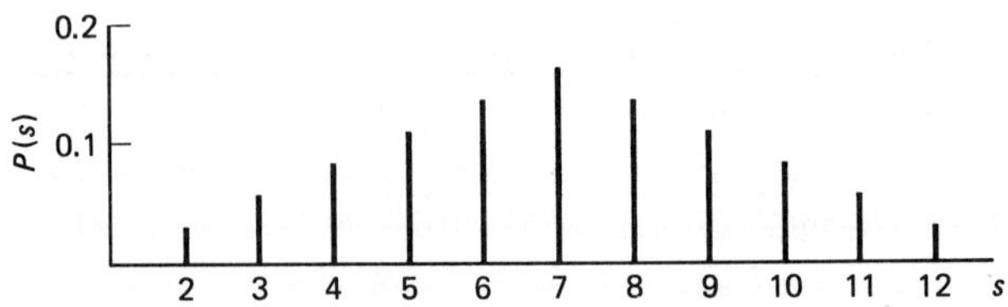

Figure 1.6-1 Discrete stochastic representation.

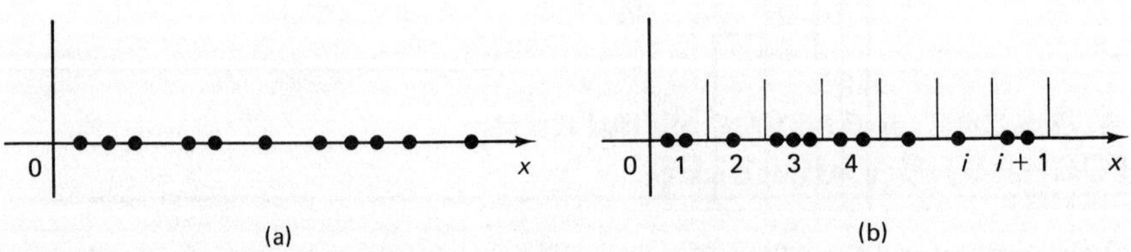

Figure 1.6-2 Distribution of points along a line.

a line, as shown in Fig. 1.6-2. Such might be the case of a gas in a container that has a fine slit in it. The gas atoms escape from the container into a vacuum, which is maintained by a pump. What is the probability that a gas atom is traveling in a certain direction? To determine this, a fluorescent screen may be set up so that every time an atom hits the screen it causes the screen to scintillate. Also, in this case, there is a continuous range of possible events—namely, an atom can strike any point on the screen.

A complete description of the distribution would require the specification of the position of every point, as in Fig. 1.6-2(a). Suppose, however, that the location of a particle cannot be exactly determined, but can only be located within a range Δx, or suppose that there simply is no interest in knowing the location more exactly. The x axis could then be divided into a set of cells each of length Δx, labeled from 1 to ∞, as shown in Fig. 1.6-2(b), and a description be given in terms of the number of points in each cell.

If n_i designates the number of points in cell i, a complete representation would require a knowledge of the whole set of n_i, written as $\{n_i\}$. To determine the probability (or frequency) that an atom strikes a certain strip, the number of hits $n(x)$ in the strip centered at x is recorded and divided by the total number of atoms that strike the screen. This record of hits is then shown by the histogram of Fig. 1.6-3. The height of each bar represents the number of hits in a strip, divided by the total number of atoms observed, N.

As the strip size is reduced, the histogram becomes more continuous and can be represented by a continuous function

$$f(x) = \lim_{\Delta x \to 0} \frac{P(x, \Delta x)}{\Delta x} \tag{1.6-2}$$

where $P(x, \Delta x)$ is the probability that an atom will strike the strip of width whose center is at x. Equation (1.6-2) is usually written as

$$P(x, dx) = f(x)\, dx \tag{1.6-3}$$

where dx is understood to be an infinitesimal. The function $f(x)$ is called a

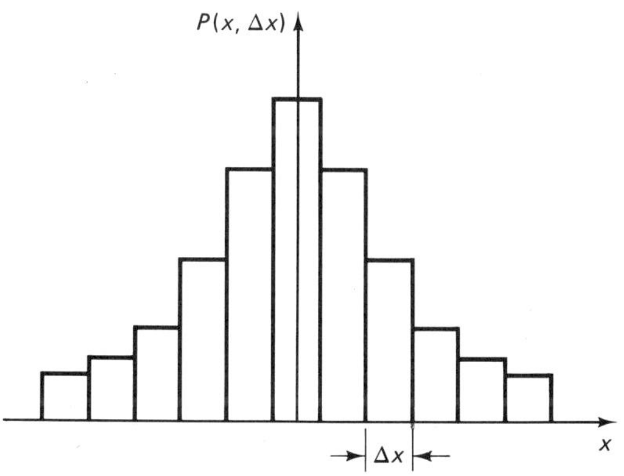

Figure 1.6-3 Probability histogram.

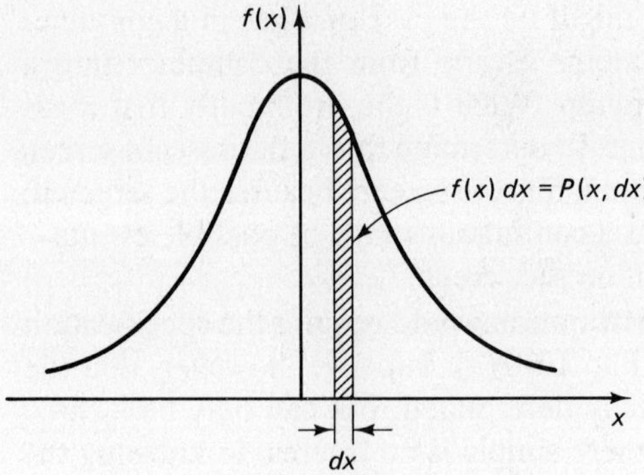

Figure 1.6-4 Probability distribution function.

probability density distribution function, or simply a *distribution function.* It represents the distribution of probability densities, so that $f(x)\,dx$ is the quantity that yields the probability $P(x)$. This is illustrated in Fig. 1.6-4.

Using the fact that the sum of the probabilities must equal unity, that is, $\Sigma P(x, \Delta x) = 1$, the normalization of Eq. (1.6-3) gives

$$\int_{-\infty}^{\infty} f(x)\,dx = 1 \tag{1.6-4}$$

A function that satisfies (1.6-4) is said to be normalized to 1.

If N is the total number of atoms, the function

$$F(x)\,dx = Nf(x)\,dx \tag{1.6-5}$$

represents a function that is normalized to N, for

$$\int_{-\infty}^{\infty} F(x)\,dx = N\int_{-\infty}^{\infty} f(x)\,dx = N \tag{1.6-6}$$

Such functions are often used in statistical mechanics and are also referred to as distribution functions. Whereas $f(x)$ represents the distribution of probability densities, $F(x)$ represents the distribution of the probable number densities. The essential difference between these distribution functions is only their normalized values (1 and N).

■ EXAMPLE 1.6-1

Draw a bar chart and a histogram for the following situations:

(a) Consider that the number of cosmic particles striking an apparatus in 40 consecutive periods of 1 min was recorded as follows:

3	0	0	1	0	2	1	0	1	1
0	3	4	1	2	0	2	0	3	1
1	0	1	2	0	2	1	0	1	2
3	1	0	0	2	1	0	3	1	2

(b) Consider the case for continuous data, such as illustrated by the following example: the heights of 100 students are measured to the nearest inch and tabulated in Table 1.6-1.

TABLE 1.6-1 STUDENT HEIGHTS

Height (in.)		Number of students
60–62		6
63–65		15
66–68		40
69–71		30
72–74		9
	Total	100

Draw the bar chart and the histogram for the respective cases.

Solution

(a) The first step is to find the *frequency* with which each value occurs. In Table 1.6-2, the values in the right-hand column form a frequency distribution. For example, it is found that "zero" occurs 13 times, but that "four" occurs only once. The frequency distribution can be plotted as in Fig. 1.6-5, where the height of each line is proportional to the frequency with which the value occurs. This diagram is called a bar chart (or line graph or frequency diagram) and is a very useful visual aid in analyzing discrete data.

TABLE 1.6-2 FREQUENCY DISTIBUTION

Number of cosmic particles		Frequency
0		13
1		13
2		8
3		5
4		1
	Total	40

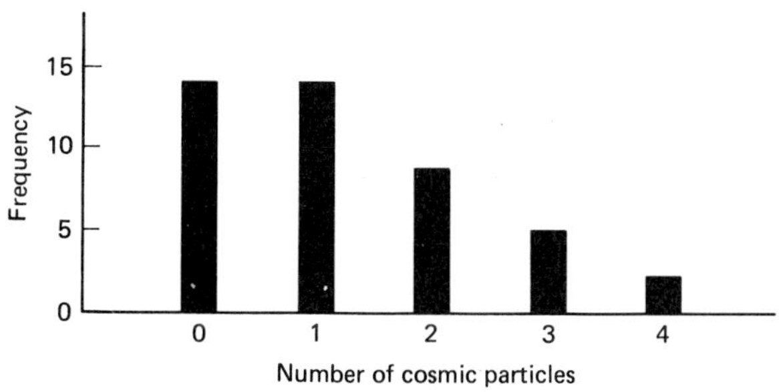

Figure 1.6-5 Bar chart of data from Table 1.6-1.

(c) The heights are divided into five groups as shown and the frequencies with which the different groups occur form a frequency distribution. The data can be plotted as a histogram in which rectangles have areas proportional to the frequency of each group.

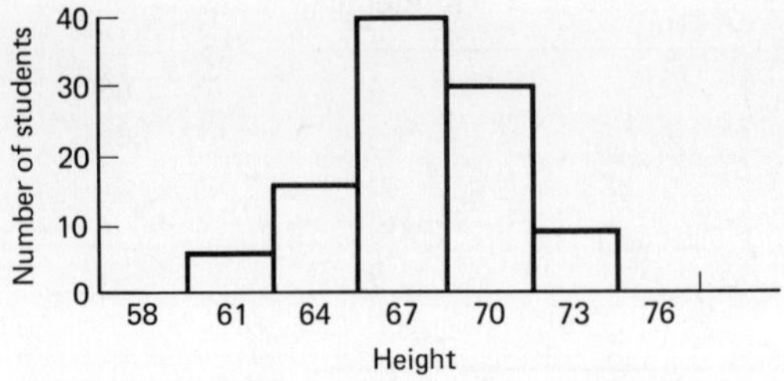

Figure 1.6-6 Histogram of data from Table 1.6-2. ■

1.7 CALCULATION OF PROPERTIES

A distribution function has many applications. For example, the distribution function $F(x)$ can be used to compute the number of atoms that strike the screen between two points x_1 and x_2:

$$n(x_1 < x < x_2) = \int_{x_1}^{x_2} F(x)\, dx \tag{1.7-1}$$

Similarly, the distribution function $f(x)$ can be used to compute the probability that an atom will strike between two points x_1 and x_2. This is just a compound event—namely, the collection of all those events (x, dx) for which $x_1 \leq x \leq x_2$. The probability of this compound event is [according to the definition $P(A) = \sum_{i \subset A} P_i$],

$$\sum_{x_2 \geq x \geq x_1} P(x, dx) = \int_{x_1}^{x_2} f(x)\, dx \tag{1.7-2}$$

Note that, if x_2 is very close to x_1, then $f(x)$ is essentially constant over the range of integration, so

$$\int_{x_1}^{x_2} f(x)\, dx \simeq f(x)(x_2 - x_1) = f(x)\, \Delta x \qquad (x_2 \geq x \geq x_1) \tag{1.7-3}$$

and this returns to the result $P(x, dx) = f(x)\, dx$ (when $\Delta x \rightarrow dx$).

For yet another example, consider the speed distribution of gaseous mercury atoms in a beam as obtained from a flight-of-time analyzer. This is an apparatus in which a collimated beam meets a pair of rotating disks which, together, constitute a speed selector. The number of atoms that get through is measured by letting the beam strike a suitable recording detector. Changing the speed of the motor that drives the disks lets through atoms with different, but known speeds. The data obtained is plotted in the form of the histogram shown in Fig. 1.7-1.

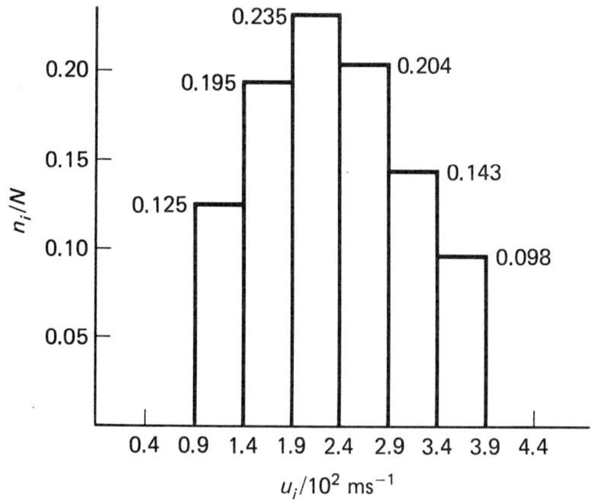

n_i/N	$u_i/10^2\,(\mathrm{m\,s^{-1}})$
0.125	1.15
0.195	1.65
0.235	2.15
0.204	2.65
0.143	3.15
0.098	3.65

Figure 1.7-1 Distribution in speeds of gaseous mercury atoms in a beam.

Suppose that it is desired to find the mean square speed $\langle u^2\rangle$ and hence the mean kinetic energy $\frac{1}{2}m\langle u^2\rangle$ of the gas atoms. Now, if out of a total of N gas atoms n_i have speed u_i, then, by definition, the mean square speed $\langle u^2\rangle$ of these atoms is

$$\langle u^2\rangle = \frac{\sum_i n_i u_i^2}{\sum_i n_i} = \frac{\sum_i n_i u_i^2}{N} = \left(\frac{n_1}{N}\right)u_1^2 + \left(\frac{n_2}{N}\right)u_2^2 + \cdots \qquad (1.7\text{-}4)$$

But the n_i/N represent the probability $P(u)$ as N becomes large, so that Eq. (1.7-4) can be written as

$$\langle u^2\rangle = \sum_u u^2 P(u) \qquad (1.7\text{-}5)$$

or, in the case of a continuous variable,

$$\langle u^2\rangle = \int_{-\infty}^{\infty} u^2 f(u)\,du \qquad (1.7\text{-}6)$$

In general, the average of a property $g(x)$ is given by

$$\langle g(x)\rangle = \sum_x g(x)\,P(x) \qquad (1.7\text{-}7)$$

for discrete variables, and

$$\langle g(x)\rangle = \int_x g(x)\,f(x)\,dx \qquad (1.7\text{-}8)$$

for continuous variables.

Returning to the example under discussion, Fig. 1.7-1 gives the fractions (n_1/N), (n_2/N), and so on with speeds u_1, u_2, and on, which may be taken to be the speeds at the midpoint of each block of the histogram, namely $1.15 \times 10^2\ \mathrm{m\,s^{-1}}$, $1.65 \times 10^2\ \mathrm{m\,s^{-1}}$, and so on. The calculation of $(n_1/N)u_1^2$, $(n_2/N)u_2^2$, is

TABLE 1.7-1 ANALYSIS OF DATA IN FIG. 1.7-3

n_i/N	$u_i/10^2\ (\mathrm{m\,s^{-1}})$	$u_i^2/10^4\ (\mathrm{m\,s^{-2}})$	$(n_i u_i^2/N)/10^4\ (\mathrm{m^2\,s^{-2}})$
0.125	1.15	1.32	0.165
0.195	1.65	2.72	0.530
0.235	2.15	4.62	1.086
0.204	2.65	7.02	1.432
0.143	3.15	9.92	1.418
0.098	3.65	13.30	1.303

set out in Table 1.7-1. Adding up the numbers in the last column of this table gives $\langle u^2 \rangle = 5.93 \times 10^4\ \mathrm{m^2\,s^{-2}}$. The mean kinetic energy of the mercury atoms in the beam is therefore given by $\frac{1}{2}m\langle u^2 \rangle = \frac{1}{2}(3.32 \times 10^{-25}\ \mathrm{kg})(5.93 \times 10^4\ \mathrm{m^2\,s^{-1}}) = 9.84 \times 10^{-21}\ \mathrm{J}$, where 3.32×10^{-25} kg is the mass of a mercury atom.

1.8 AVERAGE, VARIANCE, AND STANDARD DEVIATION

The average value (or mean value) of a set of N values $x_1, x_2, \ldots, x_N$ of x is denoted by either $\bar{x}$ or $\langle x \rangle$, and is given by

$$\bar{x} = \langle x \rangle = \frac{x_1 + x_2 + \cdots + x_N}{N} = \frac{1}{N}\sum_{j=1}^{N} x_j \tag{1.8-1}$$

The summation is over all the N values of x_j. For example, if the values of x_j are 6, 7, 6, 7, 8, 9, 7, 5, 8, the average value of x is

$$\bar{x} = \langle x \rangle = (6 + 7 + 6 + 7 + 8 + 9 + 7 + 5 + 8)/10$$

$$\bar{x} = 7$$

Since there are one 5, two 6's, four 7's, two 8's and one 9, the expression for $\bar{x}$ is

$$\bar{x} = (1 \times 5 + 2 \times 6 + 4 \times 7 + 2 \times 8 + 1 \times 9)/10.$$

This can be rewritten in the form

$$\bar{x} = \frac{1}{10} \times 5 + \frac{2}{10} \times 6 + \frac{4}{10} \times 7 + \frac{2}{10} \times 8 + \frac{1}{10} \times 9 = 7.0$$

or

$$\bar{x} = \langle x \rangle = \sum x_i P_i \tag{1.8-2}$$

Generalizing, if each member of an ensemble (a collection of identical systems) has a value g_i of the characteristics g, the average of all these values, known as the *ensemble average*, $\langle g \rangle$, is then

$$\langle g \rangle = \sum_g g P(g) \tag{1.8-3}$$

where $P(g)$ is the probability. Similarly, the average value of the square of g is given by

$$\langle g^2 \rangle = \sum_g g^2 P(g) \tag{1.8-4}$$

which, in the above case, is

$$\begin{aligned}\sum_g g^2 P(g) &= \frac{1}{10} \times 5^2 + \frac{2}{10} \times 6^2 + \frac{4}{10} \times 7^2 + \frac{2}{10} \times 8^2 + \frac{1}{10} \times 9^2 \\ &= 50.2\end{aligned}$$

The probabilities 1/10, 2/10 and so forth, for the values of x are distributed as shown in Fig. 1.8-1. The plot is centered around the average value $\langle x \rangle = 7.0$.

If the outcome of an experiment is not discrete, but continuous, a distribution function $f(x)$ must be used. In the same fashion as above, the average of a function $g(x)$ is given by

$$\langle g(x) \rangle = \int_{x_1}^{x_2} g(x)\, f(x)\, dx \tag{1.8-5}$$

where the limits on the integral extend over all possible values of x.

The average value of a quantity is not the only information that is useful. Of particular interest is how much the observed value of the quantity differs in any *particular* experiment from its average value. For example, in the present case, the average value of the number that turns up is 7.0. How much of a difference from 7.0 can be expected in any particular experiment? If this difference is large, then clearly the average value is not a good indication of the value that will be observed in a particular experiment. On the other hand, if this difference is small, then the average value is very significant for the particular experiment.

Generally, the average value of the difference $g - \langle g \rangle$ is

$$\begin{aligned}\langle (g - \langle g \rangle) \rangle &= \sum (g - \langle g \rangle)\, P(g) = \sum g P(g) - \langle g \rangle \sum P(g) \\ &= \langle g \rangle - \langle g \rangle = 0\end{aligned} \tag{1.8-6}$$

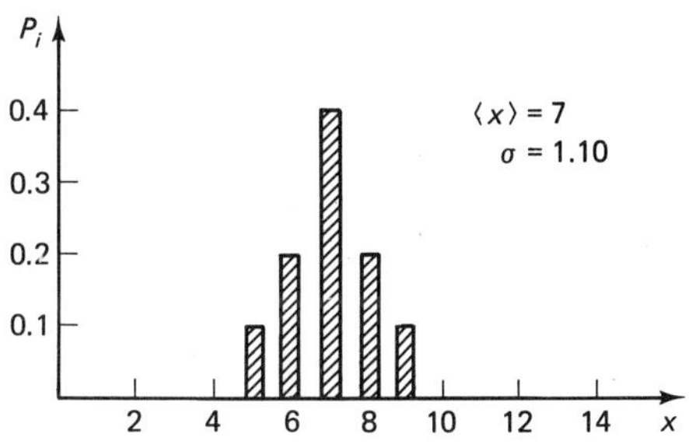

Figure 1.8-1 Probability distribution.

In other words, the average value of $g - \langle g \rangle$ vanishes, so it cannot be used to determine the average deviation in a particular experiment.

A more useful quantity is $(g - \langle g \rangle)^2$. Its average does not go to zero, and thus can be taken as an indication of the deviation from the mean:

$$\begin{aligned}\langle (g - \langle g \rangle)^2 \rangle &= \langle (g^2 - 2g\langle g \rangle + \langle g \rangle^2) \rangle \\ &= \langle g^2 \rangle - 2\langle (g\langle g \rangle) \rangle + \langle (\langle g \rangle^2) \rangle \\ &= \langle g^2 \rangle - (\langle g \rangle)^2 = [\sigma(g)]^2 = \mathrm{var}(g) \qquad (1.8\text{-}7)\end{aligned}$$

The quantity $[\sigma(g)]^2$, or var(g), is called the *variance* or *dispersion* of g. Its square root is called the *standard deviation:*

$$\sigma(g) = [\langle (g - \langle g \rangle)^2 \rangle]^{1/2} \qquad (1.8\text{-}8)$$

In the case of a continuous variable, the variance is similarly defined:

$$\mathrm{var}(x) = \int_{x_1}^{x_2} (x - \langle x \rangle)^2 f(x)\, dx \qquad (1.8\text{-}9)$$

In the above example, the variance is

$$\begin{aligned}\mathrm{var}(x) &= \sigma^2 = \langle (x_i - \langle x \rangle)^2 \rangle \\ &= \frac{1}{N}\sum_{i=1}^{N} (x_i - \langle x \rangle)^2 = \sum_i (x_i - \langle x \rangle)^2 P(x_i) \\ &= \frac{1}{10}(5-7)^2 + \frac{2}{10}(6-7)^2 + \frac{2}{10}(8-7)^2 + \frac{1}{10}(9-7)^2 \\ &= 1.2 \qquad (1.8\text{-}10)\end{aligned}$$

and the standard deviation is thus

$$\sigma = 1.1$$

The standard deviation is a good measure of the width of the probability distribution. In the present case, eight of the ten values of x lie between $\langle x \rangle - \sigma$ and $\langle x \rangle + \sigma$.

■ **EXAMPLE 1.8-1**

The raw data for a set of test scores in a class of 30 students is

53	87	76	73	62	99
78	93	82	69	65	93
92	92	78	82	89	65
63	49	88	87	94	73
85	77	98	59	93	82

Determine the statistics, namely, the mean (average), the median, the mode, and the standard deviation.

Solution. The grouping of the scores into intervals of ten and the frequency diagram are as shown in Fig. 1.8-2.

Interval	Number of scores in interval
41–50	1
51–60	2
61–70	5
71–80	6
81–90	8
91–100	8

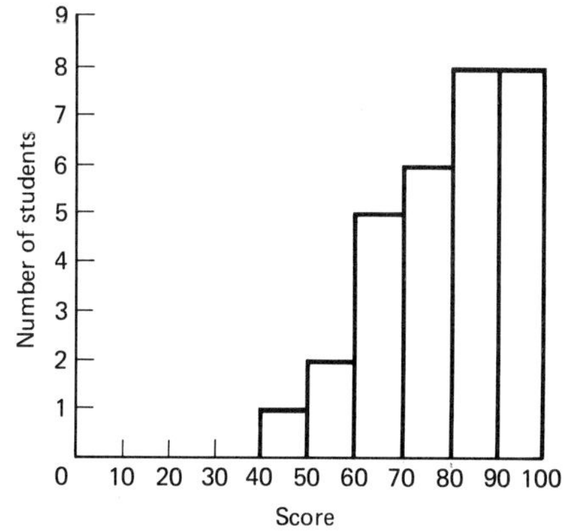

Figure 1.8-2 Frequency diagram.

A statistic is a single member that is used to summarize the properties of a large group of numbers. The *average* (also called the *mean*) is an example of a statistic:

$$\langle x \rangle = \frac{x_1 + x_2 + \cdots + x_n}{n} = \frac{2376}{30} = 79.2$$

Another statistic is the *median.* To compute this, the scores must be placed in order:

49	53	59	62	63	65
65	69	73	73	76	77
78	78	82	82	82	85
87	87	88	89	92	92
93	93	93	94	98	99

The *median* is the halfway point of the data. Half of the numbers are above it and half are below it. If there is an odd number of scores, the median is the middle score. If there is an even number of scores, then the median is the average of the two middle scores. Here, the two middle scores are 82 and 82, so the median is $(82 + 82)/2 = 82$. The median is occasionally used instead of the mean, particularly when the histogram of the observations is skewed.

Another statistic is the *mode.* The *mode(s)* is (are) the data that occur most frequently. In the above sample, there are two modes, 82 and 93. Each mode occurs three times. None of the other scores occurs three or more times.

A measure of how spread out the scores are is given by the *standard deviation:*

$$s_1 = \sqrt{\sum_{i=1}^{n} \frac{(x_i - \overline{x})^2}{n}}$$

However, it is usually easier to calculate s_1 from the formula:

$$s_1 = \sqrt{\overline{x^2} - \overline{x}^2}$$

where $\overline{x^2}$ is equal to the average value of the squares of all of the x's. There are also times when it is desirable to use an alternate standard deviation measure, called s_2:

$$s_2 = \sqrt{\frac{n}{n-1}}s_1$$

For the present scores, $\overline{x^2} = 6452$, so $s_1^2 = 6452 - 79.2^2 = 179.36$. Therefore, $s_1 = 13.39$ and $s_2 = 13.62$. ■

1.9 GAUSSIAN FUNCTIONS AND INTEGRALS

An important distribution in statistical mechanics is the Gaussian distribution function

$$f(x) = \sqrt{\frac{\alpha}{\pi}}e^{-\alpha x^2} \qquad (-\infty \leq x \leq +\infty) \tag{1.9-1}$$

centered at the origin, and

$$f(x) = \sqrt{\frac{\alpha}{\pi}}e^{-\alpha(x-x_0)^2} \tag{1.9-2}$$

centered at x_0. In both cases, α is a constant, and $\alpha/\pi^{1/2}$ is arrived at by the normalization $\int f(x)\,dx = 1$. The functions are shown in Fig. 1.9-1.

The evaluation of various integrals involving the Gaussian function will now proceed.

1. First, to evaluate the integral $I = \int \exp(-x^2)\,dx$, multiply by I to give

$$I^2 = \int_{-\infty}^{+\infty} \exp(-x)^2\,dx \int_{-\infty}^{+\infty} \exp(-y^2)\,dy$$

$$= \int_{-\infty}^{+\infty}\int_{-\infty}^{+\infty} \exp[-(x^2+y^2)]\,dx\,dy \tag{1.9-3}$$

This double integral extends over the whole of the xy plane. To evaluate it,

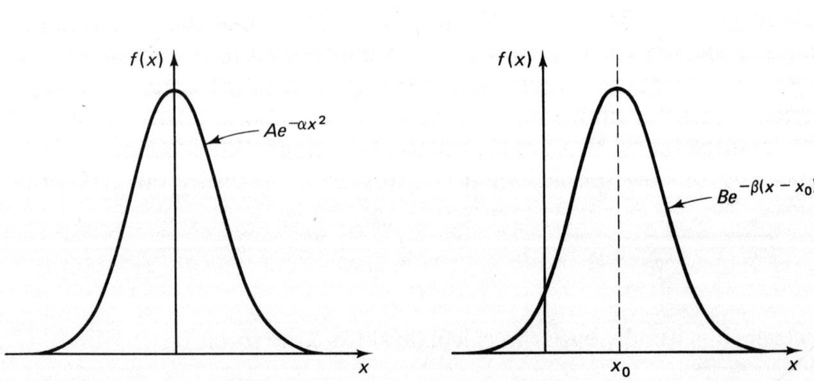

Figure 1.9-1 Gaussian function centered at origin and at x_0.

introduce polar coordinates r and θ, where $r^2 = (x^2 + y^2)$. The element of area in polar coordinates is $r\,dr\,d\theta$. Equation (1.9-3) can be rewritten in the form

$$I^2 = \int_0^\infty \int_0^{2\pi} \exp(-r^2)\, r\, dr\, d\theta = 2\pi \int_0^\infty \exp(-r^2)\, r\, dr \qquad (1.9\text{-}4)$$

If $r^2 = u$, $du = 2r\,dr$, and

$$I^2 = \pi \int_0^\infty e^{-u}\, du = \pi(-e^{-u})_0^\infty = \pi$$

Hence,

$$I = \int_{-\infty}^{+\infty} \exp(-x^2)\, dx = \pi^{1/2} \qquad (1.9\text{-}5)$$

Since the function $\exp(-x^2)$ is symmetrical about $x = 0$, it follows that

$$\int_0^\infty \exp(-x^2)\, dx = \frac{\pi^{1/2}}{2} \qquad (1.9\text{-}6)$$

2. To evaluate the integral $I_0 = \int \exp(-\alpha v^2)\, dv$, let $w = \alpha^{1/2} v$, so that $dw = \alpha^{1/2}\, dv$. Using Eq. (1.9-6), this gives

$$I_0 = \frac{1}{\alpha^{1/2}} \int_0^\infty \exp(-w^2)\, dw = \frac{1}{2}\left(\frac{\pi}{\alpha}\right)^{1/2} \qquad (1.9\text{-}7)$$

3. To evaluate $I_1 = \int v \exp(-\alpha v^2)\, dv$, let $z = \alpha v^2$, so that $dz = 2\alpha v\, dv$ and

$$I_1 = \frac{1}{2\alpha} \int_0^\infty e^{-z}\, dz = \frac{1}{2a}\left(-e^{-z}\right)_0^\infty = \frac{1}{2\alpha} \qquad (1.9\text{-}8)$$

$$I_n = \int_0^\infty v^n \exp(-\alpha v^2)\, dv = -\frac{1}{2\alpha} \int_0^\infty v^{(n-1)} \left\{\frac{d[\exp(-\alpha v^2)]}{dv}\right\}$$

where n is an integer. Integrating by parts:

$$I_n = -\frac{1}{2\alpha}\left\{[v^{n-1} \exp(-\alpha v^2)]_0^\infty - \int_0^\infty \exp(-\alpha v^2)\left(\frac{dv^{(n-1)}}{dv}\right) dv\right\}$$

Since the integrated term vanishes at both limits, this gives

$$I_n = \frac{1}{2\alpha} \int_0^\infty (n-1)\, v^{n-2} \exp(-\alpha v^2)\, dv = \left(\frac{n-1}{2\alpha}\right) I_{n-2} \qquad (1.9\text{-}9)$$

which is a recurrence relation enabling the determination of any of the I_n in terms of either I_0 or I_1. Summarizing,

$$I_0 = \int_0^\infty \exp(-\alpha v^2)\, dv = \frac{1}{2}\left(\frac{\pi}{\alpha}\right)^{1/2} \qquad (1.9\text{-}10)$$

$$I_1 = \int_0^\infty v \exp(-\alpha v^2)\, dv = \frac{1}{2\alpha} \qquad (1.9\text{-}11)$$

$$I_2 = \int_0^\infty v^2 \exp(-\alpha v^2)\, dv = \frac{\pi^{1/2}}{4\alpha^{3/2}} \qquad (1.9\text{-}12)$$

$$I_3 = \int_0^\infty v^3 \exp(-\alpha v^2)\, dv = \frac{1}{2\alpha^2} \tag{1.9-13}$$

$$I_4 = \int_0^\infty v^4 \exp(-\alpha v^2)\, dv = \frac{3\pi^{1/2}}{8\alpha^{5/2}} \tag{1.9-14}$$

$$I_5 = \int_0^\infty v^5 \exp(-\alpha v^2)\, dv = \frac{1}{\alpha^3} \tag{1.9-15}$$

$$I_6 = \int_0^\infty v^6 \exp(-\alpha v^2)\, dv = \frac{15\pi^{1/2}}{16\alpha^{7/2}} \tag{1.9-16}$$

For an application of the above, consider the Gaussian distribution function

$$f(x) = \sqrt{\frac{\alpha}{\pi}}\, e^{-\alpha x^2} \qquad (-\infty \leq x \leq +\infty) \tag{1.9-17}$$

The average value of x is

$$\langle x \rangle = \sqrt{\frac{\alpha}{\pi}} \int_{-\infty}^{\infty} x e^{-\alpha x^2}\, dx = 0 \tag{1.9-18}$$

because the integrand is an odd function of x.

The average value of x^2 is

$$\langle x^2 \rangle = \sqrt{\frac{\alpha}{\pi}} \int_{-\infty}^{\infty} x^2 e^{-\alpha x^2}\, dx$$

$$= \left(\frac{\alpha}{\pi}\right)^{1/2} \frac{\pi^{1/2}}{2\alpha^{3/2}} = \frac{1}{2\alpha} \tag{1.9-19}$$

The variance of x is

$$\langle x^2 \rangle = (\langle x \rangle)^2 = \frac{1}{2\alpha} \tag{1.9-20}$$

and the standard deviation is

$$\sigma(x) = \frac{1}{\sqrt{2\alpha}} \tag{1.9-21}$$

To appreciate these results, $f(x)$ is plotted for two different values of α in Fig. 1.9-2. The areas under both curves are equal, because the total probability is

$$\int_{-\infty}^{\infty} f(x)\, dx = 1$$

However, the smaller the value of α, the fatter is the curve, and the larger the value of $\langle x^2 \rangle$ as well as the standard deviation σ. In such a case, the value observed in the experiment differs appreciably from the average value. Anal-

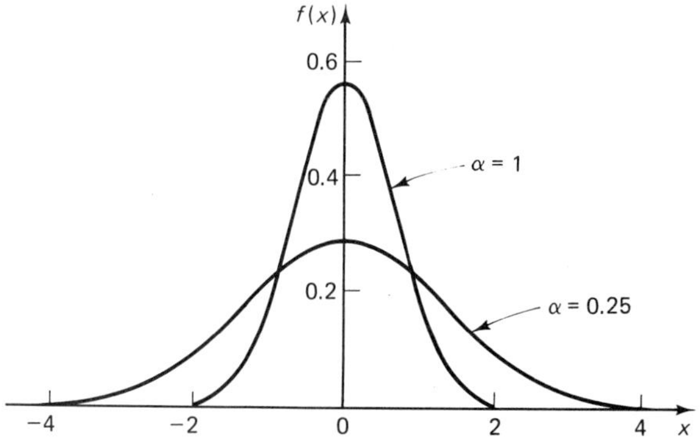

Figure 1.9-2 Distribution function $f(x) = \sqrt{\alpha/\pi}\, e^{-\alpha x^2}$ for two values of α.

ogous to the integrals considered above, but with finite rather than infinite limits, is the *error function* defined by

$$\operatorname{erf} x = \frac{1}{\sqrt{\pi}} \int_{-x}^{x} e^{-u^2}\, du = \frac{2}{\sqrt{\pi}} \int_{0}^{x} e^{-u^2}\, du \tag{1.9-22}$$

which can be evaluated by numerical methods. It is tabulated in Table 1.9-1. The constant is chosen so that, by virtue of (1.9-7), erf $\infty = 1$. Also, erf $0 = 0$.

TABLE 1.9-1 ERROR FUNCTION $\operatorname{erf} x = \frac{2}{\sqrt{\pi}} \int_0^x e^{-t^2}\, dt$

x	0	1	2	3	4	5	6	7	8	9
0.0	0.0000	0.0113	0.0226	0.0338	0.0451	0.0564	0.0676	0.0789	0.0901	0.1013
0.1	0.1125	0.1236	0.1348	0.1459	0.1569	0.1680	0.1790	0.1900	0.2009	0.2118
0.2	0.2227	0.2335	0.2443	0.2550	0.2657	0.2763	0.2869	0.2974	0.3079	0.3183
0.3	0.3236	0.3389	0.3491	0.3593	0.3694	0.3794	0.3893	0.3992	0.4090	0.4187
0.4	0.4234	0.4380	0.4475	0.4569	0.4662	0.4755	0.4847	0.4937	0.5027	0.5117
0.5	0.5205	0.5292	0.5379	0.5464	0.5549	0.5633	0.5716	0.5798	0.5879	0.5959
0.6	0.6039	0.6117	0.6194	0.6270	0.6346	0.6420	0.6494	0.6566	0.6638	0.6708
0.7	0.6778	0.6847	0.6914	0.6981	0.7047	0.7112	0.7175	0.7238	0.7300	0.7361
0.8	0.7421	0.7480	0.7538	0.7595	0.7651	0.7707	0.7761	0.7814	0.7867	0.7918
0.9	0.7969	0.8019	0.8068	0.8116	0.8163	0.8209	0.8254	0.8299	0.8342	0.8385
1.0	0.8427	0.8468	0.8508	0.8548	0.8586	0.8624	0.8661	0.8698	0.8733	0.8768
1.1	0.8802	0.8835	0.8868	0.8900	0.8931	0.8961	0.8991	0.9020	0.9048	0.9076
1.2	0.9103	0.9130	0.9155	0.9181	0.9205	0.9229	0.9252	0.9275	0.9297	0.9319
1.3	0.9340	0.9361	0.9381	0.9400	0.9419	0.9438	0.9456	0.9473	0.9490	0.9507
1.4	0.9523	0.9539	0.9554	0.9569	0.9583	0.9597	0.9611	0.9624	0.9637	0.9649
1.5	0.9661	0.9673	0.9684	0.9695	0.9706	0.9716	0.9726	0.9736	0.9745	0.9755
1.6	0.9763	0.9772	0.9780	0.9788	0.9796	0.9804	0.9811	0.9818	0.9825	0.9832
1.7	0.9838	0.9844	0.9840	0.9856	0.9861	0.9867	0.9872	0.9877	0.9882	0.9886
1.8	0.9891	0.9895	0.9899	0.9903	0.9907	0.9911	0.9915	0.9918	0.9922	0.9925
1.9	0.9928	0.9931	0.9934	0.9937	0.9939	0.9942	0.9944	0.9947	0.9949	0.9951

■ EXAMPLE 1.9-1

The density function for a continuous random variable x is given by

$$f(x) = \begin{cases} 1 & \text{for} \quad 0 < x < 1 \\ 0 & \text{otherwise} \end{cases}$$

Check that this is a proper probability density function.

Solution. The graph of this density function is as shown in Fig. 1.9-3. The probability that x falls within an interval (a,b) is the area under the curve $f(x)$ from a to b. It is proportional to the length of the interval (a,b).

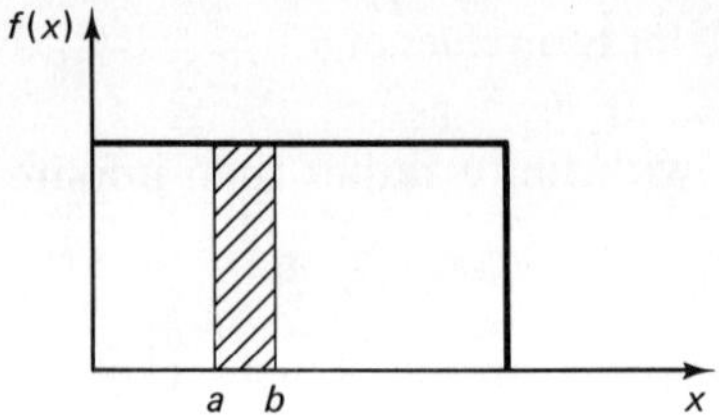

Figure 1.9-3 Density function.

The conditions imposed on $f(x)$ are that $f(x)$ be positive and that the total area between $f(x)$ and the x axis be equal to 1. Here, the total area under $f(x)$ is $1 \times (1.0) = 1$. So $f(x)$ is a proper probability density function.

Note that when the probability is defined in terms of an area under a curve, the probability that a continuous random variable is equal to a particular value is $P(x = a) = 0$, because the area of a vertical line is zero. Thus,

$$P(a < x < b) = P(a \leq x \leq b)$$

The probability that x is between a and b is

$$P(a \leq x \leq b) = 1x(b - a) = b - a$$

The probability that x is less than k is

$$P(a \leq k) = 1 \times (k - 0) = k$$

where $0 < k < 1$. ■

1.10 REPEATED TRIALS AND BINOMIAL THEOREM

To discuss the origins of the Gaussian distribution, consider the toss of a coin which gives, with equal *a priori* probability, heads or tails. One toss of the coin will have two possible results: H, T. Two tosses of the coin twice will have four possible results: HH, HT, TH, and TT. Three tosses of the coin will result in eight possibilities: HHH, HHT, HTH, THH, TTH, THT, HTT, and TTT. The toss of a coin is a good example of a random event: There is no accurate way to predict what the outcome will be in a single toss. However, if the coin is tossed more than once, some interesting predictions can be made as to what will happen. This is a general feature of random events, and it is one of the foundations of probability.

Denoting by $P(p,q)$ the *a priori* probability that $p + q$ tosses will result in p heads and q tails, irrespective of their sequence, the situation can be described by

$$N = 1: \qquad P(\mathrm{H},0) = P(0,\mathrm{T}) = \frac{1}{2}$$

$$N = 2: \qquad P(2\mathrm{H},0) = P(0,2\mathrm{T}) = \frac{1}{4},\ P(\mathrm{H},\mathrm{T}) = \frac{1}{2}$$

$$N = 3: \qquad P(3\mathrm{H},0) = P(0,3\mathrm{T}) = \frac{1}{8},\ P(2\mathrm{H},\mathrm{T}) = P(\mathrm{H},2\mathrm{T}) = \frac{3}{8}$$

The general result is

$$P(p,q) = \frac{(p+q)!}{p!q!}\left(\frac{1}{2}\right)^{p+q} \tag{1.10-1}$$

which can be deduced as follows. Each toss gives two possibilities. Therefore, $p + q$ tosses give 2^{p+q} possible sequences of tosses. Since each sequence is equally likely, its probability is $(1/2)^{p+q}$.

To determine how many different sequences give p heads and q tails, consider that there are $p + q$ objects to be put into a like number of boxes, one to a box. The first object can be put into any of the $p + q$ boxes. The second object can be put in any of the remaining $p + q - 1$ unfilled boxes. Similarly, the third object can be put into any of $p + q - 2$ boxes, and so on until the last object, which can be put only into the one remaining box. The total number of ways of arranging the $p + q$ objects, one per box, is given by the product of individual possibilities, that is, $(p+q)\ (p+q-1)\ (p+q-2)\cdots 1 = (p+q)!$. In other words, the number of possible permutations of $(p + q)$ objects is $(p + q)!$.

Now the number of permutations among the p objects alone is equal to $p!$. Similarly, the number of permutations among the q objects alone is $q!$. These do not result in a distinguishable change. Thus, the number of distinguishable ways in which a combination of p objects of one type and q objects of another type can be permuted is

$$W(p,q) = \frac{(p+q)!}{p!q!} \tag{1.10-2}$$

which, incidentally, is equal to the pth or qth binomial coefficient of order $p + q$.

Equation (1.10-2) gives the different sequences of tosses, each with probability $(1/2)^{p+q}$. The probability of a result having p heads and q tails is thus

$$P(p,q) = W(p,q)\left(\frac{1}{2}\right)^{p+q} = \frac{(p+q)!}{p!q!}\left(\frac{1}{2}\right)^{p+q} \tag{1.10-3}$$

In Fig. 1.10-1, the distribution of heads and tails in 10 tosses is represented

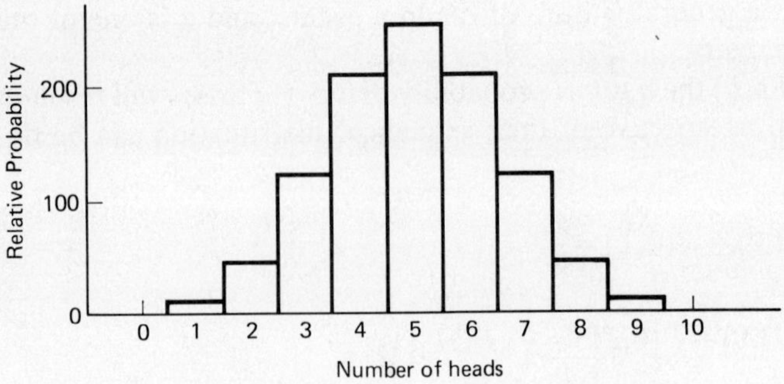

Figure 1.10-1 Distribution of results for 10 coin tosses.

in the form of a histogram. There are $2^{10} = 1024$ possible sequences. By (1.10-3), the relative probabilities of 0, 1, 2, . . . , 10 heads occurring in ten tosses are, respectively, 1, 10, 45, 120, 210, 252, 210, 120, 45, 10, and 1, which are the binomial coefficients for $n = p + q = 10$. The most probable result is five heads ($p = q = 5$), and the probability distribution is symmetric about that value. In the limit $(p + q) \to \infty$, Eq. (1.10-3) approaches a Gaussian distribution.

To summarize, if there are n tosses of a coin, all possible combinations of heads or tails can be expressed by $(p + q)$ raised to the nth power. For example, in three tosses,

$$(p + q)^3 = p^3 + 3p^2q + 3pq^2 + q^3 \tag{1.10-4}$$

The exponent of each p and q symbol is the number of heads or tails for the particular combination represented by each term of (1.10-4). The sum of the exponents in each term is equal to 3 (the number of trials). The number of terms is equal to the number of different combinations of heads and tails that are possible, four in the above example, and the coefficient of each term is equal to the number of ways that a particular combination can occur. Thus, the combination or sequence of 3 heads can result only in one way, 2 heads and 1 tail in three ways, 1 head and 2 tails in three ways, and 3 tails in one way. By the binomial theorem, the coefficient of each term in (1.10-4) is given by the expression

$$\frac{n!}{(n - m)!m!} \tag{1.10-5}$$

where m is the number of successes and $n - m$ is the number of failures in n tosses.

■ EXAMPLE 1.10-1

The probability of getting h heads in n tosses of a coin is

$$\left[\frac{n!}{(n - h)!}\right]\left(\frac{1}{2}\right)^n$$

where 1/2 is the probability of a head in a coin having two aspects (heads or tails). What are the results for the case of $n = 5$?

Solution. The number of arrangements (thermodynamic probability, in statistical mechanics) is first calculated:

for $h = 0$ (no heads) $\quad \dfrac{n!}{h!(n-h)!} = \dfrac{5!}{0!5!} = 5$

for $h = 1$ $\quad \dfrac{n!}{h!(n-h)!} = \dfrac{5!}{1!4!} = 5$

for $h = 2$ $\quad \dfrac{n!}{h!(n-h)!} = \dfrac{5!}{2!3!} = 10$

Each arrangement or thermodynamic probability must be multiplied by $(1/2)^5 = 1/32$ to give the probability for the particular number of heads. The results are summarized below in Table 1.10-1.

TABLE 1.10-1 PROBABILITY RESULTS

h	Number of combinations with h heads		Probability of h heads
0	$\dfrac{5!}{0!5!} = 1$	TTTTT	$\dfrac{1}{32} = 0.031$
1	$\dfrac{5!}{1!4!} = 5$	HTTTT, THTTT, TTHTT	$\dfrac{5}{32} = 0.156$
2	$\dfrac{5!}{2!3!} = 10$	HHTTT, HTHTT, HTTHT, HTTTH, THHTT, THTHT, THTTH, TTHHT, TTHTH, TTTHH	$\dfrac{10}{32} = 0.313$
3	$\dfrac{5!}{3!2!} = 10$	TTHHH, THTHH, THHTH, THHHT, HTTHH, HTHTH, HTHHT, HHTTH, HHTHT, HHHTT	$\dfrac{10}{32} = 0.313$
4	$\dfrac{5!}{4!1!} = 5$	HHHHT, HHHTH, HHTHH, HTHHH, THHHH	$\dfrac{5}{32} = 0.156$
5	$\dfrac{5!}{0!5!} = 1$	HHHHH	$\dfrac{1}{32} = 0.031$

■

1.11 RANDOM-WALK PROBLEM

A most important application of the preceding discussion is the so-called "random-walk problem." In its simplest form, the problem is formulated in the following way: A drunk starts out from a lamppost located on a street. Each step he takes is of equal length l. The man is, however, so drunk that the direction of each step—whether it is to the right or to the left—is completely *independent* of the preceding step. All that can be said is that each time the man takes a step, the probability of its being to the right is p, while the probability of its being to the left is $q = 1 - p$.

Let x be along the street such that $x = 0$ is the position of the lamppost. Since each step is of length l, the location of the man along the x axis must be of the form $x = ml$, where m is an integer, positive (to the right), negative (to the left). The question of interest is then the following: After the man has taken N steps, what is the probability of his being located at the position $x = ml$?

Figure 1.11-1 Random walk in one dimension.

More generally, the above is an illustration of the multidimensional problem of adding N vectors of equal length but of random directions and then asking for the probability that their resultant vector sum has a certain magnitude and direction. An example of this is diffusion in a gas: A given molecule travels in three dimensions a mean distance l between collisions with other molecules. How far is it likely to have gone after N collisions?

For simplicity, consider the random-walk problem to be one-dimensional, and think of a particle performing successive steps or displacements, each of length l. After a total of N such steps, each of length l, the particle is located at

$$x = ml \tag{1.11-1}$$

where m is an integer such that

$$-N \leq m \leq N \tag{1.11-2}$$

To calculate the probability $P(ml) = P(x)$ of finding the particle at the position $x = ml$ after N steps, let n_1 denote the number of steps to the right and n_2 the number of steps to the left. The total number of steps N is

$$N = n_1 + n_2 \tag{1.11-3}$$

Now, the many different ways of taking N steps, so that n_1 of them are to the right and n_2 are to the left, is given by

$$\frac{N!}{n_1!n_2!} \tag{1.11-4}$$

This is a well-known result which can be explained as follows: The N steps are likened to objects that can be arranged in different orders into two boxes, of which n_1 are in the first box, and n_2 are in the second, subject to $n_1 + n_2 = N$. The possible ways of doing this is given by the permutation.

$$N(N-1)(N-2)\cdots(1) = N! \tag{1.11-5}$$

However, such an enumeration considers each object as distinguishable. Since the n_1 objects of the first type are indistinguishable (e.g., all are right steps), all the $n_1!$ permutations of these objects among themselves lead to the same situation. The same condition applies to all the $n_2!$ permutations of objects of the second type among themselves. Hence, by dividing the total number $N!$ of arrangements of the objects by the number $n_1!n_2!$ of irrelevant permutations of objects of each type, Eq. (1.11-4) is obtained. This is the number of distinct ways in which N objects can be arranged if n_1 are in one box and n_2 are in another box.

The number of possible sequences, Eq. (1.11-4), multiplied by the probability of this sequence, $p^{n_1}q^{n_2}$, gives the probability $P(n_1)$ of taking (in a total of N steps) n_1 steps to the right and $n_2 = N - n_1$ steps to the left, in any order. Thus,

$$P(n_1) = \frac{N!}{n_1!n_2!}p^{n_1}q^{n_2} = \frac{N!}{n_1!(N-n_1)!}p^{n_1}q^{N-n_1} \tag{1.11-6}$$

This is precisely the binomial distribution, with

$$\frac{N!}{n_1!(N-n_1)!}$$

being the typical term encountered in expanding $(p+q)^N$ according to the binomial theorem

$$(p+q)^N = \sum_{n=0}^{N} \frac{N!}{n!(N-n)!}p^n q^{N-n} \tag{1.11-7}$$

Now, if it is known that a particle has proceeded n_1 steps to the right in a total of N steps, then its net displacement ml from the origin can be determined. The probability $P(ml) = P(x)$ that the particle is found at position x after N steps is given by Eq. (1.11-6), with the substitution

$$n_1 = (1/2)(N+m), \qquad n_2 = (1/2)(N-m) \tag{1.11-8}$$

Thus,

$$P(x) = \frac{N!}{n_1!n_2!} p^{n_1} q^{n_2}$$

$$= \frac{N!}{[(N+m)/2]![(N-m)/2]!} p^{(N+m)/2} (1-p)^{(N-m)/2} \tag{1.11-9}$$

which, in the case where $p = q = 1/2$ assumes the symmetrical form

$$P(x) = \frac{N!}{[(N+m)/2]![(N-m)/2]!} \left(\frac{1}{2}\right)^N \tag{1.11-10}$$

For example, if $N = 3$, the possible numbers of steps to the right are $n_1 = 0, 1, 2$, or 3; the corresponding displacements are $x = ml = -3l, -l, +l$, or $3l$; and the corresponding probabilities are $P(x) = 1/8$, 3/8, 3/8, and 1/8. Figure 1.11-2 shows the binomial distribution for the case of total number of steps $N = 20$. The envelope of these discrete values of $P(x)$ is a bell-shaped Gaussian curve. The physical significance of this is interesting, if expected. After N random steps, the probability of the particles being a distance of N steps away from the origin is very small, while the probability of its being located in the vicinity of the origin is largest.

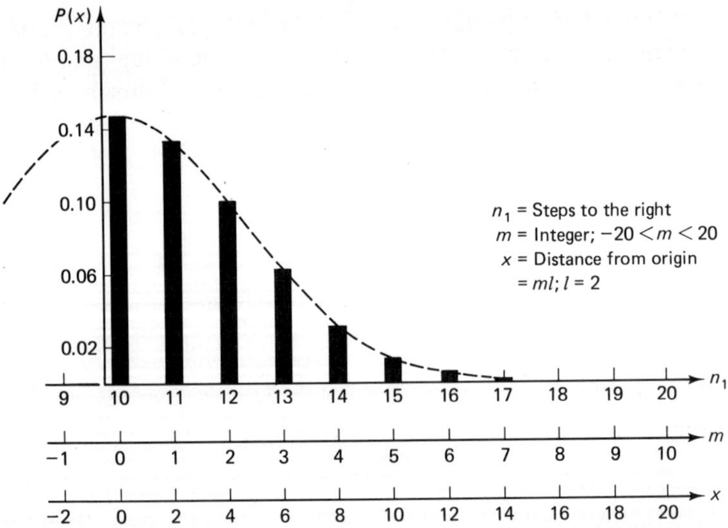

Figure 1.11-2 Random walk distribution for $p = q = \frac{1}{2}$ when $N = 20$ steps.

■ EXAMPLE 1.11-1

A plant has thirteen machines in operation. The probability that at the end of a day a machine is still in operation is 0.6. If all the machines are operating independently, what is the most probable number of machines in operation at the end of the day?

Solution. Let X = the number of machines in operation at the end of the day. With the assumption of independencc, X is binomially distributed with parameters $n = 13$ and $p = 0.6$:

$$P(x = i) = i = \binom{n}{i} p^i(1-p)^{n-i}$$

The calculations and the graph of the probability distribution are in Fig. 1.11-3 and Table 1.11-1.

The highest probability is $P(x = 8) = 0.21737$, and thus eight is the most probable number of machines that will be operating at the end of the day. The probability of that happening is 0.21737.

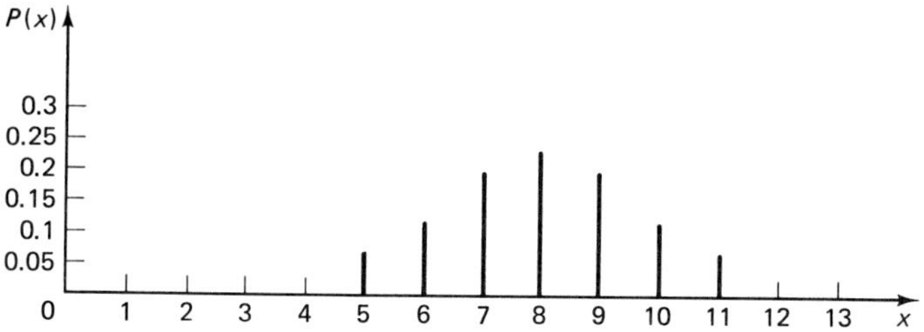

Figure 1.11-3 Probability distribution.

TABLE 1.11-1 CALCULATIONS OF PROBABILITY DISTRIBUTION

Number of machines in operation, X	Probability, $P(X = n)$	Number of machines in operation, X	Probability, $P(X = n)$
0	$\binom{13}{0}(0.6)^0(0.4)^{13} = 6.6 \times 10^{-6}$	7	$\binom{13}{7}(0.6)^7(0.4)^6 = .1932289$
1	$\binom{13}{1}(0.6)^1(0.4)^{12} = .0001287$	8	$\binom{13}{8}(0.6)^8(0.4)^5 = .21737$
2	$\binom{13}{2}(0.6)^2(0.4)^{11} = .0011544$	9	$\binom{13}{9}(0.6)^9(0.4)^4 = .1811095$
3	$\binom{13}{3}(0.6)^3(0.4)^{10} = .0063606$	10	$\binom{13}{10}(0.6)^{10}(0.4)^3 = .10868$
4	$\binom{13}{4}(0.6)^4(0.4)^9 = .0238609$	11	$\binom{13}{11}(0.6)^{11}(0.4)^2 = .04446$
5	$\binom{13}{5}(0.6)^5(0.4)^8 = .06435$	12	$\binom{13}{12}(0.6)^{12}(0.4)^1 = .0113178$
6	$\binom{13}{6}(0.6)^6(0.4)^7 = .1287$	13	$\binom{13}{13}(0.6)^{13}(0.4)^0 = .001306$

■

1.12 PERMUTATIONS AND COMBINATIONS

A major concern of statistical mechanics is the manner in which the particles of an assembly are distributed over their available quantum states. To determine this distribution, it is necessary to know the answers to several questions raised when dealing with permutations and combinations.

The first question is: What is the number of ways of arranging N distinguishable objects in an ordered sequence? The answer is

$$N(N-1)(N-2)\cdots(1) = N! \tag{1.12-1}$$

which is the number of permutations of these objects. To prove this, consider three distinguishable particles labeled a, b, and c, respectively. The three particles can be arranged in the following six ways: (a, b, c), (a, c, b), (b, a, c), (b, c, a), (c, a, b), and (c, b, a). The first particle can be chosen in three ways, namely, a or b or c. That leaves two choices for the second particle for each of the three first choices. The first two particles chosen can be arranged in 3×2 ways, namely, (a, b), (a, c), (b, a), (b, c) (c, a) and (c, b). There is only one choice left for the third particle in each of these cases. Hence, the total number of ways of arranging the three particles is $3 \times 2 \times 1$, which is three factorial and is denoted 3!. In the general case where there are N objects, the first can be chosen in N ways. Having chosen one object, there are $(N-1)$ left, which are possibilities for the next choice. Hence, the first two objects can be chosen in $N(N-1)$ different ways. After these two objects, the third can be chosen in $(N-2)$ ways, and so on. The overall result is the permutation of N objects, given by Eq. (1.12-1).

The next question is: what is the number of distinguishable arrangements of N objects among m boxes such that n_1 objects are in box 1, n_2 objects in box 2, and n_m objects in box m? The distribution is characterized by the set of occupation numbers $\{n_1, n_2, \ldots, n_m\}$, and the result is

$$C(n_1, n_2, \ldots, n_m) = \frac{N!}{n_1! n_2! \cdots n_m!} = \frac{N!}{\prod_{i=1}^{m} n_i!} \tag{1.12-2}$$

where $\sum_{i=1}^{m} n_i = N$.

To prove (1.12-2), consider first the case of N objects and two distinguishable boxes (actually a single box suffices: the inside is box 1 and the outside is box 2). Let $N = 5$, and let the objects be labeled a, b, c, d, and e. Suppose that three objects are to be selected for box 1, and two objects for box 2. One possible arrangement would be to put a, b, and c in box 1 and d and e in box 2. Having made this selection, the a, b, and c objects in box 1 can be arranged in 3! ways, namely, (a, b, c), (a, c, b), (b, a, c), (b, c, a), (c, a, b), and (c, b, a). The two objects from box 2 are then placed after the particles from box 1, and arranged in 2! ways, namely, (d, e) and (e, d). Each of the 3! ways of arranging the three objects from box 1 can be combined with each of the 2! ways of arranging the two objects from box 2 to give $3! \times 2!$ arrangements of the type for the selection of a, b, and c for box 1 and d and e for box 2. The $3! \times 2!$

arrangements obtained in this way give the 5! permutations of *a, b, c, d,* and *e*. Thus, the number of ways three distinguishable objects can be put in box 1 and two distinguishable objects in box 2 is

$$\frac{5!}{3!2!} = 10$$

In the general case of N distinguishable objects, the number of ways of selecting n_1 objects for box 1 and $n_2 = (N - n_1)$ objects for box 2 is given by

$$C(n, N - n) = \frac{N!}{n_1!n_2!} = \frac{N!}{n_1!(N - n_1)!} \qquad (1.12\text{-}3)$$

This is the number of combinations of N objects taken n_1 at a time. For the general case of N distinguishable objects placed in m distinguishable boxes—with n_1 in box 1, n_2 in box 2, . . . , n_m in box m—Eq. (1.12-2) is obtained.

Yet another question of interest in statistical mechanics is equivalent to calculating the number of ways X indistinguishable energy units can be assigned to N distinguishable particles (localized), without restriction on how many of the energy units are assigned to the distinguishable particles. To start with, if all the energy units are to be assigned to any one of the particles, a particle must be picked first. This can be done in N ways leaving $(N - 1)$ particles and X energy units. Another particle can then be chosen, which would leave the first particle without energy, or an energy unit can be assigned to the first particle. If the energies are distinguishable this choice can be made in $(N + X - 1)$ ways. The next choice can be made in $(N + X - 2)$ ways, and so on until there is either an energy unit to the last particle or a particle is without energy. Thus, the total number of ways X distinguishable energy units can be dispersed among N distinguishable particles is $N(N + X - 1)!$.

Now, for each choice of the order of the particles, there are $X!$ ways of selecting the order in which the distinguishable energies are dispersed among the particles. If the energies are indistinguishable, the order they are distributed to the particles does not matter, and $N(N + X - 1)!$ must be divided by $X!$ Also, since there are $N!$ ways of selecting N distinguishable particles, but the order of selection is of no interest, division by $N!$ is required. Hence, the number of ways of assigning X indistinguishable energy units to N distinguishable particles is

$$g = \frac{N(N + X - 1)!}{N!X!} = \frac{(N + X - 1)!}{(N - 1)!X!} \qquad (1.12\text{-}4)$$

■ EXAMPLE 1.12-1

Illustrate the concept of "thermodynamic probability" by means of putting balls into boxes.

Solution. Many problems in statistical mechanics can be expressed in the language of putting balls into boxes. For example, in tossing a coin, "heads" is box 1, and "tails" is box 2; in tossing a die, there are six "boxes." In putting letters into envelopes, the letters are the balls, and the envelopes are the boxes. In dealing cards, the cards are the balls, and the players who receive them are the boxes. In an alpha-scattering experiment, the alpha particles are the balls, and the boxes are elements of area on the detecting screen which the particles hit after they are scattered.

Consider the particular case in which there are fifteen balls and six boxes, with the number of balls to be put into the various boxes to be as follows:

Number of balls:	3	1	4	2	3	2
In box number:	1	2	3	4	5	6

The number of ways of doing this is given by Eq. (1.12-2):

$$C(n_1 n_2 \cdots) = \frac{N!}{n_1! n_2! \cdots n_m!} \tag{1.12-5}$$

This is known as the *multinomial* formula. When $m = 2$, the multinomial formula becomes the same as the binomial formula. Here,

$$C(3, 1, 4, 2, 3, 2) = \frac{15!}{3!1!4!2!3!2!} = 75{,}675{,}600$$

This is known as the *thermodynamic probability* (it is always greater than 1). The actual probability, assuming that each ball has the same probability (namely, 1/6) of being put into any one box, is

$$\frac{15!}{3! \cdot 1! \cdot 4! \cdot 2! \cdot 3! \cdot 2!} \cdot \left(\frac{1}{6}\right)^{15} = 0.0001609$$ ■

■ EXAMPLE 1.12-2

(a) A student is to take a 20-question multiple choice exam in a subject he knows absolutely nothing about. Each question has five choices. What is the probability that he will be able to get all of the answers by just guessing?

(b) A license plate consists of three letters followed by three digits such as ABC 123. What is the total number of available plates if each possible letter combination is matched with each digit combination to give a valid license plate?

(c) How many permutations are possible of the letters in the word BANANA?

Solution

(a) This is a case of sampling 20 times (with replacement) from a population size of 5. Each question has five possible answers so the total number of possible ways of choosing the answers is $5^{20} = 9.5 \times 10^{13}$. There is only one possible outcome in which to select all of the right answers, so the probability of getting all of the answers right by pure guessing is $1/(9.5 \times 10^{13}) = 10^{-14}$.

(b) First, determine how many different possibilities there are for selecting 3 letters out of 26 letters of the alphabet. This is the same as sampling three times with replacement from a population of 26, so there are $26^3 = 17{,}576$ ways of selecting the three letters. Since there are 10 possible digits, there are $10^3 = 1000$ ways of selecting the three digits. Each possible letter combination can be matched with each possible digit combination to give a valid license plate, so the total number of license plates is $17{,}576 \times 1000 = 17{,}576{,}000$.

(c) The number of permutations P of n things taken at a time, of which n_1 are alike, n_2 others are alike, n_3 others are alike, and so forth, is

$$P = \frac{n!}{n_1!n_2!n_3!\cdots} \qquad \text{with } n_1 + n_2 + n_3 + \cdots = n \tag{1.12-6}$$

In the given problem there are six letters ($n = 6$), of which two are alike, (there are two N's so that $n_1 = 2$), three others are alike (there are three A's, so that $n_2 = 3$), and one is left (there is one B, so $n_3 = 1$). Notice that $n_1 + n_3 = 2 + 3 + 1 = 6 = n$; thus,

$$P = \frac{6!}{2!3!1!} = \frac{6 \times 5 \times \not{4} \times 3!}{\not{2}! \times \not{3}! \times 1!} = 60$$

There are 60 permutations of letters in the word BANANA. ■

■ EXAMPLE 1.12-3

(a) How many different ways are there of shuffling a deck of 52 cards?

(b) Twenty students show up for class. What is the probability that they will arrive in alphabetical order?

(c) How many ways are there of selecting ten different shirts to wear during a week (seven days)?

Solution

(a) This is the case of having n objects (52 cards), and putting them in different order. There are 52 possibilities for the top card, 51 possibilities for the second card, and so on, so that altogether there are $52! = 8.07 \times 10^{67}$ ways of shuffling the deck. Generally, there are $n!$ ways of putting n objects in different order.

(b) Since there are 20 students, there are $20! = 2.43 \times 10^{18}$ different possible orders in which they can arrive. Since there is only one way of putting them in alphabetical order, the probability that they will arrive in alphabetical order is therefore $1/(2.43 \times 10^{18}) = 4.12 \times 10^{-19}$.

(c) There are ten choices for shirts to wear on Sunday, then nine choices for Monday, eight choices for Tuesday, and so on down to four choices on Saturday. So the total number of choices is

$$10 \times 9 \times 8 \times 7 \times 6 \times 5 \times 4 = 604{,}800$$

This may be rewritten as

$$10 \times 9 \times 8 \times 7 \times 6 \times 5 \times 4 = \frac{10 \times 9 \times 8 \times 7 \times 6 \times 5 \times 4 \times 3 \times 2 \times 1}{3 \times 2 \times 1}$$

$$= \frac{10!}{3!} = \frac{10!}{(10-7)!}$$

In general, there are

$$\frac{n!}{(n-j)!} \tag{1.12-7}$$

ways of selecting j objects from a population of n objects. Each way to selecting the objects is called a permutation of the objects. The formula $n!/(n-j)!$ gives the number of permutations of n things taken j at a time. ■

■ EXAMPLE 1.12-4

How many distinct 5-card hands can be drawn from a deck of 52 cards, not counting different orderings?

Solution. There are 52 possibilities for the first card drawn, then 51 possibilities for the second card, and so on. This is an example of choosing a sample of size 5 without replacement from a population of 52 cards. (Once a card is drawn, that card cannot be drawn again.) So, there are

$$\frac{52!}{(52-5)!} = \frac{52!}{47!}$$

$$= 52 \times 51 \times 50 \times 49 \times 48$$

$$= 311{,}875{,}200$$

ways of drawing the cards. Suppose that the cards drawn are

5♣ 8♦ 6♥ A♥ A♦

For all practial purposes, this hand is exactly the same as if the cards were like this:

8♦ 5♣ 6♥ A♥ A♦

In many card games, the order in which the cards are drawn does not matter; it is only which cards are drawn that is important. In counting the hands by using the formula

$$\frac{n!}{(n-j)!}$$

the two possibilities have been counted as separate hands, because they were drawn in different orders. There are, of course, a lot of different orders in which this hand could be drawn.

How many different orderings of these cards are there? Since there are 5 cards, there are $5! = 120$ different ways of arranging the cards in different orders. So, the result should be divided by that number, and the total number of distinct 5-card hands that can be drawn from a deck of 52 cards (not counting different orderings) is

$$\frac{52!}{(52-5)!5!} = 2{,}598{,}960$$

In general, to select j items without replacement from a population of n items without regard to their order, the number of possibilities is

$$\frac{n!}{(n-j)!j!}$$

This number of arrangements without regard to order is called the number of combinations. It is the number of combinations of n things taken j at a time, and is often written as

$$\binom{n}{j} = \frac{n!}{(n-j)!j!} \tag{1.12-8}$$

This expression is called the *binomial coefficient* because it occurs in the mathematical formula of the binomial theorem. ■

■ EXAMPLE 1.12-5

Calculate the probabilities of the various hands that can occur in a poker game.

Solution. The number of various 5-card hands that can occur in a poker game is calculated from

$$\binom{52}{5} = \frac{52!}{(52-5)!5!} = 2{,}598{,}960$$

Each hand is equally likely as long as the deck is well shuffled. The next step is to figure out the number of possible ways of getting each type of hand.

Royal Flush. A royal flush must contain the cards 10, J, Q, K, A, and they must all be of the same suit. Since there are four suits to choose from there are four possible royal flushes, so the probability of a royal flush is $4/2{,}598{,}960 = 1.54 \times 10^{-6}$.

Straight Flush. A straight flush must contain five cards in sequence, and they must all be of the same suit. The rules allow the ace to be the low card in a straight as well as the high card. So for each suit there are nine possible straights other than the royal straight. (The nine possible straights start with A, 2, 3, 4, 5, 6, 7, 8, or 9 as can be verified by actual counting.) There are four choices for the suit, so there are $4 \times 9 = 36$ possible straight flushes. The probability for a straight flush is $36/2{,}598{,}960 = 1.39 \times 10^{-5}$.

Four of a Kind. In drawing four cards of the same kind, there are thirteen possible choices for the type of card. This is coupled with any one of the 48 remaining cards, so altogether there are $13 \times 48 = 624$ possible hands. The probability for a four-of-a-kind hand is $624/2{,}598{,}960 = 2.4 \times 10^{-4}$.

Full House. A full house has three of one type of card and two of another. The three cards from the four suits may be chosen in $\binom{4}{3} = 4$ ways, or altogether $13 \times 4 = 52$ ways of choosing the three-of-a-kind cards. The two-of-a-kind type of cards may be chosen from the four suits in $\binom{4}{2} = 6$ ways, or altogether $12 \times 6 = 72$ ways. This means that in all, there are $52 \times 72 = 3744$ hands that are full houses, and the probability is $3744/2{,}598{,}960 = 1.44 \times 10^{-3}$.

Flush. A flush is a hand in which all five cards have the same suit. If the hand is, say, all hearts, there are five cards to choose from the thirteen hearts, so there are $\binom{13}{5} = 1287$ ways of doing this. Since a flush can be any one of the four suits, there are $4 \times 1287 = 5148$ total flushes. However, this total includes 36 straight flushes and 4 royal flushes, so the total number of plain flushes is 5108, resulting in a probability of $5108/2{,}598{,}960 = 1.97 \times 10^{-3}$.

Straight. A straight consists of any five cards in sequence starting with A, 2, 3, 4, . . . of a given suit. It can be verified that there are ten such possibilities, not counting a royal flush. For each of these possibilities, there are $4^5 = 1024$ ways of selecting the suits for the cards, so altogether there are 10,240 possible straights. Subtracting the 36 straight flushes and 4 royal flushes gives 10,200 plain straights or a probability of $10{,}200/2{,}598{,}960 = 3.92 \times 10^{-3}$.

Three of a Kind. There are thirteen ways to choose the rank of a card. Then there are four ways of selecting the cards of that rank. There are 48 choices for the fourth card in in the hand, and 44 choices for the fifth card. However, this must be divided by two because the order of drawing the fourth and fifth cards is immaterial. So altogether there are

$$13 \times 4 \times 48 \times 44 \times \tfrac{1}{2} = 54{,}912$$

possible hands with three of a kind. The probability is $54{,}912/2{,}598{,}960 = 21.1 \times 10^{-3}$.

Two Pairs. There are $\binom{13}{2} = 78$ ways of choosing the ranks of the cards that will make up the pairs. For each pair there are $\binom{4}{2} = 6$ ways of choosing the cards of that rank. As for the number of ways of choosing the remaining three cards, which can be any nonmatching cards: There are 48 choices for the third card, 44 choices for the fourth card, and 40 choices for the fifth card. This must be divided by $3! = 6$, because the order of drawing the last three cards is immaterial. Therefore, there are

$$13 \times 6 \times 48 \times 44 \times 40 \times \tfrac{1}{6} = 1{,}098{,}240$$

hands with one pair in them. The probability is $1{,}098{,}240/2{,}598{,}960 = 0.4226$.

Nothing. The above totaling 1,296,420 are the hands with something that counts. The rest, $2{,}598{,}960 - 1{,}296{,}420 = 1{,}302{,}540$, are hands with nothing. They are the majority, with $1{,}302{,}540/2{,}598{,}960 = 0.5012$ probability. The probabilities of the various hands in a poker game are summarized in Table 1.12-1.

TABLE 1.12-1 PROBABILITIES OF HANDS IN POKER

Type of hand	Number of possible hands	Probability
Royal flush	4	1.54×10^{-6}
Straight flush	36	1.39×10^{-3}
Four of a kind	624	2.40×10^{-4}
Full house	3,744	.00144
Flush	5,108	.00197
Straight	10,200	.00392
Three of a kind	54,912	.02113
Two pairs	123,552	.04754
Pair	1,098,240	.42257
Nothing	1,302,540	.50118

■

1.13 STIRLING'S APPROXIMATION

By definition, factorial $N!$ is given by

$$N! = 1 \times 2 \times 3 \times \cdots \times N \tag{1.13-1}$$

Hence,

$$\ln N! = \ln 1 + \ln 2 + \ln 3 + \cdots + \ln N \tag{1.13-2}$$

It is desirable to have an expression for $N!$ that can be differentiated and manipulated like any ordinary function. This is provided by Stirling's formula for $\ln N!$, derived below.

Consider Fig. 1.13-1 in which $\ln x$ is plotted against x; in addition, there are plotted steps, both above and below the curve, having heights equal to ln 2, ln 3, and so on. The area under the smooth curve from $x = 1$ to $x = N$ is

$$\int_1^N \ln x \, dx = N \ln N - N + 1 \tag{1.13-3}$$

where integration has been obtained by parts. This area can be approximated

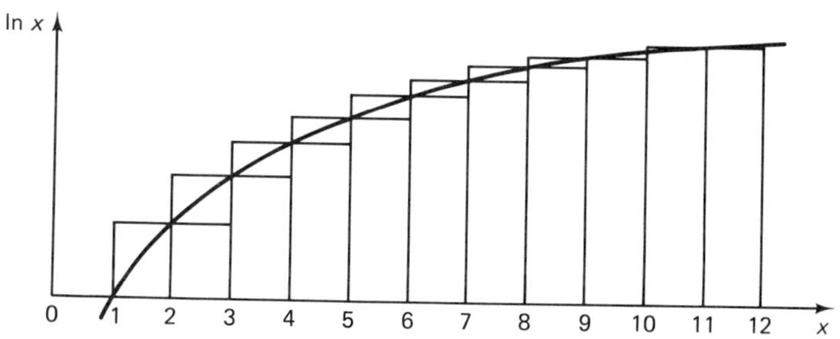

Figure 1.13-1 Approximation for area under $\ln x$.

by adding the areas of the rectangles whose tops lie above the curve: ln 2, ln 3, and so forth. This gives

$$\ln 2 + \ln 3 + \cdots + \ln N = \ln N! \tag{1.13-4}$$

On the other hand, if the areas of the rectangles whose tops lie below the curve (ln 1, ln 2, etc.) are chosen, the result is

$$\begin{aligned} \ln 1 + \ln 2 + \cdots + \ln(N-1) &= \ln(N-1)! \\ &= \ln N! - \ln N \end{aligned} \tag{1.13-5}$$

Clearly, the answer given by Eq. (1.13-3) must lie between the answers given by Eqs. (1.13-4) and (1.13-5). In other words,

$$\ln N! - \ln N < \mathrm{N} \ln N - N + 1 < \ln N!$$

Subtracting $\ln N! + N \ln N - N + 1$ and reversing signs gives

$$N \ln N - N + 1 < \ln N! < (\mathrm{N}+1) \ln N - N + 1 \tag{1.13-6}$$

If N is very large, the 1 can be neglected in Eq. (1.13-6), so that the upper and lower limits for $\ln N!$ become identical, and consequently

$$\ln N! = N \ln N - N = N \ln(N/e) \tag{1.13-7}$$

This is Stirling's approximation for $\ln N!$. It is used throughout statistical mechanics. Equation (1.13-7) is in fact quite accurate, even for small N. Table 1.13-1 shows how the difference between $\ln N!$ and $\mathrm{N} \ln N - N$ decreases as N increases.

TABLE 1.13-1 STIRLING'S APPROXIMATION

N	$\ln N!$	$(N \ln N - N)/(\ln N!)$	$\epsilon \equiv 1 - (\ln N - N)/(\ln N!)$
5	4.787	0.636	0.364
8	10.605	0.814	0.186
10	15.104	0.862	0.138
15	27.899	0.918	0.082
20	42.336	0.942	0.058
50	148.478	0.981	0.019
100	363.739	0.991	0.009
1000	5,912.128	0.999	0.001

1.14 METHOD OF UNDETERMINED MULTIPLIERS

A common problem in differential calculus involves determining the extremum of a continuous function, when one or more supplementary constraints exist.

Consider the continuous function $F(x_1, x_2, \ldots, x_n)$ of the n independent variables $x_1, x_2, \ldots, x_n$. The condition for which this function assumes a maximum or minimum is given by the expression

$$dF = \left(\frac{\partial F}{\partial x_1}\right)_{x_2,\ldots,x_n} dx_1 + \cdots + \left(\frac{\partial F}{\partial x_n}\right)_{x_1,\ldots,x_{n-1}} dx_n = 0 \tag{1.14-1}$$

Since the variables $x_1, x_2, \ldots, x_n$ are independent, the differentials $dx_1, dx_2, \ldots, dx_n$ are also independent and Eq. (1.14-1) may be satisfied only if

$$\left(\frac{\partial F}{\partial x_1}\right)_{x_2,\ldots,x_n} = \cdots = \left(\frac{\partial F}{\partial x_n}\right)_{x_1,\ldots,x_{n-1}} = 0 \tag{1.14-2}$$

This expression provides n equations from which the n unknowns ($x_1, x_2, \ldots, x_n$) may be determined.

Consider now a situation for which the variables $x_1, x_2, \ldots, x_n$ are not mutually independent. In particular, assume that these variables must satisfy two constraints given by the functional relations

$$G(x_1, x_2, \ldots, x_n) = \text{const} \tag{1.14-3}$$

$$H(x_1, x_2, \ldots, x_n) = \text{const} \tag{1.14-4}$$

It is impossible to alter any one variable, while holding all others constant. In this case *two* of the variables may no longer be altered independently, and Eqs. (1.14-2), by themselves, may no longer be used to determine the extremum.

A technique for determining an extremum, when constraints such as Eqs. (1.14-3) and (1.14-4) exist, is known as Lagrange's *method of undetermined multipliers*. The difficulty created by the constraints is the fact that there exist n unknowns $(x_1, x_2, \ldots, x_n)$ which must satisfy $n+2$ equations (1.14-2) to (1.14-4). Use of the method of undetermined multipliers to solve such a problem involves the introduction of two more unknowns, α and β. The unknown quantities α and β are each multiplied by the differential forms of the constraining equations (1.14-3) and (1.14-4), respectively. The resulting equations are

$$\alpha\, dG = \alpha \sum_{i=1}^{n} \frac{\partial G}{\partial x_i} dx_i = 0 \tag{1.14-5}$$

$$\beta\, dH = \beta \sum_{i=1}^{n} \frac{\partial H}{\partial x_i} dx_i = 0 \tag{1.14-6}$$

These equations are then added to the orginal equation (1.14-1) to obtain

$$\sum_{i=1}^{n} \left(\frac{\partial F}{\partial x_i} + \alpha \frac{\partial G}{\partial x_i} + \beta \frac{\partial H}{\partial x_i}\right) dx_i = 0 \tag{1.14-7}$$

Note that, since G and H are constant, the set of $x_1, x_2, \ldots, x_n$ that maximizes F must be identical to the set that maximizes $F + \alpha G + \beta H$.

Now, if the two variables that may not be altered arbitrarily are chosen to be x_1 and x_2, any changes in these variables must depend upon variations in

$x_3, \ldots, x_n$. That is, the differentials dx_1 and dx_2 become dependent. However, it is possible to select values of α and β such that

$$\frac{\partial F}{\partial x_1} + \alpha \frac{\partial G}{\partial x_1} + \beta \frac{\partial H}{\partial x_1} = 0 \qquad (1.14\text{-}8)$$

and

$$\frac{\partial F}{\partial x_2} + \alpha \frac{\partial G}{\partial x_2} + \beta \frac{\partial H}{\partial x_2} = 0 \qquad (1.14\text{-}9)$$

Having thus chosen the values of α and β, Eq. (1.14-7) may then be satisfied for arbitrary variations dx_i, where $3 \leq i \leq n$, by setting the coefficients equal to zero. Therefore,

$$\frac{\partial F}{\partial x_i} + \alpha \frac{\partial G}{\partial x_i} + \beta \frac{\partial H}{\partial x_i} = 0 \qquad (3 \leq i \leq n) \qquad (1.14\text{-}10)$$

But since any two of the variables may be selected as dependent and since Eqs. (1.14-8) and (1.14-9) are of precisely the same form as (1.14-10), the following equation must apply in general:

$$\frac{\partial F}{\partial x_i} + \alpha \frac{\partial G}{\partial x_i} + \beta \frac{\partial H}{\partial x_i} = 0 \qquad (1 \leq i \leq n) \qquad (1.14\text{-}11)$$

Solving for the n + 2 unknowns now becomes a matter of solving the n equations (1.14-11) for $x_1, \ldots, x_n$ and then substituting into Eqs. (1.14-3) and (1.14-4) to determine the required form of the multipliers α and β. The above discussion has allowed for the existence of two constraints. However, the method is readily extended to problems involving any number of constraints, it being necessary only to introduce an equivalent number of multipliers.

■ EXAMPLE 1.14-1

In statistical mechanics, the following function is encountered:

$$W = \frac{N!}{n_1! n_2! \cdots} = \frac{N!}{\prod_i n_i!} \qquad (1.14\text{-}12)$$

Maximize this function with respect to n_i, subject to the constraints

$$\sum_i n_i = N \qquad (1.14\text{-}13)$$

$$\sum_i n_i \, \varepsilon_i = E \qquad (1.14\text{-}14)$$

Solution. For convenience, it is easier to maximize ln W instead of W. Therefore, take the logarithm of (1.14-12):

$$\ln W = \ln N! - \sum_i \ln n_i! \qquad (1.14\text{-}15)$$

Setting the variation of (1.14-15) equal to zero gives

$$\delta \ln W = -\sum_i \delta \ln n_i! = 0 \tag{1.14-16}$$

or, using Stirling's approximation,

$$-\sum_i \delta(n_i \ln n_i - n_i) = 0 \tag{1.14-17}$$

This expands and simplifies to

$$-\sum_i \ln n_i \, \delta n_i = 0 \tag{1.14-18}$$

Similarly, Eqs. (1.14-13) and (1.14-14) give

$$\sum_i \delta n_i = 0 \tag{1.14-19}$$

$$\sum_i \varepsilon_i \, \delta n_i = 0 \tag{1.14-20}$$

Equations (1.14-18), (1.14-19), and (1.14-20) are now solved by Lagrange's method of undetermined multipliers. Accordingly, (1.14-19) is multiplied by α and (1.14-20) by β, and subtracted from (1.14-18) to give

$$\sum_i (-\ln n_i - \alpha - \beta\varepsilon_i)\delta n_i = 0 \tag{1.14-21}$$

or

$$n_i = e^{-\alpha - \beta\varepsilon_i} \tag{1.14-22}$$

This is known as the Boltzmann or *canonical ensemble distribution*. The multipliers α and β will be determined in the next chapter. ■

1.15 APPROXIMATION OF SUMS BY INTEGRATION

In many applications of statistical mechanics, it is necessary to evaluate summations of certain functions over all possible energy levels available to a molecular system. In some of these applications, the levels are so closely spaced as to allow for the assumption of a continuous distribution of the function. For example, statistical mechanics gives the classical partition function for an ideal gas composed of N monatomic molecules as

$$Z_{\text{class}} = \frac{1}{h^{3N}H!} \int \cdots \int \exp\left(\sum_i \frac{p_i^2}{2mkT}\right) dp_1 \cdots dp_{3N}\, dr_1 \cdots dr_{3N} \tag{1.15-1}$$

where the r's are the space coordinates and the p's are the conjugate momenta.

Equation 1.15-1 may be rewritten as

$$Z_{\text{class}} = \frac{1}{h^{3N}N!}\left[\int_{-\infty}^{+\infty} \exp\left(-\frac{p^2}{2mkT}\right) dp\right]^{3N} \left(\int_v dr_1\, dr_2\, dr_3\right)^N = \frac{z^N}{N!} \qquad (1.15\text{-}2)$$

where

$$Z_{\text{class}} = \frac{1}{h^3}\left[\int_{-\infty}^{+\infty} \exp\left(-\frac{p^2}{2mkT}\right) dp\right]^3 \left(\int_v dr_1\, dr_2\, dr_3\right)^N \qquad (1.15\text{-}3)$$

is the molecular partition function. It is the partition function for a single molecule, and is sometimes labeled z_1.

The integration in Eq. (1.15-3) gives

$$Z = \frac{V}{h^3}\left[\int_{-\infty}^{+\infty} \exp\left(-\frac{p_x^2}{2mkT}\right) dp_2\right]\left[\int_{-\infty}^{+\infty} \exp\left(-\frac{p_y^2}{2mkT}\right) dp_y\right]$$
$$\times\left[\int_{-\infty}^{+\infty} \exp\left(-\frac{p_z^2}{2mkT}\right) dp_z\right] = \left(\frac{2\pi mkT}{h^2}\right)^{3/2} V \qquad (1.15\text{-}4)$$

where $\int \exp(-\alpha x^2)\, dx\, dx = (\pi/\alpha)^{1/2}$ as previously established. Thus,

$$Z_{\text{cl}} = \frac{1}{N!}\left(\frac{2\pi mkT}{h^2}\right)^{3N/2} V^N \qquad (1.15\text{-}5)$$

for the classical case.

Now, quantum mechanics gives the translational energies of a particle in a cubic box of volume V as

$$\varepsilon = \frac{h^2}{8mV^{2/3}}(n_x^2 + n_y^2 + n_z^2) \qquad (1.15\text{-}6)$$

where each quantum number n_x, n_y, n_z goes from 1 to ∞, independently of the others. The translational partition function in quantum statistical mechanics is

$$Z = \frac{1}{N!}\left\{\sum_{n_x}\sum_{n_y}\sum_{n_z} \exp\left[-\frac{h^2(n_x^2 + n_y^2 + n_z^2)}{8mV^{2/3}\, kT}\right]\right\}^N = \frac{z^N}{N!} \qquad (1.15\text{-}7)$$

where

$$z = \sum_{n_x}\sum_{n_y}\sum_{n_z} \exp\left[-\frac{h^2}{8mV^{2/3}\, kT}(n_x^2 + n_y^2 + n_z^2)\right] \qquad (1.15\text{-}8)$$

is the molecular partition function.

The sums in (1.15-8) can be evaluated exactly, but not profitably. Instead, a replacement technique that is simple yet accurate is used. Each sum is replaced by an integral. This procedure is justified if the terms in the sum change very little from one term to the next, as shown below.

Consider $\sum_{n=0}^{\infty} f(n) \approx f(n+1)$ for all n. Then,

$$\sum_{n=0}^{\infty} f(n) = f(0) + f(1) + \cdots = f(0)\int_0^1 dn + f(1)\int_1^2 dn + \cdots$$
$$= \int_0^1 f(0)\, dn + \int_1^2 f(1)\, dn + \cdots \qquad (1.15\text{-}9)$$

Since the function $f(n)$ varies very slowly with n, let $f(0) \approx f(n)$ for n in the range 0 to 1 and $f(1) \approx f(n)$ for n in the range 1 to 2, and so on. Therefore,

$$\sum_{n=0}^{\infty} f(n) \approx \int_0^1 f(n)\,dn + \int_1^2 f(n)\,dn + \cdots = \int_0^{\infty} f(n)\,dn \tag{1.15-10}$$

Because $f(n)$ varies slowly, $f(0) \approx f(1) \approx f(2) \approx \cdots$. Hence, many terms contribute substantially to the sum, and the relative contribution of any one term is small and can be neglected. Thus, the sum can start at $n = 1$ instead of $n = 0$. Likewise, the integration can start at $n = 1$ instead of $n = 0$. Therefore,

$$\sum_{n=0}^{\infty} f(n) \approx \sum_{n=1}^{\infty} f(n) \approx \int_0^{\infty} f(n)\,dn \approx \int_1^{\infty} f(n)\,dn \tag{1.15-11}$$

provided $|[f(n+1) - f(n)]/f(n)| << 1$. Whether the lower limit is taken as 0 or 1 is simply a matter of which integral is easier to evaluate.

To check that the approximation (1.15-11) is applicable to the sums in (1.15-8), consider the n_x sum

$$\frac{f(n_x+1) - f(n_x)}{f(n_x)} = \frac{e^{\beta\varepsilon(n_x+1)} - e^{-\beta\varepsilon(n_x)}}{e^{-\beta\varepsilon(n_x)}} = e^{-\Delta\varepsilon/kT} - 1 \tag{1.15-12}$$

where $\varepsilon(n_x) = (h^2/8ma^2)n_x^2$ and $\Delta\varepsilon$ is the spacing between adjacent levels of x translational energy. The translational levels are extremely close together, and for typical conditions, $\Delta\varepsilon/kT \approx 10^{-9}$. Therefore,

$$e^{-\Delta\varepsilon/kT} - 1 = e^{-10^{-9}} - 1 \approx -10^{-9} \tag{1.15-13}$$

where the Taylor series $e^x = 1 + x + \cdots$ has been used. Thus, f decreases by 1 part in 10^9 for each increase in 1 in n_x, and replacement of the sum by an integral is quite justified. More generally, the molecular-partition function sum $\Sigma e^{-\varepsilon_s/kT}$ can be replaced by an integral whenever $\Delta\varepsilon_s >> kT$, where $\Delta\varepsilon_s$ is the spacing between adjacent levels of the kind of energy being considered (translational, rotational, etc.).

Applying the above to Eq. (1.15-7), in the high-temperature limit of large n_x, n_y, n_z, the summations may be replaced by integrals:

$$Z_{\text{qu}} = \frac{1}{N!}\left\{\int_0^{\infty}\int\int \exp\left[-\frac{h^2(n_x^2+n_y^2+n_z^2)}{8mV^{2/3}\,kT}\right] dn_x\,dn_y\,dn_z\right\}^N$$

$$= \frac{1}{N!}\left[\int_0^{\infty} \exp\left(-\frac{h^2n_x^2}{8mV^{2/3}\,kT}\right) dn_x\right]^{3N} = \frac{1}{N!}\left(\frac{2\pi mkT}{h^2}\right)^{3N/2} V^N \tag{1.15-14}$$

which agrees with the classical result (1.15-5).

In Fig. 1.15-1, the quantum-molecular partition function z is the sum of the rectangular areas shown [each rectangle being of height $\exp(-\varepsilon_i/kT)$ and of unit width]. Here i is the quantum state; for example, any combination of values of n_x, n_y, and n_z for the ideal monatomic gas. For a gas at normal temperatures a very large number of translational states are occupied, usually $\langle n_x\rangle \sim 10^8$. The area corresponding to the partition function may then be taken as that beneath the dotted curve shown; that is, the summation in z is replaced by integration.

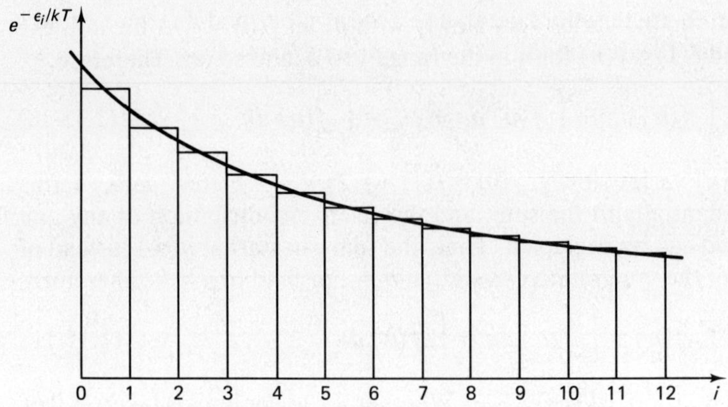

Figure 1.15-1 Approximation of summation by integration.

■ EXAMPLE 1.15-1

Evaluate the function

$$f(j) = e^{-mj} \qquad j = 0, 1, 2, \ldots \tag{1.15-15}$$

using the Euler–Maclaurian summation formula.

Solution. The Euler–Maclaurian summation formula is

$$\sum_{n=a}^{\infty} f(n) = \int_a^{\infty} f(n)\,dn + \frac{f(a)}{2} - \frac{f'(a)}{12} + \frac{f'''(a)}{720} - \frac{f^{(V)}a}{30{,}240} + \cdots$$

where $f(a)$ is $f(n)$ evaluated at $n = a$, $f'(a)$ is the first derivative evaluated at $n = a, \ldots, f^{(V)}$ is the fifth derivative of $f(n)$ evaluated at $n = a$.

First, $\int_a^\infty f(n)\,dn$ is evaluated. Letting $a = 0$, this gives

$$\int_0^{\infty} e^{-mx}\,dx = \frac{1}{m} \tag{1.15-16}$$

Then $f(0)$ and the successive derivatives at $n = 0$ are evaluated:

$$\begin{aligned} f(0) &= l \\ f'(0) &= -m \\ f''(0) &= m^2 \\ f'''(0) &= -m^2 \end{aligned} \tag{1.15-17}$$

Thus,

$$\sum_{j=0}^{\infty} e^{-mj} = \frac{1}{m} + \frac{1}{2} + \frac{m}{12} - \frac{m^3}{720} + \cdots \tag{1.15-18}$$

■

■ EXAMPLE 1.15-2

In statistical thermodynamics, the rotational partition function is given by

$$F(x) \equiv F(J) = (2J + 1)e^{-J(J+1)\Theta_{\text{rot}}/T} \tag{1.15-19}$$

where Θ_{rot} is the characteristic temperature for rotation, and J is zero or an integer. Show that the fractional error in replacing sums by integration is approximately

$$\frac{\Theta_{\text{rot}}}{2T} \tag{1.15-20}$$

Solution. Let the sum

$$\sum_{x=x_0}^{x=x_n} F(x) = F(x_0) + F(x_0+1) + F(x_0+2) + F(x_0+3) + \cdots + F(x_n)$$

be taken for all integer values of x from x_0 to x_n. Examples of such sums are the partition functions for rotation, vibration, and translation. The sums are to be replaced by integrals, that is,

$$\sum_{x=x_0}^{x=x} F(x) \approx \int_{x_0}^{x_n} F(x)\,dx \tag{1.15-21}$$

This is the idea behind the Euler–Maclaurian summation formula.

Consider now Fig. 1.15-2, which shows the curve for $F(J)$:

The shaded area (1) is $F(0) \times 1 = F(0)$, numerically.
The shaded area (2) is $F(1) \times (2-1) = F(1)$, numerically.
The shaded area (3) is $F(2) \times (3-2) = F(2)$, numerically.

Hence, the total shaded area is the sum

$$F(0) + F(1) + F(2) + \cdots \equiv \sum_{x=x_0}^{x_n} F(x) \tag{1.15-22}$$

while the area under the curve between $x = x_0$ and $x = x_n$ is the integral

$$\int_{x_0}^{x_n} F(x)\,dx \tag{1.15-23}$$

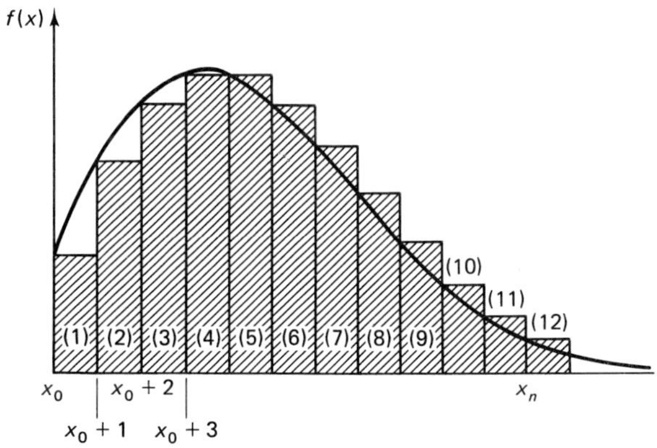

Figure 1.15-2 Area under a curve by approximation.

The difference

$$\sum_{x=x_0}^{x_n} F(x) - \int_0^{x_n} F(x)\,dx \tag{1.15-24}$$

is the sum of the areas bounded by the curve and the steps between x_0 and x_n in the figure. As a first approximation, these areas may be taken as triangles, the area of each triangle being $\frac{1}{2}$ (base × height) $= \frac{1}{2}[F(x+1) - F(x)]$. The sum of the areas of the triangles gives, altogether,

$$\tfrac{1}{2}[F(x_0) + F(x_n)] \tag{1.15-25}$$

and so, the fractional error when replacing summation by integration is

$$\frac{\frac{1}{2}[F(x_0) + F(x_n)]}{\displaystyle\int_{x_0}^{x_n} F(x)\,dx} \tag{1.15-26}$$

Applying this to the rotational partition function (with energy levels J and degeneracy $2J+1$), the following integral is first evaluated:

$$\int_0^{\infty} (2J+1)e^{-(\Theta_{\mathrm{rot}}/T)J(J+1)}\,dJ = \int_0^{\infty} e^{-(\Theta_{\mathrm{rot}}/T)w}\,dw \tag{1.15-27}$$

where the change of variable $w = J(J+1) = J^2 + J$, $dw = (2J+1)\,dJ$ has been made. Since $\int e^{-bw}\,dw = b^{-1}e^{-bw}$, this gives

$$\int_0^{\infty} F(x)\,dx = \frac{T}{\Theta_{\mathrm{rot}}} \tag{1.15-28}$$

Now, for $x_0 = J_0 = 0$, $F(0) = e^0 = 1$, and for $x = \infty$, $F(\infty) = 0$. The fractional error is thus

$$\frac{1}{2}\left(\frac{1}{T/\Theta_{\mathrm{rot}}}\right) = \frac{\Theta_{\mathrm{rot}}}{2T} \tag{1.15-29}$$

In most cases, this is less than 0.3%, when $\Theta_{\mathrm{rot}} < 2$ K. ■

1.16 SUMMARY

The distinct or mutually exclusive results of an experiment are called simple events. These events are represented as single points in sample space. The probability of an event i is

$$P_i = \lim_{N\to\infty} \frac{n_i(N)}{N} \tag{1.16-1}$$

where n_i/N is the fraction of the N experiments that result in event i. The ergodic hypothesis enables events to be interpreted in two different ways: A single physical system can be chosen, and the same experiment performed on it N times over a period of time. N identical systems (an ensemble) can be considered and the same experiment performed on each member of the ensemble. The fundamental assumption is that either method yields the same result. In other words, the time average of a dynamical quantity x is equal to the ensemble average or expectation value:

$$\bar{x} = \lim_{t \to \infty} x_i(t) = \langle x \rangle \lim_{N \to \infty} x_i(N) \tag{1.16-2}$$

The probability of a compound event A is

$$P(A) = \sum_{i \subset A} P_i \tag{1.16-3}$$

where the sum is over all the points in the event, indicated by $i \subset A$, which reads "all the i's contained in A."

The union of B and C, denoted by $B \cup C$, is the compound event consisting of all points that are either in B or in C (or in both B and C). The probability $P(B \cup C)$ is given by

$$P(B \cup C) = P(B) + P(C) \tag{1.16-4}$$

for disjoint events (having no points in common), and by

$$P(B \cup C) = P(B) + P(C) - P(BC) \tag{1.16-5}$$

for events that are not disjoint, that is, events that have points in common. The intersection of two events is denoted by $B \cap C$ or BC.

Conditional probability is defined by

$$P(X|Y) = \frac{P(XY)}{P(Y)} \tag{1.16-6}$$

where $P(X|Y)$ is the probability that event X occurs given the fact that event Y occurs. X and Y are independent if $P(XY) = P(X)P(Y)$, that is, if $P(X|Y) = P(X)$.

The probability density function for a discrete random variable is given by

$$f(a) = P(x = a) \tag{1.16-7}$$

This is just the probability that the random variable will equal the number a. For continuous events, the distribution function $f(x)$ is related to the occurrence of events by

$$P(x, dx) = f(x)\, dx \tag{1.16-8}$$

This is the probability of occurrence of the event in the infinitesimal region dx about the value x.

The average or expected value of a property $g(x)$ is

$$\overline{g(x)} = \langle g(x) \rangle = \int g(x) f(x)\, dx \tag{1.16-9}$$

The variance is

$$\text{var}(x) = \overline{(x \cdot \langle x \rangle)^2} = \langle x^2 \rangle \cdot (\langle x \rangle)^2 \tag{1.16-10}$$

Its square root, or standard deviation, is

$$\sigma(x) = (\langle x^2\rangle \cdot \langle x\rangle^2)^{1/2} \tag{1.16-11}$$

The most probable value (mode) is given by

$$\left[\frac{df(x)}{dx}\right]_{x_{\text{mode}}} = 0 \tag{1.16-12}$$

The root mean square (rms) value is

$$(\langle x^2\rangle)^{1/2} = \left[\int x^2 f(x)\, dx\right]^{1/2} \tag{1.16-13}$$

A Gaussian distribution function (centered at the origin) is

$$f(x) = Ae^{-\alpha x^2} \tag{1.16-14}$$

The evaluation of various integrals involving the Gaussian function results in

$$I_n(\alpha) = \int_0^\infty x^n e^{-\alpha x^2}\, dx \tag{1.16-15}$$

$$I_0 = \frac{1}{2}\sqrt{\frac{\pi}{\alpha}} \qquad I_3 = \frac{1}{2\alpha^2} \tag{1.16-16}$$

$$I_1 = \frac{1}{2\alpha} \qquad I_4 = \frac{3}{8}\sqrt{\frac{\pi}{\alpha^5}} \tag{1.16-17}$$

$$I_2 = \frac{1}{4}\sqrt{\frac{\pi}{\alpha^3}} \qquad I_5 = \frac{1}{\alpha^3} \tag{1.16-18}$$

The binomial distribution of a random variable x for the number of successes in n independent trials, with each trial having a probability P of success, is

$$\binom{n}{i} p^i (1-p)^{n-i} = \frac{n!}{i!(n-i)!} p^i (1-p)^{n-i} \tag{1.16-19}$$

The binomial theorem is

$$(x+y)^n - \sum_{i=0}^{n} \binom{n}{i} x^i y^{n-i} \tag{1.16-20}$$

The permutation (ordered arrangement) of N distinguishable objects is

$$N(N-1)(N-2)\cdots(1) = N! \tag{1.16-21}$$

The number of ways n objects can be selected from a set of N different objects, which is equivalent to putting n objects in box 1 and $N-n$ objects in box 2, is called the number of combinations of N objects taken n at a time, and is

$$\binom{N}{n} = \frac{N!}{n!(N-n)!} \tag{1.16-22}$$

The number of arrangements of N objects among m boxes such that n

objects are in box 1, n_2 objects in box 2, and n_m objects in box m is given by

$$C(n_1, n_2, \ldots, n_m) = \frac{N!}{n_1!n_2! \cdots n_m!} \tag{1.16-23}$$

This is known as the multinomial formula. When $m = 2$, the multinomial formula becomes the binomial formula.

The number of ways X indistinguishable objects (e.g., apples, energy units, etc.) can be distributed among N distinguishable boxes, without restriction on how many of the indistinguishable objects are in the distinguishable boxes, is

$$g = \frac{(N + x - 1)!}{(N - 1)!x!} \tag{1.16-24}$$

Stirling's approximation for $\ln N!$ is

$$\ln N! = N \ln N - N = N \ln(N/e) \tag{1.16-25}$$

It is used throughout statistical mechanics in Lagrange's method of undetermined multipliers. This is a method for maximizing the function $F(x_1\, x_2, \ldots, x_n)$ of the n independent variables x_1, x_2, x_n, subject to the constraints

$$G(x_1, x_2, \ldots, x_n) = \text{const} \tag{1.16-26}$$

$$H(x_1, x_2, \ldots, x_n) = \text{const} \tag{1.16-27}$$

Taking the variation of F and adding it to the variations of G and H, respectively multiplied by α and β, results in the n equations

$$\frac{\partial F}{\partial x_i} + \alpha \frac{\partial G}{\partial x_i} + \beta \frac{\partial H}{\partial x_i} = 0 \tag{1.16-28}$$

for the unknown x_i's. Substituting into the constraining equations then determines the multipliers α and β.

SYMBOLS FOR CHAPTER 1

A^c	A complement: probability that event A will not occur
$A \cap B$, AB	A intersect B
$A \cup B$	A union B
F_i	Frequency of event i
$F(x), f(x)$	Probability distribution function and probability density function normalized to N and to 1, respectively.
H, h	Heads
$i \subset A$	All the i's contained in event A
m	Mass
N	Total number of experiments; total number of systems (molecules, atoms, particles, objects, . . .)
n_i	Number of times event i occurs
P_i	Probability of event i
$P(X \mid Y)$	Probability that event x occurs given that event y occurred
s	Probability space
T, t	Tails; temperature, time
u	Velocity or speed in x direction

PROBLEMS

1.1. A random variable is a variable that takes on a specified value when a particular random event occurs. Random variables that can only have isolated values are called discrete. Those that can assume any number in a particular range are called continuous. Give and discuss examples of discrete and continuous random variables.

1.2. **(a)** For the experiment of drawing a card from a deck of cards, what are the simple events, and how many points are there in sample space?

(b) Using $P(X) = \sum_{i \subset X} P_i$, what are the probabilities of the compound events A = hearts, B = number three card, C = one-eyed jack?

(c) Which of the events A, B, C are disjoint?

1.3. Using $P(X \cup Y) = P(X) + P(Y) - P(XY)$, what is the probability that a card drawn from a deck of cards is either a number three or a one-eyed jack? What is the probability that a card drawn is either a heart or a number three card?

1.4. **(a)** If V is the set of vowels and C is the set of consonants, what is the intersect $V \cap C$?

(b) If A is the set of hands with five cards in sequence and B is the set of hands with five cards of the same suit, what is $A \cap B$?

(c) If A is the event of getting a head on the first toss of a coin and B is the event of getting a head on the second toss, what is $A \cap B$?

1.5. Using the conditional probability $P(X|Y) = P(XY)/P(Y)$, what is the probability that a heart is drawn if it is known that it is a number three card? What is the probability that a card is a heart if it is a one-eyed jack?

1.6. **(a)** Given a probability density function $f(x)$, the probability is then $P(x,dx) = f(x)\,dx$. What is the probability at a given x? What is the probability for which $x_1 \leq x \leq x_2$?

(b) What is the expected value of x when $f(x) = 2x$, with $0 < x < 1$?

(c) What is the expected value of x when $f(x) = 1/2\sqrt{x}$, with $0 < x < 1$?

1.7. Calculate the expected value for the continuous random variable x with probability density function

$$f(x) = \frac{x^2}{2} e^{-x} \qquad 0 < x < \infty$$

1.8. The probability that a monatomic gas atom such as helium has a velocity with an x component at x in the range dv_x is $P(v_x, dv_x) = f(v_x)\,dv_x$. The distribution function $f(v_x)$ is of the Gaussian type

$$f(v_x) = \left(\frac{\alpha}{\pi}\right)^{1/2} \exp(-\alpha v_x^2)$$

Why is there a factor $(\alpha/\pi)^{1/2}$?

1.9. A typewriter has 26 keys, one for each letter of the English alphabet. If a typist types 26 letters completely at random and then starts doing another complete set in 30 s, how long would it take the typist to type the 26 letters of the alphabet in their correct sequence?

1.10. **(a)** Show that the mean or expected value $\langle k \rangle$ of a die rolled a large number of times is

$$\sum_{k=1}^{6} kP(k) = 3.5$$

(b) Show that the average value $\langle k \rangle$ of a pair of dice rolled a large number of times is

$$\sum_{k=2}^{12} kP(k)$$

1.11. A student takes a 20-question multiple choice exam. Each question has four possible answers, so the probability is 0.25 that the answer can be arrived at by guessing.

(a) What is the probability of getting at least 10 questions right by guessing?

(b) What is the chance of getting less than three answers correct?

1.12. In a 100-question true/false exam, what is the probability of guessing randomly at least 75 answers correctly?

1.13. For the continuous distribution function

$$f(x) = \sqrt{\frac{\beta}{\pi}}e^{-\beta x^2} \qquad -\infty \le x \le \infty$$

where β is a constant, determine **(a)** the average value of x, **(b)** the average of x^2, and **(c)** the average of the magnitude of x.

1.14. Show that in any single roll of a die, the value that results can deviate considerably from the average value.

1.15. Write a computer program for the calculation of $n!$.

1.16. Write a computer program for a binomial random variable that reads in the number of coin tosses (n), the number of heads (h), and then prints a table showing the probability of getting h heads, for all values of k from $h = 0$ to $h = n$.

1.17. Find the variance and the standard deviation for the Gaussian distribution

$$f(x) = \sqrt{\frac{\beta}{\pi}}e^{-\beta x^2}$$

and discuss the effect of β on $\sigma(x)$.

1.18. The fraction of molecules of a gas having free paths (distances traveled between collisions) x or greater, is given by

$$\frac{n(x)}{n_0} = e^{-\alpha x}$$

where α is a constant.

(a) Show that the mean free path, λ, is $1/\alpha$.

(b) What fraction of molecules has free paths greater than 2λ?

1.19. The probability density for the speed s of a car on a highway is given by

$$f(s) = Ase^{-s/s_o}, \; o \le s \le \infty$$

where A and s_o are constants.

(a) Determine A in terms of s_o.

(b) Determine the average speed.

(c) Determine the most probable speed.

1.20. **(a)** How many ways are there of selecting 10 T-shirts for daily wear (without repetition) in a 7-day week?

(b) While attending an 8-horse race, what is the probability of guessing the order of the top three finishers without knowing anything about the horses involved?

(c) How many 5-card hands can be dealt from a deck of 52 cards, counting the different orderings?

(d) Same as **(c)**, but not counting the different orderings.

1.21. An artist is painting a still life that consists of three apples, five oranges, two bananas, and one plum, all arranged in a row. How many different ways can these fruits be arranged?

1.22. **(a)** What is the probability of getting an 8 on a pair of dice when a 5 has been obtained on the first roll?

(b) What is the probability of getting a royal flush when A♥, K♥ have been drawn on the first two cards?

1.23. The binomial distribution of random variable is represented by

$$P(x = i) = \binom{n}{i} p^i (1-p)^{n-i}$$

Plot the density for $n = 10$ and **(a)** $p = 0.5$, **(b)** $p = 0.4$.

1.24. **(a)** How many permutations are there of the word "algebra"? Note that two of the letters are the same.

(b) How many baseball teams of nine members can be chosen from among twelve players, without regard to the position assigned to each member?

1.25. An industrial process manufactures items that have a 1% probability of being defective. If a random sample of 100 of these is shipped out, what is the probability that the shipment contains no defectives?

1.26. A package in the mail can either be delivered, lost, or damaged while being delivered. If the probability of delivery is 0.7, the probability of loss is 0.2, and the probability of damage is 0.1, what is the probability that out of ten packages sent, six arrive safely, two are lost, and two are damaged?

1.27. The probability of h heads in n tosses a coin is given by

$$\frac{n!}{h!(n-h)!} p^h (1-p)^{n-h}$$

Design an experiment to test whether a coin is honest or not.

1.28. The probability that a monatomic gas atom has a speed between v and $v + dv$ is given by

$$P(v)\, dv = Kv^2 \exp(-mv^2/kT)\, dv$$

where K is a constant dependent on temperature. Determine: **(a)** the value of K, **(b)** the average speed, **(c)** the rms speed, and **(d)** the most probable speed.

1.29. The speed distribution of a gas is given by

$$\frac{dN(v)}{N} = \left(\frac{2}{\pi}\right)^{1/2} \left(\frac{m}{kT}\right)^{3/2} v^2 e^{-mv^2/2kT}\, dv$$

What fraction of oxygen molecules at 300 K have a speed within ±1% of the rms speed? The mass of the oxygen molecule is $m = 5.31 \times 10^{-26}$ kg.

1.30. What is the percentage error entailed in Stirling's approximation formula for $n = 60$?

1.31. Classical *random walk* is described by the following: A particle suspended in a fluid is acted on by random irregular forces. Show that the average value of the square of the particle's absolute displacement in a time interval t is proportional to t. (Assume that each time a force acts on the particle it moves a distance equal to its mean free path L in a random direction.)

1.32. A molecule has an average speed of 454 m/s and moves randomly a distance equal to its mean free path every time it undergoes a collision. What is its rms displacement after 10 s if its mean free path is 7.4×10^{-8} m (−740 Å)?

Chapter 2

Mechanics: Quantum and Statistical

The aim of statistical mechanics (also known as statistical thermodynamics) is to deduce the macroscopic properties of matter from the properties of the molecules composing a system. Typical macroscopic properties are entropy, internal energy, heat capacity, surface tension, dielectric constant, viscosity, electrical conductivity, and chemical reaction rate. Molecular properties include molecular masses, molecular geometries, and intermolecular forces. Because of the large number of molecules in a macroscopic system, statistical methods are used instead of attempting to consider the motion of each molecule in the system.

Statistical mechanics treats matter from a microscopic viewpoint. Classical thermodynamics (commonly called thermodynamics), on the other hand, is concerned with the equilibrium macroscopic properties of matter. It deals primarily with the description of physical and chemical phenomena, with the "hows" of the behavior of matter and energy. The underlying mechanisms—the "whys"—are beyond its scope. And, indeed, one of its strong points is precisely such independence of detailed models of the structure of matter.

Statistical thermodynamics, by comparison, probes more deeply than thermodynamics; it seeks to explain, as well as describe. Rather than postulating (on the basis of experiment) the laws governing macroscopic matter, it attempts to deduce them from the known facts of atomic and molecular structures. The mechanics of atomic systems, supplemented by the laws of probability, provide the fundamental basis for statistical mechanics. The word statistical is appropriate, since it is the average over the behavior of many molecules that is used to calculate the behavior of macroscopic systems.

This chapter gives an overview of statistical mechanics and its uses for predicting thermodynamic properties but does not dwell on all of its refinements or applications. Instead, the development proceeds to a familiar and plausible result—Boltzmann's distribution or canonical ensemble distribution—and obtains some working equations as quickly as possible. Quantum theory is used from the outset, but no more than is necessary for the understanding of statistical mechanics. The more detailed presentation of quantum effects occurs in later chapters.

Present discussion is restricted to noninteracting particles. Moreover, the term "particle" is used in a generic sense to represent any dynamical entity that has a well-defined energy ϵ. Thus, a particle may be an atom, a molecule, an electron, a sound wave in a solid, or a light wave. For the case of non-interacting particles, the total energy E of a system is very simply the sum of the energies of the particles.

2.1 QUANTUM STATES AND COMPLEXIONS

Two works written in the first years of the twentieth century have determined the development of the subject of thermal physics. Planck, in Berlin, in 1901 wrote a historic article on the distribution of energy in thermal radiation (the radiation emitted from a hot body). His ideas led to the theory of quanta, from which quantum mechanics developed. In the same year, Gibbs, in New Haven, wrote an extraordinarily perceptive treatise entitled *Elementary Principles in Statistical Mechanics*. The word "elementary" indicates the modesty of the author rather than the simplicity of the subject.

The study of statistical mechanics is best approached from the viewpoint of quantum mechanics than from the viewpoint of classical mechanics. This is because quantum mechanics gives a more valid description of nature, whereas classical mechanics gives a description that is incomplete at the atomic level. So, the following is the outline of the initial approach to the subject: Many different energies are possible for the individual molecules of a substance; at any instant in time, it may happen that a certain number n_1 of molecules have the same energy ϵ_1, n_2 have the energy ϵ_2, n_3 the energy ϵ_3, and so on. Elementary theory of permutations and combinations is used to determine the most probable, or average, values of $n_1, n_2, \ldots$, and it will be found that they obey a simple equation, called the Maxwell–Boltzmann distribution law. Later, it will be shown how the values so obtained are related to the energy E and entropy S of the substance; it is then found that E and S, and hence all the other thermodynamic properties, can be expressed in terms of a function Z, known as the partition function. Finally, the calculation of Z in terms of atoms and molecular parameters will be discussed. For an isolated molecule in a confined space, the Schrödinger equation

$$\hat{H}\psi = \epsilon\psi \tag{2.1-1}$$

is satisfied for certain functions $\psi_1, \psi_2, \ldots, \psi_j, \ldots$ (eigenfunctions) and for certain values of ϵ, say $\epsilon_1, \epsilon_2, \ldots, \epsilon_i \ldots$ (eigenvalues). Thus, such a molecule can exist in certain quantum states, each of which is characterized by a wave function ψ_j. The discrete values ϵ_i are referred to as energy levels, so that reference is often made to "a molecule in the level ϵ_i," meaning that the molecule is in a quantum state for which the energy is ϵ_i. It may happen that a number, say g_i with $g_i > 1$, of states (i.e., wave functions) have the same energy ϵ_i, and it is said that this energy level is g_i-fold degenerate. A particular distribution of molecules among the energy levels is said to be a complexion or arrangement.

To translate the idea of Gibbs into the language of quantum mechanics, consider the quantum states and energy levels of several atomic systems. The simplest atom is hydrogen, with one electron and one proton. States of the hydrogen atom are associated with the motion of the electron and the proton. The low-lying energy levels of hydrogen are shown in Fig. 2.1-1(a), where the number of quantum states (the degeneracy) belonging to the same energy level have been put in parentheses. The zero of energy is taken as the energy of the lowest energy level. The positions of the energy levels can be determined spectroscopically by measuring the wavelength λ of the quanta emitted from excited atoms. From the relation $\lambda\nu = c$, the speed of light, the frequency ν is determined. The energy is then given by $h\nu = \epsilon$, where h is Planck's constant.

A quantum state of a system is a state of all the particles of the system. If

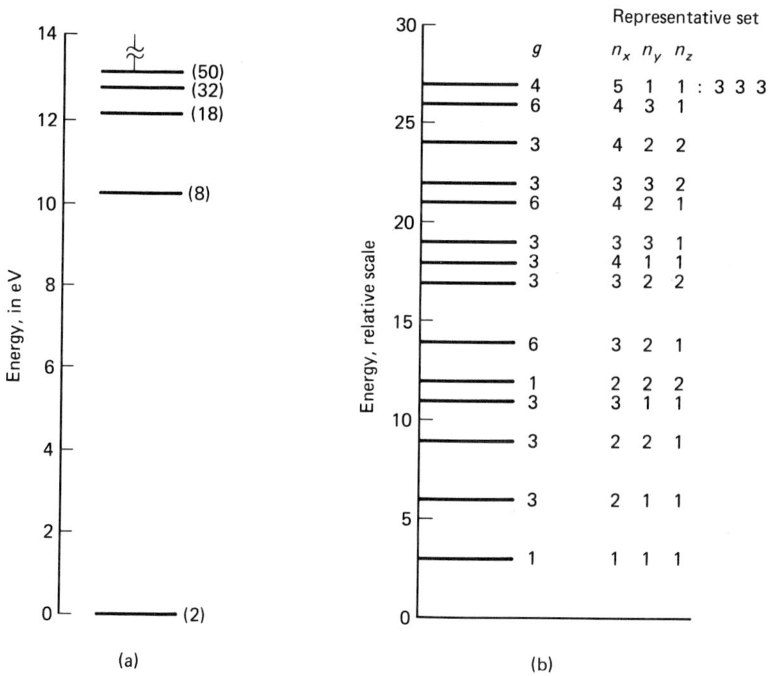

Figure 2.1-1 (a) Low-lying energy levels of atomic hydrogen. (The energies are given in electron volts, with 1 eV = 1.60×10^{-19} J.) (b) Energy levels, degeneracies, and quantum numbers n_x, n_y, n_z of a particle of spin zero confined to a cubical box.

there is cause to speak of the state of one particle, the term *orbital* is often used in place of state.

Statistical mechanics is concerned with the properties of physical systems of many different types. To describe the statistical properties of a system of N particles it is essential to know the set of values of an energy $\epsilon_1(N)$, where the notation means that ϵ is the energy of the quantum state 1 of the N particle system. The Greek letter epsilon, ϵ, is used to denote energy. The indices such as 1 may be assigned to the quantum states in any convenient arbitrary way, but two states should not be assigned the same index. Before attempting to calculate the energies $\epsilon_i(N)$ of a given system, the nature and procedure of quantum mechanics will be reviewed.

■ EXAMPLE 2.1-1

Quantum mechanics provides the result that some physical quantities may take on only selected values, but not all values. The allowed values of a physical quantity are the eigenvalues (characteristic values). Explain how this may come about through the use of operators.

Solution. The mathematical problem of finding the eigenvalues of a quantity is an eigenvalue problem; it is usually cast in the form of an equation called an eigenvalue equation. An eigenvalue equation for a physical quantity Q has the appearance

$$\hat{Q}f = qf \tag{2.1-2}$$

where f is a function, called the eigenfunction of the quantity Q, with an eigenvalue q. The element $\hat{Q}$ is called an operator, and the statement $\hat{Q}f$ says to change the function f into a new function, according to a set of instructions

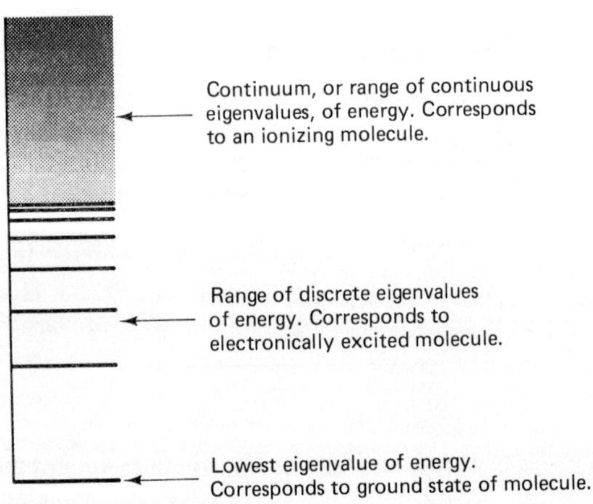

Figure 2.1-2 Eigenvalues for the energy of a molecule.

implicit in the definition of the operator $\hat{Q}$. Figure 2.1-2 shows the eigenvalues for the energy of a molecule. The task of quantum mechanics is to form operators for the physical quantities that are to be measured and solve the eigenvalue equation for the state functions and energy values. For example, the solutions for the equation $df/dx = kf$ are all functions $\{e^{kf}\}$.

Quantum mechanics itself evolved from two different points of view, which represent two analogous mathematical formulations of eigenvalue problems. The first of these points of view is the *wave mechanics* of Schrödinger. In wave mechanics, operators are differential expressions, such as the operator d/dx referred to above, and the eigenvalue equation then takes the form of a differential equation. The second formulation is the *matrix mechanics* of Heisenberg, in which operators are represented by algebraic entities called matrices.

Many problems of classical mechanics may be expressed in a simple and meaningful way as eigenvalue problems. Among these are the problems of the vibrations and rotations of a mechanical system, such as a molecule. These physical problems are of importance in molecular motion and spectroscopy. In vibrations, the normal modes and frequencies of oscillation appear as eigenvectors and eigenvalues; in rotations, the principal axes and moments of inertia emerge from an eigenvalue problem. (It should be noted that a correct description of these systems on the molecular level nearly always requires quantum mechanics, not classical mechanics.) In Chapter 3, the study of classical mechanics will be conducted to see how the vibrations of a mechanical system, such as a molecule, may be formulated as an eigenvalue problem. An attempt will be made to present Newtonian mechanics in a way that the connection to quantum mechanics can be established. ■

2.2 QUANTIZATION OF ENERGY

Mechanics is the science of motion of material bodies. Its development began with observations of the macroscopic world, but, when investigations reached the atomic and subatomic level, the old laws failed to explain the experimental results. This failure led to the development of a more general science of motion, *quantum mechanics,* of which classical mechanics is now recognized as a special case.

The fundamental concept in classical mechanics is that, if the laws governing the motion of particular particle of mass are known, and its state (i.e., position and momentum) at a given instant determined, then the state of the particle can be calculated exactly at any other time, which is equivalent to saying that the future mechanical behavior of the particle can be predicted completely. This concept also applies to a system of particles if the nature of the interactions between the particles is known.

By the end of the nineteenth century, a wide range of phenomena had been explained by classical physics. However, there was a hard core of results, such as the spectral distribution of thermal radiation from a blackbody and the low-temperature specific heats of solids, which, it was felt, should have been understood on the basis of classical physics but was not. In 1900, Planck succeeded in explaining the blackbody spectrum by assuming that electromagnetic radiation was emitted in discrete quanta each of which contained an energy

$$\epsilon = h\nu \tag{2.2-1}$$

where ν is the frequency of the radiation and h is Planck's constant. A few y. ars later Einstein explained the emission of electrons due to light falling on a metal surface by assuming that electromagnetic radiation was absorbed in discrete quanta of magnitude $h\nu$. In 1907, Einstein and, in 1912, Debye were able, by using Planck's hypothesis, to explain the specific heat of solids.

Quantum theory came into being when Bohr was able to predict the spectrum of hydrogen by postulating that the hydrogen atom consisted of a positive nucleus and an orbiting electron that could only exist in certain states for which the orbital angular momentum was an integral multiple of $h/2\pi$. Later, Wilson and Sommerfeld were able to extend Bohr's procedure to explain a wide variety of atomic systems. However, in certain more complicated problems, such as the rotational spectra of diatomic molecules, the old quantum theory did not give quantitatively correct results and was finally replaced by what is now known as *quantum mechanics*.

The dualism of the wave and particle aspects of radiation led de Broglie to suggest that particles exhibited a similar dualism, namely, that a particle of linear momentum $p = mv$ is associated with a wavelength λ given by

$$p = mv = \frac{h}{\lambda} \qquad \lambda = \frac{h}{p} \tag{2.2-2}$$

This was confirmed by the experiments of Davisson and Germer who observed the diffraction of electrons by a nickel crystal, and by those of Thomson who observed the diffraction of electrons by extremely thin metallic films. These results became more significant when Heisenberg developed the uncertainty principle, according to which it is impossible, because of the interaction of a particle with the measuring system, to determine simultaneously with precision the values of two conjugate variables (e.g., the momentum and position) of a particle. The product of the uncertainties Δp and Δq in the determination of their values is given by

$$\Delta p \, \Delta q \geq h \tag{2.2-3}$$

where h is Planck's constant: $h = 6.6256 \times 10^{-34}$ J s.

In attempting to correlate quantum and classical descriptions of matter, the clear-cut conclusions of classical physics become blurred. A degree of uncertainty is introduced. Thus, according to quantum mechanics, while the energy of a system may be sharp, its position is not determinate.

■ EXAMPLE 2.2-1

Determine the wavelengths of (a) a photon and (b) an electron, each having kinetic energy of 1 eV.

Solution

(a) The energy of a photon is given by

$$E = h\nu \tag{2.2-4}$$

where h = Planck's constant = 6.626×10^{-34} J s = 4.136×10^{-15} eV s and ν = frequency (Hz). The frequency is defined as c/λ, where c = speed of light and λ = wavelength. Therefore,

$$E = h\frac{c}{\lambda} \tag{2.2-5}$$

$$\lambda = \frac{hc}{E} \tag{2.2-6}$$

Solving the equation with E given as

$$1 \text{ eV} = 1.6021 \times 10^{-19} \text{ J}$$

yields

$$\lambda = \frac{(6.6256 \times 10^{-34} \text{ J s})(2.9979 \times 10^{8} \text{ m s}^{-1})}{1.6021 \times 10^{-19} \text{ J}}$$

$$= 12.398 \times 10^{-7} \text{ m} = 1239.8 \text{ nm}$$

(b) The energy of an electron is

$$E = \tfrac{1}{2} mv^2 \tag{2.2-7}$$

where m = mass of electron and v = velocity. Solving the equation for v gives

$$v = \left(\frac{2E}{m}\right)^{1/2} = \left[\frac{(2)(1.6021 \times 10^{-19} \text{ J})}{9.110 \times 10^{-31} \text{ kg}}\right]^{1/2}$$

$$= 5.930 \times 10^{5} \text{ m s}^{-1}$$

The wavelength of an electron is given by $\lambda = h/mv$,

$$\lambda = \frac{6.6256 \times 10^{-34} \text{ J s}}{(9.110 \times 10^{-31} \text{ kg})(5.930 \times 10^{5} \text{ m s}^{-1})}$$

$$= 1.226 \times 10^{-9} \text{ m}$$

$$= 1.226 \text{ nm}$$

■

■ EXAMPLE 2.2-2

Compare the de Broglie wavelength of an electron whose energy is 150 eV with that of a 250-g baseball traveling at 60 mi/hr.

Solution. For the electron

$$\text{eV} = \tfrac{1}{2} mv^2 \tag{2.2-8}$$

$$v^2 = \frac{2 \text{ eV}}{m} = \frac{2 \times 1.6 \times 10^{-19} \times 150}{9.11 \times 10^{-31}}$$

$$v = \sqrt{53} \times 10^{6} = 7.3 \times 10^{6} \text{ m s}^{-1}$$

Then

$$\lambda = \frac{h}{mv} = \frac{6.63 \times 10^{-34}}{9.11 \times 10^{-31} \times 7.3 \times 10^{6}} \tag{2.2-9}$$

$$= 10^{-10} \text{ m}$$

For the baseball

$$60 \text{ mi/hr} = 88 \text{ ft/s} = 88 \times 3.05 \times 10^{-1} = 26.8 \text{ m s}^{-1}$$

$$\lambda = \frac{h}{mv} = \frac{6.63 \times 10^{-34}}{250 \times 10^{-3} \times 26.8} = \frac{6.63 \times 10^{-34}}{2.50 \times 2.68} \tag{2.2-10}$$

$$= \frac{6.63 \times 10^{-34}}{6.70} = 0.99 \times 10^{-34} \text{ m}$$

The de Broglie wavelength for the baseball is insignificant compared with that of the electron. Although the latter is also extremely small by ordinary standards, it is in the X-ray range. Light is both wavelike and corpuscular. Matter is both corpuscular and wavelike. However, they cannot be both at the same time. Under certain circumstances the one aspect prevails over the other, and vice versa. ■

2.3 RULES OF QUANTUM MECHANICS

Modern quantum theory in its elementary formulation takes two different but nevertheless equivalent forms known, respectively, as *matrix mechanics* and *wave mechanics*. Because the latter is considered by a substantial majority of writers to be more suitable as a starting point for beginners, quantum theory will be introduced by a set of postulates. In quantum mechanics, the idea of a state is replaced by a state function.

POSTULATE I. For every given mechanical system at a given instant of time, there exists a function $\Psi(x,y,z,t)$, written hereafter $\Psi(q,t)$, which describes the state of that system as completely as possible. The function Ψ, which is called the state function or wave function, has the following properties: it may be complex; it is a continuous, single-valued, and finite function of q and t; its first derivatives with respect to q are single valued, continuous, and finite; and the probability of finding the system somewhere in space is

$$\int \Psi^*\Psi \, dq = 1 \tag{2.3-1}$$

where Ψ^* is the complex conjugate of Ψ. Equation (2.3-1) is called the *normalizing* condition. Ψ must go to zero at infinity. In general, $\Psi = \psi e^{i\omega t}$, where ψ (small psi) is a function of x, y, and z only.

POSTULATE II. With every measurable property of a system there is associated an operator $\hat{a}$ which can operate on the state function. The operator $\hat{a}$ has the following properties: (1) it is linear; (2) it is Hermitian; (3) there exists a set of functions ϕ_n that satisfy the equation $\hat{a}\phi_n = a_n\phi_n$, where the a_n are constants, and the ϕ_n constitute a complete set of functions with respect to the class of functions made up of all acceptable wave functions Ψ for the system. The form of the operator may be found by means of the following rules:

If a is one of the coordinates q_j or the time t, the operator is the variable itself, that is, $\hat{x} = x$.

If a is one of the momenta p_j, then the operator is $(\hbar/i)\ \partial/\partial q_j$, where q_j is conjugate to p_j and $\hbar = h/2\pi$.

For example, the operator corresponding to the x component of momentum is

$$\hat{p}_x = \frac{\hbar}{i}\frac{\partial}{\partial x} \tag{2.3-2}$$

The operator for the potential energy V is

$$\hat{V} = V \tag{2.3-3}$$

The potential for the total energy E (kinetic plus potential) is

$$\hat{E} = -\frac{\hbar}{i}\frac{\partial}{\partial t} \tag{2.3-4}$$

Because only real quantities are measurable and quantum mechanics predicts only measurable quantities, a certain restriction must be imposed on the operators. This restriction, called the *Hermitian condition* will now be explained.

For an operator $\hat{\alpha}$, the predicted measurable quantity α must be real. This means that $\alpha = \alpha^*$. In other words, any predicted measurable quantity α equals its complex conjugate. To find the implications of this statement, consider

$$\hat{\alpha}\psi = \alpha\psi \tag{2.3-5}$$

The operator $\hat{\alpha}$ corresponds to some measurement, and α is its eigenvalue, the unique result of that measurement. The complex conjugate equation must also hold:

$$\hat{\alpha}^*\psi^* = \alpha^*\psi^* = \alpha\psi^* \tag{2.3-6}$$

[As an example, the complex conjugate of the momentum operator $(\hbar/i)(\partial/\partial x)$ is $(-\hbar/i)(\partial/\partial x)$.] Making the substitution α for α^*, Eq. (2.3-5) is now multiplied by ψ^* and Eq. (2.3-6) by ψ. Since $\hat{\alpha}$ and $\hat{\alpha}^*$ are operators, and the action is just multiplication, the new factors must be placed to the left of the operator symbols. By convention, operators operate on those functions to their right. Thus, since α is a constant,

$$\psi^*(\hat{\alpha}\psi) = \psi^*(\alpha\psi) = \alpha\psi^*\psi \tag{2.3-7}$$

and

$$\psi(\hat{\alpha}^*\psi^*) = \psi(\alpha^*\psi^*) = \alpha^*\psi\psi^* = \alpha\psi\psi^* \tag{2.3-8}$$

One consequence of this is that for α to be real,

$$\psi^*(\hat{\alpha}\psi) - \psi(\hat{\alpha}^*\psi^*) = 0 \tag{2.3-9}$$

A sufficient restriction, then, in order to ensure the prediction of real quantities, is that this averages to zero when integrated over all space:

$$\int (\psi^*\hat{\alpha}\psi - \psi\hat{\alpha}^*\psi^*)\, d\tau = 0 \tag{2.3-10}$$

or

$$\int \psi^*\hat{\alpha}\psi\, d\tau - \int \psi\hat{\alpha}^*\psi^*\, d\tau = 0 \tag{2.3-11}$$

In a more general way,

$$\int \phi^*\hat{\alpha}\psi\, d\tau - \int \psi\hat{\alpha}^*\phi^*\, d\tau = 0 \tag{2.3-12}$$

where ϕ is any eigenfunction of $\hat{\alpha}$ (ϕ may be equal to Ψ). This integral form is known as the *Hermitian condition.*

POSTULATE III. If a series of measurements of the observable corresponding to the operator $\hat{A}$ is carried out on a set of identical systems all in the state $\Psi(q)$, the average value of these measurements will be given by

$$\langle\hat{\alpha}\rangle = \int \psi^*\hat{\alpha}\psi\, dq \tag{2.3-13}$$

The quantity $\langle\hat{\alpha}\rangle$ is called the expectation value. In other words, when a large number of values near some central value are measured, the average value is the one of interest and it is known as the expectation value.

In Postulate I, the state function $\Psi(q,t)$ is introduced. Postulates II and III provide the connection between the state function and the measurable properties of a system. The only thing needed now is the equation of motion that will enable the prediction of the state of a system at some future time if its state at some initial time is known. This equation will allow Ψ to be solved as function of q and t if Ψ is known as a function of q at time t_0. It is provided by the next postulate.

POSTULATE IV. The state function $\Psi(q,t)$ obeys the following equation

$$\hat{H}\Psi(q,t) = i\hbar\frac{\partial\Psi(q,t)}{\partial t} \tag{2.3-14}$$

where $\hat{H}$ is the operator corresponding to the classical Hamiltonian. The latter equation is called the *time-dependent Schrödinger equation.*

Equation (2.3-14) governs the time development of a quantum system. When $\hat{H}$ does not contain the time, $\Psi(q,t)$ is factorable into $\Psi(q,t) = \psi(q)T(t)$, giving a separation of time and space variables. Equation (2.3-14) then becomes

$$\frac{i\hbar\dot{T}}{T} = \frac{\hat{H}\psi}{\psi} \tag{2.3-15}$$

Each term in (2.3-15) must therefore equal a constant, say E. The equation for $T(t)$ is then

$$i\hbar\dot{T} - ET = 0 \tag{2.3-16}$$

Its solution is

$$T(t) = \text{const } e^{-iEt/\hbar} \tag{2.3-17}$$

The right side of (2.3-15) becomes

$$\hat{H}\psi(q) = E\psi(q) \tag{2.3-18}$$

which is the *time-independent Schrödinger equation*. The constant E is identified as the energy eigenvalue. A quantum state for which (2.3-14) reduces to (2.3-18) is known as a *stationary state*. In such a state, there is conservation of energy. All previous considerations have been concerned with stationary states. Because of the particular way in which time dependence enters into the wave function for a stationary state,

$$\Psi(q,t) = \psi(q)e^{-iEt/\hbar} \tag{2.3-19}$$

that is, the time factor can be disregarded. For example, since the absolute value of $e^{-iEt/\hbar}$ is unity,

$$\Psi^*\Psi = \psi^*\psi \tag{2.3-20}$$

POSTULATE V. An electron has a spin that can be in either of two states (designated with positive or negative signs). Two states otherwise identical but with different spins are different. Thus "spin" should be considered as part of the state function, although it is not a function of time or position. All of the eigenfunctions in the expansion of a state function of a single particle must be of the same spin (as the particle itself).

POSTULATE VI. No two electrons can occupy the same state (i.e., have the same state function). Thus, when the lowest two levels are occupied, additional electrons must go into states of higher energy. This is known as Paul's *exclusion principle*.

Before going into the details of quantum mechanics, a few remarks concerning classical and quantum situations are in order. When classical mechanics constructs equations for waves in a string, rod, or lattice, or for electromagnetic waves, it is able to derive the wave equation from a knowledge of the medium in which the waves propagate. In each case the velocity of the wave is closely linked to such physical parameters as tension, density, forces, spatial distributions, dielectric constant, and magnetic permeability.

When it comes to the question of constructing a wave equation for matter waves, there is only the experimental knowledge that matter often exhibits wavelike properties, such as in electron diffraction. There is no knowledge of a medium and its properties. The classical approach is not possible, and only a formal equation that meets very general requirements is constructed. The physical significance of its components are then interpreted to test its predictions against physical measurements.

In the matter wave equation, the common features between the particle and the wave is the energy E. Also, the momentum p of the particle is to be equated to $h\lambda$ or $\hbar k$ of the wave. Furthermore, the wave equation consists of relationships between the spatial and temporal derivatives of a displacement, and so harmonic wave solutions are sought of this wave equation.

The final choice of a wave equation is made (not derived) by the following line of argument. The energy of a particle is given by

$$E = \frac{p^2}{2m} + V \tag{2.3-21}$$

It is required that the wave equation be valid and independent of the specific energy or momentum of the particle; this means that the wave equation must contain those derivatives of the displacement that yield only E or p^2.

Suppose that the harmonic solutions of the wave equation have the form in one dimension of

$$\Psi = A \exp[i(kx - \omega t)] \tag{2.3-22}$$

where $k = 1/\lambda$ is the wave number of the traveling wave, and $\omega = 2\pi\nu$ is the circular frequency.

This means that

$$E = \hbar\omega = \frac{i\hbar}{\Psi}\frac{\partial\Psi}{\partial t} \tag{2.3-23}$$

and

$$\frac{p^2}{2m} = \frac{\hbar^2 k^2}{2m} = \frac{\hbar^2}{2m\Psi}\frac{\partial^2\Psi}{\partial x^2} \tag{2.3-24}$$

whereupon substitution into (2.3-21) gives the Schrödinger time-dependent wave equation

$$i\hbar\frac{\partial\Psi}{\partial t} = -\frac{\hbar^2}{2m}\frac{\partial^2\Psi}{\partial x^2} + V\Psi \tag{2.3-25}$$

or, generalized to three dimensions,

$$i\hbar\frac{\partial\Psi}{\partial t} = \frac{\hbar^2}{2m}\nabla^2\Psi + V\Psi \tag{2.3-26}$$

This is the equation that must be used to describe processes in which dynamic transitions involving a change in energy occur, such as in optical absorption and electron scattering. However, in the simple case of what energy states are allowed in the presence of a particular potential energy $V(x,y,z)$, the solution can be expressed as

$$\Psi(x,y,z,t) = \psi(x,y,x) \exp(-i\omega t) \tag{2.3-27}$$

that is, the dependence of Ψ on the coordinates can be separated from the dependence of Ψ on time. When this is substituted into (2.3-26), the *Schrödinger time-independent* wave equation is obtained:

$$E\psi = -\frac{\hbar^2}{2m}\nabla^2\psi + V\psi \tag{2.3-28}$$

In this equation, $\psi(x,y,z)$ and $V(x,y,z)$ are functions only of the coordinates, and E is a constant, giving the value of allowed energy for a particular ψ and V.

To find the allowed energies of an electron in the presence of a potential energy V, the problem is in principle quite simple. All that is needed is to obtain the appropriate solutions of the wave equation (2.3-28) that are mathematically well behaved and satisfy the boundary conditions. The elegance of this formulation is that the same equation can deal with any problem, the only difference from one problem to the next being in the specific form given to the potential energy V.

One final note: Schrödinger's equation is not the only differential wave equation for material particles that satisfies the equations

$$E = \hbar\omega, \qquad p = \hbar k, \qquad \hbar\omega = \frac{\hbar^2 k^2}{2m} + V \tag{2.3-29}$$

Many other differential equations embodying these dynamical relations can be constructed. But the various forms of the Schrödinger equation, besides being the mathematically simplest to satisfy the requirements, have other properties that cause them to be preferred over all other possibilities:

1. They are linear, so that if Ψ_1 and Ψ_2 are solutions to one of these equations, then any linear combination of them is a solution of the same equation. This property of *superposition* is one of the most basic properties of waves.
2. Their solutions are perfectly suited to the interpretation of Ψ as a probability amplitude.

Above all, the Schrödinger equations work; they provide the basis for a correct analysis of all kinds of molecular, atomic, and nuclear systems. Whatever questionable features there may be in the manner of their formulation are swept away in the evidence of their manifest success—a success to be seen many times in statistical mechanics.

■ EXAMPLE 2.3-1

(a) For $\Psi(x,y,z,t) = \psi(x,y,z)e^{-i\omega t}$, show that the probability of finding the particle in an infinitessimal volume dV is independent of time.

(b) If $\psi = A r e^{-r/2a_0}$ in spherical coordinates, what is the value of A from normalization?

Solution

(a) The probability is proportional to

$$\Psi^*\Psi\, dV = (\psi e^{-i\omega t})^*(\psi e^{-i\omega t})\, dV$$
$$= (\psi^* e^{+i\omega t})(\psi e^{-i\omega t})\, dV$$
$$= \psi^*\psi e^{+i\omega t - i\omega t}\, dV = \psi^*\psi e^0\, dV$$
$$= \psi^*\psi\, dV \tag{2.3-30}$$

which is independent of time.

(b) The normalization condition is

$$\int_v \psi^*\psi\, dV = 1$$

Here, $dV = 4\pi r^2\, dr$ and $\psi^*\psi = \psi^2 = A^2 r^2 e^{-r/a_0}$, so

$$A^2 \int_0^\infty r^2 e^{-r/a_0}\, 4\pi r^2\, dr = 1 \tag{2.3-31}$$

Using the integral $\int_0^\infty x^n e^{-ax}\, dx = n!/a^{n+1}$ from Chapter 1, this gives

$$4\pi A^2 24 a_0^5 = 1$$

or

$$A = (96 a_0)^{1/2}$$ ■

■ EXAMPLE 2.3-2

Let $\psi = Ae^{-ar^3 - ib}$, where a and b are real. Normalize $\Psi = \psi e^{-i\omega t}$ and find the probability that the particle is between $r = 1$ and $r = 3$. Take $a = 1$.

Solution. The normalization condition $\int_V \Psi^*\Psi\, dV = 1$ yields

$$\int_0^\infty A^2 e^{-2r^3}\, 4\pi r^2\, dr = 1 \tag{2.3-32}$$

or

$$A^2\left(\frac{4\pi}{6}\right)\int_0^\infty e^{-2r^3}\, 6r^2\, dr = 1$$

Using the substitution $u = 2r^3$, this gives

$$A^2\left(\frac{2\pi}{3}\right)\int_0^\infty e^{-u}\, du = A^2\left(\frac{2\pi}{3}\right)(-e^{-u})\Big|_0^\infty \tag{2.3-33}$$

$$= A^2\left(\frac{2\pi}{3}\right)[e^{-\infty} - (-e^0)] = A^2\left(\frac{2\pi}{3}\right)(0 + 1) = 1$$

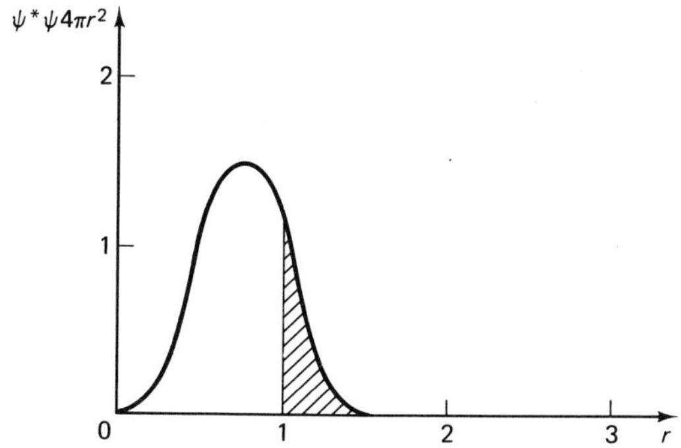

Figure 2.3-1 Probability of finding a particle at $r \geq 1$.

or

$$A = \left(\frac{3}{2\pi}\right)^{1/2} \tag{2.3-34}$$

Thus, $\psi^*\psi = (3/2\pi)e^{-2r^3}$, so the probability of finding the particle between $r = 1$ and $r = 3$ is

$$P = \int_1^3 \left(\frac{3}{2\pi}\right) e^{-2r^3} 4\pi r^2\, dr \tag{2.3-35}$$

$$= \left(\frac{3}{2\pi}\right)\left(\frac{4\pi}{6}\right) \int_1^3 e^{-2r^3} 6r^2\, dr$$

To evaluate the integral, let $u = 2r^3$ (with $u = 2$ at $r = 1$ and $u = 54$ at $r = 3$), so that

$$P = \int_2^{54} e^{-u}\, du = -e^{-u}\Big|_2^{54} = -e^{-54} - (-e^{-2}) \tag{2.3-36}$$

In other words, there is a 13.5 percent chance that the particle is between $r = 1$ and $r = 3$. The probability of finding the particle between 0 and 1 is $1 - 0.135 = 0.865$; the probability of finding the particle at $r \geq 3$ is vanishingly small. This is shown in Fig. 2.3-1. ■

2.4 SCHRÖDINGER WAVE EQUATION

In 1926, Schrödinger developed an equation equivalent to the matrix formulation published by Heisenberg a year earlier. Schrödinger combined de Broglie's matter waves, the second-order differential equation for wave motion, and other classical concepts into an equation for the total energy of a system (particle) in the manner shown below. Starting with classical mechanics, the total energy (kinetic + potential) of a system is

$$E = \tfrac{1}{2}mv^2 + V(x,y,z,t) \tag{2.3-37}$$

or, more conveniently, in terms of momentum ($p = mv$),

$$E = \frac{p^2}{2m} + V \tag{2.4-1}$$

Equation (2.4-1) is the classical expression for total energy. To obtain the quantum equation, the variables are replaced with operators, namely, using the symbol $\rightarrow$ to mean replace,

$$\begin{aligned} E &\longrightarrow -\frac{h}{2\pi i}\frac{\partial}{\partial t} \\ p_x &\longrightarrow \frac{h}{2\pi i}\frac{\partial}{\partial x} \\ p_y &\longrightarrow \frac{h}{2\pi i}\frac{\partial}{\partial y} \\ p_z &\longrightarrow \frac{h}{2\pi i}\frac{\partial}{\partial z} \end{aligned} \tag{2.4-2}$$

Therefore, Eq. (2.4-1) becomes

$$-\frac{h}{2\pi i}\frac{\partial}{\partial t} = \frac{1}{2m}\left[\left(\frac{h}{2\pi i}\right)^2\frac{\partial^2}{\partial x^2} + \left(\frac{h}{2\pi i}\right)^2\frac{\partial^2}{\partial y^2} + \left(\frac{h}{2\pi i}\right)^2\frac{\partial^2}{\partial z^2}\right] + V$$

or, upon simplification,

$$\begin{aligned} -\frac{h}{2\pi i}\frac{\partial}{\partial t} &= -\frac{h^2}{8\pi^2 m}\left(\frac{\partial^2}{\partial x^2} + \frac{\partial^2}{\partial y^2} + \frac{\partial^2}{\partial z^2}\right) + V \\ &= -\frac{h^2}{8\pi^2 m}\nabla^2 + V \end{aligned} \tag{2.4-3}$$

where the Laplacian operator $\nabla^2 \equiv \partial^2/\partial x^2 + \partial^2/\partial y^2 + \partial^2/\partial z^2$ has been used. The right-hand side of Eq. (2.4-3) is known as the *Hamiltonian* operator:

$$\hat{H} = -\frac{h^2}{8\pi^2 m}\nabla^2 + V \tag{2.4-4}$$

Introducing Ψ in Eq. (2.4-3) then yields

$$-\frac{h^2}{8\pi^2 m}\nabla^2\Psi + V\Psi = -\frac{h}{2\pi i}\frac{\partial}{\partial t}\Psi \tag{2.4-5a}$$

or

$$\hat{H}\Psi = -\frac{h}{2\pi i}\frac{\partial}{\partial t}\Psi \tag{2.4-5b}$$

Equation (2.4-5) is the *three-dimensional, time-dependent, nonrelativistic Schrödinger wave equation*. Equation (2.4-5a) is often written as

$$-\frac{\hbar^2}{2m}\nabla^2\Psi + V\Psi = i\hbar\frac{\partial}{\partial t}\Psi \tag{2.4-5c}$$

where $\hbar$ is $h/2\pi$. Note that Schrödinger's equation is not a derived equation but has been verified by the correct results it gives.

If the potential energy V does not depend directly on time, but changes only with position, Ψ can be written as $\Psi = \psi(x,y,z)e^{-i\omega t}$, and Eq. (2.4-5c) becomes

$$-e^{-i\omega t}\left(\frac{\hbar^2}{2m}\right)\nabla^2\psi + V\omega e^{-i\omega t} = i\hbar(-i\omega)\psi e^{-i\omega t}$$

$$= \hbar\omega\psi e^{-i\omega t} \tag{2.4-6}$$

But $\hbar\omega = h\nu = E$. So if (2.4-6) is divided by $e^{-i\omega t}$, it yields

$$-\frac{\hbar^2}{2m}\nabla^2\psi + V\psi = E\psi \tag{2.4-7}$$

Equation (2.4-7) is the *three-dimensional, time-independent, nonrelativistic Schrödinger wave equation*. Not having the time dependence of Ψ is a desirable simplification. If Ψ is ever wanted, the relation $E = \hbar\omega$ can be used to obtain

$$\Psi = \psi e^{-i(E/\hbar)t} \tag{2.4-8}$$

Usually, Ψ is not needed anyway, because $\Psi^*\Psi = \psi^*\psi$.

2.5 PARTICLE IN ONE-DIMENSIONAL SPACE

Consider a few simple solutions of Schrödinger's time-independent equation for systems of frequent occurrence in statistical mechanics, beginning with a particle constrained to move in the one-dimensional potential

$$\begin{aligned} V &= \infty \qquad x<0 \\ V &= 0 \qquad 0<x<L \\ V &= \infty \qquad x>L \end{aligned} \tag{2.5-1}$$

The classical Hamiltonian is

$$H = \frac{p^2}{2m} + V(x) \tag{2.5-2}$$

The corresponding quantum-mechanical operator is

$$\hat{H} = -\frac{\hbar^2}{2m}\frac{d^2}{dx^2} + V(x) \tag{2.5-3}$$

Schrödinger's equation for the region $0 \le x \le L$ is

$$-\frac{\hbar^2}{2m}\frac{d^2\psi n}{dx^2} = E_n\psi_n \tag{2.5-4}$$

Outside the region the wave function vanishes, and therefore

$$\psi_n(0) = \psi_n(L) = 0 \tag{2.5-5}$$

Equation (2.5-4) can be rewritten

$$\psi''(x) + k^2\psi(x) = 0 \tag{2.5-6}$$

having defined

$$k = \left(\frac{2mE}{\hbar^2}\right)^{1/2} \tag{2.5-7}$$

The differential equation (2.5-6) is often encountered, and its solution is straightforward. What is needed is a function whose second derivative gives the same function back again, but multiplied by a constant. Two functions that behave this way are the sine function and the cosine function, so the solution of (2.5-6) is of the form

$$\psi(x) = A \sin kx + B \cos kx \tag{2.5-8}$$

as can be verified by substitution.

The boundary conditions (2.5-5) will now be applied to (2.5-8). By the first of them,

$$\psi(0) = B = 0 \tag{2.5-9}$$

since $\sin 0 = 0$, $\cos 0 = 1$. Thus, the second term in the solution (2.5-8) must be excluded. By the second boundary condition (2.5-5):

$$\psi(a) = A \sin ka = 0 \tag{2.5-10}$$

It is assumed that $A \neq 0$, for otherwise $\psi(x)$ would equal zero everywhere and, in view of (2.5-8), the particle would be nowhere. The condition is therefore

$$\sin ka = 0 \tag{2.5-11}$$

Since a sine function vanishes at $0, \pm\pi, \pm 2\pi, \ldots$, Eq. (2.5-11) implies that

$$ka = n\pi \tag{2.5-12}$$

where n is an integer, positive, negative, or zero. The case $n = 0$ must be excluded, for then $k = 0$, and again $\psi(x)$ would vanish everywhere. Eliminating k between (2.5-7) and (2.5-12) and solving for the energy, yields

$$E_n = \frac{\hbar^2\pi^2}{2ma^2}n^2 = \frac{h^2}{8ma^2}n^2 \qquad n = 1, 2, 3, \ldots \tag{2.5-13}$$

These are the only values of the energy that allow solution of the Schrödinger equation (2.5-4) subject to the boundary conditions (2.5-5). The integer n, employed as a subscript on E, serves to designate the allowed *energy levels*. It is called a *quantum number*. Negative values of n add nothing new because n occurs in (2.5-13) as the square. Figure 2.5-1 shows part of the energy level diagram for the particle in a wire.

The occurrence of discrete or quantized energy levels is a characteristic behavior of a bound system (that is, one restricted to a finite region in space) in quantum mechanics. Note, however, that even though the energy levels are discrete, there is an infinite number of them.

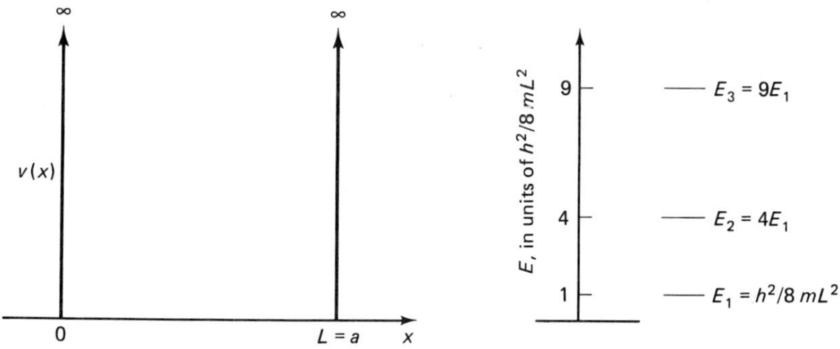

Figure 2.5-1 Potential well and energy levels for particle along a wire.

There is no accounting in classical mechanics for the discreteness of allowed energy levels. For classical systems there is a continuum of possible values for the energy. Bohr, in his theory of the hydrogen atom, put in quantization as an ad hoc assumption. This is now superseded by quantum mechanics, in which discreteness of the energy levels arises automatically.

The wave functions for the particle along a wire are given by (2.5-8) with $B = 0$ and $k = n\pi/a$:

$$\psi_n(x) = A \sin\frac{n\pi}{a}x \qquad n = 1, 2, \ldots \tag{2.5-14}$$

These, like the energy levels, can be labeled by the quantum number n. The constant A, thus far arbitrary, can be adjusted so that $\psi_n(x)$ is normalized. The normalization condition is, in this case,

$$\int_0^a [\psi_n(x)]^2\, dx = 1 \tag{2.5-15}$$

the integration being over the domain of the particle, $0 \le x \le a$. Substituting (2.5-14) into (2.5-15) gives

$$A^2 \int_0^a \sin^2\frac{n\pi}{a}x\, dx = 1 \tag{2.5-16}$$

Setting $\theta = (n\pi/a)x$,

$$\int_0^a \sin^2\frac{n\pi}{a}x\, dx = \frac{a}{n\pi}\int_0^{n\pi} \sin^2\theta\, d\theta = \frac{a}{n\pi}\frac{n\pi}{2}. \tag{2.5-17}$$

Thus, from (2.5-16) and (2.5-17),

$$A = \left(\frac{2}{a}\right)^{1/2} \tag{2.5-18}$$

for all values of n. The normalized wave functions

$$\psi_n(x) = \left(\frac{2}{a}\right)^{1/2} \sin\frac{n\pi}{a}x \tag{2.5-19}$$

and the corresponding probability densities $\Psi^*\Psi = \Psi^2$

$$\rho_n(x) = \frac{2}{a}\sin^2\frac{n\pi}{a}x \qquad (2.5\text{-}20)$$

are shown in Fig. 2.5-2.

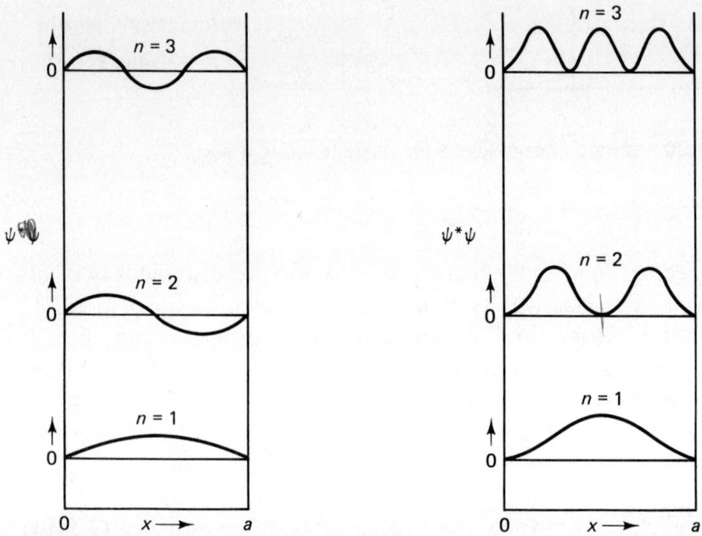

Figure 2.5-2 Eigenfunctions and probability densities for particle along a wire.

■ EXAMPLE 2.5-1

What is the ground-state energy for an electron that is confined to a potential well with a width of 0.2 nm?

Solution

$$E = \frac{h^2n^2}{8ma^2} \qquad (2.5\text{-}21)$$

$$= \frac{(6.626 \times 10^{-34}\ \text{J s})^2(1)^2}{8(9.110 \times 10^{-31}\ \text{kg})(0.2 \times 10^{-9}\ \text{m})^2}$$

$$= 1.506 \times 10^{-18}\ \text{J}$$

$$= \frac{(1.506 \times 10^{-18}\ \text{J})(6.022 \times 10^{23}\ \text{mol}^{-1})}{(10^3\ \text{J kJ}^{-1})}$$

$$= 907\ \text{kJ mol}^{-1}$$ ■

■ EXAMPLE 2.5-2

Calculate the expectation value of position for a particle in a one-dimensional infinite square wall.

Solution. The probability of finding the particle in the region dx near x is $|\psi(x)|^2\,dx$. The average or expectation value of position is then the sum of x times this probability. For a continuous distribution the summation becomes an integral, and the result is

$$\langle x\rangle = \int_{-\infty}^{+\infty} x|\psi(x)|^2\,dx \qquad (2.5\text{-}22)$$

It is important that the wave function used in this calculation be normalized in order that $|\psi(x)|^2\,dx$ be the appropriate absolute probability. Consider any one of the wave functions for an infinite square well,

$$\psi_n(x) = \sqrt{\frac{2}{L}}\sin\left(\frac{n\pi x}{L}\right) \qquad 0<x<L$$
$$= 0 \qquad \text{elsewhere} \qquad (2.5\text{-}23)$$

It can be guessed that the expectation value of x for these states will be $L/2$, since the probability density $|\psi_n(x)|^2$ is symmetric about $L/2$ for all these states. To check this, use the trigonometric identity $2\sin^2 z = 1 - \cos 2z$ to get

$$\langle x\rangle = \frac{2}{L}\int_0^L x\sin^2\left(\frac{n\pi x}{L}\right)dx$$
$$= \frac{1}{L}\int_0^L x\,dx - \frac{1}{L}\int_0^L x\cos\left(\frac{2n\pi x}{L}\right)dx \qquad (2.5\text{-}24)$$

The first integral is just $x^2/2$, which gives the first term the value $L/2$. The second integral is zero, as can be seen by setting $z = 2n\pi x/L$ and using integration by parts:

$$\frac{1}{L}\int_0^L x\cos\left(\frac{2n\pi x}{L}\right)dx = \frac{L}{(2n\pi)^2}\int_0^{2n\pi} z\cos z\,dz$$
$$= \frac{L}{(2n\pi)^2}(z\sin z + \cos z)_0^{2n\pi} \qquad (2.5\text{-}25)$$
$$= 0, \qquad \text{for integral values of } n$$

Therefore, the expectation value of x is just $L/2$ from the first term, as expected:

$$\langle x\rangle = \frac{L}{2} \qquad (2.5\text{-}26)$$

A few remarks are in order: Quite apart from the restriction of the energy to discrete (quantized) values, the wave-mechanical solution to the problem contrasts strongly with the classical description of a bound particle. In Newtonian mechanics a particle of given energy, confined between rigid walls but otherwise free, would be equally likely to be found at any point between $x = 0$ and $x = L$. In quantum mechanics, this is not so. The wave-mechanical solution leads to a probability, proportional to the square of $\psi(x)$, that varies with x and (except for the lowest state, $n = 1$) is actually zero at certain points between $x = 0$ and $x = L$, as shown in Fig. 2.5-3.

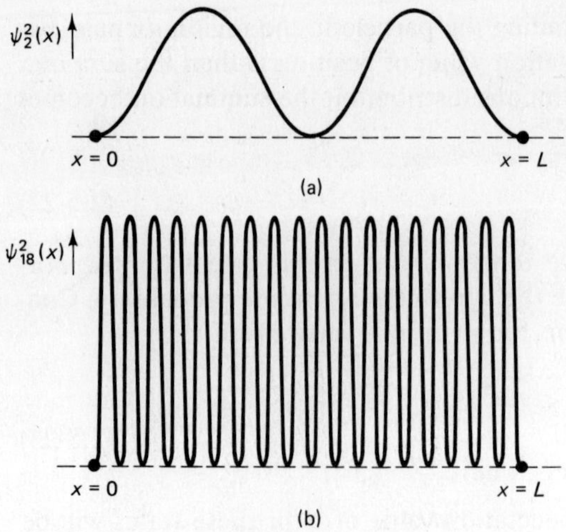

Figure 2.5-3 Probability for $n = 2$ and $n = 18$. ■

■ EXAMPLE 2.5-3

Verify that when the potential energy is zero, the function

$$\psi = Ae^{i(8\pi^2 mE)^{1/2}x/h}$$

is a solution of the time-independent Schrödinger equation for a free particle moving in one dimension.

Solution. The time-independent Schrödinger equation $\hat{H}\psi = E\psi$ in one dimension is

$$-\frac{\hbar^2}{8\pi^2 m}\frac{d^2\psi}{dx^2} = E\psi \tag{2.5-27}$$

Before substituting the given ψ in this equation, let the constants in the exponent of ψ be represented by a single constant b. Thus, $\psi = Ae^{bx}$, where $b = i(8\pi^2 mE)^{1/2}/h$. The question is

$$-\frac{h^2}{8\pi^2 m}\frac{d^2 Ae^{bx}}{dx^2} \stackrel{?}{=} EAe^{bx} \tag{2.5-28}$$

Taking the first derivative of Ae^{bx} gives Abe^{bx}. Taking the second derivative gives Ab^2e^{bx}. Substituting in the Schrödinger equation then yields

$$-\frac{h^2}{8\pi^2 m}Ab^2e^{bx} \stackrel{?}{=} EAe^{bx} \tag{2.5-29}$$

With $b = i(8\pi^2 mE)^{1/2}$, this becomes

$$\frac{h^2}{8\pi^2 m}A\frac{8\pi^2 m}{h^2}Ee^{bx} \stackrel{?}{=} EAe^{bx} \tag{2.5-30}$$

Upon simplification and canceling on each side, the answer is "yes." Thus, the wave function $\psi = Ae^{i(8\pi^2 mE)^{1/2}x/h}$ is a solution of the Schrödinger equation. ■

■ EXAMPLE 2.5-4

(a) The diameter of a hydrogen atom is about 1 Å. If the electron jumps from the second level ($n = 2$) to the first ($n = 1$), emitting a photon, what is the wavelength of the emitted radiation?

(b) A 10-g ball in a one-dimensional box, 1 m long, jumps from the second level ($n = 2$) to the first. What is the wavelength of the emitted radiation?

Solution

(a) The energies for a particle in a one-dimensional box are given by

$$E_n = \frac{n^2h^2}{8mL^2} \tag{2.5-31}$$

$$E_n = \frac{n^2(6.626 \times 10^{-34}\text{ J) s)}^2}{8(9.109 \times 10^{-31}\text{ kg})(10^{-10}\text{ m})^2} = (6.025 \times 10^{-18}\text{ J})\, n^2$$

For $n = 1$, $E_1 = 6.025 \times 10^{-18}$ J
For $n = 2$, $E_2 = (6.025 \times 10^{-18})2^2 = 24.10 \times 10^{-18}$ J

In emission, the final level has the lower n, and the frequency and spectral wavelength is given by

$$\nu = \frac{1}{h}|E_f - E_i| = \frac{c}{\lambda} \tag{2.5-32}$$

or

$$\lambda = \frac{hc}{|E_f - E_i|} \tag{2.5-33}$$

Here,

$$|E_f - E_i| = |E_1 - E_2| = |(6.03 - 24.10) \times 10^{-18}|\text{J}$$
$$= 18.07 \times 10^{-18}\text{ J}$$

so

$$\lambda = \frac{(6.626 \times 10^{-34}\text{ J s})(2.998 \times 10^8\text{ m s}^{-1})}{18.07 \times 10^{-18}\text{ J}} \tag{2.5-34}$$
$$= 11.07 \times 10^{-9}\text{ m} = 10.99\text{ nm} = 1.099\text{ Å}$$

This is in the X-ray region of radiation and does not match the shortest line of the hydrogen spectrum. The particle in a box model is not an accurate one for hydrogen, but it does illustrate some concepts.

(b) For a 10-g ball in a one-dimensional box, 1 m long,

$$E_n = \frac{n^2h^2}{8mL^2} = \frac{n^2(6.626 \times 10^{-34}\text{ J s})}{8(10 \times 10^{-3}\text{ kg})(1\text{ m})^2} \tag{2.5-35}$$
$$= (5.49 \times 10^{-66}\text{ J})\, n^2$$

$$\text{For } n = 1,\ E_1 = 5.49 \times 10^{-66}\ \text{J}$$
$$\text{For } n = 2,\ E_2 = 4E_1 = 22.0 \times 10^{-66}\ \text{J} \tag{2.5-36}$$

$$|E_f - E_i| = |E_1 - E_2| = 3\,E_1$$
$$= 16.5 \times 10^{-66}\ \text{J}$$

The wavelength is

$$\lambda = \frac{hc}{DE} = \frac{(6.626 \times 10^{-34}\ \text{J s})(2.998 \times 10^{8}\ \text{m s}^{-1})}{16.5 \times 10^{-66}\ \text{J}}$$

The ΔE is so small and the λ so large for a 10-g ball that they are impossible to measure. Thus quantization is not observed for large bodies, and their behavior may be calculated using classical (Newtonian) mechanics. ■

2.6 PARTICLE IN A BOX

This is the three-dimensional analogue of the problem worked out in the preceding section. In the statistical theory of the perfect gas, it serves as a model for a gas molecule in a containing vessel. A particle is presumed to move freely within a rectangular box of dimensions $a \times b \times c$ with impenetrable walls. In terms of potential energy, in analogy with (2.5-1),

$$V(x,y,z) = \begin{cases} 0 & \text{inside box} \\ \infty & \text{outside box} \end{cases} \tag{2.6-1}$$

Again the wave function $\psi(x,y,z)$ must vanish everywhere outside the box. If continuity is assumed, the wave function must also vanish on the surface of the box. Orienting the box so that its edges are parallel to the Cartesian axes, with one corner at (0,0,0), the following boundary conditions result:

$$\psi(x,y,z) = 0 \tag{2.6-2}$$

when $x = 0,\ x = a,\ y = 0,\ y = b,\ z = 0,\ z = c$. Inside the box, where the potential energy is everywhere zero, the Hamiltonian is simply the kinetic energy operator obtained by replacing p_i by $-i\hbar\ \partial/\partial x_i$. The Schrödinger equation is

$$-\frac{\hbar^2}{2m}\nabla^2\psi(x,y,z) = E\psi(x,y,z) \tag{2.6-3}$$

subject to the boundary conditions (2.6-2).

The second-order partial differential equation (2.6-3) is separable in Cartesian coordinates. To show this, a solution of the following form is tried:

$$\psi(x,y,z) = X(x)Y(y)Z(z) \tag{2.6-4}$$

The boundary conditions (2.6-2) can then be reexpressed as

$$X(0) = X(a) = 0$$
$$Y(0) = Y(b) = 0 \tag{2.6-5}$$
$$Z(0) = Z(c) = 0$$

Substituting (2.6-4) into (2.6-3) gives

$$YZ\frac{d^2X}{dx^2} + ZX\frac{d^2Y}{dy^2} + XY\frac{d^2Z}{dz^2} + \frac{2mE}{\hbar^2}XYZ = 0 \qquad (2.6\text{-}6)$$

Dividing by $\psi(x,y,z)$ and using the prime notation for derivatives,

$$\frac{X''}{X} + \frac{Y''}{Y} + \frac{Z''}{Z} + \frac{2mE}{\hbar^2} = 0 \qquad (2.6\text{-}7)$$

Each of the first three terms in (2.6-7) depends on one variable only, independent of the other two. This is possible only if each term separately equals a constant. Denoting these constants by $-\lambda^2$, $-\mu^2$, and $-\nu^2$, respectively, (2.6-7) is transformed into three ordinary second-order differential equations.

$$\begin{aligned} X'' + \lambda^2 X &= 0 \\ Y'' + \mu^2 Y &= 0 \\ Z'' + \nu^2 Z &= 0 \end{aligned} \qquad (2.6\text{-}8)$$

subject to the boundary conditions (2.6-5). The constants are related by

$$\frac{2mE}{\hbar^2} = \lambda^2 + \mu^2 + \nu^2 \qquad (2.6\text{-}9)$$

Each of the equations (2.6-8), with its associated boundary conditions (2.6-5), is equivalent to (2.5-6), for the particle in a wire. The normalized solutions $X(x)$, $Y(y)$, $Z(z)$ of (2.6-8) can be written down in analogy with the one-dimensional result, (2.5-19), viz.,

$$X_{n_1}(x) = \left(\frac{2}{a}\right)^{1/2} \sin\frac{n_1\pi}{a}x \qquad n_1 = 1, 2, \ldots \qquad (2.6\text{-}10a)$$

$$X_{n_2}(y) = \left(\frac{2}{a}\right)^{1/2} \sin\frac{n_2\pi}{b}y \qquad n_2 = 1, 2, \ldots \qquad (2.6\text{-}10b)$$

$$Z_{n_3}(z) = \left(\frac{2}{c}\right)^{1/2} \sin\frac{n_3\pi}{c}z \qquad n_3 = 1, 2, \ldots \qquad (2.6\text{-}10c)$$

The constants λ, μ, and ν are given by [cf. (2.5-12)]

$$\lambda = \frac{n_1\pi}{a}, \qquad \mu = \frac{n_2\pi}{b}, \qquad \nu = \frac{n_3\pi}{c} \qquad (2.6\text{-}11)$$

Equation (2.6-9) gives the energy eigenvalues (energy levels) as

$$E_{n_1,n_2,n_3} = \frac{h^2}{8m}\left(\frac{n_1^2}{a} + \frac{n_2^2}{b^2} + \frac{n_3^2}{c^2}\right) \qquad n_1, n_2, n_3 = 1, 2, \ldots \qquad (2.6\text{-}12)$$

For a cubical box of sides $a = b = c$, this becomes

$$E_{n_1,n_2,n_3} = \frac{h^2}{8ma^2}(n_1^2 + n_2^2 + n_3^2) = \frac{h^2}{8mV^{2/3}}(n_1^2 + n_2^2 + n_3^2) \qquad (2.6\text{-}13)$$

where $V = a^3$ is the volume of the box. Three quantum numbers are required to specify the state of the particle in a box. The corresponding eigenfunctions are

$$\psi_{n_1,n_2,n_3}(x, y, z) = \left(\frac{8}{V}\right)^{1/2} \sin\frac{n_1\pi x}{a} \sin\frac{n_2\pi y}{b} \sin\frac{n_3\pi z}{c} \tag{2.6-14}$$

The quantum numbers determine the form of the wave function which, by convention is written $\psi_{n_x n_y n_z}$. The most important consequence of these three quantum numbers is that several combinations can yield the same value of energy. For example, suppose one of the quantum numbers is equal to two, the others being unity. Three possible combinations of the quantum numbers lead to the same value of energy E: (1) $n_1 = 1$, $n_2 = 1$, and $n_3 = 2$; (2) $n_1 = 1$, $n_2 = 2$, and $n_3 = 1$; (3) $n_1 = 2$, $n_2 = 1$, and $n_3 = 1$. The corresponding wave functions are then

$$\psi_{112} = A \sin\left(\frac{\pi x}{L}\right) \sin\left(\frac{\pi y}{L}\right) \sin\left(\frac{2\pi z}{L}\right) \tag{2.6-15}$$

$$\psi_{121} = A \sin\left(\frac{\pi x}{L}\right) \sin\left(\frac{2\pi y}{L}\right) \sin\left(\frac{\pi z}{L}\right) \tag{2.6-16}$$

$$\psi_{211} = A \sin\left(\frac{2\pi x}{L}\right) \sin\left(\frac{\pi y}{L}\right) \sin\left(\frac{\pi z}{L}\right) \tag{2.6-17}$$

but in each instance the energy is exactly the same; that is,

$$\epsilon = \frac{6h^2}{8mL^2} \tag{2.6-18}$$

The corresponding energy level is said to be threefold degenerate. On the basis of this model, the level in which all three quantum numbers equal 2 would be nondegenerate. Table 2.6-1 has the energy level diagram for a

TABLE 2.6-1 ENERGY LEVELS AND STATES OF A PARTICLE IN A BOX

Energy levels	Energy states	Degeneracy
$3\left(\frac{h^2}{8mV^{2/3}}\right)$	$\psi_{1,1,1}$	$g_1 = 1$
$6\left(\frac{h^2}{8mV^{2/3}}\right)$	$\psi_{2,1,1}$ $\psi_{1,1,2}$ $\psi_{1,1,2}$	$g_2 = 3$
$9\left(\frac{h^2}{8mV^{2/3}}\right)$	$\psi_{2,2,1}$ $\psi_{1,2,2}$ $\psi_{2,1,2}$	$g_3 = 3$
$11\left(\frac{h^2}{8mV^{2/3}}\right)$	$\psi_{3,1,1}$ $\psi_{1,3,1}$ $\psi_{1,1,3}$	$g_4 = 3$
$12\left(\frac{h^2}{8mV^{2/3}}\right)$	$\psi_{2,2,2}$	$g_5 = 1$

particle in a three-dimensional cubic box for the ground state and some excited states, along with the degrees of degeneracy and quantum numbers.

■ EXAMPLE 2.6-1

(a) Calculate the first three energy levels of a helium gas atom in a cubical box of side 0.282 m and volume 0.0224 m^3 at STP (273 K and 1 atm).

(b) Estimate the quantum numbers for a helium gas atom of energy $\frac{3}{2}\,kT$ at 273 K.

Solution

(a) The energy eigenvalues for a particle of mass m, in a cubical box of side L and volume V, are given by

$$\epsilon = \frac{\hbar^2\pi^2}{2mV^{2/3}}\left(n_x^2 + n_y^2 + n_z^2\right) = C\left(n_x^2 + n_y^2 + n_z^2\right) \tag{2.6-19}$$

where

$$C = \frac{\hbar^2\pi^2}{2mV^{2/3}} \tag{2.6-20}$$

Here, $\hbar$ is equal to $h/2\pi$ where Planck's constant h is equal to 6.626×10^{-34} J s. The quantum numbers n_x, n_y, and n_z can have the positive integer values 1, 2, 3, ... The lowest energy eigenvalue, which is the ground state, has the quantum numbers $n_x = n_y = n_z = 1$. The energy of this state is

$$\epsilon_{1,1,1} = (1^2 + 1^2 + 1^2)C = 3C \tag{2.6-21}$$

At STP, which is 1 atm and 273 K, a volume of 0.0224 m^3 contains 1 mole of helium atoms. The mass of a helium atom is 6.65×10^{-27} kg. This gives

$$C = 1.04 \times 10^{-40}\ \text{J} \tag{2.6-22}$$

Then,

$$\epsilon_{1,1,1} = 3C = 3.12 \times 10^{-40}\ \text{J} \tag{2.6-23}$$

Now, one electron volt (1 eV) is equal to 1.602×10^{-19} J. Hence,

$$\epsilon_{1,1,1} = 1.95 \times 10^{-21}\ \text{eV} \tag{2.6-24}$$

The lowest energy level, specified by $n_x = n_y = n_z = 1$ is not degenerate. The next set of quantum states have quantum numbers n_x, n_y, and n_z equal to (2,1,1), (1,2,1), and (1,1,2), respectively, corresponding to three independent solutions of Schrödinger's equation having the same energy eigenvalue. This energy level is threefold degenerate, and has energy $(2^2 + 1^2 + 1^2)C = 6C$. The energy level with quantum numbers (2,2,1), (2,1,2), and (1,2,2) are also threefold degenerate, the energy eigenvalues being equal to $9C$.

(b) According to the kinetic theory of ideal gases, the mean energy of a helium atom in a helium gas at an absolute temperature T is $\frac{3}{2}\,kT$, where k is Boltzmann's constant. At 273 K, the mean energy of a helium gas atom is

$$\bar{\epsilon} = \tfrac{3}{2}kT = \tfrac{3}{2} \times 1.38 \times 10^{-23} \times 273 = 5.65 \times 10^{-21}\ \text{J}$$

This is 1.81×10^{19} times larger than the value of the energy of the ground state. Putting the energy equal to $\frac{3}{2}kT = 5.65 \times 10^{-21}$ J, to represent a typical helium gas atom at 273 K, gives

$$\begin{aligned} n_x^2 + n_y^2 + n_z^2 &= \frac{3kT}{2C} \\ &= \frac{5.65 \times 10^{-21}}{1.04 \times 10^{-40}} \\ &= 5.43 \times 10^{19} \end{aligned} \tag{2.6-25}$$

This shows that a helium gas atom in a cubical box of volume 0.0224 m³ kept at 273 K is likely to be in a single-particle quantum state having a huge range of quantum numbers. ■

2.7 DENSITY OF STATES

For a cubical box of volume V, quantum mechanics gives the translational energy of a molecule as

$$\epsilon_{\text{tran}} = \frac{h^2}{8mV^{2/3}}(n_x^2 + n_y^2 + n_z^2) \tag{2.7-1}$$

It is of interest to calculate the number of translational states whose energy is less than some maximum value ϵ. For $\epsilon_{\text{tran}} \le \epsilon$, the translational quantum numbers must satisfy the inequality

$$n_x^2 + n_y^2 + n_z^2 \le 8mV^{2/3}h^{-2}\epsilon \tag{2.7-2}$$

Figure 2.7-1 shows axes whose coordinates are the quantum numbers n_x, n_y, and n_z.

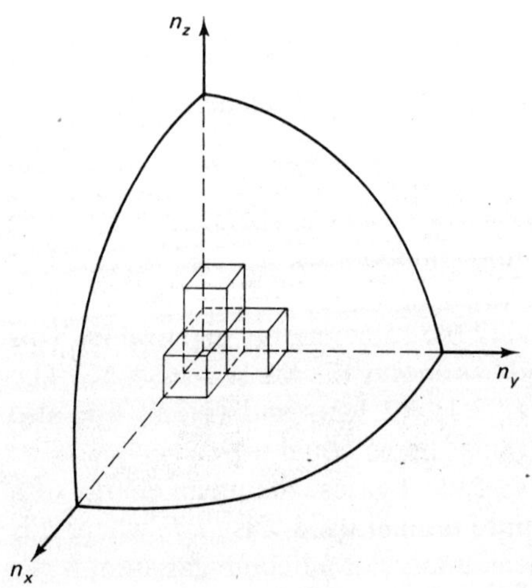

Figure 2.7-1 Translational energy states in quantum space.

Imagine a dot placed at each point whose coordinates are integers. The distance of a dot from the origin is $r = (n_x^2 + n_y^2 + n_z^2)^{1/2}$. The equation

$$n_x^2 + n_y^2 + n_z^2 = \frac{8\epsilon m V^{2/3}}{h^2} \tag{2.7-3}$$

is the equation of a sphere in the coordinate space (n_x, n_y, n_z) having a radius $(8\epsilon m V^{2/3}/h^2)^{1/2}$. Since the translational quantum numbers must be positive integers, only the positive octant of the sphere may be considered.

The allowed translational energy states can be represented by points in the three-dimensional quantum number space. Dividing the sphere into small cubes of unit volume gives one point (one allowed translational energy state) in each small cube. Provided the spherical octant is "not too small," the number of translational energy states between energy 0 and energy ϵ is then

$$\begin{aligned} g(\epsilon) &= \text{volume of the octant} \\ &= \tfrac{1}{8}(\tfrac{4}{3}\pi)\,(\text{radius})^3 \\ &= \tfrac{1}{8}\tfrac{4}{3}\pi[(8mV^{2/3}\epsilon h^{-2})^{1/2}]^3 \end{aligned}$$

that is,

$$g(\epsilon) = \frac{4\pi}{3}(2m\epsilon)^{2/3} V h^{-3} \tag{2.7-4}$$

To give some idea of the numerical size of this quantity and of the value of the translational quantum numbers that describe translational energy states, consider a helium atom of mass $6 \cdot 7 \times 10^{-27}$ kg in a container of 1 liter such that the atom is moving with a velocity of 10^3 m s^{-1}, this being a reasonable value for a helium atom if it were in a container full of helium gas at STP.

The translational energy of the helium atom is

$$\tfrac{1}{2}mv^2 = \tfrac{1}{2}(6 \cdot 7 \times 10^{-27}) \times (10^3)^2 = 3 \cdot 35 \times 10^{-21}\ \text{J} \tag{2.7-5}$$

so, from (2.7-3),

$$\begin{aligned} n^2 = n_x^2 + n_y^2 + n_z^2 &= 8mV^{2/3}\epsilon h^{-2} \\ &= \frac{8 \times 6 \cdot 7 \times 10^{-27} \times (0 \cdot 1)^{2/3} \times 3.35 \cdot 4 \times 10^{-21}}{(6 \cdot 6 \times 10^{-34})^2} \end{aligned}$$

whence

$$n = 10^9 \tag{2.7-6}$$

It is seen that the helium atom is characterized by a very high translational quantum number. From Eq. (2.7-4) it can be deduced that

$$\begin{aligned} g(\epsilon) &= \frac{4\pi}{3} \cdot \frac{(2 \times 6 \cdot 7 \times 10^{-27} \times 3 \cdot 4 \times 10^{-21})^{2/3}}{(6 \cdot 6 \times 10^{-34})^3} \\ &= 10^{28} \end{aligned} \tag{2.7-7}$$

that is, there is an enormous number of translational energy states lying between energies of (essentially) zero and $3 \cdot 35 \times 10^{-21}$ J. Equation (2.7-4) can be used to find the number of translational energy states between energy ϵ and energy $\epsilon + d\epsilon$. It is

$$g(\epsilon)\, d\epsilon = 2\pi(2m)^{3/2} V h^{-3} \epsilon^{1/2}\, d\epsilon \tag{2.7-8}$$

The value $g(\epsilon)$ is known as the *density of states* function.

Equation (2.7-8) gives the number of translational energy states within a given energy range $d\epsilon$. If $d\epsilon$ is taken small enough, it can be said that all the states in the group have the same energy. Also, it is seen that the number of states increases parabolically with the square root of the energy; that is, the degeneracy of translational quantum levels depends parabolically on the energy. These characteristics will surface when it comes time to discuss Fermi–Dirac particles (fermions). For the present, consider that the average energy separation between successive translational energy levels is given by

$$\frac{1}{g(\epsilon)} = \frac{h^3}{2\pi(2m)^{1/2}\, V \epsilon^{1/2}} \tag{2.7-9}$$

For the helium atom discussed above, this shows that the average energy separation between translational levels is about 10^{-48} J, which is very small compared with the translational energy of the atom itself ($3 \cdot 4 \times 10^{-21}$ J).

■ EXAMPLE 2.7-1

In statistical mechanics, it will be seen that for a system of N indistinguishable (nonlocalized), noninteracting particles, the canonical partition function is given by

$$Z = \frac{z^N}{N!} \tag{2.7-10}$$

where

$$z = \sum_r e^{-\beta\epsilon_r} \tag{2.7-11}$$

is the molecular partition function. Equation (2.7-10) holds if the average number of molecules in the molecular quantum state r is very much less than one, i.e., $\langle N_r \rangle << 1$. This insures that there is little probability for two molecules to be in the same molecular state. Show that the inequality $\langle N_r \rangle << 1$ reads

$$\frac{1}{60}\left(\frac{h^2}{mkT}\right)^{2/3} \frac{N}{V} << 1 \tag{2.7-12}$$

and discuss its applicability to various systems.

Solution. To show that the assumption $\langle N_r \rangle << 1$ holds, consider an ideal gas. The molecular energy is the sum of translational, rotational, vibrational, and electronic energies. Consider only the translational states (inclusion of rotational, vibrational, and electronic states would only strengthen the result).

It is known that the average translational energy per molecule is $3kT/2$, and that most molecules have ϵ_{tran} within two or three times this average value. Therefore, take $\epsilon_{max} = 3kT$ as the "maximum" ϵ_{tran} for a typical molecule. The number of quantum states with $\epsilon_{tran} \leq \epsilon_{max}$ equals the number of dots in one-eighth of a sphere whose radius r_{max} is given by Eq. (2.7-2):

$$r_{max} = (8mV^{2/3}h^{-2}\epsilon_{max})^{1/2} = (24mV^{2/3}h^{-2}kT)^{1/2} \tag{2.7-13}$$

Correspondingly, the number of available molecular translational states is then

$$\frac{1}{8}\left(\frac{4\pi r_{max}^3}{3}\right) = \frac{\pi}{6}(24mV^{2/3}h^{-2}kT)^{3/2} \approx 60\left(\frac{mkT}{h^2}\right)^{3/2} V \tag{2.7-14}$$

The condition $\langle N_r \rangle << 1$ requires that $60(mkT/h^2)^{3/2}V >> N$, where N is the number of molecules. This inequality reads the same as (2.7-12).

A typical system is 1 mol of O_2 at 0°C and 1 atm, for which $m = [32/(6 \times 10^{23})]$ g, $N = 6 \times 10^{23}$, and $V = 22{,}400$ cm^3. The left side of Eq. (2.7-12) is then 4×10^{-8}. Hence, $\langle N_r \rangle << 1$ is well satisfied. The translational levels are spaced extremely close together, so there are far more translational states available than there are molecules in the gas.

The inequality (2.7-12) does not hold for (a) extremely low temperatures, (b) extremely high densities, and (c) extremely small particle masses. Physical examples of these conditions are (a) liquid helium at 2 or 3 K, (b) white-dwarf stars and neutron stars (pulsars), and (c) conduction electrons in metals.

Liquid helium is a boson gas. White dwarfs (typical radius 10,000 mi, compared with 1,000,000 mi for the sun) do not contain atoms but are composed of atomic nuclei (mainly He^{2+}) and free electrons. Their density is typically 10^6 g cm^{-3}. A neutron star (typical radius 10 mi) is produced when gravitational collapse proceeds to the point where the electrons and protons are forced to unite to form neutrons. The density of a neutron star is about the same as the density of matter in an atomic nucleus, 10^{14} g cm^{-3}.

The conduction electrons in a metal are relatively free to move through the metal and can be approximately treated as a "gas" of electrons. Assuming one conduction electron per metal atom and a molar volume of 20 cm^3, it is found that the left side of (2.17-12) equals 700 at 0°C. The assumption $\langle N_r \rangle << 1$ fails, mainly because of the small mass of the electrons and, to a lesser extent, the higher density of the electron gas compared to an ordinary gas. ■

■ EXAMPLE 2.7-2

Determine the effect of a volume change on the energy eigenvalues of a particle in a box.

Solution. The energy eigenvalues for a particle of mass m, in a cubical box of volume V, are given by

$$\epsilon = \frac{\hbar^2\pi^2}{2mV^{2/3}}(n_x^2 + n_y^2 + n_z^2) \tag{2.7-15}$$

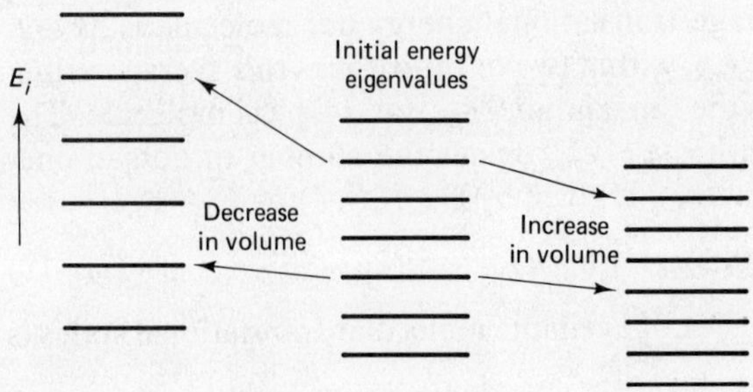

Figure 2.7-2 Effect of volume change on the energy eigenvalues of a particle in a box.

According to this, the values of the energy eigenvalues of the single-particle states for a particle in a cubical box are all inversely proportional to $V^{2/3}$, where V is the volume of the box. This dependence on V comes in via the boundary conditions used to solve Schrödinger's equation. If the volume of the box is increased, Eq. (2.7-15) shows that, for fixed values of n_x, n_y, and n_z, the values of all the energy eigenvalues decrease as illustrated in Fig. 2.7-2.

If the volume is increased, the energy eigenvalues are all decreased, whereas if the volume is decreased, the energy eigenvalues are all increased and are further apart. There is no upper limit to the integral values of n_x, n_y, and n_z. Hence, there is no upper bound to the energy of a particle in a box. ■

2.8 OPERATORS

Quantum mechanics is most conveniently formulated in terms of operators. An operator is a rule for transforming a given function into another function. A mathematical operator operates on a function and either leaves it unchanged or changes it into another function. For example, if operator $\hat{A}$ converts the function $f(x)$ into the function $g(x)$,

$$\hat{A}f(x) = g(x) \tag{2.8-1}$$

An operator is written with a caret over it.

There are many possible operators. The operator for multiplying by a constant c is represented by $\hat{c}$. The operator for multiplying by x is represented by $\hat{x}$. The operator for differentiation with respect to x is represented by D_x or d/dx. If $\hat{A}$ is the operator d/dx, then $g(x) = f'(x)$. If $\hat{A}$ is the operator "multiplication by $3x^2$," then $g(x) = 3x^2 f(x)$. If $\hat{A} = \log$, then $g(x) = \log f(x)$.

The sum of two operators $\hat{A}$ and $\hat{B}$ is defined by $(\hat{A} + \hat{B})f(x) = \hat{A}f(x) + \hat{B}f(x)$. For example, $(\log + d/dz)f(x) = \log f(x) + f'(x)$. The square of an operator is defined by $\hat{A}^2 f(x) = \hat{A}[\hat{A}f(x)]$. For example,

$$\left(\frac{d}{dz}\right)^2 f(x) = \frac{d}{dz}\left[\frac{d}{dx} f(x)\right] = \frac{d}{dx}[f'(x)] = f''(x) = \frac{d^2}{dx^2} f(x)$$

Therefore, $(d/dx)^2 = d^2/dx^2$. The product of two operators is defined by $(\hat{A}\hat{B})f(x) = \hat{A}[\hat{B}f(x)]$.

The operators occurring in quantum mechanics are all linear operators. Linear operators are such that

$$\hat{A}[f(x) + g(x)] = \hat{A}f(x) + \hat{A}f(x) \qquad \hat{A}[cf(x)] = c\hat{A}f(x) \tag{2.8-2}$$

Taking the square root is an example of an operator that is not linear.

Operators are important in quantum mechanics because every observable physical quantity has a linear operator associated with it. The operator that corresponds to p_x, the x component of momentum of a particle, is postulated to be $(h/i)\,(\partial/\partial x)$, with similar operators for p_y and p_x. Thus,

$$\hat{p}_x = \frac{\hbar}{i}\frac{\partial}{\partial x}, \qquad \hat{p}_y = \frac{\hbar}{i}\frac{\partial}{\partial y}, \qquad \hat{p}_x = \frac{\hbar}{i}\frac{\partial}{\partial z} \tag{2.8-3}$$

where $\hat{p}_x$ is the quantum-mechanical operator for the property p_x and $i \equiv \sqrt{-1}$. The operator that corresponds to the x coordinate of a particle is multiplication by x, and the operator that corresponds to $f(x,y,z)$, where f is any function, is multiplication by that function. For example,

$$\hat{x} = x x, \qquad \hat{y} = y x, \qquad \hat{z} = z x, \qquad \hat{f}(x,y,z) = f(x,y,z)\,x \tag{2.8-4}$$

To find the operator that corresponds to any other physical property, the classical–mechanical expression for that property is first expressed as a function of the Cartesian coordinates and corresponding momenta and then the coordinates and momenta replaced by their corresponding operators. An important operator in quantum mechanics is that for energy. For conservative systems, that is, systems for which the potential energy depends only on the coordinates, the total energy H is the sum of the kinetic energy T and the potential energy V. The symbol H is used for the energy to indicate that the independent variables in the function are positions and momenta (not positions and velocities, for example). The use of momenta instead of velocities arises naturally in an alternative form of classical mechanics developed by Hamilton, which is fully equivalent to Newtonian mechanics but has advantages in adapting to curvilinear coordinate systems or other descriptions more complicated than Cartesian. The function

$$H = T + V \tag{2.8-5}$$

is known as the Hamiltonian function. The Hamiltonian function for a particle of mass m is

$$H = \frac{1}{2m}(p_x^2 + p_y^2 + p_z^2) + V(x, y, z) \tag{2.8-6}$$

where $V(x, y, z)$ is the potential energy as a function of x, y, and z. Since $\hat{p}_x = (\hbar/i)\,(\partial/\partial x)$, $\hat{p}_x^2 = -\hbar^2\partial^2/\partial x^2$. Converting the classical Hamiltonian H to the quantum-mechanical Hamiltonian operator $\hat{H}$ yields

$$\hat{H} = -\frac{\hbar^2}{2m}\left(\frac{\partial^2}{\partial x^2} + \frac{\partial^2}{\partial y^2} + \frac{\partial^2}{\partial z^2}\right) + V(x, y, z) \tag{2.8-7}$$

$$= -\frac{\hbar^2}{2m}\nabla^2 + V(x, y, z)$$

For a many-particle system, $\hat{p}_{x,1} = (h/i)\ \partial/\partial x_1$ for particle 1, and the Hamiltonian operator is readily found to be

$$\hat{H} = -\frac{\hbar^2}{2m}\nabla_1^2 - \frac{\hbar^2}{2m_2}\nabla_2^2 - \cdots - \frac{\hbar^2}{2m_n}\nabla_n^2 + V(x_1, \ldots, z_n, t) \tag{2.8-8}$$

$$\nabla_1^2 \equiv \frac{\partial^2}{\partial x_1^2} + \frac{\partial^2}{\partial x_1^2} + \frac{\partial^2}{\partial z_1^2} \tag{2.8-9}$$

with similar definitions for $\nabla_2^2, \ldots, \nabla_n^2$. The terms on the right side of (2.8-8) are the operators for the kinetic energy of particle 1, the kinetic energy of particle 2, . . . , the kinetic energy of particle n, and the potential energy of the system.

In operator form, it is seen that the time-dependent Schrödinger equation can be written as

$$-\frac{h}{i}\frac{\partial \Psi}{\partial t} = \hat{H}\Psi \tag{2.8-10}$$

and the time-independent Schrödinger equation as

$$\hat{H}\psi = E\psi \tag{2.8-11}$$

Since there is a whole set of allowed stationary-state wave functions and energies, it is just as well to write (2.8-11) as

$$\hat{H}\psi_j = E_j\psi_j \tag{2.8-12}$$

where the subscript j refers to the various wave functions and energies.

When an operator $\hat{B}$ applied to the function f gives the function back again but multiplied by the constant c, that is, when $\hat{B}f = cf$, it is said that f is an eigenfunction of $\hat{B}$ with eigenvalue c. The wave functions ψ in (2.8-11) are eigenfunctions of the Hamiltonian operator, the eigenvalues being the allowed energies E.

Expectation Values Since $|\psi^2| = \psi^*\psi$ is the probability density for finding the particle between x and $x + dx$, quantum mechanics postulates that the average or expectation value of any physical property M in a system whose state function is Ψ is given by

$$\langle M \rangle = \int \Psi^* \hat{M} \Psi \, d\tau \tag{2.8-13}$$

where $\hat{M}$ is the operator for the property M and the integral goes over all space. In (2.8-13), $\hat{M}$ operates on Ψ to produce the result $\hat{M}\Psi$, which is a function; the function $\hat{M}\Psi$ is then multiplied by Ψ^*, and the resulting function $\Psi^*\hat{M}\Psi$ is integrated over the full range of the spatial coordinates of the system. For example, in the case of the momentum p_x, the operator is $p_x = (\hbar/i)\,\partial/\partial x$, and the average values of p_x for a one-particle, three-dimensional system whose state function is Ψ is

$$\langle p_x \rangle = (\hbar/i)\int_{-\infty}^{\infty}\int_{-\infty}^{\infty}\int_{-\infty}^{\infty} \Psi^* \frac{\partial \Psi}{\partial x}\, dx\, dy\, dz$$

The average value of M is the average of the results of a very large number of measurements of M made on identical systems, each of which is in the same state Ψ just before the measurement. For a stationary state, Ψ equals $e^{iEt/\hbar}\psi$. Since $\hat{M}$ does not affect the $e^{iEt/\hbar}$ factor, $\Psi^*\hat{M}\Psi = e^{iEt/\hbar}\psi^*\hat{M}e^{-iEt/\hbar}\psi = e^{iEt/\hbar}e^{-iEt/\hbar}\psi^*\hat{M}\psi = \psi^*\hat{M}\psi$, and so on, Thus,

$$\langle \hat{M} \rangle \doteq \int \psi^* \hat{M} \psi\, d\tau \tag{2.8-14}$$

For example, for a particle in a one-dimensional box, the stationary-state wave functions are $\psi = (2/a)^{1/2}\sin(n\pi x/a)$, and the average value of x^2 for the stationary state with quantum number n is given by (2.8-14) as

$$\langle x^2 \rangle = \frac{2}{a}\int_0^a x^2 \sin\left(\frac{n\pi x}{a}\right) dx$$

since the operator x^2 is simply multiplication by x^2.

The expectation value is also equal to the eigenvalue. This can be seen from the fact that if the application of an operator $\hat{F}$ to a function ψ yields a constant F_n times ψ,

$$\hat{F}\psi = F_n\psi \tag{2.8-15}$$

the expectation value is then equal to

$$\begin{aligned}\langle F \rangle &= \int x^*\hat{F}\psi\, dx\, dy\, dz = \int \psi^* F_n \psi\, dx\, dy\, dz \\ &= F_n \int \psi^*\psi\, dx\, dy\, dz = F_n\end{aligned} \tag{2.8-16}$$

This means that repeated measurements of F will always yield the same value; F is a constant of the motion.

■ EXAMPLE 2.8-1

(a) Consider the operator $(\partial/\partial x)$ and show that it is equal to

$$1 + x\frac{\partial}{\partial x} \tag{2.8-17}$$

(b) The commutator of two operators $\hat{A}$ and $\hat{B}$ is defined as

$$[\hat{A}, \hat{B}] = \hat{A}\hat{B} - \hat{B}\hat{A} \tag{2.8-18}$$

It is the difference between operating first with $\hat{B}$ and then with $\hat{A}$. Show that, in general,

$$[\hat{A}, \hat{B}] \neq 0 \tag{2.8-19}$$

Solution

(a) For any function $f(x)$,

$$\left(\frac{\partial}{\partial x} x\right) f(x) = f(x) + x\frac{\partial f(x)}{\partial x} = \left(1 + x\frac{\partial}{\partial x}\right) f(x)$$

Since this is valid for any $f(x)$, the following operator equation holds:

$$\frac{\partial}{\partial x}x = 1 + x\frac{\partial}{\partial x} \tag{2.8-20}$$

(b) Take the simple case of $\hat{A} = x$ and $\hat{B} = \partial/\partial x$. Then, for any $f(x)$,

$$\left(x, \frac{\partial}{\partial x}\right) = x\frac{\partial f(x)}{\partial x} - \frac{\partial}{\partial x}xf(x)$$

$$= x\frac{\partial}{\partial x} - \left(1 + x\frac{\partial}{\partial x}\right) f(x) = -f(x)$$

Thus,

$$\left(x, \frac{\partial}{\partial x}\right) = -1 \tag{2.8-21}$$

In other words, the operators $\hat{A} = x$ and $\hat{B} = \partial/\partial x$ do not commute. ■

■ EXAMPLE 2.8-2

Show that if all eigenfunctions of two operators $\hat{A}$ and $\hat{B}$ are the same functions, $\hat{A}$ and $\hat{B}$ commute with each other.

Solution. The eigenvalues of $\hat{A}$ and $\hat{B}$ are represented by a and b and the eigenfunctions are ψ, so that

$$\hat{A}\psi = a_i\psi_i \quad \text{and} \quad \hat{B}\psi_i = b_i\psi_i \tag{2.8-22}$$

The eigenfunctions of the operator $\hat{A}\hat{B}$ are obtained as follows.

$$\hat{A}\hat{B}\psi_i = \hat{A}(\hat{B}\psi_i) = \hat{A}b_i\psi_i = b_i\hat{A}\psi_i = b_i a_i\psi_i \tag{2.8-23}$$

The operator $\hat{B}\hat{A}$ has eigenvalue $a_i b_i$, as may be seen from

$$\hat{B}\hat{A}\psi_i = \hat{B}(\hat{A}\psi_i) = \hat{B}a_i\psi_i = a_i\hat{B}\psi_i = a_i b_i\psi_i \tag{2.8-24}$$

Since $a_i b_i$, $\hat{A}$ and $\hat{B}$ commute with each other.

When operators $\hat{A}$ and $\hat{B}$ do commute, the observables they represent can in principle be measured simultaneously and precisely. If not, simultaneous measurement is impossible beyond a certain level of precision. The failure of x

and p_x to commute leads, for example, to the Heisenberg uncertainty principle $\Delta x\, \Delta p_x \geq \hbar/2$.

■

■ EXAMPLE 2.8-3

(a) What are the eigenfunctions of the operator d/dx?

(b) Verify that $\psi = (2/a)^{1/2} \sin(n\pi x/a)$ is an eigenfunction of the equation $\hat{H}\psi = E\psi$.

Solution

(a) The eigenvalue equation for a function $f(x)$ being operated on by d/dx is

$$\frac{d}{dx} f(x) = kf(x) \tag{2.8-25}$$

or

$$\frac{df(x)}{f(x)} = k\,dx \tag{2.8-26}$$

Integration gives

$$\ln f(x) = kx + c; \qquad f(x) = e^c e^{kx} = c' e^{kx}$$

where c and c' are constants. For each different value of k there is an eigenfunction $c'e^{kx}$. Or, to put it another way, the eigenfunction $c'e^{kx}$ has the eigenvalue k.

(b) For the one-dimensional case, the Hamiltonian operator is

$$\hat{H} = -\frac{\hbar^2}{2m}\frac{d^2}{dx^2} \tag{2.8-27}$$

where $\hbar = h/2\pi$. Therefore,

$$\hat{H} = -\frac{\hbar^2}{2m}\frac{d^2}{dx^2}\left[\left(\frac{2}{a}\right)^{1/2} \sin\frac{n\pi x}{a}\right] \tag{2.8-28}$$

$$= -\frac{h^2}{4\pi^2(2m)}\left(\frac{2}{a}\right)^{1/2}\left(-\frac{n^2 x^2}{a^2}\right) \sin\frac{n\pi x}{a} = \frac{n^2h^2}{8ma^2}\left(\frac{2}{a}\right)^{1/2} \sin\frac{n\pi x}{a} = E\psi$$

where $E = n^2h^2/8ma^2$.

■

■ EXAMPLE 2.8-4

Physical properties are real (not imaginary), so eigenvalues must be real, which means that operators in quantum mechanics must be of the type called *Hermitian*. An operator $\hat{o}$ is Hermitian if the following condition holds:

$$\int u_m^* \hat{o} u_n \, dv = \int u_n \hat{o}^* u_m^* \, dv \tag{2.8-29}$$

where u_m and u_n are two eigenfunctions, the asterisk designates the complex conjugate, and dv is the volume element $(dx\ dy\ dz)$. The relationship also holds for $m = n$.

(a) Show that the operator $-i\hbar\ d/dx$ is Hermitian for the case $m = n$, using the eigenvalue $u_m = Be^{ikx/\hbar}$. (This is the solution to an actual eigen-equation for a one-dimensional case.) B is a constant.

(b) The average or expectation value of a property A is the value to be obtained when the property is measured. It is given by

$$\langle A \rangle = \frac{\int \psi^* \hat{A} \psi\ dv}{\int \psi^* \psi\ dv} = \int \psi^* \hat{A}\ dv \tag{2.8-30}$$

when $\psi^*\psi$ is normalized. Show that the expectation value (eigenvalue) obtained for an eigenfunction ψ is real if the operator $\hat{A}$ is Hermitian.

Solution

(a) Let the complex conjugates be written, replacing i by $-i$:

$$\hat{o}^* = \left(-i\hbar \frac{d}{dx}\right)^* = i\hbar \frac{d}{dx} \qquad u_m^* = Be^{-ikx/\hbar} \tag{2.8-31}$$

To verify that $-i\hbar\ d/dx$ is Hermitian, apply the test

$$\int Ae^{-ikx/\hbar}(-i\hbar)\frac{d}{dx}Ae^{ikx/\hbar}\ dx \stackrel{?}{=} \int Ae^{ikx/\hbar}(i\hbar)\frac{d}{dx}Ae^{-ikx/\hbar}\ dx$$

Now,

$$\frac{d}{dx}Ae^{ikx/\hbar} = \frac{ik}{\hbar}Ae^{ikx/\hbar} \qquad \frac{d}{dx}Ae^{-ikx/\hbar} = \frac{ik}{\hbar}Ae^{-ikx/\hbar} \tag{2.8-32}$$

Thus, the test becomes

$$\int Ae^{-ikx/\hbar}(-i\hbar)\left(\frac{ik}{\hbar}\right)Ae^{ikx/\hbar}\ dx \stackrel{?}{=} \int Ae^{ikx/\hbar}(i\hbar)\left(-\frac{ik}{\hbar}\right)Ae^{-ikx/\hbar}\ dx$$

After canceling $\hbar$, it is seen that the two integrals are equal. Thus, the operator $\hat{o} = i\hbar\ d/dx$ is Hermitian.

(b) For the case of $m = n$, the Hermitian condition is

$$\int \psi^* \hat{A} \psi\ dv = \int \psi \hat{A}^* \psi^*\ dv \tag{2.8-33}$$

Now, the expectation value is $\langle A \rangle = \int \psi^* \hat{A} \psi\ dv$. Its complex conjugate is $\langle A \rangle^* = \int \psi \hat{A}^* \psi^*\ dv$. The Hermitian property says that these two integrals are equal. Therefore, $\langle A \rangle = \langle A \rangle^*$, and for this to be true, $\langle A \rangle$ must be real. ■

2.9 HARMONIC OSCILLATOR

The quantum-mechanical treatment of a harmonic oscillator is important to the understanding of the vibration of molecules. To provide a background, the harmonic oscillator will be considered first from a classical point of view.

A particle of mass m moves in one dimension and is attracted to the coordinate origin by a force proportional to its displacement from the origin

$$F = -kx \tag{2.9-1}$$

where k is called the force constant. (When x is positive, the force is in the $-x$ direction and when x is negative, F is in the $+x$ direction.) From the relation between force and potential energy,

$$F = -\frac{dV}{dx} \tag{2.9-2}$$

or

$$-\frac{dV}{dx} = -kx \tag{2.9-3}$$

integration gives

$$V = \tfrac{1}{2}kx^2 + c = \tfrac{1}{2}kx^2 \tag{2.9-4}$$

the choice of zero of potential energy being taken at the origin $x = 0$. A plot of potential energy V versus displacement x is parabolic for a simple harmonic oscillator, as shown in Fig. 2.9-1(a). A small object sliding without friction in a parabolic well would execute harmonic motion as it slides through the minimum. As it slides up the other side, kinetic energy is converted to potential energy. When the particle is at either one of its two highest points in the oscillation, the velocity is zero, and its energy is entirely potential energy. Newton's second law, $F = ma$, gives

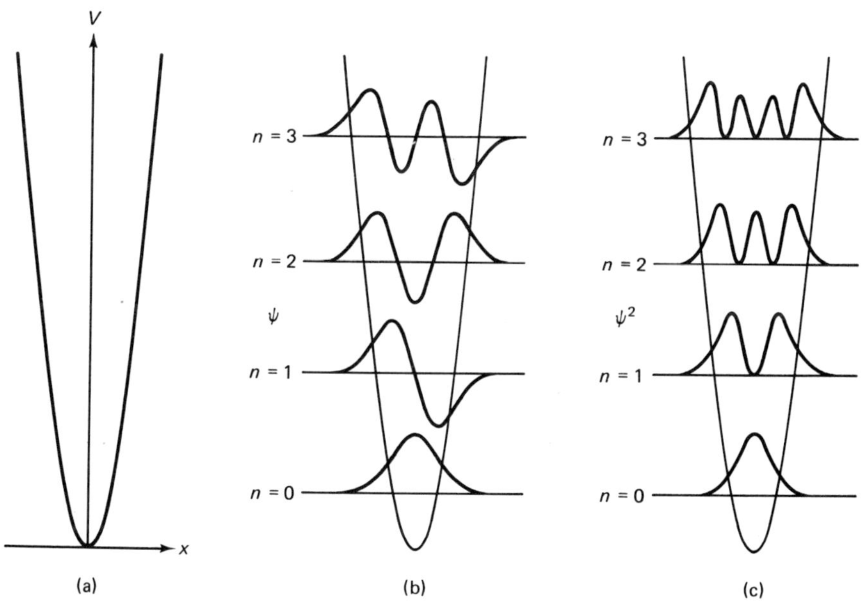

Figure 2.9-1 (a) Potential energy curve, (b) allowed energy levels, wave functions, and (c) probability density functions for harmonic oscillator.

$$m\frac{d^2x}{dt^2} = -kx \tag{2.9-5}$$

The solution to this differential equation is of the form

$$x = A\sin[(k/m)^{1/2}t + b] \tag{2.9-6}$$

as can be verified by substitution in the equation. A and b are integration constants. The maximum and minimum values of the sine function are $+1$ and -1, so the particle's x coordinate oscillates back and forth between $+A$ and $-A$. The period τ of the oscillator is the time required for one complete cycle of oscillation. For one cycle of oscillation, the argument of the sine function in (2.9-6) must increase by 2π, since 2π is the period of a sine function. Hence, the period satisfies $(k/m)^{1/2}\tau = 2\pi$, and $\tau = 2\pi(m/k)^{1/2}$. The fundamental frequency ν_0 is the reciprocal of the period and equals the number of vibrations per second ($\nu_0 = 1/\tau$),

$$\nu_0 = \frac{1}{2\pi}\left(\frac{k}{m}\right)^{1/2} \tag{2.9-7}$$

The solution of (2.9-5) is therefore

$$x - a\sin\left(\frac{k}{m}\right)^{1/2} t = a\sin 2\pi\nu_0 t \tag{2.9-8}$$

if $x = 0$ when $t = 0$.

The energy of the harmonic oscillator is

$$\begin{aligned} E = T + V &= \frac{m}{2}v_x^2 + \frac{1}{2}kx^2 \\ &= \frac{1}{2}kA^2 \end{aligned} \tag{2.9-9}$$

Equation (2.9-9) shows that the classical energy can have any nonnegative value. As the particle oscillates, its kinetic energy and potential energy continually change, but the total energy remains constant at $\frac{1}{2}kA^2$. Classically, the particle is limited to the region $-A \le x \le A$. When the particle reaches $x = A$ or $x = -A$, its speed is zero (since it reverses its direction of motion at $+A$ and $-A$) and its potential energy is a maximum, being equal to $\frac{1}{2}kA^2$. If the particle were to move beyond $x = \pm A$, its potential energy would increase above $\frac{1}{2}kA^2$. This is impossible for a classical particle: The total energy is $\frac{1}{2}kA^2$ and the kinetic energy is nonnegative, so the potential energy ($V = E - T$) cannot exceed the total energy.

Now for the quantum-mechanical treatment: From the classical Hamiltonian,

$$H = \frac{p_x^2}{2m} + \frac{1}{2}kx^2 = \frac{1}{2}kA^2 \tag{2.9-10}$$

the Hamiltonian operator $\hat{H}$ for the harmonic oscillator is obtained by substituting $(\hbar/i)(d/dx)$ for p_x in (2.9-10):

$$\hat{H} = \frac{\hbar^2}{2m}\frac{d^2}{dx^2} + \tfrac{1}{2}kx^2 \tag{2.9-11}$$

The Schrödinger equation for the harmonic escillator is then

$$-\frac{\hbar^2}{2m}\frac{d^2\psi(x)}{dx^2} + \frac{k}{2}x^2\psi(x) = E\psi(x) \tag{2.9-12}$$

or

$$\frac{d^2\psi(x)}{dx^2} + \frac{2m}{\hbar^2}\left(E - \frac{kx^2}{2}\right)\psi(x) = 0 \tag{2.9-13}$$

This is a linear differential equation with nonconstant coefficients, and some effort must be expended on its behalf. The range of the variable x must be taken from $-\infty$ to ∞, there being no finite cutoff as in the case of the particle in a one-dimensional box. A logical first step is to consider an asymptotic solution (as x tends to $\pm\infty$) of the form

$$\psi(x) = H(\gamma x)e^{-\gamma x^2/2} \tag{2.9-14}$$

where

$$\gamma = \frac{2\pi}{h}\sqrt{mk} \tag{2.9-15}$$

and $H(x)$ is a polynomial in x that terminates after a finite number of terms in order to keep $\psi(x)$ finite and square integrable. If $H(x)$ did not terminate, or if it contained infinitely high powers of x, $\psi(x)$ would not be well behaved. It is this requirement that $H(x)$ terminate after a finite number of terms, rather than any geometric boundary conditions, which leads to the quantization results found for the harmonic oscillator.

Substitution of (2.9-14) into the original differential equation results in Hermite's differential equation, whose solution is in terms of a Hermite polynomial. The condition for the polynomial to terminate with the nth power term is

$$\frac{2mE}{\hbar^2\gamma} - 1 - 2n = 0 \tag{2.9-16}$$

or

$$E_n = (n + \tfrac{1}{2})\hbar\omega_0 = (n + \tfrac{1}{2})h\nu_0 \tag{2.9-17}$$

where

$$\omega_0 = 2\pi\nu_0 = \sqrt{\frac{k}{m}} \tag{2.9-18}$$

is the natural *circular frequency* of the oscillator. The eigenfunctions and eigenvalues of Eq. (2.9-13) are given below:

$$\psi_0 = \frac{\gamma^{1/2}}{\pi^{1/4}}\exp\left(-\frac{\gamma^2x^2}{2}\right) \qquad E_0 = \frac{1}{2}h\nu_0$$

$$\psi_1 = \frac{(2\gamma)^{1/2}}{\pi^{1/4}} \gamma x \exp\left(-\frac{\gamma^2 x^2}{2}\right) \qquad E_1 = \frac{3}{2} h\nu_0$$

$$\psi_2 = \left(\frac{1}{\pi^{1/4}}\right)\left(\frac{\gamma}{2}\right)^{1/2} (-1 + 2\gamma^2 x^2) \exp\left(-\frac{\gamma^2 x^2}{2}\right) \qquad E_2 = \frac{5}{2} h\nu_0$$

$$\psi_n = \left(\frac{\gamma}{\pi^{1/2} 2^a n!}\right)^{1/2} H_n(\gamma x) \exp\left(-\frac{\gamma^2 x^2}{2}\right) \qquad E_n = \left(n + \frac{1}{2}\right) h\nu_0$$

where $\nu_0 = (1/2\pi)\sqrt{k/m}$, is the classical frequency for the harmonic oscillator, and H_n is the Hermite polynomial. The quantity γ is the natural parameter for the harmonic oscillator because it has the dimensions of inverse length, so that $\gamma^2 x^2$ is dimensionless.

The energy levels are equally spaced as shown in Fig. 2.9-1(b), where the normalized wave functions ψ_0, ψ_1, and ψ_3 have also been shown. The quantum mechanical treatment of the harmonic oscillator yields rather different results from the classical treatment. According to classical mechanics, the oscillator may have any energy but, according to quantum mechanics, the possible energy levels are given by $E_n = (n + \frac{1}{2})h\nu_0$, where n is 0, 1, 2, ... According to classical mechanics, the oscillator may be at rest and have zero energy but, according to quantum mechanics, the lowest energy level permitted is $E = \frac{1}{2}h\nu_0$, called the zero point energy. (For a collection of harmonic oscillators in thermal equilibrium, all the oscillators will fall to the ground state as the temperature goes to absolute zero; hence the name zero-point energy.)

The probability densities ψ^2 are plotted in Fig. 2.9-1(c) versus x for the first four energy levels. The probability that the x coordinate of the harmonic oscillator is between x and $x + dx$ is given by $\psi^2\,dx$. If a large number of identically prepared systems are examined, the fraction having coordinates between x and $x + dx$ is equal to this probability. In the ground state ($n = 0$) the most probable distance occurs at the position of the minimum in the potential well. This is in distinct contrast with that for a classical simple harmonic oscillator, which would spend the longest times at the turning points. As the quantum number increases, however, the quantum-mechanical probability density function approaches that for the classical harmonic oscillator. This complies with Bohr's "corresponding principle" according to which the quantum-mechanical result must approach the classical result in the limit of large quantum numbers.

Finally, notice that there is a certain probability that the particle will be outside of the classically allowed region that is indicated by the intersection of the parabola with the quantum-mechanical energy level. The probability that the particle will be outside these limits is given by the area of the plot of ψ^2 beyond the classical turning points. This and the passage through a classically forbidden zone (known as *tunneling*) are characteristic of quantum mechanics. Tunneling is more likely to occur when the mass of the particle is small. It is important in the chemistry for electrons, protons, and H atoms, because it influences the reaction rates involving these species.

The broad relevance of the results for the harmonic oscillator is the fact that wave motion, such as used to characterize the vibration of atoms in a crystal or such as electromagnetic radiation, can be represented as a collection of oscillators, each oscillator having one of the normal mode frequencies of the total system. Both crystal vibrations and light can be described in terms of classical wave formalism, and also in terms of a quantum formalism. The quantum for crystal vibrations is called the *phonon;* the quantum for light is called the *photon*. Whenever energy is given to or taken from an harmonic oscillator, the energy difference involved must be an integral multiple of $\hbar\omega_0$, that is, $\Delta E = n\hbar\omega_0$. Following this, it means that every energy interchange with crystal vibrations requires the exchange of an integral number of phonons, $n\hbar\omega_{\text{phonon}}$, and every energy interchange with light requires the exchange of an integral number of photons, $n\hbar\omega_{\text{photon}}$.

EXAMPLE 2.9-1

(a) In an harmonic oscillator, the lowest energy level is not zero, which means that some vibration is occurring even at $T = 0$ K. What principle says that this is to be expected?

(b) The wave function of the first excited state of the one-dimensional oscillator is

$$\psi_1 = \left(\frac{a}{4\pi}\right)^{1/4} 2\sqrt{a}\,xe^{-ax^2/2} \tag{2.9-19}$$

Find the constant a and the value of the energy.

Solution

(a) The principle of uncertainty indicates that if the vibrating oscillator or molecule were to come to a complete rest, its position and momentum would become precisely known. This is impossible.

(b) The first derivative of ψ is

$$\frac{d\psi_1}{dx} = \left(\frac{a}{4\pi}\right)^{1/4} 2\sqrt{a}\,(e^{-ax^2/2} - ax^2e^{-ax^2/2}) \tag{2.9-20}$$

The second derivative is

$$\frac{d^2\psi_1}{dx^2} = \left(\frac{a}{4\pi}\right)^{1/4} 2\sqrt{a}\,(-ax - 2ax + a^2x^3)\,e^{-ax^2/2}$$

$$= (-3a + a^2x^2)\psi_1 \tag{2.9-21}$$

Substituting into the Schrödinger wave equation for the harmonic oscillator

$$-\frac{\hbar^2}{2m}\frac{d^2\psi}{dx^2}n + \frac{1}{2}m\omega^2x^2\psi_n = E\psi_n \tag{2.9-22}$$

yields

$$-\frac{\hbar^2}{2m}(-3a + a^2x^2)\psi_1 + \frac{1}{2}m\omega^2x^2\psi_1 = E_1\psi_1 \tag{2.9-23}$$

or, upon division by ψ_1,

$$\left(\frac{1}{2}m\omega^2 - \frac{\hbar^2 a^2}{2m}\right)x^2 = E_1 - \frac{\hbar^2 3a}{2m} \tag{2.9-24}$$

which must be true for all x. This is possible only if both sides are zero. Thus,

$$\frac{1}{2}m\omega^2 - \frac{\hbar^2 a^2}{2m} = 0 \tag{2.9-25}$$

giving

$$a = \frac{m\omega}{\hbar} \tag{2.9-26}$$

and

$$E_1 - \frac{\hbar^2 3a}{2m} = 0 \tag{2.9-27}$$

or

$$E_1 = \frac{\hbar^2}{2m}3a = \frac{\hbar^2}{2m}3\left(\frac{m\omega}{\hbar}\right) = \frac{3}{2}\hbar\omega \tag{2.9-28}$$

which is in agreement with

$$E_n = \left(n + \frac{1}{2}\right)h\nu = \left(n + \frac{1}{2}\right)\hbar\omega \tag{2.9-29}$$

for $n = 1$. ■

2.10 ANGULAR MOMENTUM

A successful theory for the structure of hydrogenic atoms was first proposed by Bohr in 1913 when he postulated that the electron moves in discrete orbits around the nucleus for which the orbital angular momentum is quantized. The electron may emit or absorb energy only by undergoing a transition to an orbit of lower or higher energy, respectively. Only those orbits are allowed for which the angular momentum is an integral multiple of $\hbar$. These discrete orbits are the stationary state of the atom.

Angular momentum is an important property because it is conserved when a change takes place in an isolated system. Examples of angular momentum are: (1) a rotating molecule, (2) an orbiting electron, and (3) intrinsic (spin) angular momentum of electrons and certain nuclei. If a particle of mass m is revolving around a point at distance r, the angular momentum L is

$$L = rmv = mr^2\omega = I\omega \tag{2.10-1}$$

where v is the linear velocity, ω is the angular velocity ($\omega = v/r$), and $I = mr^2$ is the moment of inertia. The kinetic energy E of the particle is

$$E = \tfrac{1}{2}mv^2 = \tfrac{1}{2}mr^2\omega^2 = \tfrac{1}{2}I\omega^2 \tag{2.10-2}$$

The kinetic energy of the particle can also be written in terms of the angular momentum L:

$$E = \frac{L^2}{2I} \tag{2.10-3}$$

In three dimensions the angular momentum is represented by a vector **L**. If a mass m is rotating about a fixed point P with a linear velocity **v**, the angular momentum **L** is given by

$$\mathbf{L} = \mathbf{r} \times m\mathbf{v} = \mathbf{r} \times \mathbf{p} \tag{2.10-4}$$

where **r** is the vector from the fixed point P to the point mass and **p** is the linear momentum vector. The vector **L** is perpendicular to the plane defined by **r** and **p**. The components of the classical angular momentum vector for a single particle are

$$L_x = yp_z - zp_y \tag{2.10-5a}$$

$$L_y = zp_x - xp_z \tag{2.10-5b}$$

$$L_z = xp_y - yp_x \tag{2.10-5c}$$

The square of the angular momentum is given by the scalar product of **L** with itself:

$$\mathbf{L} \cdot \mathbf{L} = \mathbf{L}^2 = L_x^2 + L_x^2 + L_z^2 \tag{2.10-6}$$

This is a scalar. If no torque acts on a particle, its angular momentum is constant in classical mechanics. In classical mechanics all possible values of L and E are permitted.

In quantum mechanics the angular momentum is represented by an operator. The quantum-mechanical operators for the angular momentum are obtained by replacing the quantities in Eq. (2.10-5) with their corresponding quantum-mechanical operators; specifically $\hat{p}_x = (\hbar/i)(\partial/\partial x)$, and so on.

$$\hat{L}_x = -i\hbar\left(y\frac{\partial}{\partial z} - z\frac{\partial}{\partial y}\right) \tag{2.10-7a}$$

$$\hat{L}_y = -i\hbar\left(z\frac{\partial}{\partial x} - x\frac{\partial}{\partial z}\right) \tag{2.10-7b}$$

$$\hat{L}_z = -i\hbar\left(x\frac{\partial}{\partial y} - y\frac{\partial}{\partial x}\right) \tag{2.10-7c}$$

The operator for the square of the angular momentum is given by

$$\hat{L}^2 = |\hat{\mathbf{L}}|^2 = \hat{L} \cdot \hat{L} = \hat{L}_x^2 + \hat{L}_y^2 + \hat{L}_z^2 \tag{2.10-8}$$

It is often more convenient to use the angular momentum operators given in Eq. (2.10-7) in spherical polar coordinates (r, θ, ϕ) as defined in Fig. 2.10-1,

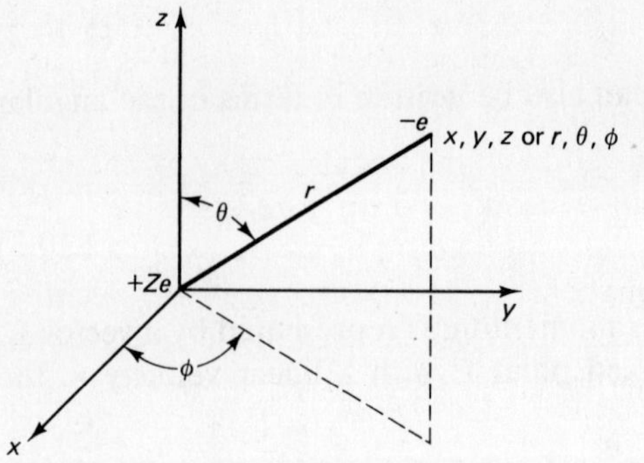

Figure 2.10-1 Orbiting electron relative to a nucleus.

which shows an electron relative to the nucleus *Ze*. (Note that math books usually interchange θ and ϕ.)

In spherical polar coordinates (*r*, θ, ϕ), the operators are

$$\hat{L}_x = i\hbar\left(\sin\phi\frac{\partial}{\partial\theta} + \cot\theta\cos\phi\frac{\partial}{\partial\phi}\right) \tag{2.10-9a}$$

$$\hat{L}_y = i\hbar\left(\cos\phi\frac{\partial}{\partial\theta} + \cot\theta\sin\phi\frac{\partial}{\partial\phi}\right) \tag{2.10-9b}$$

$$\hat{L}_z = i\hbar\frac{\partial}{\partial\theta} \tag{2.10-9c}$$

$$\hat{L}^2 = -\hbar^2\left[\frac{1}{\sin\theta}\frac{\partial}{\partial\theta}\left(\sin\theta\frac{\partial}{\partial\theta}\right) + \frac{1}{\sin^2\theta}\frac{\partial^2}{\partial\phi^2}\right] \tag{2.10-9d}$$

It can be readily shown that $\hat{L}_x$, $\hat{L}_y$, and $\hat{L}_z$, do not commute with each other, but $\hat{L}_x$, $\hat{L}_y$, and $\hat{L}_z$ all commute with L^2. Therefore, the total angular momentum and one, but only one, of its components can be measured precisely. In other words, if the magnitude of the total angular momentum

$$|\mathbf{L}| = \sqrt{L^2} = \sqrt{L_x^2 + L_y^2 + L_z^2} \tag{2.10-10}$$

is measured and L_z is measured, it is not possible to measure L_x or L_y. This is an essential difference between classical and quantum-mechanical systems and is in accord with the Heisenberg uncertainty principle.

Since $\hat{L}^2$ and $\hat{L}_z$ commute, they have the same eigenfunction. These eigenfunctions are called *spherical harmonics*, and are represented by Y_{l,m_l}. Thus,

$$\hat{L}^2 Y_{l,m_l}(\theta,\phi) = l(l+1)\hbar^2 Y_{l,m_l}(\theta,\phi) \qquad l = 0, 1, 2, \ldots \tag{2.10-11}$$

$$\hat{L}_z Y_{l,m_l}(\theta,\phi) = m_l\hbar Y_{l,m_1}(\theta,\phi) \qquad m_l = -l, -l+1, \ldots, l-1, l \tag{2.10-12}$$

The eigenvalues of the square of the angular momentum L^2 and of the z component L_z of the angular momentum are

$$L^2 = l(l+1)\hbar^2 \qquad l = 0, 1, 2, \ldots \tag{2.10-13}$$

$$L_z = m_l\hbar \qquad m_l = -l, -l+1, \ldots, l-1, l \tag{2.10-14}$$

where the quantum number l is referred to as the angular momentum quantum number (or *azimuthal* quantum number) and m_l is referred to as the *magnetic* quantum number. There are two quantum numbers because there are two coordinates, θ and ϕ. Equations (2.10-13) and (2.10-14) occur a number of times in quantum mechanics. They will be used in the next section on the two-particle rigid rotor, and will appear again in connection with the hydrogen atom.

The possible orientations of the angular momentum vector **L** with respect to a particular direction are shown in Fig. 2.10-2 for $l = 2$. The z direction might be determined by an applied magnetic field. Since the L_x and L_y components are unknown, **L** can be only be described as being on the surface of a cone, as illustrated in the figure.

Not only is the magnitude of the angular momentum quantized but its possible orientations are quantized. The magnitude of the orbital angular momentum **L** is

$$[(l+1)]^{1/2}\hbar \tag{2.10-15}$$

and the maximum component L_z in a particular direction is $1\hbar$. Thus, the magnitude of the angular momentum **L** is greater than its z component so that the angular momentum vector cannot point in the direction of an applied magnetic field. This is in agreement with the Heisenberg uncertainty principle because if the angular momentum vector pointed in the direction of the field, it would require that the orbital electron be confined to a plane perpendicular to this direction. In the absence of an electric or magnetic field, there is a degeneracy of $2l+1$ since, for any magnitude of the angular momentum vector, there are $2l+1$ values of m_l.

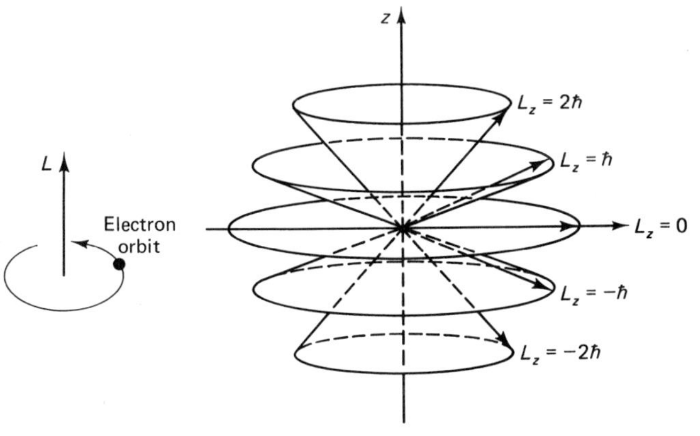

Figure 2.10-2 Quantized orientations of angular momentum (case of $l = 2$).

2.11 TWO-PARTICLE RIGID ROTOR

To a first approximation, a diatomic molecule is comprised of two nuclei which execute rotational and vibrational motion relative to each other. Section 2.9 determined the vibrational energy. Here, the rotational energy will be calculated.

A diatomic molecule with a fixed distance between the two nuclei can be represented by a two-particle rigid rotor with masses m_1 and m_2 separated by distance $r_1 + r_2$ as in Fig. 2.11-1. The translational and rotational motion of a two-particle rigid rotor can be reduced to two separate one-body problems. The translational motion of the system as a whole is treated using the total mass of the two particles, and the rotational motion of the particles is obtained by considering a hypothetical particle of reduced mass μ.

The expression for the reduced mass μ is derived as follows. In Fig. 2.11-1, the center of mass is located at the origin of the Cartesian coordinate system, so that

$$m_1 r_1 = m_2 r_2 \tag{2.11-1}$$

The distance between the two particles is represented by R_e, the equilibrium internuclear distance:

$$r_1 + r_2 = R_e \tag{2.11-2}$$

The moment of inertia I with respect to an axis that passes through the center of mass is

$$I = m_1 r_1^2 + m_2 r_2^2 \tag{2.11-3}$$

and is perpendicular to the internuclear vector. Eliminating r_1 and r_2 between Eqs. (2.11-1), (2.11-2), and (2.11-3) yields

$$I = \frac{R_e^2}{(1/m_1) + (1/m_2)} = \mu R_e^2 \tag{2.11-4}$$

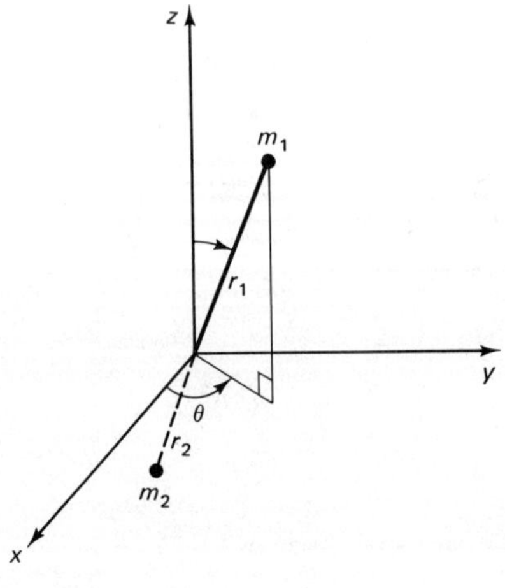

Figure 2.11-1 Coordinates for the rigid rotor. (Center of mass is defined by $m_1 r_1 = m_2 r_2$ and is located at the origin.)

where μ is the reduced mass:

$$\mu = \frac{1}{(1/m_1) + (1/m_2)} \tag{2.11-5}$$

The kinetic energy of a rigid rotor is given, classically, by $E = L^2/2I$. Thus, the classical expression for the kinetic energy of a rotating diatomic molecule is

$$E = \frac{L^2}{2\mu R_e^2} \tag{2.11-6}$$

Section 2.10 stated that in quantum mechanics the magnitude of the angular momentum L is quantized so that

$$L^2 = J(J+1)\hbar^2 \qquad J = 0, 1, 2, \ldots \tag{2.11-7}$$

where Eq. (2.10-13) has been used with the rotational quantum number J replacing the angular momentum quantum number l. The rotational energy of a diatomic molecule is given by

$$E = \frac{\hbar^2}{2I} J(J+1) \tag{2.11-8}$$

Note there is no zero-point energy for rotation. Unlike the harmonic oscillator the uncertainty principle can be satisfied even though the rotational energy is zero. The eigenfunctions corresponding to the eigenvalues in (2.11-8) are the spherical harmonics Y_{l,m_l} referred to in Section 2.10. In treating rotational energy levels the magnetic quantum number m_l is replaced with M. The energy depends only on the quantum number J. Since the values of M range from $-J$ to J, the rotational levels are $(2J+1)$-fold degenerate. The degeneracy corresponds to the different possible orientations of the angular momentum vector as seen in Fig. 2.10-2.

■ EXAMPLE 2.11-1

The reduced mass and the internuclear distance of HCl is 1.62668×10^{-27} kg and 127.5 pm, respectively. What is the moment of inertia, and the values of L, Lz, and E for the state with $J = 1$?

Solution

$$I = \mu R_e^2 = (1.626 \times 10^{-27}\ \text{kg})(127.5 \times 10^{-12}\ \text{m})^2 \tag{2.11-9}$$

$$= 2.644 \times 10^{-47}\ \text{kg m}^2$$

$$L = \sqrt{J(J+1)}\hbar = \frac{\sqrt{2}(6.626\ 176 \times 10^{-34}\ \text{J s})}{2\pi} \tag{2.11-10}$$

$$= 1.491\ 414 \times 10^{-34}\ \text{J s}$$

$$L_z = -0 \text{ or } \pm\frac{1}{2\pi}(6.626\ 176 \times 10^{-34}\ \text{J s}) \tag{2.11-11}$$

$$= 0 \text{ or } \pm 1.054\,589 \times 10^{-34}\ \text{J s}$$

$$E = \frac{\hbar^2}{2I} J(J+1) = \frac{(6.626\,176 \times 10^{-34}\ \text{J s})^2(2)}{8\pi^2(2.644 \times 10^{-47}\ \text{kg m}^2)} \tag{2.11-12}$$

$$= 4.206 \times 10^{-22}\ \text{J}$$ ■

■ EXAMPLE 2.11-2

The bond length of an O_2 molecule is 1.21 Å. The molecule is a rigid rotor with this distance as its fixed length. What are the energies of the first two rotational energy levels? Consider both atoms to be ^{16}O.

Solution. The energy levels are given by

$$E_J = J(J+1)\left(\frac{\hbar^2}{8\pi^2 I}\right) \qquad J = 0, 1, 2, \ldots \tag{2.11-13}$$

The moment of inertia I must first be determined before solving for E_0 and E_1. Actually, only E_1 need be calculated, since E_0 clearly equals zero. The moment of inertia is $I = \mu r_0^2$, where μ is the reduced mass:

$$I = \mu r_0 = \frac{m_1 m_2}{m_1 + m_2} r_0^2 \tag{2.11-14}$$

But $m_1 = m_2 = m$, so

$$I = \frac{m^2}{2m} r_0^2 = \frac{m}{2} r_0^2 \tag{2.11-15}$$

$$= \left[\frac{16.0 \times 10^{-3}\ \text{kg mol}^{-1}}{2(6.02 \times 10^{23}\ \text{mol}^{-1})}\right](1.21 \times 10^{-10}\ \text{m})^2$$

$$= 1.95 \times 10^{-46}\ \text{kg m}^2$$

Thus,

$$E_1 = 1(1+1)\left(\frac{h^2}{8\pi^2 I}\right) \tag{2.11-16}$$

$$= 2\left[\frac{(6.626 \times 10^{-34}\ \text{J s})^2}{8\pi^2(1.95 \times 10^{-46}\ \text{kg m}^2)}\right] = 5.70 \times 10^{-23}\ \text{J}$$ ■

2.12 HYDROGENLIKE ATOM

The hydrogen atom, consisting of a single electron bound to a single proton nucleus by a coulomb attractive force, is the simplest of all atoms in the universe. The Schrödinger equation can be solved exactly for this case, although such an exact solution cannot be obtained for all other atoms with more than one electron because of coulomb repulsion force between electrons.

In hydrogenlike atoms the electron and the nucleus move about their mutual center of mass. This can be provided for in the quantum-mechanical

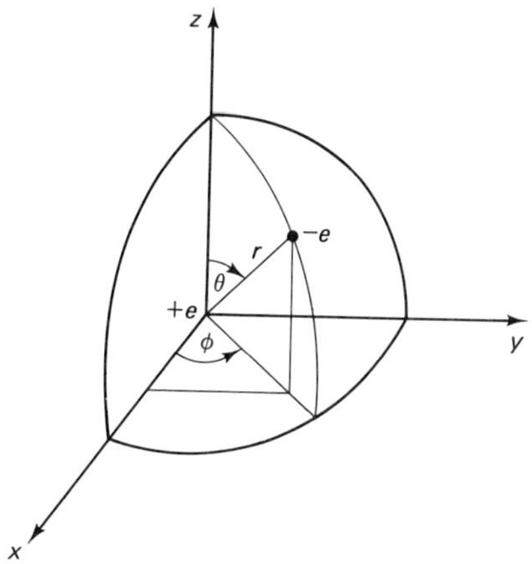

Figure 2.12-1 Spherical polar coordinates (r, θ, ϕ) of an electron $(-e)$.

treatment by using the reduced mass $\mu = m_e m_{nuc}/(m_{nuc} + m_e)$ of the electron and the nucleus. Since $m_{nuc} >> m_e$, the reduced mass of an atom is approximately equal to m_e. For a hydrogen atom, μ and m_e differ by 0.05%, and the approximation is better for more massive atoms. Thus, the mass m_e of the electron will be used in the Schrödinger equation rather than the reduced mass of the atom.

In a hydrogenlike atom the electron, shown in Fig. 2.12-1, moves under the influence of the coulomb potential.

$$V(r) = \frac{Ze^2}{4\pi\epsilon_0 r} = \frac{-Ze^2}{4\pi\epsilon_0 \sqrt{x^2 + y^2 + z^2}} \tag{2.12-1}$$

where Z is the number of positive charges in the nucleus, and ϵ_0 is the value of the dielectric constant in free space. The potential energy of an electron at infinite distance from the nucleus is zero. The potential energy of an electron in an atom is a negative quantity because it is at a lower potential than the reference state. The classical Hamiltonian H of the hydrogenlike atom is equal to the energy of the system.

$$H = \frac{1}{2m_e}(p_x^2 + p_y^2 + p_z^2) + V(r) \tag{2.12-2}$$

where p_x, p_y, and p_z are the components of the linear momentum of the electron. To convert the classical Hamiltonian H to the Hamiltonian operator $\hat{H}$, the dynamical quantities p_x, p_y, and p_z are replaced with their associated differential operators to obtain

$$\hat{H} = -\frac{\hbar^2}{2m_e}\nabla^2 + V(r) \tag{2.12-3}$$

The Hamiltonian operator $\hat{H}$ is to be used in the Schrödinger equation

$$\hat{H}\psi = E\psi \tag{2.12-4}$$

to obtain both the energy eigenvalue E and the corresponding eigenfunctions ψ for the hydrogenlike atom. To separate Eq. (2.12-4) into a set of three ordinary differential equations, each involving only one coordinate, the Laplacian operator ∇^2 in (2.12-3) is expressed in polar spherical coordinates as

$$\nabla^2 = \frac{1}{r^2}\frac{\partial}{\partial r}\left(r^2\frac{\partial}{\partial r}\right) + \frac{1}{r^2\sin^2\theta}+\frac{\partial^2}{\partial\phi^2}+\frac{1}{r^2\sin\theta}\frac{\partial}{\partial\theta}\left(\sin\theta\frac{\partial}{\partial\theta}\right) \tag{2.12-5}$$

so that

$$\frac{1}{r^2}\frac{\partial}{\partial r}\left(r^2\frac{\partial\psi}{\partial r}\right) + \frac{1}{r^2\sin\theta}\frac{\partial}{\partial\theta}\left(\sin\theta\frac{\partial\psi}{\partial\theta}\right) + \frac{1}{r^2\sin\theta}\frac{\partial^2\psi}{\partial\phi^2} + \frac{2m}{h^2}[E - V(r)]\psi = 0 \tag{2.12-6}$$

where m stands for the mass of the electron.

Solutions of (2.12-6) are of the form

$$\psi(r,\theta,\phi) = R(r)\Theta(\theta)\Phi(\phi) \tag{2.12-7}$$

where each constituent function depends on a single variable only. Substitution of (2.12-7) into (2.12-6) yields (after some rearrangement)

$$\frac{1}{\Phi}\frac{d^2\Phi}{d\Phi^2} = -\frac{\sin^2\theta}{R}\frac{d}{dr}\left(r^2\frac{dR}{dr}\right) - \frac{\sin\theta}{\Theta}\frac{d}{d\theta}\left(\sin\theta\frac{d\Theta}{d\theta}\right) - \frac{2m_e}{h^2}r^2\sin^2\theta[E - V(r)] \tag{2.12-8}$$

Since the left side of this equation does not depend on r or θ, and the right side does not depend on ϕ, each side is equal to a constant that is conveniently represented by $-m_l^2$. In this way,

$$\frac{d^2\Phi}{d\phi^2} = -m_l^2\Phi \tag{2.12-9}$$

$$\frac{1}{R}\frac{d}{dr}\left(r^2\frac{dR}{dr}\right) + \frac{2m_e r^2}{h^2}[E - V(r)] = \frac{m_l^2}{\sin^2\theta} - \frac{1}{\Theta\sin\theta}\frac{d}{d\theta}\left(\sin\theta\frac{d\Theta}{d\theta}\right) \tag{2.12-10}$$

In the second equation the left side does not depend on θ, and the right side does not depend on r. Thus, each side may be set equal to a constant; this constant is conveniently written as $l(l + 1)$. This yields

$$-\frac{1}{\sin\theta}\frac{d}{d\theta}\left(\sin\theta\frac{d\Theta}{d\theta}\right) + \frac{m_l^2\Theta}{\sin^2\theta} = l(l+1)\Theta \tag{2.12-11}$$

and

$$\frac{1}{r^2}\frac{d}{dr}\left(r^2\frac{dR}{dr}\right) + \frac{2m_e}{h^2}[E - V(r)]R = l(l+1)\frac{R}{r^2} \tag{2.12-12}$$

Now the problem of solving the Schrödinger equation has been reduced to the solution of ordinary differential equations (2.12-9), (2.12-11), and (2.12-12). Equation (2.12-9) has the solution

$$\Phi(\phi) = e^{im_l\phi} = \cos m_l\phi + i \sin m_l\phi \tag{2.12-13}$$

In order for this eigenfunction to satisfy the requirements of quantum mechanics it must be single valued. This means that it must have the same value at $\phi = 0$ and 2π. Consequently $|m_l|$ must have integer values as seen from

$$e^{im_l 0} = e^{im_l 2\pi} \tag{2.12-14}$$

$$1 = \cos m_l 2\pi + i \sin m_l 2\pi \tag{2.12-15}$$

$$m_l = 0, \pm1, \pm2, \pm3, \ldots \tag{2.12-16}$$

The quantum number m_l is referred to as the *magnetic quantum number* ($m_l h$ is the z component of the orbital angular momentum). Instead of working with the imaginary wave function (2.12-13), it is often more convenient to use real wave functions formed by taking linear combinations of complex wave functions. The justification of this procedure is that any linear combination of eigenfunctions of a degenerate energy level is also an eigenfunction of the Hamiltonian with the same eigenvalue. Thus

$$\Phi(\phi) = \sin m_l\phi \quad \text{or} \quad \cos m_l\phi \tag{2.12-17}$$

As for Eq. (2.12-11), this is the Legendre equation and it has acceptable solutions only if

$$l = |m_l|, \qquad |m_l| + 1, \qquad |m_l| + 2, \qquad |m_l| + 3, \ldots \tag{2.12-18}$$

The quantum number 1 is called the *azimuthal quantum number* (it determines the value of the orbital angular momentum). Letter symbols are associated with values of 1 as follows:

$l =$	0	1	2	3
Symbol	*s*	*p*	*d*	*f*

When these values of l are substituted in Eq. (2.12-12), it is found that this equation has acceptable solutions only for certain values of E, the total energy of the hydrogenlike atom:

$$E = -\frac{m_e e^4 Z^2}{2(4\pi_0)^2 \hbar^2 n^2} \tag{2.12-19}$$

where $n = 1, 2, 3, \ldots$ is the principal quantum number. The energy of the hydrogenlike atom is measured with respect to the nucleus and the electron separated by an infinite distance. The energy is negative because the electron in a hydrogenlike atom has less energy than when it is free. Since the lowest energy corresponds with $n = 1$, this state is referred to as the *ground state*. For $n = 2, 3, \ldots$ the hydrogenlike atom is said to be in an *excited state*.

Since the principal quantum number n can only have values

$$n = l + 1, l + 2, l + 3, \ldots \tag{2.12-20}$$

the permitted values of the quantum numbers are

$$n = 1, 2, 3, \ldots \tag{2.12-21}$$

$$l = 0, 1, 2, \ldots, (n-1) \tag{2.12-22}$$

$$m_1 = 0, \pm 1, \pm 2, \ldots, \pm 1 \tag{2.12-23}$$

This means that for a given value of the principal quantum number n there are generally several possible values of l and m_l. In other words, for a given value of the energy (eigenvalue), there are several functions (eigenfunctions). Thus, as for the particle in a three-dimensional box, the energy levels may be degenerate. In the absence of a magnetic or electric field the number of degenerate wave functions for the hydrogenlike atom is 1 for $n = 1$, 4 for $n = 2$, 9 for $n = 3, \ldots$, or, in general, n^2. The label attached to a state consists of the n quantum number followed by the l quantum number, and so on. The ground state is the 1s state.

■ EXAMPLE 2.12-1

(a) Calculate the energy required to completely remove the electron from a hydrogen atom, thus making the atom into a singly charged positive ion (a proton).

(b) What is the wavelength corresponding to the energy emitted when an electron drops from energy level $n = 3$ to energy level $n = 2$?

Solution

(a) The amount of energy called for here is simply the total energy of the electron, for this is what must be reduced to zero before the electron can be freed:

$$E = -\frac{4\pi^2 k^2 Z^2 e^4 m}{2\, n^2 h^2}$$

where Z and n are both unity, and $k = 1/4\pi\epsilon_0 = 8.987 \times 10^9$ N m^2 c^{-2} is the coulomb constant. Thus,

$$E = \frac{4\pi^2 \times (9 \times 10^9)^2 \times (1.6 \times 10^{-19})^4 \times 9.11 \times 10^{-31}}{2 \times (6.63 \times 10^{-34})^2}$$

$$= \frac{19.1 \times 10^{-85}}{8.8 \times 10^{-67}} = 2.17 \times 10^{-18} \text{ J}$$

$$= \frac{2.17 \times 10^{-18}}{1.6 \times 10^{-19}} = 13.6 \text{ V} \quad \textit{Ans.} \tag{2.12-24}$$

Electrons can be excited (as in discharge tubes) into higher and higher orbits from which they return to lower orbits, but only from allowable ones to other allowable ones, and they emit energy in quantum amounts between orbits. This leads to so-called energy level diagram to describe the structure of atoms, the one for hydrogen being the following:

______________ $n = \infty$ where $E = 0$

- - - - - - - - - -

______________ $n = 4$ where $E = 0.85 = \frac{13.6}{16}$

______________ $n = 3$ where $E = -1.51 = \frac{13.6}{9}$

______________ $n = 2$ where $E = -3.40 = \frac{13.6}{4}$

Ground state $n = 1$ where $E = 13.6$ eV

(b) Energy differences between levels can be calculated from

$$E_n - E_{n'} = \frac{4\pi^2 k^2 Z^2 e^4 m}{2h^2}\left(\frac{1}{n'^2} - \frac{1}{n^2}\right)$$

where n is the higher of the two energy states, that is, the "less negative" energy state. Also, by the quantum theory,

$$E_n - E_{n'} = h\nu = \frac{c}{\lambda}$$

Consequently there is a wavelength λ associated with each quantized energy difference, such that

$$\frac{1}{\lambda} = \frac{\nu}{c} = \frac{E_n - E_{n'}}{hc} = \frac{4\pi^2 k^2 Z^2 e^4 m}{2h^3 c}\left(\frac{1}{n'^2} - \frac{1}{n^2}\right)$$

where $4\pi^2 k^2 Z^2 e^4 m/2h^2$ is 13.6 eV or 2.17×10^{-18} J, for hydrogen. Thus,

$$\frac{1}{\lambda} = \frac{\nu}{c} = \frac{2.17 \times 10^{-18}}{hc}\left(\frac{1}{n'^2} - \frac{1}{n^2}\right) \tag{2.12-25}$$

The wavelength corresponding to the energy emitted when an electron drops from energy level $n = 3$ to energy level $n = 2$ for the hydrogen atom is

$$\frac{1}{\lambda} = \frac{2.17 \times 10^{-18}\left(\frac{1}{4} - \frac{1}{9}\right)}{6.63 \times 10^{-34} \times 3 \times 10^{8}} = \frac{2.17 \times 0.139 \times 10^{-18}}{19.9 \times 10^{-26}}$$

$$\frac{1}{\lambda} = 0.152 \times 10^{7}\ \text{m}^{-1}$$

or

$$\lambda = \frac{1}{0.152 \times 10^{7}} = 6.58 \times 10^{-7} = 6580 \times 10^{-10}\ \text{m}$$

■

■ EXAMPLE 2.12-2

The Schrödinger wave equation for an atom or ion having one electron (hydrogenlike) is solved exactly by converting to spherical coordinates and separating into three parts to yield

$$\psi(r, \theta, \phi) = [R_{n,l}(r)][\Theta_{l,m_l}(\theta)][\Phi_{m_l}(\phi)]$$

The complete wave function is the product of the three parts and is called an *orbital*.

(a) Write the wave function Φ for the first three values of m_l.

(b) Write the total wave function for a hydrogen atom in the ground state.

Solution

(a) The solution of the Φ part of the spherical Schrödinger equation is $\Phi = Ne^{im_l\phi}$, where m_l is the magnetic quantum number having values $m_l = 0$, $\pm 1, \pm 2, \ldots$. The normalization constant N is $1\sqrt{2\pi}$. With $m_l = 0$, 1, and -1, this gives

$$\Phi_0 = \frac{1}{\sqrt{2\pi}}, \qquad \Phi_1 = \frac{1}{\sqrt{2\pi}} e^{i\phi}, \qquad \Phi_{-1} = \frac{1}{\sqrt{2\pi}} e^{-i\phi}$$

The last two expressions are complex. They are "unreal" and have no physical significance. However, linear combinations may be made in order to obtain real wave functions. For example,

$$\Phi_x = \frac{1}{\sqrt{2}}(\Phi_1 + \Phi_{-1}) = \frac{1}{\sqrt{2\pi}}\left(\frac{e^{i\phi}}{\sqrt{2\pi}} + \frac{e^{-i\phi}}{\sqrt{2\pi}}\right) \tag{2.12-26}$$

$$= \frac{1}{1\sqrt{\pi}}(e^{i\phi} + e^{-i\phi}) = \frac{\cos\phi}{\sqrt{\pi}}$$

$$\Phi_y = \frac{1}{\sqrt{2}}(\Phi_1 - \Phi_{-1}) = \frac{1}{\sqrt{2}}\left(\frac{e^{i\phi}}{\sqrt{2\pi}} - \frac{e^{-i\phi}}{\sqrt{2\pi}}\right) \tag{2.12-27}$$

$$= \frac{1}{2\sqrt{\pi}}(e^{i\phi} - e^{-i\phi}) = \frac{\sin\phi}{\sqrt{\pi}}$$

(b) In the ground state, the quantum numbers are the smallest possible: $n = 1$, $l = 0$, $m_l = 0$, so that

$$\psi_{n,l,m_l} = \psi_{1,0,0} = [R_{1,0}(r)][\Theta_{0,0}(\theta)][\Phi_0(\phi)]$$

Since

$$\Phi_0 = \frac{1}{\sqrt{2\pi}}, \quad \Phi_{0,0} = \frac{1}{2}\sqrt{2}, \quad \text{and} \quad R_{1,0} = 2\left(\frac{Z}{a_0}\right)^{3/2} e^{-Zr/a_0},$$

this gives

$$\psi_{1,0,0} = 2\left(\frac{Z}{a_0}\right)^{3/2} e^{-Zr/a_0}\left(\frac{1}{\sqrt{2}}\right)\left(\frac{1}{\sqrt{2\pi}}\right) \tag{2.12-28}$$

or, with $Z = 1$ for the hydrogen atom,

$$\psi_{1,0,0} = \left(\frac{1}{\sqrt{\pi}}\right)\left(\frac{1}{a_0}\right)^{3/2} e^{-r/a_0}$$

where a_0 is the Bohr radius. ■

■ EXAMPLE 2.12-3

Discuss the solution of the R equation (spherical part of the Schrödinger wave equation).

Solution. The acceptance of Schrödinger's equation rests in large part on its ability to successfully solve the hydrogen problem. The solution of the R part of the spherical Schrödinger equation, however, requires considerable manipulation to put it in the form of an associated Laguerre equation, with solutions in terms of associated Laguerre polynomials. These expressions include both the quantum number l and a new parameter n, the principal quantum number, restricted to values $n = 1, 2, \ldots \infty$. The azimuthal quantum number l is restricted to $l = 0, 1, 2, \ldots, (n - 1)$. Some normalized values of $R_{n,l}$ are

$$R_{1,0} = 2\left(\frac{Z}{a_0}\right)^{3/2} e^{-Zr/a_0}$$

$$R_{2,0} = \frac{1}{2\sqrt{2}}\left(\frac{Z}{a_0}\right)^{3/2}\left(2 - \frac{Zr}{a_0}\right) e^{-Zr/2a_0}$$

$$R_{2,1} = \frac{1}{2\sqrt{6}}\left(\frac{Z}{a_0}\right)^{3/2} \frac{Zr}{a_0} e^{-Zr/2a_0}$$

The result of the solution of the R equation is an expression for the allowed energy levels (determined by the quantum number n):

$$E_n = -\frac{2\pi^2 \mu Z^2 e^4}{(4\pi\epsilon_0)^2 h^2 n^2} = -\frac{Z^2}{n^2} E_0$$

where μ is the reduced mass, ϵ_0 is the permittivity of free space, e is the charge of the electron, Ze is the charge of the nucleus, and E_0 is the ground-state energy of the hydrogen atom. Two useful values for E_0 are 2.1798×10^{-18} J and 13.605 eV. ■

2.13 ELECTRON SPIN

Complete definition of an electron state requires the recognition of one more property of an electron that has not yet been mentioned. In addition to the orbital angular momentum described by l, as already discussed previously, the electron also has an intrinsic angular momentum called electron spin. This is not related to the spatial coordinates of the particle as is orbital angular momentum. It is an intrinsic or internal property. Other intrinsic properties of a particle are charge, mass dipole moment, moment of inertia, and so forth. Values of such parameters comprise the internal degrees of freedom of a particle.

The optical spectra of hydrogen atoms and alkali atoms show a fine structure that is not accounted for by the quantum theory described so far. To explain these effects, Goudsmit and Uhlenbeck proposed in 1925 that an electron has an intrinsic angular momentum S and accompanying magnetic dipole moment μ_s. They further proposed that the z component S_z of this spin angular momentum is equal to $m_s\hbar$, where m_s is a fourth quantum number that can have the values $+\frac{1}{2}$ or $-\frac{1}{2}$. The complete wave function can then be designated by four quantum numbers:

$$\psi_{n,l,m_l,m_s} \tag{2.13-1}$$

In 1928 Dirac showed that spin arises in the relativistic quantum theory for a one-electron system. In the nonrelativistic quantum theory for a one-electron system, electron spin must be treated as an additional postulate. The term "electron spin" is used, although it is not really correct to think of the intrinsic angular momentum of the electron as being due to a spinning motion of the electron mass on its axis. Since the spin angular momentum of an electron has no analog in classical mechanics, the spin angular momentum operator cannot be constructed by first writing the classical Hamiltonian. However, it turns out that the treatment of spin angular momentum is closely analogous to the treatment of orbital angular momentum. The magnitudes of the spin angular momentum S and its z component S_z are given in terms of two quantum numbers, s and m_s:

$$\begin{aligned} S &= \sqrt{s(s+1)}\hbar = \frac{\sqrt{3}}{2}\hbar \\ S_z &= m_s\hbar: \qquad m_s = \pm\frac{1}{2} \end{aligned} \tag{2.13-2}$$

where the spin quantum number s (generally referred to simply as the spin) of the electron has the single value 1/2, and the quantum number m_s for the z component has two possible eigenvalues $\pm 1/2$. These two quantum numbers are known as *spin up* and *spin down.*

As in the case of orbital angular momentum there is a series of electron spin operators: $\hat{S}^2$, $\hat{S}_x$, $\hat{S}_y$, $\hat{S}_z$. When these operators are applied to the spin functions, they yield eigenvalues. Since the spin eigenfunctions do not involve other kinds of coordinates, the two possible spin functions for an electron are simply represented by α and β. These spin functions may be operated on by the operator $\hat{S}_z$ for the z component of the spin angular momentum:

$$\hat{S}^2\alpha = \tfrac{1}{2}(\tfrac{1}{2}+1)\hbar^2\alpha = \tfrac{3}{4}\hbar^2\alpha \tag{2.13-3}$$

$$\hat{S}^2\beta = \tfrac{1}{2}(\tfrac{1}{2}+1)\hbar^2\beta = \tfrac{3}{4}\hbar^2\beta \tag{2.13-4}$$

$$\hat{S}_z\alpha = +\tfrac{1}{2}\hbar\alpha \tag{2.13-5}$$

$$\hat{S}_z\beta = +\tfrac{1}{2}\hbar\beta \tag{2.13-6}$$

The operators $\hat{S}^2$ and $\hat{S}_z$ commute with the Hamiltonian operator $\hat{H}$, $\hat{L}^2$, and $\hat{L}_z$ so that the magnitude of the spin, the z component of the spin, the energy and the magnitude of the orbital angular momentum, and the z component of the orbital angular momentum can all have definite values simultaneously. A complete wave function for a hydrogenlike atom must indicate the spin state of the electron. Since there are two spin functions, there are twice as many wave functions for the hydrogen atom as indicated earlier, ψ_α and ψ_β for each of the previous ψ's. Thus, a complete state specification for the hydrogen atom requires the four quantum numbers n, l, m_l, and m_s. This increases the degeneracy of the energy levels from n^2 to $2n^2$.

The description of electron behavior in atoms, called the *vector model,* in terms of the four quantum numbers,

principal number	$n = 1, 2, \ldots, \infty$	(2.13-7)
orbital number	$l = 0, 1, 2, \ldots, n - 1$	(2.13-8)
magnetic number	$m_l = 0, \pm 1, \pm 2, \ldots, \pm 1$	(2.13-9)
spin number	$m_s = \pm \frac{1}{2}$	(2.13-10)

provides an explanation of the periodic table of elements, which is the basis of the whole of chemistry. For every electron in an atom, there is a unique set of four numbers that completely specifies the state of that electron. They can be regarded as "coordinates" for the electron, and one and only one electron can have a given set of coordinates. This is called the *Pauli exclusion principle,* from which it is possible to work out the electronic configuration of any atom. By systematically assigning sets of quantum numbers to successive electrons, a picture of the arrangement of electrons in atoms can be constructed. It is found that electronic shells and subshells exist which, when they contain a certain number of electrons, are considered to be full or closed. A shell exists for each value of n, and a subshell for each value of l associated with an n value. Thus, for $n = 3$, l can have the three values 0, 1, and 2, and there are three subshells each of which can accommodate $2(2l + 1)$ electrons. An s subshell ($l = 0$) can contain therefore a maximum of two electrons, a p subshell ($l = 1$) six electrons, a d subshell ($l = 2$) ten electrons, and so on. On this basis it can be seen that the first shell ($n = 1$) can contain at most two electrons; this is designated the K shell. Similarly, an L shell ($n = 2$) can contain $2 + 6 = 8$ electrons, and an M shell can contain $2 + 6 + 10 = 18$ electrons. In general, therefore, a shell contains $2n^2$ electrons.

A summary of the electron configurations for the lighter elements is shown in Table 2.13-1. It can be seen from this table that all atoms consist of sets of closed shells outside of which are situated one or more valence electrons. Of particular interest are the elements helium and neon which have completely filled shells and subshells. These are extremely stable since their "closed" shells are reluctant to interact with other atoms. The term "valence" is applied because it is these electrons that govern the chemical valency of the element, and, more importantly, it is these electrons that are active when atoms of the same element or dissimilar elements are brought together to form solids. The configurations shown represent the ground states of the atoms in question, and provide a convenient shorthand notation to describe the atom, rather than drawing an energy level diagram.

Minor irregularities may occur because of specific situations, but the overall pattern is maintained. This pattern is observed because electrons obey

TABLE 2.13-1 THE ELECTRON CONFIGURATION OF THE LIGHTER ELEMENTS

		K shell $n = 1$	L shell $n = 2$		M shell $n = 3$			
Z	Element	1s subshell	2s subshell	2p subshell	3s subshell	3p subshell	3d subshell	Electron configuration
1	Hydrogen	1						$1s^1$
2	Helium	2						$1s^2$
3	Lithium	2	1					$1s^2 2s^1$
4	Beryllium	2	2					$1s^2 2s^2$
5	Boron	2	2	1				$1s^2 2s^2 2p^1$
6	Carbon	2	2	2				$1s^2 2s^2 2p^2$
7	Nitrogen	2	2	3				$1s^2 2s^2 p^3$
8	Oxygen	2	2	4				$1s^2 2s^2 2p^4$
9	Flourine	2	2	5				$1s^2 2s^2 2p^5$
10	Neon	2	2	6				$1s^2 2s^2 2p^6$
11	Sodium	2	2	6	1			$1s^2 2s^2 2p^6 3s^1$
12	Magnesium	2	2	6	2			$1s^2 2s^2 2p^6 3s^2$
13	Aluminum	2	2	6	2	1		$1s^2 2s^2 2p^6 3s^2 3p^1$
14	Silicon	2	2	6	2	2		$1s^2 2s^2 2p^6 3s^2 3p^2$

the Pauli exclusion principle: only one particle per state. Other types of particles, for example, phonons and photons, do not obey the Pauli exclusion principle; in these cases more than one particle can occupy a specific energy state.

■ EXAMPLE 2.13-1

Consider one or more particles of mass m moving in a straight line and confined between the points 0 and a, that is, particles in a one-dimensional box of length a. Discuss the electron configuration of elements when electrons are added one by one until there is a total of Z electrons around a nucleus of charge Ze.

Solution. The problem of a single particle in a one-dimensional box was solved previously in Section 2.5. It was found that the energy of the particle could have only the discrete values given by

$$E_n = n^2 \frac{h^2}{8ma^2} \qquad n = 1, 2, 3, \ldots \tag{2.13-11}$$

Figure 2.13-1 shows these energy levels. Now, take the particle in the box to be an electron with intrinsic spin. The state of the system is then specified by the pair of quantum numbers (n, m_s). In each energy level, the electron can be in a state with spin up ($m_s = +\frac{1}{2}$). With two electrons, the ground (lowest energy) state of the system will have both electrons in the $n = 1$ energy level, one with spin up $(1, +\frac{1}{2})$ and one with spin down $(1, -\frac{1}{2})$, as shown in Fig. 2.13-1(a). Note that the two electrons do not have the same set of quantum numbers (n, m_s).

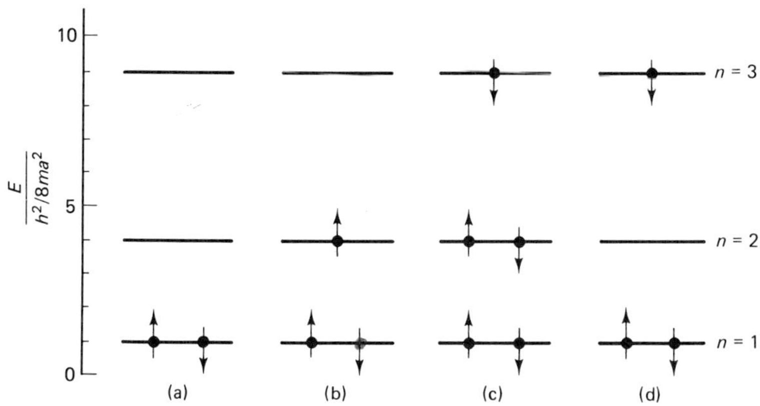

Figure 2.13-1 Placement of electrons in energy levels.

Now consider what happens when a third electron is added to the system. The Pauli exclusion principle prohibits this electron from occupying the $n = 1$ energy level; for if it were in the $n = 1$ level, two of the three electrons would have the same set of quantum numbers (n, m_s). The third electron must therefore go to a different energy level, the $n = 2$ level if the system is in its ground state, as shown in Fig. 2.13-1(b) for a spin up configuration.

A similar line of reasoning shows that a fourth electron can be put into the $n = 2$ level, but when a fifth is added it must go into the $n = 3$ level, as shown in Fig. 2.13-1(c) for a spin down configuration. Note that the Pauli exclusion principle has the effect of increasing the total energy of the ground state of the system to a much higher value than it would have if all the electrons occupied the $n = 1$ energy level.

In general, when electrons are added one by one until there is a total of Z electrons around a nucleus of charge Ze, electrons with the same value of n will occupy the same electron shell. The various shells are designated by capital letters according to the following scheme.

value of n:	1	2	3	4	...	(2.13-12)
shell letter:	K	L	M	N	...	

Within a given shell different values of l are possible; each value of l defines a subshell (which is thus equivalent to an orbit). For a given n the integer values that the quantum number l can take on are

$$l = 0, 1, 2, \ldots, n - 1 \tag{2.13-13}$$

The value of l is designated by a lowercase letter according to the following scheme:

value of l:	0	1	2	3	4	5	...	
letter symbol:	s	p	d	f	g	h	...	(2.13-14)

The elements with closed shells present a tightly bound structure relatively immune to the influence of other atoms. These are the inert or noble gases, He, Ne, A, Kr, and Xe, with high ionization potentials. The atoms with closed shells plus one extra electron are the alkali metals, Li, Na, K, Rb, and Cs, which have low ionization potentials, readily sharing the outermost loosely bound "valence" electron with other atoms. They are especially ready to share the electron with hydrogen or the halogens F, Cl, Br, and I, which have an atomic structure consisting of a closed shell minus one electron.

The electron configuration for the lighter elements is summarized in Table 2.13-1. It can be seen from this table that all atoms consist of sets of closed shells outside of which are situated one or more valence electrons. Of particular interest are the elements helium and neon (and indeed other elements) which have completely filled shells and subshells. These are extremely stable since their "closed" shells are reluctant to interact with other atoms.

The convention for specifying the number of electrons in a particular orbit, defined by the quantum numbers (n, l), is to give n, followed by the letter symbol for l, with the number of electrons as a postsuperscript. Various orbits are sequentially written one after the other, thereby defining an electron configuration. As an example, the configuration for the five electrons in the ground state of boron is $1s^2 2s^2 2p^1$.

Without the Pauli principle the ground state of the atomic system would clearly be the state in which each of the Z electrons was in the ground state of energy $-Z^2 \times 13.6$ eV. For $Z > 2$ this is not possible, however, according to the Pauli principle. With $n = 1$, l can only be zero, and hence $m = 0$, so that according to the Pauli principle only two electrons, corresponding to the two possible values of m_s, can occupy this level.

For larger values of n, the energy levels can hold more than two electrons. With the 2 values of m_s and the $2l + 1$ values of m, it follows that $2(2l + 1)$ electrons with the same value of l can be put in a common energy level. Since for given n, l can take on the values $0, 1, \ldots, n - 1$, the total number of different combinations of the three quantum numbers, l, m_l, and m_s for a given n must be equal to the sum of n terms:

$$\sum_{l=0}^{n-1} 2(2l + 1) = 2(1 + 3 + 5 + \cdots + 2n - 1) \tag{2.13-15}$$

$$= 2\frac{n[1 + (2n - 1)]}{2} = 2n^2 \tag{2.13-16}$$

In the $n = 1$ level, it is possible to have 2 electrons, in the $n = 2$ level 8, in the $n = 3$ level 18, in the $n = 4$ level 32, and so forth. This scheme would mean that the filled levels or shells are at $Z = 2$, $Z = 2 + 8 = 10$, $Z = 10 + 18 = 28$, $Z = 28 + 32 = 60$, and so forth.

Atoms with Z one to two higher than the filled-shell atoms are quite different. The additional one or two electrons would have to go into the next higher shell (higher n), and the binding energy of the outermost electron, or the ionization potential, would be much smaller than for the closed-shell atoms. The atoms with closed shells plus one electron, such as those with $Z = 3$ and $Z = 11$, display similar chemical properties and similar optical spectra, since the outermost electrons are the ones largely responsible for their properties. The atoms with closed shells plus two electrons would also be expected to be much alike, leading to a certain periodicity among the elements.

Ground States of Atoms: Periodic Table. It is possible to discuss, qualitatively, the wave functions and energy levels for atoms more complicated than the hydrogen atom. These wave functions are just those obtained for the hydrogen atom except that Z is not equal to 1.

HELIUM **($Z = 2$)** If the interaction of the electrons is neglected, the Schrödinger equation for the helium atom can be separated into two equations, one for each electron. Each equation is identical to that for the hydrogen atom except that $Z = 2$. The energy eigenvalues are given by

$$E = -\frac{Z^2E_1}{n_1^2} - \frac{Z^2E_1}{n_2^2} \qquad \text{where } E_1 = 13.6 \text{ eV} \tag{2.13-17}$$

The lowest energy, $E_0 = -2(2)^2E_1 \approx -108.8$ eV, occurs for $n_1 = n_2 = 1$. For this case, $l_1 = l_2 = 0$. The total wave function is antisymmetric if the total spin for the two electrons is zero. If the approximate ground state wave function

$$\psi_{100}(r_1, \theta_1, \phi_1)\psi_{100}(r_2, \theta_2, \phi_2) \tag{2.13-18}$$

is used to calculate the expectation value of the interaction energy of the two electrons, the result is $E_{\text{int}} \approx +34$ eV. With this correction, the ground-state energy is $E \approx -108.8 + 34 = -74.8$ eV.

The experimental value of the energy needed to remove both electrons is about 79 eV. The discrepancy between this result and the value 74.8 eV is due to the inaccuracy of the approximation used to calculate E_{int}, as indicated by the rather large value of the correction (about 30%). The helium ion He^+, formed by removing one electron, is identical to the hydrogen atom except that $Z = 2$; thus, the ground state energy is

$$-Z^2(13.6) = -54.4 \text{ eV} \tag{2.13-19}$$

The energy needed to remove the first electron from the helium atom is 24.6 eV. This value, 24.6 V, is called the *first ionization potential* of the atom.

The configuration of the ground state of the helium atom is written $1s^2$. The 1 signifies $n = 1$, the s signifies $l = 0$, and the 2 signifies that there are two electrons in this state. Since l can only be zero for $n = 1$, the two electrons completely fill the K shell ($n = 1$).

LITHIUM **($Z = 3$)** Lithium has three electrons. Two of them are in the K shell ($n = 1$), but the third cannot have $n = 1$ because of the exclusion principle. The next lowest energy state for this electron is $n = 2$. The possible l values for $n = 2$ are $l = 0$ or $l = 1$. The configuration of the lithium atom is thus $1s^22s$. The total angular momentum of the electrons in this atom is $\frac{1}{2}$ due to the spin of the outer electron. The first ionization potential for lithium is only 5.39 V. In the quantum-mechanical model of the lithium atom, the outer electron is shielded by the inner core of the two 1*s* electrons. The effective nuclear charge is greater for the 2*s* state, and this state has a lower energy.

BERYLLIUM (Z = 4) The fourth electron has the least energy in the 2*s* state. The electron configuration of beryllium is $1s^2 2s^2$. The first ionization potential is 9.32 V. This is greater than that for lithium because of the greater value of *Z*.

■ EXAMPLE 2.13-2

(a) Write symbols for the two possible wave functions of a hydrogen atom in the ground state when spin is included.

(b) What are the possible combinations for the spin functions of two electrons?

Solution

(a) The wave function of a hydrogenlike atom is the product of three parts:

$$\psi(r,\theta,\phi) = [R_{n,l}(r)][\Theta_{l,m_l}(\theta)][\Phi_{m_l}(\phi)]$$

A fourth degree of freedom is obtained by the introduction of the spin wave function to give

$$\psi_{n l m_l m_s} = R_{nl}\Theta_{l m_l}\Phi_{m_l}\alpha \qquad (2.13\text{-}20)$$

and

$$\psi_{n l m_l m_s} = R_{nl}\Theta_{l m_l}\Phi_{m_l}\beta \qquad (2.13\text{-}21)$$

where α and β are used for positive m_s and negative m_s, respectively. They are known as *spin up* and *spin down*. For the hydrogen atom, its electron is a 1*s* electron, so the two designations are simply $1s\alpha$ and $1s\beta$. The two wave functions can be written as $\psi_{1s}(1)\alpha(1)$ and $\psi_{1s}(1)\beta(1)$.

(b) For two electrons, the following combinations are possible:

$$\alpha(1)\alpha(2) \quad \beta(1)\beta(2) \quad \alpha(1)\beta(2) \quad \alpha(2)\beta(1)$$

The last two functions, however, require that the electrons be distinguishable, which isn't allowed by the Heisenberg uncertainty principle. Spin functions that don't require distinguishing between electrons can be obtained by the linear combinations

$$\frac{1}{\sqrt{2}}[\alpha(1)\beta(2)+\alpha(2)\beta(1)] \qquad \frac{1}{\sqrt{2}}[\alpha(1)\beta(2)-\alpha(2)\beta(1)]$$

where the $1/\sqrt{2}$ is a normalization constant. ■

■ **EXAMPLE 2.13-3**

(a) Discuss the Aufbau principle for constructing the lowest energy electron configuration of an atom of atomic number Z.

(b) Write the electron configurations for ^{15}P, ^{27}Co, and ^{29}Cu, with appropriate comments.

Solution

(a) The Aufbau principle (Aufbau being German for building up) is a generalization of the electronic configuration of lithium: $1s^2 2s^1$, a central nucleus surrounded by a complete K-shell of two electrons, and around that a more diffuse $2s$-electron. The rules are as follows:

1. Place the added electron in the lowest available energy level, that is, the ground state.
2. Place only two electrons in one orbital, because of the Pauli exclusion principle.
3. Place the added electron in an empty orbital whenever permitted (Hund's rule). In other words, don't pair electrons until necessary.
4. Assign the same spin to simple electrons in different equivalent orbits.

(b) For ^{15}P, the electron configuration is

$$1s^2 2s^2 2p^6 3s^2 3p^3$$

The last term could be broken up to show the three p orbitals: $3p_x^1\ 3p_y^1\ 3p_z^1$.

For ^{27}Co:

$$1s^2 2s^2 2p^6 3s^2 3p^6 4s^2 3d^7$$

The $4s$ orbital receives electrons before the $3d$ orbital does. To show the d electron arrangement, write

$$3d_{xy}^2 3d_{yz}^2 3d_{xz}^1 3d_{x^2y}^1 3d_{z^2}^1$$

Since there are more than five d electrons, the sixth and seventh form pairs according to rule 1, that is, they go into the lowest energy level available.

For ^{29}Cu:

$$1s^2 2s^2 2p^6 3s^2 3p^6 4s^2 3d^9$$

The configuration beyond $3p^6$ is actually $4s^1 3d^{10}$ which is predicted from the general rule about filled and half-filled subshells.

The complete list of orbitals in order of increasing energy, with the maximum number of electrons permitted in each orbital is

$$1s^2 2s^2 2p^6 3s^2 3p^6 4s^2 3d^{10} 4p^6 5s^2 4d^{10} 5p^6 6s^2 4f^{14} 5d^{10} 6p^6 7s^2 5f^{14}$$ ■

2.14 STATISTICAL MECHANICS

The aim of statistical mechanics is to deduce the behavior of matter in bulk, from a knowledge of the quantal laws that govern the individual molecules. In consequence of this more fundamental approach, statistical mechanics leads to a deeper insight into the significance of the laws of thermodynamics, and permits the calculation, in terms of atomic and molecular parameters, of quantities, such as heat capacity and entropy, which, in classical thermodynamics, are left to be determined experimentally. This particular aspect of statistical mechanics is called *statistical thermodynamics.*

Having established in the preceding sections the many different energies possible for the individual molecules of a substance, it is now appropriate to consider *statistical mechanics.* At any instant in time, a certain number n_1 of molecules will have the same energy ϵ_1, n_2 will have the energy ϵ_2, and so on. The elementary theory of permutations and combinations are used to determine the most probable or average, values of $n_1, n_2, \ldots$, and it will be found that they are given by some simple distribution law. It will then be shown how the values so obtained are related to the energy E and entropy S of the substance. Then E and S, and hence all the other thermodynamic properties, can be expressed in terms of a function z, known as the partition function.

Statistical mechanics begins with the concept of microstates and macrostates. Consider, as in previous discussions, the translational energy levels of a single particle in a container forming an array of energies

$$\epsilon_0 < \epsilon_1 < \epsilon_2 < \epsilon_3 \cdots$$

Each of these translational energy levels is likely degenerate and is composed of many translational energy states. The array of allowed translational energy levels can be pictures as a vertical "ladder" with each "rung" representing one of the allowed translational energies ϵ. Figure 2.14-1 illustrates this.

Each horizontal rung of the energy ladder represents an allowed translational energy. Since each allowed translational energy is described by the quantum number integers n_x, n_y, n_z, and since there is no limit to the numerical values that these may take, there will be an infinite number of rungs on the

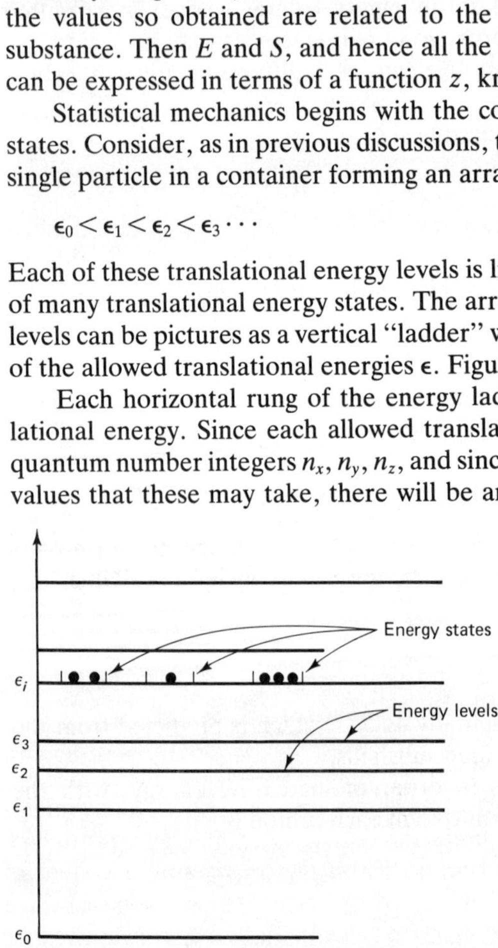

Figure 2.14-1 Translational energy levels of a particle moving in a container.

ladder, that is, an infinity of allowed translational energy levels. Each energy level of the array is such that the energy of any particular level is greater than the energy of the level immediately below it. Furthermore, the energy separation between these adjacent, successive levels, is proportional to $\epsilon^{-1/2}$ so that as the numerical value of the energy of the levels increases, their energy separation decreases. In other words, the rungs of the translational ladder are not equidistant but get closer and closer together the higher the energy of the level.

To allow for translational energy level degeneracy, each rung of the energy ladder is subdivided into several energy states (like boxes into which molecules may be placed). Figure 2.14-1 shows six particles all of the same translational energy ϵ_i, all occupying the same energy level ϵ_i and distributed amongst the three energy states (all of energy ϵ_i) that comprise the energy level ϵ_i. Thus, $g_i = 4$ for the translational energy level ϵ_i. For a gas of noninteracting particle in the absence of any external fields, the microstate is specified by the single-particle quantum states occupied by each gas atom. Each single-particle state is specified by four quantum numbers, one of which specifies the component of the spin of the particle. Let these states be labeled from 1 to ∞ for convenience. A typical microstate is schematically represented by

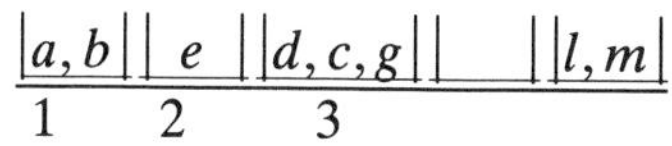

energy level i

In this microstate, particles a and b are in state 1, particle e is in state 2, and so on.

Often, such a detailed specification is unnecessary for purposes of calculating the macroscopically observed thermodynamic properties. Knowing only the number of particles rather than which of the particles are in each quantum state suffices. A specification of the number of particles in each single-particle energy level, without regard to which particles are involved, is called a *macrostate*. At each energy level, this is schematically represented by

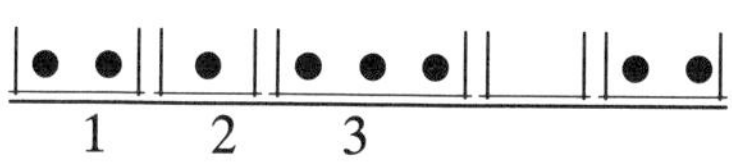

energy level i

The state of the gas is characterized by the set of occupation numbers $\{n_i\}$ at each energy level ϵ_i. With each macrostate, there are many possible microstates. The number of microstates associated with a given macrostate is called the thermodynamic probability. Note that a macrostate is an allocation of particles among the various energy levels without considering the fine detail of their degeneracies (only the number n_i is important).

■ EXAMPLE 2.14-1

Large groups of particles in a thermodynamic system can exist in many microstates, but will move toward an equilibrium or most probable macrostate with the largest number of corresponding microstates. Considering a system of three distinguishable particles a, b, and c, with a total system energy of three energy units, sketch the possible microstates and find the distribution or macrostate with the greatest thermodynamic probability.

Solution. A macrostate is defined by a set of distribution numbers of particles in energy states, and here the microstates follow from the assumption of distinguishability of particles. Also, the energy levels will be considered nondegenerate (one energy state per energy level).

The distribution of energy among atoms in a crystal is a good example. The atoms in a crystal are perfect one-dimensional harmonic oscillators so that an atom may have energies ϵ_n given by $nh\nu$, where n is the vibrational quantum number, h is Planck's constant, and ν is the vibrational frequency. (This amounts to taking the zero-point energy as the reference and measuring energy with respect to the zero-point level.) The atoms, labeled a, b, c, are distinguishable because they have different locations x_a, x_b, and x_c. The microstates and macrostates can be obtained by inspection as in Fig. 2.14-2. Each macrostate A, B, and C has a system energy of three units as required. Macrostate A has one microstate or complexion, B has three microstates, and C has six microstates. Macrostate C with six microstates has the highest thermodynamic probability, and this macrostate is the most probably occurring (60% of the time). The thermodynamic probabilities for the various macrostates are

$$W_A = \frac{3!}{0!3!0!} = 1 \qquad W_B = \frac{3!}{2!0!1!} = 3 \qquad \text{and} \quad W_C = \frac{3!}{1!1!1!} = 6$$

The total number of arrangements or microstates is obtained from the general combinatorial formula (see Chapter 1) for distributing a total of E energy units among n distinguishable particles:

$$\frac{(n+E-1)!}{(n-1)!E!} = \frac{(3+3-1)!}{(3-1)!3!} = 10. \qquad \blacksquare$$

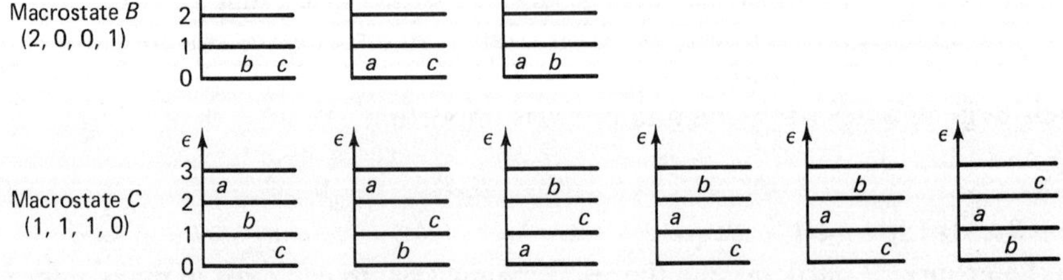

Figure 2.14-2 Macrostates and microstates of a three-particle system having total energy of three units.

2.15 STATISTICAL MODELS

So far the discussion has been concerned with distinguishable particles. The various microstates and macrostates of Example 2.14-1 have been obtained by inspection, but it is desirable to have some general rules governing them. The problem of placing N noninteracting particles into k energy levels such that the first level contains n_1 particles, the second n_2 particles, and so on, is illustrated in Fig. 2.15-1.

In Fig. 2.15-1, the large boxes correspond to the energy levels and the small (dotted) boxes to the quantum states at each level, the degeneracy of a level being the number of small boxes at that level. The balls are the non-interacting particles, of which there are three kinds, obeying different kinds of statistics: (a) identical distinguishable particles (Maxwell–Boltzmann); (b) identical indistinguishable particles of integral spin (Bose–Einstein); and (c) identical indistinguishable particles of half-integral spin (Fermi–Dirac).

Identical Distinguishable Particles (Maxwell–Boltzmann)

Consider the N molecules to be distinguishable (by their labels a, b, etc.). Let the N molecules be placed in an ordered sequence, and from this sequence place the first n_1 molecules in level 1, the next n_2 in level 2, and so on. Since there are $N!$ ways of obtaining the ordered sequence (i.e., permutations), the above procedure can be carried out in $N!$ ways. But the order within the various levels is immaterial, and so $N!$ must be divided by a factor that is the number of equivalent ways in which each combination can be achieved. The number of ways in which a given n_i molecules can be arranged in sequence in the ith level is $n_i!$, and therefore the number of ways (differing only in the ordering within the levels) of obtaining one combination is the product of such terms:

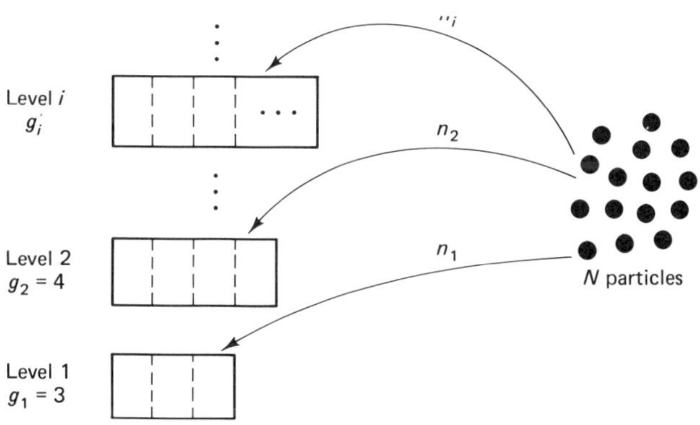

Figure 2.15-1 Placement of molecules into energy levels.

$$n_1!n_2!\cdots n_k! = \prod_{i=1}^{k} n_i! \tag{2.15-1}$$

The number of nonequivalent arrangements, that is, the number of combinations, is thus

$$C(n_i) = \frac{N!}{\prod_i n_i!} \tag{2.15-2}$$

This is the thermodynamic probability which, in statistical mechanics, is more often designated as W, the statistical weight.

More precisely, n_i designates the number of balls in the ith large box, the number of possible ways the N_i balls may be arranged in the g_i small boxes contained in each large box. For each ball there are g_i choices, so that for all balls in the ith large box there are the following number of possibilities:

$$g_i \cdot g_i \cdot g_i \cdots = g_i^{N_i} \tag{2.15-3}$$

In other words, there are $g_i^{N_i}$ possible ways to arrange the N_i balls in the g_i small boxes inside the ith large box. This means that Eq. (2.15-2) must be modified in the following way to account for all possible arrangements:

$$W = \frac{N!}{\prod_i N_i!} g_1^{N_1} \cdot g_2^{N_2} \cdot g_3^{N_3} \cdots g_i^{N_i} = N! \Pi \frac{(g_i)^{n_i}}{n_i!} \tag{2.15-4}$$

where the product is taken over all possible energy levels.

Identical Indistinguishable Particles of Integral Spin (Bose–Einstein)

The particles or balls are assumed to be indistinguishable, so that the permutations of a number of balls inside a box do not alter the basic arrangement. There are g_i small boxes and a total of N_i indistinguishable balls distributed among these boxes. The boxes can be regarded as $(g_i - 1)$ partitions, which can be permuted with the n_i balls in the manner shown in Fig. 2.15-2.

In general, if a level is g fold degenerate, there will be $(g_i - 1)$ partitions. If the distribution number is n_i, the partitions and particles can be permuted in $(n_i + g_i - 1)!$ ways. The number of ways in which the n_i particles can be placed in the ith level is, therefore,

$$\frac{(n_i + g_i - 1)!}{n_i!(g_i - 1)!} \tag{2.15-5}$$

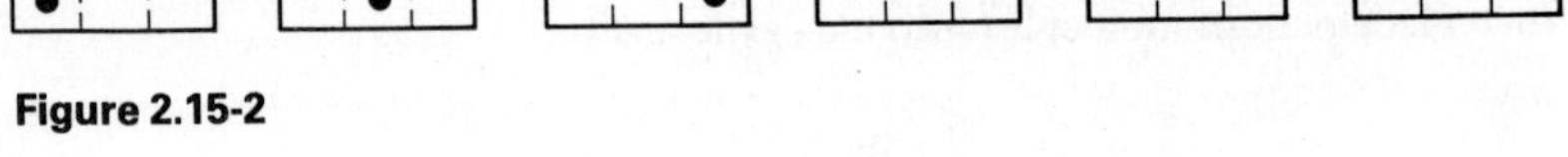

Figure 2.15-2

the factors $n_i!$ and $(g_i - 1)!$ allowing for, respectively, the equivalence of the particles and the equivalence of the partitions. The number of complexions associated with a given distribution is the product of such terms over all the energy levels:

$$W = \prod_i \frac{(n_i + g_i - 1)!}{n_i!(g_i - 1)!} \tag{2.15-6}$$

Identical Indistinguishable Particles of Half-Integral Spin (Fermi–Dirac)

This is the same as the case for Bose–Einstein, except that no more than one particle can be placed in the same single particle state. This means $n_i \leq g_i$; otherwise there would be more than one ball in each small box in Fig. 2.15-1. Consider the ith energy level, with its g_i-fold degeneracy. In the case of particles with an odd sum of numbers of protons, neutrons, and electrons of which one molecule, or other particle is composed, no two of the n_i particles that occupy this level can occupy the same quantum state. Thus, n_i quantum states will each be occupied by one particle, and therefore $(g_i - n_i)$ states will be unoccupied. The number of ways in which this can be achieved is given by the combinatorial rule, according to the following scheme:

$$g_i \begin{array}{l} \nearrow \; n_i \text{ occupied} \\ \searrow \; (g_i - n_i) \text{ unoccupied} \end{array} \tag{2.15-7a}$$

which gives

$$\frac{g_i!}{n_i!(g_i - n_i)!} \tag{2.15-7b}$$

Hence, the number of complexions corresponding to a particular distribution is

$$W = \prod_i \frac{g_i!}{n_i!(g_i - n_i)!} \tag{2.15-8}$$

Equations (2.15-4), (2.15-6), and (2.15-8) embody, respectively, the *Maxwell–Boltzmann, Bose–Einstein,* and *Fermi–Dirac* statistics that all particles must obey. The corresponding particles are known, respectively, as *boltzons, bosons,* and *fermions.* The Maxwell–Boltzmann model has been found to describe reasonably well the behavior of gas molecules, except for those at temperatures close to absolute zero and also for those electrons at extremely high temperature. The Bose–Einstein model applies well to photons of radiation as well as to gas molecules at temperatures close to absolute zero. The Fermi–Dirac model, similar to the Bose–Einstein model except that no two particles can occupy the same energy state, describes the behavior of electrons in metals except at very high temperatures.

Actually, nature consists of only two kinds of particles: bosons and fermions. Some examples of bosons are H, H_2, and He^4 (or any particle with an integer value of spin). Examples of fermions (particles with half-integer spin) are electrons, protons, neutrons, H^+, and He^3. In the limiting case where the occupation number (average number of particle in a quantum state) is much less than unity, both bosons and fermions behave like boltzons. It is well to keep in mind, however, that all particles are in reality either bosons or fermions, and that there are no boltzons. The characteristics of the particles are summarized in Table 2.15-1.

TABLE 2.15-1 SUMMARY OF STATISTICAL MODELS

Name of particle	Statistics obeyed	Type	Number of particles per quantum state
Boltzon	Maxwell–Boltzmann (MB)	Distinguishable	Any number
Boson	Bose–Einstein (BE)	Indistinguishable	Any number
Fermion	Fermi–Dirac (FD)	Indistinguishable	One

■ EXAMPLE 2.15-1

Consider an idealized system in which the single-particle energy levels are nondegenerate, are equally spaced, and have the energy eigenvalues 0, 1, 2, 3, . . . energy units. (This would be true of a one-dimensional harmonic oscillator in which the zero-point energy $\frac{1}{2}h\nu$ is ignored and in which $h\nu$ is put equal to one energy unit.) Assume initially that the system consists of three distinguishable particles of spin zero, which can be distinguished by their locations in space (labeled a, b, and c, respectively) and list the microstates and macrostates of the system. Do the same thing for the case of three bosons (identical particles of spin zero) and the case of three fermions (identical particles of spin $\pm\frac{1}{2}$).

Solution. For the case of three boltzons (identical distinguishable particles a, b, c), the macrostates are shown in Fig. 2.15-3. The macrostate (1, 1, 1, 0, . . .) has six microstates, the macrostate (2, 0, 0, 1, . . .) has three microstates, and the macrostate (0, 3, 0, 0, . . .) has one microstate. The total

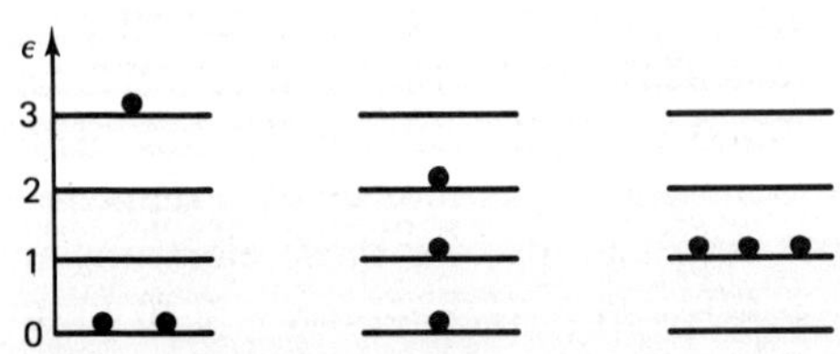

Figure 2.15-3 Macrostates for three identical bosons of spin zero having total energy of three units. For boltzons, with each dot labeled as a, b, or c, the macrostate (2, 0, 0, 1) has 3 microstates, the macrostate (1, 1, 1, 0) has 6 microstates, and the macrostate (0, 3, 0, 0) has 1 microstate.

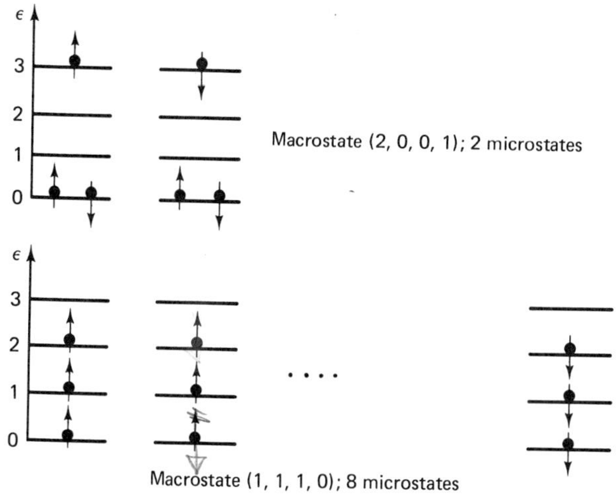

Figure 2.15-4 Macrostates for three identical fermions.

number of microstates consistent with the total energy of the system to be three energy units is ten.

For the case of three bosons (identical particles of spin zero) having a total energy of three units, the microstates are reduced to those shown in Fig. 2.15-3. There is no limit to the number of bosons that can be in a particular single particle quantum state with a given set of single particle quantum numbers, but it cannot be said which of the bosons is in which particular state. For three identical bosons of spin zero, the total number of accessible microstates is three.

For the case of fermions, let the energy of the single-particle state be specified by the quantum number n, which is equal to the number of energy units, and let the component of the spin of the fermion be denoted by the spin quantum number s, which for fermions of spin $\frac{1}{2}$ can have the values $\pm\frac{1}{2}$. There can be two, but only two, fermions of spin $\frac{1}{2}$ with a given value of n, but these must have opposite spins. The macrostates for three identical fermions of spin $\frac{1}{2}$ whose total energy is three units are shown in Fig. 2.15-4. The macrostate (2, 0, 0, 1) has two different microstates, because the fermion occupying three energy units can have spin quantum $\pm\frac{1}{2}$. Similarly, the macrostate (1, 1, 1, 0) has $2^3 = 8$ different microstates, because each fermion can have spin quantum number $\pm\frac{1}{2}$.

Originally, in the nineteenth century, it was assumed that, in principle, the atoms of a monatomic gas are distinguishable. However, this led to the difficulty of the Gibbs paradox, which will be discussed in Chapter 3. The approach from the outset is to assume that no distinction can be made between nonlocalized identical particles, such as the atoms of a monatomic gas in a box. These particles can be anywhere in the box and obey either Bose–Einstein or Fermi–Dirac statistics. ■

2.16 ENSEMBLE

The goal of statistical mechanics is to calculate thermodynamic properties of matter from information about the individual molecules that make it up. This is done by using a model for the molecules. From a thermodynamic viewpoint a macroscopic system can be completely described by specifying only several parameters like pressure, volume, and temperature. But from a microscopic viewpoint an enormous number of parameters have to be specified in order to describe the system completely. For a quantum-mechanical system of N gas molecules, the quantum states of all the molecules would have to be specified.

There are many sets of specifications for a model system that are consistent with the thermodynamic description of the macroscopic system. The approach of statistical mechanics to calculating a mechanical property is to calculate that mechanical property for each of the quantum states that has the desired macroscopic parameters. The mechanical property of the macroscopic system is then obtained by averaging over all of the possible quantum states. In calculating these averages, it is useful to think in terms of probabilities. The probability P_i of a specified result is the fraction of the number of times that result is obtained. For example, if a variable u can assume M discrete values $u_1, u_2, \ldots, u_M$ with probabilities of $P(u_1), P(u_2), \ldots, P(u_M)$, then the average value of u is

$$\bar{u} = \frac{\sum_{i=1}^{M} u_1 P(u_1)}{\sum_{i=1}^{M} P(u_1)} \tag{2.16-1}$$

where the denominator is unity because of the definition of the probability. If some quantity $f(u)$ is a function of u, then the average value of this quantity is given by

$$\overline{f(u)} = \sum_{i=1}^{M} f(u_1)P(u_1) \tag{2.16-2}$$

The variable u in these equations is a discrete random variable.

In situations where the random variable u is continuous, rather than discrete, $P(u)\,du$ is the probability that the random variable lies between u and $u + du$. In this case the average value of u and the mean of any function of u are given by

$$\bar{u} = \int uP(u)\,du \tag{2.16-3}$$

$$\overline{f(u)} = \int f(u)P(u)\,du \tag{2.16-4}$$

where the integrations are carried over the entire range of the possible values of u. The quantity $P(u)$ is referred to as a probability density, and it has the dimensions of $1/u$.

In statistical mechanics averages are first calculated for mechanical properties, such as pressure and energy, by averaging over all possible quantum states. Then the nonmechanical properties such as entropy and Gibbs energy are calculated by use of the equations of thermodynamics. These calculated average values correspond to the experimentally measured thermodynamic properties. The averaging process is facilitated by the concept of an *ensemble* (first introduced by Gibbs). An ensemble is an imaginary collection of a very large number of systems, each constructed to have the thermodynamic (macroscopic) properties of the system of interest. There are several different kinds of ensembles, depending upon the thermodynamic variables held constant; but the most commonly used ensemble is the *canonical ensemble*. In a canonical ensemble each individual system has the same number of molecules N, the same volume V, and the same temperature T. (Other principal types of ensembles are the *microcanonical* and *grand canonical*. In the microcanonical ensemble the systems are isolated systems each having the same N, V, and E. In the grand canonical ensemble the systems each have the same V, T, and chemical potential μ.)

The canonical ensemble for a single component, which is illustrated in Fig. 2.16-1, may be imagined to be set up as follows. Each of the systems, containing N molecules, of the ensemble is enclosed in a container of volume V with walls that are heat conducting, but do not permit molecules to escape. Each of these systems is placed in a heat bath at temperature T and allowed to come to equilibrium. Then each of these systems is removed from the heat bath and surrounded by insulation so that it is isolated. When this is done each system in the ensemble has a particular energy E_i that is independent of time. Since the ensemble contains a very large number of systems, all possible values of the energy consistent with N, V, T, are represented. The total energy E of the ensemble is a constant because the ensemble is an isolated system. However, it cannot be assumed a priori, that systems identical in volume, temperature, and molecular composition necessarily have the same energy. An individual system within the canonical ensemble may have an energy corresponding with any one of the energy eigenvalues for the N molecule system. The possible energy eigenvalues may be represented by E_1, E_2, E_3, . . . , with the lowest energy represented by E_1 and with any particular energy E_i repeated according to its degeneracy $W(E_i)$.

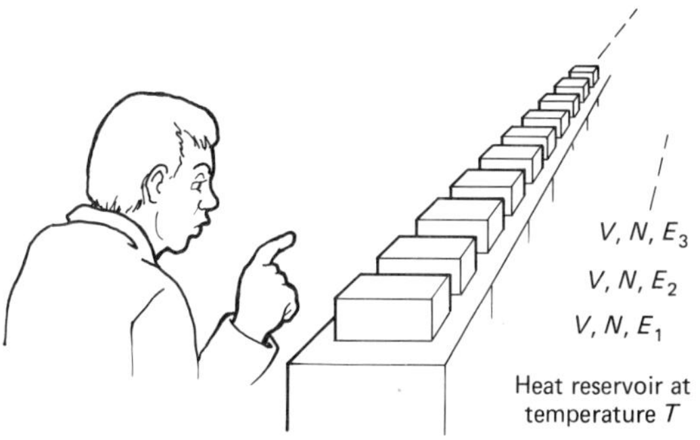

Figure 2.16-1 An inquisitive look at a canonical ensemble.

The existence of thermal contact between the systems implies that the systems can exchange energy with one another. In consequence, the energies E_i of the individual systems fluctuate about a mean value (as do the energies ϵ_i of the individual molecules of a gas). It will later be found that significant probable fluctuations in the E_i are, in contrast with those of the molecular energies, very small (too small to be observed), but this in no way detracts from the mathematical treatment.

Each ensemble member will fluctuate among many possible states. If the ensemble is observed at some instant of time, the various members will be distributed over all the possible quantum states. If the number of ensemble members is sufficiently large, this instantaneous distribution will be representative of the entire set of possible states. Since the ensemble is imaginary, there is no difficulty in supposing that the number of members approaches infinity. All quantum states will then be well represented, and the probability of observing an ensemble member in state i at a given instant is simply the fraction of members in that state:

$$P_i = \frac{N_i}{\sum_i N_i} \tag{2.16-5}$$

where N_i is the number of members in state i and the summation is over all states. Note that P_i is normalized, because $\Sigma_i P_i = 1$. The ensemble average of any dynamical property such as the average system energy can be obtained from

$$\langle E \rangle = \sum_i E_i P_i \tag{2.16-6}$$

Now, if the ensemble is truly representative of the real physical system, the ensemble average energy should be the same as the time average energy for the system, so that

$$U = \lim_{\tau \to \infty} \frac{1}{\tau} \sum_i E_i \, \Delta t_i = \lim_{N \to \infty} \sum_i E_i P_i \tag{2.16-7}$$

Equation (2.16-7) formally states the equivalence of time and ensemble averaging; similar equations may be written for other dynamical properties. The above equivalence implies that the time Δt_i spent in quantum state i is proportional to P_i. In Eq. (2.16-7), N is the total number of ensemble members, and the limit $N \to \infty$ ensures that all quantum states are properly represented. Thus, although time averaging cannot be performed, the ensemble average can be calculated once P_i is known, depending on the thermodynamic constraints on the real physical system.

2.17 EQUILIBRIUM DISTRIBUTION

The concept of an ensemble is suited to the general case of interacting particles. It is also useful to the case of independent particles. By independent, it is meant that the forces of interaction between the particles are not explicitly considered when the total energy is calculated. Thus, the total energy is simply the sum of the energies of the individual molecules.

Considering the particles themselves as systems, in an ensemble at constant V and T, there are n_1 systems having energy ϵ_1, n_2 systems having energy ϵ_2, n_i systems having energy ϵ_i, and so on:

State number	1	2	3	...
Energy	ϵ_1	ϵ_2	ϵ_3	...
Occupation number	n_1	n_2	n_3	...

The set of occupation numbers ($n_1, n_2, n_3, \ldots$) is called a distribution. The total energy of the ensemble is

$$\sum_i n_i \epsilon_i = \epsilon \tag{2.17-1}$$

and the total number of systems in the ensemble is

$$\sum_i n_i = n \tag{2.17-2}$$

The number n can be as large as imaginable. There are many possible distributions $n_1, n_2, n_3, \ldots$ that satisfy Eqs. (2.17-1) and (2.17-2). According to the basic postulate of statistical mechanics, every distribution that satisfies the conditions that the number of systems in the ensemble be fixed and the total energy of the ensemble be fixed must be given equal weight in calculating an average property for the ensemble.

The most probable distribution for an ensemble in equilibrium will be that with the greatest thermodynamic probability as discussed in Section 2.15. To find this, a statistical model is first chosen. Since nature essentially consists of bosons and fermions, the former (identical, indistinguishable particles of integer spin) is selected, because it is not restricted to an occupation number of one particle per quantum state. Thus, from Eq. (2.15-6),

$$w = \prod_i \frac{(n_i + g_i - 1)!}{(n_i! g_i - 1)!} \tag{2.17-3}$$

This is to be maximized with respect to n_i, but it is more convenient to work with ln w, so

$$\ln w = \sum_i [\ln(n_i + g_i - 1)! - \ln n_i! - \ln(g_i - 1)!] \tag{2.17-4}$$

If both g_i and n_i in (2.17-4) are large compared to unity, ones may be neglected and Stirling's approximation ($\ln x! = x \ln x - x$) may be applied to obtain

$$\ln w = \sum_i [(n_i + g_i)\ln(n_i + g_i) - (n_i + g_i) - n_i \ln n_i + n_i - g_i \ln g_i + g_i] \tag{2.17-5}$$

The maximization process is carried out by setting

$$\frac{\partial(\ln w)}{\partial n_i} = 0 \tag{2.17-6}$$

This means that the variation in ln w is zero for a small variation in n_i from the equilibrium distribution, or

$$\delta(\ln w) = 0 = \frac{\partial(\ln w)}{\partial n_i}\,\delta n_i \tag{2.17-7}$$

Using Eq. (2.17-4), this condition gives

$$\frac{\partial(\ln w)}{\partial n_i} = 1 + \ln(n_i + g_i) - 1 - \ln n_i = \ln\left(\frac{n_i + g_i}{n_i}\right) = 0 \tag{2.17-8}$$

Equation (2.17-8) is subject to the restrictions

$$n = \sum_i n_i, \qquad \text{constant number of particles} \tag{2.17-9}$$

or

$$\sum_i \delta n_i = 0 \tag{2.17-10}$$

and

$$u = \epsilon = \sum_i \epsilon_i n_i \qquad \text{constant total internal energy} \tag{2.17-11}$$

where ϵ_i designates the energy of each particle occupying the ith energy level. The constant energy condition gives

$$\delta u = 0 = \frac{\partial u}{\partial n_i}\,\delta n_i$$

or

$$\sum_i \epsilon_i\,\delta n_i = 0 \tag{2.17-12}$$

Also, Eq. (2.17-8) can be rewritten as

$$\sum_i \ln\frac{n_i + g_i}{n_i}\,\delta n_i = 0 \tag{2.17-13}$$

Using Lagrange's method of undetermined multipliers, as explained in Section 1.14, Eq. (2.17-12) is multiplied by a constant, $-\beta$, which is a function of the total system energy, and Eq. (2.17-10) is multiplied by a constant, $-\alpha$, which is a function of the total number of particles. The resulting equations added to Eq. (2.17-13) give

$$\sum_i \left[\ln\left(1 + \frac{g_i}{n_i}\right) - \beta\epsilon_i - \alpha\right]\delta n_i = 0 \tag{2.17-14}$$

In effect, the constants β and α have incorporated the energy and number of particles restriction into Eq. (2.17-13) so that the n_i's may now be treated as independent. Setting all the coefficients of δn_i in Eq. (2.17-14) equal to zero gives

$$\ln\left(1+\frac{g_i}{n_i}\right)-\beta\epsilon_i-\alpha=0 \tag{2.17-15}$$

Hence,

$$e^{\alpha}e^{\beta\epsilon_i}=1+\frac{g_i}{n_i} \tag{2.17-16}$$

or

$$n_i=\frac{g_i}{e^{\alpha}e^{\beta\epsilon_i}-1} \tag{2.17-17}$$

and so, the most probable average population per quantum state at level ϵ_i is

$$\langle n_i\rangle=\frac{n_i}{g_i}=\frac{1}{e^{\alpha}e^{\beta\epsilon_i}-1}=\frac{1}{Ae^{\beta\epsilon_i}-1} \tag{2.17-18}$$

where A stands for e^{α}. The constants α and β are called undetermined multipliers. They play a role similar to constants of integration in the solution of differential equations in that their values must be determined from appropriate physical boundary conditions. These constants will be determined subsequently. Equation (2.17-18) is known as the *Bose–Einstein distribution law*.

A similar law, called the Fermi–Dirac (FD) distribution, is obtained in an analogous way, using the FD model, Eq. (2.15-8):

$$w=\prod_i\frac{g_i!}{n_i!(g_i-n_i)!} \tag{2.17-19}$$

Taking logarithms and applying Stirling's approximation,

$$\begin{aligned}\ln w&=\sum_i \ln_i g_i!-\sum_i \ln_i n_i!-\sum_i \ln(g_i-n_i)!\\&=\sum_i\left[g_i\ln g_i-n_i\ln n_i-(g_i-n_i)\ln(g_i-n_i)\right]\end{aligned} \tag{2.17-20}$$

Hence,

$$\frac{\partial\ln w}{\partial n_i}=\ln(g_i-n_i)-\ln n_i=\ln\frac{g_i-n_i}{n_i} \tag{2.17-21}$$

Proceeding as before yields

$$n_i=\frac{g_i}{e^{\alpha}e^{\beta\epsilon_i}+1} \tag{2.17-22}$$

or

$$\langle n_i\rangle=\frac{n_i}{g_i}=\frac{1}{e^{\alpha}e^{\beta\epsilon_i}+1}=\frac{1}{Ae^{\beta\epsilon_i}+1} \tag{2.17-23}$$

This is the *Fermi–Dirac distribution law*.

The quantity $n_i(\epsilon)$ in Eqs. (2.17-18) and (2.17-23) is called the *occupancy* of the orbital (a one-particle quantum state). For bosons, $n(\epsilon)$ is not the same as the probability that an orbital be occupied. For fermions, the occupancy and the probability are the same, because only 0 or 1 particle may occupy an orbital.

■ EXAMPLE 2.17-1

Plot and discuss the Bose–Einstein and Fermi–Dirac distribution functions in terms of $(\epsilon - \mu)/kT$, where μ is the chemical potential.

Solution. Equations (2.17-18) and (2.17-23), the distribution functions for indistinguishable particles, can be represented by the single equation

$$\langle n_i \rangle = \frac{n_i}{g_i} = \frac{1}{Ae^{\beta\epsilon_i} \mp 1} \tag{2.17-24}$$

where -1 in the denominator is for bosons, and $+1$ is for fermions. $\langle n_i \rangle$ is the average occupancy of an orbital. It will later be shown that (2.17-24) can be rewritten (after A and β have been determined) as

$$\langle n_i \rangle = \frac{1}{e^{(\epsilon_i - \mu)/kT} \mp 1} \tag{2.17-25}$$

where μ is the chemical potential per particle and k is the Boltzmann constant. To apply this expression to a particular system, the permitted energy levels ϵ_i and the chemical potential μ must be known. For the present, it is instructive to plot and compare the two distribution functions in terms of the dimensionless quantity $(\epsilon_i - \mu)/kT$. This is shown in Fig. 2.17-1. The ordinate is the occupation number at those abscissas at which the energy ϵ_i has one of its permitted values. (The energy increases toward the right.)

When $\epsilon_i = \mu$, the occupation value $\langle n_i \rangle$ in the Bose–Einstein (BE) model becomes infinite, and for levels in which ϵ_i is less than μ, it is negative and hence meaningless. Thus, for bosons, the chemical potential must be less than the energy of the lowest permitted energy level. The particles like to concentrate in levels for which ϵ_i is only slightly greater than μ. For fermions, on the other hand, all levels are populated down to the lowest and as ϵ_i decreases, $\langle n_i \rangle$

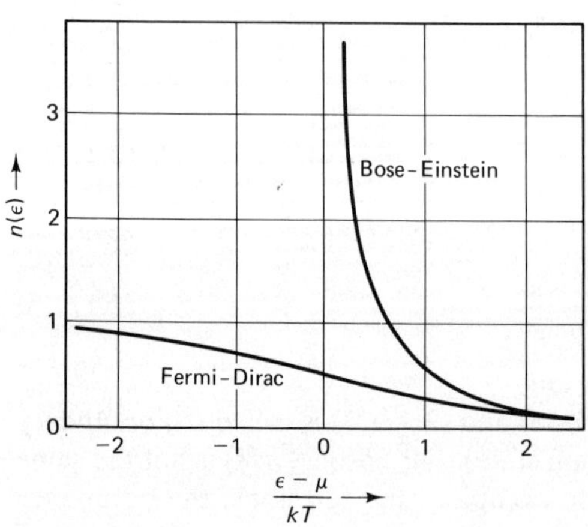

Figure 2.17-1 Comparison of Bose–Einstein and Fermi–Dirac distribution functions.

approaches 1. That is, the low-energy levels are very nearly uniformly populated with one particle per state.

When $\langle n_i \rangle = n_i/g_i$ is very small, the BE and FD distributions very nearly coincide, and both reduce to what is known as the classical regime, attained for $(\epsilon_i - \mu)kT >> 1$. It will be seen in Chapter 4 that in the degenerate regime at low temperature, the chemical potential μ for a FD distribution is positive, and changes to negative at high temperatures. The value $\mu(0)$ of the chemical potential at the absolute zero of temperature is called the *Fermi energy* (ϵ_F). The chemical potential of a BE distribution is always negative if the zero of energy is chosen to coincide with the energy of the lowest orbital. ■

2.18 CLASSICAL LIMIT: BOLTZMANN DISTRIBUTION

In many systems of indistinguishable particles, the average number of particles n_i/g_i in a level is very much less than the number of states g_i in the level, and so the average number of particles per state, $\langle n_i \rangle$, is very small. The denominator in the distributions (2.17-18) and (2.17-23)

$$\langle n_i \rangle = \frac{n_i}{g_i} = \frac{1}{Ae^{\beta\epsilon_i} \mp 1} \tag{2.18-1}$$

becomes very large, and the 1 can be neglected. Both the BE and FD distributions then reduce to

$$\langle n_i \rangle = \frac{1}{Ae^{\beta\epsilon_i}} \tag{2.18-2}$$

which is the classical *Maxwell–Boltzmann distribution*. This is the distribution law associated with many applications of statistical mechanics when quantum effects can be neglected without significant error.

It is instructive to independently derive the distribution law for classical (distinguishable) particles, starting with Eq. (2.15-4):

$$w = N! \prod_i \frac{g_i^{n_i}}{n_i!} \tag{2.18-3}$$

Taking logarithms and going through the same procedure as before,

$$\begin{aligned} \ln w &= \ln N! + \sum_i (n_i \ln g_i - \ln n_i) \\ &= N \ln N - N + \sum_i (n_i \ln g_i - n_i \ln n_i + n_i) \\ &= N \ln N - \sum_i (n_i \ln g_i - n_i \ln n_i) \end{aligned} \tag{2.18-4}$$

Setting the variation of this equal to zero gives

$$\begin{aligned} \delta \ln w &= \sum_i [(\ln g_i)\delta n_i - (\ln n_i)\delta n_i - \delta n_i] \\ &= \sum_i (\ln g_i - \ln n_i - 1)\delta n_i = 0 \end{aligned} \tag{2.18-5}$$

Likewise, the variations of $\Sigma_i n_i = n$ and $\Sigma_i \epsilon_i n_i = \epsilon_i = u$ are

$$\sum_i \delta n_i = 0 \tag{2.18-6}$$

$$\sum_i \epsilon_i \delta n_i = 0 \tag{2.18-7}$$

Multiplying Eq. (2.18-6) by γ, and Eq. (2.18-7) by β, and subtracting the results from Eq. (2.18-5) then yields

$$\sum_i (\ln g_i - \ln n_i - 1 - \gamma - \beta\epsilon_i)\delta n_i = 0$$

Letting $\alpha = 1 + \gamma$,

$$\ln g_i - \ln n_i - \alpha - \beta\epsilon_i = 0 \tag{2.18-8}$$

and, solving for n_i,

$$n_i = g_i e^{-\alpha - \beta\epsilon_i} \tag{2.18-9}$$

or

$$\langle n_i \rangle = \frac{n_i}{g_i} = \frac{1}{e^{\alpha} e^{\beta\epsilon_i}} = \frac{1}{A e^{\beta\epsilon_i}} \tag{2.18-10}$$

where $A = e^{\alpha}$. Equation (2.18-10) is the *Maxwell–Boltzmann distribution law.* It is the canonical ensemble distribution pertaining to the individual molecules of a macroscopic system. It applies exactly for a perfect gas, and is a good approximation for real gases under ordinary conditions. In modified form it can also be applied to crystals, in which case the weakly interacting systems are not molecules, but modes of vibration. As stated in Example 2.17-1, it will later be shown that (2.18-10) can be rewritten in terms of the chemical potential μ, as

$$\langle n_i \rangle = \frac{1}{e^{(\epsilon_i - \mu)/kT}} \qquad \text{MB} \tag{2.18-11}$$

along with the quantum distributions

$$\langle n_i \rangle = \frac{1}{e^{(\epsilon_i - \mu)/kT} - 1} \qquad \text{BE} \tag{2.18-12}$$

$$\langle n_i \rangle = \frac{1}{e^{(\epsilon_i - \mu)/kT} + 1} \qquad \text{FD} \tag{2.18-13}$$

A comparison of the MB, BE, and FD distribution functions is shown in Fig. 2.18-1. Note that the reference energy is zero for the MB and BE distributions, but is the Fermi energy ϵ_F for the FD distribution. For large values of ϵ_i, all distributions become identical with the MB distribution. [The latter has no significance except when $(\epsilon_i - \mu)/kT$ is large.]

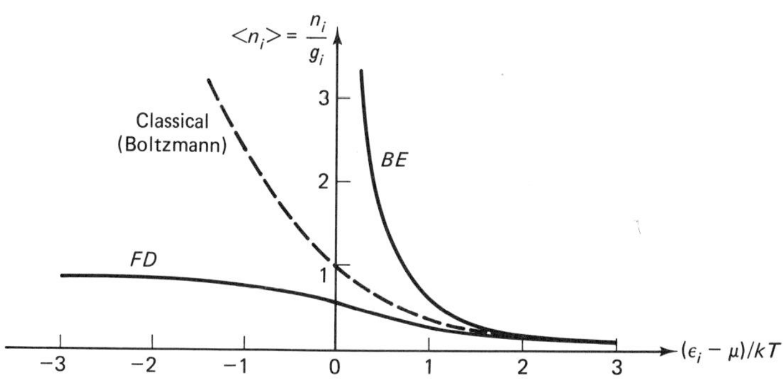

Figure 2.18-1 Comparison of the *MB*, *BE*, and *FD* distribution functions.

Some final comments: The quantum nature of the statistical model can be removed in the following way in classical statistical mechanics. g_i is taken to be 1 for all energy levels, and the energy is assumed to have a continuous distribution so that no quantum restrictions are placed on the energy levels. In this case each energy level can be slightly different from others and there is no need to be concerned about different energy states (since $g_i = 1$ for each energy level). There is, in effect, an infinite number of possible energy states for a continuous distribution. There is no degeneracy, and the number of particles having energy ϵ_i is simply given by Eq. (2.18-10).

The MB model does, however, have limitations: certain types of microscopic particles just are not distinguishable. It is not possible, for example, to tell one oxygen molecule from another, or one electron from another. Also, a truly classical theory cannot involve h. Nevertheless, many problems are more easily solved by staying with classical mechanics. In the following sections, the classical distribution function will be used to explore the thermal properties of the ideal gas, beginning with the monatomic gas.

An additional way of bringing out the similarities and differences of the three distributions is to plot the quotient n_i/g_i, the *occupation index*, versus energy for various temperatures, as shown in Fig. 2.18-2 and listed in Table 2.18-1.

TABLE 2.18-1 SUMMARY OF CLASSICAL AND QUANTUM STATISTICS

Statistics	W	N_i
Maxwell–Boltzmann	$W = N! \prod_i \frac{g_i^{N_i}}{N_i!}$	$N_i = g_i e^{-\lambda - \epsilon_i/kT}$
Bose–Einstein	$W = \prod_i \frac{(N_i + g_i - 1)!}{N!(g_i - 1)!}$	$N_i = \frac{g_i}{e^{\lambda + \epsilon_i/kT} - 1}$
Fermi–Dirac	$W = \prod_i \frac{g_i!}{N_i!(g_i - N_i)!}$	$N_i = \frac{g_i}{e^{\lambda + \epsilon_i/kT} + 1}$

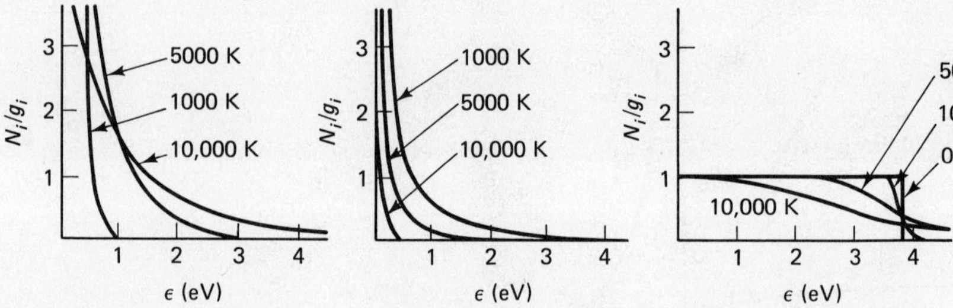

(a) Maxwell-Boltzmann. (b) Bose-Einstein. (c) Fermi-Dirac.

Figure 2.18-2 *MB* energy distribution for two values of β (inverse temperature).

■ EXAMPLE 2.18-1

The classical and quantum-mechanical distribution functions have one thing in common: they all depend upon the energy of the state and not upon the quantities that define the state. Because of this, it is possible, for example, to write the particle distribution in energy in the form

$$dn(\epsilon) = Fg(\epsilon)\,d\epsilon \tag{2.18-14}$$

where F is a fundamental distribution function that describes how the particles are, on the average, distributed among the permitted energy states, and $g(\epsilon)$ is the density of states with respect to energy. F maintains the same form for a given statistics (e.g., $F = F_{MB}$ for classical statistics). The distribution function is the fraction of the particles with energy ϵ within a range $d\epsilon$. In other words,

$$F = \frac{1}{n}\frac{dn(\epsilon)}{g(\epsilon)\,d\epsilon} \tag{2.18-15}$$

Equation (2.18-8), the Maxwell–Boltzmann distribution, states that the average number of particles with energy ϵ_i is

$$\langle n_i \rangle = Ce^{-\beta\epsilon_i} \tag{2.18-16}$$

where C is a constant.

(a) Plot the MB distribution with respect to energy for values of β_1 and $\beta_1/2$.

(b) For a value of $\beta = 38.6$, what are the relative numbers of molecules in a gas having translational energies of 10^{-2} eV, and 10^{-1} eV?

(c) Suppose that ϵ_1 is proportional to a quantity $J_i(J_i + 1)$, that is, $\epsilon_i = BJ_i(J_i + 1)$, and g_i is proportional to a quantity $(2J_i + 1)$, that is, $g_i = C(2J_i + 1)$. What form does the MB distribution take?

Solution

(a) The MB distribution for translational energy is sketched for β_1 and $\beta_1/2$. The curves are normalized to contain the same area under the curve from $\epsilon_i = 0$ to $\epsilon_i = \infty$.

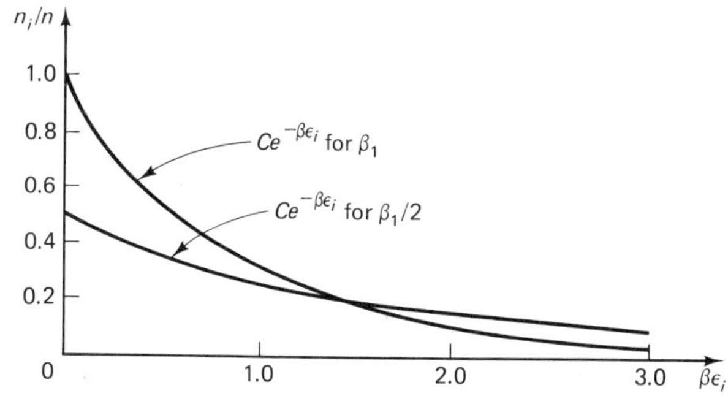

Figure 2.18-3 Form of the Maxwell–Boltzmann distribution for $e_i \propto J_i(J_i+1)$ and $g_i \propto (ZJ_i+1)$.

The most characteristic feature of the distribution is that the number of particles in a given state decreases exponentially with increasing energy of the state for a given β. On the other hand, as β (shown in the next section to be equal to $1/kT$) decreases, $\epsilon_i\beta$ becomes smaller and smaller and so $\exp(-\epsilon_i\beta)$ increases. The net result is that more particles are found in higher energy states when β is decreased (temperature increased).

(b) With $\beta = 38.6\ \text{eV}^{-1}$, Boltzmann's distribution gives
For $\epsilon_1 = 10^{-2}$ eV,

$$n_1 = C\exp(-38.6\times10^{-2}) = C\exp(-0.386) = 0.680C$$

For $\epsilon_2 = 5\times10^{-2}$ eV,

$$n_2 = C\exp(-38.6\times5\times10^{-2}) = C\exp(-1.93) = 0.145C$$

For $\epsilon_3 = 10^{-2}$ eV,

$$n_3 = C\exp(-38.6\times10^{-1}) = C\exp(-3.86) = 0.021C$$

A comparison of n_2 and n_3 to n_1 shows that

$$\frac{n_2}{n_1} = \frac{0.145}{0.680} = 0.213 \qquad \frac{n_3}{n_1} = \frac{0.021}{0.680} = 0.031$$

The number of molecules having 10^{-1} eV energy is only 3.1% of the number having an energy of 10^{-2} eV. For an energy of 1 eV, the number drops to 1.7×10^{-15}% of the number of molecules having an energy of 10^{-2} eV.

(c) When the degeneracies of the energy levels are not all the same, the Maxwell–Boltzmann distribution takes the form

$$n_i = Cg_i\exp(-\beta\epsilon_i) \tag{2.18-17}$$

Here, $\epsilon_i = BJ_i(J_i+1)$, and $g_i = C(2J_i+1)$. Thus,

$$n_i = C(2J_i+1)\exp[-\beta BJ_i(J_i+1)] \tag{2.18-18}$$

To show the form of this curve, let $\beta B = 0.1$, and $C = 1$. Then,

$$n_i = (2J_i+1)\exp[-0.1J_i(J_i+1)] \tag{2.18-19}$$

TABLE 2.18-2 VALUES OF N_i AS A FUNCTION OF J_i

J_i	$2J_i+1$	$J(J_i+1)$	$\exp[-BJ(J_i+1)]$	N_i
0	1	0	1.000	1.000
1	3	2	0.819	2.457
2	5	6	0.549	2.745
3	7	12	0.301	2.107
4	9	20	0.135	1.215
5	11	30	0.050	0.550
6	13	42	0.015	0.195

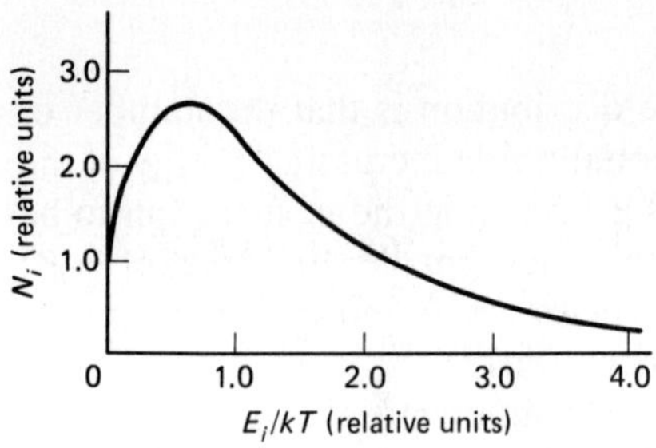

Figure 2.18-4

The calculated values of n_i are given in Table 2.18-2 for several values of J_i and plotted in Fig. 2.18-4.

■

2.19 IDENTIFICATION OF α AND β LAGRANGIAN MULTIPLIERS

The distribution laws, Eqs. (2.18-10), (2.18-12), and (2.18-13), all contain undetermined multipliers (constants), and these must be evaluated before the calculation of n_i is possible. In principle, the unknowns α and β can be determined by substituting the most probable distribution into the boundary conditions

$$\sum_i n_i = n \tag{2.19-1}$$

$$\sum_i \epsilon_i n_i = \epsilon \tag{2.19-2}$$

and then solving the resulting equations. In practice, however, this is easily done only for the MB distribution, and so the discussion will focus on this model, leaving the quantum models to Chapter 4. Thus, starting with,

$$n_i = g_i e^{-\alpha} e^{-\beta\epsilon_i} \tag{2.19-3}$$

substitution into (2.19-1) gives

$$\begin{aligned} n &= g_1 e^{-\alpha} e^{-\beta\epsilon_1} + g_2 e^{-\alpha} e^{-\beta\epsilon_2} + \cdots + g_i e^{-\alpha} e^{-\beta\epsilon_i} + \cdots \\ &= e^{-\alpha}(g_1 e^{-\beta\epsilon_1} + g_2 e^{-\beta\epsilon_2} + \cdots + g_i e^{-\beta\epsilon_i} + \cdots) \\ &= e^{-\alpha} \sum_i g_i e^{-\beta\epsilon_i} \end{aligned}$$

Hence,

$$e^{-\alpha} = \frac{n}{\sum_i g_i e^{-\beta\epsilon_i}} \tag{2.19-4}$$

The quantity appearing in the denominator on the right-hand side of Eq. (2.19-4) is denoted by z:

$$z = \sum_i g_i e^{-\beta\epsilon_i} \tag{2.19-5}$$

and is called the *partition function*, the symbol z being from German: *zustandssumme,* or sum-over-states. The lowercase z is used to denote the molecular partition function to avoid confusion with Z, the partition function of an N particle system. Equation (2.19-3) gives

$$n_i = n \frac{g_i e^{-\beta\epsilon_i}}{\sum_i g_i e^{-\beta\epsilon_i}} = n \frac{g_i e^{-\beta\epsilon_i}}{z} \tag{2.19-6}$$

The evaluation of β is a less straightforward matter. However, it is possible to draw the following conclusions about the properties of this parameter:

1. β must be positive because, with a negative or zero value of β, z, as given by Eq. (2.19-5), would diverge. Thus, if β were zero, z would be the sum of an infinite number of terms g_i; if β were negative, the situation would be even worse, for then f would be the sum of an infinite number of progressively increasing terms.

2. β is related to the temperature of the system. Consider two assemblies of, respectively, n' and n'' particles, having total energies ϵ' and ϵ'', the most probable distribution numbers being correspondingly

$$n_i' = g_i' e^{-\alpha'} e^{-\beta'\epsilon_i'} \tag{2.19-7}$$

$$n_i'' = g_i'' e^{-\alpha''} e^{-\beta''\epsilon_i''} \tag{2.19-8}$$

Suppose that the assemblies are placed in thermal contact (but are otherwise isolated), so that they can interchange energy until equilibrium is reached. In the new equilibrium state, the energies of the assemblies need no longer be ϵ' and ϵ''; in fact, all that is known a priori is that the total energy of the combined system is $\epsilon' + \epsilon''$. Initially,

$$\sum_i n_i' = n' \qquad \sum_i n_i'' = n'' \tag{2.19-9}$$

$$\sum_i n_i' \epsilon_i' = \epsilon' \qquad \sum_i n_i'' \epsilon_i'' = \epsilon'' \tag{2.19-10}$$

Now, in the final state of equilibrium, $\epsilon' + \epsilon''$. The method of undetermined multipliers, applied once more, gives, for the average occupation,

$$\langle n_i' \rangle = g_i' e^{-\alpha'} e^{-\beta \epsilon_i'} \tag{2.19-11}$$

$$\langle n_i'' \rangle = g_i'' e^{-\alpha''} e^{-\beta \epsilon_i''} \tag{2.19-12}$$

where β has the same value for both systems. Hence, if any two assemblies for which the Maxwell–Boltzmann law is valid can exchange energy and are in equilibrium, then they must have the same value of β. Thermodynamically, any two systems that can exchange energy have the same temperature when they are in equilibrium. Thus, β in statistical mechanics has the same role as T in thermodynamics: or, mathematically, β is some universal function of T.

3. A decrease in β is associated with an increase in temperature. This can be shown by taking the ratio of any two most-probable distribution numbers of a system,

$$\frac{\langle n_k \rangle}{\langle n_l \rangle} = \frac{g_k e^{-\beta \epsilon_k}}{g_e e^{-\beta \epsilon_l}} = \frac{g_k}{g_l} e^{-\beta(\epsilon_k - \epsilon_l)} \tag{2.19-13}$$

If $\epsilon_k > \epsilon_l$, that is, if ϵ_k is a higher energy level than ϵ_l, then when β decreases, the ration $\langle n_k \rangle / \langle n_l \rangle$ must increase. Hence, a decrease in the value of β causes an increase in the numbers of molecules in the higher energy levels, at the expense of lower levels, which must mean that the total energy of the system is increased. Thermodynamically, such an increase in energy would be associated with a rise in temperature. The quantitative relationship between β and T is discussed in the next section.

2.20 TEMPERATURE PARAMETER

To further evaluate β, the results of statistical mechanics will be compared with the combined first and second laws of thermodynamics for a closed system with only pressure–volume work, that is,

$$du = T\,ds - p\,dv \tag{2.20-1}$$

In statistical mechanical terms, if v and T are changed reversibly at constant n for a system, the differential of u may be calculated from

$$u = \frac{1}{z} e^{-\beta \epsilon_i} = \sum_i P_i \epsilon_i \tag{2.20-2}$$

where

$$P_i = \frac{e^{-\beta \epsilon_i}}{z} \tag{2.20-3}$$

is the probability that a system in the ensemble is in the ith quantum state. Equation (2.20-2) gives

$$du = \epsilon_i \, dP_i + \sum_i P_i \, d\epsilon_i \tag{2.20-4}$$

The first term is due to changes that do not affect the energy levels but involve a change in the populations of the various energy levels; thus, it comes from thermal changes. The second term does involve a change in energy levels, and this suggests that it corresponds to work done on the system.

From the Boltzmann distribution, Eq. (2.20-3), upon taking the logarithm,

$$\ln P_i + \ln z = -\beta\epsilon_i$$

or

$$\epsilon_i = -\frac{1}{\beta}(\ln P_i + \ln z) \tag{2.20-5}$$

Substituting this expression into the first term of Eq. (2.20-4) and recognizing in the second that the energy states ϵ_i at constant n can only be changed by changing the volume,

$$du = -\frac{1}{\beta}\sum_i (\ln P_i + \ln z) dP_i + \sum_i P_i \left(\frac{\partial \epsilon_i}{\partial v}\right)_n dv \tag{2.20-6}$$

This equation may be simplified by noting that $\Sigma_i \ln P_i \, dP_i = \ln P_i \, \Sigma_i \, dP_i = 0$ since $\Sigma_i \, dP_i = 0$ from $\Sigma_i \, P_i = 1$. Furthermore,

$$d\left(\sum_i P_i \ln P_i\right) = \sum_i dP_i + \sum_i \ln P_i \, dP_i = \sum_i \ln P_i \, dP_i \tag{2.20-7}$$

so that Eq. (2.20-6) becomes

$$du = -\frac{1}{\beta} d\left(\sum_i P_i \ln P_i\right) + \sum_i P_i \left(\frac{\partial \epsilon_i}{\partial v}\right)_n dv \tag{2.20-8}$$

Now, $-p_i \, dv = d\epsilon_i$ is the work that has to be done on the system to change the energy of the ith state by $d\epsilon_i$. Thus,

$$du = -\frac{1}{\beta} d\left(\sum_i P_i \ln P_i\right) - \sum_i P_i p_i \, dv \tag{2.20-9}$$

But $\Sigma_i \, P_i p_i = p$, the pressure of the system, so that (2.20-9) becomes

$$du = -\frac{1}{\beta} d\left(\sum_i P_i \ln P_i\right) - p \, dv \tag{2.20-10}$$

Comparing this with Eq. (2.20-1) from thermodynamics indicates that

$$-\frac{1}{\beta} d\left(\sum_i P_i \ln P_i\right) = T \, ds \tag{2.20-11}$$

Multiplying both sides by a constant k gives

$$\frac{1}{\beta}\left[-d\left(k\sum_i P_i \ln P_i\right)\right] = kT\,ds \tag{2.20-12}$$

Equation (2.20-12) enables the following identifications

$$\frac{1}{\beta} = kT \tag{2.20-13}$$

$$ds = -d\left(k\sum_i P_i \ln P_i\right) \tag{2.20-14}$$

where the Boltzmann constant

$$k = R/N_{\text{Avogadro}}$$

is used to bring statistical mechanics into agreement with thermodynamics. Integration of (2.20-14) yields

$$s = -k\sum_i P_i \ln P_i \tag{2.20-15}$$

The constant of integration is omitted because it often cancels out in any calculation of Δs. This equation embodies the statistical version of the third law of thermodynamics. If there is a single state (e.g., a perfect crystal) at $T = 0$, then $s = 0$. Thus, the entropy depends only on the probabilities P_i of the various states accessible to a system. Since the probabilities are fractions, the entropy is necessarily a positive quantity. Equation (2.20-15) in fact yields Boltzmann's famous relation. This can be seen as follows: If all the members of the ensemble have the same energy ϵ_i, they all have, classically, the same degeneracy $w_i = w$. Since every state of a member of the ensemble has equal a priori probability, $P_i = 1/w$. Therefore, the summation in (2.20-15) consists of w identical terms $w(1/w)\ln(1/w)$, and becomes

$$s = k \ln w \tag{2.20-16}$$

This is the bridge between statistical mechanics and thermodynamics. With the value of β determined, the most probable distribution, Eq. (2.19-6), can now be written as

$$n_i = n\frac{g_1 e^{-\epsilon_i/kT}}{\sum_i g_i e^{-\epsilon_i/kT}} = n\frac{g_i e^{-\epsilon_i/kT}}{z} \tag{2.20-17}$$

This is the final result. It will be shown that all the thermodynamic properties of a system can be obtained once the partition function z is calculated.

The entropy must be a logarithmic function of the thermodynamic probability. This can be seen by starting with the entropy S of a macroscopic system as some function of the number of configurations or arrangements W,

$$S = f(W) \tag{2.20-18}$$

The function $f(W)$ must be a single-valued, monotonically increasing function of its argument W. To determine the nature of the function f, let two systems A and B be brought together. System A is in a state having the entropy S_A with W_A configurations, and system B is in a state having the entropy S_B with W_B configurations. It is known from the extensive character of the entropy that the combined entropy of the two systems is the sum of the entropies of each,

$$S_{\text{total}} = S_A + S_B \tag{2.20-19}$$

On the other hand, the number of configurations accessible to the combined system is the product of the number of configurations of its parts, because each configuration of one part can be combined with any of the configurations of the other part to give one configuration of the combined system,

$$W_{\text{total}} = W_A W_B \tag{2.20-20}$$

Thus,

$$S_{\text{total}} = f(W_{\text{total}})$$

or

$$S_A + S_B = f(W_A) + f(W_B) = f(W_A W_B)$$

The only functional relationship existing between S and W that can satisfy the property

$$f(W_A W_B) = f(W_A) + f(W_B) \tag{2.20-21}$$

is

$$S = k \ln W \tag{2.20-22}$$

where k is Boltzmann's constant. It will be shown to be the gas constant per molecule. Equation (2.20-22) is known as *Boltzmann's relation,* although it was first written by Planck in 1906. It appears on Boltzmann's tombstone.*

■ EXAMPLE 2.20-1

A simple method of identifying Boltzmann's constant k is to consider the calculation of the entropy change which occurs when n moles of an ideal gas are transferred isothermally from a box with a volume of V_1 to one with a volume of V_2. Thermodynamically, for an infinitesimal, isothermal, reversible process,

$$(dQ)_{\text{rev}} = p\, dV = \frac{nRT}{V}\, dV$$

---

* . . . so it stands, carved out on Boltzmann's memorial in the central cemetery in Vienna, floating in the clouds over his majestic bust. (From *Thermodynamics and Statistical Mechanics,* Volume 5 of Lectures on Theoretical Physics, by Arnold Sommerfeld, Academic Press, New York, 1956.)

and the entropy change is therefore

$$\Delta S = \int_{V_1}^{V_2} \frac{(dQ)_{\text{rev}}}{T} = \int_{V_1}^{V_2} \frac{nR\,dv}{V} = nR \ln \frac{V_2}{V_1} \tag{2.20-23}$$

Show that Boltzmann's relation gives the same result.

Solution. Consider the ratios of configurations W_2/W_1 and probabilities P_1/P_2. Suppose, first, that the gas consisted of only one molecule. Then the ratio P_2/P_1 is equal to V_2/V_1, because if the molecule is able to move freely by means of a passageway between two boxes having the volumes V_1 and V_2, the probability of finding it in one of the boxes is proportional to the volume of that box. Let two molecules be present, and compare the probability that both are in V_1 with the probability that both are in V_2. The laws of probability state that if the probabilities of two separate events are P_a and P_b, then the probability that both events will occur simultaneously is the product, $P_a P_b$. Therefore, the ratio of the probability that both molecules will be found in volume V_2 to the probability that both are in volume V_1 is

$$\frac{P_2}{P_1} = \left(\frac{V_2}{V_1}\right)^2 \tag{2.20-24}$$

Similarly, if there are N molecules in the gas, the ratio of the probability that they are all in V_2 to the probability that they are all in V_1 is

$$\frac{P_2}{P_1} = \left(\frac{V_2}{V_1}\right)^N \tag{2.20-25}$$

Thus, from the Boltzmann relation,

$$\Delta S = k \ln \frac{P_2}{P_1} = k \ln \left(\frac{V_2}{V_1}\right)^N \tag{2.20-26}$$

If n moles of gas are present, then $N = nN_A$, where N_A is Avogadro's number, and (2.20-26) may be written

$$\Delta S = n(N_A k) \ln \frac{V_2}{V_1} \tag{2.20-27}$$

Comparison of (2.20-27) with (2.20-23) shows that if the gas constant R is identified with the product $N_A k$, the same result is obtained from the Boltzmann idea as from a thermodynamic treatment. The constant k in Boltzmann's relation is thus the gas constant per molecule, or $k = R/N_A$. Its value is 1.36×10^{-23} J K^{-1}. ■

■ EXAMPLE 2.20-2

The notion of distinguishable and indistinguishable particles (systems) occurs throughout statistical mechanics. When are particles distinguishable, and how does the counting of microstates depend on this fact?

Solution. Statistical mechanics seeks to evaluate the properties of an assembly that consists of a large number N of weakly interacting identical particles. There are two types of assembly.

One type is a gaseous assembly, in which the identical particles are the gas molecules themselves. In quantum mechanics, it is recognized that the molecules are not only identical, but they are also indistinguishable. (It is not possible to paint a label on a particular molecule and follow its history.)

Another type of assembly is one in which the particle is distinguishable. The physical example is that of a solid. It is made up of N identical atoms, although the atoms themselves are indistinguishable. The solid can be thought of as a set of N lattice sites, in which each lattice contains an atom. The particle is distinguished not by the identity of the atom, but by the location of each lattice site. The solid is an assembly of localized particles, and it is this locality that makes the particles distinguishable. This means that the particles can be counted just as macroscopic objects. A microstate will thus specify the state for each distinct particle. The number of possible arrangements of N objects into piles with n_i in a typical pile is then given by $N!/\prod_i n_i!$ ■

2.21 DISTRIBUTION OF MOLECULAR SPEEDS

The parameter β was originally introduced as an undetermined multiplier. It was then proved that $\beta = 1/kT$. The same conclusion can be arrived at by first obtaining the Maxwellian distribution of translational energies of a thermally perfect gas and comparing it with the result derived from kinetic theory.

The molecules of a gas have a distribution of speeds. The information about the distribution of speeds is described by a distribution function $f(v)$, which is defined so that $f(v)\,dv$ is the fraction of the molecules having speeds in the range v to $v + dv$. Maxwell derived the speed distribution function $f(v)$ for a perfect gas in 1859 using classical mechanics. However, having already considered quantum mechanics and statistical mechanics the canonical ensemble can be used to derive the expression for $f(v)$. To this end, the following model is adopted: The molecules of a perfect gas do not interact with each other, and so a single molecule in a volume V can be chosen as the system, the ensemble being a collection of N molecules in a volume V. The molecules are indistinguishable, and the average number of molecules in each of the possible energy states is extremely small. The proper statistics is therefore the Maxwell–Boltzmann statistics.

The energy of one molecule in a cubic box of volume V has been determined earlier in Section 2.6. If $n^2 = n_x^2 + n_y^2 + n_z^2$, then

$$\epsilon_n = \frac{\pi^2\hbar^2 n^2}{2mV^{2/3}} \tag{2.21-1}$$

The probability of finding a replica system, that is, one molecule, with energy ϵ_n, is

$$P_n = \frac{g_n \exp(-\beta\epsilon_n)}{\sum_n g_n \exp(-\beta\epsilon_n)} \tag{2.21-2}$$

according to Eq. (2.19-6).

The degeneracy of each energy level is represented by g_n. As explained in Section 2.7, the translational energy state of a molecule is determined by the three quantum numbers n_x, n_y, and n_z. Thus, each state may be represented by a point in the space determined by three axes n_x, n_y, and n_z. There is one state per unit volume because n_x, n_y, and n_z can change only by integer values. The points corresponding to a particular energy ϵ_n lie on a spherical surface. Since the quantum numbers n_x, n_y, and n_z are very large numbers, they can be treated as continuous variables. The number of states with energies in the range ϵ_n to $\epsilon_n + d\epsilon$ is equal to the volume of the spherical shell of radius n and thickness dn. The surface area of this shell is $4\pi n^2$, and its volume is $4\pi n^2 dn$. Since the quantum numbers are positive, only the states in one octant of the space need be considered. Thus, the number g_n of quantum states in the range ϵ_n to $\epsilon_n + d\epsilon$ is

$$g_n = \frac{1}{8} \cdot 4\pi n^2\, dn = \frac{\pi}{2} n^2\, dn \tag{2.21-3}$$

This is the "density of states" mentioned in Section 2.7.

The denominator of Eq. (2.21-2) is now evaluated by replacing the summation by an integral:

$$\begin{aligned} \sum_n g_n \exp(-\beta\epsilon_n) &= \sum_n \frac{\pi}{2} n^2 \exp(-\beta\epsilon_n)\, dn \\ &= \frac{\pi}{2} \int_0^\infty n^2 \exp\left(-\frac{\beta\pi^2\hbar^2 n^2}{2mV^{2/3}}\right) dn \end{aligned} \tag{2.21-4}$$

This integral is evaluated by making use of

$$\int_0^\infty x^2 e^{-ax^2}\, dx = \frac{\pi^{1/2}}{4a^{3/2}}$$

Thus, Eq. (2.21-4) gives

$$\sum_n g_n \exp(-\beta\epsilon_n) = \frac{\pi}{n} \frac{\pi^{1/2}}{4} \left(\frac{2mV^{2/3}}{\beta\pi^2\hbar^2}\right)^{3/2} = \left(\frac{m}{2\pi\hbar^2\beta}\right)^{3/2} V \tag{2.21-5}$$

In order to obtain the speed distribution from Eq. (2.21-2), it is necessary to express n in terms of the molecular speed. Equation (2.21-1) may be written

$$n^2 = \frac{2mV^{2/3}}{\pi^2\hbar^2} \epsilon n = \frac{2mV^{2/3}}{\pi^2\hbar^2} \left(\frac{1}{2} mv^2\right) \tag{2.21-6}$$

where the energy is entirely kinetic. From Eq. (2.21-6),

$$n = \frac{mV^{1/3} v}{\pi\hbar} \tag{2.21-7}$$

$$dn = \frac{mV^{1/3}}{\pi\hbar} dv \tag{2.21-8}$$

Equation (2.21-3) thus gives

$$\begin{aligned} g_n &= \frac{\pi}{2} n^2\, dn \\ &= \frac{\pi}{2}\left[\frac{2mV^{2/3}}{\pi^2\hbar^2}\left(\frac{1}{2}mv^2\right)\right]\left(\frac{mV^{1/3}}{\pi\hbar^2}\right) dv \\ &= \frac{m^3Vv^2\, dv}{2\pi^2\hbar^2} \end{aligned} \qquad (2.21\text{-}9)$$

so that Eq. (2.21-2) becomes

$$\begin{aligned} P(n) &= \frac{g_n \exp(-\beta\epsilon_n)}{\sum\limits_n g_n \exp(-\beta\epsilon_n)} \\ &= \frac{\dfrac{m^3Vv^2\, dv}{2\pi^2\hbar^2}\exp(-\beta\epsilon_n)}{\left(\dfrac{m}{2\pi\hbar^2\beta}\right)^{3/2} V} \end{aligned}$$

which simplifies to

$$P(n) = \frac{4\pi v}{2^{3/2}\pi^{3/2}}(\beta m)^{3/2} \exp(-\beta\epsilon_n)\, dv$$

or

$$P(n) = (4\pi v^2)\left(\frac{\beta m}{2\pi}\right)^{3/2} \exp\left(-\frac{\beta m v^2}{2}\right) dv \qquad (2.21\text{-}10)$$

Comparing this with the standard result obtained from kinetic theory,

$$P(n) = f(v)\, dv - (4\pi v^2)\left(\frac{m}{2\pi kT}\right)^{3/2} \exp\left(-\frac{mv^2}{2kT}\right) dv \qquad (2.21\text{-}11)$$

shows, once again, that

$$\beta = \frac{1}{kT} \qquad (2.21\text{-}12)$$

A plot of $f(v)$ versus the magnitude of the molecular speed v is shown in Fig. 2.21-1 for oxygen at 100, 300, 500, and 1000 K. The probability that a molecule has its speed between any two values is given by the area under the curve between these two values. The probability of a zero speed is zero. Also, $f(v)$ approaches zero as the speed v approaches infinity. At higher speeds the probability decreases toward zero because the exponential term decreases much more rapidly than v^2 increases. Thus, very few molecules have very high or very low speeds. The fraction of the molecules having speeds greater than 10 times the most probable speed is 9×10^{-42} at any temperature. (The Avogadro constant times this fraction is a very small number.) As the temperature is increased, the most probable speed (given by the maximum in the plot)

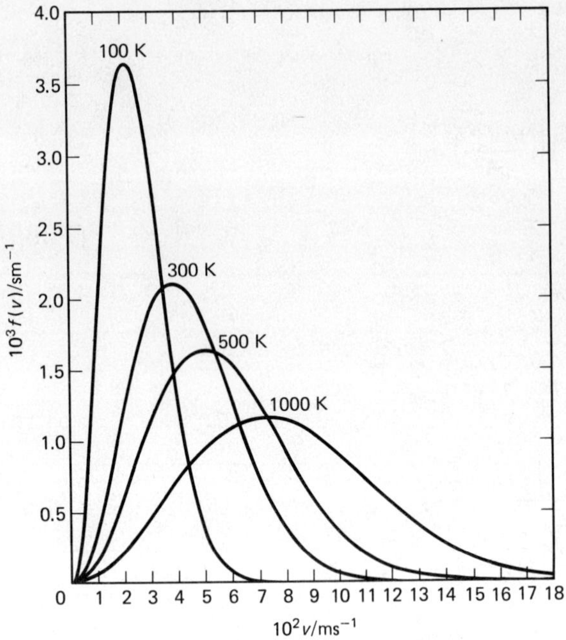

Figure 2.21-1 Probability of speeds v for oxygen at 100, 300, 500, and 1000 K.

moves to higher speeds and the distribution of speeds also becomes broader, but the relative shape is independent of temperature.

Since there is a distribution of speeds, various types of averages may be considered. The most probable speed v_p is obtained by setting $df(v)/dv$ equal to zero.

$$\frac{df(v)}{dv} = \left(\frac{m}{2\pi kT}\right)^{3/2} e^{-mv^2/2kT}\left[8\pi v + 4\pi v^2\left(\frac{-mv}{kT}\right)\right] \tag{2.21-13}$$

This derivative vanishes at $v = v_p$ (often called v_{mode}), where

$$v_p = \left(\frac{2kT}{m}\right)^{1/2} = \left(\frac{2RT}{M}\right)^{1/2} \tag{2.21-14}$$

The mean speed $\langle v\rangle$ is the summation of each possible speed times the probability $P(v)$ of that speed, that is, $\langle v\rangle = \Sigma v\, P(v)$. For the case of a continuous speed distribution function, the probability that a molecule has a speed between v and $v + dv$ is represented by $f(v)\, dv$, so that

$$\langle v\rangle = \int_0^\infty v f(v)\, dv \tag{2.21-15}$$

Integration is used instead of summation because the number of molecules in a macroscopic sample is so large that the speed distribution may be taken to be continuous. Substituting the Maxwell distribution,

$$\langle v \rangle = 4\pi\left(\frac{m}{2\pi kT}\right)^{3/2} \int_0^\infty e^{-mv^2/2kT}\, v^3\, dv \tag{2.21-16}$$

This gives, with the aid of

$$\int_0^\infty x^3\, e^{-ax^2}\, dx = \frac{1}{2a^2}$$

the result

$$\langle v \rangle = \left(\frac{8kT}{\pi m}\right)^{1/2} = \left(\frac{8RT}{\pi M}\right)^{1/2} \tag{2.21-17}$$

The root-mean-square (rms) speed is

$$\langle v^2 \rangle = \left(\frac{1}{N} \sum_{i=1}^{N} v_i^2\right)^{1/2} \tag{2.21-18}$$

Since the speed distribution is continuous, the rms speed is obtained by multiplying each speed squared by the probability of that speed, integrating over all speeds, and taking the square root:

$$\langle v^2 \rangle^{1/2} = \left[\int_0^\infty v^2 f(v)\, dv\right]^{1/2} \tag{2.21-19}$$

Substituting the Maxwell distribution and using the appropriate definite integral

$$\int_0^\infty x^4\, e^{-ax^2}\, dx = \frac{3}{8}\pi^{1/2}$$

gives

$$\langle v^2 \rangle^{1/2} = \left(\frac{3kT}{m}\right)^{1/2} = \left(\frac{3RT}{M}\right)^{1/2} \tag{2.21-20}$$

It can be seen that at a given temperature each of these average speeds is inversely proportional to the square root of the molar mass. Lighter molecules move more rapidly, so that their average kinetic energies are exactly equal to those for the heavier molecules.

■ EXAMPLE 2.21-1

Calculate the most probable speed v_{mode}, the average speed $\langle v \rangle$, and the rms speed $\langle v^2 \rangle^{1/2}$ for hydrogen molecules at 0°C.

Solution

$$v_p = \left(\frac{2RT}{M}\right)^{1/2} = \left[\frac{(2)(8.314\ \text{J K}^{-1}\ \text{mol}^{-1})(273\ \text{K})}{2.016 \times 10^{-3}\ \text{kg mol}^{-1}}\right]^{1/2}$$

$$= 1.50 \times 10^3\ \text{m s}^{-1}$$

$$\langle v \rangle = \left(\frac{8RT}{\pi M}\right)^{1/2} = \left[\frac{(8)(8.314\ \text{J K}^{-1}\ \text{mol}^{-1})(273\ \text{K})}{(3.146)(2.016) \times 10^{-3}\ \text{kg mol}^{-1})}\right]^{1/2}$$

$$= 1.69 \times 10^3\ \text{m s}^{-1}$$

$$\langle v^2 \rangle^{1/2} = \left(\frac{3RT}{M}\right)^{1/2} = \left[\frac{(3)(8.314 \text{ J K}^{-1} \text{ mol}^{-1})(273.15 \text{ K})}{2.016 \times 10^{-3} \text{ kg mol}^{-1}}\right]^{1/2}$$

$$= 1.84 \times 10^3 \text{ m s}^{-1}$$

The rms speed of a hydrogen molecule at 0°C is 6620 km hr^{-1}, but at ordinary pressures it travels only an exceedingly short distance before colliding with another molecule and changing direction. ■

2.22 THERMODYNAMIC PROPERTIES AND THE PARTITION FUNCTION

The value of a mechanical property for a system can be obtained by summing the probability P_i times the mechanical property of state i over all of the states of the ensemble. For example, in the case of the internal energy,

$$u = \sum_i P_i \epsilon_i = \frac{1}{z} \sum_i \epsilon_i \, e^{-\beta \epsilon_i} \tag{2.22-1}$$

Since the numerator is the negative derivative of the denominator with respect to β, this becomes

$$u = -\frac{1}{z}\left(\frac{\partial z}{\partial \beta}\right)_{v,n} = -\left(\frac{\partial \ln z}{\partial \beta}\right)_{v,n} \tag{2.22-2}$$

the differentiation being carried out at constant volume because the energy levels ϵ_i depend on the volume. With the identification of $\beta = 1/kT$, Eq. (2.22-2) may be written as

$$u = kT^2 \left(\frac{\partial \ln z}{\partial T}\right)_{v.n} \tag{2.22-3}$$

The expression for the entropy in terms of the partition function may be derived from the expression for the entropy in terms of the probabilities of the various states,

$$s = -k \sum_i P_i \ln P_i \tag{2.22-4}$$

from

$$P_i = \frac{e^{-\epsilon_i /kT}}{z} \tag{2.22-5}$$

taking the logarithm

$$\ln P_i = -\frac{\epsilon_i}{kT} - \ln z \tag{2.22-6}$$

and substituting in (2.22-4),

$$s = -k \sum_i P_i \epsilon_i + k \ln z \sum_i P_i \tag{2.22-7}$$

Since $\Sigma_i P_i \epsilon_i = u$ and $\Sigma_i P_i = 1$, this gives

$$s = \frac{u}{T} + k \ln z \tag{2.22-8}$$

From $a = u - Ts$, the Helmholtz function becomes

$$a = -kT \ln z \tag{2.22-9}$$

Since $p = -(\partial a/\partial v)_{T,n}$, the pressure is

$$p = kT\left(\frac{\partial \ln z}{\partial v}\right)_{T,n} \tag{2.22-10}$$

Since $h = u + pv$, the enthalpy is

$$h = kT^2\left(\frac{\partial \ln z}{\partial v}\right)_{v,n} + pv \tag{2.22-11}$$

Since $g = h - Ts = a + pv$, the Gibbs function is

$$g = kT \ln z + pv \tag{2.22-12}$$

These relations are perfectly general and apply to any collection of molecules. They show that any thermodynamic property can be calculated once the partition function is known. Before applying them, however, the evaluation of z will be discussed. The partition function is the bridge between thermodynamics, spectroscopy, and statistical and quantum mechanics. For a perfect gas, the energy of the system is given by the sum of the energies of the individual molecules. For an imperfect gas, intermolecular forces will have to be taken into account. First, the partition function for a localized subsystem will be considered, that is, one in which the independent particles can be distinguished from each other. An example of a localized subsystem would be a perfect crystal in which the individual atoms may be identified by their location in the crystal lattice. Then, nonlocalized subsystems of identical particles, for example, a perfect gas, will be discussed.

If a system is made up of independent and distinguishable particles a, b, $c, \ldots$ each in its own energy level j, k, $l, \ldots$, the energy of the system is represented by

$$E_i = \epsilon_{a_j} + \epsilon_{b_k} + \epsilon_{c_l} + \cdots \tag{2.22-13}$$

Substituting this expression into the canonical ensemble partition function gives

$$\begin{aligned} Z &= \sum_i e^{-E_i/kT} = \sum_{j,k,\ldots} e^{-(\epsilon_{a_j} + \epsilon_{b_k} + \epsilon_{c_l})/kT} \\ &= \sum_j e^{-\epsilon_{a_j}/kT} \sum_k e^{-\epsilon_{b_k}/kT} \sum_l e^{-\epsilon_{c_l}/kT} \ldots \\ &= z_a z_b z_c \end{aligned} \tag{2.22-14}$$

where the molecular partition function for a molecule such as a is given by

$$z_a = \sum_j e^{-\epsilon_{a_j}/kT} \tag{2.22-15}$$

Equation (2.22-15) is an important result because it shows that if the particles are independent but distinguishable, the canonical ensemble partition function Z is equal to the product of the partition functions of the independent particles.

If molecules $a, b, \ldots$ are identical their molecular partition functions are identical so that for N independent molecules the canonical ensemble partition function is given by

$$Z = z^N \qquad \text{(localized subsystems)} \tag{2.22-16}$$

This relation is limited to localized subsystems (molecules) because it has been assumed that molecules can be identified as $a, b, \ldots$. (If molecules are in a crystal they are localized and may be identified by position.)

Now, the molecules of a gas are not localized and, according to quantum mechanics, they are *indistinguishable.* In order to see how Eq. (2.22-16) has to be modified for gases, consider a gas in which one molecule has quantum numbers n_1', n_2', n_3' and another has quantum number n_1'', n_2'', n_3''. In deriving Eq. (2.22-16), it was assumed that the state of the system is different if the two molecules are interchanged so that the first has quantum numbers $n_1'', n_2'\ n_3'$, and the second has quantum numbers n_1', n_2', n_3'. However, this is not so, because the molecules are identical and the state of the system is the same in the two cases. Fortunately it is easy to make a correction because, in a gas, the number of microstates is far greater than the number of molecules so that it is very unlikely that there will be more than one molecule in a quantum state. Therefore, the right side of Eq. (2.22-16) is divided by $N!$, the number of ways of permuting N molecules.

$$Z = \frac{z^N}{N!} \qquad \text{(nonlocalized subsystems)} \tag{2.22-17}$$

When the number of available molecular states is much larger than the number of particles, Eq. (2.22-17) prevails. For example, the internal energy of the system is

$$\begin{aligned} U &= kT^2 \frac{\partial}{\partial T}\left[\ln \frac{\left(\sum_i e^{-\epsilon_i/kT}\right)^N}{N!}\right] \\ &= kT^2 \frac{\partial}{\partial T}\left[N \ln\left(\sum_i e^{-\epsilon_i/kT}\right) - \ln N!\right] \\ &= N \frac{\sum_i \epsilon_i e^{-\epsilon_i/kT}}{\sum_i e^{-\epsilon_i/kT}} \end{aligned} \tag{2.22-18}$$

The average energy $\langle u \rangle$ of a single molecule is

$$\frac{U}{N} = \frac{\sum_i \epsilon_i e^{-\epsilon_i/kT}}{\sum_i e^{-\epsilon_i/kT}}$$

and the probability P_i that a single molecule is in the ith energy state is

$$P_i = \frac{e^{-\beta\epsilon_i}}{z} \tag{2.22-19}$$

This can be compared to the probability $e^{-\beta E_i}/Z$ for the N body system.

Factorization of the Partition Function

The energy of a molecule is the sum of contributions from its different modes of motion, translational and internal, which includes rotation, vibration, electronic, nuclear, thus giving

$$\epsilon_i = \epsilon_{\text{tran}} + \epsilon_{\text{rot}} + \epsilon_{\text{vib}} + \epsilon_{\text{nuc}} \tag{2.22-20}$$

This separation is only approximate (except in the case of translation) because the modes are not perfectly independent, but in most cases it is satisfactory (the separation of the electronic and nuclear motions, for example, is justified by the Born–Oppenheimer approximation). The advantage of the separation is that the partition function factorizes into contributions from each mode, as will be seen shortly.

With the energy of a molecule as a sum of terms as in Eq. (2.22-20), it is relatively easy to calculate the molecular partition function z from data obtained from spectroscopy. Substituting (2.22-20) into (2.22-14) and using $e^{a+b+c+\cdots} = e^a e^b e^c \cdots$, gives the factored partition function

$$z = \sum_j e^{-\epsilon_{\text{tran}_j}/kT} \sum_k e^{-\epsilon_{\text{rot}_k}/kT} \sum_j e^{-\epsilon_{\text{vib}_l}/kT} \sum_m e^{-\epsilon_{\text{elec}_m}/kT} \tag{2.22-21}$$

or

$$z = z_{\text{tran}} z_{\text{rot}} z_{\text{vib}} z_{\text{elec}} \tag{2.22-22}$$

The canonical partition function for an N particle system is then

$$Z = \frac{1}{N!}(z_{\text{tran}} z_{\text{rot}} z_{\text{vib}} z_{\text{elec}})^N \tag{2.22-23}$$

$$= Z_{\text{tran}} Z_{\text{rot}} Z_{\text{vib}} Z_{\text{elec}} \tag{2.22-24}$$

where

$$Z_{\text{tran}} = \frac{(z_{\text{tran}})^N}{N!} \tag{2.22-25}$$

$$Z_{\text{rot}} = (z_{\text{rot}})^N \tag{2.22-26}$$

$$Z_{\text{vib}} = (z_{\text{vib}})^N \tag{2.22-27}$$

$$Z_{\text{elec}} = (Z_{\text{elec}})^N \tag{2.22-28}$$

Note that the $N!$ is part of the translational factor as per Eq. (2.22-17). Each of the factors in (2.22-22) is a sum of the general form

$$1 + e^{-\theta_1/T} + e^{-\theta_2/T} + \cdots \tag{2.22-29}$$

where

$$\theta_1/T = \epsilon_1/kT, \qquad \theta_2/T = \epsilon_2/kT, \quad \text{etc.} \tag{2.22-30}$$

and the θ's are *characteristic temperatures* for the particular energies involved, such as rotational, vibrational, electronic, and nuclear. In the case of nuclear levels, the excited states of the nucleus have such high energies that θ_{nuc} is of the order of millions of degrees and is of significance only in thermonuclear systems. In the case of electronic levels, θ_{elec} is much lower, but still so large that z_{elec} usually reduces to the degeneracy value g of the lowest or ground electronic state.

Translational Partition Function

Lowest of all the characteristic temperatures is θ_{tran}. The quantum evaluation of the translational partition function parallels the classical treatment in that summation is replaced by integration. This is permissible if the energy difference between adjacent quantum states $\Delta\epsilon$ is smaller than kT (usually the case above 50 K).

Consider an ideal gas of N monatomic molecules all in the same electronic state. The quantum mechanical energy levels for a molecule in a container with dimensions $a \times b \times c$ are those given in Section 2.6. Substituting the equation for the energies of the allowed quantum states of a molecule of mass m into (2.22-15) yields

$$z = \sum_{n_1=1}^{\infty} \sum_{n_2=1}^{\infty} \sum_{n_3=1}^{\infty} \exp\left[-\frac{h^2}{8mkT}\left(\frac{n_1^2}{a^2} + \frac{n_2^2}{b^2} + \frac{n_3^2}{b^2}\right)\right] \tag{2.22-31}$$

$$= \sum_{n_1=1}^{\infty} \exp\left(-\frac{h^2 n_1^2}{8ma^2kT}\right) \sum_{n_2=1}^{\infty} \exp\left(\frac{h^2 n_2^2}{8mb^2kT}\right) \sum_{n_3=1}^{\infty} \exp\left(-\frac{h^2 n_3^2}{8mc^2kT}\right) \tag{2.22-32}$$

To evaluate these sums, it is noted that for macroscopic containers the exponents are very small, unless T is very small. Since the successive terms in the summations differ by only small amounts, the summations may be replaced by integrations, considering the quantum number n as a continuous variable that is not restricted to integer values,

$$\int_0^{\infty} e^{-h^2n^2/8ma^2kT}\, dn = \frac{(2\pi mkT)^{1/2} a}{h} \tag{2.22-33}$$

since

$$\int_0^{\infty} e^{-\alpha x^2}\, dx = \frac{1}{2}\left(\frac{\pi}{\alpha}\right)^{1/2} \tag{2.22-34}$$

Substituting integrals obtained in this way into (2.22-32) and noting that $abc = V$, the following expression for the translational partition function is obtained:

$$z = \left(\frac{2\pi mkT}{h^2}\right)^{3/2} V \qquad (2.22\text{-}35)$$

Thus, the translational partition function for a molecule is proportional to the volume V in which it moves.

Vibrational Partition Function

The vibrational levels of diatomic molecules have been discussed in Section 2.9. For many purposes, it is sufficient to use the harmonic oscillator approximation because only lower levels are occupied. For the purpose of calculating the partition function, vibrational energy is usually measured from the ground state ($n = 0$) rather than from the bottom of the potential energy curve. This simplifies the equations, because the zero-point energy, $h\omega_0/2$, where ω_0 is the fundamental vibration frequency, does not appear. In the harmonic oscillator approximation,

$$\epsilon_v = nh\omega \qquad n = 0, 1, 2, \ldots \qquad (2.22\text{-}36)$$

so that

$$z = \sum_{n=0}^{\infty} e^{-nh\omega/kT} = 1 + x + x^2 + x^3 + \cdots \qquad (2.22\text{-}37)$$

where $x = e^{-h\omega/kT}$. This is a geometric series. When $x < 1$,

$$\frac{1}{1-x} = 1 + x + x^2 + x^3 + \cdots \qquad (2.22\text{-}38)$$

Since $e^{-h\omega/kT} < 1$,

$$z = \frac{1}{1 - e^{-h\omega/kT}} = \frac{1}{1 - e^{-\theta_v/T}} \qquad (2.22\text{-}39)$$

where θ_v is the characteristic vibrational temperature; $\theta_v = h\omega/k = hc\overline{\omega}/k$. The characteristic temperature of a vibration θ_v is the temperature at which the energy of the vibrational quantum is equal to kT.

Rotational Partition Function

The expression for the rotational energy of a rigid diatomic molecule was found in Section 2.11 to be

$$\epsilon_r = \frac{J(J+1)h^2}{8\pi^2 I} \qquad (2.22\text{-}40)$$

J is the rotational quantum number, and I is the moment of inertia of the diatomic molecule. Although the rotational energy depends only on J, the state of a rigid rotator is specified by the quantum number J and an additional quantum number M, where M can have integral values between $-J$ and $+J$. Thus, there are $2J + 1$ values of M for each value of J; in other words, the rotational levels are $(2J + 1)$-fold degenerate. The rotational partition function is

$$z = \sum_{J=0}^{x} (2J+1) e^{-J(J+1)h^2/8\pi^2 IkT} \qquad (2.22\text{-}41)$$

For molecules with large moments of inertia, the energy levels are ordinarily so close together that they may be considered to be continuous above 10 to 100 K, so that the summation may be replaced by an integration:

$$z = \int_0^x (2J+1)e^{-J(J+1)h^2/8\pi^2 IkT}\, dJ \tag{2.22-42}$$

Since $(2J+1)\, dJ$ is the differential of $J(J+1) = J^2 + J$, this equation may be integrated by multiplying and dividing through by $8\pi^2 IkT/h^2$ to obtain

$$z = \frac{8\pi^2 Ik}{h^2}\int_0^x e^{-u}\, du = \frac{8\pi^2 Ik}{h^2} = \frac{T}{\theta_r} \tag{2.22-43}$$

where θ_r is the characteristic rotational temperature defined by

$$\theta_r = \frac{h^2}{8\pi^2 Ik} \tag{2.22-44}$$

Equation (2.22-44) holds for a heteronuclear molecule. For a homonuclear diatomic molecule (like Cl_2), calculation has to take into account the fact that rotation by 180° interchanges two equivalent nuclei. Since the new orientation is indistinguishable from the first, it is necessary to divide by 2 so that the indistinguishable orientations are counted only once. For a heteronuclear diatomic molecule (like HCl), a 180° rotation produces a distinguishable orientation. The effect of symmetry is taken into account by introducing a symmetry number, σ, that has the value 2 for a *homonuclear molecule* and the value 1 for a *heteronuclear molecule.* In general then

$$z = \frac{T}{\sigma\theta_r} \tag{2.22-45}$$

Also, the behavior of H_2 at low temperatures is complicated by the quantum-mechanical symmetry requirements applicable to homonuclear molecules. H_2 is a mixture of two molecular species. Hydrogen with a symmetric nuclear wave function and J odd is called *orthohydrogen.* Hydrogen with an antisymmetric nuclear wave function and J even is called *parahydrogen.* The rotational partition function for the equilibrium mixture is

$$z = \sum_{J\text{ even}} (2J+1)\exp\left[\frac{J(J+1)\theta_r}{T}\right] + 3\sum_{J\text{ odd}} (2J+1)\exp\left[-\frac{J(J+1)\theta_r}{T}\right] \tag{2.22-46}$$

Electronic Partition Function

As for the electronic contribution to the thermodynamic properties of a perfect gas, the electronic partition function is given by

$$z = \sum_i e^{-\epsilon_i/kT} \tag{2.22-47}$$

where the summation is carried over all electronic states. This partition function is more generally written as a sum over levels

$$z = \sum_j g_j e^{-\epsilon_j/kT} \tag{2.22-48}$$

instead of states. Thus, if the degeneracy of the ground state is g_0 and the degeneracy of the first excited state is g_1.

$$z = g_0 + g_1 e^{-\epsilon_1/kT} + \cdots \tag{2.22-49}$$

where the energies of the excited states are measured with respect to the energy of the ground state. If the energies of the excited states are all large with respect to T, only the degeneracy g_0 of the ground state has to be taken into account in calculating thermodynamic properties. For most cases, the electronic energy separations from the ground state are very large, and so $z = 1$. An important exception is the alkali metal atoms, which have doubly degenerate ground states (corresponding to the two orientations of their electron spin), and so $z = 2$. Since $Z_{\text{elec}} = z_{\text{elec}}^N$, the degeneracy g_0 of the ground state of an atom adds $R \ln g_0$ to the entropy, $-RT \ln g_0$ to the Helmholtz and Gibbs energies, but nothing to C_p° and H°.

The various equations relating the thermodynamic properties to the canonical partition function are listed in Table 2.22-1. They will be applied extensively in Chapters 3 and 5. For the present, it is sufficient to state that the theory is based on the assumption that the Hamiltonian operator is separable into independent parts, one involving only center-of-mass motion (translation), the other involving only internal degrees of freedom (vibrational, rotational, etc.) This procedure is usually valid for monatomic fluids in both the gaseous and liquid states, but may lead to errors for polyatomic fluids, especially at higher densities. It is not valid for very low temperature solids or liquid hydrogen and helium. The overall partition function is the product of each contribution accordng to the rule set by (2.22-24).

TABLE 2.22-1 THERMODYNAMIC PROPERTIES AND THE PARTITION FUNCTION Z

$Z = \sum_i \exp\left(\frac{-\epsilon_i/kT}{kT}\right)$	$C_p = \left(\frac{\partial H}{\partial T}\right)_{p,N}$
$U = kT^2\left(\frac{\partial \ln Z}{\partial T}\right)_{V,N}$	$S = k \ln Z + \frac{U}{T}$
$p = kT\left(\frac{\partial \ln Z}{\partial V}\right)_{T,N}$	$-A = kT \ln Z$
$H = kT^2\left(\frac{\partial \ln Z}{\partial T}\right)_{V,N} + kTV\left(\frac{\partial \ln Z}{\partial V}\right)_{T,N}$	$G = -kT \ln Z + kTV\left(\frac{\partial \ln Z}{\partial V}\right)_{T,N}$
$C_v = \left(\frac{\partial U}{\partial T}\right)_{V,N}$	$\mu = -kT\left(\frac{\partial \ln Z}{\partial N}\right)_{T,V}$

■ EXAMPLE 2.22-1

Show that the partition function gives an indication of the number of states that are accessible to a system at the temperature of interest. In particular, consider $T \to 0$ and $T \to \infty$.

Solution. Some insight into the significance of the partition function can be obtained by considering the molecular partition function

$$z = \sum_j e^{-\epsilon_j/kT}$$

where the summation is over states j. When the temperature of the system is close to absolute zero, every term except one in the summation is of the form e^{-x}, with $x \to \infty$, and so they all diappear. The exception is the term with $\epsilon_j = 0$, because then $(\epsilon_j/kT) = 0$ for all values of T, including $T = 0$. As there is only one term that survives at $T = 0$, and its value is 1, it follows that at $T = 0$, $z = 1$.

At the other extreme, consider the case when T is so large that for every term in the sum $(\epsilon_j/kT) \to 0$. Since $e^{-x} = 1$ when $x = 0$, every term in the sum contributes the value 1. Then the sum is equal to the total number of molecules, which, in general, is so large that it is infinite. Thus, as $T \to \infty$, $z \to \infty$.

The partition function is seen to give an indication of the number of states that are accessible to the system at the temperature of interest. At absolute zero only the ground state is accessible, and $z = 1$; at the highest temperatures, virtually all the states are accessible, and so z approaches infinity. The proportion of molecules in the energy state ϵ_j is $(1/z)\exp(-\epsilon_j/kT)$. ■

■ EXAMPLE 2.22-2

In the absence of electronic excitation, the molecular partition function for a monatomic gas was shown to be

$$z = \frac{(2\pi mkT)^{2/3}\, V}{h^3}$$

From this, obtain the canonical ensemble partition function and derive expressions for the translational contributions to the thermodynamic properties of a perfect gas.

Solution. The canonical ensemble partition function for an assembly of indistinguishable particles is

$$z = \frac{z^N}{N!} = \left(\frac{2\pi mkT}{h^2}\right)^{3N/2} \frac{V^N}{N!} = \left(\frac{2\pi mkT}{h^2}\right)^{3N/2} \frac{V^N e^N}{N^N} \tag{2.22-50}$$

where the last form has been obtained using Stirling's approximation $N! = N^N/e^N$. This relation may be used to calculate the translational contributions to the thermodynamic properties of a perfect gas. Before applying Table 2.22-1, it is convenient to have the logarithm of Eq. (2.22-50) as

$$\ln Z = N \ln\left(\frac{2\pi mkT}{h^2}\right)^{3N/2} \frac{Ve}{N} \tag{2.22-51}$$

or

$$\ln Z = \frac{3N}{2}\left[\ln\left(\frac{2\pi mk}{h^2}\right) + \ln T\right] + N\ln\left(\frac{Ve}{N}\right) \tag{2.22-52}$$

With these expressions for ln Z, Table 2.22-1 yields

$$p = kT\left(\frac{\partial \ln Z}{\partial V}\right)_{T,N} = \frac{NkT}{V} \tag{2.22-53}$$

which reproduces the ideal gas law. Note that this particular result applies as well to a perfect polyatomic gas since the rotational, vibrational, and electronic factors in the molecular partition function are independent of V. Other expressions for the translational contributions are

$$U = kT^2\left(\frac{\partial \ln Z}{\partial T}\right)_{V,N} = \frac{3}{2}NkT \tag{2.22-54}$$

$$H = U + pV = \frac{3}{2}NkT + NkT = \frac{5}{2}NkT \tag{2.22-55}$$

$$C_V = \left(\frac{\partial U}{\partial T}\right)_V = \frac{3}{2}Nk; \qquad C_p = \left(\frac{\partial H}{\partial T}\right)_p = \frac{5}{2}Nk \tag{2.22-56}$$

$$S = k\ln Z + \frac{U}{T} = Nk\ln\left[\left(\frac{2\pi mkT}{h^2}\right)^{3/2}\frac{V}{N}\right] + \frac{5}{2}Nk \tag{2.22-57}$$

$$A = kT\ln Z = -NkT\ln\left[\left(\frac{2\pi mkT}{h^2}\right)^{3/2}\frac{V}{N}\right] - NkT \tag{2.22-58}$$

$$G = A + pV = -NkT\ln\left[\left(\frac{2\pi mkT}{h^2}\right)^{3/2}\frac{V}{N}\right] \tag{2.22-59}$$

These equations may be used to calculate molar thermodynamic properties by letting N equal Avogadro's constant N_A. To calculate thermodynamic quantities for the standard state, the value of V corresponding to the standard state is used. It is therefore convenient to rewrite theses equations for calculating molar thermodynamic properties at the standard state pressure of $p° = 10^5$ Pa or 1 bar. Substituting N_A for N in Nk and replacing V/N_A by $kT/p°$, the above equations may be rewritten for 1 mol:

$$p = \frac{RT}{V} \tag{2.22-60}$$

$$U° = \frac{3}{2}RT; \qquad H° = \frac{5}{2}RT \tag{2.22-61}$$

$$C_V^\circ = \frac{3}{2}R; \qquad C_p^\circ = \frac{5}{R} \tag{2.22-62}$$

$$S° = R\ln\left[\left(\frac{2\pi mkT}{h^2}\right)^{3/2}\frac{kT}{p°}\right] + \frac{5}{2}R \tag{2.22-63}$$

$$A° = -RT\left\{1 + \ln\left[\left(\frac{2\pi mkT}{h^2}\right)^{3/2} \frac{kT}{p°}\right]\right\} \tag{2.22-64}$$

$$G° = -RT \ln\left[\left(\frac{2\pi mkT}{h^2}\right)^{3/2} \frac{kT}{p°}\right] \tag{2.22-65}$$

The translational $U°$, $H°$, $A°$, and $G°$ calculated with these equations approach zero as the thermodynamic temperature approaches zero. Therefore, these properties are relative to a standard state of a hypothetical perfect gas at absolute zero. The translational value of $S°$ approaches $-\infty$ as T approaches zero, but, of course, the gas condenses before this happens. The entropy of a perfect crystal approaches zero as the thermodynamic temperature approaches zero. ■

■ EXAMPLE 2.22-3

(a) What is the molecular translational partition function and the de Broglie wavelength of a hydrogen atom in a mole volume at 300 K and 1 bar?

(b) What are the translational contributions to $C_p°$, $H°$, $S°$, and $G°$ for $O_2(g)$ at 298.15 K?

Solution

(a) The volume that would contain 1 mol of perfect gas at 3000 K and 1 bar pressure is 0.249 m^3. The molecular partition for translational motion of the hydrogen atom is

$$z = \left(\frac{2\pi mkT}{h^2}\right)^{3/2} V \tag{2.22-66}$$

$$= \left[\frac{2\pi(1.0080 \times 10^{-3}\ \text{kg mol}^{-1})(1.3806 \times 10^{-23}\ \text{J K}^{-1})(3000\ \text{K})}{(6.022 \times 10^{23}\ \text{mol}^{-1})(6.626 \times 10^{-34}\ \text{J s})^2}\right]^{3/2}$$

$$\times\, 0.2494\ \text{m}^3$$

$$= 7.791 \times 10^{30}$$

Note that the molecular partition function is a dimensionless quantity and may be interpreted approximately as the number of energy levels accessible to the single hydrogen atom in this volume. The criterion for perfect gas conditions is that the Boltzmann limit $z/N \gg 1$. This requirement is met by most gases at STP, for

$$\frac{z}{N} = \left(\frac{2\pi mkT}{h^2}\right)^{3/2} \frac{V}{N} \approx 10^6 \gg 1 \tag{2.22-67}$$

In fact, only for liquid helium does this fail.

The quantity $(h^2/2\pi mkT)^{1/2}$ is called the thermal de Broglie wavelength and is represented by Λ. Thus,

$$z = \frac{V}{\Lambda^3} \tag{2.22-68}$$

or

$$\Lambda = \left(\frac{V}{z}\right)^{1/2} = \left(\frac{0.2492\ \text{m}^3}{7.791 \times 10^{30}}\right)^{1/3} = 3.175 \times 10^{-11}\ \text{m} \tag{2.22-69}$$

The condition for the applicability of Boltzmann statistics is that the thermal de Broglie wavelength must be small compared with the dimensions of the container.

(b) For O_2

$$m = \frac{M}{N_A} = \frac{31.9988 \times 10^{-3}\ \text{kg mol}^{-1}}{6.022\ 045 \times 10^{23}\ \text{mol}^{-1}} \tag{2.22-70}$$

$$= 5.313\ 61 \times 10^{-26}\ \text{kg} \quad \left(\frac{2\pi mkT}{h^2}\right)^{3/2} \frac{kT}{p^\circ}$$

$$= \left[\frac{2\pi(5.313\ 61 \times 10^{-26}\ \text{kg})(1.380\ 662 \times 10^{-23}\ \text{J K}^{-1})(298.15\ \text{K})}{(6.626\ 176 \times 10^{-34}\ \text{J s})^2}\right]^{3/2}$$

$$\times \frac{(1.380\ 662 \times 10^{-23}\ \text{J K}^{-1})(298.15\ \text{K})}{10^5\ \text{N m}^{-2}} = 7.208\ 93 \times 10^6$$

The translational contributions are

$$C_p^\circ = \tfrac{5}{2}R = \tfrac{5}{2}(8.314\ 41\ \text{J K}^{-1}\ \text{mol}^{-1}) = 20.786\ \text{J K}^{-1}\ \text{mol}^{-1}$$

$$H^\circ = \tfrac{5}{2}RT = \tfrac{5}{2}(8.314\ 41\ \text{J K}^{-1}\ \text{mol}^{-1})(298.15\ \text{K}) = 6.197\ \text{kJ mol}^{-1}$$

$$S^\circ = (8.314\ 41\ \text{J K}^{-1}\ \text{mol}^{-1})(2.5 + \ln 7.208\ 93 \times 10^6)$$

$$= 152.077\ \text{J K}^{-1}\ \text{mol}^{-1}$$

$$G^\circ = -(8.314\ 41\ \text{J K}^{-1}\ \text{mol}^{-1})(298.15\ \text{K}) \ln 7.208\ 93 \times 10^6$$

$$= -39.145\ \text{kJ mol}^{-1}$$

■

■ EXAMPLE 2.22-4

(a) Develop expressions for the rotational contributions to the thermodynamic properties of perfect gases.

(b) What are the rotational contributions to C_p°, S°, H°, and G° for $O_2(g)$ at 298.15 K? The moment of inertia for $O_2(g)$ is $I = 1.9373 \times 10^{-46}$ kg m^2.

Solution

(a) To determine the rotational contributions, the canonical ensemble partition function for rotation, Z, must first be calculated. Since $Z_{\text{rot}} = z_{\text{rot}}^N$, the equations $U = kT^2(\partial \ln Z/\partial T)_{V,N}$ and $S = k \ln Z + (U/T)$ lead to the following expressions for the rotational contributions to the molar thermodynamic quantities for diatomic molecules:

$$U_r^\circ = kT^2 \left(\frac{\partial \ln z_r^N}{\partial T}\right)_{V,N} = RT \tag{2.22-71}$$

$$(C_V^\circ)_r = R \tag{2.22-72}$$

$$H_r^\circ = RT \tag{2.22-73}$$

$$(C_p^\circ)_r = R \tag{2.22-74}$$

$$S_r^\circ = k \ln z \frac{N}{r} + \frac{U_r}{T} = R \ln\left(\frac{eT}{\sigma\theta_r}\right) \tag{2.22-75}$$

$$G_r^\circ = H_r - TS_r = -RT \ln\left(\frac{T}{\sigma\theta_r}\right) \tag{2.22-76}$$

These results confirm the classical principle of equipartition of energy, namely, the expectation that the two components of rotational motion of a rigid linear molecule each contribute $R/2$ to C_V° or C_p°.

(b) The characteristic rotation temperature and rotational contributions to the thermodynamic properties of O_2(g) at 298.15 K are as follows:

$$\theta_r = \frac{h^2}{8\pi^2 Ik} = \frac{(6.626\ 18 \times 10^{-34}\ \text{J s})^2}{8\pi^2(1.9373 \times 10^{-46}\ \text{kg m}^2)(1.380\ 66 \times 10^{-23}\ \text{J K}^{-1})}$$

$$= 2.079\ \text{K}$$

$$(C_p^\circ)_r = R = 8.314\ \text{J K}^{-1}\ \text{mol}^{-1}$$

$$S_r^\circ = R \ln\left(\frac{eT}{\sigma\theta_r}\right) = (8.314\ 41\ \text{J K}^{-1}\ \text{mol}^{-1}) \ln\left[\frac{(2.7183)(298.15\ \text{K})}{(2)(2.079\ \text{K})}\right]$$

$$= 43.838\ \text{J K}^{-1}\ \text{mol}^{-1}$$

$$H_r^\circ = RT = (8.314\ 41\ \text{J K}^{-1}\ \text{mol}^{-1})(298.15\ \text{K}) = 2.479\ \text{kJ mol}^{-1}$$

$$G_r^\circ = RT \ln\left(\frac{T}{\sigma\theta_r}\right)$$

$$= -(8.314\ 41\ \text{J K}^{-1}\ \text{mol}^{-1})(298.15\ \text{K}) \ln\left[\frac{298.15\ \text{K}}{2 \times 2.079\ \text{K}}\right]$$

$$= -10.591\ \text{kJ mol}^{-1}$$ ■

■ EXAMPLE 2.22-5

(a) Develop expressions for the vibrational contributions to the thermodynamic properties of perfect gases.

(b) What are the vibrational contributions to C_p°, S°, H°, and G° for O_2(g) at 298.15 K? The vibrational frequency is 1580.246 cm^{-1}.

Solution (a) It has been shown earlier that the vibrational partition function (molecular z_{vib}) is given by

$$z_{\text{vib}} = \frac{1}{1 - e^{-\theta_{\text{vib}}/T}} \tag{2.22-77}$$

where θ_{vib} is the characteristic vibrational temperature. Since $Z_{\text{vib}} = z_{\text{vib}}^N$, the vibrational contributions to the various thermodynamic quantities for diatomic molecules are as follows:

$$U^\circ_{vib} = kT^2\left(\frac{\partial \ln z^N_{vib}}{\partial T}\right)_{N,V} = RT^2\left(\frac{\partial \ln z_{vib}}{\partial T}\right)_{N,V} = RT\frac{\theta_{vib}/T}{e^{\theta_{vib}/T}-1} \tag{2.22-78}$$

$$(C^\circ_V)_{vib} = \frac{\partial U^\circ_{vib}}{\partial T} = R\left(\frac{\theta_{vib}}{T}\right)^2 \frac{e^{\theta_{vib}/T}}{(e^{\theta_{vib}/T}-1)^2} \tag{2.22-79}$$

$$H^\circ_{vib} = RT\frac{\theta_{vib}/T}{e^{\theta_{vib}/T}-1} \tag{2.22-80}$$

$$(C^\circ_p)_{vib} = \frac{\partial U^\circ_{vib}}{\partial T} = R\left(\frac{\theta_{vib}}{T}\right)^2 \frac{e^{\theta_{vib}/T}}{(e^{\theta_{vib}/T}-1)^2} \tag{2.22-81}$$

$$S^\circ_{vib} = k\ln z^N_{vib} + \frac{U^\circ_{vib}}{T} = R\left[\frac{\theta_{vib}/T}{e^{\theta_{vib}/T}-1} - \ln(1-e^{-\theta_{vib}/T})\right] \tag{2.22-82}$$

$$G^\circ_{vib} = H^\circ_{vib} - TS^\circ_{vib} = RT\ln(1-e^{-\theta_{vib}/T}) \tag{2.22-83}$$

(b) The characteristic vibrational temperature and the vibrational contributions to the thermodynamic properties of $O_2(g)$ at 298.15 K are

$$\theta_{vib} = \frac{hc}{k} \tag{2.22-84}$$

$$= \frac{(6.26\ 176\times10^{-34}\ \text{J s})(2.997\ 925\times10^{8}\ \text{m s}^{-1})(1.580\ 246\times10^{5}\ \text{m}^{-1})}{1.380\ 662\times10^{-23}\ \text{J K}^{-1}}$$

$$= 2273.64\ \text{K}$$

$$(C^\circ_p)_{vib} = (8.314\ 41\ \text{J K}^{-1}\ \text{mol}^{-1})\left(\frac{2273.64}{298.15}\right)^2 \frac{e^{2273.64/298.15}}{(e^{2273.64/298.15}-1)^2}$$

$$= 0.236\ \text{J K}^{-1}\ \text{mol}^{-1}$$

$$S^\circ_{vib} = (8.314\ 41\ \text{J K}^{-1}\ \text{mol}^{-1})\left[\frac{2273.64/298.15}{e^{2273.64/298.15}-1} - \ln(1-e^{-2273.64/298.15})\right]$$

$$= 0.035\ \text{J K}^{-1}\ \text{mol}^{-1}$$

$$H^\circ_{vib} = (8.314\ 41\ \text{J K}^{-1}\ \text{mol}^{-1})(298.15\ \text{K})\frac{2273.64\ \text{K}/298.15\ \text{K}}{e^{2273.64/298.15}-1}$$

$$= 0.009\ \text{kJ mol}^{-1}$$

$$G^\circ_{vib} = (8.314\ 41\ \text{J K}^{-1}\ \text{mol}^{-1})(298.15\ \text{K})\ln(1-e^{-2273.64/298.15})$$

$$= -0.001\ \text{kJ mol}^{-1}$$ ■

■ EXAMPLE 2.22-6

What is the electronic partition function and the electronic contributions C°_p, H°, S°, and G° for $O_2(g)$ at 298.15 K? The electronic ground state is a triplet, and the spectroscopic dissociation energy D_0 is 491.888 kJ mol^{-1}. (The energy of the first excited electronic state is very high and does not have to be considered.)

Solution. The electronic partition function is more generally written as a sum over levels instead of over states. Thus, if the degeneracy of the ground state is g_0 and the degeneracy of the first excited state is g_1,

$$z_e = g_0 + g_1 e^{-\epsilon_1/kT} + \cdots \tag{2.22-85}$$

where the energies of the excited states are measured with respect to the energy of the ground state. If the energies of the excited states are all large with respect to kT, only the degeneracy g_0 of the ground state has to be taken into account in calculating thermodynamic properties. The electronic contribution to the canonical ensemble partition function Z_{elec} is z_e^N.

In calculating the electronic partition function for a diatomic molecule it is convenient to measure energy with respect to the atoms in a perfect gas state at absolute zero. When this is done the calculated thermodynamic properties of the molecule and its constituent atoms are relative to the same reference state. On this basis the energy of the electronic ground state of the molecule is $-D_0$, where D_0 is the spectroscopic dissociation energy. The energy of the molecule in its first excited state is then $-D_0 + \epsilon_1$, where ϵ_1 is the energy with respect to the ground state. The electronic partition function is therefore written

$$\begin{aligned} z_e &= g_0 e^{D_0/kT} + g_1 e^{(D_0 - \epsilon_1)/kT} + \cdots \\ &= e^{D_0/kT}(g_0 + g_1 e^{-\epsilon_1/kT} + \cdots) \end{aligned} \tag{2.22-86}$$

If $\epsilon_1 \gg kT$, the contributions to the thermodynamic functions of a diatomic molecule are

$$(C_p^\circ)_e = 0 \quad \text{and} \quad H_l^\circ = -D_0 \tag{2.22-87}$$

The contribution to the entropy is given by

$$S_e^\circ = R \ln(g_0 + g_1 e^{-\epsilon_1/kT} + \cdots) \tag{2.22-88}$$

The contribution to the Gibbs function is

$$G_e^\circ = -D_0 - RT \ln(g_0 + g_1 e^{-\epsilon_1/kT} + \cdots) \tag{2.22-89}$$

However, these equations for diatomic molecules are not a good approximation if an excited electronic level is appreciably populated. For $O_2(g)$ at 298.15 K, the electronic contributions are

$$(C_p^\circ)_e = 0$$

$$H_e^\circ = -D_0 - 491.888 \text{ kJ mol}^{-1}$$

$$S_e^\circ = (8.314\,41 \text{ J K}^{-1}) \ln 3 = 9.134 \text{ J K}^{-1} \text{ mol}-1$$

$$G_e^\circ = -491.888 \text{ kJ mol}^{-1}$$
$$- (8.314\,41 \times 10^{-3} \text{ kJ K}^{-1} \text{ mol}-^{1})(298.15 \text{ K}) \ln 3$$
$$= -494.611 \text{ kJ mol}^{-1}$$

Having calculated the translational, rotational, vibrational, and electronic contributions in the foregoing illustrations, the results are added up to give

$$C_p^\circ = (C_p^\circ)_{\text{tr}} + (C_p^\circ)_{\text{rot}} + (C_p^\circ)_{\text{vib}} + (C_p^\circ)_e$$
$$= (20.786 + 8.314 + 0.236 + 0) \text{ J K}^{-1} \text{ mol}^{-1} = 29.336 \text{ J K}^{-1} \text{ mol}^{-1}$$
$$S^\circ = S_{\text{tr}}^\circ + S_{\text{rot}}^\circ + S_{\text{vib}}^\circ + S_e^\circ = (152.077 + 43.838 + 0.035 + 9.134) \text{ J K}^{-1} \text{mol}^{-1}$$
$$= 205.085 \text{ J K}^{-1} \text{ mol}^{-1}$$
$$H^\circ = H_{\text{tr}}^\circ + H_{\text{rot}}^\circ + H_{\text{vib}}^\circ + H_e^\circ = (6.197 + 2.479 + 0.009 - 491.888) \text{ kJ mol}^{-1}$$
$$= -483.203 \text{ kJ mol}^{-1}$$
$$G^\circ = G_{\text{tr}}^\circ + G_{\text{rot}}^\circ + G_{\text{vib}}^\circ + G_e^\circ$$
$$= (-39.145 - 10.591 - 0.001 - 494.611) \text{ kJ mol}^{-1}$$
$$= -544.349 \text{ kJ mol}^{-1}$$

■

2.23 STATISTICAL INTERPRETATION OF HEAT AND WORK

Classical thermodynamics treats the concepts of heat and work as primitive (phenomenological). Statistical thermodynamic affords a different visualization of these quantities.

Consider the change in energy of a system resulting from the following processes: First, when a system is heated, the energy levels are unchanged, but the populations of the levels are changed. This is shown in Fig. 2.23-1(a). The levels in this case can be regarded as the particle-in-a-box energy levels of Section 2.6. On the other hand, when work is done on a system, the energy levels themselves are changed. (The levels depend on the size of the box and move apart as its volume is decreased.) This is shown in Fig. 2.23-1(b).

These results can be verified from the fact that the internal energy of a system is given by

$$U = \Sigma N_i \epsilon_i \tag{2.23-1}$$

so that the change in internal energy can be written

$$dU = \sum_i \epsilon_i \, dN_i + \sum_i N_i \, d\epsilon_i \tag{2.23-2}$$

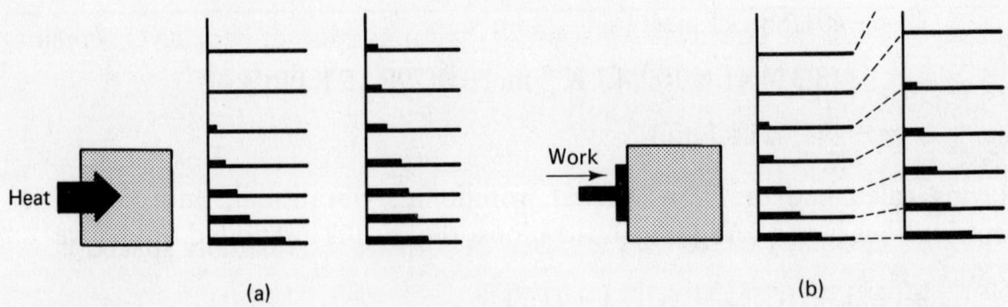

Figure 2.23-1 Microscopic viewpoint of the transfer of heat and work. (a) Change of population due to heat transfer. (b) Change of energy levels due to work transfer.

When V is fixed the allowable energy states are fixed. Taking $\epsilon_i = f(V)$, Eq. (2.23-2) becomes

$$dU = \sum_i \epsilon_i \, dN_i + \sum_i N_i \left(\frac{d\epsilon_i}{dV}\right) dV \tag{2.23-3}$$

If this equation is applied to a quasistatic process for a simple compressible substance, it is seen that for zero volume change the work is zero and the internal energy must increase because of the heat interactions. This identifies the first term in (2.23-3) as

$$dQ = \sum_i \epsilon_i \, dN_i \tag{2.23-4}$$

For zero heat transfer, but with work addition at the boundary, the second term on the right-hand side of Eq. (2.23-4) is identified as

$$dW = \sum_i N_i \left(\frac{d\epsilon_i}{dV}\right) dV - \sum_i N_i \, d\epsilon_i \tag{2.23-5}$$

From this discussion, it is possible to make the following microscopic appraisal of the quasistatic process. A constant volume heat addition is evidence by a change in the population of the energy states (N_i), but the allowable energy levels are not changed. In contrast, an adiabatic compression process causes a

TABLE 2.23-1 SUMMARY OF THERMODYNAMIC LAWS

	Classical	Statistical
First law	$dU = dQ - dW$	$dU = \Sigma\epsilon_i \, dN_i + \Sigma N_i \, dt_i$
Second law	$dS = \left(\frac{dQ}{T}\right)_{\text{rev}}$	$dS = kd \ln W$
Third law	$S_0 =$ at absolute zero for perfect crystal	$S_0 = k \ln W$; for perfect crystal, $W = 1$, and $S_0 = 0$

shift upward in the energy levels ϵ_i while the population of each level remains the same. A summary of thermodynamic laws from both a classical and a statistical point of view is shown in Table 2.23-1.

■ EXAMPLE 2.23-1

Find the partition function for a molecule with a set of equally spaced nondegenerate energy levels.

Solution. Consider a molecule with a set of equally spaced nondegenerate energy levels. (This can be the case of the vibrational energy levels of a diatomic molecule.) If the separation between levels is ϵ, the partition function is

$$z = \sum_i e^{-\epsilon_i/kT} = 1 + e^{-\epsilon/kT} + e^{-2\epsilon/kT} + e^{-3\epsilon/kT} + \cdots \tag{2.23-6}$$

This is a geometric series

$$a + ar + ar^2 + \cdots + ar^{n-1} \tag{2.23-7}$$

whose sum is

$$\frac{a(1-r^n)}{1-r} \tag{2.23-8}$$

for $r < 1$. When the series is infinite, the sum is

$$\frac{a}{1-r} \tag{2.23-9}$$

Two versions frequently encountered in statistical mechanics are

$$1 - x + x^2 + \cdots = \frac{1}{1+x} \tag{2.23-10}$$

$$1 - x + x^2 + \cdots = \frac{1}{1-x} \tag{2.23-11}$$

The second of these special cases applies here, so Eq. (2.23-6) gives

$$z = \frac{1}{1 - e^{-\epsilon/kT}}$$

Numerical values can be obtained for various temperatures, as depicted in Fig. 2.23-2. It is seen that a very low temperatures z is close to unity and only the lowest state is significantly populated. As the temperature is raised, the population breaks out of the lowest state, and the upper states become progressively more populated. At the same time, the partition function (which gives an indication of the range of states populated) rises from near unity to progressively higher values. ■

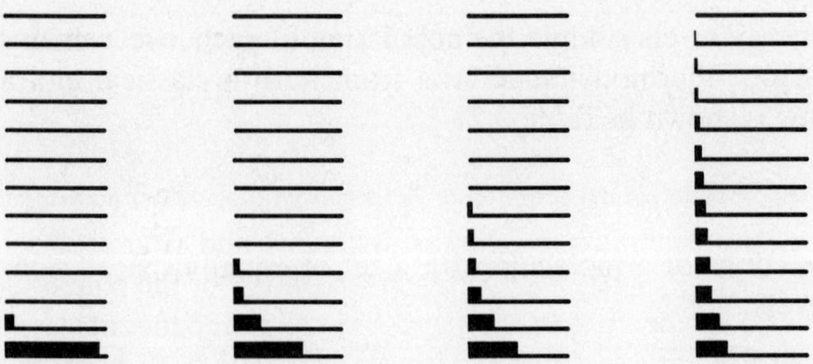

Figure 2.23-2 Populations of energy levels and partition function at different temperatures.

2.24 INFORMATION THEORY AND ALTERNATE DERIVATIONS

Lagrange's method of undetermined multipliers is not the only one for obtaining the Boltzmann or canonical distribution. There are other ways of deriving the equilibrium distribution; they are now discussed.

In a canonical ensemble (systems in thermal equilibrium at the same temperature), a system can exchange energy with the reservoir as well as with other systems, so that the energy of the system is no longer constant. For a system of fixed volume, composition, and temperature, all microstates that have the same energy are equally probable. This assumption, along with the postulate of equality of ensemble averages and time averages, discussed in Section 1.2, is the basis of all of statistical mechanics. This means that if microstates i and k of the (macroscopic) system have the same energy ($E_i = E_k$), then the probability P_i equals the probability P_k. Therefore, P_j can depend only on the energy of state j:

$$P_j = f(E_j) \tag{2.24-1}$$

To find the function f, consider two macroscopic systems I and II, which are in thermal equilibrium with the same heat reservoir. All the systems labeled I are macroscopically identical to one another. All the systems labeled II are macroscopically identical. However, systems I and II are not necessarily identical to each other; they can differ in volume and composition. In other words,

$$P_{\mathrm{I},j} = f(E_{\mathrm{I},j}) \tag{2.24.2}$$

$$P_{\mathrm{II},k} = g(E_{\mathrm{II},k}) \tag{2.24-3}$$

where $P_{\mathrm{I},j}$ is the probability that system I is in the microstate j (whose energy is $E_{\mathrm{I},j}$), $P_{\mathrm{II},k}$ is the probability that system II is in the microstate k (whose energy is $E_{\mathrm{II},k}$), and f and g are (at this point) unknown functions. Since systems I and II may differ, f and g are not necessarily the same function.

Now, I and II can also be considered to form a single composite system (I + II) for which

$$P_{\mathrm{I+II},i} = h(E_{\mathrm{I+II},i}) \qquad (2.24\text{-}4)$$

where $P_{\mathrm{I+II},i}$ is the probability that system I + II is in the microstate i (whose energy is $E_{\mathrm{I+II},i}$) and h is some function. The subsystems I and II are independent of each other, so $E_{\mathrm{I+II},i} = E_{I,j} + E_{\mathrm{II},k}$. The microstate i of the composite system I + II is defined by the microstates j and k of the independent subsystems I and II. The probability that two independent events both happen is the product of the probabilities of each event, so the probability that system I + II is in microstate i is the product of the independent probabilities $P_{\mathrm{I},j}$ and $P_{\mathrm{II},k}$, that is,

$$P_{\mathrm{I+II},i} = P_{\mathrm{I},j}\, P_{\mathrm{II},k} \qquad (2.24\text{-}5)$$

or

$$h(E_{\mathrm{I},j} + E_{\mathrm{II},k}) = f(E_{\mathrm{I},j})g(E_{\mathrm{II},k}) \qquad (2.24\text{-}6)$$

which has the form

$$h(x+y) = h(z) = f(x)g(y) \qquad (2.24\text{-}7)$$

where $x = E_{\mathrm{I},j}$, $y = E_{\mathrm{II},k}$, $z = x + y$. Equation (2.24-7) is solved by noting that for any function $h(x+y)$ which depends on x and y only through their sum, the derivative of $h(x+y)$ with respect to x equals the derivative of $h(x+y)$ with respect to y. In other words, $dh/dz = f'(x)g(y) = f(x)g'(y)$. This gives

$$\frac{f'(x)}{f(x)} = \frac{g'(y)}{g(y)} = -\beta \qquad (2.24\text{-}8)$$

where β is a constant. From the first relation,

$$\frac{df(x)}{f(x)} = -\beta\, dx \qquad (2.24\text{-}9)$$

integration yields

$$\ln f = -\beta x + \text{const}$$

or

$$f(x) = ae^{-\beta x}$$

where a is a constant. Similarly, $g'(y)/g(y)$ yields

$$g(y) = ce^{-\beta y} \qquad (2.24\text{-}10)$$

where c is another constant. The use of two systems, however, is merely a temporary device, so going back to the ensemble with a single system in a heat bath,

$$P_j = ae^{-\beta E_i} \qquad (2.24\text{-}11)$$

where β is a function of T and a is a function of T, V, and composition. The identification of a and β then proceeds as in Sections 2.19 and 2.20.

Information Theory

Yet another way of obtaining the canonical ensemble distribution function is by means of *information theory*. The basic concept is that information is a physical quantity that can be generated and measured (in an information source), transmitted or communicated via a physical medium (cable, radio link, magnetic tape, etc.), and received (by a receiver which may be an instrument or a human being). Let the uncertainty of the receiver about the outcome of an event (before the message has been sent) be defined by

$$-k \ln P_i \tag{2.24-12}$$

where k is a constant while P_i is the probability of the outcome of the event about which the message is being sent before receipt of the message. Since P_i is always less than unity, its logarithm (whatever the base) is negative and the initial uncertainty always positive.

Now, suppose the message is sent such that after its receipt by the receiver, the latter is still not completely certain about the outcome of the event. (This is the case of a noisy channel.) Let the uncertainty of the receiver about the outcome of the event (after the message is sent) be defined by

$$-k \ln P_f \tag{2.24-13}$$

where P_f is the (final) probability of the outcome of the event after receipt of the message. The information contained in the message is thus

$$\begin{aligned} I &= \text{decrease in uncertainty of the receiver} \\ &= -(\text{final uncertainty} - \text{initial uncertainty}) \\ &= -[-k \ln P_f - (-\mathrm{k} \ln P_i)] = -k \ln \frac{P_i}{P_f} \end{aligned} \tag{2.24-14}$$

For an ideal loss-less, noise-free communication channel, the receiver would be in no doubt, that is, the receiver would be completely certain about the outcome of the event. Hence, the probability of the event after receipt of the message is unity ($P_{\mathrm{i}} = 1$) because there are no alternatives. In this case,

$$I = -k \ln \frac{P_i}{1} = -k \ln P_i \tag{2.24-15}$$

Extended to statistical mechanics, (2.24-15) leads to the uncertainty of an event whose outcomes have probabilities $P_1, P_2, \ldots, P_n$, as

$$H(P_1, P_2, P_3, \ldots, P_n) = -k \sum_i P_i \ln P_i \tag{2.24-16}$$

It can be verified that H readily fulfills the following requirements:

1. The uncertainty must be a symmetric continuous function of the probabilities of the various outcomes.
2. If the probabilities are equal, that is $P = 1/n$ where n is the number of possible outcomes, then the uncertainty must be a monotonically increasing function of n.
3. If the possible outcomes of a particular experiment depend upon the

possible outcomes of n subsidiary experiments, the uncertainty for the experiment is the sum of the uncertainties of the n subsidiary experiments.

Consider now the situation where the mean value $\langle x_i \rangle$ of a particular variable x_i is given. In this case,

$$H = -\sum_i P_i \ln P_i \tag{2.24-17}$$

where k in (2.24-16) has been chosen equal to unity for simplicity reasons. Equation (2.24-17) is to be maximized, subject to

$$\sum_i P_i = 1 \tag{2.24-18}$$

and

$$\sum_i P_i x_i = \langle x_i \rangle \tag{2.24-19}$$

The technique for handling this problem is again Lagrange's method of undetermined multipliers, which calls for the maximization of the function,

$$-\sum_i (P_i \ln P_i + \lambda P_i + \beta P_i' x_i) \tag{2.24-20}$$

where the undetermined multipliers λ and β are determined from the constraints (2.24-18) and (2.24-19). Differentiating (2.24-20) with respect to each of the P_i, and equating the derivative to zero, leads to

$$\ln P_i + 1 + \lambda + \beta x_i = 0 \tag{2.24-21}$$

or

$$P_i = \exp(-1 - \lambda) \exp(-\beta x_i) \tag{2.24-22}$$

The constraint (2.24-18) yields

$$\exp(1 + \lambda) = \sum_i \exp(-\beta x_i) \tag{2.24-23}$$

hence,

$$P_i = \frac{\exp(-\beta x_i)}{\sum_i \exp(-\beta x_i)} \tag{2.24-24}$$

and the average value of x is thus

$$\langle x_i \rangle = \frac{\sum_i x_i \exp(-\beta x_i)}{\sum_i \exp(-\beta x_i)} = -\frac{\partial}{\partial \beta}\left\{\ln\left[\sum_i \exp(-\beta x_i)\right]\right\} \tag{2.24-25}$$

The partial derivative indicates that any parameters on which the x_i depend, are to be kept constant. Equation (2.24-24) is the same as the result previously obtained in Section 2.18. The evaluation of β proceeds exactly as before.

■ EXAMPLE 2.24-1

A three-level system has allowable particle energies $\epsilon = 0$, $\epsilon = 1$, and $\epsilon = 2$, respectively. Given that the average energy of the system is known to be 0.3 total, what is the most likely distribution?

Solution. Let P_i be the probability that the system is in the quantum state *i*. Then the given information is

$$\sum_i P_i = 1 \qquad \text{the system is in some state} \tag{2.24-26}$$

$$\sum_i P_i \epsilon_i = \langle E \rangle \qquad \text{the system has energy} \tag{2.24-27}$$

The question is this: Given the value $\langle E \rangle$, what values should be assigned to the P_i, the probability of the *i*th state? The answer, if prejudice is to be avoided, is that P_i should be assigned values that make the uncertainty a maximum. Thus,

$$H = -\sum_i P_i \ln P_i \tag{2.24-28}$$

is to be maximized, subject to (2.24-26) and (2.24-27). Using Lagrange's method once again, this means working with

$$\sum_i (P_i \ln P_i + \lambda P_i + \beta P_i \epsilon_i) \tag{2.24-29}$$

the undetermined multipliers λ and β being determined from the constraints (2.24-26) and (2.24-27). Differentiating Eq. (2.24-29) with respect to each of the P_i, and equating the derivative to zero, leads to

$$\ln P_i + 1 + \lambda + \beta\epsilon_i = 0$$

or

$$P_i = \exp(-1 - \lambda)\exp(-\beta\epsilon_i) \tag{2.24-30}$$

The constraint (2.24-26) gives

$$\exp(1 + \lambda) = \sum_i \exp(-\beta\epsilon_i) \tag{2.24-31}$$

Hence, from (2.24-30),

$$P_i = \frac{\exp(-\beta\epsilon_i)}{\sum_i \exp(-\beta\epsilon_i)} \tag{2.24-32}$$

β can be determined from

$$0.3 = \frac{\sum_i \epsilon_i \exp(-\beta\epsilon_i)}{\sum_i \exp(-\beta\epsilon_i)} = -\frac{\partial}{\partial\beta}\left[\ln\left(\sum_i \exp(-\beta\epsilon_i)\right)\right] \tag{2.24-33}$$

that is,

$$0.3 = \frac{0 + e^{-\beta} + 2e^{-2\beta}}{1 + e^{-\beta} + 2e^{-2\beta}} \tag{2.24-34}$$

which is solved to give $\beta = 1.34$, and hence $P_0 = 0.751$, $P_1 = 0.197$, and $P_2 = 0.052$. ■

2.25 FLUCTUATIONS

In statistical mechanics, once an ensemble (a replication of systems) has been established, all the thermodynamic properties can be evaluated by taking the average over the ensemble using the most probable distribution. This dominating configuration is overwhelmingly the most probable, and it dominates the properties of the system virtually completely. The probability that a system of fixed volume, temperaturre, and composition is in quantum state j is given by

$$P_j = \frac{e^{-E_j/kT}}{Z} \tag{2.25-1}$$

where Z, the partition function, is a constant at fixed T, V, and composition. Equation (2.25-1) gives the probability of a system being in a single quantum state. However, as seen earlier, the quantum energy levels of a system with a large number of molecules are highly degenerate, with many different quantum states having the same energy. Let g_i be the number of quantum states that have energy E_i; that is, g_i is the degeneracy of the level E_i. The probability of the system having energy E_i is thus

$$P(E_i) = \frac{g_i e^{-E_i/kT}}{Z} \tag{2.25-2}$$

The degeneracy or density of states (number of states per unit energy range), is a sharply increasing function of the system's energy E_i. The exponential function $e^{-E_i/kT}$, on the other hand, is a sharply decreasing function of E_i, especially since Boltzmann's constant k is a very small number. The product of a sharply increasing function and a sharply decreasing function gives a probability $P(E_i)$ that is peaked extremely narrowly about a single value, which is the observed thermodynamic internal energy U, as shown in Fig. 2.25-1. The probability of a significant fluctuation in the thermodynamic internal energy is extremely small. Most members of the ensemble have an energy very close to the mean.

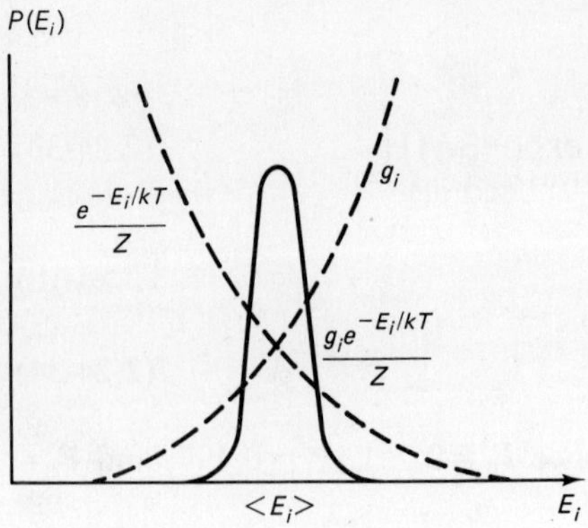

Figure 2.25-1 Value of the internal energy.

To complete the discussion, expressions for the departures from the mean values of observed properties will be derived, specifically, the probability that the value of the energy may differ from the mean energy $\langle E \rangle$. Consider, for example, the ensemble average of the mean square deviation from the average energy:

$$\langle(\Delta E)^2\rangle = \langle(E - \langle E \rangle)^2\rangle = \langle E^2 \rangle - \langle E \rangle^2 \tag{2.25-3}$$

By definition is the average energy

$$\langle E \rangle = \frac{\sum_i E_j\, e^{-\beta E_i}}{\sum_i e^{-\beta E_i}} = \frac{\sum_i E_i\, e^{-\beta E_i}}{Z} \tag{2.25-4}$$

where β has been used instead of $1/kT$ to keep the writing simple. This may be rearranged into

$$\langle E \rangle Z = \sum_i E_i e^{-\beta E_i} \tag{2.25-5}$$

Differentiating (2.25-5) with respect to β,

$$\frac{\partial\langle E \rangle}{\partial\beta} Z + \langle E \rangle \frac{\partial Z}{\partial\beta} = -\sum_i E_i^2\, e^{-\beta E_i} \tag{2.25-6}$$

Dividing by Z,

$$\frac{\partial\langle E \rangle}{\partial\beta} + \langle E \rangle \frac{\partial Z}{Z\,\partial\beta} = \frac{\sum_i E_i\, e^{-\beta E_i}}{Z} \tag{2.25-7}$$

But $\partial Z/Z\,\partial\beta = -\langle E \rangle$ by (2.25-4). Also, the right-hand side of (2.25-7) is recognized as $\langle E^2 \rangle$. Thus, Eq. (2.25-7) is

$$\frac{\partial\langle E \rangle}{\partial\beta} - \langle E \rangle^2 = -\langle E^2 \rangle \tag{2.25-8}$$

and, by (2.25-3),

$$\frac{\partial\langle E\rangle}{\partial\beta} = -\langle\Delta E)^2\rangle \qquad (2.25\text{-}9)$$

Replacing β by its value, $1/kT$, Eq. (2.25-9) becomes

$$-kT^2\frac{\partial\langle E\rangle}{\partial T} = -\langle(\Delta E)^2\rangle \qquad (2.25\text{-}10)$$

or, with $\partial\langle E\rangle/\partial T = \langle C_v\rangle$,

$$kT^2\langle C_v\rangle = \langle(\Delta E)^2\rangle \qquad (2.25\text{-}11)$$

whereupon

$$\frac{\langle(\Delta E)^2\rangle^{1/2}}{E} = \frac{T\langle kC_v\rangle^{1/2}}{E} \qquad (2.25\text{-}12)$$

This is the relative rms error in E. For a typical thermodynamic system, $E = NkT$, therefore, $C_v = Nk$. Using this in Eq. (2.25-12) gives

$$\frac{\langle(\Delta E)^2\rangle^{1/2}}{E} \approx \frac{1}{\sqrt{N}} \qquad (2.25\text{-}13)$$

Equation (2.25-13) says that the relative fluctuation of the energy of a system in a heat bath is inversely proportional to the square root of its size. Herein lies the essential characteristics of the various systems, notwithstanding their similarities. For a macroscopic system, $N \simeq 10^{24}$. Therefore, fluctuations amount to about 1 part in 10^{12}, exceedingly minute on an observational level. The most probable value is, therefore, overwhelmingly the most probable. For a single molecule, however, $N = 1$, so that $\langle(\Delta E)^2\rangle^{1/2} \simeq E$; fluctuations in energy are comparable in magnitude to the energy itself.

Similarly, the thermal motions of the individual molecules of a gas lead to density fluctuations, and thus to pressure fluctuations, in the gas. The density fluctuations in a system of constant T, V, and μ, however, will normally be very small. Only when the number density of a system becomes small will fluctuations have to be reckoned with. Example 2.25-1 shows that the same inverse root law, (2.25-13), also applies to the density fluctuations, that is, the rms fluctuation in volume is given by

$$\frac{\langle(\Delta V)^2\rangle^{1/2}}{V} = \frac{1}{\sqrt{N}} \qquad (2.25\text{-}14)$$

By way of example, there are, on average, around 3×10^{16} molecules in a volume of 10^{-9} m^3 (1 mm^3) of air at STP. Thus, (2.25-14) gives the fractional rms fluctuation in their volume as approximately 6×10^{-9}, which is insignificant. On the other hand, a volume of 2×10^{-19} m^3 (the volume of a cube of side equal to the wavelength of yellow light) contains some 6×10^6 molecules, so that the fractional rms fluctuation in volume (or density) is now around 4×10^{-4}, that is, 0.04%. Such density fluctuations produce fluctuation in the index of refraction of the air and these in turn lead to scattering of light as it passes through a gas.

An interesting example of the role of density fluctuations is in the appearance of blue sky. If there were no density fluctuations in the air, sunlight would not be scattered and the sky would be black except in the direct path to the sun. Actually, very small adjacent regions of greatly different densities act as scattering centers, deflecting light from its straight path. To have appreciable scattering requires that a significant density fluctuation occur over a region with dimension of the order of the wavelength of sunlight. In a gas, the molecules are far apart and most of the light incident on the gas passes through without being absorbed. However, the oxygen and nitrogen molecules of the air do "scatter" the light (absorb and reemit it) in all directions. Since red light has longer wavelength than blue, and significant fluctuations in density are more likely to occur in smaller regions, blue light is scattered more than red. This is what gives air its bluish color when the sky is looked at in any direction other than directly at the sun.

By the same token, the glancing rays that penetrate through a very deep layer of air at sunrise and sunset tend to be red. (The blues tend to be scattered out before they can be seen.) This is why dawn and dusk are characteristically rich in red hues. The light that reaches the earth is reddish orange and this is the color seen reflected from clouds and dust near the surface of the earth at sunset.

Figure 2.25-2 Red hues in the sunset.

■ **EXAMPLE 2.25-1**

The presence of volume fluctuations in a small amount of gas containing a fixed number N of molecules implies corresponding fluctuations in the number $n = N/V$ of molecules per unit volume, and thus also in the mass density. Show that for a system in which fluctuations occur due to the thermal motions of the individual molecules, the rms fluctuation in volume is given by Eq. (2.25-14).

Solution. Consider the energy required to produce a local increase in pressure from p_0 (the mean value) to $p_0 + \delta p$ within a small region of the gas. This can readily be done by picturing a portion of the gas, containing N molecules and originally occupying a volume V, as being compressed in the piston-and-cylinder arrangement shown in Fig. (2.25-3), until its volume is $V - \delta V$.

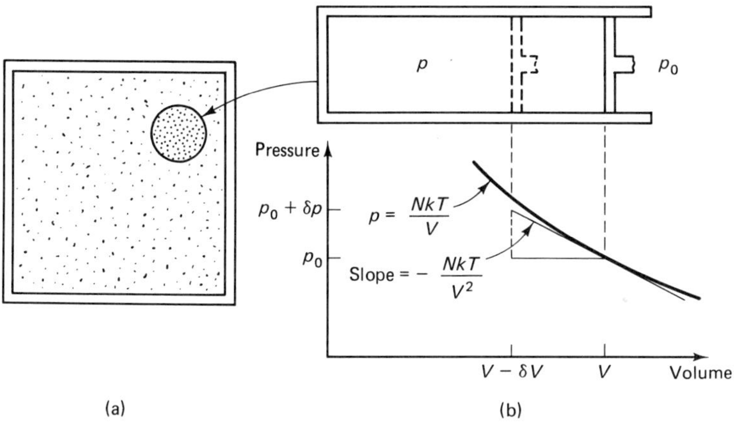

Figure 2.25-3 Work done in density fluctuation.

The amount of work (required) to produce a compression, is equal to the shaded area on the p–V diagram, which is $\frac{1}{2}$ (base) times height, or $\frac{1}{2}(\delta V)(\delta p)$. Now, for a perfect gas, its pressure p and volume V are related to the total number of atoms present, N, and the temperature T by

$$p = \frac{NkT}{V} \tag{2.25-15}$$

or

$$\delta p = \frac{NkT}{V^2}\,\delta V \tag{2.25-16}$$

Thus, the work done is

$$-\frac{1}{2}(\delta V)(\delta p) = \frac{1}{2}(\delta V)^2\left(\frac{NkT}{V^2}\right)$$

In other words, averaged over a period of time during which δV fluctuates erratically as a displacement in the x direction, the mean potential energy of the small volume of gas is

$$\frac{1}{2}NkT\frac{\langle(\delta V)^2\rangle}{V} \tag{2.25-17}$$

Setting this equal to $\frac{1}{2}kT$ (in the one direction) gives

$$\frac{1}{2}NkT\frac{\langle(\delta V)^2\rangle}{V} = \frac{1}{2}kT \tag{2.25-18}$$

from which the rms fluctuation in volume is

$$\frac{\langle(\delta V)^2\rangle^{1/2}}{V} = \frac{1}{\sqrt{N}} \tag{2.25-19}$$

This is Eq. (2.15-14). It is instructive to note that the volume occupied by the N molecules fluctuates between

$$V + \frac{V}{\sqrt{N}} = V(1 + N^{-1/2})$$

and

$$V - \frac{V}{\sqrt{N}} = V(1 - N^{-1/2})$$

It follows that the number density of the molecules fluctuates between

$$\frac{N}{V(1 + N^{-1/2})} \tag{2.25-20}$$

and

$$\frac{N}{V(1 - N^{-1/2})} \tag{2.25-21}$$

Writing n for the mean number density N/V and expanding $(1 + N^{-1/2})^{-1}$ and $(1 - N^{-1/2})^{-1}$ by the binomial theorem shows that the number density of the molecules fluctuates between

$$n\left(1 - \frac{1}{N^{1/2}}\right) \tag{2.25-22}$$

and

$$n\left(1 + \frac{1}{N^{1/2}}\right) \tag{2.25-23}$$

■

■ EXAMPLE 2.25-2

For 10^{22} molecules in a container, what is the probability that at some instant, there are 5.000001×10^{21} molecules in one half of the container and 4.99999×10^{21} molecules in the second half?

Solution. Consider an ideal gas of N molecules in a container, and assume an imaginary partition dividing the volume into equal parts. What is the probability of finding N_1 molecules on the left side of the partition and N_2 molecules on the right side, where $N_1 + N_2 = N$? The situation is shown in Fig. 2.25-4.

The problem is equivalent to the number of distinct ways in which the molecules can be placed into the box to give the distribution (N_1, N_2). In general, for N molecules, the distribution (N_1, N_2) can be realized in

$$\frac{N!}{N_1!(N - N_1)!} = \frac{N!}{N_1!N_2!} \tag{2.25-24}$$

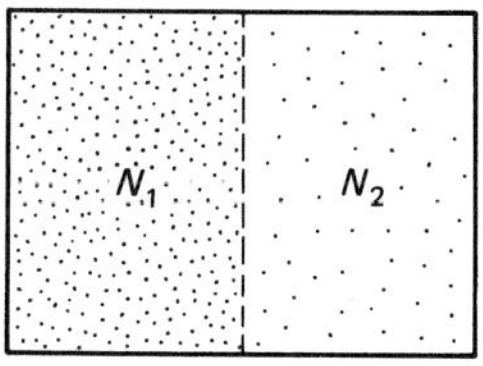

Figure 2.25-4 Ideal gas with N_1 and N_2 particles across an imaginary partition.

In statistical physics, this quantity is the statistical weight, or thermodynamic probability W of the distribution (N_1, N_2). For the distribution (2,1), the statistical weight is

$$W = \frac{3!}{2!1!} = 3 \qquad (2.25\text{-}25)$$

Now, the a priori probability for finding a specific molecule in either compartment is 0.5. The probability of finding N_1 specific molecules on the left side and N_2 specific molecules on the right side is then

$$(0.5)^{N_1}(0.5)^{N_2} \qquad (2.25\text{-}26)$$

This is the a priori probability of the distribution (N_1, N_2). It is the same for all distributions. Note that there is an implicit assumption that the molecules are distinguishable, that is, they can be tagged in some way and are known where they are at all times. The absolute probability is

$$P(N_1, N_2) = (0.5)^{N_1}(0.5)^{N_1}(0.5)^{N_2}\frac{N!}{N_1!N_2!} \qquad (2.25\text{-}27)$$

P is tabulated for all possible distributions for $N = 4$ and $N = 5$ in Table 2.25-1. Clearly, the distribution of maximum probability is that for which there are equal numbers of molecules on the two sides for N even, and for N_2 and N_1 differing by only one molecule for N odd.

TABLE 2.25-1 VALUES OF $P(N_1,N_2)$ FOR $N = 4$ AND $N = 5$

$N = 4$			$N = 5$		
N_1	N_2	P	N_1	N_2	P
4	0	$\frac{1}{16}$	5	0	$\frac{1}{32}$
3	1	$\frac{4}{16}$	4	1	$\frac{5}{32}$
2	2	$\frac{6}{16}$	3	2	$\frac{10}{32}$
1	3	$\frac{4}{16}$	2	3	$\frac{10}{32}$
0	4	$\frac{1}{16}$	1	4	$\frac{5}{32}$
			0	5	$\frac{1}{32}$

Let

$$N_1 = \frac{N}{2} + n, \qquad N_2 = \frac{N}{2} - n \qquad (2.25\text{-}28)$$

where n is a measure of the deviation from the most probable value. Then

$$P(n) = (0.5)^N \frac{N!}{[(N/2)+n]![(N/2)-n]!} \tag{2.25-29}$$

For very large N and for $N \gg n$ this expression reduces to

$$P(n) = \left(\frac{2}{\pi}\right)^{1/2} \left(\frac{1}{N}\right)^{1/2} \exp\left(-\frac{2n^2}{N}\right) \tag{2.25-30}$$

This distribution function is of the *Gaussian* or *normal distribution* form

$$(n) = a\exp(-bn^2) \tag{2.25-31}$$

where a and b are constants. In Fig. 2.25-5, the total probability $P(N_1, N_2)$ for $N = 4$ is plotted as a function of N_1; a normal distribution of the same area is plotted for comparison. It is easy to see that $P(N_1, N_2)$ should approach a normal distribution as N becomes large.

The width of the normal distribution at one-half the peak value may be shown to be $b^{1/2}$; in the present case

$$b^{1/2} = \left(\frac{2}{N}\right)^{1/2} \approx \left(\frac{1}{N}\right)^{1/2} \tag{2.25-32}$$

for a large N. For a real system of molecules, N is usually extremely large, so that the normal curve is extremely narrow and it is overwhelmingly more probable to find equal numbers of molecules in the two sections of the container. For 10^{22} molecules in a container,

$$n = N_1 - \frac{N}{2} = (5.000001 - 5.000000) \times 10^{21} = 10^{15}$$

Therefore,

$$P(n) = \left(\frac{2}{\pi N}\right)^{1/2} \exp\left(-\frac{2n^2}{N}\right) = \frac{1}{10^{11}} \exp(-2 \times 10^8) = 10^{-108}$$

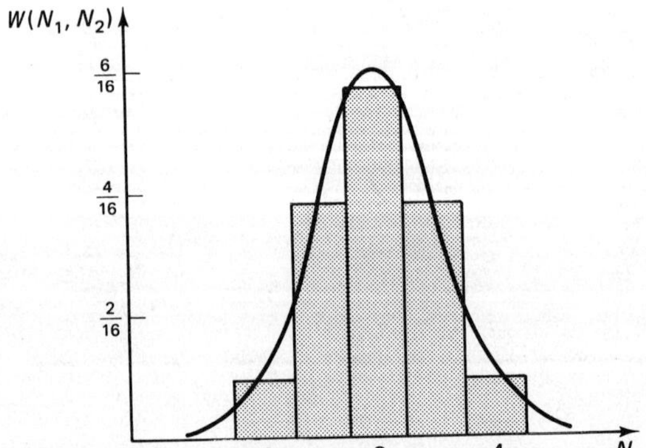

Figure 2.25-5 Total probability $P(N_1, N_2)$ for $N = 4$ as function of N.

which is an extraordinarily small number! Even fluctuations of one part in one million are extremely unlikely. ■

Example 2.25-2 shows that the state of a real, macroscopic system in thermal equilibrium is determined by specifying a very small number of parameters (such as pressure, volume, energy, etc.) derived from a study of the statistical properties of the microscopic behavior. Any system left to itself will approach a state of thermal equilibrium, and, for a given initial total energy, the same state is reached (independent of any other initial conditions imposed upon the constituent particles). This results from the fact that there is a large number of particles contained in any macroscopic system. Organized motions tend to become randomized, and the probability that a macroscopic parameter will be found to have a value different from the most probable value is quite negligible.

2.26 SUMMARY

Many different energy states are possible for the individual molecules of a substance. At any instant in time, it may happen that a certain number n_1 of molecules have the same energy ϵ_1, n_2 have the energy ϵ_2, n_3 the energy ϵ_3, and so on. For an isolated molecule in a confined space, the *Schrödinger equation*

$$H\psi = \epsilon\psi \tag{2.26-1}$$

is satisfied for certain function $\psi_1, \psi_2, \ldots, \psi_j$ (eigenvalues). Such a molecule can exist in certain quantum states, each of which is characterized by a wave function ψ_j. The discrete values ϵ_i are referred to as energy levels, meaning that the molecule is in a quantum state for which the energy is ϵ_i. It may happen that a number, say g_i with $g_i > 1$, of states have the same energy ϵ_i, and it then said that this energy level is g_i-fold degenerate.

The Schrödinger equation can also be applied to macroscopic systems of particles. It is convenient to use the symbols E and ψ for the energy and wave function of a macroscopic system. For example, the quantum states ψ_j and energy levels E_i of 1 mol of a substance, containing about 10^{25} particles (nuclei and electrons), would, in principle, be found by solving the equation

$$H\psi = E\psi \tag{2.26-2}$$

a differential equation involving 3×10^{25} variables (e.g., the x, y, and z coordinates for each of the particles). It is quite unnecessary (even if it were possible) for the purposes of statistical mechanics to seek a rigorous solution of this equation; the important thing is that such a solution exists.

For a *free particle* (a particle that has no force acting on it except at the walls of a container) Schrödinger's equation gives the possible values of the translational energy of the particle as

$$\epsilon_{\text{tran}} = \frac{h^2}{8m}\left(\frac{n_x^2}{L_x^2} + \frac{n_y^2}{L_y^2} + \frac{n_z^2}{L_z^2}\right) \tag{2.26-3}$$

where m is the mass of the particle and L_x L_y, and L_z are the dimensions of the container. The quantities n_x, n_y, and n_z are all positive integers, which may be the same or different. For each different set of values (n_x, n_y, n_z), it is said that the particle is in a different quantum state.

For the *rotational energy* of molecules, quantum theory shows the allowed values as

$$\epsilon_{rot} = \left(\frac{h^2}{8\pi^2 I}\right) J(J+1) \qquad J = 0, 1, 2, \ldots \tag{2.26-4}$$

where I is the moment of inertia of the molecule and J is the rotational quantum number.

Of even greater importance than the quantization of ϵ_{rot} is the restriction that quantum mechanics places on the *vibrational motion* of atoms. Schrödinger's equation gives the possible energies of a harmonic oscillator as

$$\epsilon_{vib} = \left(n + \tfrac{1}{2}\right) h\nu \qquad n = 0, 1, 2, \ldots \tag{2.26-5}$$

where ν is the characteristic frequency of the oscillator. The lowest value of the energy of a harmonic oscillator is $\frac{1}{2} h\nu$(for $n = 0$), which is known as the zero-point energy.

The distinction between *classical theory* and *quantum theory* becomes important only when a small amount of energy is exchanged with a system. Classically, it is always possible to exchange any small amount of energy with the system, whereas quantum theory prescribes a certain minimum amount of energy, namely, the energy difference between two of the quantum states of the system. The most directly compelling evidence for the quantization of energy comes from the observation of the frequencies of light absorbed and emitted by atoms and molecules. A typical atomic or molecular spectrum features radiation of discrete rather than continuous frequencies being emitted or absorbed.

The simplest systems for the application of statistical mechanics are those in which intermolecular forces can be neglected (e.g., perfect gas systems). The total energy of such a system is just the sum of the individual molecular energies; each of these will be one or other of the ϵ_i values derived from Schrödinger's equations (whose solutions are known as orbitals). For some particular quantum state of the whole system, there are n_1 molecules in the level ϵ_1, n_2 molecules in the level ϵ_2, and so on. The condition of the system which is characterized by the numbers $n_1, n_2, \ldots, n_i, \ldots$ is called a distribution (of molecules among the energy levels, or, equivalently, of energy among the molecules), and the numbers $n_1, n_2, \ldots, n_i$, are themselves called distribution numbers. If the distribution of the particles of the system among its energy states is known, the macroscopic properties of the system can be determined. Thus, a central problem of statistical mechanics is to determine the possible distributions of particles among energy levels and energy states.

An assembly or large number of replicate systems is known as an *ensemble*. The systems may be as simple as single particles, or they may themselves be identical assemblies of particles. They may be distinguishable (like the molecules in a crystal, which can be labeled by the positions they occupy in the crystal lattice), or they may be indistinguishable (like the molecules of a gas, which cannot be labeled). Whether the particles are distinguishable or not, a specification of the number of particles in each energy level defines a *macrostate* of the assembly. If the particles are distinguishable, a specification of the energy state of each particle defines a *microstate* of the assembly. That is, not only how many particles in each state, but which particles they are, is specified. The number of different microstates, for a given set of occupation numbers, is much greater if the particles are distinguishable than if they are indistinguishable.

A *microcanonical* ensemble is an assembly whose members have the same values of N, V, and E. A *canonical* ensemble is composed of members at constant N, V and T, and a *grand canonical* ensemble has members at constant μ, V, and T. In an isolated, closed system, the energy E and the total number of particles N are both constant. However, as time goes on, interactions between the particles of an isolated, closed system will result in changes in the numbers of particles occupying the energy states, and, if the particles are distinguishable, will result in changes in the energy state of each particle. Every interchange results in a change in the microstate of the assembly, but every possible microstate is equally probable. In a canonical ensemble (constant N, V, T), the energies of the systems fluctuate, but all microstates of a system that have the same energy are equally probable. This simple postulate is the basis of all of statistical mechanics.

The number of equally probable microstates that correspond to a given macrostate is called the thermodynamic probability W (Wahrscheinlichkeit). This is a number which can be very large, and is not the same as the mathematical probability. The latter, for a given distribution, is the thermodynamic probability W for that distribution divided by the sum of the values of W for all possible distributions. The general evaluation of W depends on the particular statistics obeyed by the assembly. In the *Bose–Einstein statistics,* the particles are considered indistinguishable, and there is no restriction on the number of particles that can occupy any energy state. The energy states, however, are distinguishable. The number of arrangements of n_j particles for the jth level is

$$W_j = \frac{(g_j + n_j - 1)!}{(g_j - 1)!n_j!} \tag{2.26-6}$$

For each of the possible distributions in any one level, any one of the distributions in one of the other levels is possible, so the thermodynamic probability is the product over all levels of the values of W_j for each level, or

$$W_{\mathrm{BE}} = \prod_j \frac{(g_j + N_j - 1)!}{(g_j - 1)!N_j!} \tag{2.26-7}$$

In *Fermi–Dirac statistics,* the particles are considered indistinguishable, but they obey Pauli's exclusion principle, according to which there can be no more than one particle per quantum state

$$W_j = \frac{g_j!}{(g_j - n_j)!n_j!} \tag{2.26-8}$$

and so, the thermodynamic probability of a given distribution of particles among energy levels is the product over all levels, or

$$W_{\mathrm{FD}} = \prod_j \frac{g_j!}{(g_j - N_j)!N_j!} \tag{2.26-9}$$

In the *Maxwell–Boltzmann statistics,* the particles of an assembly are considered distinguishable, but as in the Bose–Einstein statistics there is no restriction on the number of particles that can occupy the same energy or quantum state. The particles are to be distributed in some manner among the various energy levels ϵ_j, each of which has a degeneracy or number of quantum states g_j. First, the number of ways in which n distinguishable particles can be divided into groups or levels, with n_1 particles in the first level, n_2 in the second, and so on, is

$$\frac{n!}{\prod_j n_j!} \tag{2.26-10}$$

However, for each of these groups or levels, j, the number of ways that the n_j particles in that group can be placed in the g_i quantum states is

$$W_j = g_j^{n_j} \tag{2.26-11}$$

Therefore, the total number of different distributions, or the thermodynamic probability of a macrostate is

$$W_{\mathrm{MB}} = \frac{N!}{\prod_j N_j!} \prod_j g_j^{N_j} = N! \prod_j \frac{g_j^{N_j}}{N_j!} \tag{2.26-12}$$

Of the many possible distributions in a macroscopic system, the equilibrium distribution stands out as the single most probable distribution. Applying Lagrange's method of undetermined multipliers, the most probable distribution can be found. For indistinguishable particles obeying Bose–Einstein statistics this is

$$n_j = \frac{g_j}{e^{\alpha} e^{\beta\epsilon_j} - 1} \tag{2.26-13}$$

For indistinguishable particles obeying Fermi–Dirac statistics, the equilibrium distribution is

$$n_j = \frac{g_j}{e^{\alpha} e^{\beta\epsilon_j} + 1} \tag{2.26-14}$$

And for distinguishable particles obeying Maxwell–Boltzmann statistics, it is

$$n_j = g_j e^{-\alpha} e^{-\beta\epsilon_j} \tag{2.26-15}$$

These are the distribution laws for *bosons, fermions,* and *boltzons.* The ratio n_j/g_j is the average occupation number per state in any level. The probabilities

$$F_{\mathrm{BE}} = \frac{1}{e^{\alpha}e^{\beta\epsilon_j} - 1} \tag{2.26-16}$$

$$F_{\mathrm{BE}} = \frac{1}{e^{\alpha}e^{\beta\epsilon_j} + 1} \tag{2.26-17}$$

$$F_{\mathrm{MB}} = Ae^{-\beta\epsilon_j} \tag{2.26-18}$$

are known respectively as the Bose–Einstein, Fermi–Dirac, and Maxwell–Boltzmann *distribution functions.* Nature decrees that particles, whether they be photons, electrons, or atoms, are either bosons or fermions, that is they obey either Bose–Einstein or Fermi-Dirac statistics. However, for most gases at low to moderate density (relatively low pressure or high temperature), the number of quantum states available at any level is much larger than the number of particles in that level. In that case,

$$\frac{N_j}{g_j} \ll 1 \tag{2.26-19}$$

for all j. From the distribution equations, this is seen to be equivalent to the statement

$$e^{\alpha}e^{\beta\epsilon_j} \gg 1 \tag{2.26-20}$$

Therefore, under these special conditions, the -1 in the denominator of the distribution equation for Bose–Einstein statistics and the $+1$ for Fermi–Dirac statistics are negligible, and both models predict essentially the same distribution of particles among energy levels as does Boltzmann statistics. However, the thermodynamic probability for any given distribution is different by $N!$ for the Boltzmann statistics, or

$$W_{\mathrm{BE}} \approx W_{\mathrm{FD}} \approx \left(\frac{W}{N!}\right)_{\mathrm{Boltz}} \tag{2.26-21}$$

Dividing the value of W for Boltzmann statistics by $N!$ has the effect of discounting the distinguishability of the N particles. This model is known as semi-classical, or corrected Boltzmann, statistics.

The derivation for the equilibrium state or most probable distribution of particles N_j among energy levels, $N_j = g_j e^{-\alpha}e^{-\beta\epsilon_j}$, introduces two undetermined multipliers, α and β, which are determined from the constraints $N = \Sigma_j N_j$, $U = \Sigma_j N_j \epsilon_j$. The first constraint leads to

$$e^{-\alpha} = \frac{N}{\sum_j g_j e^{-\beta\epsilon_i}} = \frac{N}{Z}; \qquad Z = \sum_j g_j e^{-\beta\epsilon_i} \tag{2.26-22}$$

The second constraint leads, with the aid of $S = k \ln W$, to

$$\beta = \frac{1}{kT} \tag{2.26-23}$$

The relation $S = k \ln W$ is the connecting link between statistical and classical thermodynamics. Using the results for α and β, a thermodynamic property, for example, the internal energy, may then be obtained:

$$\langle u \rangle = \frac{U}{N} = \frac{\sum_j g_j \epsilon_j e^{-\epsilon_j/kT}}{\sum_j g_j e^{-\epsilon_j/kT}} = \frac{1}{Z} \sum_j g_j \epsilon_j e^{-\epsilon_j/kT} \tag{2.26-24}$$

The quantity

$$Z = \sum_j g_j e^{-\epsilon_j/kT} \tag{2.26-25}$$

is the *partition function.* The summation in this expression is to be made over all the j levels of energy available to the particles. In many cases, however, it proves to be more convenient to perform the summation directly over all the individual quantum states, referred to by the subscript i. Thus,

$$Z = \sum_{j \text{ levels}} g_j e^{-\epsilon_j/kT} = \sum_{i \text{ states}} e^{-\epsilon_j/kT} \tag{2.26-26}$$

The calculation of the thermodynamic properties of a system is intimately connected with the evaluation of the partition function. This in turn depends upon a knowledge of the energy levels and degeneracies for the various modes of energy. In theory, all the thermodynamic properties of a system can be calculated once the partition function is known. The thermodynamic properties of, say, a system at fixed T, V, and composition can be found by averaging over the systems of a hypothetical ensemble, each system in the ensemble having the same T, V, and composition, but not necessarily the same energy. The key statistical–mechanical formulas obtained for the calculation of thermodynamic properties in terms of $Z = Z(T, V, N_A, N_B, \ldots)$ are

$$Z = \sum_j e^{-\epsilon_j/kT} \qquad S = \frac{U}{T + k \ln Z} \tag{2.26-27}$$

$$p = kT \left(\frac{\partial \ln Z}{\partial V}\right)_{T, N_B} \qquad A = -kT \ln Z \tag{2.26-28}$$

$$U = kT^2 \left(\frac{\partial \ln Z}{\partial T}\right)_{V, N_B} \qquad \mu_B = -RT \left(\frac{\partial \ln Z}{\partial N_B}\right)_{T, V, N_{C \neq B}} \tag{2.26-29}$$

where E_j is the energy of quantum state j of the thermodynamic system, and the sum in the partition function goes over all possible quantum states of the thermodynamic system. If species B is electrically charged, μ_B is to be replaced by the electrochemical potential.

A fundamental distribution function describes how particles in a system are, on the average, distributed among the available states of the system. It describes, for each state, the probability that a particle will be found in that state. Classical and quantum-mechanical *fundamental distribution functions* have one thing in common: they all depend only on the energy of the state. For example, the particle distribution in energy is of the form

$$dn(\epsilon) = Fg(\epsilon)d\epsilon \tag{2.26-30}$$

where F is the probability of finding a particle in a given state at energy ϵ, and $g(\epsilon)$ is the density of states (with respect to energy). The *Maxwell–Boltzmann* distribution function, F_{MB}, or probability for a particle to be found in the state of velocity (v_x, v_y, v_z) is

$$F_{MB} = \left(\frac{m}{2\pi kT}\right)^{3/2} e^{-m(v_x^2+v_y^2+v_z^2)/2kT} = \left(\frac{m}{2\pi kT}\right)^{3/2} e^{-\epsilon_i/kT} \tag{2.26-31}$$

The *Bose–Einstein* distribution function, F_{BE}, giving the average number of bosons in a system in equilibrium at temperature T in a particular state i with energy ϵ_i is

$$F_{BE} = \frac{1}{e^{\alpha}e^{\epsilon_i/kT} - 1} \tag{2.26-32}$$

The *Fermi-Dirac* distribution function, F_{FD}, for the expected number of fermions in equilibrium at temperature T in a particular state i with energy ϵ_i is

$$F_{FD} = \frac{1}{e^{\alpha}e^{\epsilon_i/kT} + 1} \tag{2.26-33}$$

In classical thermodynamics, the condition of equilibrium is one for which the thermodynamic properties $U, p, H, A, \ldots$ are considered to have fixed and determinable values. On a microscopic level, these properties are not constant at all, but instead are continually fluctuating about some average values. The fractional standard deviation in internal energy of a monatomic ideal gas is inversely proportional to the square root of the number of particles in the system. For a cubic centimeter of gas at room temperature and pressure, the ratio $\sigma_U/\overline{U}$ is on the order of magnitude of 10^{-10}, a quantity much too small to be detected experimentally. For a real substance, the deviation in U is also negligibly small in nearly every region of temperature and pressure. However, in the immediate vicinity of the critical point, fluctuations in U become appreciable, regardless of the size of the system.

Similarly, for an ideal gas the fractional standard deviation in N is inversely dependent upon the average number of particles in the system and is negligibly small for normal macroscopic systems. Fluctuations become appreciable only for an extremely rarefied gas. However, near the critical point, density fluctuations become appreciable. This causes the phenomenon of critical *opalescence,* a scattering of light making the substance appear cloudy. These large-scale fluctuation make difficult the measurement of the critical density for a substance.

SYMBOLS IN CHAPTER 2

A, a	Helmholtz function
$\hat{A}, a$	Operator, Eigenvalue
β	Rotational constant
C, c	Constant
E	Energy
F, f	Distribution functions
G, g	Gibb's function, Degeneracy
H, h	Enthalpy, Planck's constant
I, i	Moment of inertia, Dummy index
J, j	Angular momentum number, Dummy index
k	Boltzmann constant; Force constant
L, l	Angular momentum, Angular momentum number
m	Mass
N, n	Number of particles
P	Probability
p	Momentum coordinate, Pressure
Q	Heat transfer
q	Space coordinate
R	Gas constant
S, s	Entropy, Spin
T	Temperature
U, u	Internal energy
V, v	Volume, specific volume, velocity
W	Thermodynamic probability
Z, z	Partition function
$\hat{A}, \hat{B}, \hat{H}, \hat{O}, \ldots$	Operators
α, β, γ	Lagrangian multipliers
E, ϵ	Energy
θ_r, θ_v	Characteristic rotational, vibrational temperature
Λ	de Broglie wavelength
ν	Frequency, Hz
Ψ	Time dependent wave function
ψ	Space dependent wave function
ω	Circular frequency, $2\pi\nu$

PROBLEMS

2.1. Discuss the quantization of the following types of motion: translation, vibration, rotation, and electronic. Also, give the relative separations of the different types of energy levels of atoms and molecules.

2.2. **(a)** What are the results of operating on the functions e^{-ax^2}, $\cos x$, and e^{ikx} with the operator d/dx? What eigenvalues are obtained?

(b) Show that the function $\psi = 8e^{5x}$ is an eigenfunction of the operator d/dx. What is the eigenvalue?

2.3. Show that e^{ax} is an eigenfunction of the operator d/dx, and find the corresponding eigenvalue. Show that e^{ax^2} is not an eigenfunction of d/dx.

2.4. The wave function of an electron in the lowest energy state of the hydrogen atom is the function

$$\psi(r) \propto e^{-r/a_0} = \left(\frac{1}{\pi a_0^3}\right)^{1/2} e^{-r/a_0}$$

where r is the distance from the nucleus and $a_0 = 52.9$ pm. What is the average value of the distance of an electron from the nucleus of the hydrogen atom? What is the value of an individual observation?

2.5. What is the de Broglie wavelength of a hydrogen atom with a translational energy corresponding to room temperature? (Room temperature is 298 K.)

2.6. What are the wavelengths of a photon and an electron, each having kinetic energy of 1 eV?

2.7. The first ionization potential of atomic hydrogen is 13.60 V. What is the wavelength of the light produced when a free electron without kinetic energy returns to the inner orbit?

2.8. **(a)** Determine the degeneracies of the following orbitals for hydrogenlike atoms: $n = 1$, $n = 2$, $n = 3$.

(b) The 1s orbital of a hydrogenlike atom is $\psi_{1s} \propto e^{-zr/r_0}$. What is the most probable distance from proton to electron?

2.9. The electron-in-a-box is a simplified model of molecular structure; it can be used for estimating rough values of transition energies. For a particle in an infinite square potential, the energies, and wavefunctions are

$$E_n = \frac{n^2h^2}{8mL^2} \qquad n = 1, 2, \ldots$$

$$\psi_n = \left(\frac{2}{L}\right)^{1/2} \sin\frac{n\pi x}{L}$$

Applying these results to an electron confined to a molecule of length 1.0 nm (about five atoms long), what is **(a)** its minimum energy and **(b)** the minimum excitation energy from that state? What is the probability of locating the electron between $x = 0$ and $x = 0.2$ nm?

2.10. For a hydrogenlike atom, the eigenvalues for the energy are given as

$$E = -\frac{m_e Z^2 e^4}{2n^2h^2(4\pi\epsilon_0)^2}$$

The negative sign indicates that the electron has less energy than when it is free.

(a) List the orbitals for hydrogenlike atoms in order of increasing energy.

(b) Calculate the energy of the hydrogen atom in its ground state ($n = 1$) using the mass of the electron rather than the reduced mass of the electron.

2.11. The ground state wave function of a hydrogen atom is the wave function corresponding to the lowest energy. It is spherically symmetrical, and decays exponentially with the distance from the nucleus:

$$\psi = \left(\frac{1}{\pi a_0^3}\right)^{1/2} e^{-r/a_0}$$

where a_0 is a collection of fundamental constants. It has the dimensions of length and a magnitude of 53 pm. It is called the Bohr radius. r is the distance between the electron and the nucleus in the ground state of the hydrogen atom. Find the expected position of the electron (mean radius r).

2.12. For rotating molecules, the angular momentum quantum number is normally denoted J, and the rotational energy is given by

$$E_{\text{rot}} = \frac{\hbar^2}{2I} J(J+1) \qquad J = 0, 1, 2, \ldots$$

Calculate the energies of the first five rotational levels of H_2 ($I = 4.603 \times 10^{-48}$ kg m^2), the corresponding magnitudes of the angular momentum, and the number of different values of m_l in each case.

2.13. In the vibrational motion of HI, the iodine atom essentially remains stationary because of its large mass. Assuming that the hydrogen atom undergoes harmonic motion and that the force constant k is 317 N m^{-1}, what is the fundamental vibration frequency ν_0?

2.14. Calculate the energy in joules per quantum, joules per mole, and electron volts of photons of wavelength 300 nm.

2.15. The vibrational frequency of a carbon monoxide molecule is measured to be 6.53×10^{13} Hz. Calculate the force constant of the molecule if the mass of the carbon atom is 2.00×10^{-26} kg and the mass of the oxygen atom is 2.66×10^{-26} kg.

2.16. An assembly is comprised of four indistinguishable particles which may be distributed among only four possible energy levels. These levels have the relative energy values ϵ_j of 0, 1, 2, 3, and the total energy of the system is equal to 3. This means that the only distributions of particles among levels permitted are those that result in the given total energy. List the possible distributions of the particles among the energy levels, and for each distribution calculate the thermodynamic probability and the mathematical probability. What are the thermodynamic probabilities if, instead, the particles are indistinquishable from one another?

2.17. An assembly comprised of four particles has permitted relative energy levels ϵ_j of 0, 1, 2, 3, ... The total energy of the system is 3, and the degeneracies of the levels are 3, 3, 4, and 4, respectively. Determine the thermodynamic probability for each of the possible distributions, assuming **(a)** Boltzmann statistics, **(b)** Bose–Einstein statistics, and **(c)** Fermi–Dirac statistics.

2.18. Determine the number of ways of distributing

(a) six distinguishable balls into four boxes such that one ball is in the first box, one ball is in the second, three balls are in the third, and one ball is in the fourth;

(b) six indistinguishable balls into four distinguishable boxes;

(c) six indistinguishable balls into eight distinguishable boxes with no more than one per box.

2.19. Statistical mechanics gives the equilibrium distribution of a particle as

$$N_j = g_j e^{-\alpha} e^{-\beta\epsilon_j}$$

where the index j refers to energy levels, and α and β are two unknown multipliers.

(a) Express α in terms of β and the system parameters g_j and ϵ_j.

(b) Making use of the kinetic theory result that internal energy of an ideal monatomic gas is $U = \frac{3}{2}NkT$, show that $\beta = 1/kT$.

2.20. **(a)** What is the most probable distribution of $N = 5$ distinguishable particles among three energy levels if the intrinsic degeneracy or statistical weights are $g_1 = g_2 = g_3 = 1/3$?

(b) If $\epsilon_1 = 0$, $\epsilon_2 = \epsilon$, and $\epsilon_3 = 2\epsilon$, what is the most probable distribution if the total energy is fixed at $\epsilon_{\text{total}} = 3\epsilon$?

2.21. From the definition of the thermodyanmic probability

$$W = N!\prod_i 2\frac{g_i^{n_i}}{n_i!}$$

and Boltzmann's relation, $S = k \ln W$, show that

$$\beta = \frac{1}{kT}$$

using the most probable distribution $n_i = (N/z)g_i e^{-\beta\epsilon_i}$.

2.22. Show that if two assemblies for which the Maxwell-Boltzmann law is valid, can exchange energy, and are in equilibrium, then they must have the same value of β.

2.23. Show that if the conditions $g_j \gg 1$ and $N_j/g_j \ll 1$ apply, the thermodynamic probabilities given by the BE and FD statistics reduce to

$$W = \prod_i \frac{g_j^{N_j}}{N_j!}$$

and consequently $W_{\text{BE}} = W_{\text{FD}} = W_{\text{MB}}/N!$.

2.24. From the following expression for the distribution of atoms in reference to energy,

$$dn(\epsilon) = ce^{-\beta\epsilon}\epsilon^{1/2}\,d\epsilon$$

plot the energy distribution function and the speed distribution function.

2.25. The NO molecule has a doubly degenerate electronic ground state and a doubly degenerate excited state 121 cm^{-1}. Calculate the electronic partition function at zero, 25°C, and very high temperatures.

2.26. For a linear rotor (such as CO_2, HCL, and C_2H_2) in which the atoms are regarded as mass points, the rotation occurs only about an axis perpendicular to the line of atoms. The energy is confined to

$$E_J = hcBJ(J+1)$$

where B is the rotational constant, expressed as a wave number (cm^{-1}). Each J level has a degeneracy of $2J+1$, and the partition function is

$$z_{\text{rot}} = \sum_J (2J+1)e^{-\beta E_J}$$

Calculate the rotational partition function of $^1\text{H}\,^{35}\text{Cl}$ at 25°C for which $B = 10.59$ cm^{-1}.

2.27. The wavenumbers of the three modes of H_2O are $\nu = 3656.7$ cm^{-1}, 1594.8 cm^{-1}, and 3755.8 cm^{-1}, respectively. Evaluate the vibrational partition function at 1500 K.

2.28 Using the Boltzmann distribution calculate the ratio of populations at 25°C of energy levels separated by **(a)** $\nu = 1000$ cm^{-1}, and **(b)** $E = 10$ kJ mol^{-1}.

2.29. The thermodynamic probability for N molecules having two equal a priori aspects is given by

$$W = \frac{N!}{(n/2)!(N/2)!}$$

Using this, and Stirling's approximation, $\ln N! \approx N \ln N - N$, calculate the statistical mechanical value of the entropy of CO attributed to the randomness of orientation of CO molecules in its crystals at absolute zero.

2.30. Evaluate the translational partition function for $H_2(g)$ at 1000 K and 1 bar.

2.31. An equation that is quoted as the first given for an absolute determination of entropy, and whose origin is wholly in statistical thermodynamics, is the well-known Sakur–Tetrode formula

$$S^\circ = R\left\{\frac{5}{2} + \ln\left[\left(\frac{2\pi mkT}{h^2}\right)^{3/2} \frac{V}{N_A}\right]\right\}$$

Use this to calculate the entropy of neon (molar mass $M = 20.179 \times 10^{-3}$ kg mol^{-1}) at 20°C and 1 bar.

2.32. Evaluate the translational partition functions of hydrogen atoms and hydrogen molecules at 500 K in a volume of 4.157×10^{-2} m^{-3}. (This is the molar volume of a perfect gas at this temperature and a pressure or 1 bar).

2.33. What are the translational partition functions for H, H_2, and H_3 at 1000 K and 1 bar? What are the rotational partition functions of H_2 and H_3 (linear) at 1000 K? H_2 has a moment of inertia 4.6054×10^{-48} kg m^2, and H_3 a moment of inertia 2.96×10^{-47}kg m^2.

2.34. The characteristic vibrational and rotational temperatures for HBr are $\theta_v = 3700$ K and $\theta_r = 12.1$ K. What is the ratio of the number of HBr molecules in state $n = 2, J = 5$ to the number in state $n = 1, j = 2$ at 1000 K?

2.35. From the expression for the vibrational partition function

$$z_{\text{vib}} = \frac{1}{1 - e^{-h\nu/kT}}$$

derive the expression for the vibrational contribution to the internal energy

$$U = \frac{RTx}{e^x - 1}$$

where $x = h\nu/kT$.

2.36. Determine the temperature at which 10% of the molecules in a system will be in the first excited electronic state if this state is 400 kJ mol^{-1} above the ground state.

2.37. Evaluate the electronic partition function of Cl(g) at 25°C and 1000 K. The ground state is twofold degenerate. The first excited state is 875.4 cm^{-1} higher in energy and is twofold degenerate.

2.38. Compare the Fermi–Dirac distribution with the classical Maxwell–Boltzmann distribution and the Bose–Einstein distribution, using ϵ_F, the Fermi energy, as reference for the first distribution, and zero as the reference for the other two distributions. Plot all three distributions.

2.39. Using the geometry of quantum space in relation to energy,

$$\left(n_x^2 + n_y^2 + n_z^2\right) = \frac{2mL^2}{\hbar^2\pi^2}\,\epsilon$$

show that the density of orbital states for electrons per unit volume is given by

$$n(\epsilon) = \frac{1}{2\pi^2}\left(\frac{2m}{\hbar^2}\right)^{3/2} \epsilon^{1/2}$$

2.40. For a system comprised of $N = 3$ distinguishable particles of spin zero and having U energy units, the number of assumable microstates is given by

$$g_T = \frac{(3 + U - 1)!}{(3 - 1)!U!} = \frac{(U + 2)(U + 1)}{2}$$

(a) Calculate the values of g_T for $N = 3$ and for selected values of U in the range 0 to 20 energy units.

(b) Do the same thing for the case of three identical bosons of spin zero (Bose–Einstein statistics).

2.41. The ground state of hydrogen has an energy $\epsilon_1 = -13.6$ eV. With two degenerate spin states for 1*s*, two independent states are available, and the statistical weight of the ground state is $g_1 = 2$. The first excited state has an energy $\epsilon_2 = \epsilon_1/2^2 = -13.6\text{ eV}/4 = -3.4$ eV. With two states for 2*s* and six states for 2*p*, eight states are available for $n = 2$, and $g_2 = 8$. In a room of typical size (50 m^3) filled with atomic hydrogen at STP, how many atoms are in the first excited state ($n = 2$)?

2.42. The quantum mechanics of the hydrogen atom show that the degeneracy of the atomic levels is $2n^2$. Thus, $g = 2$ for the ground state ($n = 1$) and $g = 8$ for the first excited state ($n = 2$).

(a) In a gas of atomic hydrogen at room temperature, what is the relative population of the first excited state at $\epsilon = 10.2$ eV? How much hydrogen would be required in order to have a reasonable probability of finding one atom in the first excited state?

(b) At what temperature would it be expected to find 10.0% of the atoms in the first excited state?

2.43. The Fermi energy of electrons in a metal is given by

$$\epsilon_F = \frac{h^2}{2m}\left(\frac{3N}{8\pi V}\right)^{2/3}$$

where V is the volume of the metal. Compute the Fermi energy ϵ_F for sodium.

2.44. **(a)** Show that in the dependence of the partition function on temperature, $z \rightarrow g_0$ as $T \rightarrow 0$, $z \rightarrow \infty$ as $T \rightarrow \infty$.

(b) Show that for a molecule with a set of equally spaced nondegenerate energy levels,

$$z = \frac{1}{1 - e^{-\epsilon/kT}}$$

2.45. HD gas has a characteristic temperature of rotation $\theta_r = 64$ K, and a characteristic temperature of vibration $\theta_v = 6100$ K. Determine:

(a) z_{rot}

(b) ϵ_{rot}

(c) ϵ_{total}, all at 96 K

2.46. The translational energies for a particle in a box are given by

$$\epsilon = \frac{h^2}{8ma^2}(n_x^2 + n_y^2 + n_z^2)$$

The molecular translational partition function is thus

$$z = \sum_{n_x=1}^{\infty}\sum_{n_y=1}^{\infty}\sum_{n_z=1}^{\infty}\exp\left[\frac{h^2}{8ma^2kT}(n_x^2 + n_y^2 + n_z^2)\right]$$

$$= \sum_{n_x=1}^{\infty}\exp\left(-\frac{h^2}{8ma^2kT}n_x^2\right)\sum_{n_y=1}^{\infty}\exp\left(-\frac{h^2}{8ma^2kT}n_y^2\right)\sum_{n_z=1}^{\infty}\exp\left(-\frac{h^2}{8ma^2kT}n_z^2\right)$$

Replacing the sums by integrals

$$\int_0^\infty e^{-ax^2}\,dx = \frac{1}{2}\sqrt{\frac{\pi}{\alpha}}$$

show that

$$z = V\left(\frac{2\pi mkT}{h^2}\right)^{3/2}$$

where $V = a^3$ is the volume. What is the molar partition function?

see p 165

Chapter 3

Classical Assemblies

The formulation of statistical mechanics corresponding to the description of atomic systems by classical and quantum mehcanics are known, respectively, as *classical* and *quantum statistical mechanics*. Quantum-statistical mechanics is the more exact formulation, but classical statistical mechanics is often easier to apply computationally. In fact, whenever integration is used instead of summation for the calculation of results, classical statistical mechanics is tacitly assumed. Thus, the classical method is used when classical and quantum results agree. However, there is no such choice in cases when incorrect results arise from neglect of quantum effects.

Classical statistical mechanics was developed mainly from the work of Boltzmann and Gibbs during the latter part of the nineteenth century, before the advent of quantum mechanics. It is a subject of great beauty and insight, expressed by Liouville's theorem of classical mechanics based on the postulate that the number of microstates is proportional to the accessible volume of phase space. Classical statistical mechanics may be introduced as a limiting case of quantum-statistical mechanics. In this chapter, however, it will be developed independently of quantum-statistical mechanics so that a comparison between the two methods may be established later. It emerges that many successful applications of statistical mechanics, although based on the atomic model of matter, are independent of quantum principles.

At sufficiently high temperatures, for molecules in states with large quantum numbers and the energy difference between adjacent quantum states much smaller than kT, classical mechanics may be used in place of quantum mechanics. Although the latter provides a more accurate description of particle behavior, there are many instances where it is easier to use classical laws with no danger of error. This is usually the case for the translational energy levels (kinetic and potential energies of the center of mass) of a system of molecules, provided neither the molecular mass nor the temperature is too small. However, the classical treatment cannot be used for the distribution of molecules among vibrational or electronic states at normal temperatures because the spacing between such levels is of the order of kT.

3.1 CLASSICAL FORMULATION

In classical statistical mechanics, the instantaneous state of motion of a particle is represented by a point in phase space. For a one-dimensional motion, the phase space is a two-dimensional space whose abscissa is the position x and whose ordinate is the momentum p_x. For a three-dimensional motion, phase space is a six-dimensional space of coordinates x, y, z, p_x, p_y, and p_z. In classical statistics, the energy distribution of a system is characterized by the number of points in each small volume of phase space, say $\Delta x\ \Delta y\ \Delta z\ \Delta p_x\ \Delta p_y\ \Delta p_z$. Such a small volume element is called a *cell* in phase space, and points in the cell have total energy between E and $E + dE$, corresponding to momentum values between p_x and $p_x + \Delta p_x$, and so on, and position values between x and $x + \Delta x$, and so on. The cell is chosen to be small enough that the average total energy of its represenative points differs little from the energy of any one of them; it is chosen large enough so that there are many points in a cell, thereby permitting the application of statistical ideas. Hence, the size of a cell is somewhat arbitrary and indefinite, but once it is chosen the cell is characterized by an average total energy and a population number. The cell then is the classical statistical analogue to the quantum state of quantum statistics. This is illustrated in Fig. 3.1-1 for the one-dimensional case.

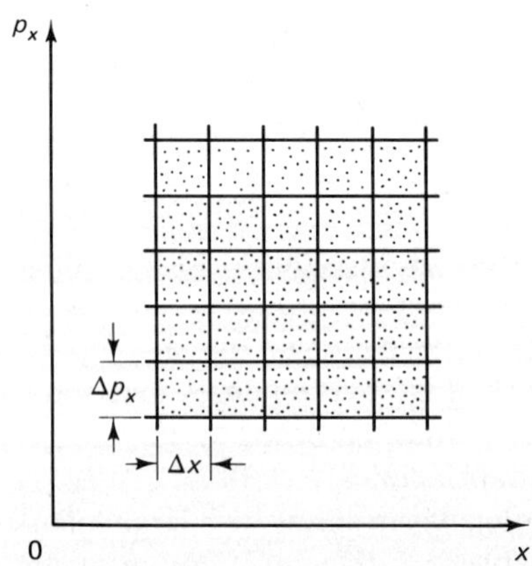

Figure 3.1-1 Phase space and representative points for a one-dimensional system.

If $p_1, p_2, \ldots, p_M$ and $q_1, q_2, \ldots, q_M$ denote the momenta and positions, respectively, of the systems of the assembly, then a typical element in phase space is obtained by specifying infinitesimal ranges for these coordinates, that is, p_1 to $p_1 + dp_1$, p_2 to $p_2 + dp_2$, and so on. The volume of such an element in phase space is

$$dV = dp_1\, dq_1\, dp_2\, dq_2 \cdots dp_M\, dq_M \tag{3.1-1}$$

In this expression M is the total number of positional coordinates required to specify the precise configuration of all the systems of the assembly. Thus, if each system contained n atoms, then for an assembly of N system $M = 3 \cdot n \cdot N$.

The energy associated with any position in phase space is a function of the p's and q's. (For an ideal gas of structureless systems it is a function only of the p's.) Since equal probabilities attach to equal volumes of phase space of the same energy, the probability of an energy state is given by

$$P(\text{state}, dV) = \frac{\exp(-E/kT)dp_1\, dq_1\, dp_2\, dq_2 \cdots dp_M\, dq_M}{\int \cdots \int_{p,q} \exp(-E/kT)\, dp_1\, dq_1 \cdots dp_M\, dq_M} \tag{3.1-2}$$

The denominator of this expression is the classical partition function, or *phase integral*

$$Z_{\text{classical}} = \int \cdots \int \exp(-E/kT)\, dp_1\, dq_1 \cdots dp_M\, dq_M \tag{3.1-3}$$

where the multiple integral is evaluated for the whole accessible region of phase space.

The classical formulation in terms of phase space can be related to the quantum formulation if the phase space is suitably divided into cells of volume h^M when the systems of the assembly are localized, and $h^M \times N!$ if the systems are nonlocalized. Each cell then corresponds to a single quantum state. The value of the classical partition function can be made to correspond with that of the quantum partition function by dividing by the relevant factor. This yields the semiclassical results

$$Z_{\text{crystal}} = h^{-M} \int \cdots \int \exp\left(-\frac{E}{kT}\right) dV \tag{3.1-4}$$

$$Z_{\text{gas}} = \frac{h^{-M}}{N!} \int \cdots \int \exp\left(-\frac{E}{kT}\right) dV \tag{3.1-5}$$

where the symbol dV is defined by Eq. (3.1-1).

When quantum effects are not important Eqs. (3.1-4) and (3.1-5) give the same numerical value of Z as the quantum formulation. If quantum effects are important they give larger values. The use of Eqs. (3.1-4) and (3.1-5) does not, of course, affect the probability that an assembly is found in a volume dV of phase space, for both numerator and denominator of Eq. (3.1-2) are divided by the same factor.

Furthermore, just as in the last chapter, developments lead to equations that relate the thermodynamic propeties of an assembly in a thermostat to the energies of the states allowed to the assembly. These equations, for the quantum partition function

$$Z = \sum_{\substack{\text{quantum}\\\text{states}}} \exp(-E/kT) \tag{3.1-6}$$

now go over to the classical formulation

$$Z = \int_{p,q} \cdots \int \exp(-E/kT)\, dp_1\, dp_2 \cdots dp_M\, dq_M \tag{3.1-7}$$

■ EXAMPLE 3.1-1

The procedure of statistical mechanics starts with, say, an assembly of N identical systems (particles) in a volume V in thermal equilibrium with a heat bath of temperature T. What are the questions asked and the decisions to be made in the analysis of the problem? How does classical statistical mechanics fit in the scheme?

Solution. The basic task of statistical mechanics (statistical thermodynamics) is to take a system in an equilibrium state and compute the various thermodynamic properties of that system from an assumed microscopic model. The questions to be asked and decisions to be made are represented by the "tree of choice" below.

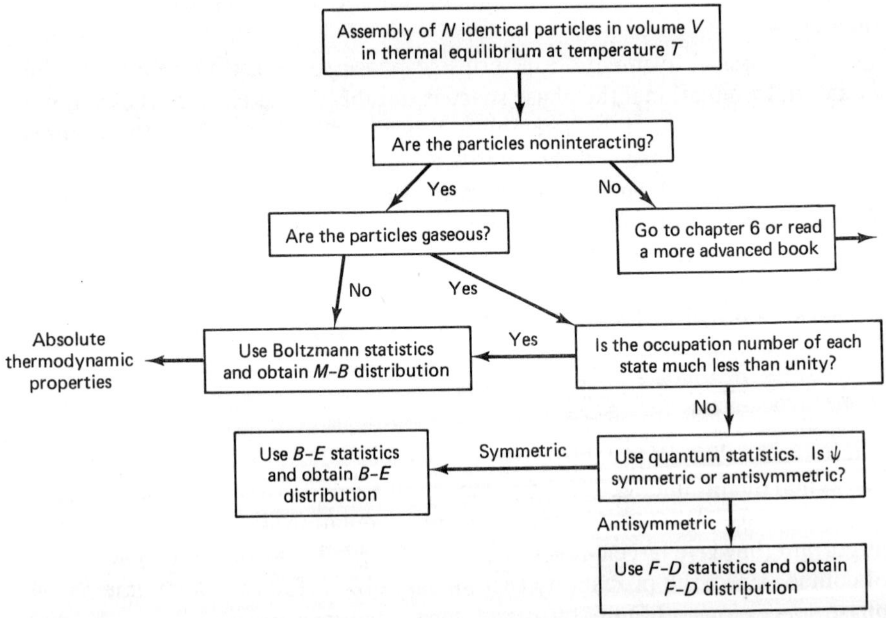

Classical statistical mechanics was developed before the advent of quantum mechanics. It is based on the assumption that particles are identical but distinguishable (as by the lattice positions in a solid). In a gas, the particles occupy a large macroscopic volume V and are indistinguishable. The methods of classical statistical mechanics, however, can still be applied to a gas, because the energy levels of a gas particle are extremely close together. ■

3.2 PERFECT CLASSICAL GAS

In previous discussions, a perfect gas has often been used to illustrate general results. In this chapter, the perfect classical gas will be studied in more detail. It will be exactly defined and its partition function obtained. The conditions under which the classical approximation holds will be stated, and derivations made for the two equations that thermodynamically characterize the perfect classical gas: the equation of state $PV = NkT$, and Joule's law $E = E(T)$, which states that the internal energy E is independent of the volume and the pressure of the gas. It will develop that classical statistical mechanics becomes the natural generalization of the treatment of a perfect gas.

Conceptually, a gas consists of molecules moving about fairly freely in space. The molecules are rather far from each other and interact only very weakly. At any instant only a small fraction are interacting strongly, that is, are colliding. Typical values for the mean separation between molecules and for the molecular diameter are of the order of 30 Å and 3 Å, respectively. At sufficiently low density, the molecules of a gas will interact very little with one another.

The perfect gas represents an idealization in which the potential energy of interaction between the molecules is negligible compared to their kinetic energy of motion. Since the interaction energy is negligible, it is an immediate consequence that the energy of a perfect gas subdivides into a sum of energies for each molecule. These energies

$$\varepsilon_1 \le \varepsilon_2 \le \cdots \le \varepsilon_r \le \cdots \tag{3.2-1}$$

correspond to the complete set of discrete quantum states, labeled

$$1, 2, \ldots, r, \ldots \tag{3.2-2}$$

in which a single molecule can exist. (In classical mechanics, the energies form a continuous set.)

Consider an enclosure of volume V, containing N molecules of gas, the enclosure being at temperature T, that is, immersed in a heat bath at temperature T. Let the state of the gas be specified by listing the number of molecules in each possible individual state; n_1 molecules in state 1, n_2 molecules in state 2, and n_r in the general state r. (n_r is called the occupation number of the rth state.) The energy of the gas in this state, determined by the occupation numbers $n_1, n_2, \ldots, n_r, \ldots$, is

$$E(n_1, n_2, \ldots, n_r, \ldots) = \sum_r n_r \varepsilon_r \tag{3.2-3}$$

where

$$N = \sum_r n_r \tag{3.2-4}$$

Since there are no interactions, the partition function for the gas may be expressed in terms of the partition function of a single molecule in the container of volume V:

$$Z_1(T, V) \equiv Z(T, V, 1) = \sum_r \exp(-\beta\varepsilon_r) \tag{3.2-5}$$

For a gas composed of N indistinguishable molecules, the correct partition function is given by

$$Z \equiv Z(T, V, N) = \frac{1}{N!}\left[\sum_r \exp(-\beta\varepsilon_r)\right]^N = \frac{1}{N!}[Z_1(T, V)]^N \tag{3.2-6}$$

where the division by $N!$ is made so that each distinct state of the N particle gas is counted exactly once, because the molecules are nonlocalized, and therefore, indistinguishable. Equation (3.2-6) is the basic result from which all properties of a perfect gas follow easily.

For a polyatomic molecule, assuming that all modes of energy storage are completely referable, the total energy of the molecule system is

$$\varepsilon = \varepsilon_i^{\text{trans}} + \varepsilon_j^{\text{rot}} + \varepsilon_k^{\text{vib}} + \varepsilon_l^{\text{elec}} + \varepsilon_m^{\text{nuc}} \tag{3.2-7}$$

and the canonical ensemble partition function is

$$Z = \frac{1}{N!}(z_{\text{trans}}\, z_{\text{rot}}\, z_{\text{vib}}\, z_{\text{elec}}\, z_{\text{nuc}})^N \tag{3.2-8}$$

$$= Z_{\text{trans}}\, Z_{\text{rot}}\, Z_{\text{vib}}\, Z_{\text{elec}}\, Z_{\text{nuc}} \tag{3.2-9}$$

where

$$Z_{\text{trans}} = \frac{z_{\text{trans}}^N}{N!} \tag{3.2-10}$$

$$Z_{\text{rot}} = z_{\text{rot}}^N \tag{3.2-11}$$

$$Z_{\text{vib}} = z_{\text{vib}}^N \tag{3.2-12}$$

$$Z_{\text{elec}} = z_{\text{elec}}^N \tag{3.2-13}$$

$$Z_{\text{nuc}} = z_{\text{nuc}}^N \tag{3.2-14}$$

Note that the division by $N!$ is included in the translational factor as seen from Eq. (3.2-6).

These results characterize the classical regime, in which the single-particle states are mostly empty and very few contain one or more molecules. Alternately, the results hold for high temperatures and low densities. This is the classical limit set by Maxwell and Boltzmann.

In classical mechanics, the state of a system (particle) has a simple interpretation. For example, a particle moving in one dimension along the x axis is described at any instant of time by its coordinate x and its momentum p_x, so that in the course of time, it traces a trajectory in the two-dimensional space with x and p_x as orthogonal axes. This concept, when generalized to motions in more than one dimension, defines the phase space and is the starting point of classical statistical mechanics.

■ EXAMPLE 3.2-1

Describe the motion and trajectory of the representative point in phase space for a particle in a one-dimensional box.

Solution. According to quantum mechanics the state of any system is completely described by the wave function for the system. Thus, while the energy of a system may be sharp, its position is indeterminate. In classical mechanics there is no restriction on the precision with which conjugate variables such as position and momentum can be specified. Indeed the proper classical specification of the state of any system requires exact information about both its position and its momentum at any instant. The classical picture of a system thus contains more information than the quantum-mechanical picture.

For the purpose of molecular thermodynamics, it is convenient to describe the classical motions of systems in terms of their momenta and spatial coordinates, rather than in terms of their velocities and positions. Furthermore when several particles or systems are considered, their combined motions are represented by the motion of a single point in phase space. This is a multidimensional space in which distances along different axes give the positional and momentum coordinates of the systems of the assembly. It is illustrated by Fig. 3.2-1.

A particle is in a one-dimensional box. Its motion is back and forth along a line of length L as shown in Fig. 3.2-1(a). The instantaneous state of the particle is specified by stating its position, q, and its momentum, p, as in Fig. 3.2-1(b). Any point in the (p,q)-plane then represents a particular position and momentum of the particle, that is, the instantaneous state of the particle. As the particle moves along the line in Fig. 3.2-1(a), its momentum remains constant until it collides with the wall at $q = L$, whereupon its momentum changes abruptly from $+q$ to $-q$. The particle then moves back along the line

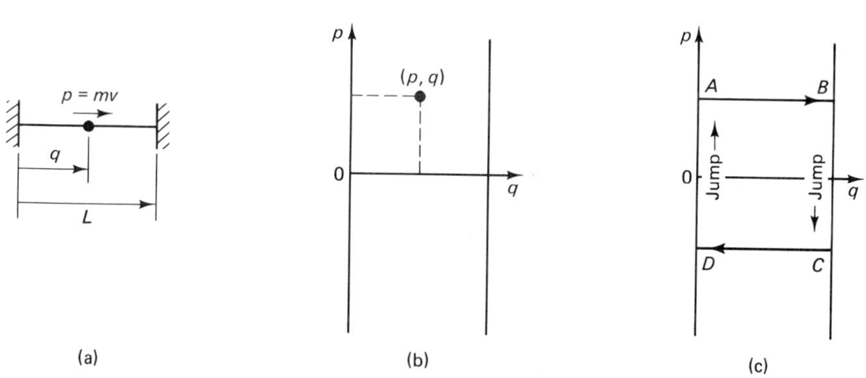

Figure 3.2-1 Phase space for motion of a particle along a line.

until $q = 0$ when the momentum again abruptly changes sign. The representative point of the system thus executes the cycle $ABCDA$ in Fig. 3.2-1(c). The motion of the point along AB and CD is regular but the motions from B to C and from D to A are jumps. The (p,q)-plane in Figs. 3.2-1(b) and 3.2-1(c) is the phase plane or the phase space for the system, and the cycle $ABCDA$ is called the trajectory of the representative point in phase space. ■

3.3 PHASE SPACE

Classical statistical mechanics begins by assigning to each particle generalized space and momentum coordinates. Consider a particle moving in one dimension, along the x axis, say. According to classical mechanics, its state of motion at any instant can be established by specifying its position coordinate x and its momentum p_x. Its motion is described by a trajectory such as is shown in Fig. 3.3-1, with x and p_x as orthogonal axes. This two-dimensional space, with x and p_x as axes, is called the *phase space* of the particle.

Any point in the phase space represents a definite state of motion. The same geometrical language is used to describe the three-dimensional motion of a particle. Phase space is then six-dimensional with space coordinates x, y, z and momentum coordinates, p_x, p_y, p_z as orthogonal axes. Again each point in phase space represents a state of motion of the particle. In classical mechanics every point represents a possible state of motion, so it is reasonable to say that the number of states in a given region of phase space is proportional to the six-dimensional volume of that region. In particular, the number of states where the particle has position and momentum coordinates within the ranges

$$\left.\begin{array}{r} x, x+dx;\ y, y+dy;\ z, z+dz; \\ p_x, p_x+dp_x;\ p_y, p_y+dp_y;\ p_z, p_z+dp_z; \end{array}\right\} \tag{3.3-1}$$

is proportional to the volume of this element of phase space, that is, to

$$(dx\, dy\, dx)(dp_x\, dp_y\, dp_z) = d^3r d^3p \tag{3.3-2}$$

This is essentially the description used by Maxwell, Boltzmann, and Gibbs in classical statistical mechanics. The constant of proportionality, however,

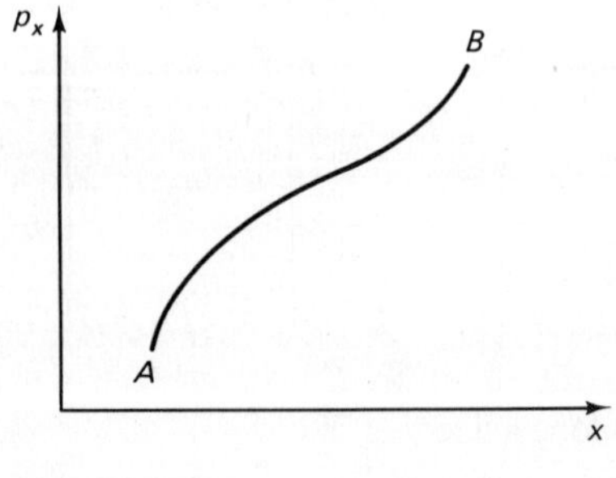

Figure 3.3-1 Phase space for a particle moving in one dimension.

they could not fix, because classically the possible states form a continuum. Quantum mechanics, on the other hand, gives the number of states for a molecule in an enclosure V having momentum between p and $p + dp$ and kinetic energy between ε and $\varepsilon + d\varepsilon$ to be

$$f(p)\,dp = V\frac{4\pi p^2\,dp}{h^3} \tag{3.3-3}$$

The constant of proportionality can be found by relating Eqs. (3.3-2) and (3.3-3). The volume $d\Gamma$ in phase space for which the particle in the enclosure V has momentum within the range p to $p + dp$ is, from Eq. (3.3-2),

$$d\Gamma = \int_V d^3r \int_p^{p+dp} d^3p \tag{3.3-4}$$

The integral in coordinate space is simply V, while the integral in momentum space is the volume of a spherical shell with radii p and $p + dp$, that is,

$$\int_p^{p+dp} d^3p = 4\pi p^2\,dp \tag{3.3-5}$$

Equation (3.3-4) gives

$$d\Gamma = V(4\pi p^2\,dp) \tag{3.3-6}$$

Comparing this with Eq. (3.3-3), it is seen that phase space can be thought of as subdivided into cells of volume h^3. Each such cell represents one quantum state, so that

$$\frac{1}{h^3}\,d^3r\,d^3p \tag{3.3-7}$$

is the number of states within a volume $d^3r\,d^3p$ of phase space. (Converting continuous classical phase space into this kind of quantum-mechanical phase space was proposed by Planck in connection with his theory of blackbody radiation at the beginning of this century, before the advent of quantum mechanics. It was unforeseen by the early masters of statistical mechanics such as Maxwell, Boltzmann, and Gibbs.)

Generally, for a particle with F degrees of freedom, the state of motion at a given time is specified by $2F$ coordinates: $q_1, \ldots, q_F$ and $p_1, \ldots, p_F$. This is a $2F$-dimensional phase space with the F generalized coordinates $q_1, \ldots, q_F$ and the F conjugate momenta $p_1, \ldots, p_F$ as coordinate axes. Each point in this phase space corresponds to a definite state of motion of the system. Hence, the number of states in a given region of phase space will be proportional to the volume of that region of phase space. The constant of proportionality cannot be obtained within the framework of classical mechanics, but with quantum mechanics foresight, the number of states of the system with coordinates and momenta lying within the intervals $q_1, q_1 + dq_1; \cdots q_F, q_F + dq_F$; $p_1, p_1 + dq_1; \cdots p_F, p_F + dp_F$; is taken as

$$\frac{1}{h^F}\,dq_1 \cdots dq_F\,dp_1 \cdots dp_F \tag{3.3-8}$$

■ **EXAMPLE 3.3-1**

Consider a one-dimensional harmonic oscillator of mass m and spring constant k, and discuss it in terms of classical mechanics. Denote the displacement coordinate of the oscillator by x and its linear momentum by p.

Solution. Phase space here is two-dimensional. The total energy E (called the Hamiltonian) of the oscillator is

$$E = \frac{p^2}{2m} + \frac{1}{2}kx^2 \tag{3.3-9}$$

where the first term on the right is its kinetic, the second term its potential energy. For a constant energy E, Eq. (3.3-9) describes an ellipse in phase space, that is, in the p-x plane. Suppose that the energy of the oscillator is known to be within the small range between E and $E + \delta E$. Then there are still many cells in phase space contained between the two ellipses corresponding to the respective energies E and $E + \delta E$; that is, many different sets of x and p are possible for the oscillators in a representative ensemble. If the only information available about the oscillator is that it is in equilibrium with an energy in the specified range, then the statistical postulate asserts that it is equally probable that the oscillator has values of x and p lying within any one of these cells.

The phase-space trajectory is an ellipse in the x-p plane

$$\frac{p^2}{2m} + \frac{kx^2}{2} = E \tag{3.3-10}$$

or

$$\frac{p^2}{a^2} + \frac{x^2}{b^2} = 1 \tag{3.3-11}$$

where $a = \pm\sqrt{2mE}$ and $b = \pm\sqrt{2E/K}$. The generalized coordinates are $q = x$ and $p = p_x$. Figure 3.3-2 shows the one-dimensional classical harmonic oscillator with energy between E and $E + \delta E$. The accessible region of phase space consists of the area lying between the two ellipses.

Note that a given interval dx corresponds to a larger number of cells (i.e.,

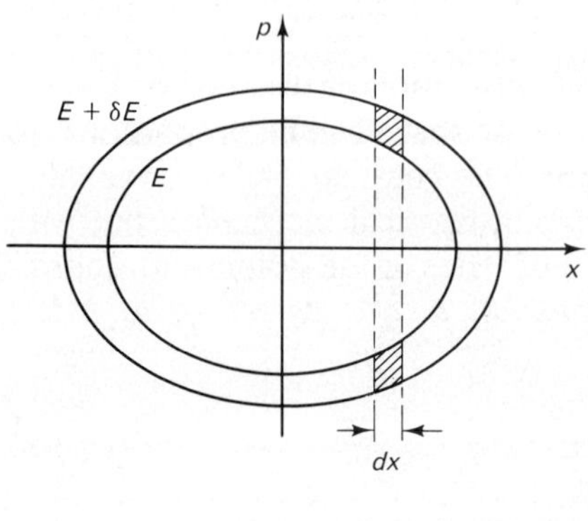

Figure 3.3-2 Phase-space trajectory for one-dimensional harmonic oscillator.

a larger area) lying between the two ellipses when $x \approx \pm a$ than when $x \approx 0$. Hence, it is more probable than an oscillator in the ensemble is found with its position x close to $\pm a$ than close to 0. This result is, of course, also obvious from the fact that near the extremes of its position, where $x \approx \pm a$, the oscillator has small velocity; hence, it spends a longer time there than near $x \approx 0$, where it moves rapidly. ■

■ EXAMPLE 3.3-2

The Hamiltonian function of a system is defined as the sum of the kinetic energy and the potential energy expressed entirely in terms of the p's and q's. In general, each particular choice of q's has its own conjugate p's. The general rule for finding the proper p's are as follows:

1. Express the kinetic energy in terms of the $\dot{q}$'s. This gives

$$\varepsilon_{\text{kin}} = \varepsilon_{\text{kin}}(\dot{q}_1, \ldots, \dot{q}_f) \tag{3.3-12}$$

2. By definition the p's are given as

$$p_i = \frac{\partial \varepsilon_{\text{kin}}}{\partial \dot{q}_i} \qquad i = 1, \ldots, f \tag{3.3-13}$$

That is, differentiate the kinetic energy partially with respect to the $\dot{q}$'s to obtain the conjugate p's.

Using this procedure, find the Hamiltonian for the following systems: (a) a free point mass in space; (b) a symmetric rotor with a fixed axis; (c) a point mass moving in space about the origin.

Solution

(a) *Free Point Mass in Space.* Let a point mass be located by the Cartesian coordinates x, y, and z. Then

$$\varepsilon_{\text{kin}} = \varepsilon_{\text{kin}}(\dot{q}_1, \dot{q}_2, \dot{q}_3) = \frac{1}{2m(\dot{x}^2 + \dot{y}^2 + \dot{z}^2)} \tag{3.3-14}$$

and

$$p_1 = \frac{\partial \varepsilon_{\text{kin}}}{\partial \dot{q}_1} = m\dot{x}$$

$$p_2 = \frac{\partial \varepsilon_{\text{kin}}}{\partial \dot{q}_2} = m\dot{y} \tag{3.3-15}$$

$$p_3 = \frac{\partial \varepsilon_{\text{kin}}}{\partial \dot{q}_3} = m\dot{z}$$

in agreement with the previous statement. Solving these for the $\dot{q}$'s and expressing the kinetic energy in terms of the p's gives

$$\varepsilon_{\text{kin}} = \frac{1}{2m}(p_1^2 + p_2^2 + p_3^2) \tag{3.3-16}$$

In the absence of any potential energy term, this is the total energy of the particle, and so

$$H = \frac{1}{2m}(p_1^2 + p_2^2 + p_3^2) \tag{3.3-17}$$

(b) *Symmetric Rotor with a Fixed Axis.* In this case the q and $\dot{q}$ coordinates are naturally taken as θ and $\dot{\theta}$. In terms of $\dot{\theta}$ the kinetic energy expression is

$$\varepsilon_{\text{kin}} = \frac{1}{2I\dot{\theta}^2} \tag{3.3-18}$$

where I is the moment of inertia about the given axis. Defining p_θ as

$$p_\theta = \frac{\partial \varepsilon_{\text{kin}}(\dot{\theta})}{\partial \dot{\theta}} = I\dot{\theta} \tag{3.3.-19}$$

yields the ordinary angular momentum of the system. The status of the system is given in a two-dimensional phase space with orthogonal axes for θ and p_θ. The Hamiltonian function for the system is found by taking the sum

$$\varepsilon^{\text{kin}}(\dot{\theta}) + \varepsilon^{\text{pot}}(\theta) \tag{3.3-20}$$

and eliminating $\dot{\theta}$ in favor of p_θ. The resulting sum is the Hamiltonian function

$$H = H(p, q) = \frac{1}{2I}p_\theta^2 + \varepsilon^{\text{pot}}(\theta) \tag{3.3-21}$$

If there is no force acting, $\varepsilon^{\text{pot}} = 0$, and H reduces to a single squared term.

(c) *Mass Point Moving in Space About the Origin.* Let the Cartesian coordinates of the mass point be replaced by the spherical coordinates, r, θ, and ϕ. These then constitute the q's,

$$\begin{aligned} x &= r \sin\theta \cos\phi \\ y &= r \sin\theta \sin\theta \\ z &= r \cos\theta \end{aligned} \tag{3.3-22}$$

The kinetic energy can always be written directly in terms of the Cartesian velocities, $\dot{x}$, $\dot{y}$, and $\dot{z}$, as

$$\varepsilon_{\text{kin}} = \frac{1}{2m}(\dot{x}^2 + \dot{y}^2 + \dot{z}^2) \tag{3.3-23}$$

On differentiating (3.3-22) with respect to time and substituting in (3.3-23), there results, after considerable reduction,

$$\varepsilon_{\text{kin}} = \varepsilon_{\text{kin}}(\dot{q}_1, \dot{q}_2, \dot{q}_3) = \frac{m}{2}(\dot{r}^2 + \dot{r}^2\dot{\theta}^2 + r^2 \sin^2\theta\dot{\phi}^2) \tag{3.3-24}$$

The moments conjugate to r, θ, and ϕ are then formed by differentiation of the expression to give

$$p_r = m\dot{r}$$

$$p_\theta = mr^2\dot{\theta}$$
$$p_\phi = mr^2 \sin^2 \theta\dot{\phi} \qquad (3.3\text{-}25)$$

The coordinates r, θ, ϕ, p_r, and p_ϕ are thus a proper set of dynamic or Hamiltonian coordinates for this problem.

To find the Hamiltonian function, Eq. (3.3-25) is solved for $\dot{r}$, $\dot{\theta}$, and $\dot{\phi}$ and the ε_{kin} expressed in terms of the momenta. On adding to this the potential function $\varepsilon^{pot}(r, \theta, \phi)$, there results

$$H = H(p_i q_i) = \frac{1}{2m} p_r^2 + \frac{1}{2mr^2} p_\theta^2 + \frac{1}{2mr^2 \sin^2 \theta} p_\phi^2 + \varepsilon^{pot}(r, \theta, \phi) \qquad (3.3\text{-}26)$$

■

■ EXAMPLE 3.3-3

Select generalized coordinates for the motion of a diatomic molecule consisting of two atoms joined together.

Solution. A diatomic molecule may be regarded as consisting of two point masses, m_1 and m_2, joined together. The kinetic energy can be expressed as the sum of terms in Cartesian coordinates, one term for each of the point masses:

$$T = \tfrac{1}{2} m_1(\dot{x}_1^2 + \dot{y}_1^2 + \dot{z}_2^2) + \tfrac{1}{2} m_2(\dot{x}_2^2 + \dot{y}_2^2 + \dot{z}_2^2)$$

Choose the following (generalized) coordinates: r is the distance between the point masses; X, Y, and Z are the coordinates of the center of mass of m_1 and m_2; θ, and ϕ are polar angles of the line joining the mass points, referred to the x, y, and z axes as shown in Fig. 3.3-3. It is found that

$$T = \frac{m_1 + m_2}{2}(\dot{X}^2 + \dot{Y}^2 + \dot{Z}^2) + \frac{r^2}{2} \frac{m_1 m_2}{m_1 + m_2}(\dot{\theta} + \sin^2 \theta\dot{\phi}^2) + \frac{1}{2} \frac{m_1 m_2}{m_1 + m_2} \dot{r}^2 \qquad (3.3\text{-}27)$$

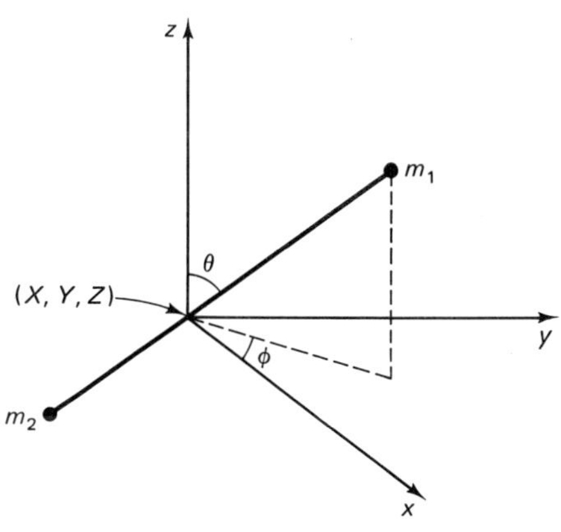

Figure 3.3-3 Coordinate system for a diatomic molecule.

This can be expressed in a simpler form, because $m_1 + m_2$ is the total mass m of the molecule, $m_1m_2/(m_1 + m_2)$ is the reduced mass μ, and μr^2 is the moment of inertia I of the molecule. Thus,

$$T = \tfrac{1}{2} m(\dot{X}^2 + \dot{Y}^2 + \dot{Z}^2) + \tfrac{1}{2} I(\dot{\theta}^2 + \sin^2\theta\dot{\phi}^2) + \tfrac{1}{2}\mu\dot{r}^2 \tag{3.3-28}$$

The three terms on the right-hand side of Eq. (3.3-28) have simple physical meanings. The first is the energy of translation of the molecule as a whole, the second is the energy of rotation, and the third is the energy of vibration.

The velocities can be eliminated by introducing the conjugate momenta. By differentiation of Eq. (3.3-28) with respect to the various generalized velocities;

$$\begin{aligned} p_x &= m\dot{X}; \quad & p_y &= m\dot{Y}; \quad & p_z &= m\dot{Z} \\ p_\theta &= I\dot{\theta}; \quad & p_\phi &= I\sin^2\theta\dot{\phi}; \quad & p_r &= \mu\dot{r} \end{aligned} \tag{3.3-29}$$

Substituting back in (3.3-28) then yields

$$T = \frac{1}{2}m(p_x^2 + p_y^2 + p_z^2) + \frac{1}{2I}\left(p_\theta^2 + \frac{1}{\sin^2\theta}p_\phi^2\right) + \frac{1}{2\mu}p_r^2 \tag{3.3-30}$$

Equation (3.3-30) will be referred to later on. ■

3.4 LIOUVILLE'S THEOREM

By use of the generalized coordinates of space and momenta as the dimensions of phase space, it is possible to derive Liouville's theorem, which states that if particles are followed in phase space, the density of the representative points does not change with respect to time.

Consider an ensemble or a system of N molecules with s degrees of freedom. As given in classical mechanics, the motion of these molecules is described by Hamilton's equations of motion,

$$\dot{q}_1 = \frac{\partial H}{\partial p_i}, \qquad \dot{p}_i = -\frac{\partial H}{\partial q_i}, \qquad i = 1, \ldots, s \tag{3.4-1}$$

which constitute $2s$ first-order differential equations in $2s$ unknowns. Once all the values of the p_i and q_i are known at any one time, the motion of these molecules is completely known for all time. Thus, in principle, the macroscopic properties of any system can be determined by a detailed investigation of the microscopic motion of the molecules constituting the system.

The behavior, in time, of the system can be graphically represented by means of a trajectory in a $2s$-dimensional space that is determined by the $2s$ coordinates $p_1, \ldots, p_s$ and $q_1, \ldots, q_s$. At any time the instantaneous state of the system is specified by the position of a representative point in this space (the phase space of the system). Sometimes a phase space is used to represent the motion of a single molecule contained in the system. Or the system could be described by plotting the orbit of each of N molecules in this phase space. The macroscopic properties of the system are then time averaged over a segment of the trajectory in phase space.

Following the procedure of Gibbs, a large number of similar systems is considered, each consisting of N molecules and s degrees of freedom, but whose representative points in phase space are suitably random so that every state accessible to the actual system in the course of time is represented by at least one system at any instant of time. For a large enough assembly of systems, there is a density $\rho(p_i, q_i, t)$ with which the representative points are distributed in phase space. Replacing time averages over a single system by ensemble averages over the ensemble at a fixed time, the average value of a function $A(p,q)$ is then

$$\langle A \rangle = \frac{\int A(p_i,q_i)\rho(p_i,q_i)\, dq_1 \cdots dp_s}{\int \rho(p_i,q_i)\, dq_1 \cdots dp_s} \tag{3.4-2}$$

In considering how $\rho(p_i,q_i,t)$ changes with time, distinction must be made between its rate of change with respect to a fixed point in phase space, which is $\partial\rho/\partial t$ since p_i and q_i will not vary, and its rate of change in a coordinate system moving with a particle through phase space, $d\rho/dt$. These two rates of change are related by

$$\frac{d\rho}{dt} = \frac{\partial \rho}{\partial t} + \sum_i \left(\frac{\partial \rho}{\partial p_i} \dot{p}_i + \frac{\partial \rho}{\partial q_i} \dot{q}_i \right) \tag{3.4-3}$$

It will be shown that $d\rho/dt = 0$. This is *Liouville's theorem* (1838).

To proceed with the proof of Liouville's theorem, the number of representative points N in a volume of δV at time t is given by

$$N = \rho(p_i,q_i,t)\, \delta V \tag{3.4-4}$$

This number will change with time because the number of representative points entering δV through any face will, in general, be different from the number leaving through the opposite face. The number of representative points entering δV during unit time through the face at q_i is

$$\rho \dot{q}_i \frac{\delta V}{\delta q_i} \tag{3.4-5}$$

and the number of points leaving δV through the opposite face is

$$\left(\rho + \frac{\partial \rho}{\partial q_i} \delta q_i\right)\left(\dot{q}_i + \frac{\partial \dot{q}_i}{\partial q_i}\right) \frac{\delta V}{\delta q_i} \tag{3.4-6}$$

where $\delta V/\delta q_i$ is the area of the face (i.e., δV divided by δq_i is an area, and is not to be confused with a partial derivative such as $\partial\rho/\partial q_i$).

The rate of change with time in the number of representative points lying within δV is thus

$$\frac{\partial N}{\partial t} = -\sum_i \left[\rho\left(\frac{\partial \dot{q}_i}{\partial q_i} + \frac{\partial \dot{p}_i}{\partial p_i} \right) + \left(\frac{\partial \rho}{\partial q_i} \dot{q}_i + \frac{\partial \rho}{\partial p_i} \dot{p}_i \right) \right] \delta V \tag{3.4-7}$$

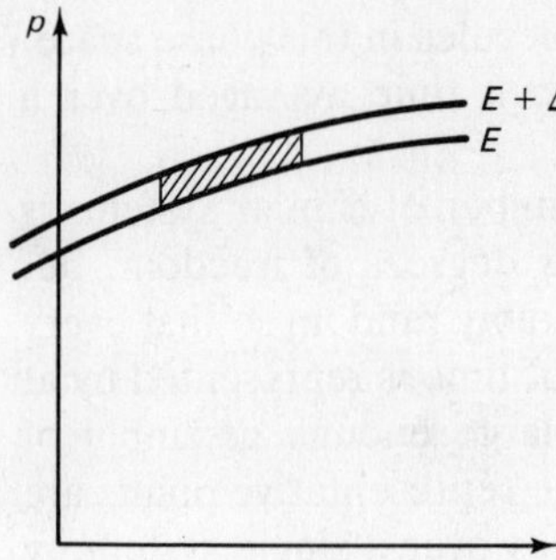

Figure 3.4-1 Incompressible flow in phase space.

neglecting second-order terms. From Hamilton's equations

$$\dot{q}_i = \frac{\partial H}{\partial p_i} \qquad \dot{p}_i = -\frac{\partial H}{\partial q_i} \tag{3.4-8}$$

it follows that

$$\frac{\partial \dot{q}_i}{\partial q_i} = \frac{\partial^2 H}{\partial q_i \, \partial p_i} = -\frac{\partial \dot{p}_i}{\partial p_i} \tag{3.4-9}$$

Hence, substituting Eq. (3.4-9) in Eq. (3.4-7) yields

$$\frac{\partial N}{\partial t} = -\sum_i \left(\frac{\partial \rho}{\partial q_i} \dot{q}_i + \frac{\partial \rho}{\partial p_i} \dot{p}_i \right) \delta V \tag{3.4-10}$$

Dividing through by δV gives

$$\frac{\partial N}{\partial t} = -\sum_i \left(\frac{\partial \rho}{\partial q_i} \dot{q}_i + \frac{\partial \rho}{\partial p_i} \dot{p}_i \right) \tag{3.4-11}$$

which combined with Eq. (3.4-1) yields

$$\frac{d\rho}{dt} = 0 \tag{3.4-12}$$

this completes the proof of Liouville's theorem.

Thus, the density in phase space remains constant in the neighborhood of a particle that is moving in phase space. An equivalent statement of this theorem is that if $\delta V'$ is the volume occupied at time $t + dt$ by the representative points that occupied a volume δV at time t, then $\delta V = \delta V'$. If the system to be represented by an ensemble is in thermal equilibrium, the ensemble averages must be independent of time, that is,

$$\frac{\partial \rho}{\partial t} = 0 \tag{3.4-13}$$

An ensemble satisfying Eq. (3.4-13) is called a *stationary ensemble*.

It is this property of conservation of volume in phase space during the motion that distinguishes phase space from any other space in which the motion of the system could be represented. Hamilton's equations, which hold for the coordinates q,p, of phase space give it this unique quality.

The situation is depicted graphically in Fig. 3.4-1, where the total energy of the gas lies between E and $E + \Delta E$. By Liouville's theorem, the distribution of representative points moves about in this region like an incompressible fluid. The density of representative points of the ensemble remains uniform for all times.

■ EXAMPLE 3.4-1

What is the trajectory in phase space for the one-dimensional free-fall motion of a point particle of mass m?

Solution. For this case the energy (kinetic + potential) is given by

$$H = \frac{p^2}{2m} + mgz = E \tag{3.4-14}$$

Substitution into the canonical equations of motion

$$\dot{p}_i = \frac{\partial H}{\partial q_i} \tag{3.4-15}$$

$$\dot{q}_i = \frac{\partial H}{\partial p_i} \tag{3.4-16}$$

leads to

$$\begin{aligned} \dot{p} &= -mg \\ p &= p_o - mgt \end{aligned} \tag{3.4-17}$$

$$\begin{aligned} \dot{z} &= \frac{p}{m} = \frac{p_o}{m} - gt \\ z &= \frac{p_o t}{m} - \frac{1}{2} gt^2 + z_o \end{aligned} \tag{3.4-18}$$

Liouville's theorem says that a phase point in the $p-z$ plane representing the system moves in the same way as the particles of a two-dimensional incompressible fluid having constant density. The trajectory of the particle for the system energy given by (3.4-14) is seen to be a parabola. The same information is given in parametric form by (3.4-17) and (3.4-18).

Consider now an infinite number of bodies, all having the same mass m and filling the energy continuum between E and $E + \Delta E$. Then all the phase points at $t = 0$ occupy the shaded rectangle at (p_o, z_o) as shown in Fig. 3.4-2.

After a time t the phase points have moved to the region determined by (p,z) and the parallelograms (no longer a rectangle) is determined by the four lines

$$p = p_o - mgt$$

$$p = p_o - mgt + \Delta p_o$$

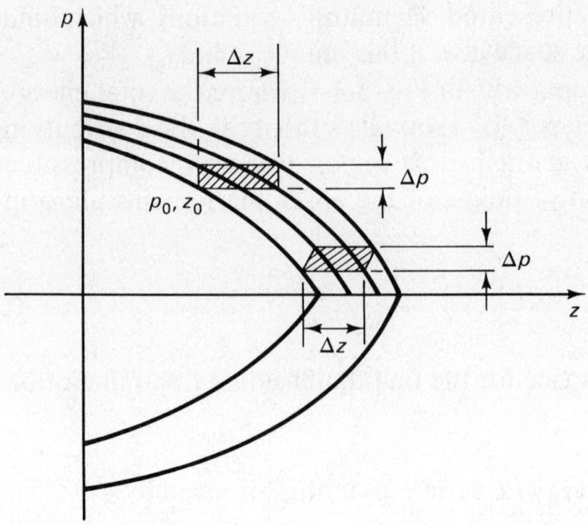

Figure 3.4-2. Motion of point particles in phase space under action of constant gravitational force.

$$z = z_o + \frac{pt}{m} + \frac{1}{2} g t^2$$

$$z = z_o + \frac{pt}{m} + \frac{1}{2} g t^2 + \Delta z_o$$

The parallelogram has a base of length Δz_o and height Δp_o. The sides are inclined against the base at an angle $\delta = \tan^{-1}(m/t)$. ■

3.5 BASIC ASSUMPTIONS

It has been shown that the dynamical state of a gas can be represented by a set of points in a phase space of generalized coordinates q and conjugate momenta p appropriate to the description of the state of a single particle. In the absence of collisions, these representative points move through phase space in the fashion of an incompressible fluid.

To add to the anonymity of different portions of phase space, the latter can be divided into regions of equal size in such a way that the time spent by a representative point in traveling any one region is the same as that spent in traversing any other. This assumption of equal a priori probability for equal volumes in phase space is the fundamental postulate of classical statistical mechanics.

This is generalized and strengthened by the ergodic hypotheses of Maxwell and Boltzmann, which say that the representative point for an isolated system (constant V, N, E) visits every accessible point in phase space before returning to its starting point. Accessibility is, in this case, restricted by the constant energy requirement, whereby

$$H(q,p) = E \qquad (3.5\text{-}1)$$

Each system of an ensemble will, in the course of a sufficiently long period of time, assume the complexion of every other member of the ensemble. Thus, the time averaging of any thermodynamic property (for example, pressure, energy, entropy, and so on) is identical to the instantaneous averaging over the whole ensemble.

Using an overbar to denote time average, the value of a dynamical quantity is given by

$$\overline{X(q,p)} = \frac{1}{\tau}\int_0^{\tau} X[q,(t),p(t)]\, dt \tag{3.5-2}$$

where τ represents some sufficiently long time to enable a meaningful measurement of the thermodynamic property under consideration. The ensemble average, on the other hand, is defined by

$$\langle X(q,p)\rangle = \int \cdots \int \rho(q,p,t) X(q,p)\, dq_1 \cdots dp_F \tag{3.5-3}$$

representing the average value of $X(q,p)$ among all the systems of the ensemble at some instant of time. In view of the above discussion,

$$\overline{X(q,p)} = \langle X(q,p)\rangle \tag{3.5-4}$$

The averaging involves the partition function, whose quantal form is

$$Z = \sum_{E_r} g(E_r) \exp(-\beta E_r) \tag{3.5-5}$$

that is, a sum over discrete energy levels. According to classical mechanics, a system does not have discrete energy levels. The energy, like all other variables characterizing the state of the system is a continuous variable. To express Eq. (3.5-5) in classical terms, a classical way of denoting energy states and counting them must be found.

For a system with F degrees of freedom, the state of motion at any given time is specified by $2F$ coordinates: $q_1, \ldots, q_F$ and $p_1, \ldots, p_F$. The number of states of the system with coordinates and momenta lying in the intervals q_1, $q_1 + dq_1$; $\cdots$ q_F, $q_F + dq_F$; p_1, $p_1 + dp_1$; $\cdots$ p_F, $p_F + dp_F$; is given by

$$\frac{1}{h^F} dq_1 \cdots dq_F\, dp_1 \cdots dp_F \tag{3.5-6}$$

Equation (3.5-6) gives the density of states in phase space. Classically each point of phase space represents a definite state. Quantum-mechanically the states have a discrete nature. The phase space can be considered as subdivided into cells of volume h^F each. Each cell represents one state. Using this density of states, the expression for the partition function, Eq. (3.5-5), can be translated into classical language. Equation (3.5-5) is a sum of Boltzmann factors $\exp(-\beta E_r)$ over all energy levels, each counted the correct number of times. This sum will be replaced by an integral over states in phase space by means of Eq. (3.5-6).

To write down the integral, use is made of the fact that each point of phase space has a definite energy associated with it, given by

$$H(q,p) \equiv H(q_1,\ldots,q_F; p_1,\ldots,p_F) = K + U \tag{3.5-7}$$

The energy function $H(q_1,\ldots,q_F;\, p_1,\ldots,p_F)$, abbreviated $H(q,p)$ is, by definition, the Hamiltonian of the system. Moreover, each point in phase space has a Boltzmann factor $\exp[-\beta H(q,p)]$ associated with it, and the partition function (3.5-5) thus goes over into the phase integral

$$Z = \frac{1}{h^F}\int \exp[-\beta H(q,p)]\, dq_1 \cdots dq_F\, dp_1 \cdots dp_F \tag{3.5-8}$$

Equation (3.5-8) is the desired result that has been obtained mainly by plausibility arguments. (It will more formally be derived in the next section.) If the system consists of N identical nonlocalized subsystems, the right-hand side of the equation must be divided by $N!$ to ensure that each state of the gas is counted only once. In deriving Eq. (3.5-8), two approximations are made. First, the degeneracies $g(E_r)$ in the quantal partition function (3.5-5) are replaced by the density of states (3.5-6). This is usually valid, except in the case of Bose–Einstein condensation. The second approximation is to replace the discrete discontinuously varying Boltzmann factors $\exp[-\beta H(q,p)]$ by continuously varying factors. This is justified if successive Boltzmann factors in the quantal partition function (3.5-5) differ sufficiently little from each other. Such is the case if all level spacings $E_{r+1} - E_r$ are sufficiently small, that is, if

$$|E_{r+1} - E_r| \ll kT \tag{3.5-9}$$

Equation (3.5-9) represents the real limitation on the validity of the classical partition function (3.5-8). If ΔE, the spacing between adjacent levels of the kind of energy being considered is not small compared to kT, deviations from classical behavior will occur.

■ EXAMPLE 3.5-1

Discuss the procedure for making the transition from the classical to the quantum model in the case of (a) a particle in a one-dimensional box, (b) a one-dimensional rotator, and (c) a harmonic oscillator.

Solution. Classical and quantum mechanics represent different models for the behavior of particles, systems, or assemblies. Since the two models give essentially the same results when applied to macroscopic bodies, although their descriptions diverge more and more as the dimensions of the bodies or particles decrease, there is a general procedure for making the transition from the classical to the quantum model.

According to quantum mechanics it is not possible to know the position and momentum of a particle precisely. The uncertainties in the two quantities are related by the *uncertainty principle*, namely,

$$\Delta p \times \Delta p \approx h \tag{3.5-10}$$

Since the quantity $\Delta p \cdot \Delta q$ represents an area in two-dimensional phase space, it is reasonable to suppose that the way to introduce the necessary uncertainty into the classical description of motions is to divide phase space into cells each of an area $\Delta p \cdot \Delta q = h$. Such a step is examined in the following cxamples.

(a) For a particle in a one-dimensional box of length L, the quantum condition is

$$p = n\left(\frac{h}{2L}\right) \tag{3.5-11}$$

where n may take any positive integral value. Since the energy of the system is entirely kinetic, $\epsilon = p^2/2m$, lines of constant momentum in the phase plane are also contours of constant total energy. Accordingly the classical phase space trajectory for one-dimensional translation is a pair of lines parallel to the q axis and equidistant from it. The line with positive p is traversed from left to right, and the line with negative p from right to left. Two trajectories corresponding to successive quantum states with quantum numbers r and $r + 1$ are shown in Fig. 3.5-1(a). The area between these trajectories (shown shaded) is made up of two equal parts whose combined area is

$$2 \times L\left[(r+1)\left(\frac{h}{2L}\right) - r\left(\frac{h}{2L}\right)\right] = h \tag{3.5-12}$$

Phase space for a particle in one-dimensional translational motion can thus be divided into cells of area h so that the cell boundaries are the classical phase space trajectories corresponding to the quantum-mechanical stationary states. Each cell is made up of two equal areas, one above and one below the q axis.

(b) For a one-dimensional rotator, quantum mechanics gives the allowable angular momenta as

$$p_\theta = I\omega = mvr = j\left(\frac{h}{2\pi}\right) \qquad j = 0, \pm 1, \pm 2, \ldots \tag{3.5-13}$$

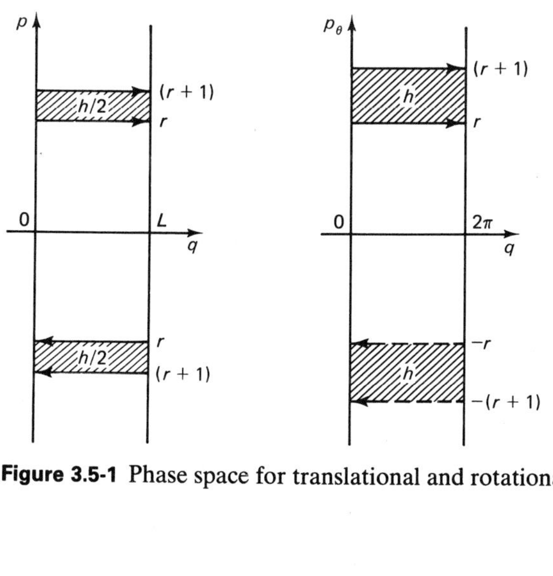

Figure 3.5-1 Phase space for translational and rotational systems.

The values of j are both positive and negative because the angular momentum can be positive or negative depending upon whether the rotation is anticlockwise or clockwise.

The classical phase space trajectory for a one-dimensional rotator is a single line parallel to the q_θ axis, a trajectory above the q_θ axis representing anticlockwise rotation and one below, clockwise rotation. Two trajectories corresponding to rotational states with successive positive quantum numbers r and $r+1$ are shown by the full lines in Fig. 3.5-1(b). The shaded area between trajectories is

$$2\pi\left(\frac{h}{2\pi}\right) = h \tag{3.5-14}$$

The phase plane may thus again be divided into cells of area h so that each cell boundary represents the classical trajectory corresponding to a quantum mechanical stationary state.

(c) In both examples above, the total energy of the system is kinetic, and lines of constant momentum correspond to contours of constant total energy. For the harmonic oscillator this is not so. The trajectories of constant energy in the phase plane are ellipses (as previously shown) whose major and minor axes are $2\sqrt{2\epsilon/k}$, and $2\sqrt{2m\epsilon}$. The area enclosed by any trajectory is then

$$\begin{aligned} &\frac{\pi}{4} \times (\text{product of major and minor axes}) \\ &= \frac{\pi}{4} \times 4 \times 2\epsilon\sqrt{\frac{m}{k}} = 2\pi\epsilon\sqrt{\frac{m}{k}} \end{aligned} \tag{3.5-15}$$

Since the frequency of the harmonic oscillator is

$$\nu = \frac{1}{2\pi}\sqrt{\frac{k}{m}} \tag{3.5-16}$$

the area enclosed by any trajectory is

$$\text{area} = \frac{\epsilon}{\nu} \tag{3.5-17}$$

Now, the energy levels allowed by quantum mechanics for the harmonic oscillator are

$$\epsilon = h\nu\left(n + \frac{1}{2}\right) \qquad \Delta\epsilon = h\nu \tag{3.5-18}$$

where $\Delta\epsilon$ is the spacing of the energy levels. If trajectories are drawn in the phase plane to represent the energies allowed by quantum mechanics, it is readily seen from (3.5-17) and (3.5-18) that the area between successive trajectories is

$$\Delta(\text{area}) = \frac{\Delta\epsilon}{\nu} = h \tag{3.5-19}$$

This result may be stated in a general way: For any type of motion in one dimension the area of phase space between classical trajectories that correspond to successive quantum-mechanical stationary states is exactly h. For motions in s dimensions which require a $2s$-dimensional phase space for their representation, one quantum state corresponds to a cell volume of h^3. Thus, if a volume V of phase space is accessible to a system or assembly, the number of quantum states that are accessible is V/h^3.

Furthermore, for assemblies of nonlocalized systems, according to quantum theory, the systems of the assembly are indistinguishable, and the assembly wave function must be constructed in such a way as to include this indistinguishability. In the classical construction of the phase space for an N particle assembly, the particles are, in fact, considered distinguishable by assigning the $6N$ coordinates required to define the phase space. Since the number of states for labeled and unlabeled particles differ by $N!$, the transition between localized and nonlocalized systems finally reads:

For assemblies of localized systems, number of quantum states corresponding to a volume of V of phase space $= V/h^3$ (3.5-20)

For assemblies of nonlocalized systems, number of quantum states corresponding to a volume V of phase space $= V/h^3N!$ (3.5-21)

■

■ **EXAMPLE 3.5-2**

The limitation on the validity of classical statistical mechanics is that $\Delta\epsilon$, the spacing between adjacent energy levels of a particle, be small compared to kT. Investigate this for the case of ^{4}He gas contained in a 10-cm cube at a temperature of 300 K.

Solution. To show that a very large number of energy levels are available for occupation, consider the energies of the particle to be given by

$$\epsilon_i = \frac{h^2}{8mV^{2/3}}(n_x^2 + n_y^2 + n_z^2)$$

The energy spacing

$$\Delta\epsilon_i = \frac{h^2}{8mV^{2/3}}$$

is of the order of 10^{-40} J. This is very small compared to kT which equals 4×10^{-21} J at 300 K. Clearly, $\Delta\epsilon_i \ll kT$, and classical statistical mechanics holds.

The energy states are so numerous for a gas that they are, in classical statistical mechanics, treated in groups, the ith group taken to contain g_i states of energy ϵ_i. A test for the grouping is to consider the average number of particles per state, n_i/g_i. Since the energy level spacings is about 10^{-19} times kT, it is quite possible to use groups, say 10^{10} states each, while maintaining an energy level spacing $\Delta\epsilon_i$ which is still minute compared to the energy scale kT. ■

3.6 CLASSICAL MICROSTATES AND MACROSTATES

In discussing the distribution of representative points in phase space, it has been stated that the phase space is divided into cells of equal size and the density is the number of particles in each cell. This is analogous to the notion of a distribution function previously introduced. Figure 3.6-1 shows a typical distribution of points in the phase space q–p, which has been divided into cells of equal size.

The cells are labeled from 1 to ∞ in some arbitrary fashion, or from 1 to M if the accessible phase space is limited to a finite number of cells. A distribution can be defined by giving the set of occupation numbers $\{n_i\}$, where n_i is the number of representative points in the cell i. To know the set of numbers is to know the the macroscopic properties of the gas that is being represented, at least within the degree of accuracy determined by the cell size. The set of numbers constitute a *macrostate*.

Now, if in addition to specifying how many representative points are in each cell, it is specified which ones $(a, b, \ldots)$ are in each cell, the arrangement represents a *microstate* of the gas. This is shown in Fig. 3.6-2.

According to the fundamental postulate of classical statistical mechanics, equal a priori probabilities must be assigned to equal volumes in phase space. It follows that if the phase space is divided into M cells of equal size, then the a priori probability of each microstate is the same. For each microstate there is a particular macrostate. But for each macrostate there are many possible microstates, the number depending on the particular macrostate. The probabilities of different macrostates are therefore not the same.

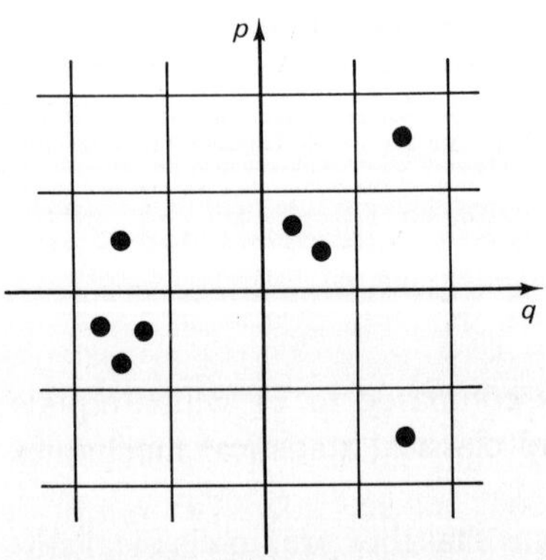

Figure 3.6-1 Macrostate in phase space.

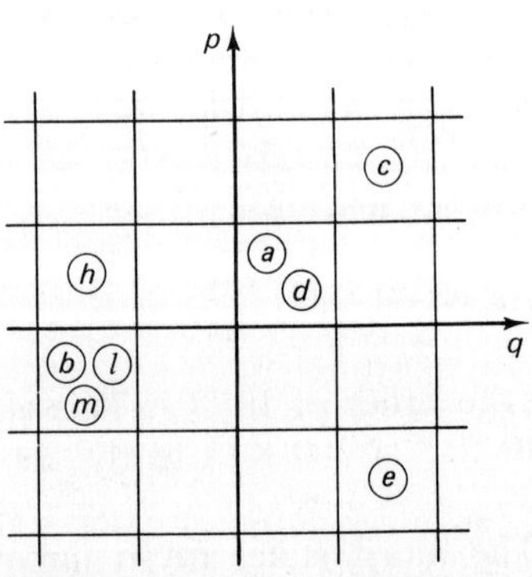

Figure 3.6-2 Microstate in phase space.

The number of microstates associated with the macrostate $\{n_i\}$ is known as its *thermodynamic probability* $W(n_i)$. To determine this, it is necessary to know the number of ways N particles may be arranged among M cells in such a way that there are a certain fixed number of particles in each cell. In Chapter 1, this number was found to be

$$W(n_1, n_2, \ldots) = \frac{N!}{\Pi n_i!} \tag{3.6-1}$$

Equation (3.6-1) can be verified by first letting the N molecules (or their labels a, b, etc.) be placed in an ordered sequence, and from this sequence place the first n_1 molecules in cell 1, the next n_2 in cell 2, and so on. Since there are $N!$ ways of obtaining the ordered sequence (i.e., permutations), the above procedure can be carried out in $N!$ ways. Now, in determining the number of combinations, the order within each cell is immaterial, so $N!$ must be divided by a factor that is equivalent to the ways in which each combination can be achieved.

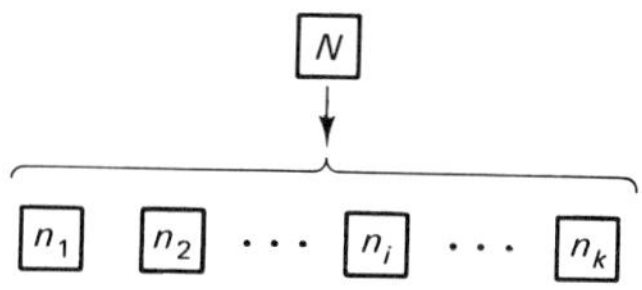

The number of ways in which a given n_i molecules can be arranged in sequence in the ith cell is $n_i!$, and therefore the number of ways (differing only in the ordering within the cells) of obtaining one combination is the product of such terms:

$$n_1!n_2! \cdots n_k! = \prod_{i=1}^{k} n_i! \tag{3.6-2}$$

The number of nonequivalent arrangements, that is, the number of combinations (thermodynamic probability) is thus

$$W(n_1, n_2, \ldots) = \frac{N_1}{n_1!n_2! \cdots} = \frac{N!}{\Pi n_i!} \tag{3.6-3}$$

This is the combinatorial rule for obtaining W.

3.7 ENSEMBLE AND DISTRIBUTIONS

It has been stated that the properties of an assembly of systems, for example, a gas, could be derived by replacing the time average of the assembly by an average over a large number of identical assemblies which are in thermal contact but which are otherwise isolated. Such a collection of assemblies is called an ensemble.

For the sake of completeness, some of the notions previously introduced will be reviewed and extended. Consider the properties of a gas of molecules. It is convenient to refer to a single molecule in the gas as a system, and to the gas as a whole as an assembly. An ensemble is then a collection of a large number of assemblies which are essentially independent of one another but which have been made macroscopically as identical as possible. By being independent of one another, the interactions between molecules can be ignored.

By being macroscopically identical, each assembly is characterized by the same value of some set of macroscopic parameters that uniquely determines the equilibrium state of the assembly; for example, if each assembly has the same temperature and volume, and is made up of the same number and kind of systems, or if each assembly has the same energy and pressure and is made up of the same number and type of systems. For each set of parameters chosen, there is a different type of ensemble. The three most commonly used ensembles are the *microcanonical,* the *canonical,* and the *grand canonical* ensemble. They are shown in Fig. 3.7-1.

The microcanonical ensemble is a collection of essentially independent assemblies having the same energy E, volume V, and number N of systems (for simplicity only one type of system). The individual assemblies are separated by rigid, impermeable, and well-insulated walls.

The macroscopic energy of the assembly cannot be specified exactly. The best that can be done is to specify the energy within some narrow range. Let the energy fall between $E - \Delta E$ and E. Let P_i be the probability that an assembly is in the state i and E_i be the corresponding eigenvalue. Then for the microcanonical ensemble,

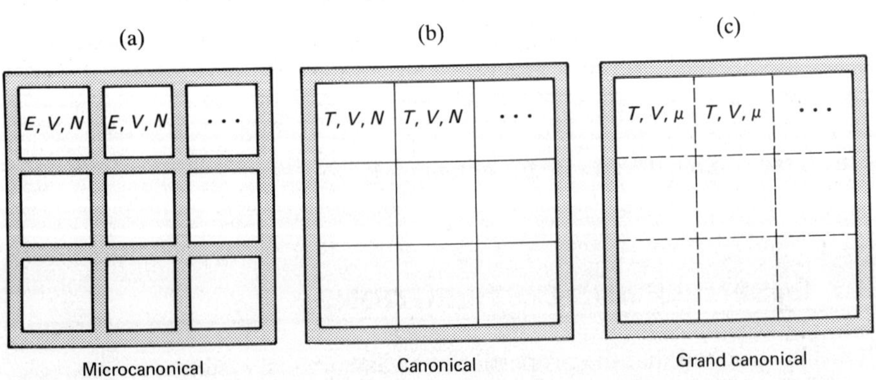

Figure 3.7-1 Types of ensembles. (For another look at ensembles, see Fig. 2.9-2.)

$$\begin{aligned} P_i &= \frac{1}{Z} \qquad E - \Delta < E \leq E_i \leq E \\ P_i &= 0 \qquad \text{otherwise} \end{aligned} \tag{3.7-1}$$

Quantum mechanically, Z is the number of assembly eigenstates between $E - \Delta E$ and E. Classically, Z is

$$Z(q,p) \int \cdots \int_{H(q,p)=E} dq_1 \cdots dp_N = \text{const} \tag{3.7-2}$$

Equation (3.7-1) is the *microcanonical distribution*. All of the equilibrium properties of the assembly can be calculated if Z is known as a function of E, V, and N. The integral (or summation) Z is the microcanonical partition function.

The canonical ensemble is a collection of essentially independent assemblies having the same temperature T, volume V, and number of identical systems N. Each system is in thermal contact with a large heat reservoir at temperature T. Alternatively, all the assemblies are in thermal contact with one another. The number of assemblies is large, and each assembly is effectively immersed in a heat bath. In Fig. 3.7-1(b), the individual members of a canonical ensemble are shown to be separated by rigid, impermeable, but diathermic walls. Energy can be exchanged among the members, so they all arrive at a common temperature.

The distribution function for a canonical ensemble, by far one of the most important results in classical statistical mechanics, will be shown in the following sections to be

$$P_i = \frac{\exp(-E_i/kT)}{\sum\limits_i \exp(-E_i/kT)} = \frac{\exp[-H_i(q,p)/kT]}{Z} \tag{3.7-3}$$

where

$$Z = \sum_i \exp\left(-\frac{E_i}{kT}\right) \tag{3.7-4}$$

$$= \int \cdots \int \exp\left[\frac{-H_i(q,p)}{kT}\right] dq_1 \cdots dp_N \tag{3.7-5}$$

is the canonical partition function or phase integral. Equation (3.7-3) is the *canonical distribution*. It has been employed extensively, and practically exclusively, in the preceding chapter.

The grand canonical ensemble is a collection of essentially independent assemblies having the same temperature T, volume V, and chemical potential μ. In Chapter 4, the chemical potential will be discussed in greater detail. For the present, it is sufficient to know that if two gases of the same species are allowed to exchange molecules freely with each other, their chemical potentials will equalize. The grand canonical ensemble is in essence a collection of assemblies, each occupying a separate volume V, but which can exchange energy and molecules with one another.

3.8 CANONICAL ENSEMBLE DISTRIBUTION

Consider a collection of identical molecules which are placed in a rigid, impermeable, and well-insulated container. The dynamical state of each molecule is represented by a point in phase space. These points move along a trajectory determined by the equations of motion. Since the molecules in a box periodically collide with one another, the point representing a particular molecule will not follow a continuous trajectory but will jump from trajectory to trajectory with each collision. The gas therefore passes rapidly from one macrostate to another. However, there will be a most probable macrostate.

More generally, take N representative members of a canonical ensemble (each corresponding to a thermodynamic system of specified V, N, and T) and stack them together into a large box having thermally insulating walls. The systems are allowed to exchange energy freely, and each system, which has specified V and N, can be considered to be immersed in a large thermostat, made up of the other $N-1$ systems.

A convenient mode of description is based on the division of phase space into cells numbered 1, 2, . . . , V. The extension of each cell is shrunken to zeros as $V \rightarrow \infty$. A description of the state of the ensemble would specify the distribution: "there are N_1 systems in cell 1, N_2 systems in cell 2, . . ." By the fundamental postulate, every arrangement $\{N_i\}$ consistent with

$$N_1 + N_2 + \cdots = \Sigma N_i = N \tag{3.8-1}$$

$$N_1E_1 + N_2E_2 + \cdots = \Sigma N_i H(q_i, p_i) = E \tag{3.8-2}$$

is equally probable. Therefore, the distribution or macrostate admitting the largest number of arrangements must be the most probable distribution.

The number of arrangements or microstates consistent with a distribution $\{N_i\}$ is given by the combinatorial formula of Section 3.6, which is

$$W(N_1N_2, \ldots) = \frac{N!}{N_1!N_2! \cdots N_2!} = \frac{N!}{\pi_i N_i!} \tag{3.8-3}$$

It is required to maximize (3.8-3) with respect to variations in the N_i subject to the constraints imposed by (3.8-1) and (3.8-2). It is more convenient, however, to maximize the logarithm of (3.8-3), that is,

$$\begin{aligned} \ln W &= \ln N! - \sum_i \ln N_i! \\ &= N \ln - N - \Sigma(N_i \ln N_i - N_i) \end{aligned} \tag{3.8-4}$$

Setting the variation of Eq. (3.8-4) to zero gives

$$\delta\left[N \ln N - N - \sum_i (N_i \ln N_i - N_i)\right] = 0$$

or, since $N \ln N - N$ is a constant,

$$-\sum_i \delta(N_i \ln N_i - N_i) = 0 \tag{3.8-5}$$

Expanding Eq. (3.8-5) yields

$$-\sum \ln N_i \, \delta N_i - \sum_i N_i \, \delta \ln N_i + \sum_i \delta N_i = 0$$

or

$$-\sum_i \ln N_i \, \delta N_i - \sum_i N_i \frac{\delta N_i}{N_i} + \sum_i \delta N_i = 0$$

which simplifies to

$$-\sum_i \ln N_i \, \delta N_i = 0 \tag{3.8-6}$$

The variations of Eqs. (3.8-1) and (3.8-2) are, respectively,

$$\sum_i \delta N_i = 0 \tag{3.8-7}$$

$$\sum_i H_i \, \delta N_i = 0 \tag{3.8-8}$$

According to Lagrange's method of undetermined multipliers, Eq. (3.8-7) is multiplied by $-\alpha$, Eq. (3.8-8) by $-\beta$, and both added to Eq. (3.5-6):

$$-\sum_i \ln N_i \, \delta N_i - \sum_i \alpha \, \delta N_i - \sum_i \beta E_i \, \delta N_i = 0$$

or

$$\sum_i (-\ln N_i - \alpha - \beta H_i) = 0 \tag{3.8-9}$$

Setting each coefficient equal to zero then gives

$$-\ln N_i - \alpha - \beta H_i = 0$$

or

$$N_i = e^{-\alpha - \beta H(q_i, p_i)} \tag{3.8-10}$$

This is the number of points in a cell in phase space centered at (q, p). The distribution function is thus

$$\rho(q, p) = e^{-\alpha} e - \beta H(q, p) \tag{3.8-11}$$

The value of the constant $e^{-\alpha}$ is determined from the normalization condition that the integration of $\rho(q, p)$ over all q and p must give the total number N. The constant β will be shown to equal $1/kT$ by computing the average kinetic energy of molecules in an ideal gas. Equation (3.8-11) is the *canonical ensemble distribution.* It is of general validity, because H is not restricted to the energy of a monatomic molecule, and it is applicable to polyatomic molecules for which the number of degrees of freedom is greater than 3.

■ EXAMPLE 3.8-1

Show that for a system of s degrees of freedom, phase space is subdivided into volume elements h^s for the transition from summation to integration.

Solution. From the preceding development, it would be expected that the transition from quantum-statistical mechanics to classical statistical mechanics can be made by replacing the sum in the partition function by an integral and making the substitution

$$\varepsilon_i \rightarrow H(p_i, q_i) \tag{3.8-12}$$

However, this does not settle the question of the volume element of integration that should be used. It cannot simply be

$$dp_1 \cdots dp_s\, dq_1 \cdots dq_s \tag{3.8-13}$$

where s is the number of degrees of freedom of the system, because this would yield a partition function with the dimensions of (J s)^s, whereas the partition function for a system of N molecules,

$$Z_N = \sum_i \exp(-\beta\varepsilon_i) \tag{3.8-14}$$

is dimensionless.

To investigate the situation, consider the case of a one-dimensional harmonic oscillator, defined by the potential

$$V = \tfrac{1}{2} aq^2 \tag{3.8-15}$$

The Hamiltonian of the oscillator is

$$H(p, q) = \frac{p^2}{2m} + \frac{1}{2} aq^2 \tag{3.8-16}$$

and so the curves of constant energy are ellipses in phase space (i.e., $p-q$ space) as illustrated in Fig. 3.8-1.

The area of the ellipse is given by $\oint p\,dq$, the integration being carried out over one complete period of the oscillator. Classically any ellipse in phase space would represent a state of the oscillator, but in quantum mechanics, the Wilson–Sommerfeld quantization rule requires that

$$\oint p\,dq = nh \tag{3.8-17}$$

where h is Planck's constant and n is some integer. The allowed energy surfaces then subdivide phase space into regions of area h.

For a system of s degrees of freedom, the energy surfaces corresponding to allowed states subdivide phase space into regions of volume h^s. In other words, each quantum state is associated with a volume h^s. Therefore, in making the transition to an integral (classical case),

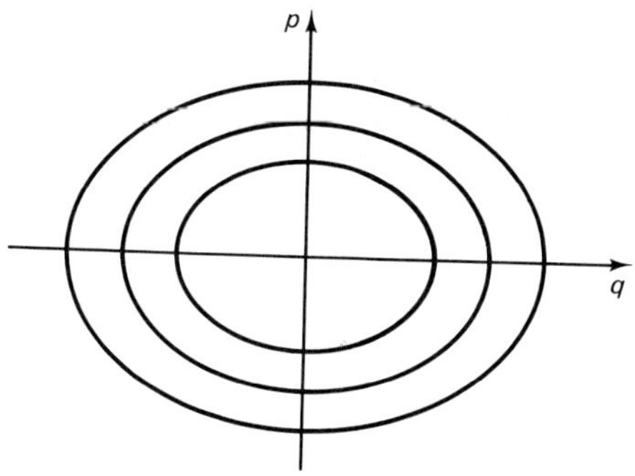

Figure 3.8-1 Constant energy paths for one-dimensional harmonic oscillator.

$$\varepsilon_i \rightarrow H(q_i, p_i) \tag{3.8-18}$$

$$\varepsilon_i \rightarrow \frac{dq_i \cdots dp_s}{h^s} \tag{3.8-19}$$

The partition function becomes

$$Z_N = \sum_i g_i \exp(-\beta\varepsilon_i) \qquad \text{(quantum)} \tag{3.8-20}$$

$$Z_N = \frac{1}{h^s} \int \exp(-\beta H)\, dq_i \cdots dp_s \qquad \text{(classical)} \tag{3.8-21}$$

and is dimensionless in both cases. ■

■ **EXAMPLE 3.8-2**

Show that the classical replacement for the degeneracy g_i of an energy level is

$$g_i = \frac{dp_i \cdots dp_n\, dq_1 \cdots dq_n}{h^n}$$

State the condition for the validity of this approximation.

Solution. In relating classical statistical to quantum statistical mechanics, consider the single-particle quantum mechanical partition function:

$$z = \sum_j g_j e^{-\beta\varepsilon_j} \tag{3.8-22}$$

The fractional number of molecules with energy ε_j is

$$\frac{N_j}{N} = \frac{g_j e^{-\beta\varepsilon_j}}{z} \tag{3.8-23}$$

For certain systems, the energy levels are closely spaced, so that the functions $e^{-\beta\varepsilon_j}$ and $e^{-\beta\varepsilon_{j+1}}$ differ but slightly. Under these circumstances, it is possible to evaluate the partition function sum (3.8-22) as an integral. In effect, ε_j is regarded, not as an exact number, but as a small energy interval, over which the the quantity $e^{-\beta\varepsilon_j}$ is almost constant and in which the number of states g_j is reasonably large. Accordingly, g_j is taken to be a volume element of phase space divided by h^n:

$$g_j = \frac{dp_1 \cdots dp_n \, dq_1 \cdots dq_n}{h^n} \tag{3.8-24}$$

The partition function then becomes

$$z = \frac{1}{h^n} \int_{p_1} \cdots \int_{q_n} e^{-\beta\varepsilon} \, dp_1 \cdots dp_n \, dq_1 \cdots dq_n \tag{3.8-25}$$

For such a classical system, the energy is identical with the Hamiltonian $H(p_1, \ldots, q_n)$, and so

$$z = \frac{1}{h^n} \int_{p_1} \cdots \int_{q_n} e^{-\beta H(p_1, \ldots, q_n)} \, dq_1 \cdots dq_n \tag{3.8-26}$$

while the classical phase integral is

$$\int_{p_1} \cdots \int_{q_n} e^{-\beta H(p_1, \ldots, q_n)} \, dq_1 \cdots dq_n \tag{3.8-27}$$

It is seen from (3.8-26) that the classical approximation consists of dividing the phase integral by h^n to obtain the quantum-mechanical partition function. The necessary condition for the validity of this approximation is that the spacing of the adjacent discrete energy levels ε_j and ε_{j+1} be small compared with kT. The ratio of the Boltzmann factors is then very close to unity,

$$\frac{e^{-\varepsilon_{j+1}/kT}}{e^{-\varepsilon_j/kT}} = e^{-(\varepsilon_{j+1} - \varepsilon_j)/kT} \approx 1$$

and the sum in (3.8-22) can be replaced by an integral. Such a substitution of the integral for a sum is made in evaluating the translational and rotational partition function, although, in order to discuss the infrared rotational spectrum of a diatomic molecule, it is necessary to use quantum-mechanical calculation of the energy levels. The term classical statistical mechanics loosely describes the approximation in which the phase integral is divided by h^n to obtain the partition function. The early Maxwell–Boltzmann distribution law based on purely classical mechanics does not contain the quantum of action, h. ■

3.9 IDENTIFICATION OF THE MULTIPLIERS

As was done in Chapter 2, the Lagrange multiplier α is classically determined from the normalization condition

$$\int \cdots \int \rho(q, p) \, dq_1 \cdots dp_N = 1 \tag{3.9-1}$$

With $\rho(q,p)$ from Eq. (3.8-11), this gives

$$\int\cdots\int e^{-\alpha}e^{-\beta H(q,p)}\,dq_1\cdots dp_N = 1 \tag{3.9-2}$$

and solving for $e^{-\alpha}$,

$$e^{-\alpha} = \frac{1}{\int\cdots\int e^{-\beta H(q,p)}\,dq_1\cdots dp_N} = \frac{1}{Z} \tag{3.9-3}$$

where

$$Z = \int\cdots\int e^{-\beta H(q,p)}\,dq_1\cdots dp_N \tag{3.9-4}$$

is the classical partition function (phase integral). At this point, the canonical ensemble distribution may be written

$$\rho(q,p) = \frac{e^{-\beta H(q,p)}}{Z} \tag{3.9-5}$$

It remains to identify β. As mentioned in Chapter 2, it governs the partitioning of energy among systems, and so has to do with temperature. β will be determined to be $1/kT$ by comparing the statistical and thermodynamic results for the energy of a monatomic ideal gas. It will then be shown that $\beta = 1/kT$ even if the gas is polyatomic.

Statistically, the interval energy (E in statistical mechanics, U in thermodynamics) is the ensemble average of the Hamiltonian:

$$E = U = \langle H(q,p)\rangle = \int\cdots\int H(q,p)\rho(q,p)\,dq_1\cdots dp_N$$

or, with Eq. (3.9-4),

$$E = \frac{1}{Z}\int\cdots\int H(q,p)e^{-\beta H(q,p)}\,dq_1\cdots dp_N \tag{3.9-6}$$

Noting that the integral can be obtained from the differentiation of Z with respect to β, Eq. (3.9-6) can be written as

$$E = -\frac{1}{Z}\frac{\partial Z}{\partial\beta} = -\frac{\partial \ln Z}{\partial\beta} \tag{3.9-7}$$

To proceed further, Z must be evaluated. Consider a perfect monatomic gas: there are no intermolecular interactions, hence no potential energy terms in $H(q,p)$, and the latter is simply

$$H(q,p) = \frac{p_{1x}^2}{2m} + \frac{p_{1y}^2}{2m} + \cdots + \frac{p_{Nz}^2}{2m} = \sum_{i=1}^{3N}\frac{p_i^2}{2m} \tag{3.9-8}$$

Using this in (3.9-4), the phase integral becomes

$$Z = \int\cdots\int \exp\left[-\frac{\beta}{2m}(p_1^2 + \cdots + p_{3N}^2)\right] dq_1\cdots dq_{3N}\,dp_1\cdots dp_{3N} \tag{3.9-9}$$

Since $\exp[-\beta/2m(p_1^2+\cdots+p_{3N}^2)]$ is independent of the q's, the integration in (3.9-9) will be carried out separately, noting that

$$\iiint dq_1\,dq_2\,dq_3=\iiint dx_1\,dy_1\,dz_1=V \tag{3.9-10}$$

where V is the volume of the container, and

$$\int\cdots\int dq_1\cdots dq_{3N}=V^N \tag{3.9-11}$$

Equation (3.9-9) can now be written as

$$Z=V^N\left(\int dp_1e^{-\beta p_1^2/2m}\right)\cdots\left(\int dp_3e^{-\beta p_{3N}^2/2m}\right) \tag{3.9-12}$$

The $3N$ integrals in (3.9-12) are all identical. The limits of integration are $-\infty$ to ∞, even though the actual momentum of each particle will have an upper limit determined by the total energy. However, $e^{-\beta p^2/2m}$ will be negligibly small for large p (assuming $\beta>0$). Using the results on Gaussian integrals, each integral in (3.9-12) contributes a factor

$$\int_{-\infty}^{\infty} dpe^{-\beta p^2/2m}=\left(\frac{2\pi m}{\beta}\right)^{1/2} \tag{3.9-13}$$

Thus, Eq. (3.9-12) gives

$$Z=V^N\left(\frac{2\pi m}{\beta}\right)^{3N/2} \tag{3.9-14}$$

This is the translational partition function for a system of N molecules (classically distinguishable). If the molecules are indistinguishable, and quantum mechanics taken into account, the right-hand side of (3.9-14) should be divided by $N!$ and h^{3N} to give

$$Z=\frac{V^N}{N!h^{3N}}\left(\frac{2\pi m}{\beta}\right)^{3N/2} \tag{3.9-15}$$

The result (3.9-15) is known as the semiclassical partition function. For a system composed of a single molecule, the molecular partition function is

$$Z_1=\frac{V}{h^3}\left(\frac{2\pi m}{\beta}\right)^{3/2} \tag{3.9-16}$$

Proceeding with the determination for the multiplier β, taking the logarithm of (3.9-14), while isolating the term that contains β from those terms that do not,

$$\ln Z=-\frac{3N}{2}\ln\beta+\text{terms independent of }\beta \tag{3.9-17}$$

Differentiating (3.9-17) with respect to β gives

$$\frac{\partial\ln Z}{\partial\beta}=-\frac{3N}{2\beta} \tag{3.9-18}$$

Now, according to thermodynamics, the internal energy of a perfect gas is

$$E = \tfrac{3}{2} NkT \tag{3.9-19}$$

where k is Boltzmann's constant. Substituting (3.9-18) into (3.9-7) and setting the statistical and thermodynamic results for the internal energy equal to each other gives

$$\frac{3N}{2\beta} = \frac{3}{2} NkT \quad \text{or} \quad \beta = \frac{1}{kT} \tag{3.9-20}$$

Having identified β, the canonical ensemble distribution function is therefore given, in the general case, by

$$\rho(q,p) = \frac{e^{-H(q,p)/kT}}{Z} \tag{3.9-21}$$

where

$$Z(V,N,T) = \int \cdots \int dq_1 \cdots dp_n e^{-H(q,p)/kT} \tag{3.9-22}$$

■ EXAMPLE 3.9-1

Summarize and further comment on the evaluation of $\beta = 1/kT$ for a monatomic perfect gas by letting the Hamiltonian for a molecule of such a gas be

$$H(q,p) = \frac{p_x^2}{2m} + \frac{p_y^2}{2m} + \frac{p_z^2}{2m} \tag{3.9-23}$$

Solution. The average translational energy of the molecule is

$$\langle \varepsilon \rangle = \frac{\iiint \left(\frac{p_x^2}{2m} + \frac{p_y^2}{m} + \frac{p_z^2}{2m} \right) \exp\left[-\frac{\beta}{2m} (p_x^2 + p_y^2 + p_z^2) \right] dp_x \, dp_y \, dp_z}{\iiint \exp\left[-\frac{\beta}{2m} (p_x^2 + p_y^2 + p_z^2) \right] dp_x \, dp_y \, dp_z} \tag{3.9-24}$$

where each component of momentum can have values from $-\infty$ to $+\infty$.

Because the components are all independent of one another, the triple integrals of (3.9-24) are separable into individual integrals of the form

$$\int_{-\infty}^{\infty} x^2 e^{-ax^2} \, dx = \frac{1}{2} \sqrt{\frac{\pi}{a^3}} \tag{3.9-25}$$

$$\int_{-\infty}^{\infty} e^{-ax^2} \, dx = \sqrt{\frac{\pi}{a}} \tag{3.9-26}$$

Performing the indicated integrations, the remarkably simple result is obtained:

$$\langle \varepsilon \rangle = \frac{3}{2} \cdot \frac{1}{\beta} \tag{3.9-27}$$

From the kinetic theory of gases, it is known that the average translational energy of a gas molecule is $\frac{3}{2}kT$. Hence,

$$\frac{3}{2}\frac{1}{\beta} = \frac{3}{2}kT \quad \text{or} \quad \beta = \frac{1}{kT} \tag{3.9-28}$$

Having shown this for a monatomic perfect gas, it will now be shown that $\beta = 1/kT$ even if the gas is polyatomic. To this effect, consider a composite system consisting of two gases separated by a diathermal wall and contained within a rigid, impermeable, and well-insulated container.

The two gases are completely independent of one another except for the fact that energy can be passed from one gas to another. Let $\{N_i'\}$ represent a particular macrostate of one gas and $\{N_i''\}$ represent a particular macrostate of the second gas, and $W'(N_i')$ and $W''(N_i'')$ the respective thermodynamic probabilities. Since the two systems are independent, a macrostate for the composite system will be given by the product of W' and W''. Thus,

$$W(N_i'N_i'') = W'(N_i')W''(N_i'') = \frac{N'!}{\prod_i N_i'!}\frac{N''!}{\prod_i N_i''!} \tag{3.9-29}$$

where N' nd N'' are the total number of molecules, respectively, of the gases.

Now maximize ln W subject to the constraints

$$\Sigma N_i' = N' \tag{3.9-30}$$

$$\Sigma N_i'' = N'' \tag{3.9-31}$$

$$\Sigma N_i' + \Sigma N_i''E_i'' = E \tag{3.9-32}$$

This is equivalent to maximizing the function

$$\begin{aligned} F(N_i', N_i'', \alpha', \alpha'', \beta) &= \ln W(N_i', N_i'') \\ &+ \alpha'(\Sigma N_i' - N') + \alpha''(\Sigma N_i'' - N'') \\ &- \beta(\Sigma N_i'E_i' - \Sigma N_i''E_i'' - E) \end{aligned} \tag{3.9-33}$$

Taking the partial derivatives of F with respect to N_i' and N_i'' and setting them equal to zero, gives

$$N_i' = \exp(\alpha' - \beta E_i') \tag{3.9-34}$$

$$N_i'' = \exp(\alpha'' - \beta E'') \tag{3.9-35}$$

It is seen that all gases in thermal contact will have the same value of β, which, because it is $1/kT$ for a monatomic gas, will therefore be $1/kT$ for a polyatomic gas as well. ■

3.10 BOLTZMANN DISTRIBUTION

By applying the canonical ensemble distribution to the individual molecules of a macroscopic system, the Maxwell–Boltzmann distribution law can be obtained. Indeed, the simple (but restricted) approach would have been to consider the molecule as the system, and the ensemble as N molecules. (The more general ensemble method is not needed until Chapter 4.)

Assuming that the molar Hamiltonian can be expressed as a sum of individual molecular contributions

$$H(q,p) = \sum_{i=1}^{N} h_i(q_i, p_i) \tag{3.10-1}$$

the Boltzmann distribution formula can be derived in exactly the same way as in Section 3.8, making the substitution

system $\rightarrow$ molecule

The result is

$$\rho_{\text{molec}}(q,p) = \frac{e^{-h(q,p)/kT}}{z} \tag{3.10-2}$$

where z is the molecular phase integral, defined by

$$z = \int \cdots \int dq_i \cdots dp_i \, e^{-h_i(q_i,p_i)/kT} \tag{3.10-3}$$

the integration being over the degrees of freedom appropriate to a single molecule. The system partition function (for N molecules) is then

$$z = z^N \tag{3.10-4}$$

The method is applicable, even in cases when there exist intermolecular interactions, so that $H(q,p)$ contains terms of the form h_{ij} and h_{jk}, which involve the coordinates of two or more molecules. Equation (3.10-1) provides good approximation if

$$h_i \gg h_{ij},\ h_{ijk}, \quad \text{and so on} \tag{3.10-5}$$

so that the one-particle terms dominate. The additivity condition (3.10-1) holds exactly for a perfect gas. It is a good approximation for real gases under ordinary conditions. It can also be applied to crystals, in which case the weakly interacting subsystems are not molecules but modes of vibration.

Consider now a perfect monatomic gas enclosed in a volume V. The energy of one atom (call it particle 1) of this perfect monatomic gas is given by

$$H_1 = \frac{p_x^2 + p_y^2 + p_z^2}{2m} + U_1 \tag{3.10-6}$$

where

$$\begin{aligned} U_1 &= 0 \quad \text{inside } V \\ U_1 &= \infty \quad \text{outside } V \end{aligned} \tag{3.10-7}$$

The partition function becomes

$$Z_1 = \frac{1}{h^3} \int_{-\infty}^{\infty} \exp\left[-\beta \frac{(p_x^2 + p_y^2 + p_z^2)}{2m}\right] dp_x\, dp_y\, dp_z \int_V dx\, dy\, dz \tag{3.10-8}$$

$$= \frac{V}{h^3} \int_{-\infty}^{\infty} \exp\left(\frac{-\beta p_x^2}{2m}\right) dp_x \int_{-\infty}^{\infty} \exp\left(\frac{-\beta p_y^2}{2m}\right) dp_y \int \exp\left(\frac{-\beta p_z^2}{2m}\right) dp_z \tag{3.10-9}$$

With

$$\int_{-\infty}^{\infty} \exp(-ax^2)\, dx = 2\int_{-\infty}^{\infty} \exp(-ax^2)\, dx = \sqrt{\frac{\pi}{a}} \tag{3.10-10}$$

Equation (3.10-9) gives

$$Z_1 = \frac{(2\pi mkT)^{3/2} V}{h^3} \tag{3.10-11}$$

For a macroscopic system of N molecules, the partition function is

$$Z_N = Z_1^N \tag{3.10-12}$$

for localized molecules, and

$$Z_N = \frac{1}{N!} Z_1^N \tag{3.10-13}$$

for nonlocalized molecules.

Having determined the partition function, all the thermodynamic properties may be obtained in terms of its partial derivatives. For example,

$$\langle E \rangle = NkT^2 \frac{\partial}{\partial T} \ln\left(\frac{2\pi mkT}{h^2}\right)^{3/2} \frac{V}{N!} = \frac{3}{2} NkT \tag{3.10-14}$$

$$C_v = \frac{3}{2} Nk \tag{3.10-15}$$

$$p = kT\left(\frac{\partial \ln Z_N}{\partial V}\right) = \frac{NkT}{V} \tag{3.10-16}$$

$$S = \frac{5}{2} Nk + Nk \ln\left[\left(\frac{2\pi mkT}{h^2}\right)^{3/2} \frac{V}{N}\right] \tag{3.10-17}$$

These relations were derived in Section 2.15. The Z_N used is that with the $1/N!$ present. Equation (3.10-17) is commonly known as the *Sackur–Tetrode equation*. It gives the entropy of an ideal monatomic gas in terms of N, V, and T.

■ EXAMPLE 3.10-1

From the definition of the phase integral,

$$\iint e^{-\varepsilon/kT}\, dq\, dp$$

evaluate the partition function of a diatomic molecule by means of classical mechanics.

Solution. The partition function of a diatomic molecule consists of the translational partition function, the vibrational partition function, and the rotational partition function.

(a) *The Translational Partition Function.* If x, y, and z are the coordinates of the center of gravity of the molecule, and m is the molecular mass, then

$$\varepsilon_{\text{trans}} = \frac{1}{2} m(\dot{x}^2 + \dot{y}^2 + \dot{z}^2)$$

$$= \frac{1}{2m}(p_x^2 + p_y^2 + p_z^2) \tag{3.10-18}$$

The translational factor in the phase integral for a molecule moving in a rectangular box of dimensions a, b, and c, is therefore

$$\int_{-\infty}^{\infty}\int_{-\infty}^{\infty}\int_{-\infty}^{\infty}\int_0^a\int_0^b\int_0^c \exp\left[\frac{-(p_x^2+p_y^2+p_z^2)}{2mkT}\right] dx\, dy\, dz\, dp_x\, dp_y\, dp_z \tag{3.10-19}$$

Since

$$\exp\left[\frac{-(p_x^2+p_y^2+p_z^2)}{2mkT}\right] = \exp\left(\frac{-p_x^2}{2mkT}\right)\exp\left(\frac{-p_y^2}{2mkT}\right)\exp\left(\frac{-p_z^2}{2mkT}\right) \tag{3.10-20}$$

Equation (3.10-19) can be written in the form

$$\int_{-\infty}^{\infty} e^{-p_x^2/2mkT}\, dp_x \int_{-\infty}^{\infty} e^{-p_y^2/2mkT}\, dp_y \int_{-\infty}^{\infty} e^{-p_z^2/2mkT}\, dp_z \int_0^a dx \int_0^b dy \int_0^c dz \tag{3.10-21}$$

Each of the first three integrals is of the form:

$$\int_{-\infty}^{\infty} e^{-ax^2}\, dx = \left(\frac{\pi}{a}\right)^{1/2} \tag{3.10-22}$$

that is,

$$\int e^{-p_x^2/2mkT}\, dp_x = (2\pi mkT)^{1/2} \tag{3.10-23}$$

Therefore, the translational phase integral is

$$(2\pi mkT)^{3/2}\, abc = (2\pi mkT)^{3/2}\, V \tag{3.10-24}$$

Note that, strictly speaking, the classical phase integral does not contain Planck's constant h. However, to account for Heisenberg's uncertainty principle in quantum mechanics, the phase integral is divided by h^F in order to obtain the partition function. F is the degrees of freedom which here is equal to 3. Thus, the classical value of the partition function is

$$z_{\text{trans}} = \frac{\text{phase integral}}{h^3}$$

$$= \frac{(2\pi mkT)^{3/2}\, V}{h^3} \tag{3.10-25}$$

which is identical with the result obtained by quantum mechanics.

It may be remarked that, for the space part of the phase integral, the result

$$\iiint dx\,dy\,dz = V \tag{3.10-26}$$

is true whatever the shape of the vessel containing the gas, so the above derivation is of general validity.

(b) *The Vibrational Partition Function.* According to classical mechanics, the vibrational energy of a diatomic molecule is given by

$$\varepsilon_{\text{vib}} = \frac{1}{2\mu} p_r^2 + \frac{1}{2} k(r - r_0)^2 \tag{3.10-27}$$

where μ is the reduced mass $m_1 m_2/(m_1 + m_2)$, k is the force constant, and r is the internuclear distance (of value r_0 at equilibrium). The phase integral for vibration is thus

$$\int_{-\infty}^{\infty}\int_{-\infty}^{\infty} \exp\left\{-\frac{1}{kT}\left[\frac{p_r^2}{2\mu} + \frac{k(r - r_0)^2}{2}\right]\right\} dr\,dp_r$$

$$= \int_{-\infty}^{\infty} \exp\left(\frac{-p_r^2}{2\mu kT}\right) dp_r \int_{-\infty}^{\infty} \exp\left[\frac{-k(r - r_0)^2}{2kT}\right] dr \tag{3.10-28}$$

(The limits of integration may seem unrealistic. If, however, r differs greatly from r_0, the potential energy, and hence ε_{vib} is large, and $\varepsilon^{-\varepsilon_{\text{vib}}/kT}$ is small, consequently, the integrand is negligible for such values of r, and the limits can be chosen to be arbitrarily large, or small, respectively.) Both of the integrals in (3.10-28) can be expressed in the form (3.10-22), the second by setting $x = r - r_0$. The result is

$$(2\pi\mu kT)^{1/2}\left(\frac{2\pi kT}{k}\right)^{1/2} = 2\pi kT\left(\frac{\mu}{k}\right)^{1/2} \tag{3.10-29}$$

Now,

$$\left(\frac{\mu}{k}\right)^{1/2} = \frac{1}{2\pi\omega} \tag{3.10-30}$$

where ω is the vibrational frequency. This gives, for the vibrational phase integral,

$$\frac{kT}{\omega} \tag{3.10-31}$$

Since ε_{vib} is a function of one generalized coordinate and its conjugate momentum, the degrees of freedom is $F = 1$, and the corresponding classical partition function is

$$Z_{\text{vib}} = \frac{kT}{h\omega} = \frac{T}{\theta_{\text{vib}}} \tag{3.10-32}$$

This is not the same as the quantal result, but is an approximation to it at high temperatures.

(c) *The Rotational Partition Function.* Classical mechanics gives the rotational kinetic energy as

$$\varepsilon_{\text{rot}} = \frac{1}{2I}\left(p_\theta^2 + \frac{p_\phi^2}{\sin^2\theta}\right) \tag{3.10-33}$$

Therefore, the rotational phase integral is

$$\int_0^{2\pi}\int_0^{\pi}\int_{-\infty}^{\infty}\int_{-\infty}^{\infty} \exp\left[-\frac{1}{2IkT}\left(p_\theta^2 + \frac{p_\psi^2}{\sin^2\theta}\right)\right] dp_\theta\, dp_\phi\, d\theta\, d\phi \tag{3.10-34}$$

All spatial configurations are covered if θ runs from 0 to π and ϕ from 0 to 2π, hence, the limits of integration in Eq. (3.10-34).

Because of the term $p_\phi^2/\sin^2\theta$, complete factorization is not immediately possible. However, (3.10-34) can be written as

$$\int_0^{2\pi} d\phi \int_{-\infty}^{\infty} \exp\left(\frac{-p_\theta^2}{2IkT}\right) dp_\theta \int_0^{\pi}\int_{-\infty}^{\infty} \exp\left(\frac{-p_\phi^2}{2IkT/\sin^2\theta}\right) dp_\phi\, d\theta \tag{3.10-35}$$

The first two integrals in this expression are immediately solved, giving, respectively, 2π and $(2\pi IkT)^{1/2}$. The double integral is solved by integrating with respect to p_ϕ and then with respect to θ:

$$\int_0^{\pi}\int_{-\infty}^{\infty} \exp\left(\frac{-p_\phi^2}{2IkT\,\sin^2\theta}\right) dp_\phi\, d\theta = \int_0^{\pi} [(2\pi IkT)^{1/2}\sin\theta]\, d\theta$$

$$= (2\pi IkT)^{1/2} \times 2 \tag{3.10-36}$$

Thus, (3.10-34) simplifies to

$$8\pi^2 IkT \tag{3.10-37}$$

With the degrees of freedom $F = 2$, the classical rotational partition becomes

$$Z_{\text{rot}} = \frac{8\pi^2 IkT}{h^2} \tag{3.10-38}$$

This is identical with the quantal result. ■

■ EXAMPLE 3.10-2

Derive the thermodynamic rotational and vibrational properties of a classical diatomic gas.

Solution. It is instructive to discuss in more detail how the quantization of the internal states (rotational and vibrational) of a diatomic molecule can be handled with some precision. Consider a molecule in an electronic state that gives rise to a typical attractive potential energy. In such a diatomic molecule the relative motion of one of the atoms about the other is the same as if one of the atoms had an infinite mass, and hence acted as a center for the other, which has a mass $\mu = m_1 m_2/(m_1 + m_2)$, where m_1 and m_2 are the true separate masses of the atoms, μ being the reduced mass. The total energy of the system is the sum of the potential energy (which includes the electronic energy) and the kinetic energy, and is given in classical mechanics by

$$\epsilon = \epsilon_k + \epsilon_p = \tfrac{1}{2}\,\mu v^2 + \epsilon_p \tag{3.10-39}$$

here v is the relative velocity.

When polar coordinates are used, v is conveniently resolved at any given instant into three mutually perpendicular components. One of these, the radial component, is along the line joining the two atoms and is equal to $\dot{r}$; the second is along a circle of longitude in a sphere whose center is at the origin and is equal to $r\dot{\theta}$, where θ is the polar angle; and the third is along a circle of latitude and is equal to $r\dot{\phi}\sin\theta$, where ϕ is the longitudinal angle, $r\sin\theta$ being the radius of the circle of latitude. Thus,

$$\epsilon = \tfrac{1}{2}\mu\dot{r}^2 + \tfrac{1}{2}\mu r^2\dot{\theta}^2 + \tfrac{1}{2}\mu r^2\dot{\phi}\sin^2\theta + \epsilon_p \tag{3.10-40}$$

From the form of the generalized momentum as $p_i = \partial\epsilon_k/\partial\dot{q}_i$, it is seen that

$$p_r = \mu\dot{r} \tag{3.10.41}$$

$$p_\theta = \mu r^2\dot{\theta} \tag{3.10-42}$$

$$p_\phi = \mu r^2\dot{\phi}\sin^2\theta \tag{3.10-43}$$

and therefore,

$$\epsilon = \frac{p_r^2}{2\mu} + \frac{p_\theta^2}{2\mu r^2} + \frac{p_\phi^2}{2\mu r^2\sin^2\theta} + \epsilon_p \tag{3.10-44}$$

Equation (3.10-44) is a general expression for the energy of a diatomic molecule; the second and third terms on the right-hand side of this equation give the rotational part of the energy. If the vibration is not too highly excited, so that r does not change too greatly from r_e, the equilibrium value r_e can replace r in (3.10-44), and with the moment of inertia $I = \mu r_e^2$, the following expression for the rotational energy is obtained:

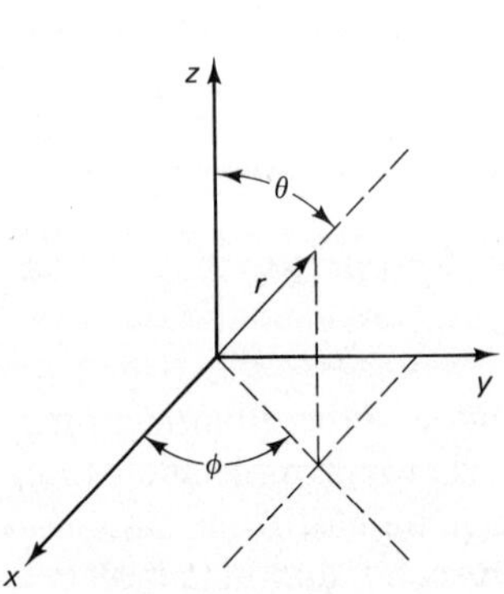

Figure 3.10-1 Coordinate system for a diatomic molecule.

$$\varepsilon_{\text{rot}} = \frac{p_\theta^2}{2I} + \frac{p_\phi^2}{2I \sin^2 \theta} \tag{3.10-45}$$

This may now be substituted into

$$z_{\text{rot}} = \frac{1}{h^n} \int e^{-\epsilon/kT} \, dq \, dp$$

where $n = 2$ and the element of volume in phase space is $dp_\theta \, dp_\phi \, d\theta \, d\phi$. Hence,

$$z_{\text{rot}} = h^{-2} \int_{\phi=0}^{2\pi} \int_{\theta=0}^{\pi} \int_{p_\phi=-\infty}^{\infty} \int_{p_\theta=-\infty}^{\infty} \exp\left(-\frac{p_\theta^2}{2IkT} - \frac{p_\phi^2}{2I(\sin^2 \theta)\, kT}\right) dp_\theta \, dp_\phi \, d\theta \, d\phi \tag{3.10-46}$$

Integrating with respect to ϕ, p_θ, and p_ϕ gives

$$z_{\text{rot}} = 2\pi h^{-2}(2\pi IkT)^{1/2} \int_0^{\pi} [2\pi I(\sin^2 \theta)\, kT]^{1/2} \, d\theta$$

$$= \frac{8\pi^2 IkT}{h^2} \tag{3.10-47}$$

If the two atoms in a diatomic molecule are alike, rotating the molecule through an angle of π (taking as an axis a perpendicular bisector of the line of centers) will carry it into a state indistinguishable from the initial state. Such a rotation amounts to changing ϕ to $\phi + \pi$ and θ to $\pi - \theta$. The integral in Eq. (3.10-46) covers both positions. To compensate for the duplication, the rotational partition function is generally written as

$$z_{\text{rot}} = \frac{8\pi^2 IkT}{h^2 \sigma} \tag{3.10-48}$$

where the symmetry factor is $\sigma = 2$ if the two atoms are alike and $\sigma = 1$ if the two atoms are different.

The thermodynamic functions associated with rotation are readily calculated from (3.10-48). They are given, per mole, by

$$E_{\text{rot}} = RT \tag{3.10-49}$$

$$C_{\text{rot}} = R \tag{3.10-50}$$

$$S_{\text{rot}} = R \ln\left(\frac{8\pi^2 eIkT}{h^2 \sigma}\right) \tag{3.10-51}$$

Consider now the evaluation of vibrational partition function for a diatomic molecule. If the vibrational energy of the molecule does not become too high (essentially, if the temperature is not too high), the vibrations of the molecule are harmonic, and the energy levels are given by

$$\epsilon_{\text{vib}} = (n + \tfrac{1}{2})\, h\nu \tag{3.10-52}$$

where n can take the values 0, 1, 2, 3, ..., and ν is the frequency of the oscillator. The partition function is then

$$z_{\text{vib}} = \sum_{\nu=0}^{\infty} \exp\left[\frac{-(n+\frac{1}{2})h\nu}{kT}\right]$$

$$= \exp\left(\frac{-\frac{1}{2}h\nu}{kT}\right) \sum_{\nu=0}^{\infty} \exp\left(\frac{-nh\nu}{kT}\right) \tag{3.10-53}$$

The last sum is a geometric series and is equal to $[1-\exp(-h\nu/kT)]^{-1} = \exp(h\nu/kT)/[\exp(h\nu/kT)-1]$. Hence,

$$z_{\text{vib}} = \frac{\exp(-\frac{1}{2}h\nu/kT)\exp(h\nu/kT)}{\exp(h\nu/kT)-1} \tag{3.10-54}$$

From this the thermodynamic functions are readily found, and the following relations for the molal energy, specific heat, and entropy obtained:

$$E_{\text{vib}} = \frac{Nh\nu}{2} + \frac{Nh\nu}{\exp(h\nu/kT)-1} \tag{3.10-55}$$

$$C_{\text{vib}} = R\left(\frac{h\nu}{kT}\right)^2 \frac{\exp(h\nu/kT)}{[\exp(h\nu/kT)-1]^2} \tag{3.10-56}$$

$$S_{\text{vib}} = R\left(\frac{h\nu}{kT}\right)\left[\exp\left(\frac{h\nu}{kT}\right)-1\right]^{-1} - R\ln\left[1-\exp\left(\frac{-h\nu}{kT}\right)\right] \tag{3.10-57}$$

Here, N is the Avogadro number. The quantity $h\nu/k$, designated as θ, is the characteristic temperature of the oscillator. In the expressions for C_{vib} and S_{vib} the quantities ν and T appear only in the combination $h\nu/kT = \theta T$. ■

3.11 PERFECT GAS IN THE CLASSICAL REGIME

In this section, the perfect classical gas will be defined and studied in more detail. Two equations that thermodynamically characterize the perfect classical gas will be derived: the equation of state $PV = NkT$, and Joule's law $E = E(T)$ which states that the internal energy E is independent of the volume and the pressure of the gas. Also, the conditions under which the classical approximations hold will be studied.

A gas consists of molecules moving about freely in space. The molecules are separated rather far from one another and interact only very weakly. At any instant only a small fraction are interacting strongly, that is, are colliding. Typical values for the mean separation between molecules under ordinary conditions and for the molecular diameter are on the order of 30 and 3 Å, respectively. The force between molecules, at separations large compared with the molecular diameter, is the weak van der Waals attraction which decreases rapidly with separation (like the seventh power of the intermolecular distance). At sufficiently low density, the molecules of a gas will interact very little with one another.

From an empirical macroscopic viewpoint, the ideal gas is characterized by the equation of state

$$pV = NRT \tag{3.11-1}$$

and by the dependence of its internal energy on temperature alone,

$$E = E(T) \quad \text{or} \quad \left(\frac{\partial E}{\partial V}\right)_T = 0 \tag{3.11-2}$$

From a microscopic viewpoint, the perfect gas is characterized by negligible intermolecular forces or by additivity of the energy according to

$$H(q,p) = \sum_i h_i(q_i,p_i) \tag{3.11-3}$$

The molecular Hamiltonian h_i can be further separated into parts referring to individual degrees of freedom within a molecule, that is, the translational motion of the center of mass can be separated from internal motions (rotations and vibrations of the molecule, as well as electronic excitation):

$$h = h_{\text{trans}} + h_{\text{int}} \tag{3.11-4}$$

and correspondingly,

$$\varepsilon = \varepsilon_{\text{trans}} + \varepsilon_{\text{int}} \tag{3.11-5}$$

Since the translational degrees of freedom are fully excited ($\Delta\varepsilon << kT$), the partition function can be evaluated from the corresponding phase integral:

$$\frac{1}{h^3}\int \cdots \int e^{-(p_x^2 + p_y^2 + p_z^2)/2mkT}\, dx\, dy\, dz\, dp_x\, dp_y\, dp_z$$

$$= \frac{1}{h^3}\left(\int_{-\infty}^{\infty} e^{-p^2/2mkT}\, dp\right)^3 V = (2\pi mkT)^{3/2}\, V \tag{3.11-6}$$

and, in the absence of electronic excitation, the molecular partition function in the monatomic case is

$$z = \frac{(2\pi mkT)^{3/2}\, V}{h^3} \tag{3.11-7}$$

A criterion for perfect gas conditions is that

$$\langle n_i \rangle = \frac{Ne^{-\varepsilon i/kT}}{z} << 1 \tag{3.11-8}$$

or, equivalently, that the Boltzmann limit

$$\frac{z}{N} >> 1 \tag{3.11-9}$$

To see whether this criterion holds, consider liquid helium at STP:

$$\frac{z}{N} = \frac{(2\pi mkT)^{3/2}\, V}{h^3 N} \sim 10^6 >> 1 \tag{3.11-10}$$

In fact, only for liquid helium does this fail. This gives the molar partition function in the Boltzmann limit for an assembly of independent indistinguishable particles as

$$Z = \frac{z^N}{N!} \tag{3.11-11}$$

so that

$$Z = \frac{(2\pi mkT)^{3N/2}\, V^N}{h^{3N} N!} \tag{3.11-12}$$

Once the canonical ensemble partition function is known, all the thermodynamic functions for a perfect monatomic gas can be derived. Noting that

$$\ln (N!) = N \ln N - N = N \ln \left(\frac{N}{e}\right) \tag{3.11-13}$$

this gives

$$\begin{aligned} \ln Z &= \ln \frac{z^N}{N!} = N \ln z - \ln N! \\ &= N \ln z - N \ln \left(\frac{N}{e}\right) = N \ln \left(\frac{ze}{N}\right) \\ &= N \ln \left[\frac{(2\pi mkT)^{3/2} Ve}{h^3 N}\right] \end{aligned} \tag{3.11-14}$$

where (3.11-7) has been used for z.

The Helmholtz function, $A = -kT \ln Z$, is then

$$A = -NkT \ln \left[\frac{(2\pi mkT)^{3/2}\, Ve}{h^3 N}\right] \tag{3.11-15}$$

The pressure, $p = -(\partial A / \partial V)_{N,T}$, is

$$p = NkT \left\{\frac{\partial}{\partial V} \ln \left[\frac{(2\pi mkT)^{3/2}\, Ve}{h^3 N}\right]\right\}_{N,T} \tag{3.11-16}$$

Since the logarithmic term in (3.11-16) consists of $\ln V$ + terms independent of V, Eq. (3.11-16) gives

$$p = \frac{NkT}{V} \tag{3.11-17}$$

which reproduces the *ideal gas law*. This particular result applies, as well, to a perfect polyatomic gas since the rotational, vibrational, and electronic factors in the molecular partition function are independent of V.

The Gibbs function $G = A + pV$ is, using (3.11-15) and (3.11-17),

$$G = A + pV = -NkT \ln \left[\frac{(2\pi mkT)^{3/2} V}{h^3 N} \right] \tag{3.11-18}$$

noting that $pV = NkT = NkT \ln e$.

The internal energy has already been established as

$$E = kT^2 \left(\frac{\partial \ln Z}{\partial T} \right)_{V,N} = \frac{3}{2} NkT \tag{3.11-19}$$

and thus

$$C_V = \left(\frac{\partial E}{\partial T} \right)_{V,N} = \frac{3}{2} Nk \tag{3.11-20}$$

in agreement with the equipartition principle.

The chemical potential per molecule, $\mu = (\partial A / \partial N)_{V,T}$, is

$$\mu = -kT \left(\frac{\partial \ln Z}{\partial N} \right)_{V,T} \tag{3.11-21}$$

$$= -kT \ln \left[\frac{(2\pi mkT)^{3/2} V}{h^3 N} \right] \tag{3.11-22}$$

where (3.11-15) has been used, and in taking partial derivatives, only those parts containing the appropriate variable need be considered.

Finally, the entropy, $S = -(\partial A / \partial T)_{V,N}$, of a perfect monatomic gas is

$$S = \left(\frac{\partial kT \ln Z}{\partial T} \right)_{V,N} = \frac{3}{2} Nk + Nk \ln \left[\frac{(2\pi mkT)^{3/2} Ve}{h^3 N} \right] \tag{3.11-23}$$

where (3.11-15) has been used. This relation is one form of the *Sakur–Tetrode equation.*

A gas need not be monatomic to be ideal. For polyatomic molecules there are rotational, vibrational, and electronic contributions to the partition function in addition to the translational contribution already computed. The Boltzmann limit remains applicable and is, in fact, further enhanced, since internal degrees of freedom serve to increase the number of quantum states available to a given number of molecules. The molecular partition function becomes

$$Z = \frac{1}{N!} (z_{\text{trans}} z_{\text{rot}} z_{\text{vib}} z_{\text{elec}})^N \tag{3.11-24}$$

This can be expressed as

$$Z = Z_{\text{trans}} Z_{\text{rot}} Z_{\text{vib}} Z_{\text{elec}} \tag{3.11-25}$$

with

$$Z_{\text{trans}} = \frac{z_{\text{trans}}^N}{N!} \qquad Z_{\text{rot}} = z_{\text{rot}}^N \qquad Z_{\text{vib}} = z_{\text{vib}}^N \qquad Z_{\text{elec}} = z_{\text{elec}}^N \tag{3.11-26}$$

Translational contributions in polyatomic gases are identical to those already derived for the monatomic gas. (Contributions from the internal degrees of freedom will be considered in Chapter 5.)

3.12 VELOCITY AND SPEED DISTRIBUTION

Although applications in depth of statistical mechanics will not occur until Chapter 5, it should be mentioned here that some of the most important applications of classical statistical mechanics coincide with the subject of kinetic theory. In many of these phenomena, quantum effects can be neglected without significant error. The distribution function for molecular velocities and speeds was first obtained by Maxwell in 1859.

The problem can be stated as follows: Given N molecules of a gas confined to some volume V, what is the probability or number of molecules with velocity components v_x in dv_x, v_y in dv_y, and v_z in dv_z? Historically, the Maxwell–Boltzmann velocity distribution was first derived for a perfect gas from the kinetic theory point of view. Its validity will now be shown, in classical statistical mechanics, to extend to imperfect gases and liquids as well.

Consider the Hamiltanian function

$$H(q,p) = \sum_{i=1}^{n} \frac{p_i^2}{2m} + V(q_1 \cdots q_n) \tag{3.12-1}$$

where the last term on the right takes account of intermolecular interactions that are present in nonideal gases and liquids. Substituting (3.12-1) into the canonical ensemble distribution function

$$\rho(q,p) = \frac{e^{-H(q,p)/kT}}{z} \tag{3.12-2}$$

gives

$$\rho(q,p) = \frac{e^{-p_1^2/2mkT}\, e^{-p_2^2/2mkT} \cdots e^{-p_n^2/2mkT}\, e^{-V(q)/kT}}{\int \cdots \int dq_1 \cdots dp_n\, e^{-p_1^2/2mkT} \cdots e^{-p_n^2/2mkT}\, e^{-V(q)/kT}} \tag{3.12-3}$$

This distribution function will now be reduced to the momentum distribution of a single particle (call it particle 1). Accordingly, (3.12-3) is integrated over all the variables except $p_1 = p_{1x}$, $p_2 = p_{1y}$, $p_3 = p_{1z}$ to give

$$\rho(p_{1x}, p_{1y}, p_{1z}) = \frac{e^{-(p_{1x}^2 + p_{1y}^2 + p_{1z}^2)/2mkT}}{\int_{-\infty}^{\infty}\iint e^{-(p_{1x}^2 + p_{1y}^2 + p_{1z}^2)/2mkT}\, dp_{1x}\, dp_{1y}\, dp_{1z}} \tag{3.12-4}$$

The simplification occurs because every other factor in the numerator integrand cancels its counterpart in the phase integral. In particular, dependence on $V(q, \ldots, q_n)$ has dropped out. Applying the result of the integral

$$\int_{-\infty}^{\infty} e^{-ax^2}\, dx = \left(\frac{\pi}{a}\right)^{1/2} \tag{3.12-5}$$

to (3.12-4) gives

$$\rho(p_{1x}, p_{1y}, p_{1z}) = \left(\frac{1}{2\pi mkT}\right)^{3/2} e^{-(p_{1x}^2 + p_{1y}^2 + p_{1z}^2)/2mkT} \tag{3.12-6}$$

or, in shorter notation,

$$\rho(p_1) = \left(\frac{1}{2\pi mkT}\right)^{3/2} e^{-p_1^2/2mkT} \tag{3.12-7}$$

Since this result applies equally well to any particle, the subscript 1 can be dropped. Thus, the momentum distribution function of any particle is

$$\rho(\mathbf{p}) = \left(\frac{1}{2\pi mkT}\right)^{3/2} e^{-\mathbf{p}^2/2mkT} \tag{3.12-8}$$

This relation is completely independent of the form of intermolecular interactions, of internal degrees of freedom such as rotations and vibrations, and of density, pressure, and the like. Equation (3.12-8) is the *probability distribution for momentum.*

The *velocity distribution* follows easily. Since $p_x = mv_x$, $p_y = mv_y$, $p_z = mv_z$, the two distributions are related by noting that

$$\rho(\mathbf{p})\, dp_x\, dp_y\, dp_z = \rho(\mathbf{p})\, m^3\, dv_x\, dv_y\, dv_z \tag{3.12-9}$$

or

$$\rho(\mathbf{v}) = m^3\rho(\mathbf{p}) = \left(\frac{m}{2\pi kT}\right]^{3/2} e^{-mv^2/2kT} \tag{3.12-10}$$

Thus, Maxwell–Boltzmann's velocity distribution law has been reproduced, but with greater generality, as part of the canonical ensemble distribution. The subsequent reduction to the Maxwell–Boltzmann of molecular speeds will follow. But first, consider (3.12-10) in more detail:

$$\rho(\mathbf{v}) = \rho(v_x, v_y, v_z) = \left(\frac{m}{2\pi kT}\right)^{3/2} e^{-m(v_x^2 + v_y^2 + v_z^2)/2kT} \tag{3.12-11}$$

The one-dimensional version of this equation is

$$\rho(\mathbf{v}_x) = \left(\frac{m}{2\pi kT}\right)^{1/2} e^{-mv_x^2/2kT} \tag{3.12-12}$$

It is plotted in units of $(kT/m)^{1/2}$ by setting

$$w = v_x/(kT/m)^{1/2} \tag{3.12-13}$$

Equation (3.12-12) then becomes

$$P(v_x)\, dv_x = dv_x \left(\frac{m}{2\pi kT}\right)^{1/2} \exp\left(\frac{-mv_x^2}{2kT}\right) \tag{3.12-14}$$

$$= \frac{1}{\sqrt{2\pi}}\, dw \exp\left(-\frac{1}{2}w^2\right) \equiv F(w)\, dw \tag{3.12-15}$$

where $P(v_x)\, dv_x$ is the probability that a molecule should have an x component of velocity in the interval $(kT/m)^{1/2}(v_x + dv_x)$.

Now, to reduce the Maxwell velocity distribution to a speed distribution, Eq. (3.12-11), the velocity distribution function, can be rewritten as

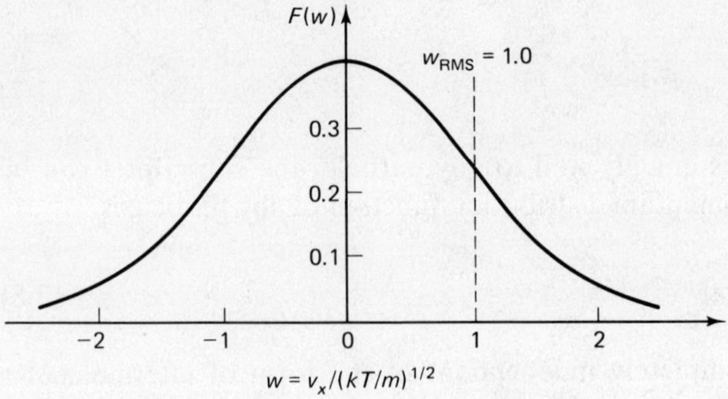

Figure 3.12-1 One-component velocity distribution function.

$$\rho(v, \theta, \phi) = \left(\frac{m}{2\pi kT}\right)^{3/2} e^{-mv^2/2kT} \tag{3.12-16}$$

by transformation to spherical coordinates in which the Cartesian components v_x, v_y, and v_z are replaced by v, θ, and ϕ, given by

$$v^2 = v_x^2 + v_y^2 + v_y^2 + v_z^2, \qquad \cos\theta = \frac{v_z}{v}, \qquad \tan\phi = \frac{v_y}{v_x} \tag{3.12-17}$$

as seen in Fig. 3.12-2.

In spherical polar coordinates, the volume element in Fig. 3.12-2 is given by

$$dr = v^2 \sin\theta \, dv \, d\theta \, d\phi \tag{3.12-18}$$

The normalization condition for (3.12-16) is

$$\int_0^\infty \int_0^\pi \int_0^{2\pi} v^2 \sin\theta \rho(v, \theta, \phi) \, dv \, d\theta \, d\phi = 1 \tag{3.12-19}$$

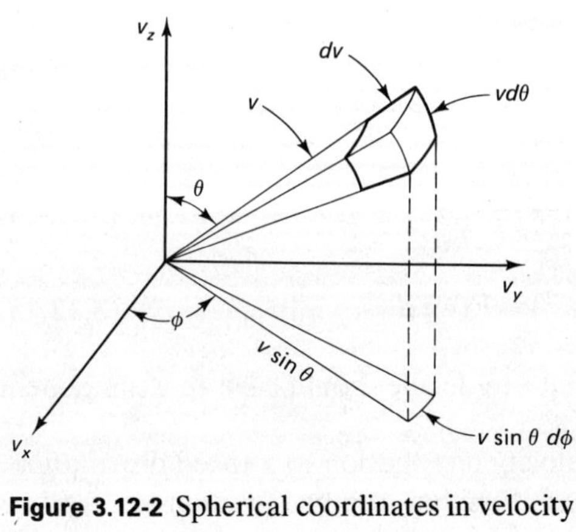

Figure 3.12-2 Spherical coordinates in velocity space.

Since $\rho(v, \theta, \phi)$ is independent of θ and ϕ, integration over the angles simply produces a factor 4π. Consequently,

$$4\pi \int_0^\infty v^2 \rho(v, \theta, \phi)\, dv = \int_0^\infty \rho(v)\, dv = 1 \tag{3.12-20}$$

which shows that the distribution for speeds is

$$\rho(v) = 4\pi v^2 \rho(v, \theta, \phi) \tag{3.12-21}$$

or, making use of (3.12-16),

$$\rho(v) = 4\pi v^2 \left(\frac{m}{2\pi kT}\right)^{3/2} e^{-mv^2/2kT} \tag{3.12-22}$$

Equation (3.12-21) is the Maxwell–Boltzmann distribution law for *molecular speeds.* Note that $\rho(v) = 0$ when $v = 0$, whereas the distribution function for each velocity component has a maximum at the origin. The reason, as seen from Eq. (3.12-21), is that the distribution function for speeds is the product of two quantities: (a) the Boltzmann factor $\rho(v, \theta, \phi)$, and (b) the density of states factor $4\pi v^2$, as shown in Fig. 3.12-3. These two factors respectively decrease and increase with energy, so that $\rho(v)$ has a maximum. The function $\rho(v)$ is an indication of the distribution of translational kinetic energy of one molecule in a gas at temperature T. This energy changes with every collision of the molecule.

The above discussion pertains to the distribution at a given temperature. The speed distribution at various temperatures is plotted in Fig. 3.12-4. It is seen that the most probable speed (labeled v_{mode}) increases as the temperature is increased.

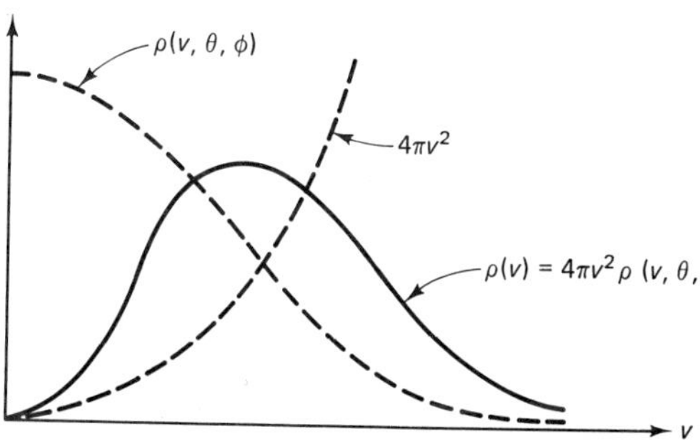

Figure 3.12-3 Factors in the speed distribution.

■ EXAMPLE 3.12-1

The Maxwell–Boltzmann distribution is rather sharply peaked about the most probable value, v_{mode}, so that most of the molecules will have speeds fairly close to v_{mode}. Only a small fraction of molecules have higher speeds. These exceptionally fast molecules can, however, play an important role in chemical reactions. Calculate the fraction of molecules having speeds in excess of some $v = v_0$.

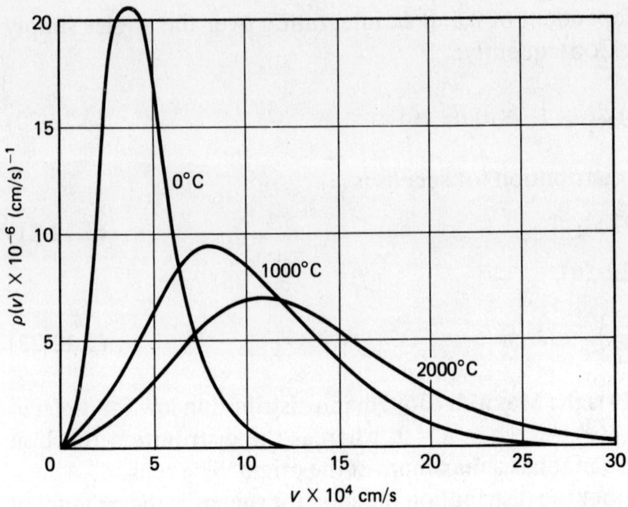

Figure 3.12-4 Maxwell–Boltzmann speed distribution for N_2 at 0C, 1000C, and 2000C.

Solution. The probability of a molecule having speeds in excess of v_0 is

$$P(v \geq v_0) = \int_{v_0}^{\infty} \rho(v)\, dv = 4\pi \left(\frac{m}{2\pi kT}\right)^{3/2} \int_{v_0}^{\infty} v^2 e^{-mv^2/2kT}\, dv \tag{3.12-23}$$

Changing variables to

$$x = \left(\frac{mv^2}{2kT}\right)^{1/2} \tag{3.12-24}$$

gives

$$P(v \geq v_0) = \frac{4}{\sqrt{\pi}} \int_{x_0}^{\infty} x^2 e^{-x^2}\, dx \tag{3.12-25}$$

The integral will be expressed in terms of the error function, but first, it is necessary to integrate by parts. Noting that

$$\frac{d}{dx} e^{-x^2} = -2xe^{-x^2} \tag{3.12-26}$$

and that $\lim_{x\to\infty} xe^{-x^2} = 0$,

$$P(v \geq v_0) = \frac{4}{\sqrt{\pi}} \int_{x_0}^{\infty} x d\left(-\frac{1}{2} e^{-x^2}\right) = \frac{4}{\sqrt{\pi}}\left(-\frac{x}{2} e^{-x^2}\Big]_{x_0}^{\infty} + \frac{1}{2}\int_{x_0}^{\infty} e^{-x^2}\, dx\right)$$

$$= \frac{2}{\sqrt{\pi}} x_0 e^{-x_0^2} + \frac{2}{\sqrt{\pi}} \int_{x_0}^{\infty} e^{-x^2}\, dx \tag{3.12-27}$$

The last integral is found from

$$\frac{2}{\sqrt{\pi}} \int_{x_0}^{\infty} e^{-x^2}\, dx = \frac{2}{\sqrt{\pi}} \int_{0}^{\infty} e^{-x^2}\, dx - \frac{2}{\sqrt{\pi}} \int_{0}^{x_0} e^{-x^2}\, dx = 1 - \operatorname{erf} x_0 \tag{3.12-28}$$

where

$$\int_0^\infty e^{-ax^2}\,dx = \frac{1}{2}\sqrt{\frac{\pi}{a}}$$

has been used, and erf x is given by Table 1.9-2.

Finally, returning to the original variable,

$$P(v \ge v_0) = \frac{2}{\sqrt{\pi}}\left(\frac{mv_0^2}{2kT}\right)^{1/2} e^{-mv_0^2/2kT} + 1 - \operatorname{erf}\left[\left(\frac{mv_0^2}{2kT}\right)^{1/2}\right] \qquad (3.12\text{-}29)$$

For $v_0 = 2v_{\text{mode}} = 2(2kT/m)^{1/2}$, $P(v \ge v_0) = 0.26$. For $v_0 = 3v_{\text{mode}}$, $P(v \ge v_0) = 0.10$. For $v_0 = 6v_{\text{mode}}$, the fraction of molecules having that speed drops to less than 0.01. ■

3.13 AVERAGE, MOST PROBABLE, AND RMS SPEEDS

In a speed distribution, several values acquire a special significance. It is instructive to determine what they are.

The average molecular speed $\langle v \rangle$ is

$$\begin{aligned} \langle v \rangle &= \int_0^\infty v\rho(v)\,dv \\ &= 4\pi\left(\frac{m}{2\pi kT}\right)^{3/2}\int_0^\infty v^3 e^{-mv^2/2kT}\,dv \end{aligned} \qquad (3.13\text{-}1)$$

where ρ from (3.12-22) has been used. With

$$\int_0^\infty x^3 e^{-\alpha x^3}\,dx = \frac{1}{2\alpha^2} \qquad (3.13\text{-}2)$$

Eq. (3.13-1) gives

$$\langle v \rangle = \left(\frac{8kT}{\pi m}\right)^{1/2} \qquad (3.13\text{-}3)$$

In terms of the molecular weight $M = N_0 m$ and ideal gas constant $R = N_0 k$, where N_0 is the Avogadro number, the average speed is

$$\langle v \rangle = \left(\frac{8RT}{\pi M}\right)^{1/2} \qquad (3.13\text{-}4)$$

Similarly, the average of the square speed, $\langle v^2 \rangle$, is given by

$$\int_0^\infty v^2\,\rho(v)\,dv$$

or

$$\langle v^2 \rangle = 4\pi\left(\frac{m}{2\pi kT}\right)^{3/2}\int_0^\infty v^4 e^{-mv^2/2kT}\,dv \qquad (3.13\text{-}5)$$

This can be evaluated by knowing that

$$\int_0^\infty x^4 e^{-\alpha x^2}\, dx = \frac{3\pi^{1/2}}{8\alpha^{5/2}} \tag{3.13-6}$$

The result is

$$\langle v^2 \rangle = \frac{3kT}{m} \qquad \langle \varepsilon \rangle = \frac{1}{2} m \langle v^2 \rangle = \frac{3}{2} kT \tag{3.13-7}$$

The square root of $\langle v^2 \rangle$, known as the *root-mean-square (RMS) speed* is

$$v_{\mathrm{RMS}} = (\langle v^2 \rangle)^{1/2} = \sqrt{\frac{3kT}{m}} \tag{3.13-8}$$

Finally, the most probable speed, also known as v_{mode}, is that which maximizes the distribution function $\rho(v)$. Setting the derivative equal to zero gives

$$\frac{d\rho}{dv} = 4\pi \left(\frac{m}{2\pi kT}\right)^{3/2} \left(2ve^{-mv^2/2kT} - v^2 \frac{mv}{kT} e^{-mv^2/2kT}\right) = 0 \tag{3.13-9}$$

The two roots of this equation are

$$v = 0 \quad \text{and} \quad v = \sqrt{\frac{2kT}{m}} \tag{3.13-10}$$

To determine which root corresponds to a maximum in $\rho(v)$, the second derivative must be evaluated:

$$\frac{d^2\rho}{dv^2} = 4\pi \left(\frac{m}{2\pi kT}\right)^{3/2} \left(2 - \frac{5m}{kT} v^2 + \frac{m^2}{k^2T^2} v^4\right) e^{-mv^2/2kT} \tag{3.13-11}$$

Now

$$\left(\frac{d^2\rho}{dv^2}\right)_{v=0} = 4\pi \left(\frac{m}{2\pi kT}\right)^{3/2} \times 2 > 0 \tag{3.13-12}$$

and

$$\left(\frac{d^2\rho}{dv^2}\right)_{v=(2kT/m)^{1/2}} = 4\pi \left(\frac{m}{2\pi kT}\right)^{3/2} (2 - 10 + 4)\, e^{-2} < 0 \tag{3.13-13}$$

Thus, only the second value corresponds to a maximum, and the most probable speed is

$$v_{\mathrm{mode}} = \sqrt{\frac{2kT}{m}} = \left(\frac{2}{3}\right)^{1/2} v_{\mathrm{RMS}} = 0.816\, v_{\mathrm{RMS}} \tag{3.13-14}$$

The three quantities, the *most probable speed*, the *average speed*, and the *root-mean-square speed*, are in constant ratio to one another, irrespective of molecular weight or temperature, that is,

$$v_{\mathrm{mode}} : \langle v \rangle : v_{\mathrm{RMS}} = \sqrt{2} : \sqrt{8/\pi} : \sqrt{3} \approx 1.414 : 1.596 : 1.732 \tag{3.13-15}$$

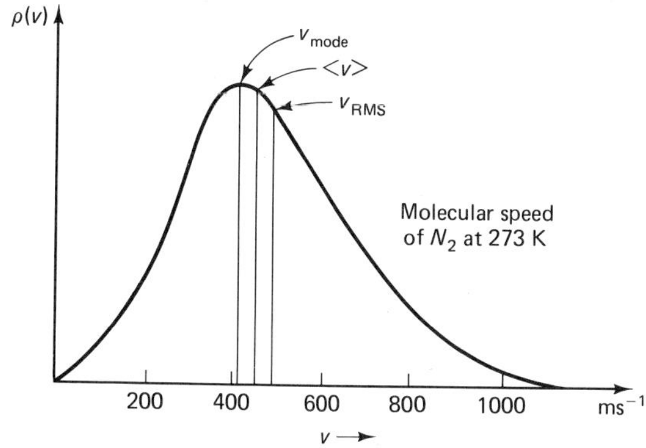

Figure 3.13-1 Average, most probable, and RMS speeds for Maxwell–Boltzmann distribution.

This relationship is shown graphically in Fig. 3.13-1. For N_2 molecules at 0°C, v_{mode}, $\langle v \rangle$, and v_{RMS} are, respectively, 4.02, 4.54, and 4.93×10^4 cm/s. These speeds are of the order of 1000 mi/hr.

■ EXAMPLE 3.13-1

Determine the number of molecules and the average speed of a molecule for a perfect gas composed of O_2 at $p = 1.013 \times 10^5$ N/m^2 (1 atm), $T = 300$ K, and $V = 10^2$ cm^3. Also, what is the probable number of O_2 molecules that have a speed in the range $10^{-6}\, v_{\mathrm{mode}}$ of the value $v = v_{\mathrm{mode}}$?

Solution. The number of molecules in the system is given by the perfect gas relation $pV = NkT$, with $k = 1.38 \times 10^{-23}$ J K^{-1},

$$N = \frac{1.013 \times 10^5 \times 10^{-4}}{1.38 \times 10^{-23} \times 300} = 2.45 \times 10^{21}$$

To obtain the average speed of an O_2 molecule, its mass must be first determined. The molecular weight of an O_2 molecule is 32, and the value of one atomic mass unit is 1.66×10^{-27} kg. Hence,

$$m = 32 \times 1.66 \times 10^{-27} = 5.31 \times 10^{-26} \text{ kg}$$

From Eq. (3.13-3), the average speed is

$$\langle v \rangle = \left(\frac{8kT}{\pi m}\right)^{1/2} = \left(\frac{8 \times 1.38 \times 10^{-23} \times 300}{3.14 \times 5.31 \times 10^{-26}}\right)^{1/2}$$

$$= 455 \text{ m s}^{-1}$$

This is quite fast, at least compared with ordinary experience (it is roughly 1000 mi/hr). From these values, it is seen that there is a great deal of activity going on in a little bit of gas.

To find the probable number of O_2 molecules that have a speed in the range $10^{-6}\,v_{\text{mode}}$ of the value $v = v_{\text{mode}}$, the distribution function for molecular speeds is used:

$$\rho(v) = 4\pi\left(\frac{m}{2\pi kT}\right)^{3/2} v^2 e^{-mv^2/2kT}$$

The number of molecules having the speed v_{mode} within a range of $10^{-6}\,v_{\text{mode}}$ is then

$$N\rho(v = v^{\text{mode}})(10^{-6}\,v_{\text{mode}})$$

$$= 4\pi N\left(\frac{m}{2\pi kT}\right)^{3/2} v_{\text{mode}}^2\, e^{-mv_{\text{mode}}^2/2kT}\,(10^{-6}\,v_{\text{mode}})$$

Using the fact that $v_{\text{mode}} = (2kT/m)^{1/2}$, this reduces to

$$\frac{4}{\sqrt{\pi}}\,Ne^{-1} \times 10^{-6} = 0.83 \times 10^{-6} N = 0.83 \times 10^{-6} \times 2.45 \times 10^{21} \simeq 2.04 \times 10^{15}$$

Thus, even in this small range of speeds ($10^{-6}\,v_{\text{mode}} \simeq 0.04$ cm s^{-1}) there is a large number of molecules. ■

■ EXAMPLE 3.13-2

What fraction of the total number of particles has speeds above a given speed v_0 in a system of particles described by the Maxwell–Boltzmann distribution? Sketch the graph of $N(>v_0)$ versus $(m/2kT)^{1/2}\,v$. Select v_0 to be $2v_m$, where $v_m = (2kT/m)^{1/2}$.

Solution. The fraction or probability of molecules having speeds in excess of some $v = v_0$ is given by

$$P(v \geq v_0) = \int_{v_0}^{\infty} \rho(v)\,dv = 4\pi\left(\frac{m}{2\pi kT}\right)^{3/2}\int_{v_0}^{\infty} v^2 e^{-mv^2/2kT}\,dv \tag{3.13-16}$$

Changing variables to

$$x = \left(\frac{mv^2}{2kT}\right)^{1/2} \tag{3.13-17}$$

$$P(v \geq v_0) = \frac{4}{\sqrt{\pi}}\int_{x_0}^{\infty} x^2 e^{-x^2}\,dx \tag{3.13-18}$$

The integral can be expressed in terms of the error function

$$\text{erf}\,x = \frac{1}{\sqrt{\pi}}\int_{-x}^{x} e^{-u^2}\,du = \frac{2}{\sqrt{\pi}}\int_{0}^{x} e^{-u^2}\,du \tag{3.13-19}$$

First, it is necessary to integrate by parts. Noting that

$$\frac{d}{dx}\,e^{-x^2} = -2xe^{-x^2}$$

or

$$d\left(-\frac{1}{2}e^{-x^2}\right) - xe^{-x^2} \tag{3.13-20}$$

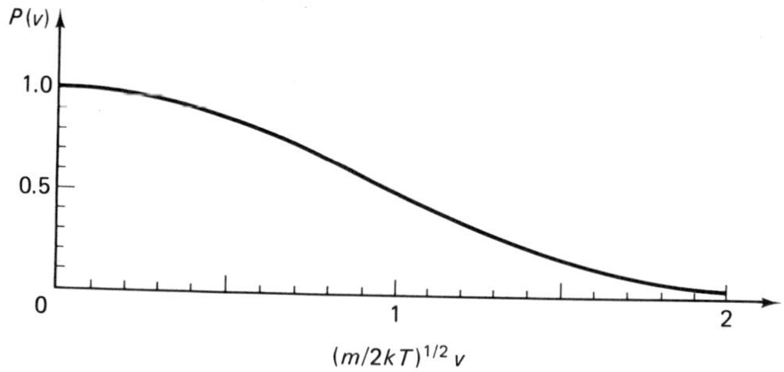

Figure 3.13-2

and that $\lim_{x\to\infty} xe^{-x^2} = 0$, Eq. (3.13-18) becomes

$$P(v \geq v_0) = \frac{4}{\sqrt{\pi}} \int_{x_0}^{\infty} xd\left(-\frac{1}{2}e^{-x^2}\right) = \frac{4}{\sqrt{\pi}}\left(-\frac{x}{2}e^{-x^2}\right]_{x_0}^{\infty} + \frac{1}{2}\int_{x_0}^{\infty} e^{-x^2}\,dx\right)$$

$$= \frac{2}{\sqrt{\pi}} x_0 e^{-x_0^2} + \frac{2}{\sqrt{\pi}} \int_{x_0}^{\infty} e^{-x^2}\,dx \tag{3.13-21}$$

The last integral is evaluated as

$$\frac{2}{\sqrt{\pi}}\int_{x_0}^{\infty} e^{-x^2}\,dx = \frac{2}{\sqrt{\pi}}\int_{x_0}^{\infty} e^{-x^2}\,dx - \frac{2}{\sqrt{\pi}}\int_{0}^{x_0} e^{-x^2}\,dx = 1 - \operatorname{erf} x_0 \tag{3.13-22}$$

Finally, returning to the original variable,

$$P(v \geq v_0) = \frac{2}{\sqrt{\pi}}\left(\frac{mv_0^2}{2kT}\right)^{1/2} e^{-mv_0^2/2kT} + 1 - \operatorname{erf}\left[\left(\frac{mv_0^2}{2kT}\right)^{1/2}\right] \tag{3.13-23}$$

For $v_0 = 2v_m = 2(2kT/m)^{1/2}$, $P(v \geq 2v_m) \approx 0.26$. Similarly, for $v_0 = 3v_m$, $P(v \geq 3v_m) \approx 0.10$. For $v \geq 6v_m$, the fraction drops to less than 0.01. The graph of $P(v \geq v_0)$ versus $(m/2kT)^{1/2}\, v$ is shown in Fig. 3.13-2. ■

3.14 COLLISIONS WITH A SURFACE OR OPENING

The Maxwellian distribution of molecular speeds occurs in a variety of problems that involve low-density gaseous systems (whenever the intermolecular forces are negligible compared with the kinetic energy of the molecules). A few examples of such problems are now being discussed.

In studying the collision of a gas with a wall, it is necessary to calculate the rate with which molecules of the gas strike a plane surface. The same equation gives the rate with which molecules will pass through a small hole into a vacuum (*effusion*). The equation arrived at applies to a hole so small that the leakage of molecules does not affect the equilibrium speed distribution in the bulk phase. Furthermore, it is assumed that the mean free path of the molecules is long compared with the diameter of the hole so that molecular collisions in the neighborhood of the hole can be neglected.

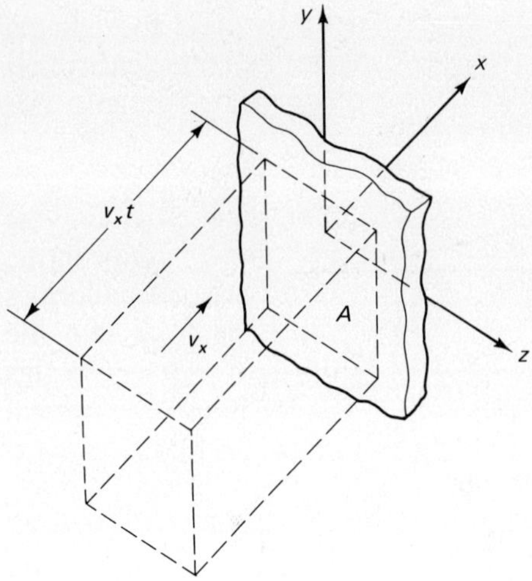

Figure 3.14-1 Molecular effusion.

Consider a plane surface of area A with a coordinate system in which the x axis is perpendicular to the surface. All the molecules having an x component of velocity v_x which are within a distance $v_x t$ of the surface will strike it in time t. It is not necessary to consider the velocity components v_y and v_z because they do not move molecules toward the surface.

The volume containing molecules that strike the area A in time t is $v_x tA$, and the number of molecules striking the surface is

$$\int_0^\infty \rho v_x f(v_x)\, tA\, dv_x \tag{3.14-1}$$

where $\rho = N/V$ is the number density of gas molecules and $f(v_x)\, dv_x$ is the probability that a molecule has a velocity between v_x and $v_x + dv_x$. The integration goes from zero to infinity, since only those molecules moving toward the surface from one side are involved. Thus, the flux Γ_n, which is the ratio of the number of molecules passing through a hole to the product of hole area and time, is

$$\Gamma_n = \int_0^\infty \rho v_x f(v_x)\, dv_x \tag{3.14-2}$$

With

$$f(v_x)\, dv_x = \left(\frac{m}{2\pi kt}\right)^{1/2} \exp\left(\frac{mv_x^2}{2kT}\right) dv_x \tag{3.14-3}$$

Eq. (3.14-2) gives

$$\Gamma_n = \rho\left(\frac{kT}{2\pi m}\right)^{1/2} = \rho\frac{\langle v\rangle}{4} \tag{3.14-4}$$

where the integral has been evaluated from expressions given in Section 1.9, and Eq. (3.13-3) has been used for $\langle v\rangle$. It is seen that there is a very simple relationship among the flux Γ_n, the average velocity $\langle v \rangle$, and the number density ρ of molecules. As previously described, the number flux density may be considered to be the number of molecules striking a surface per unit area per unit time or the number of molecules that will escape through a hole into a vacuum per unit area per unit time.

The result given by Eq. (3.14-4) is useful, for example, in estimating the rate at which gas will escape from a container that has a small pinhole in the wall (this process is called *molecular effusion*). In this case all molecules striking the pinhold will escape, instead of being reflected back into the interior. Such an estimate of the rate of effusion assumes, of course, that this rate is sufficiently slow for the remaining gas to be able to stay in thermal equilibrium [otherwise $f(v)$ would not be Maxwellian]. If the pinhole area is small enough, this is a good approximation.

To illustrate, consider a perfect gas composed of O_2 molecules. Assume that the pressure is 1.013×10^5 N m^{-2} (1 atm), $T = 300$ K, and $V = 10^2$ cm^3. The number of molecules in the system is given by $pV = NkT$, and since $k = 1.38 \times 10^{-23}$ J deg^{-1},

$$N = \frac{1.013 \times 10^5 \times 10^{-4}}{1.38 \times 10^{-23} \times 300} = 2.45 \times 10^{21}$$

To obtain the average speed of an O_2 molecule, its mass is first determined. The molecular weight of an O_2 molecule is 32, and the value of one atomic mass unit is 1.66×10^{-27} kg. Hence,

$$m = 32 \times 1.66 \times 10^{-27} = 5.31 \times 10^{-26} \text{ kg}$$

Using Eq. (3.13-3), the average speed is

$$\langle v \rangle = \left(\frac{8kT}{\pi m}\right)^{1/2} = \left(\frac{8 \times 1.38 \times 10^{-23} \times 300}{3.14 \times 5.31 \times 10^{-26}}\right)^{1/2}$$

$$= 445 \text{ m-s}^{-1}$$

In this case, the probable number of molecules that would escape through a pinhole with an area of 10^{-8} cm^2 would be

$$\frac{1}{4}\left(\frac{2.45 \times 10^{21}}{10^2}\right) \times 4.45 \times 10^4 \times 10^{-8} = 2.7 \times 10^{15}$$

which is about one out of every million molecules.

The rate of change of the number of molecules N in a container of volume V, the molecular effusion, is given by

$$\frac{dN}{dt} = -\left(\frac{N}{V}\right)\left(\frac{kT}{2\pi m}\right)^{1/2} A \qquad (3.14\text{-}5)$$

It is noted that dN/dt is proportional to $m^{-1/2}$, so that molecules of larger mass escape more slowly. In fact, at one time, measurements of the rates of effusion were used to determine the molecular weights of gases.

A more recent application involves the problem of separating ^{235}U from ^{238}U. The uranium, when combined with fluorine, forms the gas UF_6 (uranium

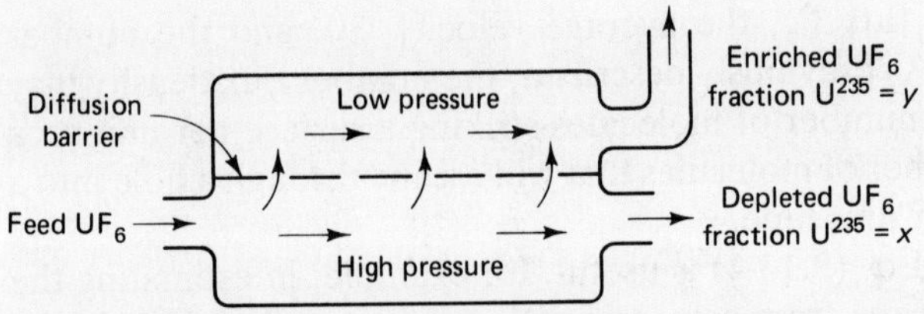

Figure 3.14-2 Diffusion isotope separation.

hexafluoride). When this gas effuses through a small hole, the molecules of UF_6 that contain the lighter isotope ^{235}U effuse faster than the others. In this way (by repeating the process several thousand times!) it is possible to achieve a high degree of separation of the isotopes, resulting in enriched ^{235}U. Figure 3.14-2 shows the gaseous diffusion process currently employed in the United States to separate uranium isotopes. The isotopes diffuse through a porous barrier at different rates, depending on their molecular weights.

The amount of separation obtained in one barrier is equal to the square root of the ratio of the molecular weights of the gases, that is,

$$\alpha = \frac{y(1-x)}{x(1-y)} = \sqrt{\frac{M_{UF_6^{238}}}{M_{UF_6^{235}}}} = \sqrt{\frac{352}{349}} = 1.00429 \tag{3.14-6}$$

3.15 PERFECT GAS IN AN EXTERNAL FIELD

Consider the effect of an external field acting on a perfect gas. The ever-present gravitational field is one example of such an external field. (Another possibility is an electric field on charged particles.)

For the Earth's atmosphere, one of the air molecules is treated as the system and the rest of the atmosphere as the heat reservoir of absolute temperature T. According to the Boltzmann distribution, the probability that at thermal equilibrium the gas molecule is in a single particle state with total energy (kinetic energy plus potential energy) is

$$\frac{1}{z}e^{-\beta\varepsilon} \tag{3.15-1}$$

Here,

$$\varepsilon = \tfrac{1}{2}mv^2 + mgz \tag{3.15-2}$$

where mgz is the potential energy due to the external field. This makes the probability (3.15-1) depend additionally on the position of the particle. The molecular partition function z will be calculated by considering a column of gas as shown in Fig. 3.15-1.

Let A be the base area and z the height. Then

$$z = A\int\int \exp(-\beta\varepsilon)\,dv\,dz$$

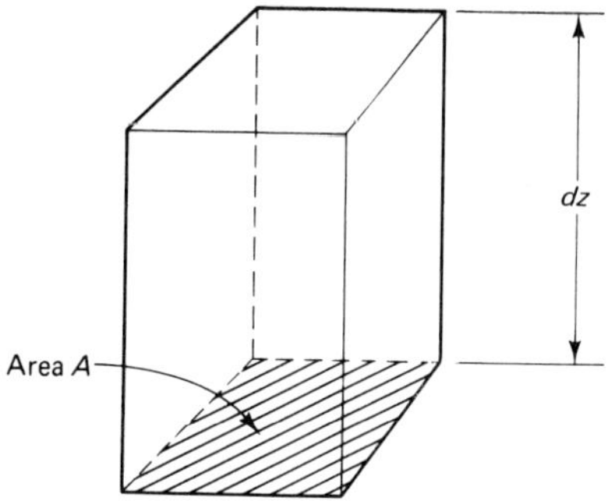

Figure 3.15-1 Fluid column in a gravitational field.

$$= A \int_0^z e^{-\beta mgz}\, dz \int_{-\infty}^{\infty} e^{-\beta m v^2/2}\, dv_x\, dv_y\, dv_z \tag{3.15-3}$$

where the integration on z is from 0 to z, and the integration on the velocity is from $-\infty$ to ∞. Equation (3.15-3) gives

$$z = \left(\frac{2\pi}{m\beta}\right)^{3/2} \frac{A}{\beta mg}\left(1 - e^{-\beta mgz}\right) \tag{3.15-4}$$

The partition function increases as the height z increases. Since the probability decreases as z increases, this means that the particle is more likely to be found near the ground ($z = 0$) than at great heights ($z >> 0$). Also, if βmgz is small, then

$$e^{-\beta mgz} \simeq 1 - \beta mgz \tag{3.15-5}$$

and (3.15-4) becomes

$$z \simeq \left(\frac{2\pi}{m\beta}\right)^{3/2} Az = \left(\frac{2\pi}{m\beta}\right)^{3/2} V \tag{3.15-6}$$

Thus, if z is small, the gravitational field has no effect.

For an isothermal atmosphere ($\beta = 1/kT =$ constant), the $\exp(-\beta m v^2/2)$ term in (3.15-3) is the same at all altitudes. Hence, the probability that the air molecule is at an altitude between z and $(z + dz)$ is given by

$$P(z)\, dz \propto \exp\left(\frac{-mgz}{kT}\right) dz \tag{3.15-7}$$

The above equation is applicable to each of the air molecules, and so the variation of the number of molecules per cubic meter with altitude in an isothermal atmosphere is given by

$$n = n_0 \exp\left(\frac{-mgz}{kT}\right) \tag{3.15-8}$$

where n_0 is the number of molecules per cubic meter at sea level. Now, the pressure of an ideal gas is proportional to the concentration. Therefore,

$$p = p_0 \exp\left(\frac{-mgz}{kT}\right) \tag{3.15-9}$$

where p_0 is the pressure at sea level. The pressure of an isothermal atmosphere decreases exponentially with height. Equation (3.15-9) is known as the *barometric pressure equation*. It gives the dependence of the pressure on the height z in an isothermal atmosphere of a single chemical species.

From Eq. (3.15-9) it is seen that there is a characteristic height $z_{c.h.}$ at which the atmospheric pressure decreases by the fraction $e^{-1} \simeq 0.37$. To estimate this characteristic height, substitute the numerical values, $k = 1.38 \times 10^{-23}$ J K^{-1}, $T = 300$ K, $g = 9.8$ m s^{-2}, and $m = 4.9 \times 10^{-26}$ kg to represent an average air molecule (80% nitrogen, 20% oxygen). This gives

$$z_{c.h.} = \frac{kT}{mg} = \frac{(1.38 \times 10^{-23})(300)}{(4.9 \times 10^{-26})(9.8)} = 8.6 \text{ km}$$

This is approximately 5 miles. According to Eq. (3.15-9), the pressure of an isothermal atmosphere at 300 K is reduced to 0.37 of its sea level value at an altitude of 8.6 km. In practice $z_{c.h.}$ is about 8 km near sea level. The difference arises from the significant decrease of temperature with altitude in the lower atmosphere. For example, at $z = 10$ km, the temperature is down to 223 K. A lower average value of temperature should be used in the equation for $z_{c.h.}$ to estimate the scale height near sea level. Also, lighter molecules such as H_2 and He will extend farther up, but these have largely escaped from the atmosphere in the course of time. The pressure calculated for N_2 at 290 K is plotted against height in Fig. 3.15-2. Experimental values of the total pressure as found from rocket observations are also shown. (The actual atmosphere of the earth is not exactly isothermal.)

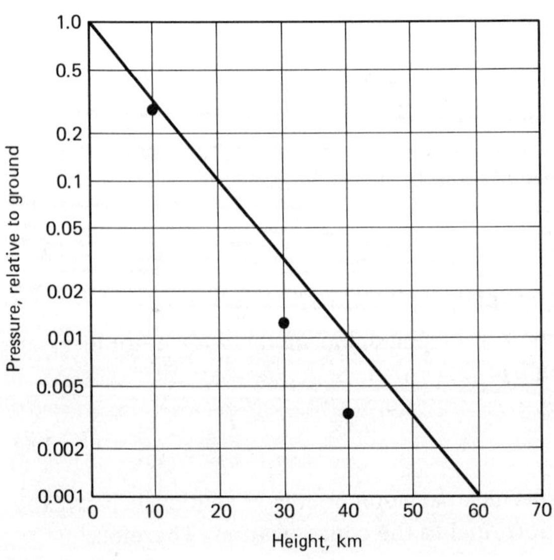

Figure 3.15-2 Pressure–altitude relation for atmospheric air.

■ EXAMPLE 3.15-1

Derive the barometric pressure formula by means of a simple balance of forces.

Solution. It is instructive to verify Eq. (3.15-9) by means of a simple balance of forces. The pressure p at an altitude z equals the weight of fluid above z divided by the area A. The relationship between p and z is found by considering an infinitesimal slab of fluid between z and $z + dz$. The corresponding pressures are denoted p and $p + dp$, respectively. At equilibrium, the net upward force, $pA - (p + dp)A$, must exactly balance the weight of the slab, $\rho(p)gA\,dz$:

$$dp = -\rho(p)g\,dz \tag{3.15-10}$$

where $\rho(p)$ is the density of the fluid when under a pressure p. The minus sign reflects the decrease in pressure with increasing altitude. When (3.15-10) is applied to a gas, the dependence of density on pressure must be taken into account. Assuming that the gas is ideal,

$$\frac{dp}{p} = -\frac{g}{RT}\,dz \tag{3.15-11}$$

$$\ln\frac{p(z)}{p(0)} = -\frac{gz}{RT}$$

$$p(z) = p(0)\,e^{-gz/RT} \tag{3.15-12}$$

where $p(0)$ represents the pressure at sea level. This is the same as Eq. (3.15-9). ■

■ EXAMPLE 3.15-2

An important application of the barometric equation is that it can be used to directly determine the value of Boltzmann's constant. This method was used by Jean B. Perrin and his coworkers in the early part of this century. They suspended colloidal particles of gamboze (a yellowish gum resin) in a liquid and observed the number of these particles at various heights z after the particles obtained an equilibrium distribution. These particles behave like a perfect gas, their number density being given by

$$n = n_0 \exp\left(\frac{-mgz}{kT}\right) \tag{3.15-13}$$

The effect of the liquid is to reduce the mass of the particles (by giving them a buoyancy force). If ρ is the mass density of gamboze, and v_0 is the volume of a colloidal particle, then the effective mass of a colloidal particle is

$$m = v_0(\rho - \rho_{\text{liquid}}) \tag{3.5-14}$$

From Fig. 3.15-3, it can be seen that the number of particles in dz about z_1 is to the number of particles in dz about z_2 as

$$\frac{n(z_1)}{n(z_2)} = e^{-mg(z_1 - z_2)/kT} \tag{3.15-15}$$

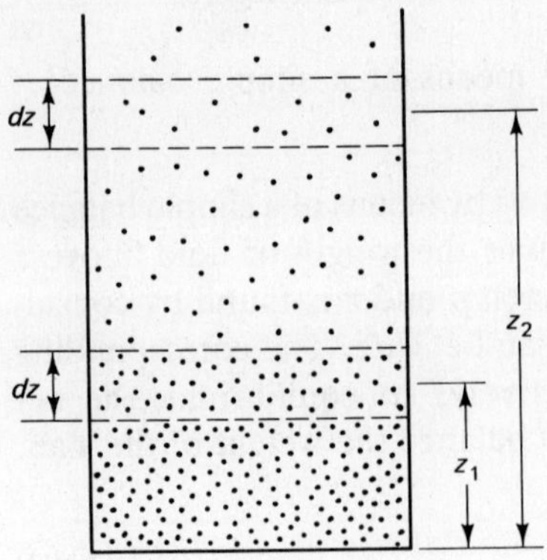

Figure 3.15-3 Colloidal particles in gamboze.

By making a large number of observations using a microscope with a very limited field of vision, Perrin was able to obtain several hundred records at each of two levels z. The fixed volume of observation, determined by the field of vision and depth of the focal plane of the microscope, was small enough to make the number of visible particles small (less than five or six), so that they could be readily counted. In this way, the average number of particles at z_1 and z_2 was obtained, and the ratio gives an experimental value for the left side of Eq. (3.15-15). Knowing z_1, z_2, m, g, and T_1, Perrin determined the value of the Boltzmann constant k from the right side of Eq. (3.15-15). Using an experimental value of the gas constant R, Perrin was then able to obtain a value for Avogadro's number N_0, where $N_0 = R/k$. The experiment is all the more remarkable when it is considered that, by observing less than six particles at a time, it is possible to accurately predict a number on the order of 10^{23}! ■

3.16 FACTORIZATION OF THE PHASE INTEGRAL

Just as the calculation of the partition function in Chapter 2 is simplified by expressing it as the product of simpler sums, so the calculation of the phase integral is simplified by expressing it as the product of simpler integrals. Quite generally, if $f_1(a)$ and $f_2(b)$ are two functions of independent variables a and b, then

$$\int_{a_1}^{a_2}\int_{a_1}^{b_2} f_1(a)\, f_2(b)\, db\, da = \int_{a_1}^{a_2} f_1(a)\, da \int_{a_1}^{b_2} f_2(b)\, db \tag{3.16-1}$$

Multiple integrals can be handled in the same way. It is often possible to factorize the integrand of the phase integral into independent functions, that is,

$$\int_{\text{phase}} = \int_{(2n)} \cdots \int f_1(q_1)\, f_2(q_2) \cdots f_n(q_n)\, f_{n+1}(p_1)\, f_{n+2}(p_2) \cdots f_{2n}(p_n)\, dq_1\, dq_2 \cdots dq_n\, dp_1\, dp_2 \cdots dq_n \tag{3.16-2}$$

Hence,

$$\int_{\text{phase}} = \int f_1(q_1)\,dq_1 \int f_2(q_2)\,dq_1 \cdots \int f_n(q_n)\,dq_n$$
$$\times \int f_{n+1}(p_1)\,dp_1 \int f_{n+2}(p_2)\,dp_2 \cdots \int f_{2n}(p_n)\,dp_n \tag{3.16-3}$$

and the problem of evaluating a $2n$-fold integral simplifies to one of evaluating $2n$ simple integrals.

The instantaneous value of the kinetic energy of a diatomic molecule (treated as a pair of point masses) is

$$\epsilon_{\text{trans}} + \epsilon_{\text{rot}} + \epsilon_{\text{vib}(k)} \tag{3.16-4}$$

where

$$\epsilon_{\text{trans}} = \frac{1}{2m}(p_x^2 + p_y^2 + p_z^2) \text{ is the translational energy} \tag{3.16-5}$$

$$\epsilon_{\text{rot}} = \frac{1}{2I}\left(p_\theta^2 + \frac{1}{\sin^2\theta}p_\varphi^2\right) \text{ is the rotational kinetic energy} \tag{3.16-6}$$

$$\epsilon_{\text{vib}(k)} = \frac{1}{2\mu}p_r^2 \text{ is the kinetic energy of vibration} \tag{3.16-7}$$

Thus, the total energy is, to a good degree of approximation, the sum of three independent terms:

$$\epsilon = \epsilon_{\text{trans}} + \epsilon_{\text{rot}} + \epsilon_{\text{vib}} \tag{3.16-8}$$

where ϵ_{trans} and ϵ_{rot} are as defined above, and

$$\epsilon_{\text{vib}} = \epsilon_{\text{vib}(k)} + \tfrac{1}{2}k\,(r - r_0)^2 \tag{3.16-9}$$

Therefore,

$$\begin{aligned} \epsilon^{-\epsilon/kT} &= e^{-(\epsilon_{\text{trans}} + \epsilon_{\text{rot}} + \epsilon_{\text{vib}})/kT} \\ &= e^{-\epsilon_{\text{trans}}/kT}\, e^{-\epsilon_{\text{rot}}/kT}\, e^{-\epsilon_{\text{vib}}/kT} \end{aligned} \tag{3.16-10}$$

that is, $e^{-\epsilon/kT}$ factorizes into three independent terms. The phase integral can be written as the product of three factors, corresponding respectively to translation, rotation, and vibration:

$$\int_{\text{phase}} = \int_{\text{trans}} \int_{\text{rot}} \int_{\text{vib}} \tag{3.16-11}$$

where

$$\int_{\text{trans}} = \iiiint\!\!\iint e^{-\epsilon_{\text{trans}}/kT}\,dp_x\,dp_y\,dp_z\,dx\,dy\,dz$$

$$\int_{rot} = \iiiint e^{-\epsilon_{\text{rot}}/kT}\,dp_\theta\,dp_\phi\,d\theta\,d\phi$$

$$\int_{\text{vib}} = \iint e^{-\epsilon_{\text{vib}}/kT}\,dp_r\,dr$$

It will be seen that each of these multiple integrals is itself expressible as the product of simpler factors.

Translational Partition Function

If x, y, and z are the coordinates of the center of gravity of the molecule, and m is the molecular mass, then

$$\epsilon_{\text{trans}} = \frac{1}{2} m (\dot{x}^2 + \dot{y}^2 + \dot{z}^2)$$

$$= \frac{1}{2m} (p_x^2 + p_y^2 + p_z^2) \qquad (3.16\text{-}12)$$

The translational factor in the phase integral for a molecule moving in a rectangular box of dimension a, b, and c, is

$$\int_{\text{trans}} = \int_{-\infty}^{\infty} \int_{-\infty}^{\infty} \int_{-\infty}^{\infty} \int_0^a \int_0^b \int_0^c \exp\left[\frac{-(p_x^2 + p_y^2 + p_z^2)}{2mkT}\right] dx\, dy\, dz\, dp_x\, dp_y\, dp_z \qquad (3\text{-}16.13)$$

Since

$$\exp\left[\frac{-(p_x^2 + p_y^2 + p_z^2)}{2mkT}\right] = \exp\left(\frac{-p_x^2}{2mkT}\right) \exp\left(\frac{-p_y^2}{2mkT}\right) \exp\left(\frac{-p_z^2}{2mkT}\right)$$

Eq. (3.16-13) can be written in the form

$$\int_{\text{trans}} = \int_{-\infty}^{\infty} e^{-p_x^2/2mkT}\, dp_x \int_{-\infty}^{\infty} e^{-p_y^2/2mkT}\, dp_y \int_{-\infty}^{\infty} e^{-p_z^2/2mkT}\, dp_z \int_0^a dx \int_0^b dy \int_0^c dz$$

Each of the first three integrals is of the form

$$\int_{-\infty}^{\infty} e^{-ax^2}\, dx = \left(\frac{\pi}{a}\right)^{1/2} \qquad (3.16\text{-}14)$$

that is

$$\int e^{-p_x^2/2mkT}\, dp_x = (2\pi mkT)^{1/2}$$

Hence, the translational phase integral is

$$\int_{\text{trans}} = (2\pi mkT)^{3/2} abc$$

$$= (2\pi mkT)^{3/2}\, V \qquad (3.16\text{-}15)$$

and the classical value of the partition function is, upon division by h^3, since $n = 3$,

$$z_{\text{trans}} = \frac{(2\pi mkT)^{3/2}\, V}{h^3} \qquad (3.16\text{-}16)$$

This is the same result as that obtained Chapter 2 by quantum mechanics. For the space part of the integral, the result

$$\iiint dx\, dy\, dz = V \tag{3.16-17}$$

is true whatever the shape of the vessel containing the gas, so the above derivation is of general applicability.

Vibrational Partition Function

According to previous discussion, the vibrational energy of a diatomic molecule is

$$\epsilon_{\text{vib}} = \frac{1}{2\mu} p_r^2 + \frac{1}{2} k (r - r_0)^2 \tag{3.16-18}$$

Hence,

$$\int_{\text{vib}} = \int_{-\infty}^{\infty} \int_{-\infty}^{\infty} \exp\left\{-\frac{1}{kT}\left[\frac{p_r^2}{2\mu} + \frac{k(r-r_0)^2}{2}\right]\right\} dr\, dp_r$$

$$= \int_{-\infty}^{\infty} \exp\left(\frac{-p_r^2}{2\mu kT}\right) dp_r \int_{-\infty}^{\infty} \exp\left[\frac{-k(r-r_0)^2}{2kT}\right] dr \tag{3.16-19}$$

Both of the integrals can be expressed in the form (3.16-14) (the second by putting $x = r - r_0$). Thus,

$$\int_{\text{vib}} = (2\pi\mu kt)^{1/2} \left(\frac{2\pi kT}{k}\right)^{1/2} = 2\pi kT \left(\frac{\mu}{k}\right)^{1/2} \tag{3.16-20}$$

Now, for an oscillator,

$$\left(\frac{\mu}{k}\right)^{1/2} = \frac{1}{2\pi\omega} \tag{3.16-21}$$

where ω is the circular frequency. The vibrational phase integral is

$$\int_{\text{vib}} = \frac{kT}{\omega} \tag{3.16-22}$$

Since ϵ_{vib} is a function of one generalized coordinate and its conjugate momentum, the corresponding classical partition function is

$$z_{\text{vib}} = \frac{kT}{h\omega} = \frac{T}{\theta_{\text{vib}}} \tag{3.16-23}$$

This is an approximation to the quantal result at high temperatures.

Rotational Partition Function

From previous development,

$$\epsilon_{\text{rot}} = \frac{1}{2I}\left(p_\theta^2 + \frac{p_\phi^2}{\sin^2\theta}\right) \tag{3.16-24}$$

and

$$\int_{\text{rot}} = \int_0^{2\pi}\int_0^{\pi}\int_{-\infty}^{\infty}\int_{-\infty}^{\infty} \exp\left[-\frac{1}{2IkT}\left(p_\theta^2 + \frac{p_\varphi^2}{\sin^2\theta}\right)\right] dp_\theta\, dp_\phi\, d\theta\, d\varphi \tag{3.16-25}$$

Examination of Fig. 3.3-3 shows that all spatial configurations are covered if θ runs from 0 to π and ϕ from 0 to 2π. Because of the term $p_\phi^2/\sin^2\theta$, complete factorization is not immediately possible. However,

$$\int_{\text{rot}} = \int_0^{2\pi} d\phi \int_{-\infty}^{\infty} \exp\left(\frac{-p_\theta^2}{2IkT}\right) dp_\theta \int_0^{\pi}\int_{-\infty}^{\infty} \exp\left(\frac{-p_\varphi^2}{2IkT/\sin^2\theta}\right) dp_\phi\, d\theta \tag{3.16-26}$$

The first two integrals in this expression are immediately solved, giving, respectively, 2π and $(2\pi IkT)^{1/2}$. The double integral is most easily solved by integrating first with respect to p_ϕ and then with respect to θ:

$$\int_0^{\pi}\int_{-\infty}^{\infty} \exp\left(\frac{-p_\phi^2}{2IkT/\sin^2\theta}\right) dp_\phi\, d\theta = \int_0^{\pi} [(2\pi IkT)^{1/2}\sin\theta]\, d\theta$$

$$= (2\pi IkT)^{1/2} \times 2 \tag{3.16-27}$$

Hence, Eq. (3.16-25) simplifies to

$$\int_{\text{rot}} = 8\pi^2 IkT \tag{3.16-28}$$

Since $n = 2$, division by h^n gives

$$z_{\text{rot}} = \frac{8\pi^2 IkT}{h^2} \tag{3.16-29}$$

This was established in Chapter 2.

3.17 INADEQUACIES OF THE CLASSICAL THEORY

Historically, the inadequacy of classical statistical mechanics showed up in the failure of the equipartion of energy theorem which will now be derived. In many cases, the energy of a system can be expressed in a form that is quadratic in the position and momentum variables of the system. For example, the kinetic energy of a free particle mass m, which has velocity v and momentum p equal to mv, can be expressed in the form

$$\frac{p_x^2}{2m} + \frac{p_y^2}{2m} + \frac{p_z^2}{2m} \tag{3.17-1}$$

Similarly, the instantaneous value of the kinetic energy of a diatomic molecule is the sum of three terms:

$$\varepsilon_{\text{trans}} + \varepsilon_{\text{rot}} + \varepsilon_{\text{vib}} \tag{3.17-2}$$

where

$$\varepsilon_{\text{trans}} = \frac{1}{2m}\,(p_x^2 + p_y^2 + p_z^2) \tag{3.17-3}$$

is the translational energy,

$$\varepsilon_{\text{rot}} = \frac{1}{2I}\left(p_\theta^2 + \frac{1}{\sin^2\theta}\,p_\phi^2\right) \tag{3.17-4}$$

is the rotational energy, and

$$\varepsilon_{\text{vib}} = \frac{1}{2\mu}\,p_r^2 \tag{3.17-5}$$

is the kinetic energy of vibration. In addition, there is also a potential energy V. If effects of external fields are neglected, V does not change with the orientation of the molecule as a whole, that is, it is independent of x, y, z, θ, and ϕ. V does, however, depend upon r, and, if Hooke's law is assumed,

$$V = \tfrac{1}{2}k(r - r_0)^2 \tag{3.17-6}$$

where k is the force constant and V is the potential energy relative to that of the equilibrium configuration. Thus, the total energy is, to a good degree of approximation, the sum of three independent terms:

$$\varepsilon = \varepsilon_{\text{trans}} + \varepsilon_{\text{rot}} + \varepsilon_{\text{vib}} \tag{3.17-7}$$

where $\varepsilon_{\text{trans}}$ and ε_{rot} are as defined above, and

$$\varepsilon_{\text{vib}} = \frac{1}{2\mu}\,p_r^2 + \frac{1}{2}k(r - r_0)^2 \tag{3.17-8}$$

Therefore,

$$\begin{aligned}\varepsilon^{-\varepsilon/kT} &= e^{-(\varepsilon_{\text{trans}} + \varepsilon_{\text{rot}} + \varepsilon_{\text{vib}})/kT} \\ &= e^{-\varepsilon_{\text{trans}}/kT}\, e^{-\varepsilon_{\text{rot}}/kT}\, e^{-\varepsilon_{\text{vib}}/kT}\end{aligned} \tag{3.17-9}$$

that is, $e^{-\varepsilon/kT}$ factorizes into three independent terms.

In each of the examples discussed above, it happens that the part of the energy being considered depends upon the squares of the variables concerned. The contribution to the molecular energy associated with a particular variable w (w might be a momentum, e.g., p_θ or a coordinate, e.g., r) is of the form

$$\varepsilon_w = cw^2 \tag{3.17-10}$$

where c is some parameter.

Correspondingly, the phase integral, and hence the partition function, contains the factor

$$\int_{-\infty}^{\infty} e^{-cw^2/kT}\,dw = \left(\frac{\pi kT}{c}\right)^{1/2} = z_w \tag{3.17-11}$$

The contribution to the molar energy is then

$$NkT^2\left(\frac{\partial \ln z_w}{\partial T}\right) = \frac{NkT}{2} = \frac{RT}{2} \tag{3.17-12}$$

In other words, the mean value of each contribution to the molar total energy, which is quadratic in either a position or a momentum coordinate, is $RT/2$ or $\frac{1}{2}$ kT per molecule.

The total number of quadratic terms per molecule is the sum of those due to translation, rotation, and vibration. For a molecule with N atoms, there are $3N-6$ normal modes of vibrations if the molecule is nonlinear, $3N-5$ normal modes if the molecule is linear. Thus, for a nonlinear molecule there are $3+3+2(3N-6)=6N-6$, and for a linear molecule, $3+2+2(3N-5)=6N-5$. The molar internal energy is given by

$$\overline{E} = \begin{cases} (3N-3)\,RT & \text{(nonlinear)} \\ (3N-\frac{5}{2})\,RT & \text{(linear)} \end{cases} \tag{3.17-13}$$

while the molar heat capacity is

$$\overline{C}_v = \begin{cases} (3N-3)\,R & \text{(nonlinear)} \\ (3N-\frac{5}{2})\,R & \text{(linear)} \end{cases} \tag{3.17-14}$$

These results are independent of the mass of the atoms. A comparison of predicted with experimental values of C_V at 25°C is given in Table 3.17-1.

The equipartion of energy theorem is seen to fail for all but monatomic gases. However, as the temperature is increased, C_V does approach its equipartion value in the limit $T \to \infty$ (if the molecule does not dissociate first). The error in classical statistical mechanics is resolved in the next chapter with the adoption of quantum-mechanical description of molecular systems.

TABLE 3.17-1 HEAT CAPACITIES OF GASES AT 25°C

Molecule	C_v/R (experimental)	C_v/R (equipartion)
He	1.50	1.5
Ne	1.50	1.5
H_2	2.47	3.5
N_2	2.50	3.5
O_2	3.15	3.5
I_2	3.43	3.5
CO_2	3.47	6.5
NH_3	3.29	9.00
C_2H_6	5.33	21.0

■ EXAMPLE 3.17-1

(a) Evaluate the constant A in the distribution function $Ae^{-E/kT}$ when the energy of a particle is of the form $\epsilon = \Sigma c_i q_i^2$, where the c_i are constants and the q_i are position and momentum coordinates. (b) Show that the average energy per particle is equal to $\frac{1}{2}kT$ times the number of squared position and momentum terms in the energy expression.

Solution

(a) The normalization condition for the distribution function $Ae^{-E/kT} = Ae^{-(\Sigma c_i q_i^2)/kT}$ is

$$N = A\int_{-\infty}^{\infty}\int_{-\infty}^{\infty}\cdots\int_{-\infty}^{\infty} e^{-\left(\sum_{i=1}^{n} c_i q_i^2\right)/kT}\, dq_1\, dq_2\cdots dq_n$$

$$= A\left(\int_{-\infty}^{\infty} e^{-c_1 q_1^2/kT}\, dq_1\right)\left(\int_{-\infty}^{\infty} e^{-c_2 q_2^2/kT}\, dq_2\right)\cdots\left(\int_{-\infty}^{\infty} e^{-c_n q_n^2/kT}\, dq_n\right)$$

$$= A\,(2)\left[\frac{1}{2}\left(\frac{\pi kT}{c_1}\right)^{1/2}\right]\left[\frac{1}{2}\left(\frac{\pi kT}{c_2}\right)^{1/2}\right]\cdots\left[\frac{1}{2}\left(\frac{\pi kT}{c_n}\right)^{1/2}\right]$$

$$= A\,\frac{(\pi kT)^{n/2}}{(c_1, c_2, \ldots, c_n)^{1/2}}$$

which gives

$$A = N\,\frac{(c_1, c_2, \ldots, c_n)^{1/2}}{(\pi kT)^{n/2}}$$

For three velocity coordinates with $c_1 = c_2 = c_3 = m/2$, $n = 3$, this gives

$$A = N\left(\frac{m}{2\pi kT}\right)^{3/2}$$

which leads to Maxwell's velocity distribution

$$N(v_x, v_y, v_z) = N\left(\frac{m}{2\pi kT}\right)^{3/2} e^{-m(v_x^2 + v_y^2 + v_z^2)/2kT}$$

(b) The above shows that

$$\int_{-\infty}^{\infty}\int_{-\infty}^{\infty}\cdots\int_{-\infty}^{\infty} e^{-E/kT}\, dq_1\, dq_2\cdots dq_n = \frac{(\pi kT)^{n/2}}{(c_1, c_2, \ldots, c_n)^{1/2}}$$

If both sides are differentiated with respect to kT, this gives

$$\frac{1}{(kT)^2}\int_{-\infty}^{\infty}\int_{-\infty}^{\infty}\cdots\int_{-\infty}^{\infty} Ee^{-E/kT}\, dq_1\, dq_2\cdots dq_n = \frac{\pi^{n/2}(n/2)(kT)^{n/2-1}}{(c_1, c_2, \ldots, c_n)^{1/2}} = \frac{N}{A}\frac{n/2}{kT}$$

or, transposing terms,

$$\frac{1}{N}\int_{-\infty}^{\infty}\cdots\int_{-\infty}^{\infty} E(Ae^{-E/kT})\, dq_1\cdots dq_n = n\left(\frac{1}{2}kT\right)$$

But the left-hand side is recognized as the average energy of the particle. Thus, the energy equals $\frac{1}{2}kT$ times the number of squared position and/or momentum terms (equipartition theorem). ■

■ EXAMPLE 3.17-2

A model for a diatomic molecule that is free to rotate, but not to vibrate, is a dumbbell with the moment of inertia about the axis passing through the masses approximately equal to zero. Find the molar specific heat, using the equipartition theorem.

Solution. The energy of one diatomic molecule is equal to

$$E = \frac{1}{2m}(p_x^2 + p_y^2 + p_z^2) + \frac{1}{2}(k_x x^2 + k_y y^2 + k_z z^2)$$

There are six terms of the form cq^2. By the equipartition theorem, 1 mole has an energy

$$E_{\text{tot}} = 6N_0\left(\frac{kT}{2}\right) = 3N_0 kT$$

where N_0 is Avogadro's number. The molar specific heat is thus

$$C_V = \frac{dE_{\text{tot}}}{dT} = 3N_0 k = 3R$$

which is the *Dulong–Petit law*. This result is found experimentally to be correct at high temperatures, but at low temperatures quantum effects cause a change. (Quantum effects will be discussed in Chapter 4.) ■

3.18 THE GIBBS PARADOX

Another failure of classical statistical mechanics is the Gibbs paradox. In classical statistical mechanics, it was originally assumed in the nineteenth century that, in principle, a distinction could be made between the atoms of a monatomic gas. This assumption led to the difficulty which is now discussed.

Consider two equal volumes V, both of which contain N atoms of the same ideal monatomic gas, and are separated, initially, by a solid partition as shown in Fig. 3.18-1(a). Both gases are in thermal equilibrium with a heat reservoir of absolute temperature T. The partition is then removed, and the gases are allowed to mix. If the monatomic gas is the same in both the volumes V in (a), there is no change in the total entropy on mixing. If the atoms in the two volumes V are different, there is an increase in the total entropy when the gases mix.

The Sackur–Tetrode equation gives the entropy of an ideal monatomic gas in terms of N, V, and T as

$$S = Nk\left[\ln(qV/N) + \tfrac{3}{2}\ln(mkT/2\pi\hbar^2) + \tfrac{5}{2}\right] \tag{3.18-1}$$

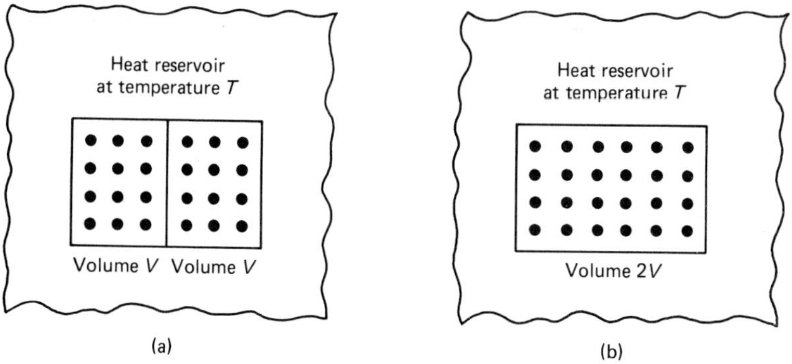

Figure 3.18-1 Entropy of mixing. The partition in (a) is removed and the gases in the two separate volumes V are allowed to mix and fill the volume $2V$ in (b). If the monatomic gas is the same in both the volumes V in (a), there is no change in the total entropy on mixing. If the atoms in the two volumes V are different, there is an increase in the total entropy, when the gases mix.

where g is the total degeneracy associated with the internal degrees of freedom. The entropy of the gas in each compartment (of volume V) is thus

$$S_1 = S_2 = Nk\,[\ln (V/N) + C_1] \tag{3.18-2}$$

with

$$C_1 = \tfrac{3}{2}\ln (mkT/2\pi\hbar^2) + \ln q + \tfrac{5}{2} \tag{3.18-3}$$

being a constant which has been separated from the volume V. The total initial entropy of the two separate volumes of gas is

$$S_i = S_1 + S_2 = 2Nk\,[\ln (V/N) + C_1] \tag{3.18-4}$$

When the partition is removed, there are $2N$ atoms in a volume $2V$, as shown in Fig. 3.18-1(b), and so the final entropy is

$$S_f = 2Nk\,[\ln (2V/2N) + C_1] \tag{3.18-5}$$

which is the same as (3.18-4). It is seen that there is no entropy change upon mixing of the same gas.

The situation would be different had classical statistical mechanics been strictly applied and the $N!$ term not included in the expression for the partition function

$$Z = g^N V^N (mkT/2\pi\hbar^2)^{3N/2}/N! \tag{3.18-6}$$

Then the Sackur–Tetrode equation for the entropy would have been

$$S = Nk\,[\ln V + \ln q + \tfrac{3}{2}\ln (mkT/2\pi\hbar^2) + \tfrac{3}{2}] \tag{3.18-7}$$

$$= Nk\,[\ln V + C_1 - 1] \tag{3.18-8}$$

and the initial and final entropies would have become

$$S_i = 2Nk\,(\ln V + C_1 - 1) \tag{3.18-9}$$

$$S_f = 2Nk\,(\ln 2V + C_1 - 1) \tag{3.18-10}$$

giving an entropy of mixing

$$S_f - S_1 = 2Nk \ln 2 \tag{3.18-11}$$

This was the *Gibbs paradox* of the nineteenth century. It was only resolved by including the $N!$ term in Eq. (3.18-6) for the N-particle partition function. The $N!$ does not arise in the classical approach when it is assumed that the particles are distinguishable.

3.19 SUMMARY

In its most exact formulation, statistical mechanics should be based on quantum mechanics. The quantal treatment is, in many ways, simpler than the classical, while at the same time being fundamentally more correct. As remarked earlier, however, certain problems are more easily solved in terms of classical mechanics. This is especially true when integration replaces summation.

The conditions under which it is permissible to replace quantum by classical mechanics are easily deduced from a general theorem known as the limiting principle (Fowler). This states that quantum mechanical calculation of the properties of molecules in states with large quantum numbers yields the same results as the corresponding calculation by classical mechanics, or

$$\lim_{n\to\infty} \text{(quantized system)} = \text{(classical system)}$$

This is true, for example, for the translational properties of molecules of a gas, since the quantum numbers become very large for the important translational energy levels.

The development of classical statistical mechanics rests on the notion of an ensemble, of which the canonical ensemble is the most representative. A canonical ensemble consists of N_s distinguishable systems which are identical in that they have the same number of molecules, N, volume V, and thermodynamic state. As N_s becomes large, the time (or observable) average of any mechanical property of a system, such as p or E, becomes equal to the instantaneous ensemble average value for the property. Nonmechanical properties, such as T and S, must be calculated using known thermodynamic relationships. The ensemble acts as a single supersystem having a volume N_sV, number of molecules $N_s N$, and a total energy of E_t. The classical state of a system of molecules is described by means of phase space. For a one-dimensional, one-particle system, for example, phase space has position coordinate $q = x$ and momentum coordinate $p = p_x$. For molecular systems involving n atoms, complete classical descriptions require $3n$ position coordinates and $3n$ momentum coordinates. The state of the system is then indicated as a phase point in $6n$-dimensional phase space.

The evaluation of the number of configurations (thermodynamic probability) for a collection of N molecules proceeds by counting the number of ways individual molecules can be distributed among cells. The notion of a cell in phase space assumes the same role in classical statistical mechanics that the notion of a quantum state assumes in quantum statistics. In general, the energy of a cell depends on its position coordinate and its momentum coordinate; thus, for a one-dimensional system,

$$\epsilon(p_x, x) = \frac{1}{2m} p_x^2 + V(x) \tag{3.19-1}$$

where $V(x)$ is the potential energy function of the system. If each cell is sufficiently small (p_x and x varying only slightly within the cell), then the energy within the cell can be considered to be constant. The basic problem in classical statistical mechanics is then the following. A given number of molecules, N (a very large number), is given a definite total energy, E. The molecules may be distributed in many different ways among the cells in phase space. If the cells are given the serial labels, 0, 1, 2, . . . , and if n_0 molecules are placed in cell 0, n_1 in cell 1, n_2 in cell 2, . . . , then the number of configurations leading to this distribution is

$$W = \frac{N!}{\Pi n_i!} \tag{3.19-2}$$

The most probable distribution of molecules among the cells is the set of values of the n_i which maximizes W subject to the conditions

$$\sum n_i = N \tag{3.19-3}$$

$$\sum n_i \epsilon_i = E \tag{3.19-4}$$

where ϵ_i is the energy of the ith cell. The problem is mathematically solved, giving the most probable distribution as

$$n_i = A \exp(-\beta\epsilon_i) \tag{3.19-5}$$

where A and β are constants determined by the constraints (on N and E).

Since the population of a cell is proportional to the area of the cell, $\Delta x\, \Delta p_x$,

$$n_i = A' \exp(-\beta\epsilon_i)\, \Delta x\, \Delta p_x \tag{3.19-6}$$

where A has been replaced by $A'\, \Delta x\, \Delta p_x$ and A' is no longer dependent on the size chosen for the cell. The condition $\Sigma n_i = N$ now becomes

$$N = A' \sum_{\text{all cells}} \exp(-\beta\epsilon_i)\, \Delta x\, \Delta p_x \tag{3.19-7}$$

If the cells are made sufficiently small, this sum can be replaced by an integral and ϵ_i can be regarded as a continuously varying function, $\epsilon_i = \epsilon(x, p_x)$, giving

$$N = A' \int \exp[-\beta\epsilon(x, p_x)]\, dx\, dp_x \tag{3.19-8}$$

so that

$$A' = \frac{N}{\int \exp[-\beta\epsilon(x, p_x)]\, dx\, dp_x} \tag{3.19-9}$$

The denominator in the above expression is known as the phase integral. The average energy of the system is

$$\langle\epsilon\rangle = \frac{E}{N} = \frac{\int \epsilon(x, p_x) \exp[-\beta\epsilon(x, p_x)]\, dx\, dp_x}{\int \exp[-\beta\epsilon(x, p_x)]\, dx\, dp_x} \tag{3.19-10}$$

$$= -\frac{\int (d/d\beta) \exp[-\beta\epsilon(x, p_x)]\, dx\, dp_x}{\int \exp[-\beta\epsilon(x, p_x)]\, dx\, dp_x} \tag{3.19-11}$$

$$= -\frac{d}{d\beta} \ln \int \exp[-\beta\epsilon(x, p_x)]\, dx\, dp_x \tag{3.19-12}$$

The constant β is shown to be equal to $1/kT$ by evaluating this integral for a one-dimensional gas confined to a container of length a, for which the thermal energy is known to be

$$\langle\epsilon\rangle = \tfrac{1}{2}kT \tag{3.19-13}$$

For such a system, the energy is entirely kinetic,

$$\epsilon(x, p_x) = \frac{p_x^2}{2m} \tag{3.19-14}$$

so that, since p_x can range from $-\infty$ to $+\infty$, and x can have values between 0 and a,

$$\begin{aligned} \int_0^a \int_{-\infty}^{\infty} \exp\left(\frac{\beta p_x^2}{2m}\right) dp_x\, dx &= a \int_{-\infty}^{\infty} \exp\left(\frac{\beta p_x^2}{2m}\right) dp_x \\ &= \left(\frac{2\pi m}{\beta}\right)^{1/2} a \end{aligned} \tag{3.19-15}$$

(Recall that $\int_{-\infty}^{\infty} \exp(-ax^2)\, dx = (\pi/a)^{1/2}$.) Thus,

$$\int \exp[-\beta\epsilon(x, p_x)]\, dx\, dp_x = \left(\frac{2\pi m}{\beta}\right)^{1/2} a$$

and

$$\ln \int \exp[-\beta\epsilon(x, p_x)]\, dx\, dp_x = -\tfrac{1}{2} \ln \beta + \tfrac{1}{2} \ln (2\pi m a^2) \tag{3.19-16}$$

so that (3.19-12) becomes

$$\langle\epsilon\rangle = -\frac{1}{2\beta} \tag{3.19-17}$$

Equating this to (3.19-13) gives

$$\beta = \frac{1}{kT} \tag{3.19-18}$$

The phase integral $\int\int \exp[-\epsilon(x,p_x)/kT]\, dx\, dp_x$ is the classical analogue of the partition function in quantum-statistical mechanics. A more direct correspondence is obtained when the classical partition function is defined as

$$z_{\text{class}} = \frac{1}{h}\int\int \exp\left(\frac{\epsilon}{kT}\right) dp_x\, dx \tag{3.19-19}$$

Then, if the temperature is sufficiently high that the intervals between successive energy levels are small compared to kT, the classical partition function will approximate the quantum partition function. For three-dimensional molecular systems involving n atoms, a $6n$-dimensional phase space is required if a phase point is to indicate the classical state of the system. Each quantum state in such a $6n$-dimensional phase space defines a cell of volume h^{3n}; the classical equivalent to the partition function is therefore

$$z_{\text{class}} = \frac{1}{h^{3n}}\int \underset{\substack{\text{all phase}\\ \text{space}}}{\cdots} \int \exp\left(-\frac{E}{kT}\right) dp_{x1}\, dx_1 \cdots dp_{zn}\, dz_n \tag{3.19-20}$$

where E is the energy expressed as a function of the $3n$ position coordinates, $x_1, y_1, z_1, \ldots, z_n$ and the $3n$ associated momentum coordinates, $p_{x1}, p_{y1}, \ldots, p_{zn}$.

For example, a monatomic gas molecule moving in a rectangular box of dimensions a by b by c (volume $= V = abc$) has the classical energy

$$E = \frac{1}{2m}(p_x^2 + p_y^2 + p_z^2) \tag{3.19-21}$$

so that the partition function is

$$\begin{aligned} z_{\text{class}} &= \frac{1}{h^3}\int_{-\infty}^{\infty}\int_{-\infty}^{\infty}\int_{-\infty}^{\infty}\int_0^c\int_0^b\int_0^a \exp\left[-\frac{(p_x^2+p_y^2+p_z^2)}{2mkT}\right] dx\, dy\, dz\, dp_x\, dp_y\, dp_z \\ &= \frac{1}{h^3}\int_{-\infty}^{\infty} \exp\left(-\frac{p_x^2}{2mkT}\right) dp_x \int_{-\infty}^{\infty} \exp\left(-\frac{p_y^2}{2mkT}\right) dp_y \\ &= \int_{-\infty}^{\infty} \exp\left(-\frac{p_z^2}{2mkT}\right) dp_z \int_0^a dx \int_0^b dy \int_0^c dz \\ &= \frac{(2\pi mkT)^{3/2}}{h^3}\, abc = \left(\frac{2\pi mkT}{h^2}\right)^{3/2} V \end{aligned} \tag{3.19-22}$$

This result is exactly the same as the partition function obtained from quantum theory. Classical application of the canonical ensemble distribution to a single particle gives the Maxwell–Boltzmann distribution function, which states that the probability of a particle being in a cell $dE = dx\, dy\, dz\, dp_x\, dp_y\, dp_z$ of phase space is

$$F(x, y, x, p_x, p_x, p_y, p_z)\, dE = Ce^{-\epsilon/kT}\, dE \tag{3.19-23}$$

where C is a constant determined by normalization. Carrying this out in terms of different variables results in

$$F(p) = \left(\frac{1}{2\pi mkT}\right)^{3/2} e^{-p^2/2mkT} \tag{3.19-24}$$

for the momentum distribution function,

$$F(v_x, v_y, v_z) = \left(\frac{m}{2\pi kT}\right)^{3/2} e^{-m(v_x^2+v_y^2+v_z^2)/2kT} \tag{3.19-25}$$

for the velocity distribution function, and

$$F\left(v = \sqrt{v_x^2 + v_y^2 + v_z^2}\right) = 4\pi v^2 \left(\frac{m}{2\pi kT}\right)^{3/2} e^{-mv^2/2kT} \tag{3.19-26}$$

for the speed distribution function. The latter also shows that the Maxwell–Boltzmann distribution has the form

$$n(v)\,dv = Av^2T^{-3/2}\exp\left(\frac{-Bv^2}{T}\right)dv \tag{3.19-27}$$

where A and B are constants and T is the absolute temperature. Such a distribution indicates that the majority of molecules have velocities close to a certain mean velocity v. If the temperature is raised, the maximum shifts to higher energies and the curve flattens out. If the temperature is reduced to zero, the mean velocity also becomes zero, indicating that all the molecules have zero energy at this temperature.

The most important assumption of classical statistics, however, is that any number of particles may have identical energies. Since it will be shown that an energy band is made up of a large number of discrete levels each of which can only be occupied by one electron at a time, it is unlikely that the energy distribution in a band will follow Maxwell–Boltzmann statistics. (Chapter 4 will discuss the Fermi–Dirac statistics for electrons.)

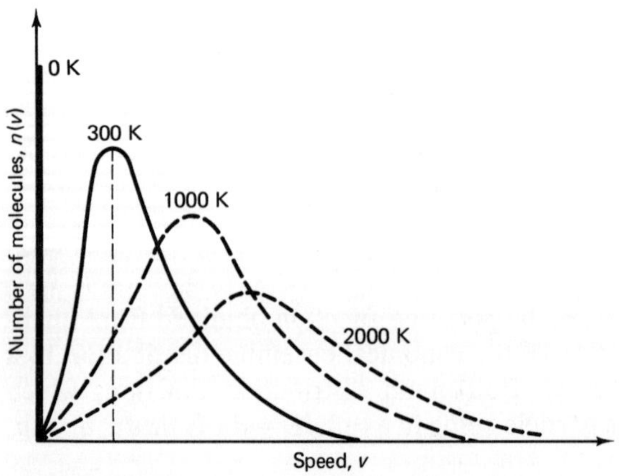

Figure 3.19-1

The contributions to the energy ϵ of an ideal gas are translational, rotational, vibrational, electronic, and nuclear, giving

$$\epsilon = \epsilon_{\text{trans}} + \epsilon_{\text{rot}} + \epsilon_{\text{vib}} + \epsilon_{\text{elec}} + \epsilon_{\text{nuc}} \tag{3.19-28}$$

$$z = z_{\text{trans}}\, z_{\text{rot}}\, z_{\text{vib}}\, z_{\text{elec}}\, z_{\text{nuc}} \tag{3.19-29}$$

or, on a mole basis,

$$Z = \frac{z^N}{N!}\, z_{\text{rot}}\, z_{\text{vib}}\, z_{\text{elec}}\, z_{\text{nuc}} \tag{3.19-30}$$

The $N!$ correction term is traditionally associated with the translational partition function because not all gases have vibrational and rotational contributions, and the nuclear and electronic contributions are negligible for most gases. Equation (3.19-30) does not contain minor contributions for rotational–vibrational interactions or for intramolecular rotation energy levels.

For a particle mass m in a three-dimensional box of side a, quantum mechanics gives the energy as

$$\epsilon_{\text{trans}} = (n_x^2 + n_y^2 + n_z^2)\frac{h^2}{8ma^2} \tag{3.19-31}$$

where the n_i are integers and h is Planck's constant. Classical mechanics gives the energy in the form of Eq. (3.19-14). The values of ϵ_{trans} are on the order of 10^{-42} J apart and at room temperature a large number of states are occupied by the molecules (because $kT \approx 4 \times 10^{-12}$ J). The quantum-mechanical partition function for translation is given by

$$z_{\text{trans}} = \sum_{n_x, n_y, n_z} e^{-(n_x^2 + n_y^2 + n_z^2)h^2/8ma^2kT}$$

$$= \left(\sum_{n_x=0}^{\infty} e^{-n_x^2 h^2/8ma^2 kT}\right)\left(\sum_{n_y=0}^{\infty} e^{-n_y^2 h^2/8ma^2 kT}\right)\left(\sum_{n_z=0}^{\infty} e^{-n_z^2 h^2/8ma^2 kT}\right)$$

Replacing the summations by integration because of the large number of states occupied gives

$$z_{\text{trans}} = \left(\int_0^{\infty} e^{-n^2h^2/8ma^2kT}\, dn\right)^3 = \left[(2\pi mkT)^{1/2}\frac{a}{h}\right]^3 = (2\pi mkT)^{3/2}\frac{V}{h^3} \tag{3.19-32}$$

which is precisely the same result as that obtained from strict classical theory.

The average translational kinetic energy of a molecule is $\frac{3}{2}kT$, which amounts to $\frac{1}{2}kT$ per degree of freedom. This is an instance of a more general principle (the equipartition of energy). For polyatomic molecules, the Hamiltonian contains several types of quadratic terms:

$$\text{const } q_i^2 \quad \text{and} \quad \text{const } p_i^2 \tag{3.19-33}$$

For example, the translational kinetic energy is

$$\frac{p_x^2}{2m} + \frac{p_y^2}{2m} + \frac{p_z^2}{2m} \tag{3.19-34}$$

The rotational kinetic energy for a nonlinear molecule is

$$\frac{J_x^2}{2I_x} + \frac{J_y^2}{2I_y} + \frac{J_z^2}{2I_z} \tag{3.19-35}$$

where the J's are components of angular momenta and the I's the associated moments of inertia. For a linear molecule, the rotational kinetic energy is

$$\frac{J_x^2}{2I} + \frac{J_y^2}{2I} \tag{3.19-36}$$

In the case of molecular vibrations, each normal mode contributes two quadratic terms

$$\frac{1}{2\mu} p^2 + \frac{K}{2} q^2 \tag{3.19-37}$$

where μ is the reduced mass for the normal mode and K the effective force constant. For a molecule with N atoms, there are $3N - 6$ normal modes of vibrations if the molecule is nonlinear, $3N - 5$ normal modes if the molecule is linear. If the Hamiltonian is composed entirely of quadratic terms,

$$H(q, p) = \sum_i a_i x_i^2 \tag{3.19-38}$$

where the x_i represents both q's and p's, the phase integral is

$$\int \cdots \int dq_1 \cdots dp_n e^{-\Sigma_i a_i x_i^2 / kT} \tag{3.19-39}$$

The internal energy is correspondingly

$$E = \frac{\int \cdots \int dq_1 \cdots dp_n \left(\sum_j a_j x_j^2 \right) e^{-\Sigma_i a_i x_i^2 / kT}}{\int \cdots \int dq_1 \cdots dp_n e^{\Sigma_i a_i x_i^2 / kT}} \tag{3.19-40}$$

where the summations run from i or $j = 1$ to $(6n - 6 \text{ or } 5)\, N$. Integrating over all variables except x_j, a typical term is

$$\frac{a \int_{-\infty}^{\infty} dx\, x^2 e^{-ax^2/kT}}{\int_{-\infty}^{\infty} dx\, e^{-ax^2/kT}} = \frac{\frac{1}{2}(\pi k^3 T^3/a^3)^{1/2} a}{(\pi k T/a)^{1/2}} = \frac{1}{2} kT \tag{3.19-41}$$

independent of the value of a or the source of the quadratic term. Therefore, the internal energy equals $\frac{1}{2}kT$ per quadratic term in the mole Hamiltonian,

$$E = (6N - 6 \text{ or } 5) \times \tfrac{1}{2} kT \tag{3.19-42}$$

and the molar internal energy is given by

$$\overline{E} = \begin{cases} (3N - 3)\, RT & \text{(nonlinear)} \\ (3N - \frac{5}{2})\, RT & \text{(linear)} \end{cases} \tag{3.19-43}$$

while the molar heat capacity is

$$\bar{C}_V = \begin{cases} (3N-3)R & \text{(nonlinear)}, \\ (3N-\frac{5}{2})R & \text{(linear)}. \end{cases} \tag{3.19-44}$$

A comparison of predicted with experimental values of C_V shows that the principle of equipartition of energy fails for all but monatomic gases. However, as the temperature is increased, C_V does approach its equipartition value. These difficulties of classical statistical mechanics are cleared up when a quantum-mechanical description is adopted for the description of molecular systems.

For a one-dimensional rotator, that is, rotation in a plane about a single axis of a mass m constrained to move at a distance r from a fixed axis, the energy is entirely kinetic and is given by

$$\epsilon = \frac{I\dot{\theta}^2}{2} = \frac{P_\theta^2}{2I} \tag{3.19-45}$$

where $\dot{\theta}$ is the angular velocity in radians per second, and p_θ is the angular momentum defined as $I\dot{\theta}$. This is equivalent to the equation $\epsilon = mv^2/2$ where, v is the orbital speed (say in $m \cdot s^{-1}$). The angular momentum may take any value from $-\infty$ to $+\infty$ and the conjugate positional coordinate q_θ can have any value between 0 and 2π, corresponding to one complete revolution. The partition function is then

$$z_{\text{rot}} = h^{-1} \int\int \exp\left(-\frac{p_\theta^2}{2IkT}\right) dp_\theta\, dq_\theta \tag{3.19-46}$$

$$= h^{-1}(2\pi) \int_{-\infty}^{\infty} \exp\left(-\frac{p_\theta^2}{2IkT}\right) dp_\theta$$

$$= h^{-1}(2\pi)(2\pi IkT)^{1/2} = \pi \left(\frac{8\pi^2 IkT}{h^2}\right)^{1/2} \tag{3.19-47}$$

The energy of a classical harmonic oscillator depends upon both the position and velocity (or momentum) of the oscillator. If the oscillator is regarded as a mass m constrained to move in a straight line according to simple harmonic motion, the energy is given by

$$\epsilon = \frac{kq^2}{2} + \frac{p^2}{2m} \tag{3.19-48}$$

where k is the force constant, and q the displacement from equilibrium. For a more complex oscillator a similar equation holds but p, q, and m are now functions of the momenta, positions and masses of the individual atoms. The classical partition function is obtained by integration over all accessible regions of phase space, that is, for both p and q between $-\infty$ and $+\infty$.

$$z_{\text{vib}} = h^{-1} \int\int \exp\left(-\frac{kq^2m + p^2}{2mkT}\right) dp\, dq \tag{3.19-49}$$

Since p and q are independently variable the integrations can be performed independently giving

$$z_{\text{vib}} = h^{-1}\left(\frac{2\pi mkT}{km}\right)^{1/2}(2\pi mkT)^{1/2}$$

$$= h^{-1}(2\pi kT)\left(\frac{m}{k}\right)^{1/2} \tag{3.19-50}$$

Solution of the equation of motion for the oscillator

$$\dot{p} + kq = 0 \tag{3.19-51}$$

shows that the frequency of oscillation in cycles per unit time is

$$\nu = \left(\frac{1}{2\pi}\right)\left(\frac{k}{m}\right)^{1/2} \tag{3.19-52}$$

Substituting this into (3.19-50) then gives

$$z_{\text{vib}} = \frac{kT}{h\nu} \tag{3.19-53}$$

This is the same result as that obtained by quantum theory when θ_{vib}/T is small. In general, the partition function can always be obtained from the quantum formulation by assuming that the energy levels are very close together compared to kT and replacing summation by integration.

SYMBOLS IN CHAPTER 3

A, a	Helmholtz function
E	Energy
F, f	Distribution functions
G, g	Gibbs function, degeneracy
H, h	Enthalpy, Hamiltonian, Planck constant
I, i	Moment of inertia, dummy index
k	Boltzmann constant; force constant
m	Mass
N, n	Number of particles
P, p	Probability, momentum coordinate, pressure
q	Space coordinate
R	Gas constant
S, s	Entropy
T	Temperature, kinetic energy
U, u	Internal energy
V, v	Volume, specific volume, velocity
Z, z	Partition function, phase integral
β	Lagrange multiplier $1/kT$
Γ	Flux
ε	Energy
θ_r, θ_v	Characteristic temperature of rotation, vibration
ν	Frequency, Hz
ρ	Density, distribution function
ω	Circular frequency $2\pi\nu$

PROBLEMS

3.1. Discuss the analytic nature the probability distribution of energy of $P(p, q)$ for a microcanonical ensemble representing a system of two independent, free particles of mass M moving in one dimension on a line segment of length L. The energy of the system is E_0.

3.1. What is the motion in phase space for: **(a)** a one-dimensional rotator, **(b)** a one-dimensional oscillator, **(c)** an n particle system?

3.3. Find the Lagrangian and Hamiltonian forms of the equation of motion for a classical harmonic oscillator in one dimension.

3.4. One of the requirements for an ensemble to satisfactorily represent a system in equilibrium is that the distribution of the ensemble be independent of time, that is, $\partial p(p, q)/\partial t$ should be equal to zero. Examine this requirement in light of Hamilton's equations and Liouville's theorem.

3.5. Discuss the phase space motion of a one-dimensional system consisting of a bead moving along a wire between two stops at $x = 0$ and $x = a$.

3.6. An object rotates about a fixed axis with angular velocity ω, and angular momentum $p_\theta = I\omega$ (I being the moment of inertia of the object with respect to the axis). Let the position of the object as it rotates about the axis be given by the angular coordinate θ. The coordinate θ is restricted to the range 0 to 2π. The energy is $\epsilon = p_\theta^2/2I$.

(a) Draw the phase space of the system and show the locus of a typical phase point.

(b) According to the quantum theory, the energy of an object rotating about an axis fixed in space can only have the values $\epsilon = j^2h^2/8\pi^2I$, where $j = 0, \pm 1, \pm 2, \ldots$. Show that quantum result divides the classical phase space into cells of area h.

3.7. Phase space can be subdivided into rectangular cells of dimensions Δx by Δp_z as shown in Fig. P3.7. Each cell has an energy and an identity, and molecules can be distributed among the cells in various ways to find the number of configurations, w, of the system.

(a) What are the dimensions of the area in phase space?

(b) By how much do areas enclosed by successive paths differ?

(c) Do phase points have preference for one region of phase space over another?

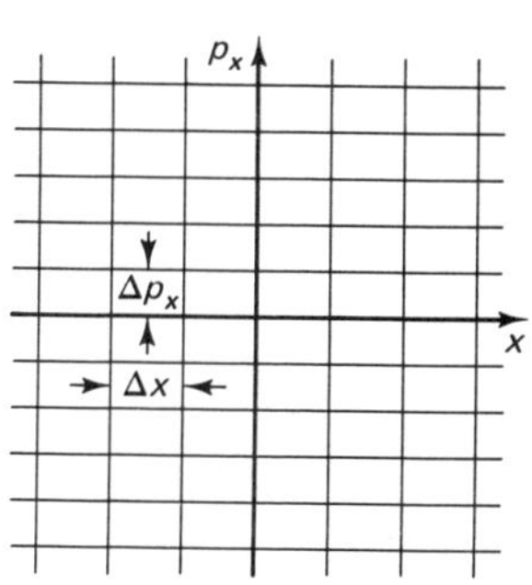

Figure P3.7

3.8. Let the phase space of some one-dimensional system be subdivided into cells of equal volume and any convenient shape (see Fig. P3.8; a rational size of cell

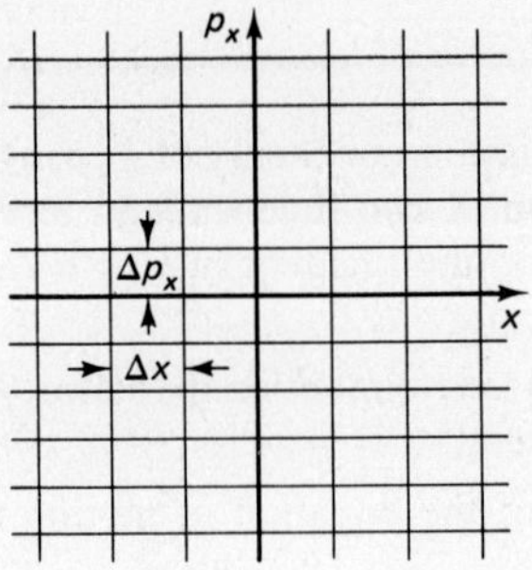

Figure P3.8

would obviously be one of area h). Then the evaluation of the thermodynamic probability, w, for a collection of N molecules can proceed by counting the number of ways individual molecules can be distributed among these cells. For example, in a one-dimensional system,

$$\epsilon(p_x, x) = \frac{1}{2m} p_x^2 + V(x)$$

where $V(x)$ is the potential energy function of the system. The basic problem in classical statistical mechanics is the following. A given number of molecules, N (a very large number), is given a definite total energy, E. The molecules may be distributed in different ways among the cells in phase space. Show that the most probable distribution is given by

$$n_i = A \exp(-\beta\epsilon_i)$$

where A and β are constants to be determined in such a way as to satisfy the conditions

$$\sum n_i = N \qquad \sum n_i \epsilon_i = E$$

3.9. For a one-dimensional system with phase space (x, p_x), the population of a cell is proportional to the area of the cell, $\Delta x\, \Delta p_x$, so that

$$n_i = A\Sigma \exp(-\beta\epsilon_i)\, \Delta x\, \Delta p_x$$

If the cells are made sufficiently small, this sum can be replaced by an integral and ϵ_i can be regarded as a continuously varying function, $\epsilon_i = \epsilon(x, p_x)$, giving

$$N = \Sigma n_i = A \int\!\!\int \exp[-\beta\epsilon(x, p_x]\, dx\, dp_x$$

where the integration is in all phase space. The constant A is thus

$$A = \frac{N}{\int\!\!\int \exp[-\beta\epsilon(x, p_x)]\, dx\, dp_x}$$

Show that the constant β can be determined by evaluating the energy of a one-dimensional gas confined to a container of length a, and setting it equal to $\frac{1}{2}NkT$.

3.10. A particle in translational phase space x, y, z, p_x, p_y, p_z has energies given by

$$\epsilon_i = \frac{p_x^2}{2m} + \frac{p_y^2}{2m} + \frac{p_z^2}{2m}$$

Because the smallest discernible volume in phase space is h^3, the degeneracy of an energy level is given by

$$g_i = \frac{dx\, dy\, dz\, dp_x\, dp_y\, dp_z}{h^3}$$

From the definition of the molecular partition function, $z = \sum_i g_i e^{-\beta\epsilon_i}$, show that

$$z = \int\int\int\int\int\int \frac{m^3}{h^3} \exp\left[-\frac{m\beta}{2}(v_x^2 + v_y^2 + v_z^2)\right] dx\, dy\, dz\, dv_x\, dv_y\, dv_z$$

$$= \frac{m^3 V}{h^3}\left(\frac{2\pi}{m\beta}\right)^{3/2}$$

where V us the volume of the container.

3.11. Taking a one-dimensional system as an example, show that if the temperature is sufficiently high for intervals between successive energy levels to be small compared with kT, the classical partition function (phase integral divided by h), defined by

$$z_{\text{class}} = \frac{1}{h}\int\int \exp\left(-\frac{\epsilon_i}{kT}\right) dp_x\, dx$$

will approximate to the partition function from quantum mechanics,

$$z_{\text{quant}} = \Sigma \exp\left(-\frac{\epsilon_i}{kT}\right)$$

3.12. From considerations of the Heisenberg principle, show that the statistical weight or degeneracy for a classical monatomic gas is

$$g(v_x, v_y, v_z) = \frac{m^3 V}{h^3}\, dv_x\, dv_y\, dv_z$$

Show that the partition function for translational energy is

$$Z = V\left(\frac{2\pi mkT}{h^2}\right)^{3/2}$$

3.13. Show that the classical and quantum partition functions tend to become equal at high temperatures for a harmonic oscillator.

3.14. Show that for $h^2/8ma^2kT << 1$, the classical partition function and the quantum partition function are the same for a particle in a one-dimensional box of length a.

3.15. Consider a molecular system involving n atoms. Since each atom requires three position coordinates to define its location in space, the specification of the spatial configuration of the molecule requires $3n$ coordinates. The complete classical state of such a system is determined by specifying $3n$ position coordinates plus $3n$ momentum coordinates. Therefore, a $6n$-dimensional phase space is required if a phase point is to indicate the classical state of the system. Discuss the procedure for applying classical statistical mechanics to obtain the partition function from the phase integral and compare the result with that obtained from quantum-statistical mechanics.

3.16. In 1859, Maxwell derived the velocity distribution in the x direction to be

$$f(v_x) = \left(\frac{m}{2\pi kT}\right)^{1/2} e^{-mv_x^2/2kT}$$

Similar expressions hold for $f(v_y)$ and $f(v_z)$ in the other directions. The complete velocity distribution is thus

$$F(v_x, v_y, v_z) = f(v_x) f(v_y) f(v_z)$$

$$= \left(\frac{m}{2\pi kT}\right)^{3/2} e^{-m(v_x^2+v_y^2+v_z^2)/2kT}$$

Derive this same result from the standpoint of classical statistical mechanics.

3.17. Given the statistical weight (number of states in a velocity element $dv_x\, dv_y\, dv_z$ at v_x, v_y, v_z) to be

$$g(v_x, v_y, v_z) = \left(\frac{m^3 V}{h^3}\right) dv_x\, dv_y\, dv_z$$

find the Maxwell–Boltzmann distribution in velocity components and in speed for a classical monatomic gas.

3.18. Given the Maxwell velocity distribution

$$f(v_x, v_y, v_z) = \left(\frac{m}{2\pi kT}\right)^{3/2} \exp\left[-\frac{m}{2kT}\left(v_x^2 + v_y^2 + v_z^2\right)\right]$$

convert it to a distribution of molecular speeds.

3.19. Develop the distribution function for molecular velocities as obtained by Maxwell in 1859. The problem can be stated as follows: Given N molecules of a gas confined to some volume V, let the number with velocity components v_x in dv_x, v_y in dv_y, and v_z and dv_z be $NF(v_x, v_y, v_z)\, dv_x\, dv_y\, dv_z$. Determine the form of $F(v_x, v_y, v_z)$ when the gas is in equilibrium at temperature T.

3.20. Given a system of particles for which the energy E can be a function of the coordinates x, y, and z and momentum p_x, p_y, and p_z, statistical mechanics gives the probability of a particle being in the cell $d\tau = dx\, dy\, dz\, dp_x\, dp_y\, dp_z$ as

$$f(x, y, z, p_x, p_y, p_z)\, d\tau = Ce^{-E/kT}\, d\tau$$

where the constant C is determined by the normalization condition

$$\int Ce^{-E/kT}\, d\tau = 1$$

Apply this to an ideal gas for the case of translational energies and obtain the momentum distribution as well as Maxwell's velocity distribution.

3.21. The classical Maxwell–Boltzmann distribution law may be written as

$$\frac{dn(v)}{N} = F_{\text{MB}}\, dv_x\, dv_y\, dv_z$$

where $n(v)$ is the number of particles with velocity components between v_x and $v_x + dv_x$, v_y and $v_y + dv_y$, v_z and $v_z + dv_z$, and F_{MB} is the distribution function. From the canonical ensemble distribution, which is proportional to $e^{-E/kT}$, find the normalized Maxwell–Boltzmann distribution.

3.22. **(a)** Given a system of particles for which the energy has a discrete set of values, the probability distribution of a particle having the energy E_i is

$$f_i = Cg_i e^{-E_i/kT} \tag{1}$$

where the constant C is determined by the normalization condition and g_i is the statistical weight (the number of quantum states having the same energy). The first excited state of the hydrogen atom is 10.2 eV above the ground state. What is the ratio of the number of atoms in the first excited state to the number in the ground state at $T = 300$ K? The statistical weights are $g_0 = 2$ and $g_1 = 8$.

(b) Express Eq. (1) in the case of the enery being continuous rather than discrete.

3.23. Maxwell first showed in 1859 that the distribution function for the x component of molecular velocity is

$$f(v_x) = \left(\frac{m}{2\pi kt}\right)^{1/2} e^{-mv_x^2/kT}$$

Similar expressions hold for $f(v_x)$ and $f(v_z)$. The complete velocity distribution is thus

$$F(v_x, v_y, v_z) = f(v_x) f(v_y) f(v_z) = \left(\frac{m}{2\pi kt}\right)^{3/2} e^{-m(v_x^2+v_y^2+v_z^2)/2kT}$$

From this, obtain: **(a)** the speed distribution; **(b)** the most probable speed; **(c)** the average and the RMS speeds. What is the most probable speed for helium at 0°C? The mass of the helium atom is 6.645×10^{-24} g.

3.24. The probability of a particle having energy E within dE is given by

$$f(E)\, dE = Cg(E)\, e^{-E/kT}\, dE \tag{1}$$

where $g(E)\, dE$ is the statistical weight or number of energy states in dE. For an ideal gas for which the energy is given by

$$E = \frac{1}{2m}(p_x^2 + p_y^2 + p_z^2) = \frac{p^2}{2m} \tag{2}$$

show that $g(E)dE \propto E^{1/2}dE$ and $f(E)dE \propto E^{1/2}e^{-E/kT}dE$. How does this compare with the speed distribution?

3.25. Consider a system in which the energy of a particle is continuous and given by

$$E = c_1x^2 + c_2y^2$$

where c_1 and c_2 are constants and x and y are any type of position or momentum coordinates. Using the Maxwell–Boltzmann distribution, show that an average energy of $\frac{1}{2}kT$ is associated with each coordinate or momentum component appearing in the energy as a squared term. (This is the equipartition theorem.)

3.26. For translational motion, find the number of classical states, $g(E)\, dE$, in the energy interval between E and $E + dE$, and thereby show that the density of states, $g(E)$, has the form $g(E) \propto E^{1/2}$. Also, evaluate $g(E)$ in quantum space.

3.27. The quantum-mechanical solution for the energy of a simple harmonic oscillator is

$$E_n = (n + \tfrac{1}{2})\, h\nu$$

where E_n is the energy of the state corresponding to the integer n $(n = 0, 1, 2, \ldots)$, h is Planck's constant, and ν is the frequency of oscillation. Determine the possible distributions among energy states of three simple harmonic oscillators such that $E_T = 9E_0$. Is the distribution $(1, 0, 1, 1, 0, \ldots)$ an acceptable answer?

3.28. The fundamental classical and quantum-mechanical distribution functions have one thing in common: they all depend only upon the energy of the state and not upon the quantities that are in the state. Because of this property, it is possible to write the particle distribution in the form

$$d\,n(q) = Fg(q)\, dq$$

where q is a dynamical quantity that depends only on the energy, F is a statistical factor, and $g(q)$ is a density-of-state factor. For classical statistics, $F = F_{\text{MB}}$. Evaluate $g(q)$ for $q = v$ and $q = E$, respectively.

3.29. The canonical ensemble distribution states that the probability of finding a molecule in the speed range v to $v + dv$ is proportional to the number of states in this range (which is proportional to v^2) times the probability of a single state (which is proportional to $e^{-\epsilon/kT}$), or

$$F_{\mathrm{MB}}\, dv = Av^2 \exp\left(-\frac{mv^2}{2kT}\right) dv$$

Find the factor A and, therefore, the Maxwell–Boltzmann distribtion function. Does the canonical ensemble treatment of the MB distribution law extend to imperfect gases and liquids as well?

3.30. From statistical mechanics, derive the barometric equation or law of atmospheres for the pressure p at a distance h above the surface of the earth. Also, show that the same result follows from hydrostatic considerations.

3.31. If the fraction of gas atoms to be found at heights between p and $h + dh$ is

$$\frac{d\,n(h)}{n} = \left(\frac{mg}{kT}\right) e^{-mgh/kT}\, dh$$

and the probability of a gas atom having component velocities in the range v_x to $v_x + dv_x$, v_y to $v_y + dv_y$, and v_z to v_z to $v_z + dv_z$ is

$$\frac{d^3n(v_x, v_y, v_z)}{n} \left(\frac{m}{2\pi kT}\right)^{3/2} e^{-m(v_x^2+v_y^2+v_z^2)/2kT}\, dv_x\, dv_y\, dv_z$$

what is the distribution function for a gas having potential energy as well as kinetic energy? What is the mean energy of the multivariable system?

3.32. **(a)** From the assumptions (1) that molecules exist in energy states, and (2) that the Boltzmann factor is correct, derive the expression for the energy of a mole of substance

$$E = RT^2 \frac{d \ln z}{dT}$$

where R is the gas constant per mole.

(b) Show that for a diatomic gas, the vibrational energy of N molecules at temperature T is given by

$$E_{\mathrm{vib}} = \frac{Nh\nu}{e^{h\nu/kT} - 1} = \frac{RT_y}{e^y - 1}$$

where $y = h\nu/kT$.

3.33. Calculate the moment of inertia for $^{14}N_2$ and $^{14}N^{15}N$. The two molecules have the same intermolecular distance of 0.1095 nm. What is the molar rotational partition function for these molecules? The symmetry number, σ, for $^{14}N_2$ is 2. For $^{14}N^{15}N$, it is 1.

3.34. What is the fraction $N(v_0)/N$ of the total number of particles N having speeds above a given speed v_0 in a system of particles described by the Maxwell–Boltzmann distribution? Sketch the graph of $N(>v_0)$ versus $(m/2kT)^{1/2}\, v$.

3.35. **(a)** A diatomic molecule that is free to rotate, but not to vibrate, is a dumbell with the moment of inertia about the axis passing through the masses approximately equal to zero. Find the molar heat capacity of a diatomic gas, using the equipartition theorem.

(b) If a crystal lattice is pictured as a system of regularly spaced atoms connected by springs, find the molar specific heat, using the equipartition theorem.

3.36. For the Maxwell–Boltzmann distribution, evaluate the relative magnitudes of the following speeds: v_{mode}, v_{av}, and v_{RMS}.

3.37. The equipartition theorem is useful in understanding the heat capacity of solids. In 1819, Dulong and Petit pointed out that the molar heat capacity of most solids was very nearly equal to 3R. This result was used by them to obtain unknown molecular weights from the experimentally determined heat capacities. Derive the Dulong–Petit law and discuss the conditions for its validity.

3.38. An important application of the equipartition theorem occurs for a system suspended so that it can rotate about a given axis, for example, the suspended coil of a moving coil galvanometer. The system performs minute irregular movements, partly due to collisions with molecules of the surrounding gas, and partly due to the thermal motion of the atoms in the suspension. Show that as a result of this, the minimum current that can be detected with a galvanometer is

$$i_{\min} = \text{const}\left(\frac{kT}{\alpha}\right)^{1/2}$$

where α is the restoring torque per unit deflection.

3.39. In the Earth's atmosphere, one of the air molecules can be treated as a small system and the rest of the atmosphere as a heat reservoir. Allowance must be made for the variation of the gravitational potential energy of the molecule with its altitude H above sea level. The sum of the potential and kinetic energy of the air molecule is

$$\epsilon = mgH + \tfrac{1}{2}mv^2$$

Substituting this in the canonical ensemble distribution, find the pressure variation with height in an isothermal atmosphere, and determine the scale height at 300 K.

3.40. In statistical mechanics, it is assumed that an energy level has an intrinsic probability g_i of occupation. Classically, regarding energy intervals, if there are twice as many states in ΔE_i as in ΔE_j, then it should be twice as probable for a particle to have an energy in the interval ΔE_i as in ΔE_j. Under this assumption, obtain an expression for the probability of finding a particular distribution of particles $\{n_1, n_2, \ldots, n_r\}$.

3.41. Statistical mechanics assumes that among all distributions of particles, the one of highest probability corresponds to equilibrium of the system. From the probability of a particular distribution $\{n_1, n_2, \ldots\}$ given by

$$P_n = \frac{N!}{n_1!\, n_2! \cdots n_r}\, g_1^{n_1} g_2^{n_2} \cdots$$

determine the most probable distribution subject to the conditions that the number of particles and total energy of the system are constant.

Chapter 4

Quantum Assemblies

Macroscopic systems are characterized according to the distribution of energy among their subsystems (particles). The distribution laws for classical and quantum subsystems are known, respectively, as classical and quantum statistics. So far, a classical system has been considered as, for example, a perfect gas of noninteracting particles under conditions such that for all single-particle states $r(=1, 2, \ldots)$ the mean occupation numbers $\langle n_r \rangle$ are very small compared to unity. The classical treatment, based on the methods of Maxwell and Boltzmann, involves quantum mechanics only in so far as the values of the allowed energy levels and their degeneracies are required in the evaluation of the partition function. As seen in the preceding chapter, most of the results could have been obtained without the use of quantum theory. (The quantal and quasiclassical expressions do, however, contain h, and this term can never appear in a truly classical result, but the powers of h always cancel when thermodynamic functions are evaluated.)

If the condition $\langle n_r \rangle \ll 1$ does not hold, quantum effects become important, as in the case of electrons in metals or atoms in liquid helium. The many cases in which quantum theory plays a vital role fall into two classes: (1) effects modifying the distribution law, and (2) effects modifying the partition function. This chapter deals with the modifications of the classical statistics to accommodate for quantum effects of indistinguishability, degeneracy, and permutational symmetry requirements. In the classical treatment, the canonical ensemble concept was used to obtain the Maxwell–Boltzmann distribution for closed systems (constant N). In this chapter, the theory will be refined and broadened by the use of the grand canonical ensemble (open system; variable N) for the derivation of the Bose–Einstein and Fermi–Dirac distributions. The thermodynamic variables U, p, S, and N will then be obtained when the grand partition function Ξ of the system is known as a function of the temperature, volume, and chemical potential of the system.

4.1 CLASSICAL AND QUANTUM FORMULATIONS

In classical mechanics, the instantaneous state of motion of a particle is represented by a point in phase space, and the energy distribution of a system is given by the number of points in each small volume of phase space, say $\Delta x\ \Delta y\ \Delta z\ \Delta p_x\ \Delta p_y\ \Delta p_z$. Points in that cell have total energy between E and $E + dE$, corresponding to momentum values between p_x and $(p_x + \Delta p_x)$, and so on, and position values between x and $(x + \Delta x)$, and so on. The cell is chosen to be small enough that the average total energy of its representative points differs little from the energy of any one of them; it is chosen large enough so that there are many points in a cell, thereby permitting the application of statistical ideas.

In particular, the number of states where the particle has position and momentum coordinates with the ranges $x, x + dx, \ldots,$ and $p_x, p_x + dp_x, \ldots,$ is proportional to the volume of this element of phase space, that is, to

$$d^3\mathbf{r}\, d^3\mathbf{p} \equiv (dx\, dy\, dz)(dp_x\, dp_y\, dp_z) \tag{4.1-1}$$

This is essentially the description used by Maxwell, Boltzmann, and Gibbs in classical statistical mechanics. The constant of proportionality they could not fix, as classically the possible states form a continuum.

The volume $d\Gamma$ phase space in which the particle lies with momentum within the range p to $(p + dp)$ is given by

$$d\Gamma = \int_V d^3\mathbf{r} \int_{p,p+dp} d^3\mathbf{p} = V \int_{p,p+dp} d^3\mathbf{p} \tag{4.1-2}$$

The integral in momentum space is simply the volume in momentum space of a spherical shell with radii p and $(p + dp)$, that is,

$$\int_{p,p+dp} d^3\mathbf{p} = 4\pi p^2\, dp \tag{4.1-3}$$

Thus, Eqs. (4.1-2) and (4.1-3) give

$$d\Gamma = V 4\pi p^2\, dp \tag{4.1-4}$$

for the volume of phase space, while quantum-mechanically, the number of states in which a molecule in an enclosure of volume V has momentum in the interval p to $(p + dp)$ [and kinetic energy in a corresponding range $\varepsilon_{\text{trans}}$ to $(\varepsilon_{\text{trans}} + d\varepsilon_{\text{trans}})$] is given by

$$f(p)\, dp = \frac{V 4\pi p^2\, dp}{h^3} \tag{4.1-5}$$

The constant of proportionality involves Planck's constant, a consequence of the quantal picture of discrete states. Thus, the volume in phase space for each discrete quantum state is h^3, and phase space can be thought to be subdivided into cells of volume h^3, so that

$$\frac{1}{h^3} d^3\mathbf{r}\, d^3\mathbf{p} \tag{4.1-6}$$

is the number of states where the particle lies within a volume element $d^3\mathbf{r}$ of (ordinary) coordinate space and within a volume element $d^3\mathbf{p}$ of momentum space. (As mentioned earlier, the term h does not appear in a truly classical derivation, but the powers of h always cancel when thermodynamic functions are evaluated.)

The quantum-mechanical description of a system is based on a single function $\Psi(q_1, \ldots, q_n, t)$ of the n generalized coordinates and time. Just as the dynamical behavior of a classical system is governed by the Hamiltonian function $H(q,p,t)$ through Hamilton's equations of motion, so the dynamical behavior of the corresponding quantum system is governed by the Hamiltonian operator $\hat{H}(q, -i\hbar\,\partial/\partial_q, t)$, and the Schrödinger equation

$$\hat{H}\psi(q) = E\psi(q), \tag{4.1-7}$$

which applies to stationary (time-independent) states. The formulation of quantum-statistical mechanics is, like the classical formulation, based on the ensemble concept. A quantum ensemble is a representative collection of microscopic states, each represented by a wave function $\psi_i(q,t)$, and each compatible with a specified macroscopic state. In quantum mechanics, the allowed values of most dynamical variables form a discrete rather than continuous set. Therefore, the number of independent microscopic states (which are not merely linear combination of other states) is no longer infinite, in contrast to the classical case. It is still extremely large, however, in most instances.

Just as classical statistical mechanics is based on the postulate that, for an isolated system, each region of phase space has equal a priori probability, so the quantum analogue is:

> The different quantum-mechanical states of an isolated system have equal a priori probabilities (and random a priori phases).[1] (4.1-8)

A second fundamental postulate of quantum-statistical mechanics is:

> The (sufficiently long) time average of every physical observable is equal to its ensemble average. (4.1-9)

The two postulates together lead to the quantum analogue of the ergodic hypothesis: Over a sufficiently long period of time, an isolated system spends equal amounts of time in each of the available quantum states. Thus, both

---

[1] Wave functions are, in general, complex functions. Ψ can be expressed in polar form as $|\Psi|e^{i\phi}$, where $\phi(q,t)$ is known as the phase. The postulate of random phases means that, in an ensemble, all values of ϕ between 0 and 2π are equally probable.

the time average $\bar{x}$ and the ensemble average $\langle x \rangle$ of an observable x can be associated with the expectation value, that is,

$$\bar{x} = \langle x \rangle = \sum_i P_i x_i = \int_\tau \Psi^* \hat{x} \Psi \, d\tau \tag{4.1-10}$$

where P_i is the probability of state i; and $\hat{x}$ is the Hamiltonian operator for the dynamical variable.

The main difference between classical and quantum-statistical mechanics is that the applicability of classical statistics rests on three assumptions:

1. The systems are, at least approximately, energetically independent. (4.1-11)
2. The systems, although identical, are distinguishable. (4.1-12)
3. Accessibility of phase space to each subsystem is independent of its prior occupation by other systems. (4.1-13)

For assemblies of molecules, the first condition essentially defines the perfect gas. Assumption (4.1-11) must be moderated, however, by the stipulation that there must be sufficient coupling among the subsystems to provide a mechanism for energy exchange and hence thermal equilibration. This is almost always fulfilled in practice. In a gas, for example, collisions among the molecules provide the necessary interaction.

Under certain conditions, analogues of (4.1-11), (4.1-12), and (4.1-13) can also apply to quantum systems. In most instances, however, modifications are required. When the subsystems are identical nonlocalized microscopic particles (for example, gas molecules), quantum mechanics demands that they be indistinguishable. In such instances, (4.1-12) is invalid. In addition, many-particle quantum systems are subject to certain permutational symmetry restrictions. This, in turn, leads to exclusion of certain quantum states, in contrast to (4.1-13). Stated another way, permutational symmetry is dynamically equivalent to an additional interaction among identical particles. (For bosons, these are effectively attractive forces; for fermions, effectively repulsive forces.)

4.2 PAULI EXCLUSION PRINCIPLE

The central notion in quantum-statistical mechanics is that identical particles are *indistinguishable*. The interchange of two identical particles in a system does not give rise to a different state of the system. For a system of two particles, say a and b, where a and b represent the coordinates of the two particles, interchanging the two particles yields the same state and thus $|\psi(b,a)| = |\psi(a,b)|$. This means that the two wave functions can only differ by a phase factor,

$$\psi(b,a) = e^{i\theta}\,\psi(a,b) \tag{4.2-1}$$

or

$$\psi(b, a) = \pm\psi(a, b) \tag{4.2-2}$$

where the wave functions of a single particle are

$$\psi_1, \psi_2, \psi_3, \ldots, \psi_j, \ldots \tag{4.2-3}$$

and the corresponding energies are

$$\varepsilon_1, \varepsilon_2, \varepsilon_3, \ldots, \varepsilon_j, \ldots \tag{4.2-4}$$

Let the two particles be isolated from each other. Each will be in a quantum state characterized by one of the wave functions, and will have the corresponding energy; for particle a, the energy will be ε_r, and the wave function will be $\psi_r(a)$; for molecule b, the energy will be ε_s, and the wave function will be $\psi_s(b)$.

Now suppose that the two particles are brought together to form part of the same system, while still being distinguishable from each other (e.g., both might occupy sites in one crystal). If intermolecular forces can be neglected, the energy of a and b together is

$$E = \varepsilon_r + \varepsilon_s \tag{4.2-5}$$

and the product function

$$\psi_1 = \psi_r(a)\psi_s(b) \tag{4.2-6}$$

is an eigenfunction of the Schrödinger equation for the system of the two particles. This wave function characterizes a complexion of the system.

Since the two particles are of the same type, the same wave functions ψ and energies ε are available to both. Therefore, another complexion of the system is the one characterized by

$$\psi_2 = \psi_r(b)\psi_s(a) \tag{4.2-7}$$

in which particle b happens to be in the quantum state associated with ψ_r and a in the state associated with ψ_s. The corresponding energy is again

$$E = \varepsilon_r + \varepsilon_s \tag{4.2-8}$$

the same as that associated with ψ_1.

If the two particles are indistinguishable, there should be no observable change when the two particles are formally interchanged. The probability density $|\psi(a, b)|^2$ must therefore be equal to the density with the particles permuted, $|\psi(b, a)|^2$, or

$$\psi(b, a) = \pm\psi(a, b) \tag{4.2-9}$$

Neither the function in (4.2-6) nor the function (4.2-7) possess the required symmetry given by (4.2-9), but the following linear combinations of ψ_1 and ψ_2 do

$$\psi_S = C(\psi_1 + \psi_2) = \frac{1}{\sqrt{2}}[\psi_r(a)\psi_s(b) + \psi_r(b)\psi_s(a)] \tag{4.2-10}$$

$$\psi_A = C(\psi_1 - \psi_2) = \frac{1}{\sqrt{2}}[\psi_r(a)\psi_s(b) - \psi_r(b)\psi_s(a)] \tag{4.2-11}$$

where C is a normalization constant equal to $1/\sqrt{2}$. Interchanging the labels a and b clearly leaves ψ_S unchanged, but changes ψ_A to

$$\frac{1}{\sqrt{2}}[\psi_r(b)\psi_s(a) - \psi_r(a)\psi_s(b)] = -\psi_A \tag{4.2-12}$$

that is, changes the signs but not the magnitude of ψ_A at every point. The value ψ_S is said to be *symmetric* in the particles, and ψ_A *antisymmetric* in the particles. The linear combination in either (4.2-10) or (4.2-11) thus achieves a representation conforming to the indistinguishablity property. No longer can any explicit correspondence between (a, b) and (r, s) be made. Only the distribution of particles among orbitals, but not the arrangement, has any physical significance.

Thus far nothing has appeared that would forbid the same system from having both symmetrical and antisymmetrical states; however, nature (experiment) has decreed that certain particles can have only symmetrical wave functions and others only antisymmetrical wave functions. Thus, electrons, protons, neutrons, and other elementary particles with *half-integral spin,* m_s, are denoted *Fermi–Dirac particles,* or *fermions,* and can occur only in antisymmetrical many-particle states. Correspondingly, photons, π mesons, and other elementary particles with *integral (or zero) spin,* which are denoted *Bose–Einstein particles,* or *bosons*, can occur only in symmetrical states.

Moreover, composite particles such as atoms and molecules comprised of an *odd* number of electrons, protons, and neutrons have a half-integer spin quantum number and are therefore fermions. Atoms and molecules made up of an *even* number of such subparticles have an integer spin quantum number and are classified as bosons. Examples of atoms and molecules that are bosons are H, H_2, ^4He, O, and N_2; a collection of H atoms would then, for example, have to be described by a symmetrical wave function. Atoms and molecules that are classified as fermions are D, ^3He, NO, and NH_3.

Consider the possibility of having two electrons (fermions) in the same state (denoted by the four quantum numbers n, l, m_l, and m_s). Since fermions are described by an antisymmetric wave function, Eq. (4.2-11) yields

$$\frac{1}{\sqrt{2}}[\psi_r(a)\psi_r(b) - \psi_r(b)\psi_r(a)] = 0 \tag{4.2-13}$$

that is, the probability of two electrons being in the same state r is identically equal to zero. This is the content of the *Pauli exclusion principle*: no two electrons can be in the same quantum state. Otherwise stated, no two electrons (fermions) can have the same set of quantum numbers (n, l, m_l, m_s). Pauli's

TABLE 4.2-1 PROPERTIES OF CLASSES OF PARTICLES

Particle type	Spin in units of $\hbar$	Examples	Occupancy, $n(\varepsilon)$, of an orbital	Distribution function
Fermion	$\frac{1}{2}, \frac{3}{2}, \frac{5}{2}, \ldots$	Electron, proton, neutron, ^{3}He atom	0, 1	Fermi–Dirac $n(\varepsilon) \equiv f_{FD} = \dfrac{1}{e^{\beta(\varepsilon-\mu)}+1}$
Boson	$0, 1, 2, \ldots$	Photon, deutron, ^{4}He atom	$0, 1, 2, 3, \ldots$	Bose–Einstein $n(\varepsilon) \equiv f_{BE} = \dfrac{1}{e^{\beta(\varepsilon-\mu)}-1}$

proof (1940), showing the connection between spin (first introduced by Uhlenbeck and Goudsmit) and based on relativistic quantum field theory, is one of the milestones of modern physics.

The Pauli principle is responsible for the shell structure of atoms, that is, for the regularities of the periodic table of elements. For example, an atom of lithium with three electrons has in the ground state, one electron in the $1s\uparrow$ orbital, and one electron in the $2s\uparrow$ orbital. Here, the number 1 or 2 is the principal quantum number of the orbital; the letter s by convention denotes a state of zero orbital angular momentum; and the arrow $\uparrow$ or $\downarrow$ denotes the direction of the electron spin assigned to the orbital.

The exclusion principle is known to be satisfied by every particle, elementary or composite, whose spin or intrinsic angular momentum is a half-integral multiple of $\hbar$. (Neutrons and neutrinos, as well as electrons, positrons, and protons, have spin $\frac{1}{2}\hbar$ and obey the exclusion principle; so does the composite atom ^{3}He, with two electrons, two protons, and one neutron.) The properties of the two classes of particles, fermions and bosons, are summarized by their distribution laws, given in Table 4.1-1.

■ EXAMPLE 4.2-1

Generalize the Pauli exclusion principle to a system of N particles instead of two particles.

Solution. Quite generally, for a system of N particles $a, b, c, d, \ldots$, with wave functions $\psi_r(a), \psi_s(b), \psi_t(c), \psi_u(d), \ldots$, the total energy is

$$E = \varepsilon_r + \varepsilon_s + \varepsilon_t + \varepsilon_u + \cdots \tag{4.2-14}$$

and the wave function for the whole system is the product

$$\psi_1 = \psi_r(a)\psi_s(b)\psi_t(c)\psi_u(d)\cdots \tag{4.2-15}$$

Other wave functions $\psi_2, \psi_3, \ldots$, and so on, having the same associated energy E are obtained by interchanging the quantum states of the individual particles. Symmetric and antisymmetric wave functions can be constructed for the general system of N particles. $N!$ products, $\psi_1, \psi_2, \psi_3, \psi_4, \ldots$, can be formed by permuting the particles among the occupied quantum states, ψ_r, $\psi_s, \ldots$. Then,

$$\psi_S = \left(\frac{1}{\sqrt{N!}}\right)(\psi_1 + \psi_2 + \psi_3 + \psi_4 + \cdots) \tag{4.2-16}$$

$$\psi_A = \left(\frac{1}{\sqrt{N!}}\right)(\psi_1 - \psi_2 + \psi_3 - \psi_4 + \cdots) \tag{4.2-17}$$

the order of $\psi_1, \psi_2, \ldots$, in the latter expression being chosen so that each ψ differs from the previous ψ by the interchange of just two labels in the product. For example, considering a system of three particles, and denoting $\psi_r(a)\psi_s(b)\psi_t(c)$ by abc, $\psi_r(b)\psi_s(a)\psi_t(c)$ by bac, and so on,

$$\psi_A = \frac{1}{\sqrt{6}}(abc - bac + bca - acb + cab - cba) \tag{4.2-18}$$

Suppose that, in a system of three molecules, the quantum state ψ_r is doubly occupied, that is, ψ_r appears twice in the product functions. The product functions are now

$$\psi_1 = \psi_r(a)\psi_r(b)\psi_t(c) \tag{4.2-19}$$

$$\psi_2 = \psi_r(b)\psi_r(a)\psi_t(c) \tag{4.2-20}$$

$$\psi_3 = \psi_r(b)\psi_r(c)\psi_t(a) \tag{4.2-21}$$

Obviously, ψ_1 and ψ_2 differ only in the order in which the terms in the product have been written down, that is,

$$\psi_1 \equiv \psi_2 \tag{4.2-22}$$

Hence, they cancel out in the expression (4.2-18) for ψ_A. The same is true for the pairs ψ_3,ψ_4 and ψ_5,ψ_6. Therefore,

$$\psi_A \equiv 0 \tag{4.2-23}$$

A generalized way of expressing the wave function of a many-fermion system is to use the Slater determinant,

$$\psi(a, b, \ldots, N) = \frac{1}{\sqrt{N!}}\begin{vmatrix} \psi_r(a) & \psi_s(a) & \cdots & \psi_i(a) \\ \psi_r(b) & \psi_s(b) & \cdots & \psi_i(b) \\ \cdots & \cdots & \cdots & \cdots \\ \psi_r(N) & \psi_s(N) & \cdots & \psi_i(N) \end{vmatrix} \tag{4.2-24}$$

where $1/\sqrt{N!}$ is the normalization constant. Should two of the particles occupy the same state, then two columns of the determinant would be equal and ψ would be zero. This is again the Pauli exclusion principle. For the case of two particles, the determinant wave function (4.2-24) reduces to

$$\psi(a,b) = \frac{1}{\sqrt{2}}\begin{vmatrix} \psi_r(a) & \psi_s(a) \\ \psi_r(b) & \psi_s(b) \end{vmatrix} = \frac{1}{\sqrt{2}}[\psi_r(a)\psi_s(b) - \psi_r(b)\psi_s(a)]$$

which is Eq. (4.2-12). ■

■ EXAMPLE 4.2-2

(a) In a given atom, can two electrons have $(1, 0, 0, \frac{1}{2})$ and $(1, 0, 0, -\frac{1}{2})$, respectively, as their quantum numbers (n, l, m_l, m_s)?

(b) Can two nearby atoms both have $(1, 0, 0, \frac{1}{2})$ electrons?

Solution

(a) Yes. The numbers agree with the rules for the quantum numbers. For $n = 1$, l must equal 0; then m_l must equal 0, and m_s can be $+\frac{1}{2}$ or $-\frac{1}{2}$. Also, the two sets of quantum numbers are different (corresponding to two different states).

(b) Yes. The Pauli exclusion principle applies to one unconnected system at a time, as for example, either atom. Therefore, two $(1, 0, 0, \frac{1}{2})$ electrons cannot be in the same atom (same system). But the two electrons can be in two different atoms (two different unconnected systems). ■

■ EXAMPLE 4.2-3

Consider two particles to be distributed in all possible ways between two energy states. Sketch and discuss the possible arrangements according to Boltzmann, Bose–Einstein, and Fermi–Dirac statistics, respectively.

Solution. The possible arrangements are shown below. In Boltzmann counting, the particles are distinguishable, and labeled a and b. In quantum statistics, the particles are indistinguishable, and are all labeled x. Additionally, in Fermi–Dirac statistics, only one particle may occupy an energy state.

Maxwell–Boltzmann		Bose–Einstein		Fermi–Dirac	
a, b	___	x, x	___	x	x
___	a, b	___	x, x		
a	b	x	x		
b	a				

Figure 4.2-1 Distribution of particles among energy states according to MB, BE, and FD statistics.

According to MB counting, there are four possible arrangements. In the BE counting, there are three possible combinations, and in the FD counting, only one arrangement is possible.

Normally, in classical physics, above a certain temperature (the characteristic temperature of the system), the difference in counting does not matter, because many more energy states are available than there are particles to place into them. In this way the excessive number of combinations given by classical statistics is not noticed (so many cells remain empty). Things change, however, at low temperature where the thermal energy is low and most of the states tend to be occupied. Since in Fermi–Dirac statistics there is only room for one particle in each state an enormous number of states will be filled even at absolute zero and this fact alone leads to a high zero-point energy. From this model Fermi calculated the specific heat of a highly degenerate gas and found that it should be proportional to the absolute temperature at low temperatures. ■

4.3 STATISTICS FOR LOCALIZED SYSTEMS

Before proceeding with the detailed treatment of nonlocalized (indistinguishable) systems, the distribution law for localized system will be reviewed. This will serve to bring out some salient points. In a perfect crystal, the lattice units (atoms and molecules) are localized in space and are therefore distinguishable. In this and similar instances of localized systems, the assumptions of classical statistics apply.

The distribution law for systems, following from a derivation analogous to that in Section 2.10, gives, for the a priori probability of the ith participle quantum state,

$$P_i = \frac{e^{-\varepsilon_i/kT}}{z} \tag{4.3-1}$$

Alternatively, in terms of the average occupation number of the ith quantum state,

$$\langle N_i \rangle = \frac{Ne^{-\varepsilon_i/kT}}{z} \tag{4.3-2}$$

Equations (4.3-1) and (4.3-2) represent the quantum-mechanical Boltzmann distribution. The molecular partition function is defined, as in Chapter 2, by

$$z = \sum_{i(\text{states})} e^{-\varepsilon_i/kT} \tag{4.3-3}$$

or, by

$$z = \sum_{j(\text{levels})} g_j e^{-\varepsilon_j/kT} \tag{4.3-4}$$

where g_j represents the degeneracy of the jth molecular energy level.

As for the molar partition function, it is expressed as

$$Z = \sum_E G(E)e^{-E/kT} \tag{4.3-5}$$

where, for every set of molecular energy levels $\varepsilon^{(1)}, \varepsilon^{(2)}, \ldots, \varepsilon^{(N)}$ adding up to E,

$$G(E) = g(\varepsilon^{(1)})g(\varepsilon^{(2)}) \cdots g(\varepsilon^{(N)}) \tag{4.3-6}$$

since the overall multiplicity equals the product of component multiplicities. The superscripts refer to levels and not quantum states. With

$$E = \sum_i \varepsilon^{(i)} \tag{4.3-7}$$

the partition function (4.3-5) becomes

$$Z = \sum_{\varepsilon^{(1)}} \sum_{\varepsilon^{(2)}} \cdots \sum_{\varepsilon^{(N)}} g(\varepsilon^{(1)}) g(\varepsilon^{(2)}) \cdots g(\varepsilon^{(N)}) e^{-(\varepsilon^{(1)} + \varepsilon^{(2)} + \cdots + \varepsilon^{(N)})/kT} \tag{4.3-8}$$

which can be rearranged to

$$Z = \left[\sum_{\varepsilon^{(1)}} g(\varepsilon^{(1)}) e^{-\varepsilon^{(1)}/kT}\right]\left[\sum_{\varepsilon^{(2)}} g(\varepsilon^{(2)}) e^{-\varepsilon^{(2)}/kT}\right] \cdots \left[\sum_{\varepsilon^{(N)}} g(\varepsilon^{(N)}) e^{-\varepsilon^{(N)}/kT}\right] \tag{4.3-9}$$

But these N factors, differing only in dummy index, are identical and all equal to (4.3-4) for a single particle. Therefore,

$$Z = z^N \qquad \text{(localized systems)} \tag{4.3-10}$$

This relationship holds for molar and molecular partition functions in classical statistical mechanics.

For the case of nonlocalized (indistinguishable) particles, the overwhelming majority of accessible quantum states for the system places (almost) all N particles in N different orbitals. Now, each distribution is associated with $N!$ possible arrangements. The multiplicity (4.3-6) will therefore be reduced by a factor $N!$, since the state of the system remains the same under permutation of particle indices. The complement to (4.3-10) for nonlocalized (indistinguishable) particles is thus

$$Z = \frac{z^N}{N!} \qquad \text{(nonlocalized systems)} \tag{4.3-11}$$

Equations (4.3-10) and (4.3-11) constitute the fundamental link between thermodynamics and molecular mechanics. On the one hand, all the thermodynamic functions can be expressed in terms of the molar partition function z, as shown in Chapter 2. On the other hand, microscopic molecular parameters, derived for the most part from spectroscopy, are directly applicable to evaluation of the molecular partition function, z. Molecules in crystals are essentially completely localized. Those in nearly perfect gases are completely mobile. Molecules in liquids and highly imperfect gases constitute an intermediate case between localization and mobility. In such cases, it is more difficult to relate the molar and molecular partition functions.

4.4 GRAND CANONICAL ENSEMBLE

So far, the formulation of statistical mechanics has been based on the canonical ensemble. Figure 4.4-1(a) shows a system of fixed volume V and given number of particles N at temperature T immersed in a heat bath at this

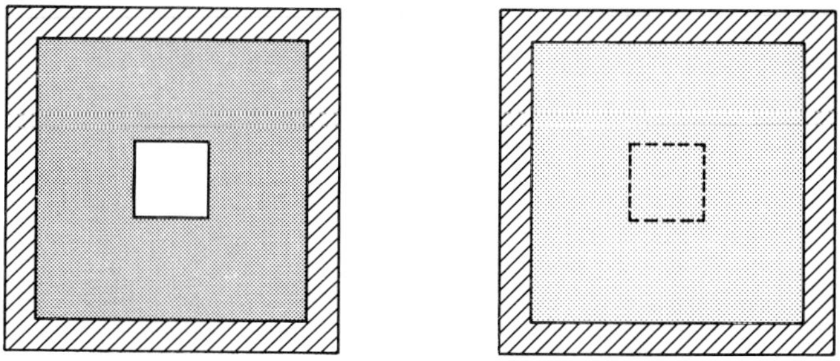

Figure 4.4-1 Canonical and grand canonical ensemble. (a) (T, V, N) fixed. E fluctuates, $F = -kT \ln Z(T, V, N)$. (b) (T, V, μ) fixed. E and N fluctuate, $\Omega = -kT \ln \Xi(T, V, \mu)$.

temperature. The statistical and thermodynamic descriptions are given by the partition function $Z(T, V, N)$ and the Helmholtz function A. These two quantities are related through

$$A = -kT \ln Z(T, V, N)$$

For such a system in a heat bath the energy is not constant; it fluctuates. However, in a macroscopic system, these fluctuations are generally extremely small so the energy of the system has, after all, a sharply defined value. In considering the properties of the system, it is then immaterial whether the system is at fixed energy or at fixed temperature.

In this chapter, another ensemble, Fig. 4.4-1(b), is selected for the formulation of the laws of statistical mechanics. This ensemble lends itself well to the derivation of the Fermi–Dirac and Bose–Einstein distributions. A *grand canonical ensemble* represents an *open isothermal system* (V, μ, and T specified). The variables N as well as E are, in concept, allowed to fluctuate. The state of each member of the ensemble is specified by N and $E_i(N)$, the latter representing the energy of the ith quantum state for an N-particle system. The volume V is fixed, but the number of particles N of the system can vary: $N = 1$, $2, \ldots$. The system possesses a sequence of states which can be arranged in order of increasing energy:

$$E(N_1) \leq E(N_2) \leq \cdots \leq E(N_i) \leq \cdots \tag{4.4-1}$$

where $E(N_i)$ is the energy of the ith state of the system.

For the system shown in Fig. 4.4-1(b) the temperature T and the volume V are given. It will be seen that for a complete description of the system, the chemical potential μ must be specified. The statistical and thermodynamic descriptions of the system will lead to two new quantities, the grand partition function $\Xi(T, V, \mu)$ and the grand potential Ω, which are related by

$$\Omega = -kT \ln \Xi(T, V, \mu) \tag{4.4-2}$$

This formulation of statistical thermodynamics is very convenient in many applications. For the problem of a gas it avoids the difficulty, encountered before, that the number of particles in the system be constant. There are many situations where the number of particles is not conserved, notably in chemical reactions, photons, and phonons, and for these the present chapter is particularly appropriate.

The most probable distribution of systems within a grand canonical ensemble with respect to particle numbers and energy states is now determined in accordance with postulate (4.1-8). It is recalled that in dealing with a particular outcome of several variables, say x_i, y_i, . . . , when the mean of all these variables, $\langle x \rangle$, $\langle y \rangle$, . . . , is known; that is, when

$$\sum_i P_i = 1$$

$$\sum_i P_i x_i = \langle x \rangle$$

$$\sum_i P_i y_i = \langle y \rangle$$

the result is given by

$$P_i = \frac{\exp(-\beta x_i - \gamma y_i \cdots)}{\sum_i \exp(-\beta x_i - \gamma y_i \cdots)}$$

Here, the conditions to be met are

$$\sum_N \sum_i P_i(N) = 1 \tag{4.4-3}$$

$$\sum_N \sum_i M_i(N) = M \tag{4.4-4}$$

$$\sum_N \sum_i M_i(N) E_i(N) = E \tag{4.4-5}$$

and

$$\sum_N \sum_i M_i(N) N = L \tag{4.4-6}$$

where $M_i(N)$ is the number of N-particle systems in the appropriate ith quantum state. Using Lagrange multipliers α, β, and γ, respectively, the maximization follows along the same lines as that in Chapters 2 and 3, and the result is

$$P_i(N) = e^{-\alpha} e^{-\beta E_i(N)} e^{-\gamma N} \tag{4.4-7}$$

The multiplier α is identified from the normalization condition applied on the a priori probability $P_i(N)$,

$$1 = e^{-\alpha} \sum_N \sum_i e^{-\beta E_i(N)} e^{-\gamma N} \tag{4.4-8}$$

so that

$$e^{\alpha} = \Xi \tag{4.4-9}$$

where

$$\Xi(V, \mu, T) \equiv \sum_N \sum_i e^{-\beta E_i(N)} e^{-\gamma N} \tag{4.4-10}$$

is defined as the *grand (canonical) partition function*, which can also be expressed as

$$\Xi(V, \mu, T) = \sum_N Z(V, N, T) e^{-\gamma N} \tag{4.4-11}$$

in terms of the canonical ensemble partition function. Furthermore, β can again be identified as kT.

It remains to identify Ξ and γ in thermodynamic terms. This can be done most easily by restricting the ensemble to fixed N, that is,

$$\Xi = Ze^{-\gamma N} \tag{4.4-12}$$

(For fixed N, $e^{-\gamma N}$ is treated as just another constant.) Taking logarithms of (4.4-12), multiplying by kT, and rearranging, gives

$$-kT \ln Z + kT \ln \Xi = -kT\gamma N \tag{4.4-13}$$

Making use of the fact that in thermodynamics, for a single one-component phase, the Gibbs function is proportional to the particle number N,

$$G \equiv H - TS = A + pV = \mu N \tag{4.4-14}$$

Eqs. (4.4-13) and (4.4-10) show that, in addition to Eq. (4.4-11),

$$pV = kT \ln \Xi \tag{4.4-15}$$

and

$$\mu = -kT\gamma \tag{4.4-16}$$

or

$$\gamma = -\frac{\mu}{kT} \tag{4.4-17}$$

Thus, the Lagrange multiplier, $-\gamma$, is the *chemical potential* in units of kT.

Incorporating the thermodynamic meaning of β and γ into (4.4-10), the grand canonical partition function becomes

$$\begin{aligned} \Xi(V, \mu, T) &= \sum_N \sum_i e^{-E_i(N)/kT} e^{\mu N/kT} \\ &= \sum_N \sum_i e^{[\mu N - E_i(N)]/kT} \end{aligned} \tag{4.4-18}$$

or, in terms of the *absolute activity*, $\lambda = e^{\mu/kT}$

$$\Xi(V, \lambda, T) = \sum_N \sum_i e^{-E_i(N)/kT} \lambda^N \tag{4.4-19}$$

The probability, as given by (4.4-7), is then

$$P_i(N) = \frac{e^{-E_i(N)/kT}\, e^{\mu N/kT}}{\Xi(V, \mu, T)} \tag{4.4-20}$$

This is the *grand canonical ensemble distribution*. It will be applied to several quantum systems of interest in this chapter.

■ EXAMPLE 4.4-1

Discuss the grand canonical ensemble distribution for the small system inside a heat and particle reservoir as shown in Fig. 4.4-2.

Solution. The probability $P(N_i, \varepsilon_i)$ that, at thermodynamic equilibrium, the small system has N_i particles and is in a microstate of energy eigenvalue $\varepsilon_i\,(V, N_i)$ is given by

$$P(N_i, \varepsilon_i) = C \exp[(\mu N_i - \varepsilon_i)/kT] \tag{4.4-21}$$

where μ is the chemical potential of the particle reservoir, N_i is the number of particles in the ith N-particle state of the small system, and ε is the energy of the N particles in the ith state of the system. The normalization condition to obtain C is that the summation over all the states i for all the values of N must be equal to unity, Thus,

$$C \sum_N \sum_i \exp[(\mu N - \varepsilon_i)/kT] = 1 \tag{4.4-22}$$

This gives

$$P(N, \varepsilon_i) = \frac{\exp[(\mu N - \varepsilon_i)/kT]}{\Xi} \tag{4.4-23}$$

where Ξ is the *grand partition function*. One way to evaluate Ξ is to sum first over all the states of the N-particle small system for a fixed value of N to give

$$\Xi = \sum_N \exp\left(\frac{\mu N}{kT}\right) Z(N)$$

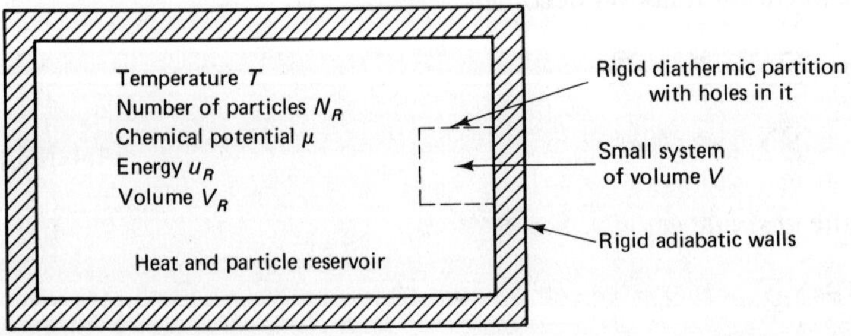

Figure 4.4-2 Open system.

with

$$Z(N) = \sum_i \exp(-\varepsilon_i/kT) \tag{4.4-24}$$

being the partition function for a fixed value of N. The partition function Ξ is then obtained by summing $\exp(\mu N/kT)Z(N)$ for all values of N. The grand sum, or grand partition function is a function of the absolute temperature T, the chemical potential μ, and the volume V of the system. ■

4.5 BOSE–EINSTEIN AND FERMI–DIRAC DISTRIBUTIONS

Particles of integral spin (0, 1, 2, . . .) obey Bose–Einstein statistics and are called *bosons*. There is no limit to the number of bosons that can be in any particular single-particle state. In Fig. 4.4-2, the bosons in the adiabatic box act as the particle reservoir for the open system. The grand partition function

$$\Xi = \sum_N \sum_i \exp[(\mu N - \varepsilon_i)/kT] \tag{4.5-1}$$

is to be summed over all states i for all numbers of particles N. Since ε_i is dependent on the number of particles populating that state, and there is no limit on the number of particles, the possible energy states are 0 for $N = 0$, ε for $N = 1$, 2ε for $N = 2$, 3ε for $N = 3$, etc. Accordingly, (4.5-1) gives

$$\Xi = \exp(0) + \exp[(\mu - \varepsilon)/kT] + \exp[(2\mu - 2\varepsilon)/kT] + \cdots \tag{4.5-2}$$

Letting $\exp[(\mu - \varepsilon)/kT] = x$, this becomes

$$\Xi = \sum_N x^N = (1 - x)^{-1} \tag{4.5-3}$$

as can be seen by using the binomial theorem. Now, the ensemble average value of any variable $a(N, i)$ is

$$\langle a \rangle = \sum_N \sum_i P(N, i)a(N, i) \tag{4.5-4}$$

$$= (1/\Xi) \sum_N \sum_i a(N, i) \exp[(\mu N - \varepsilon_i)/kT] \tag{4.5-5}$$

Putting a equal to N in (4.5-5) gives $f_{BE}(\varepsilon)$, the mean number of bosons in the chosen single-particle state:

$$f_{BE}(\varepsilon) = \langle N \rangle = (1/\Xi) \sum_N \sum_i N \exp[(\mu N - \varepsilon_i)/kT] \tag{4.5-6}$$

or, using (4.5-2),

$$f_{BE}(\varepsilon) = \langle N \rangle = (1/\Xi)\{\exp[0] + 1 \exp[(\mu - \varepsilon)/kT] + 2 \exp[(2\mu - 2\varepsilon)/kT] + 3 \exp[(3\mu - 3\varepsilon)/kT] + \cdots\} \tag{4.5-7}$$

Letting $\exp[(\mu - \varepsilon)/kT] = x$, it is seen that (4.5-7) is

$$f_{BE}(\varepsilon) = (1/\Xi) \sum_N N x^N \tag{4.5-8}$$

Now, from (4.5-3),

$$x\left(\frac{d}{dx}\right) \sum_N x^N = x \sum_N N x^{N-1} = \sum_N N x^N \tag{4.5-9}$$

Since $\sum N x^N = (1 - x)^{-1}$, this gives

$$\sum_N N x^N = x\left(\frac{d}{dx}\right)\left(\frac{1}{1-x}\right) = \frac{x}{(1-x)^2} \tag{4.5-10}$$

Substituting (4.5-10) and (4.5-3) into (4.5-8) yields

$$f_{BE}(\varepsilon) = (1 - x)\left[\frac{x}{(1-x)^2}\right] = \frac{x}{1-x} \tag{4.5-11}$$

or, remembering that $x = \exp[(\mu - \varepsilon)/kT]$,

$$f_{BE}(\varepsilon) = \frac{\exp[(\mu - \varepsilon)/kT]}{1 - \exp[(\mu - \varepsilon)/kT]} \tag{4.5-12}$$

Equation (4.5-12) gives the mean boson occupancy $b(\varepsilon)$. Multiplying the numerator and denominator by $\exp[(\varepsilon - \mu)/kT]$ puts it into the more familiar form

$$f_{BE}(\varepsilon) = \frac{1}{\exp[(\varepsilon - \mu)/kT]^{-1}} \tag{4.5-13}$$

This is the *Bose–Einstein distribution.*

To obtain the Fermi–Dirac distribution function, which applies to particles of half integral spin ($\frac{1}{2}, \frac{3}{2}, \frac{5}{2}, \ldots$), it is first observed that only one fermion of a group of identical fermions can occupy any single-particle quantum state (orbital). For example, an electron is a fermion of spin $\frac{1}{2}$, and only one of the electrons in an atom can have a given set of values of the quantum numbers n, l, m_l, and m_s. Now, consider again a system of N fermions inside a heat and particle reservoir as in Fig. (4.4-2). In the case of fermions, there can only be 0 or 1 fermions in a particular state, according to the Pauli exclusion principle. Thus, there are only two possibilities, namely, $N = 0$, energy $= 0$ and $N = 1$, energy ε. The grand partition function, Eq. (4.5-1), consists of terms

$$\Xi = \exp(0) + \exp[(\mu - \varepsilon)/kT] = 1 + \exp[(\mu - \varepsilon)/kT] \tag{4.5-14}$$

Letting α equal to N in Eq. (4.5-5) then gives $f_{FD}(\varepsilon)$, the mean (ensemble average) number of fermions in a given single-particle state of energy eigenvalue ε:

$$\begin{aligned} f_{FD}(\varepsilon) = \langle N \rangle &= (1/\Xi) \sum_N \sum_i N \exp[(\mu N - \varepsilon_i)/kT] \\ &= \frac{0 \exp(0) + 1 \exp[(\mu - \varepsilon)/kT]}{\exp(0) + 1 \exp[(\mu - \varepsilon)/kT]} \\ &= \frac{\exp[(\mu - \varepsilon)/kT]}{1 + \exp[(\mu - \varepsilon)/kT]} \end{aligned} \tag{4.5-15}$$

Multiplying (4.5-15) by $\exp[(\varepsilon - \mu)/kT]$ yields the familiar form

$$f_{FD}(\varepsilon) = \frac{1}{\exp\,[(\varepsilon - \mu)/kT] + 1} \tag{4.5-16}$$

for the *Fermi–Dirac distribution function*. Note that it has a (+1) in the denominator instead of a (−1) as in the Bose–Einstein distribution. The *chemical potential* μ is often called the *Fermi level*, ε_F, in solid-state physics. The three distributions of statistical mechanics, namely, Maxwell-Boltzmann, Fermi–Dirac, and Bose–Einstein, can all be expressed in a single compact form by writing

$$f_{FD}(\varepsilon), f_{MB}(\varepsilon), f_{BE}(\varepsilon) = \frac{1}{\exp[(\varepsilon - \mu)/kT]\ +1,\ +0,\ -1} \tag{4.5-17}$$

The quantum distributions, f_{FD} and f_{BE} will be used extensively from here on.

■ EXAMPLE 4.5-1

For purposes of comparison, let the three distribution functions be written as

$$n_{MB}(\varepsilon) = \frac{1}{e^{\alpha}e^{\varepsilon/kT}} \qquad n_{BE}(\varepsilon) = \frac{1}{e^{\alpha}e^{\varepsilon/kT} - 1} \qquad n_{FD}(\varepsilon) = \frac{1}{e^{\alpha}e^{\varepsilon/kT} + 1}$$

where $n(\varepsilon)$ is the average, or probable number, of particles of a system in equilibrium at temperature T to be found in a quantum state of energy ε. The parameter α, for a given temperature and system, is specified by the total number of particles it contains. (a) Plot the M-B distribution function versus energy for three different values of T and α. (b) Plot the B-E distribution function versus energy for three different values of T, all with $\alpha = 0$. (c) Plot the F-D distribution function versus energy for four different values of T and α. (d) Plot all three distribution functions versus ε/kT for two different values of α.

Solution

(a) The Boltzmann distribution function is a pure exponential, falling by a factor of $1/e$ with each increase kT in energy. The energy kT for each temperature is shown at the top of Fig. 4.5-1. α for any temperature T is fixed by the density of the system of particles.

(b) The Bose distribution function versus energy is plotted in Fig. 4.5-2 for three different values of T with $\alpha = 0$. The case of $\alpha = 0$, $e^{\alpha} = 1$, has been chosen, because it applies to a photon gas, for which the total number of particles, N, is not fixed. Notice that at energies small compared to kT the number of particles per quantum state is greater for the Bose distribution than for the Boltzmann distribution. At energies large compared to kT, however, the distribution approaches the exponential form characteristic of the Boltzmann distribution.

(c) The Fermi distribution function versus energy for different values of T and α has a maximum of $n(\varepsilon) = 1$ as shown in Fig. 4.5-3. (The exclusion principle sets the limit of one particle per quantum state.) The Fermi energy ε_F is shown for each curve at the bottom, and the energy kT is shown at the top. The function drops in a region of width kT centered on ε_F, becoming more gradual as the temperature increases. At high temperatures and energies, the function approaches the Boltzmann distribution function. The curve applies to a material with electron density similar to that of potassium, whose Fermi energy is 2.1 eV.

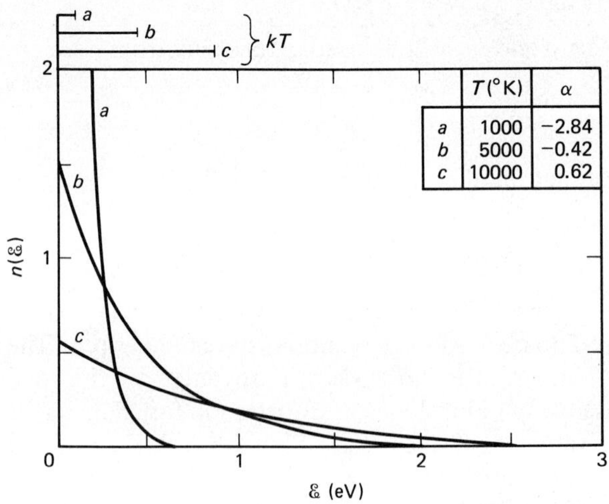

Figure 4.5-1 Boltzmann distribution for different values of T and α.

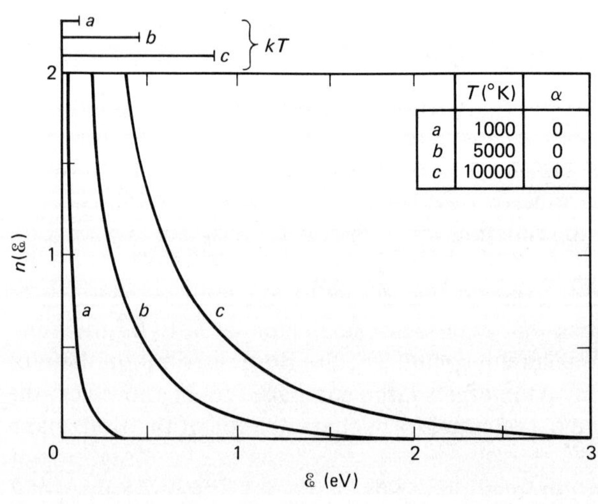

Figure 4.5-2 Bose distribution for $\alpha = 0$ and different values of T.

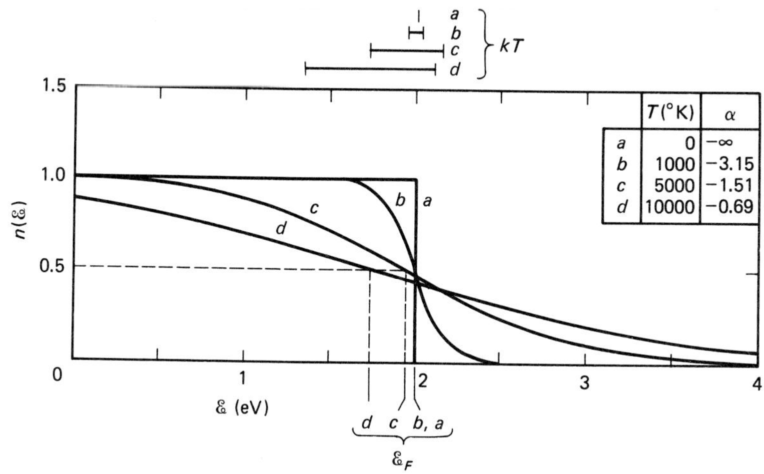

Figure 4.5-3 Fermi distribution for four different values of T and α.

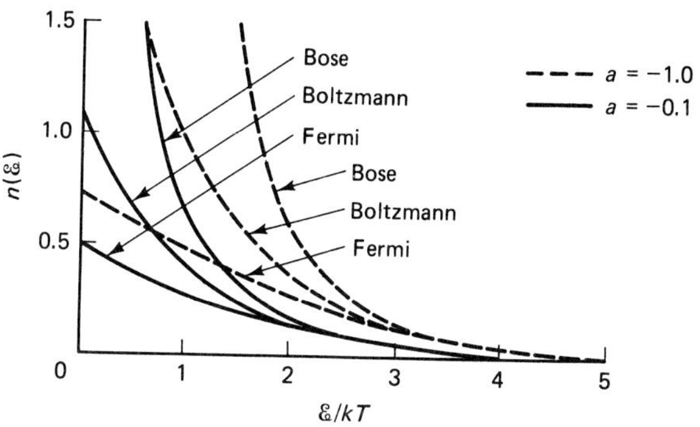

Figure 4.5-4 Boltzmann, Bose, and Fermi distributions for two different values of α.

(d) The three distribution functions are plotted versus ε/kT in Fig. 4.5-4 for values of $\alpha = -0.1$ and $\alpha = -1.0$. The dashed curves, if moved to the left 0.9 units $[-0.1-(-1.0)]$, would coincide exactly with the solid curves.

Certain general features of the three distribution functions should be cited. At high energies ($\varepsilon \gg kT$) where the number of particles per quantum state is much less than one, the quantum distributions merge with the classical distribution. At low energies ($\varepsilon \ll kT$) where the number of particles per quantum state is comparable to or larger than one, the quantum distributions fall on opposite sides of the classical distribution. Table 4.5-1 summarizes the important attributes of the three distribution functions.

TABLE 4.5-1 COMPARISON OF DISTRIBUTION FUNCTIONS

	Maxwell–Boltzmann	Bose–Einstein	Fermi–Dirac
Mathematical form	$Ae^{-\varepsilon/kT}$	$\dfrac{1}{e^{\alpha}e^{\varepsilon/kT}-1}$	$\dfrac{1}{e^{\alpha}e^{\varepsilon/kT}+1}$
Characteristics	Particles are distinguishable	Particles are indistinguishable. Any number can occupy a quantum state.	Particles are indistinguishable, but obey the Pauli exclusion principle.
Eigenfunctions of particles	No symmetry requirements	Symmetric under exchange of labels. Zero or integral spin.	Antisymmetric under exchange of labels. Add half integral spin.
Behavior versus ε/kT	Exponential	Exponential for $\varepsilon >> kT$. Lies above Boltzmann for $\varepsilon \ll kT$.	For $\varepsilon \gg kT$, exponential where $\varepsilon \gg \varepsilon_F$. If $\varepsilon_F \gg kT$, decreases abruptly near ε_F.
Examples of application	Dilute gases at practically any temperature.	Photon gas (cavity radiation); phonon gas (heat capacity); liquid helium.	Electrons in metals (electronic specific heat, contact potential, thermionic emission) protrons, neutrons, white dwarfs.

■

■ EXAMPLE 4.5-2

(a) Consider the Fermi–Dirac distribution function to be given by

$$f_{FD} = \frac{1}{e^{(\varepsilon - \mu)/kT} + 1} \tag{4.5-18}$$

where μ, the chemical potential or Fermi level is positive, but decreases and changes to negative at high temperatures. Draw a graph of the function versus energy for $T = 0$ K.

(b) Find the number of fermions $n(\varepsilon)$ with energy ε if the density of states for electrons in a metal is given by

$$g(\varepsilon) = \frac{2\pi(2m)^{3/2}\,V}{h^3}\,\varepsilon^{1/2} = \frac{1}{2\pi^2}\left(\frac{2m}{\hbar^2}\right)^{3/2}\varepsilon^{1/2}\,V$$

Plot $n(\varepsilon)$ versus ε.

(c) Consider the Maxwell–Boltzmann distribution function to be given by

$$f_{MB} = Ae^{-\varepsilon/kT}$$

Develop an expression for the density of states in terms of kinetic energy, find A by normalization, and then obtain the Maxwell speed distribution

$$n(v) = \left(\frac{m}{kT}\right)^{3/2}\sqrt{\frac{2}{\pi}}\,v^2 e^{-mv^2/2\,kT} \tag{4.5-19}$$

Plot the speed distribution function and compare it with the distribution function in (b).

Solution

(a) Near $T = 0$ K, $\mu \approx \varepsilon_F > 0$, where ε_F is the Fermi level, so the Fermi–Dirac distribution function

$$f_{\text{FD}} = \frac{1}{e^{(\varepsilon - \mu)/kT} + 1} \tag{4.5-20}$$

has a positive μ and does not change sign (as at high temperature). For $\varepsilon < \mu$ and $T \to 0$, $e^{(\varepsilon - \mu/kT)} \to 0$, and $f_{\text{FD}} = 1$. For $\varepsilon > \mu$ and $T \to 0$, $e^{(\varepsilon - \mu)/kT} \to \infty$, and $f_{\text{FD}} = 0$. Therefore, the plot of f_{FD} against ε at 0 K has the form given by the solid curve in Fig. 4.5-5. For $T > 0$ K, the curve takes on the form shown by the dashed curve.

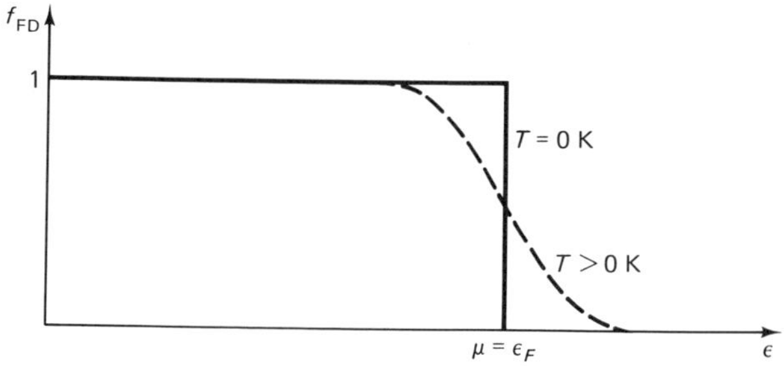

Figure 4.5-5 Fermi–Dirac distribution function at two different temperatrures.

(b) The number of fermions $n(\varepsilon)$ with energy ε is equal to the product of the density of states by the distribution function. In other words, the number of particles having energies in the interval between ε and $(\varepsilon + d\varepsilon)$ is

$$\begin{aligned} dn(\varepsilon) &= g(\varepsilon) f_{\text{FD}}\, d\varepsilon \\ &= \frac{1}{2\pi^2}\left(\frac{2m}{\hbar^2}\right)^{3/2} \frac{\varepsilon^{1/2}\, V\, d\varepsilon}{\exp[(\varepsilon - \mu)/kT] + 1} \end{aligned} \tag{4.5-21}$$

or, per unit volume, and $\mu = \varepsilon_F$,

$$n(\varepsilon) = \frac{1}{2\pi^2}\left(\frac{2m}{\hbar^2}\right)^{3/2} \frac{\varepsilon^{1/2}}{\exp[(\varepsilon - \varepsilon_F)/kT] + 1} \tag{4.5-22}$$

This is plotted in Fig. 4.5-6. The shaded area represents the occupied orbitals at absolute zero temperature. The occupation of orbitals at $T > 0$ K is shown by the dashed curve.

(c) The Maxwell–Boltzmann distribution is

$$f_{MB} = A e^{-\varepsilon/kT} \tag{4.5-23}$$

To obtain the population occupation, the distribution function must be multiplied by the density of states function. In the present case, the expression for discrete quantized energy levels

$$n(\varepsilon) = g(\varepsilon) f(\varepsilon) \tag{4.5-24}$$

must be changed to

$$n(v) = g(v) f(v) \tag{4.5-25}$$

for a continuous distribution of allowed speeds, v. To develop an expression

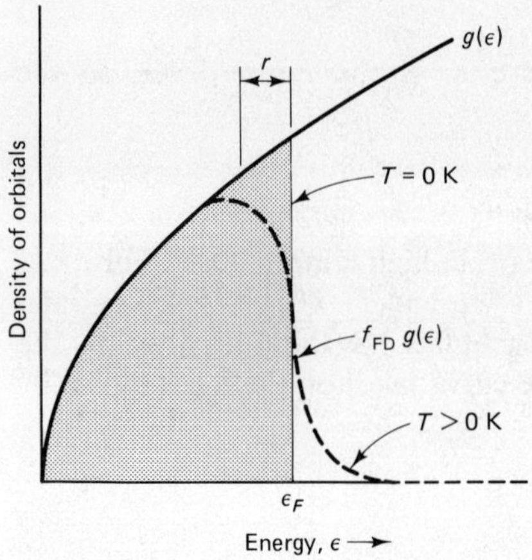

Figure 4.5-6 Density of orbitals and occupation of orbitals at two different temperatures.

for the density of states in speed, $g(v)$, consider a spherical shell of radius v and thickness dv. The volume $4\pi v^2\,dv$ gives the number of speed states between v and $v + dv$. Therefore, $g(v)$ is simply $4\pi v^2$, and the number of particles with speed between v and $v + dv$ is

$$n(v)dv = Av^2 e^{-mv^2/2\,kT}\,dv \tag{4.5-26}$$

where $\varepsilon = \frac{1}{2}mv^2$ and the constant A includes the 4π. To determine the constant A, $n(v)\,dv$ is integrated between 0 and infinity to catch all the fractions of molecules:

$$1 = A\int_0^\infty v^2 e^{-mv^2/2\,kT}\,dv \tag{4.5-27}$$

Using integral tables (from Chapter 1 or the Appendix), the result is $A = (m/kT)^{3/2}\sqrt{2/\pi}$. Thus, the Maxwell speed distribution for an ideal gas is

$$n(v) = \left(\frac{m}{kT}\right)^{3/2}\sqrt{\frac{2}{\pi}}\,v^2 e^{-mv^2/2\,kT} \tag{4.5-28}$$

This distribution is plotted in Fig. 4.5-7 for two different temperatures.

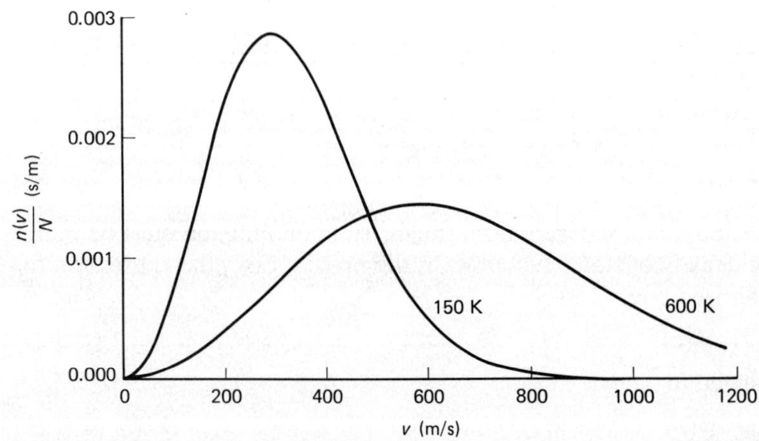

Figure 4.5-7 Maxwell speed distribution for air molecules at 150 K and 600 K. ■

4.6 CLASSICAL AND QUANTUM DESCRIPTIONS

It is worthwhile to compare the classical and quantum modes of describing a system. In Chapter 3, it was seen how the instantaneous state of motion of a classical particle can be represented by a point in *phase space*. The representation can be in terms of position coordinates and conjugate momenta, or in terms of position coordinates $(q_1, \ldots, q_\nu)$ and conjugate momenta $(p_1, \ldots, p_\nu)$. However, the latter description is chosen for reasons stated below.

Classically, a six-dimensional phase space with x, y, z, p_x, p_y, p_z as coordinate axes is introduced. (Figure 4.6-1 shows a graphical representation for a one-dimensional motion along the x axis.) For a system with ν degrees of freedom, the state of motion at a given time is specified by 2ν coordinates: $(q_1, \ldots, q_\nu)$, $(p_1, \ldots, p_\nu)$ in a 2ν-dimensional phase space. Each point in this phase space corresponds to a definite state of motion of the system. Hence the number of states in a given region of phase space is proportional to the volume of that region. The constant of proportionality, however, cannot be strictly determined within the framework of classical mechanics, since classically the states of a system form a continuum. Planck proposed to convert the continuous classical phase space into a quantum-mechanical phase space by taking the density of states to be

$$\frac{1}{h^\nu} dq_1 \cdots dq_\nu \, dp_1 \cdots dp_\nu \qquad (4.6\text{-}1)$$

This is the number of states of a system with coordinates and momenta lying within the intervals $q_1, q_1 + dq_1; \ldots q_\nu, q_\nu + dq_\nu; p_1, p_1 + dp_1; \ldots p_\nu, p_\nu + dp_\nu$. The result (4.6-1) has two properties which are essential to the success of the theory. First, a given region of phase space always contains the same number of states. Second, consider a given region Γ of phase space at time t. In the course of time, each point of the region traces out a trajectory which is completely determined by the equations of motion of the system. (The initial values of the position coordinates and momenta completely determine the subsequent motion.) At a later time t, the region Γ will have gone over into a new region Γ', but Γ and Γ' contain the same number of states, and so these

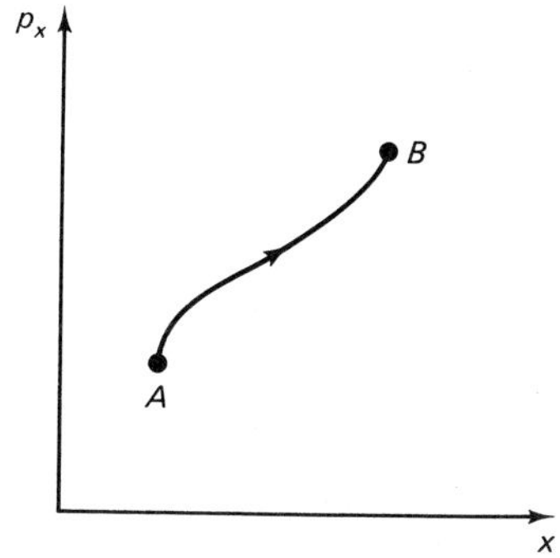

Figure 4.6-1 Phase space for particle moving in one dimension.

two regions must have the same volume in phase space. These statements are known as *Liouville's theorem*, and were shown to be true in Chapter 3. An analogous formulation of these results in a 2ν-dimensional space with position coordinates $(q_1, \ldots, q_\nu)$ and velocity coordinates $(\dot{q}_1, \ldots, \dot{q}_\nu)$ fails.

Equation (4.6-1) indicates that phase space is subdivided into cells of volume h^ν each. Each cell represents one state. Using this language, the quantum expression

$$z = \sum_i q(\varepsilon_i) e^{-\varepsilon_i/kT} \tag{4.6-2}$$

can be translated into classical language. Equation (4.6-2) is a sum of Boltzmann factors $e^{-\varepsilon_i/kT}$ over discrete energies. This sum is now replaced by an integral over phase space of Eq. (4.6-1). To write down this integral, use is made of the fact that each point of phase space [labelled by the 2ν coordinates $(q_1, \ldots, p_\nu)$] has a definite energy associated with it given by the Hamiltonian $H(q_1, \ldots, q_\nu, p_1, \ldots, p_\nu)$ of the system. Hence each point in classical phase space has a Boltzmann factor $\exp[-H(q,p)/kT]$, and the partition function (4.6-2) goes over into

$$z = \frac{1}{h^\nu} \int_q \int_p \exp[-H(q,p)/kT]\, dq\, dp \tag{4.6-3}$$

This is the end result. Finally, if the system consists of N identical nonlocalized particles, then the right-hand side of Eq. (4.6-3) must be divided by $N!$ as shown in Chapter 3. Quantum considerations indicate that there is a definite smallest size to any cell. A cell in which $\Delta x\, \Delta p_x$ is less than h would be more precise than allowed by the uncertainty principle. Thus, for a six-dimensional phase space, the smallest cell has a volume of h^3 and for a 2ν-dimensional phase space, a volume of h^ν.

These explanations show the applicability of classical statistical mechanics (characterized by easier calculations: integration instead of summation) to many quantum situations. At high temperatures, the limitation of a smallest-size cell is not important, because there is not more than one particle per state. The classical distribution function is then valid, regardless of the indistinguishability of the particles. At low temperatures, on the other hand, the number of particles per cell is more than one, and the quantum (real) world's limitation to the smallness of a cell becomes important. Classical Boltzmann's statistics no longer applies, and quantum distributions (Bose–Einstein or Fermi–Dirac) must be used.

4.7 PHOTON STATISTICS

Quantum radiation was first studied by Planck in 1900, before the advent of quantum mechanics. Consider a gas of identical particles in a volume V in equilibrium at the temperature T, and proceed as follows: Label the possible quantum states of a single particle by r. Denote the energy of a particle in state

r by ε_r. Denote the number of particles in state r by n_r. Consider the possible quantum states of the whole gas as R.

The assumption of negligibly small interaction between the particles enables the total energy of the gas, when it is in some state R where there are n_1 particles in state $r = 1$, n_2 particles in state $r = 2$, and so on, to be written

$$E_R = n_1\varepsilon_1 + n_2\varepsilon_2 + n_3\varepsilon_3 + \cdots = \sum_r n_r\varepsilon_r \tag{4.7-1}$$

where the sum extends over all the possible states r of a particle. Furthermore, if the total number of particles in the gas is known to be N, then

$$\sum_r n_r = N \tag{4.7-2}$$

In order to calculate the thermodynamic functions of the gas, it is necessary to calculate its partition function

$$Z = \sum_R e^{-\beta E_R} = \sum_R e^{-\beta(n_1e_1 + n_2e_2 + \cdots)} \tag{4.7-3}$$

where β, which stands for $1/kT$, has been kept to shorten the writing somewhat. The sum is over all possible states R of the whole gas, that is, essentially over all the various possible values of the numbers $n_1, n_2, n_3, \ldots$. Since $\exp[-\beta(n_1\varepsilon_1 + n_2\varepsilon_2 + \cdots)]$ is the relative probability of finding the gas in a particular state where there are n_1 particles in state 1, n_2 particles in state 2, and so on, the mean number of particles in a state s is

$$\langle n_s \rangle = \frac{\sum_R n_s e^{-\beta(n_1\varepsilon_1 + n_2\varepsilon_2 + \cdots)}}{\sum_R e^{-\beta(n_1\varepsilon_1 + n_2\varepsilon_2 + \cdots)}} \tag{4.7-4}$$

$$= \frac{1}{Z}\sum_R \left(-\frac{1}{\beta}\frac{\partial}{\partial \varepsilon_s}\right) e^{-\beta(n_1\varepsilon_1 + n_2\varepsilon_2 + \cdots)} = -\frac{1}{\beta Z}\frac{\partial Z}{\partial \varepsilon_s}$$

$$= -\frac{1}{\beta}\frac{\partial \ln Z}{\partial \varepsilon_s} \tag{4.7-5}$$

Thus, the mean number of particles in a given single-particle state s can be expressed in terms of the partition function Z. Other properties, for example, the entropy, can be likewise expressed.

In Bose–Einstein and photon statistics, the particles are considered indistinguishable, so that mere specification of the numbers $(n_1, n_2, n_3, \ldots)$ is enough to specify the state of the gas. It is necessary only to sum over all possible numbers of particles in each single-particle state, that is, over all possible values

$$n_r = 0, 1, 2, 3, \ldots \quad \text{for each } r \tag{4.7-6}$$

If the total number of particles is fixed, these numbers must satisfy the restriction (4.7-2). A simpler special case is when there is no requirement fixing the total number of particles. This occurs, for example, when the particles are photons enclosed in a container of volume V, since the photons can be readily absorbed and emitted by the walls. There is then no equation of constraint (4.7-2) to be satisfied, and the mean number of particles $\langle n_s \rangle$ proceeds directly from Eq. (4.7-4):

$$\langle n_s \rangle = \frac{\sum\limits_{n_1, n_2, \ldots} n_s e^{-\beta(n_1 \varepsilon_1 + n_2 \varepsilon_2 + \cdots + n_s \varepsilon_s + \cdots)}}{\sum\limits_{n_1, n_2, \ldots} e^{-\beta(n_1 \varepsilon_1 + n_2 \varepsilon_2 + \cdots + n_s \varepsilon_s + \cdots)}} \tag{4.7-7}$$

By summing first over all possible values of n_s, using the multiplicative property of the exponential function, and rearranging the order of summation, (4.7-7) can be written in the form

$$\langle n_s \rangle = \frac{\sum\limits_{n_s} n_s e^{-\beta n_s \varepsilon_s} \sum\limits_{n_1, n_2, \ldots}^{(s)} e^{-\beta(n_1 \varepsilon_1 + n_2 \varepsilon_2 + \cdots)}}{\sum\limits_{n_s} n_s e^{-\beta n_s \varepsilon_s} \sum\limits_{n_1, n_2, \ldots}^{(s)} e^{-\beta(n_1 \varepsilon_1 + n_2 \varepsilon_2 + \cdots)}} \tag{4.7-8}$$

In (4.7-8), the last sums in the numerator and denominator omit from consideration the particular state s (this is indicated by the superscript s on the summation symbol).

In photon statistics, the numbers $n_1, n_2, \ldots$, assume all values $n_r = 0, 1, 2, 3, \ldots$, for each r, without any further restriction. The sums $\Sigma^{(s)}$ in the numerator and denominator of (4.7-8) are then identical and cancel, and what is left is simply

$$\langle n_s \rangle = \frac{\sum\limits_{n_s} n_s e^{-\beta n_s \varepsilon_s}}{\sum\limits_{n_s} e^{-\beta n_s \varepsilon_s}} \tag{4.7-9}$$

The rest of the calculation is straightforward. The result (4.7-9) becomes

$$\langle n_A \rangle = \frac{(-1/\beta)(\partial/\partial \varepsilon_s)\Sigma e^{-\beta n_s \varepsilon_s}}{\Sigma e^{-\beta n_s \varepsilon_s}} = \frac{1}{\beta} \frac{\partial}{\partial \varepsilon_s} \ln(\Sigma e^{-\beta n_s \varepsilon_s}) \tag{4.7-10}$$

But the last sum is just an infinite geometric series which can be summed

$$\sum_{n_s = 0}^{\infty} e^{-\beta n_s \varepsilon_s} = 1 + e^{-\beta \varepsilon_s} + e^{-2\beta \varepsilon_s} + \cdots = \frac{1}{1 - e^{-\beta \varepsilon_s}} \tag{4.7-11}$$

using the binomial theorem $\Sigma x^n = \dfrac{1}{1-x}$.

Equation (4.7-10) gives

$$\langle n_s \rangle = \frac{1}{\beta} \frac{\partial}{\partial \varepsilon_s} \ln(1 - e^{-\beta \varepsilon_s}) = \frac{e^{-\beta \varepsilon_s}}{1 - e^{-\beta \varepsilon_s}} \tag{4.7-12}$$

or

$$\langle n_s \rangle = \frac{1}{e^{\beta \varepsilon_s} - 1} \tag{4.7-13}$$

This is the *Planck distribution*.

■ EXAMPLE 4.7-1

Photons in a cavity and phonons in a solid follow the Bose–Einstein distribution law

$$n(\varepsilon) = \frac{1}{e^{\alpha}e^{\varepsilon/kT} - 1} = \frac{1}{e^{\varepsilon/kT} - 1} \tag{4.7-14}$$

(a) Explain why α is zero for photons and phonons.

(b) The total number of phonons increases with temperature. How does the average number of phonons per energy state increase with temperature for frequencies that are less than $(k/h)T$?

(c) The density of states or number of quantum states per unit energy interval at energy ε for photons in a cavity is given by

$$g(\varepsilon) = \frac{8\pi V}{c^3}\frac{\varepsilon^2}{h^3} \tag{4.7-15}$$

Show that the energy per unit volume in the energy interval ε to $\varepsilon + d\varepsilon$ is

$$\rho(\varepsilon)\, d\varepsilon = \frac{8\pi\varepsilon^3\, d\varepsilon}{c^3h^3(e^{\varepsilon/kT} - 1)} \tag{4.7-16}$$

(d) For a system of bosons having a constant number of particles, how does α vary with temperature?

Solution

(a) The value of the parameter α is governed by the total number of particles in the system. For photons, this number is not constant, since particles are constantly emitted and absorbed by the walls. Thus, for a system of photons, the distribution function cannot contain the term e^{α}. Because the number of particles in the system varies, it is necessary that $\alpha = 0$ so that $e^{\alpha} = 1$. The same explanation holds for phonons (quanta of mechanical wave energy), which are emitted and absorbed by vibrating atoms at the lattice points in a solid.

(b) The distribution function for phonons is

$$n(\varepsilon) = \frac{1}{e^{\varepsilon/kT} - 1} \tag{4.7-17}$$

where $\varepsilon = h\nu$ gives the energy of each phonon, or $\varepsilon/kT = h\nu/kT$. For $\nu \ll kT/h$, $h\nu/kT \ll 1$, so that $e^x \simeq 1 + x$ for small x. Thus,

$$n(\varepsilon) = \frac{1}{1 + (h\nu/kT) - 1} = \frac{kT}{h\nu} \tag{4.7-18}$$

For the given frequency of $\nu = kT/h = (2.08 \times 10^{10}\ \text{Hz}/K)T$, the average number of phonons per energy state is directly proportional to the temperature of the system.

(c) Multiplying the density of states by the distribution and by the energy carried by each photon:

$$\varepsilon g(\varepsilon)n(\varepsilon) = \varepsilon\left(\frac{8\pi V}{c^3}\frac{\varepsilon^2}{\hbar^3}\right)\left(\frac{1}{e^{\varepsilon/kT}-1}\right)$$

$$= \frac{8\pi V\varepsilon^3}{c^3h^3(e^{\varepsilon/kT}-1)} \qquad (4.7\text{-}19)$$

Dividing this by the volume V of the cavity gives, for the energy per unit volume in the energy interval ε to $\varepsilon + d\varepsilon$,

$$\rho(\varepsilon)\,d\varepsilon = \frac{8\pi\varepsilon^3\,d\varepsilon}{c^3h^3(e^{\varepsilon/kT}-1)} \qquad (4.7\text{-}20)$$

Planck's spectrum law follows at once by using $\varepsilon = h\nu$ to convert from ε to ν:

$$\rho(\nu)d\nu = \frac{8\pi V^2}{c^3}\cdot\frac{h\nu\,d\nu}{e^{h\nu/kT}-1} \qquad (4.7\text{-}21)$$

In Planck's original derivation, the radiation is a set of waves confined to the cavity. Each of the standing waves is a mode of vibration that is distinguishable from all the others, just as for the lattice vibration modes in a Debye solid, so it is equally valid to apply the Boltzmann distribution to them.

(d) For systems having a constant total number of particles, α decreases to zero as the temperature is lowered. Near a particular low temperature, this behavior causes the particles to drop rapidly into the ground energy level. This is known as *Bose–Einstein condensation* and is part of the explanation for superfluidity (zero viscosity) of liquid helium and for superconductivity. ■

4.8 INTERNAL ENERGY AND ENTROPY

The quantum distribution functions found in the preceding sections and written as

$$\left(\frac{n_i}{g_i}\right)_{\substack{\text{FD}\\\text{BE}}} = \frac{1}{e^{\alpha+\beta\varepsilon_i}\pm 1} \qquad (4.8\text{-}1)$$

are the basic results. In principle, the Lagrange multipliers may be determined by substituting the most probable distribution into $\Sigma n_i = N$ and $\Sigma n_i\varepsilon_i = E$ and solving for α and β. However, this is not simple to do, and only for the Maxwell–Boltzmann distribution is the procedure easily implemented. Since fluctuations in the various ensembles are often negligible, the canonical ensemble will be selected for mathematical convenience. In the limit of a dilute assembly, $(g_i/n_i) \gg 1$, the unit term in Eq. (4.8-1) is neglected, and a common result is obtained for all three statistical models. Thus, irrespective of whether the assembly is comprised of fermions or bosons, the equilibrium distribution is provided by

$$\frac{n_i}{g_i} = e^{-\alpha} e^{-\beta \varepsilon_i} \tag{4.8-2}$$

and the probability that a member of the ensemble occupies a state with energy ε_i is

$$P_i = \frac{n_i}{N} = e^{-\alpha} e^{-\beta \varepsilon_i} \tag{4.8-3}$$

The Lagrange multiplier α is determined by normalizing the probability

$$\sum_i P_i = \sum_i e^{-\alpha} e^{-\beta \varepsilon_i} = 1 \tag{4.8-4}$$

so that

$$e^{-\alpha} = \frac{1}{\sum_i e^{-\beta \varepsilon_i}} = \frac{1}{z} \tag{4.8-5}$$

where

$$z = \sum_{i(\text{states})} e^{-\beta \varepsilon_i} \tag{4.8-6}$$

is the molecular partition function. The sum in Eq. (4.2-20) is over quantum states. As a sum over energy levels, it is

$$z = \sum_{i(\text{levels})} g_i e^{-\beta \varepsilon_i} \tag{4.8-7}$$

where g_i is the degeneracy of the ith level.

Now, the value of a mechanical property for a system can be obtained by summing the probability P_i times the mechanical property of state i over all states of the ensemble. Consider a canonical ensemble specified by V, N, and T. Its internal energy U is identified with the ensemble average, that is,

$$U = \langle E \rangle = \sum_i E_i P_i = \frac{\sum_i E_i e^{-\beta E_i}}{Z} \tag{4.8-8}$$

where P_i is the probability. Also, upon summation over quantum states,

$$Z(V, N, T) = \sum_{i(\text{states})} e^{-\beta E_i} \tag{4.8-9}$$

is the system partition function. Noting that

$$\left(\frac{\partial Z}{\partial \beta}\right)_{V,N} = -\sum_i E_i e^{-\beta E_i} \tag{4.8-10}$$

this gives, with (4.8-8),

$$U = -\frac{1}{Z}\left(\frac{\partial Z}{\partial \beta}\right)_{V,N} = -\left(\frac{\partial \ln Z}{\partial \beta}\right)_{V,N} \tag{4.8-11}$$

Equation (4.8-11) is the formal statistical representation for internal energy.

In order to evaluate β, the relations of statistical mechanics will be compared with the combined first and second laws of thermodynamics: $dU = T\,dS - p\,dV$. In statistical mechanics, a reversible process can be regarded as a continuous sequence of ensembles. Thus, if V and T are changed reversibly at constant N for a system,

$$dU = \sum_i E_i\,dP_i + \sum_i P_i\,dE_i \tag{4.8-12}$$

The first sum represents a redistribution of the systems in the ensemble among a fixed set of energy levels E_i. The second sum represents a perturbation of the set of energy eigenvalues but without changing the population distribution. The two parts of (4.8-12) are interpreted as reversible heat and work, respectively:

$$dQ = \sum_i E_i\,dP_i \tag{4.8-13}$$

$$dW = \sum_i P_i\,dE_i = \sum_i P_i\left(\frac{\partial E_i}{\partial V}\right)_N dV \tag{4.8-14}$$

where it is recognized in (4.8-14) that the energy states E_i can only be changed by changing the volume. Thus, (4.8-12) becomes

$$dU = \sum_i E_i\,dP_i + \sum_i P_i\left(\frac{\partial E_i}{\partial V}\right)_N dV \tag{4.8-15}$$

Comparison with the thermodynamic relation

$$dU = T\,dS - p\,dV \tag{4.8-16}$$

shows that

$$\sum_i P_i\left(\frac{\partial E_i}{\partial V}\right)_N dV = -p\,dV \tag{4.8-17}$$

and therefore

$$p = -\sum_i P_i\left(\frac{\partial E_i}{\partial V}\right)_N = -\left\langle\left(\frac{\partial E_i}{\partial V}\right)_N\right\rangle \tag{4.8-18}$$

The mechanical property pressure, $p = \Sigma_i p_i P_i$, is thus identified as the ensemble average of the change of E_i with respect to volume.

Now, the Boltzmann distribution states that the probability of being in the ith energy state is

$$P_i = \frac{e^{-\beta E_i}}{Z} \tag{4.8-19}$$

Solving for E_i gives

$$E_i = -\frac{1}{\beta}(\ln P_i + \ln Z) \tag{4.8-20}$$

Substituting this into (4.8-15) yields

$$dU = -\frac{1}{\beta}\sum_i (\ln P_i + \ln Z)dP_i + \sum_i P_i \left(\frac{\partial E_i}{\partial V}\right)_N dV \tag{4.8-21}$$

This equation may be simplified by noting that $\Sigma(\ln Z)dP_i = 0$ since $\Sigma dP_i = 0$ from $\Sigma P_i = 1$. The first term may be rewritten using

$$d\left(\sum_i P_i \ln P_i\right) = \sum_i dP_i + \sum_i \ln P_i\, dP_i = \sum_i \ln P_i\, dP_i$$

The second term is simply $-p\,dV$. Thus, Eq. (4.8-21) becomes

$$dU = -\frac{1}{\beta} d\left(\sum_i P_i \ln P_i\right) - p\,dV \tag{4.8-22}$$

Comparing this with $dU = T\,dS - p\,dV$ shows that

$$T\,dS = -\frac{1}{\beta} d\left(\sum_i P_i \ln P_i\right) = \frac{1}{k\beta} d\left(-k\sum_i P_i \ln P_i\right), \tag{4.8-23}$$

giving

$$T = \frac{1}{k\beta}; \qquad \beta = \frac{1}{kT} \tag{4.8-24}$$

and

$$S = -k\sum_i P_i \ln P_i \tag{4.8-25}$$

where the proportionality constant k depends on the size of the unit of temperature. The Boltzmann constant ($k = R/N_{\text{Avogadro}}$) is used because this brings statistical mechanics into agreement with the temperature scale of the second law.

Note that β is necessarily positive as can be seen from normalization of $P_i = (1/Z)\exp(-\beta E_i)$. A negative value of β would not enable normalization. Equation (4.8-25) embodies the statistical version of the third law of thermodynamics. If there is a single state (e.g., a perfect crystal) at $T = 0$, then $S = 0$. The entropy depends only on the probabilities P_i of the various states accessible to a system. Since the probabilities are fractions, the entropy is necessarily a positive quantity.

■ EXAMPLE 4.8-1

Derive expressions of thermodynamic properties in terms of the partition function Z.

Solution. The two concepts that are of particular importance in the application of statistical mechanisms to the calculation of thermodynamic properties are: (1) The evaluation of the total energy of a system as the sum of the energies of the separate particles, and the identification of this total with the thermodynamic internal energy. (2) The statistical mechanical interpretation of temperature; from this, an expression for the entropy of the system is obtained. The remaining thermodynamic properties can be deduced from known thermodynamic equations involving the energy and entropy functions.

With the identification of $\beta = 1/kT$, the internal energy U may be written in terms of the partition function using Eq. (4.8-11):

$$U = -\frac{1}{Z}\left(\frac{\partial Z}{\partial \beta}\right)_{V,N} = -\left(\frac{\partial \ln Z}{\partial \beta}\right)_{V,N}$$

$$= kT^2\left(\frac{\partial \ln Z}{\partial T}\right)_{V,N} \qquad (4.8\text{-}26)$$

The entropy in terms of the probabilities of the various states is

$$S = -k\Sigma P_i \ln P_i \qquad (4.8\text{-}27)$$

The Boltzmann distribution is

$$P_i = \frac{1}{Z}e^{-\beta E_i} = \frac{1}{Z}e^{-E_i/kT}$$

$$\ln P_i = -\ln Z - E_i/kT$$

Substitution into (4.8-27) gives

$$S = -k\sum_i P_i\left(-\ln Z - \frac{E_i}{kT}\right) = (k \ln Z)\sum_i P_i + \frac{1}{T}\sum_i P_i E_i$$

But $\Sigma_i P_i = 1$, and $\Sigma_i P_i E_i = U$. Thus,

$$S = k \ln Z + \frac{U}{T} \qquad (4.8\text{-}28)$$

Helmboltz function: $A = U - TS$. From (4.8-28),

$$TS = kT \ln Z + U$$

This gives

$$A = U - kT \ln Z - U$$

$$= -kT \ln Z \qquad (4.8\text{-}29)$$

Pressure: $p = -(\partial A/\partial V)_{T,N}$. From (4.8-29),

$$(\partial A/\partial V)_{T,N} = -kT\left(\frac{\partial \ln Z}{\partial V}\right)_{T,N}$$

Thus,

$$p - kT\left(\frac{\partial \ln Z}{\partial V}\right)_{T,N} \tag{4.8-30}$$

Enthalpy: $N = U + pV$. Using (4.8-26),

$$H = kT^2\left(\frac{\partial \ln Z}{\partial T}\right)_{V,N} + pV \tag{4.8-31}$$

Gibbs function: $G = H - TS = A + pV$. Using (4.8-29),

$$G = -kT \ln Z + pV \tag{4.8-32}$$

■

4.9 QUANTUM BEHAVIOR NEAR ABSOLUTE ZERO

It is of interest to compare the low-temperature behavior of quantum systems with that of a classical system. Equally important is the fact that at low temperatures, N-particle systems behave differently according to whether they are bosons or fermions. To bring out these features, consider the distributions

$$\frac{1}{e^{(\varepsilon_i - \mu)/kT} \pm 1} \qquad \begin{matrix}(+)\,FD \\ (-)\,BE\end{matrix} \tag{4.9-1}$$

to be written in the form

$$\zeta_i = (1 \pm e^{-\varepsilon_i/kT}\lambda)^{\pm 1} \qquad \begin{matrix}(+)\,FD \\ (-)\,BE\end{matrix} \tag{4.9-2}$$

In the case of bosons, it is implicitly assumed that

$$e^{-\varepsilon_i/kT}\lambda = e^{-(\varepsilon_i - \mu)/kT} < 1, \qquad \text{all } \varepsilon_i \tag{4.9-3}$$

This requires that all $\varepsilon_i > \mu$ and, in particular,

$$\varepsilon_0 > \mu \tag{4.9-4}$$

Therefore, the chemical potential of a many-boson system must be less than the ground-state orbital energy. (There is no similar requirement for fermions.) By virtue of (4.9-3), for bosons, as $T \to 0$,

$$e^{-(\varepsilon_0 - \mu)/kT} \gg e^{-(\varepsilon_i - \mu)/kT} \qquad \text{for } \varepsilon_i > \varepsilon_0 \tag{4.9-5}$$

$$\zeta_0 \gg \zeta_i \qquad \text{for } \varepsilon_i > \varepsilon_0$$

and

$$\lim_{T \to 0} \zeta_i = 1 \qquad \text{all } i \tag{4.9-6}$$

Now, the grand partition function

$$\Xi(V, \mu, T) = \sum_N \sum_i e^{-E_i(N)/kT} e^{\mu N/kT} \tag{4.9-7}$$

may be written in terms of the activity

$$\lambda = e^{\mu/kT} \tag{4.9-8}$$

as

$$\Xi(V, \lambda, T) = \sum_N \sum_i e^{-E_i(N)/kT} \lambda^N \tag{4.9-9}$$

Also, the probability $P_i(N)$ that a member of the ensemble contains N particles in state i is

$$P_i(N) = \frac{e^{-E_i(N)/kT} e^{\mu N/kT}}{\Xi(V, \mu, T)} \tag{4.9-10}$$

$$= \frac{e^{-E_i(N)/kT} \lambda^N}{\Xi(V, \lambda, T)} \tag{4.9-11}$$

The grand partition function is

$$\Xi(V, \lambda, T) = \zeta_0 \, \zeta_1 \, \zeta_2 \cdots = \prod_i \zeta_i \tag{4.9-12}$$

and the average number of particles occupying the ith orbital is

$$\langle N_i \rangle = \frac{1}{e^{(\varepsilon_i - \mu)/kT} \pm 1} = \frac{1}{e^{\varepsilon_i/kT} \lambda^{-1} \pm 1} \qquad \begin{matrix} (+)\ FD \\ (-)\ BE \end{matrix} \tag{4.9-13}$$

$$= \lambda \left(\frac{\partial \ln \zeta_i}{\partial \lambda} \right)_{V,T} \tag{4.9-14}$$

As the temperature nears zero,

$$\lim_{T \to 0} \langle N_0 \rangle = N \tag{4.9-15}$$

while

$$\lim_{T \to 0} \langle N_i \rangle = 0 \qquad \text{for } \varepsilon_i > \varepsilon_0 \tag{4.9-16}$$

This means that in an N-boson system at absolute zero, every particle has fallen to its lowest orbital-energy level. This is borne out in Fig. 4.9-1, which shows the Bose distribution function versus energy for three different values of T, all with $\alpha = 0$. At energies large compared to kT, the function approaches the exponential Boltzmann distributions, but at energies small compared to kT, it exceeds the Boltzmann values, tending to infinity as the energy goes to zero. The energy kT for each temperature is shown at the right of the figure.

The same characteristics do not apply to fermions, because of the exclusion principle. From (4.9-1) with the plus sign, it is seen that

$$\lim_{T \to 0} \langle N_i \rangle = \begin{cases} 1 \text{ for } \varepsilon_i < \mu \\ 0 \text{ for } \varepsilon_i > \mu \end{cases} \tag{4.9-17}$$

This defines the step function drawn in Fig. 4.9-2. For a many-fermion system at absolute zero, the particles occupy singly all orbitals lower than μ, the

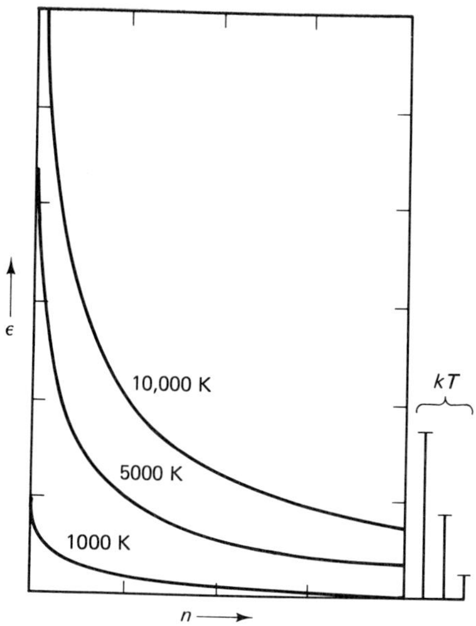

Figure 4.9-1 Bose distribution for three different temperatures.

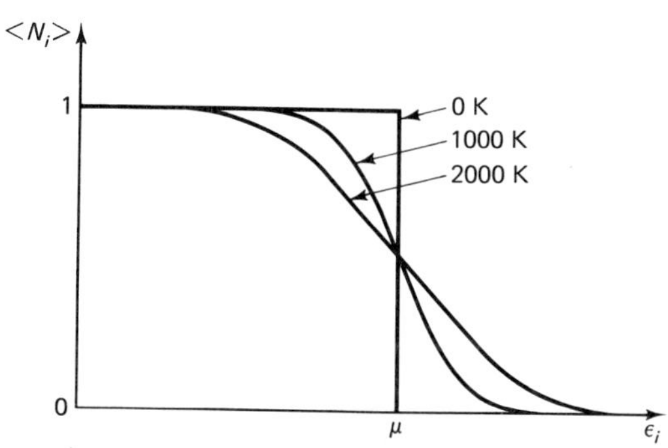

Figure 4.9-2 Fermi–Dirac distribution at zero and at temperatures above zero.

chemical potential (also called the *Fermi energy*). At higher temperatures, some particles become excited to energies above the Fermi level. The resulting distributions are as shown.

4.10 BOLTZMANN LIMIT

The transition from the quantum-mechanical distributions

$$\langle N_i \rangle = \frac{1}{e^{(\varepsilon_i - \mu)/kT} \pm 1} \qquad \begin{matrix} (+)FD \\ (-)BE \end{matrix} \tag{4.10-1}$$

to the Boltzmann distribution

$$\langle N_i \rangle = \frac{Ne^{-\varepsilon_i/kT}}{\sum_i e^{-\varepsilon_i/kT}} = \frac{Ne^{-\varepsilon_i/kT}}{z} \tag{4.10-2}$$

occurs when the chemical potential μ is sufficiently lower than the lowest orbital energy, that is, when

$$e^{(\varepsilon_i - \mu)/kT} = e^{\varepsilon_i/kT}\lambda^{-1} \gg 1, \qquad \text{all } \varepsilon_i \tag{4.10-3}$$

To a good approximation, in both the Fermi–Dirac and Bose–Einstein distributions, the term ± 1 can be neglected, giving

$$\langle N_i \rangle = \lambda e^{-\varepsilon_i/kT} \tag{4.10-4}$$

Comparing with (4.10-4) gives

$$\langle N_i \rangle = \ll 1 \tag{4.10-5}$$

and so the average occupation numbers all become very small, most often, $N_i = 0$; occasionally, $N_i = 1$; very rarely (for bosons), $N_i = 2$ or more. Evidently the number of accessible orbitals far exceeds the number of particles. This is, in turn, consistent with the low chemical potential, if the latter is construed as a *particle pressure* imposed on an open system.

Summing (4.10-4) over i gives

$$\sum_i \langle N_i \rangle = \lambda \sum_i e^{-\varepsilon_i/kT} = \lambda\zeta \tag{4.10-6}$$

But $N = \Sigma_i \langle N_i \rangle$, so that

$$\lambda = \frac{N}{\zeta} \tag{4.10-7}$$

which, when substituted into (4.10-4) gives

$$\langle N_i \rangle = \frac{Ne^{-\varepsilon_i/kT}}{\zeta} \tag{4.10-8}$$

thereby retrieving the Boltzmann distribution.

A sufficient condition for the validity of the Boltzmann limit is, from (4.10-4) and (4.10-5),

$$\lambda \ll 1 \tag{4.10-9}$$

so that, using (4.10-7),

$$\frac{\zeta}{N} \gg 1 \tag{4.10-10}$$

This is, in fact, fulfilled in most applications, some notable exceptions being the electrons in a metal, the photon gas comprising blackbody radiation, and liquid helium at very low temperatures.

Lastly, consider, in the Boltzmann limit, the molar partition function for an assembly of identical nonlocalized systems, such as an ideal gas. From the relationship with fixed N, expressed in terms of λ,

$$\Xi = \zeta\lambda^N \tag{4.10-11}$$

$$\ln Z = \ln \Xi - N \ln \lambda \tag{4.10-12}$$

Now, for a perfect gas, $pV = kT \ln \Xi = NkT$. Using N for $\ln \Xi$ and (4.10-7) for λ, (4.10-12) gives

$$\ln Z = N - N \ln N + N \ln z \tag{4.10-13}$$

And, by the factorial approximation $\ln N! = N \ln N - N$,

$$\ln Z = N \ln z - \ln(N!) \tag{4.10-14}$$

which gives

$$Z = \frac{z}{N!} \tag{4.10-15}$$

in agreement with the previous result for nonlocalized systems.

4.11 THERMODYNAMIC PROPERTIES AND THE GRAND PARTITION FUNCTION

Just as previously thermodynamic relations were obtained in terms of the canonical ensemble partition function, so can the approach to statistical thermodynamics be based on the grand partition function

$$\Xi = \sum_N \sum_i \exp\{\beta[\mu N - \varepsilon_i(V, N)]\} \tag{4.11-1}$$

and the probability

$$P_{N,i} = \frac{\exp\{\beta[\mu N - \varepsilon_i(V, N)]\}}{\Xi} \tag{4.11-2}$$

In classical equilibrium thermodynamics the grand potential Ω is defined as

$$\Omega = U - TS - \mu N \tag{4.11-3}$$

The grand potential is a function of the thermodynamic state of the system. It is an extensive variable. For infinitesimal changes,

$$d\Omega = dU - T\,dS - S\,dT - \mu\,dN - N\,d\mu \tag{4.11-4}$$

Now, according to the thermodynamic identity,

$$dU = T\,dS - p\,dV + \mu\,dN \tag{4.11-5}$$

substitution into (4.11-4) gives

$$d\Omega = -S\,dT - N\,d\mu - p\,dV \tag{4.11-6}$$

The natural variables for the grand potential Ω are T, μ, and V. Equation (4.11-6) shows that

$$S = -\left(\frac{\partial \Omega}{\partial T}\right)_{\mu, V}; \qquad N = -\left(\frac{\mu \Omega}{\partial \mu}\right)_{T, V}; \qquad p = -\left(\frac{\partial \Omega}{\partial V}\right)_{T, \mu} \tag{4.11-7}$$

For a one-component system,

$$G = \mu N \tag{4.11-8}$$

Since $G = H - TS = U + pV - TS$, Eq. (4.11-8) gives

$$\mu N = U + pV - TS \tag{4.11-9}$$

and, upon substitution into Eq. (4.11-3),

$$\Omega = -pV \tag{4.11-10}$$

In statistical mechanics, the grand potential Ω is defined in terms of the grand partition function Ξ by

$$\Omega = -kT \ln \Xi \tag{4.11-11}$$

where $\Xi(T, \mu, V)$ is given by (4.11-1) and is a function of V, the volume of the system, T, the absolute temperature of the heat reservoir, and μ, the chemical potential of the particle reservoir. Equations (4.11-10) and (4.11-11) give

$$pV = kT \ln \Xi \tag{4.11-12}$$

Substitution of the grand partition function Ξ from (4.11-1) into the expression

$$kT\left(\frac{\partial \ln \Xi}{\partial \mu}\right)_{T, V} = \frac{kT}{\Xi}\left(\frac{\partial \Xi}{\partial \mu}\right)_{T, V} \tag{4.11-13}$$

and carrying out the partial differentiation gives

$$\langle N \rangle = -\left(\frac{\partial \Omega}{\partial \mu}\right)_{V, T} = kT\left(\frac{\partial \ln \Xi}{\partial \mu}\right)_{V, T} \tag{4.11-14}$$

Similarly, substitution of Ξ from (4.11-1) into

$$kT\left(\frac{\partial \ln \Xi}{\partial V}\right)_{\mu, T} = \frac{kT}{\Xi}\left(\frac{\partial \Xi}{\partial V}\right)_{\mu, T} \tag{4.11-15}$$

gives

$$\langle p \rangle = kT\left(\frac{\partial \ln \Xi}{\partial V}\right)_{\mu, T} \tag{4.11-16}$$

Substitution of the probability, $P_{N,i}$, given by (4.11-2) into the Boltzmann expression for the entropy, namely

$$S = -k \sum_N \sum_i P_{N,i} \ln P_{N,i} \tag{4.11-17}$$

yields

$$S = -k \sum_N \sum_i P_{N,i}\left[-\left(\frac{\varepsilon_i}{kT}\right) + \left(\frac{\mu N}{kT}\right) - \ln \Xi\right] \tag{4.11-18}$$

But

$$-k \sum_N \sum_i P_{N,i}\left[-\left(\frac{\varepsilon_i}{kT}\right)\right] = \frac{U}{T} \tag{4.11-19}$$

$$-k \sum_N \sum_i P_{N,i}\left[-\left(\frac{\mu N}{kT}\right)\right] = \frac{\mu\langle N\rangle}{T} \tag{4.11-20}$$

and, since Ξ is constant if T, V, and μ are constants,

$$k \sum_N \sum_i P_{N,i} \ln \Xi(T, V, \mu) = k \ln \Xi \sum_N \sum_i P_{N,i} = k \ln \Xi \tag{4.11-21}$$

Thus, Eq. (4.11-18) becomes

$$S = \frac{U}{T} - \frac{\mu\langle N\rangle}{T} + k \ln \Xi \tag{4.11-22}$$

The above equations give the thermodynamic variables U, p, S, and N, if the grand partition function Ξ is known as a function of T (the absolute temperature of the heat reservoir), μ (the chemical potential of the particle reservoir), and V (the volume of the system). This approach to statistical thermodynamics, based on the grand partition function, will be applied to the ideal fermion gas, the ideal boson gas, and the ideal photon gas in the following sections.

■ EXAMPLE 4.11-1

From the definition of the grand potential Ω for systems of variable particle number

$$\Omega = \Omega(T, V, \mu) = -kT \ln \Xi(T, V, \mu)$$
$$= \langle E\rangle - TS - \mu\langle N\rangle \tag{4.11-23}$$

derive the relations for the chemical potential μ and other thermodynamic variables. ■

Solution. For a closed system, the fundamental thermodynamic relation is

$$dE = T\,dS - p\,dV \tag{4.11-24}$$

This comes from $S = S(E, V)$. For the case of an open system, $S = S(E, V, N)$, so that differentiation gives

$$dS = \frac{\partial S}{\partial E}dE + \frac{\partial S}{\partial V}dV + \frac{\partial S}{\partial N}dN = \frac{dE}{T} + \frac{p\,dV}{T} - \frac{\mu\,dN}{T}$$

that is,

$$dE = T\,ds - p\,dV + \mu\,dN \tag{4.11-25}$$

This is the required generalization of Eq. (4.11-24).

Equation (4.11-25) can be expressed in terms of other variables. From $A = E - TS$ and $G = E + pV - TS$ one obtains, respectively,

$$dA = -S\,dT - p\,dV + \mu\,dN \tag{4.11-26}$$

$$dG = -S\,dT + V\,dp + \mu\,dN \tag{4.11-27}$$

Equations (4.11-25) to (4.11-27) lead to many thermodynamic relations, including the following expressions for the chemical potential

$$\mu = -T\left(\frac{\partial S}{\partial N}\right)_{E,V} = \left(\frac{\partial F}{\partial N}\right)_{T,V} = \left(\frac{\partial G}{\partial N}\right)_{T,p} \tag{4.11-28}$$

This last relation leads to

$$G(T,p,N) = \mu N \tag{4.11-29}$$

which will now be shown. It follows from the extensive property of the Gibbs free energy G that for any factor λ

$$\lambda G(T,p,N) = G(T,p,\lambda N) \tag{4.11-30}$$

Differentiating this equation with respect to λ and putting $\lambda = 1$ after differentiation, gives

$$G(T,p,N) = N\left(\frac{\partial G}{\partial N}\right)_{T,p} \tag{4.11-31}$$

Elimination of $\partial G/\partial N$ between Eqs. (4.11-28) and (4.11-31) leads to Eq. (4.11-29). This last equation enables the grand potential to be written in a particularly simple form. From Eqs. (4.11-23) and (4.11-29),

$$\Omega = E - TS - \mu N = E - TS - G$$

and since $G = E + pV - TS$, the grand potential Ω reduces to

$$\Omega = -pV \tag{4.11-32}$$

Lastly, differentiation of (4.11-23) leads to

$$d\Omega = dE - T\,dS - S\,dT - \mu\,dN - N\,d\mu$$

which, on account of (4.11-25) reduces to

$$d\Omega = -S\,dT - p\,dV - N\,d\mu \tag{4.11-33}$$

From this differential form follow the thermodynamic relations with T, V, and μ as independent variables:

$$S = -\left(\frac{\partial \Omega}{\partial T}\right)_{V,\mu}, \qquad p = -\left(\frac{\partial \Omega}{\partial V}\right)_{T,\mu}, \qquad N = -\left(\frac{\partial \Omega}{\partial \mu}\right)_{T,V} \tag{4.11-34}$$

These last three relations have already been obtained earlier. ■

4.12 IDEAL FERMION GAS

Fermions are particles of half-integral spin which obey Fermi–Dirac statistics. If a composite particle contains an odd number of fermions (i.e., particles of spin $\frac{1}{2}, \frac{3}{2}, \ldots$) and any number of bosons (i.e., particles of spin 0, 1, . . .), then its spin is half-integral and it behaves like a fermion. If it contains an even number of fermions and any number of bosons, then its spin is integral and it behaves like a boson.

Consider an ideal fermion gas consisting of N identical fermions, each of mass m, in a box of volume V, kept at a constant absolute temperature T. It will be assumed that there are no interactions between the fermions, and that each fermion occupies a single-particle state. (A single-particle state is sometimes referred to as an orbital.) These single-particle states (orbitals) can be determined by solving Schrödinger's time-independent equation for a single particle in a box.

According to Chapter 2, the number of single-particle states (orbitals) for a free particle in a cubical box, which differ in at least one of the quantum numbers n_x, n_y, and n_z and spin quantum number m_s and which have energy eigenvalues in the energy range ε to $(\varepsilon + d\varepsilon)$ is

$$D(\varepsilon)\, d\varepsilon = \frac{gV}{4\pi^2}\left(\frac{2m}{\hbar^2}\right)^{3/2} \varepsilon^{1/2}\, d\varepsilon \qquad (4.12\text{-}1)$$

For particles such as electrons, g, the spin degeneracy is equal to 2. Equation (4.12-1) is the correct expression for the density of states $D(\varepsilon)$ whatever the shape of the box. It holds for both fermions and bosons. The variation of D with ε is shown in Fig. 4.12-1(a). According to Eq. (4.12-1), D is proportional to $\varepsilon^{1/2}$ at a fixed volume V, and displays a parabolic shape.

According to Section 4.5, the ensemble average number of fermions in a

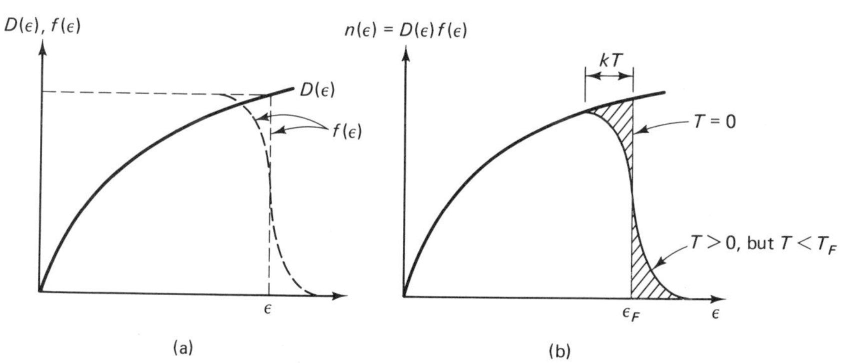

Figure 4.12-1 (a) Density of single-particle states for fermions in a box. (b) Energy distribution for fermions in a box at $T = 0$ and at $T < T_F$.

particular single-particle state, which has (single-particle) energy eigenvalue ε, is given by the Fermi–Dirac distribution function

$$\frac{n}{g} = f(\varepsilon) = \frac{1}{\exp[(\varepsilon - \mu)/kT] + 1} = \frac{1}{\exp[\beta(\varepsilon - \mu)] + 1} \tag{4.12-2}$$

where T is the absolute temperature of the heat reservoir and μ is the chemical potential of the particle reservoir, which in the present case is the fermions in the other single-particle states inside the box. (The determination of μ in terms of N, V, and T will be discussed later.)

The variation of $f(\varepsilon)$ with ε, at various absolute temperatures T, is shown in Fig. 4.12-2. If $T = 0$ and if ε is less than $\mu(0)$, the value of the distribution function at the absolute zero of temperature is $f(\varepsilon) = (e^{-\infty} + 1)^{-1} = 1$. If $T = 0$ and $\varepsilon > \mu(0)$, then $f(\varepsilon) = (e^{+\infty} + 1)^{-1} = 0$. Hence, according to Eq. (4.11-2), at the absolute zero of temperature all the single-particle states that have energy eigenvalues up to the value of the chemical potential $\mu(0)$ are occupied. All the single-particle states having energy eigenvalues greater than $\mu(0)$ are empty, as shown in Fig. 4.12-2(a). A Fermi gas at absolute zero temperature is labeled *completely degenerate.*

The value $\mu(0)$ of the chemical potential at the absolute zero of temperature is called the Fermi energy and is denoted ε_F. The Fermi temperature T_F is defined by the equation

$$\mu(0) = \varepsilon_F = kT_F \tag{4.12-3}$$

The Fermi–Dirac distribution function for an absolute temperature small compared to T_F, that is, $0 < T << T_F$ is shown in Fig. 4.12-2(b). The example chosen is $T = 0.1T_F$. When $T = 0.1T_F$ the value of the chemical potential $\mu(T)$ is $0.9918\varepsilon_F$, which is only just below the Fermi energy ε_F. It can be seen that at low values of ε, much less than ε_F, $f(\varepsilon)$ is very close to unity. For $T = 0.1T_F$, the large changes in $f(\varepsilon)$ are confined to an energy range of about kT on either side of the Fermi energy ε_F. In this energy range some of the fermions, which at $T = 0$ would have occupied single-particle states having energy eigenvalues just below the Fermi energy will, at a temperature $T << T_F$, be found in single-

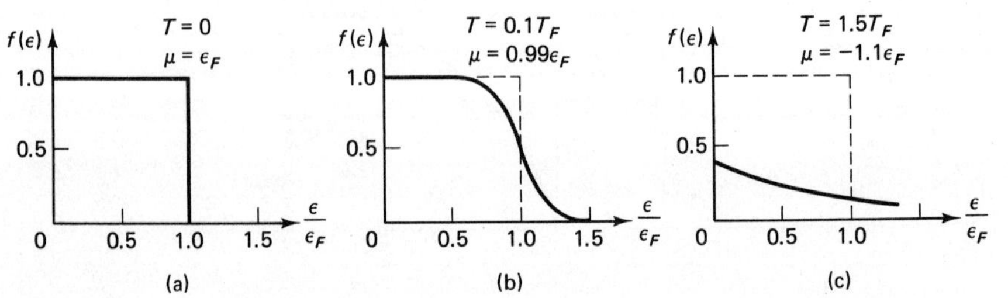

Figure 4.12-2 Fermi–Dirac distribution at different temperatures.

particle states having energy eigenvalues just above the Fermi energy. Such an electron gas is called *extremely degenerate*.

The variation of the Fermi–Dirac distribution function $f(\varepsilon)$ with ε for the case when $T = 1.5T_F$ is shown in Fig. 4.12-2(c). In this case $f(\varepsilon)$ is less than 0.5 for the single-particle ground state, and $f(\varepsilon)$ decreases continuously with increasing ε. Also, the chemical potential $\mu(T)$ is now negative. (The values of ε_F and T_F, however, do not vary with temperature if N and V are kept constant.)

The average number of fermions $n(\varepsilon)\,d\varepsilon$ in single-particle states which have energy eigenvalues between ε and $(\varepsilon + d\varepsilon)$ is given by the product of $D(\varepsilon)\,d\varepsilon$, the number of single-particle states (orbitals) having energy eigenvalues in the energy range ε to $(\varepsilon + d\varepsilon)$ and $f(\varepsilon)$, the mean number of fermions in each of these single-particle states. Using Eqs. (4.12-1) and (4.12-2),

$$n(\varepsilon)\,d\varepsilon = \frac{gV}{4\pi^2}\left(\frac{2m}{\hbar^2}\right)^{3/2} \frac{\varepsilon^{1/2}\,d\varepsilon}{\exp[(\varepsilon-\mu)/kT]+1} \tag{4.12-4}$$

With $g = 2$ for electrons, Eq. (4.12-4) gives, for the number of electrons with energy between ε and $\varepsilon + d\varepsilon$,

$$\bar{n}(\varepsilon) f(\varepsilon)\,d\varepsilon = \left[\frac{4\pi V}{h^3}(2m)^{3/2}\right]\varepsilon^{1/2}\,d\varepsilon\,\frac{1}{e^{\beta(\varepsilon-\mu)}+1} \tag{4.12-5}$$

The total number of electrons N in the electron gas is obtained by integration of Eq. (4.12-5) from $\varepsilon = 0$ to $\varepsilon = \infty$ (call it N),

$$N = \left[\frac{4\pi V}{h^3}(2m)^{3/2}\right]\int_0^\infty \frac{\varepsilon^{1/2}\,d\varepsilon}{e^{\beta(\varepsilon-\mu)}+1} \tag{4.12-6}$$

Equation (4.12-6) determines the chemical potential μ for a given number of electrons (N), a given volume (V), and a given temperature $T(= 1/\beta k)$:

$$\mu = \mu(T, V, N) \tag{4.12-7}$$

For a given piece of metal, its volume V is given. Also, N is fixed. It is determined by the atomic properties of the metal. In the case of the alkalis or the noble metals there is one conduction electron per atom. For simplicity, V and N are omitted from Eq. (4.12-7) and therefore,

$$\mu = \mu(T) \tag{4.12-8}$$

The Fermi energy ε_F is then defined as the value of the chemical potential at the absolute zero temperature

$$\varepsilon_F \equiv \mu(0) \tag{4.12-9}$$

The physical significance of ε_F will now be examined. First, note that it is positive. Figure 4.12-2(a) shows that at $T = 0$ K the system is in its state of lowest energy. But because of the Pauli principle not all N electrons can crowd into the single-particle state of lowest energy ε_1. There can be only one particle per state. Hence, with the single-particle states ordered in increasing energy

$$\varepsilon_1 \leq \varepsilon_2 \leq \cdots \tag{4.12-10}$$

the first N (lowest) single-particle states are filled. The states above these first N states are empty. The meaning of the Fermi energy ε_F is now quite clear! It is the energy of the topmost occupied level at $T = 0$ K. Figs. 4.12-1, 4.12-2, and the dotted outline in Fig. 4.12-3 show this quite clearly.

Equation (4.12-2) gives the mean occupation number of a perfect FD gas. Its energy distribution is reflected by Eq. (4.12-5). The distribution for $T = 0$ has a sharp cutoff. The sharp cutoff at $\varepsilon = \varepsilon_F$ reflects the behavior of $\bar{n}(\varepsilon)$; the parabolic rise reflects the density-of-states factor $\varepsilon^{1/2}$.

From the interpretation of ε_F as the topmost energy level occupied at $T = 0$ K, the Fermi energy ε_F can be readily found. At $T = 0$ K, Eq. (4.12-2) has the outline shown in Fig. 4.12-2(a). Hence, Eq. (4.12-6) need only be integrated to ε_F, and becomes, at $T = 0$ K,

$$N = \left[\frac{4\pi V}{h^3}(2m)^{3/2}\right]\int_0^{\varepsilon_F} \varepsilon^{1/2}\, d\varepsilon = \left[\frac{4\pi V}{h^3}(2m)^{3/2}\right]\frac{2}{3}(\varepsilon_F)^{3/2} \tag{4.12-11}$$

so that

$$\varepsilon_F = \frac{h^2}{2m}\left(\frac{3}{8\pi}\frac{N}{V}\right)^{2/3} \tag{4.12-12}$$

It is seen that the Fermi energy depends on the mass m of the particles (in most cases, the electron mass) and on their mean concentration N/V. (For monovalent metals V/N is the atomic volume, i.e., the volume per atom of the solid.)

The Fermi temperature T_F is defined by

$$\varepsilon_F \equiv kT_F \tag{4.12-13}$$

The order of magnitude of ε_F and T_F may be estimated from Eqs. (4.11-12) and (4.11-13). With $N/V \approx 5 \times 10^{22}$ cm^{-3} [which would correspond to an atomic spacing $(V/N)^{1/3} = 2.7$ Å] and one conduction electron per atom, the values $\varepsilon_F \approx 4.5$ eV, $T_F \approx 5 \times 10^4$ K are obtained.

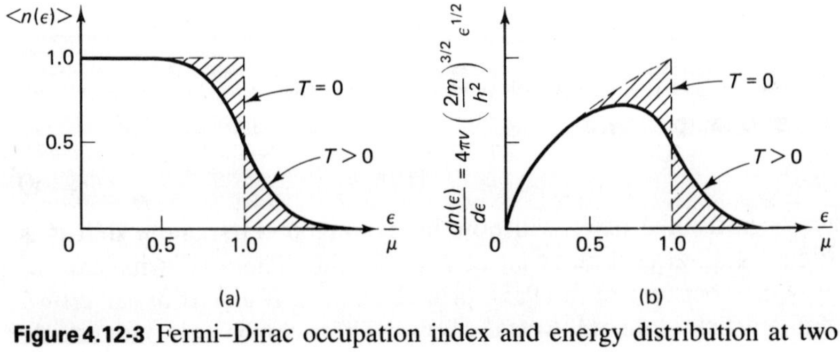

Figure 4.12-3 Fermi–Dirac occupation index and energy distribution at two different temperatures.

A Fermi gas is extremely degenerate if its temperature is very small compared with T_F. Since the Fermi temperature T_F is on the order of 10^4 to 10^5 K for metals, it follows that the conduction electrons are in an extremely degenerate state under all usual conditions. Figure 4.12-3 shows the FD mean occupation number $\bar{n}(\varepsilon)$, Eq. (4.12-2), and the FD energy spectrum, Eq. (4.12-5), for a temperature T such that $kT = 0.2\mu$, plotted against ε/μ. Also shown as dashed lines are the corresponding curves for $T = 0$ K. It is seen that as the temperature is raised from $T = 0$ K, electrons are excited from single-particle states with energy $\varepsilon < \mu$ to single-particle states with energy $\varepsilon > \mu$. Most of the transfer occurs for electrons with energy near $\varepsilon = \mu$.

The value of the chemical potential $\mu(V, N, T)$ in Eqs. (4.12-2) or (4.12-4) can be determined (in principle). The total number N of fermions in the ideal fermion gas is equal to the integral of (4.12-4) from $\varepsilon = 0$ to $\varepsilon = \infty$:

$$N = \int_0^\infty n(\varepsilon)\, d\varepsilon = \frac{gV}{4\pi^2}\left(\frac{2m}{\hbar^2}\right)^{3/2} \int_0^\infty \frac{\varepsilon^{1/2}\, d\varepsilon}{\exp[(\varepsilon - \mu)/kT] + 1} \qquad (4.12\text{-}14)$$

This is an integral equation which can be solved numerically to determine the chemical potential μ in terms of T, V, and N. If N, V, or T is changed, the chemical potential μ is also changed. Usually, V and N are fixed, so that essentially $\mu = \mu(T)$. The variation of the chemical potential μ of an ideal fermion gas with absolute temperature T is shown in Fig. 4.12-4.

At $T = 0$, the chemical potential $\mu(0)$ is equal to the Fermi energy ε_F, where ε_F is given by Eq. (4.12-12). When T is very much less than the Fermi temperature T_F, the chemical potential μ is only just below the Fermi energy. As the temperature increases further the chemical potential decreases significantly, and is zero at a temperature just below the Fermi temperature T_F. As T increases above the Fermi temperature, the chemical potential gets more and more negative. In the high-temperature limit, μ is negative.

In the low-temperature limit when $T << T_F$, the chemical potential of an ideal fermion gas is given, to a good approximation, by

$$\mu \simeq \varepsilon_F\left[1 - \frac{\pi^2}{12}\left(\frac{T}{T_F}\right)^2\right] \qquad (T << T_F) \qquad (4.12\text{-}15)$$

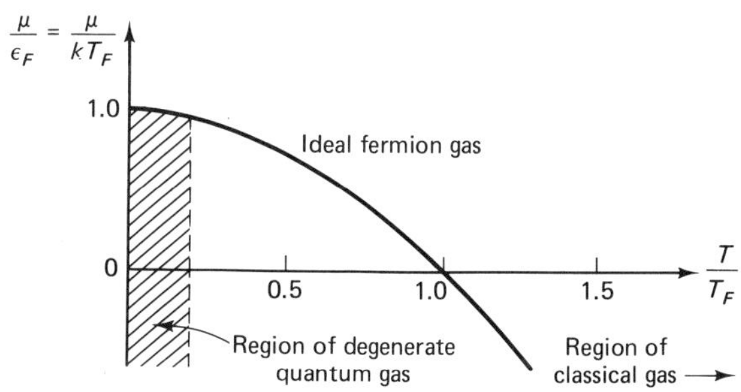

Figure 4.12-4 Variation of the chemical potential μ of a fermion gas with absolute temperature T).

For example, according to Eq. (4.12-15), if $T = 0.01T_F$, μ is equal to $0.999918\varepsilon_F$, whereas, if $T = 0.1T_F$, μ is equal to $0.9913\varepsilon_F$. Hence, in the low-temperature limit, when $T \ll T_F$, the chemical potential μ is only just fractionally less than the Fermi energy ε_F, which is the value of the chemical potential at the absolute zero.

In the high-temperature limit, when $T \gg T_F$, the ideal fermion gas approximates to an ideal classical gas. For an ideal monatomic gas in the classical limit,

$$\exp\left(\frac{-\mu}{kT}\right) = \left(\frac{gV}{N}\right)\left(\frac{mkT}{2\pi\hbar^2}\right)^{3/2} \tag{4.12-16}$$

For a fermion gas in the classical limit, consider a mole of ${}^3_2He$ gas atoms at STP. In this case, $\exp(-\mu/kT)$ is equal to 3.4×10^5, so that $\mu/kT = -12.7$. This illustrates that, in the high-temperature limit, when $T \gg T_f$, the chemical potential μ is negative, and $\exp(-\mu/kT) \gg 1$.

■ EXAMPLE 4.12-1

Show by means of sketches, the Fermi–Dirac distribution for a system of $N = 8$ fermions of spin $\frac{1}{2}$ at $T = 0$, and $0 < T \ll T_F$, respectively.

Solution. Consider, for simplicity, the free electrons to be on a wire of length L and find what orbitals are occupied in the ground state of the system. By the Pauli exclusion principle each orbital can be occupied by at most one electron. This principle applies, of course, to electrons in solids as well as to electrons in atoms. In a one-dimensional world the free-electron orbitals are of the form $\sin n\pi x/L$, the quantum number is n, and the spin quantum number is m_s.

With each value of the integer n in the wave function there is an orbital for each of the two possible orientations of the electron spin (along the $+z$ or along the $-z$ direction, with $m_s = \pm\frac{1}{2}$). Each energy level labeled by the quantum number n has two orbitals, one with the spin up and one with the spin down.

All the eight fermions in the system cannot go into the single-particle ground state at the absolute zero, but they occupy the single-particle states having the lowest energy eigenvalues. These range in energy from zero up to the value $\mu(0)$ of the chemical potential at $T = 0$, as shown in Fig. 4.12-5(a). For fermions of spin $\frac{1}{2}$ in a wire, there can be two fermions having spin quantum numbers $+\frac{1}{2}$ and $-\frac{1}{2}$, respectively, associated with each set of the quantum number n. The two different spin states with spin quantum numbers $+\frac{1}{2}$ and $-\frac{1}{2}$ are represented by arrows pointing in opposite directions in Fig. (4.12-5). The value $\mu(0)$ of the chemical potential at the absolute zero of temperature is the Fermi energy and is denoted ε_F.

(a) At $T = 0$, the ground state of the system, all the single-particle states up to ε_F are filled. All the single-particle states above ε_F are empty.

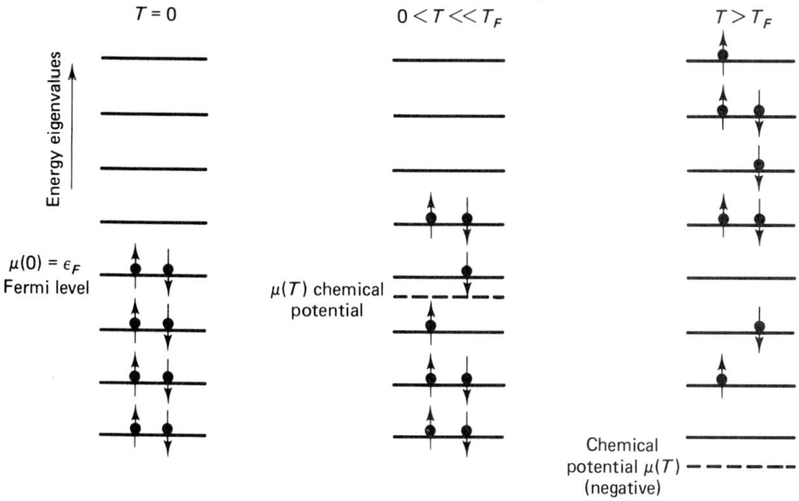

Figure 4.12-5 Fermi–Dirac distribution for $N = 8$ fermions of spin $\frac{1}{2}$.

(b) When $0 < T << T_F$, some of the fermions in single-particle states just below ε_F have moved to single-particle states just above ε_F, In this case, the chemical potential $\mu(T)$ is just below the Fermi energy ε_F.

(c) When $T > T_F$, the fermions have spread into the higher energy states, and the occupation for the single-particle ground state is less than 0.5. Also, the chemical potential $\mu(T)$ is negative. ■

Some remarks are appropriate at this point. It is possible to neglect, to a first approximation, the mutual interaction of the conduction electrons in a metal. These electrons can therefore be treated as an ideal gas. Their concentration in a metal is, however, so high that they cannot be treated at ordinary temperatures by the approximation of classical statistics. Hence, Fermi–Dirac statistics must be used to discuss conduction electrons in a metal.

The mean number of particles in state s is given by the FD distribution

$$\bar{n}_s = \frac{1}{e^{\alpha + \beta\varepsilon_s} + 1} = \frac{1}{e^{\beta(\varepsilon_s - \mu)} + 1} \tag{4.12-17}$$

where the parameter α or μ is to be determined by the condition that

$$\sum_s \bar{n}_s = \sum \frac{1}{e^{\beta(\varepsilon_s - \mu)} + 1} = N, \tag{4.12-18}$$

N being the total number of particles in the volume V. By virtue of (4.12-18), μ is then a function of the temperature. Consider the distribution function

$$f(\varepsilon) \equiv \frac{1}{e^{\beta(\varepsilon - \mu)} + 1} \tag{4.12-19}$$

in terms of ε, this energy being measured above its lowest possible value $\varepsilon = 0$. If μ is such that $\beta\mu << 0$, then $e^{\beta(\varepsilon - \mu)} >> 1$ and $f(\varepsilon)$ reduces to the Maxwell–Boltzmann distribution.

The other limit is

$$\beta\mu = \frac{\mu}{kT} >> 1 \tag{4.12-20}$$

In this case, if $\varepsilon << \mu$, then $\beta(\varepsilon - \mu) << 0$ so that $F(\varepsilon) = 1$. On the other hand, if $\varepsilon >> \mu$, then $\beta(\varepsilon - \mu) >> 0$ so that $f(\varepsilon) = e^{\beta(\mu - \varepsilon)}$ falls off exponentially like a classical Boltzmann distribution. If $\varepsilon = \mu$, then $f = \frac{1}{2}$. The transition region in which f goes from a value close to 1 to a value close to zero corresponds to an energy interval on the order of kT about $\varepsilon = \mu$, as shown in Fig. 4.12-6(a).

In the limit when $T \to 0$ (or $\beta \to \infty$) this transition region becomes infinitesimally narrow. In this case $f = 1$ for $\epsilon < \mu$ and $f = 0$ for $\varepsilon > \mu$, as illustrated in Fig. 4.12-6(b). For the absolute zero situation, the gas is in its ground state of lowest energy. Since the exclusion principle requires that there be no more than one particle per single-particle state, the lowest energy of the gas is obtained by putting all particles into the lowest available unoccupied states until all the particles are accommodated. The last particle thus added to the pile has quite a considerable energy μ, since all the lower energy states have already been used up. Consequently, a fermion gas has a large mean energy, even at absolute zero.

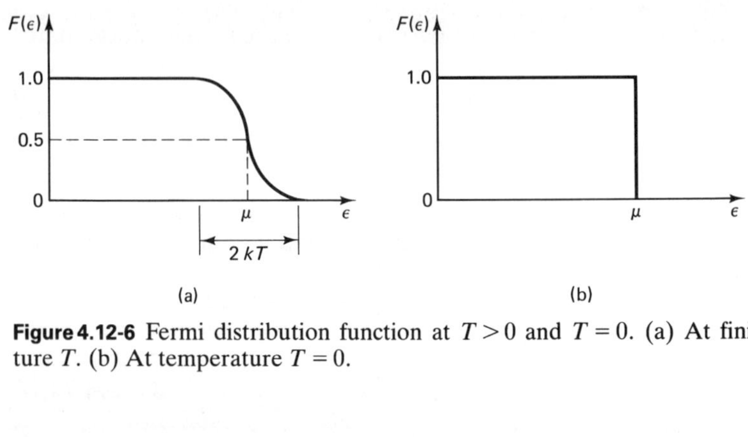

Figure 4.12-6 Fermi distribution function at $T > 0$ and $T = 0$. (a) At finite temperature T. (b) At temperature $T = 0$.

■ EXAMPLE 4.12-2

Determine the density of states for a fermion gas. This is the number of states having energy between ε and $\varepsilon + d\varepsilon$.

Solution. For a particle in a wire of length l, the energy is

$$\varepsilon_x = \left(\frac{\hbar^2\pi^2}{2mL^2}\right) n_x \tag{4.12-21}$$

In three dimensions, the energy is

$$\varepsilon = \varepsilon_x + \varepsilon_y + \varepsilon_z = \frac{\hbar^2\pi^2}{2mL^2}(n_x^2 + n_y^2 + n_z^2) \tag{4.12-22}$$

This can be rewritten as

$$(n_x^2 + n_y^2 + n_z^2) = \frac{2mL^2}{\hbar^2\pi^2} \tag{4.12-23}$$

In n-space, Eq. (4.1-23) is the equation of a sphere with radius $(2mL^2\varepsilon/\hbar^2\pi^2)^{1/2}$. As seen in Fig. 4.12-7, there is one state (defined by the triplet n_x, n_y, n_z) per unit volume of n-space.

Since n_x, n_y, and n_z take on only positive integer values, the number of states with energy less than ε is given by the volume of the positive octant of a sphere with radius $(2mL^2\varepsilon/\hbar^2\pi^2)^{1/2}$:

$$\frac{1}{8}\frac{4\pi}{3}\left(\frac{2mL^2}{\hbar^2\pi^2}\varepsilon\right)^{3/2} = \frac{V}{6\pi^2}\left(\frac{2m}{\hbar^2}\right)^{3/2}\varepsilon^{3/2} \tag{4.12-24}$$

where the volume $V = L^3$. Therefore, the number of states with energy between ε and $(\varepsilon + d\varepsilon)$ is

$$\frac{d}{d\varepsilon}\left[\frac{V}{6\pi^2}\left(\frac{2m}{\hbar^2}\right)^{3/2}\varepsilon^{3/2}\right] = \frac{V}{4\pi^2}\left(\frac{2m}{\hbar^2}\right)^{3/2}\varepsilon^{1/2} \tag{4.12-25}$$

This is the density of orbital states (one-particle solutions of Schrödinger's equation). For fermions with a spin degeneracy of 2, the density of states per unit volume is

$$g(\varepsilon) = \frac{1}{2\pi^2}\left(\frac{2m}{\hbar^2}\right)^{3/2}\varepsilon^{1/2} \tag{4.12-26}$$

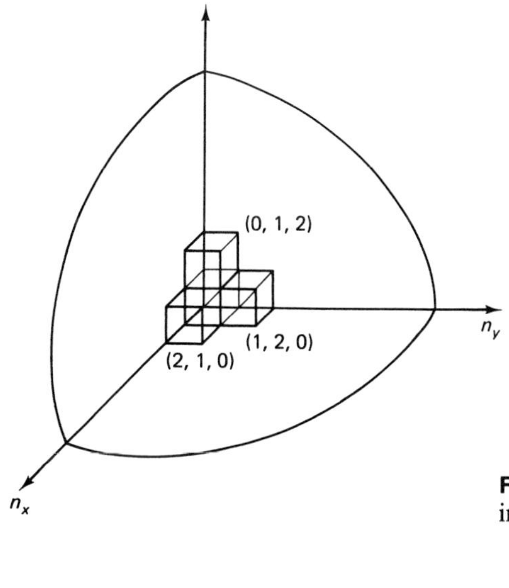

Figure 4.12-7 Representation of volume in n space.

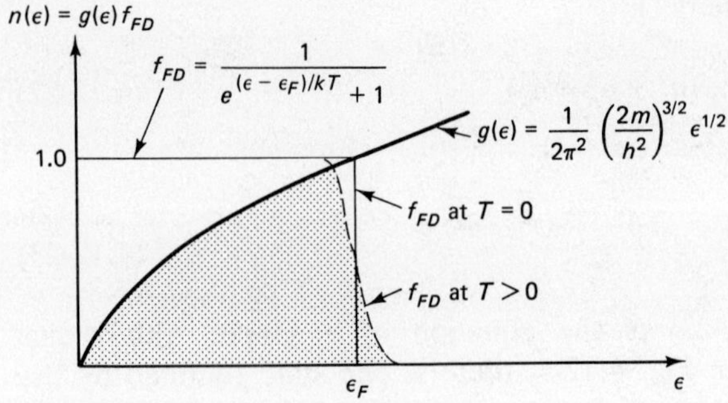

Figure 4.12-8 Occupation index for fermions at $T = 0$ and $T > 0$.

This is plotted versus ε in Fig. 4.12-8. When $g(\varepsilon)$ is multiplied by the Fermi distribution

$$f_{FD}(\varepsilon) = \frac{1}{e^{(\varepsilon - \varepsilon_F)/kT} + 1} \tag{4.12-27}$$

the result is

$$n(\varepsilon) = \frac{1}{2\pi^2}\left(\frac{2m}{\hbar^2}\right)^{3/2} \frac{\varepsilon^{1/2}}{e^{(\varepsilon - \varepsilon_F)/kT} + 1} \tag{4.12-28}$$

where $n(\varepsilon)$ is the occupation index, or number of particles occupying energy state ε. Note in Fig. 4.12-8 that at absolute zero temperature, all states are filled up to the Fermi energy ε_F. At $T > 0$, some of the electrons near the Fermi level cross over and become more energetic. ■

4.13 ENERGY, PRESSURE, AND SPECIFIC HEAT OF FERMION GAS

One of the most impressive accomplishments of the theory of degenerate fermion gas is the evaluation of the specific heat of electrons in metals. For an ideal monatomic gas, the heat capacity is $\frac{3}{2}Nk$, but for electrons in a metal, much lower values prevail. The results obtained are in excellent agreement with experimental findings.

According to the preceding development, the ensemble average energy of the ideal fermion gas is

$$U = \int_0^\infty \varepsilon\, n(\varepsilon)\, d\varepsilon = \frac{gV}{4\pi^2}\left(\frac{2m}{h^2}\right)^{1/2} \int_0^\infty \frac{\varepsilon^{3/2}\, d\varepsilon}{\exp[\beta(\varepsilon - \mu)] + 1} \tag{4.13-1}$$

The details of handling this expression are given in Example 4.13-1. The variation of U with the absolute temperature T, at constant volume V, is shown in Fig. 4.13-1(a). According to classical theory, U is equal to $3NkT/2$ for an ideal

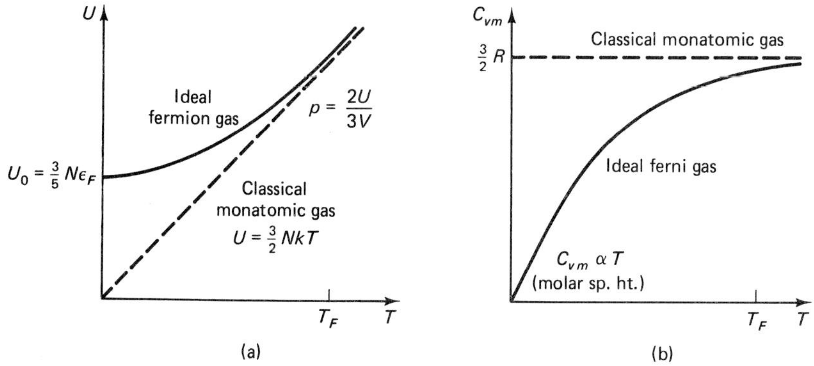

Figure 4.13-1 Energy, pressure, and specific heat of ideal fermion gas.

monatomic classical gas. This relation is shown by the dotted line in the figure. It can be seen that the energy of the ideal fermion gas is always greater than the energy an ideal monatomic classical gas would have if the volume V, the temperature T, and the number of particles N were the same. When T is greater than the Fermi temperature T_F, the behavior of the ideal fermion gas begins to approximate to the ideal monatomic classical gas laws developed in Chapter 3. When $T < T_F$, the ideal fermion gas departs markedly from the ideal classical gas laws. The ideal fermion gas is said to be *degenerate* in this temperature range. It will be shown later that, when $T = 0$, the mean energy of the fermion gas is equal to $3N\varepsilon_F/5$.

The specific heat at constant volume is

$$C_V = \left(\frac{\partial U}{\partial T}\right)_V \tag{4.13-2}$$

where U is given by Eq. (4-13-1). The variation of the molar specific heat $C_{V,\text{mol}}$ of the ideal fermion gas with temperature is shown in Fig. 4.13-1(b). At low temperatures when $T << T_F$, the variation of $C_{V,\text{mol}}$ with T is linear. When $T > T_F$, the molar heat capacity approaches the value of $\frac{3}{2}R$, which is valid for an ideal monatomic gas in the classical limit.

As for the pressure of a fermion gas, consider such a fluid contained inside an adiabatic cylinder fitted with a movable adiabatic piston of area A. The piston can move in and out without friction. Assume that the gas is in its ith microstate with an energy eigenvalue ε_i. In this microstate, the pressure exerted by the system on the piston is p_i. If the piston is given a virtual displacement dx inward, the change in the volume of the system would be equal to $A\,dx$, and the work done on the system would be

$$p_i A\,dx = -p_i\,dV \tag{4.13-3}$$

where dV would be the volume change of the system. If all the work went into increasing the energy of the microstate from ε_i to $(\varepsilon_i + d\varepsilon_i)$, then

$$d\varepsilon_i = -p_i\,dV \tag{4.13-4}$$

Hence, when the system is in its ith microstate having energy eigenvalue ε_i, according to the principle of virtual work, the pressure of the system is

$$p_i = -\frac{\partial E_i}{\partial V} \tag{4.13-5}$$

Applying Eq. (4.13-5), the mean pressure of the ideal fermion gas is

$$p = \sum P_i \left[\frac{\partial E_i(V,N)}{\partial V}\right] \tag{4.13-6}$$

where $E_i(V,N)$ is the energy eigenvalue of the ith N-particle state of the N-particle fermion system, and P_i is the probability that the N-particle system is in that state. According to quantum mechanics, for a cubical box, the energy eigenvalues of the single-particle states are given by

$$\varepsilon = \left(\frac{\hbar^2\pi^2}{2mV^{2/3}}\right)(n_x^2 + n_y^2 + n_z^2) \tag{4.13-7}$$

In an ideal fermion gas, in which there are no interactions between the fermions, the energy of the N-fermion system is equal to the sum of the energies of the N individual fermions. Hence, the energy eigenvalue $E_i(V,N)$ of the ith microstate of the N-fermion system is

$$E_i(V,N) = \left(\frac{\hbar^2\pi^2}{2mV^{2/3}}\right)\sum[(n_x)_i^2 + (n_y)_i^2 + (n_z)_i^2] = \frac{K_i}{V^{2/3}} \tag{4.13-8}$$

The summation is over the N single-particle states occupied by the N fermions, when the system is in its ith N-particle state. E_i is proportional to $V^{-2/3}$, and K_i is a constant for the ith N-particle state. Substituting for E_i in the expression for p given by Eq. (4.13-6) and carrying out the partial differentiation gives

$$p = \sum P_i\left(\frac{2K_i}{3V^{5/3}}\right) = \frac{2}{3}\sum P_i\left(\frac{E_i}{V}\right) = \frac{2U}{3V} \tag{4.13-9}$$

where U is the ensemble average energy of the N-fermion system.

Equation (4.13-9) is valid for all values of U for an ideal fermion gas, in which there is no interaction between the fermions. The same argument can be applied to derive the equation for an ideal boson gas. Since, according to Eq. (4.13-9) the pressure p of an ideal fermion gas is proportional to U, if V is constant, the variation of p with T at constant volume has the same form as the variation of U with T shown in Fig. 4.13-1(a). It can be seen that the pressure of an ideal fermion gas is always greater than the pressure an ideal classical monatomic gas would have, if V, T, and N are the same.

■ EXAMPLE 4.13-1

(a) Given the density of states for a fermion gas in a volume V to be $(V/2\pi^2)(2m/\hbar^2)^{3/2}\,\varepsilon^{1/2}$, show that the Fermi energy is

$$\varepsilon_F = \frac{\hbar^2}{2m}\left(\frac{3\pi^2N}{V}\right)^{2/3} = \frac{h^2}{8m}\left(\frac{3N}{\pi V}\right)^{2/3}$$

(b) At absolute zero, Cu has a free-electron concentration of 8.5×10^{28} m^{-3}. What is the Fermi energy and Fermi temperature?

Solution

(a) Let N be the number of fermions in a volume V. The electrons will populate the energy states from the ground up until the last electron, which has the energy ε_F. Thus,

$$N = \int_0^{\varepsilon_F} \frac{V}{2\pi^2}\left(\frac{2m}{\hbar^2}\right)^{3/2} \varepsilon^{1/2}\, d\varepsilon$$

$$= \frac{V}{3\pi^2}\left(\frac{2m}{\hbar^2}\right)^{3/2} \varepsilon_F^{3/2}$$

or

$$\varepsilon_F = \frac{\hbar^2}{2m}\left(\frac{3\pi^2 N}{V}\right)^{2/3} = \frac{\hbar^2}{8m}\left(\frac{3N}{\pi V}\right)^{2/3}$$

$$= (3.646 \times 10^{-19}\ \text{eV m}^2)\left(\frac{N}{V}\right)^{2/3}$$

(b) With $N/V = 8.5 \times 10^{28}\ \text{m}^{-3}$, the Fermi energy at absolute zero is

$$\varepsilon_F = (3.646 \times 10^{-19}\ \text{eV m}^2)(8.5 \times 10^{28}\ m^{-3})^{2/3}$$

$$= 7.0\ eV$$

The corresponding Fermi temperature is

$$T_F = \frac{\varepsilon_F}{k} = 8.2 \times 10^4\ \text{K}$$

■

4.14 COMPLETELY DEGENERATE FERMION GAS: FERMI SPHERE

It is instructive to examine the extremely low and high temperature regimes of an ideal Fermion gas. The evaluation of properties is relatively simple and straightforward. At the absolute zero, when $T = 0$, there are no fermions having energies above the Fermi energy ε_F, and the upper limit of infinity in equation for the total number N of fermions

$$N = \int_0^{\infty} n(\varepsilon)\, d\varepsilon = \frac{gV}{4\pi^2}\left(\frac{2m}{\hbar^2}\right)^{3/2} \int_0^{\infty} \frac{\varepsilon^{1/2}\, d\varepsilon}{\exp[(\varepsilon - \mu)\, kT + 1]} \qquad (4.14\text{-}1)$$

can be replaced by ε_F. Hence, at $T = 0$, Eq. (4.13-1) becomes

$$N = \frac{gV}{4\pi^2}\left(\frac{2m}{\hbar^2}\right)^{3/2} \int_0^{\varepsilon_F} \varepsilon^{1/2}\, d\varepsilon = \frac{gV}{4\pi^2}\left(\frac{2m}{\hbar^2}\right)^{3/2} \frac{2\varepsilon_F^{3/2}}{3} \qquad (4.14\text{-}2)$$

Rearranging, it is found that the Fermi energy ε_F which is also equal to kT_F, is given by

$$\varepsilon_F = kT_F = \frac{\hbar^2}{2M}\left(\frac{6\pi^2 N}{gV}\right)^{2/3} \tag{4.14-3}$$

The Fermi energy ε_F and Fermi temperature T_F are both proportional to $(N/V)^{2/3}$, so that the Fermi energy and Fermi temperature both increase, if the concentration of the fermions (N/V) is increased. The Fermi energy ε_F and Fermi temperature T_F are inversely proportional to the mass of the fermions so that, for example, for the same concentration (N/V), the Fermi energy and Fermi temperature are bigger for electrons than for a gas of $^3_2$He atoms. The ensemble average energy of a fermion in an ideal fermion gas at the absolute zero of temperature is

$$\langle\varepsilon\rangle = \frac{1}{N}\int_0^{\varepsilon_F} \varepsilon n(\varepsilon)\, d\varepsilon = \frac{gV}{4\pi^2 N}\left(\frac{2m}{\hbar^2}\right)^{3/2}\int_0^{\varepsilon_F} \varepsilon^{3/2}\, d\varepsilon \tag{4.14-4}$$

$$= \frac{gV}{10\pi^2 N}\left(\frac{2m}{\hbar^2}\right)^{3/2} \varepsilon_F^{5/2} \tag{4.14-5}$$

Using (4.14-3), it is found that at $T = 0$ the average fermion energy is

$$\langle\varepsilon\rangle = \tfrac{3}{5}\,\varepsilon_F \tag{4.14-6}$$

The total energy of an ideal fermion gas consisting of N noninteracting fermions is the product of N and the mean energy of each fermion. Hence, at $T = 0$, the total energy of an ideal fermion gas is

$$U_0 = \frac{3}{5N\varepsilon_F} \tag{4.14-7}$$

As for the pressure, using (4.14-7) in (4.12-9) results, in the case of an ideal fermion gas at the absolute zero of temperature,

$$p_0 = \frac{2}{5(N/V)\varepsilon_f} = \frac{\hbar^2}{5m}\left(\frac{6\pi^2}{g}\right)^{2/3}\left(\frac{N}{V}\right)^{5/3} \tag{4.14-8}$$

where g, the spin degeneracy has been retained in the equation. This pressure can have enormous values, as witnessed in the theory of white dwarf and neutron stars.

At $T = 0$, all the single particle states up to the Fermi energy are occupied. Since the fermions are identical, this corresponds to only one state of the N-particle system. Hence, at $T = 0$, the entropy given by

$$S = k \ln W \tag{4.14-9}$$

is equal to $k \ln 1$, which is zero, showing that the entropy of an ideal fermion gas is zero at the absolute zero, in agreement with the *third law* of thermodynamics.

Further insight can be obtained by writing the distribution function of a Fermion gas in terms of the momentum **p**, in the limit of $T \to 0$:

$$\bar{n}(\mathbf{p}) = \begin{cases} gV/h^3 & \text{if } p^2/24 < \varepsilon_F \\ 0 & \text{if } p^2/2m > \varepsilon_F \end{cases} \tag{4.14-10}$$

This is the *degenerate* case. The distribution function is entirely different from the Maxwellian distribution. The function $\bar{n}(\mathbf{p})$ in (4.14-34) is discontinuous at momenta with the value $p^2/2m = \varepsilon_F$. If $\bar{n}(\mathbf{p})$ is represented in momentum space, as previously done and repeated below, it has a uniform value of gV/h^3 inside a sphere of radius $\sqrt{2m\varepsilon_F}$ and is zero outside this sphere. This sphere is known as the *Fermi sphere,* and its radius is referred to as the *Fermi momentum:*

$$p_F \equiv \sqrt{2m\varepsilon_F} \tag{4.14-11}$$

At the other extreme,

$$\bar{n}(\mathbf{p}) \simeq \bar{N}(2\pi mkT)^{3/2}\, e^{-\beta p^2/2m} \qquad (\text{if } 1 >> e^{\beta\mu}) \tag{4.14-12}$$

This is the *nondegenerate* case. The distribution function is identical with the Maxwellian distribution. Thus, the classical result is recovered if $1 >> e^{\beta\mu}$. According to (4.14-27), this means that

$$1 >> \frac{\langle N\rangle h^3}{gV(2\pi mkT)^{3/2}} \qquad (\text{classical limit}) \tag{4.14-13}$$

Previously, in Chapter 3, a similar condition was obtained for a *nondegenerate* perfect gas, namely,

$$1 >> \frac{6}{\pi}\frac{Nh^3}{V(2mkT)^{3/2}} \tag{4.14-14}$$

This shows once again that the translational motion of a nondegenerate perfect gas can be described by a Maxwellian distribution.

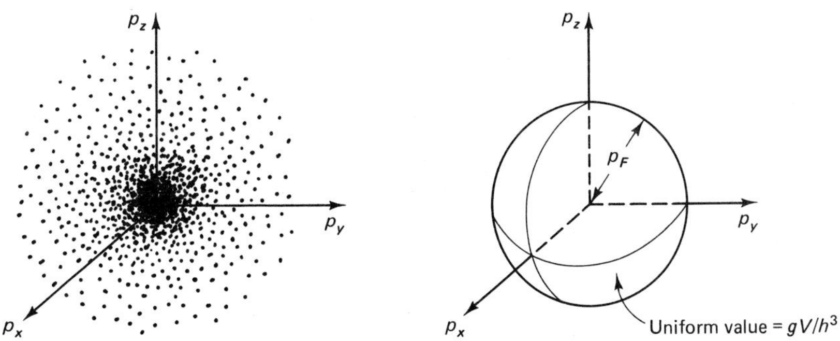

Figure 4.14-1 Fermi–Dirac distribution in the nondegenerate and the completely degenerate limit.

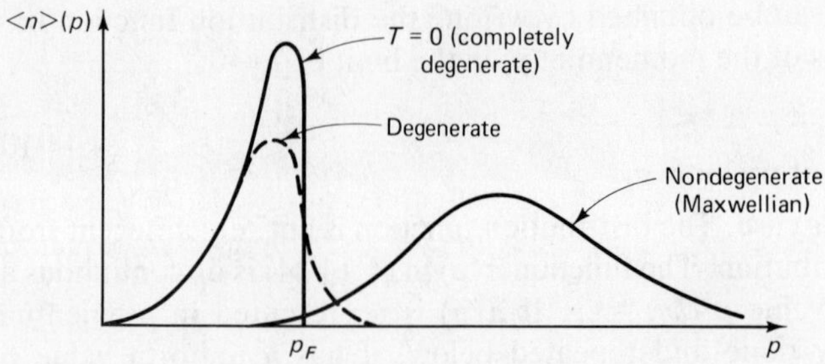

Figure 4.14-2 Degenerate and nondegenerate distribution function for fermions and boltzons.

The various cases are illustrated in Fig. 4.14-2. At $T = 0$ the distribution function has a sharp cutoff at the Fermi momentum p_F. The fact that $\bar{n}(p)$ increases in relation to p^2 is simply due to the number of available particle states $4\pi p^2\, dp$ in the region dp about p. The dotted curve represents the-distribution in a degenerate case when $T \neq 0$. Note that the particles near the Fermi surface (that have $p \simeq p_F$) are the ones first affected by the rise in temperature. At high temperatures, the nondegenerate Maxwellian distribution (which is actually much further out on the p axis) is recovered.

■ EXAMPLE 4.14-1

For an ideal monatomic gas in the classical limit,

$$\exp(-\mu/kT) = \frac{gV}{N}\left(\frac{mkT}{2\pi\hbar^2}\right)^{3/2} >> 1$$

Show, in the case of ${}^3_2$He gas atoms, that the mean occupancy of any particular single-particle state is very small in the classical limit.

Solution. For a mole of ${}^3_2$He gas atoms at STP, $\exp(-\mu/kT)$ is equal to 3.4×10^5, and $\mu/kT = 12.7$. The mean kinetic energy of a monatomic gas atom in the classical limit is $3kT/2$. ${}^3_2$He atoms are fermions. Choosing $\varepsilon = 3kT/2$ to represent a typical fermion in the classical limit, and using a value of 3.4×10^8 for $\exp(-\mu/kT)$, it is found that at STP,

$$f_{FD}\left(\varepsilon = \frac{3kT}{2}\right) = \frac{1}{\exp[(\varepsilon - \mu)/kT] + 1} = 0.66 \times 10^{-6}$$

This result shows that the mean occupancy of any particular single-particle state is very small in the classical limit.

Since the energy ε is always positive, $\exp(\varepsilon/kT)$ is always greater than unity. Hence in the classical limit, when $\exp(-\mu/kT) >> 1$,

$$e^{\varepsilon/kT} e^{-\mu/kT} >> 1$$

for all values of ε, and the Fermi–Dirac distribution function tends to

$$f(\varepsilon) = e^{\mu/kT} e^{-\varepsilon/kT}$$

The $\exp(\mu/kT)$ term is known as the *absolute activity* and is generally denoted by the symbol λ. It can be seen that, in the classical limit, the probability that there is a fermion in the single-particle state of energy eigenvalue ε is equal to

$$\lambda \exp(-\varepsilon/kT)$$

But this is precisely the classical Boltzmann distribution developed in Chapter 3. ■

4.15 FREE-ELECTRON THEORY OF METALS

In metals, some of the electrons in the outer shells of the atoms may become detached from the individual atoms. These are the conduction electrons. In solid metals, the positive ions are confined to the vicinity of the lattice, whereas the conduction electrons are free to move anywhere in the metal. In examining the aspects of electrons in a metal, only those electrons that are involved in the conduction of electrical charge (current flow) and thermal energy (heat flow) are considered. These electrons represent a small fraction of all the electrons in the metal; but they are a very important fraction. There are in fact several types of crystal binding in a solid such as covalent binding, ionic binding, and so on; however, these types of crystal contain electrons that are so tightly bound that an externally applied electric field does not readily produce a flow of current.

The simplest picture of a metal is a lattice structure of atoms held together by binding forces. The innermost electrons of the lattice atoms remain bound to their parent atom and move around the nuclei in Bohr orbits much as they would if each atom were completely isolated. However, the outermost electrons, especially those in the incomplete outermost electron shell (the valence electrons) behave very differently. Their Bohr orbits are about the same size as the distance between the nuclei, so that the electrons of neighboring atoms interact powerfully with each other. The result is that the individual electronic quantum levels of the valence electrons split into an energy band known as the valence band. This band consists of a very large number of closely spaced quantum levels. (The electrons are constrained to move in a volume of the metal, so their energies are quantized.)

There is a very close analogy between the translational energy states of a particle moving in a laboratory container and the translational energy states of a valence electron moving in the valence band of a metal since the electron can only possess translational, that is, kinetic energy. In fact, each quantum state of a valence electron when in the valence band is characterized by just two quantum numbers, n and m_s. The quantum number n is the translational quantum number; it must have an integral value and can reach very large numerical values. The quantum number m_s is the spin quantum number and can have only two values, $+\frac{1}{2}$ and $-\frac{1}{2}$.

Two valence electrons can occupy the same translational energy state characterized by the translational quantum number n and not disobey the Pauli principle, provided that they have different values of spin quantum number; this means that one of the electrons must have $m_s = +\frac{1}{2}$ while the other electron must have $m_s = -\frac{1}{2}$. A single-particle translational energy state has prescribed n and m_s values and can only be occupied by one electron at most. In the simplest version of the *free-electron* model, the electrons are thought of as being free to move within the volume of the lattice, and the potential energies of interaction are neglected. As with perfect gases, collisions, though infrequent, play an important role in maintaining thermal equilibrium. This model was used successfully long before the development of quantum mechanics to give an explanation of Ohm's law and of the Wiedemann–Franz law expressing the proportionality between electrical and thermal conductivities. It was naturally assumed then that the electron gas was describable by Maxwell–Boltzmann statistics and this led to a discrepancy with respect to the heat capacity which, classically, would be expected to add $\frac{3}{2}Nk$ to the lattice contribution. The experimental absence of the extra term represented a blow against the model, as did the behavior of the paramagnetic susceptibility of the conduction electrons. Quantum mechanics not only led to a better understanding of the motion of electrons in periodic lattices and of metallic bonding, but also explained heat capacity and paramagnetic susceptibility.

Figure 4.15-1(a) shows the Sommerfeld free electron model of a metal, consisting of a flat-bottom potential well with its array of valence band translational energy levels. Its simplicity depends entirely on the fact that the electrons have only translational, that is, kinetic energy which is readily handled, as already seen in Chapter 2. Whether it is a good model depends entirely on how well the theoretical results agree with experiment. A more realistic model, such as the Kronig–Penney potential (Chapter 6), is shown in Fig.

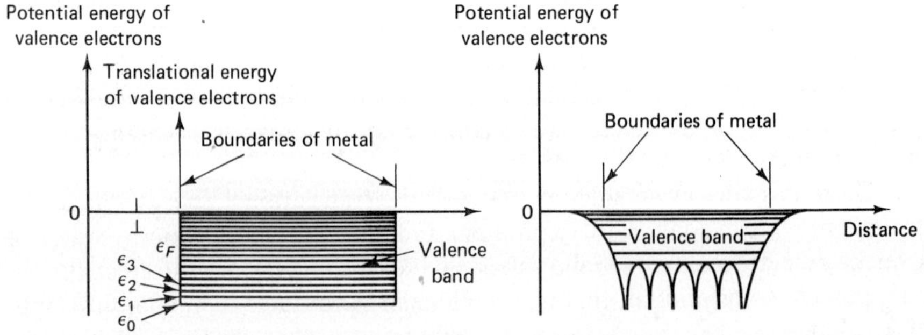

Figure 4.15-1 Potential energy of an electron as a function of its distance from the boundary of a metal; (a) simplified model, (b) actual model.

(4.15-1b), where the potential varies in the vicinity of the positive ions with the periodicity of the crystal lattice.

Experiments, for example, on the photoelectric effect, havc shown that work must be done to remove an electron from a metal. The work function Φ is the energy required to remove an electron, which has an energy equal to the chemical potential (Fermi level), to the outside of the metal. The depth of the potential well in Fig. (4.15-1a) is equal to $[\mu(T) + \Phi]$. The chemical potential of the free electrons in a metal is typically about 5 eV above the bottom of the potential well, and the work function is typically of the order of 3 eV, so the depth of a potential well is typically of the order of 8 eV. Now, the pressure of the free-electron gas at $T = 0$ is

$$p_0 = \frac{2}{3}\left(\frac{N}{V}\right)\varepsilon_F$$

In the case of the free electrons in copper, the concentration of free electrons, N/V, is 8.5×10^{28} m^{-3} at 300 K and the Fermi energy ε_F is 1.13×10^{-18} J at absolute zero temperature. If the concentration remained equal to 8.5×10^{28} at the absolute zero, the value of p_0 would be 3.8×10^{10} N m^{-2} or 3.8×10^5 atmospheres, which is considerable. At 300 K the pressure of the free electron gas would be even higher. However, the free electrons are prevented from leaving the metal by the work function. At very high temperatures, some of the electrons do have sufficient energies to leave the metal. This is known as *thermionic emission* (discussed in Chapter 6).

Returning to the heat capacity of a free electron gas, it can be shown that in the low-temperature limit when $T \ll T_f$, which is true for the electrons in a metal at room temperature, the internal energy of a fermion gas is approximately

$$U = \frac{3}{5} NE_F \left[1 + \frac{5\pi^2}{12}\left(\frac{T}{T_F}\right)^2\right] \qquad (4.15\text{-}1)$$

The heat capacity of the free electron gas is, therefore,

$$c_{\text{elec}} = \left(\frac{\partial U}{\partial T}\right)_V = \frac{\pi^2}{2} Nk \left(\frac{T}{T_F}\right)$$

$$= 4.93\ Nk \left(\frac{T}{T_F}\right) \qquad (4.15\text{-}2)$$

According to this equation, when $T \ll T_F$, the contribution of the free electrons to the total heat capacity of a metal is proportional to the absolute temperature T. This can be seen in Fig. 4.15-2(a) which shows the variation of the molar specific heat, $c_{V,m}$, with absolute temperature. When $T \ll T_f$, $c_{V,m}$ is proportional to T. At high temperatures, $c_{V,m}$ tends to the classical value of $3R/2$.

The linear dependence of the heat capacity of a fermion gas on absolute temperature, when $T \ll T_F$ can be verified by a simple order of magnitude

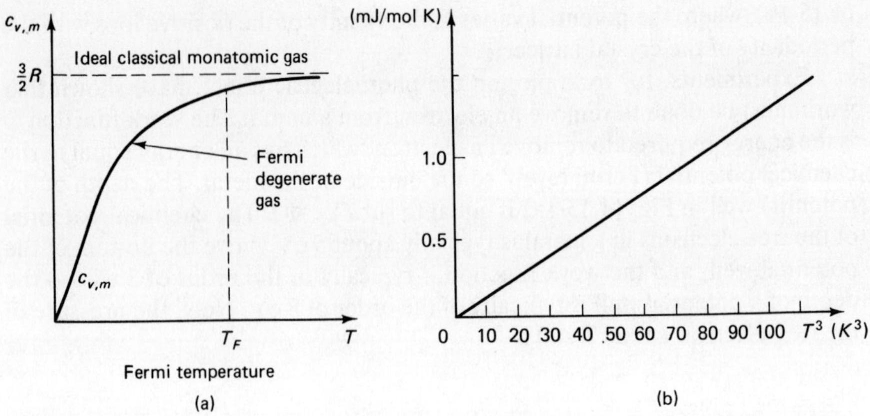

Figure 4.15-2 Dulong-Petit and Debye laws for specific heat. (a) Variation of molar specific heat, $C_{V,m}$, with absolute temperature. (b) Debye's T^3 law: specific heat of T_iO_2.

estimate. Assume that, for the free electrons in a typical metal, $T = 0.01T_F$, and $kT = 0.01\,\varepsilon_F$. From

$$\mu(T) \simeq \left[1 - \frac{\pi^2}{12}\left(\frac{T}{T_F}\right)^2\right]\varepsilon_F \qquad T << T_F \tag{4.15-3}$$

the chemical potential μ is equal to $0.999\varepsilon_F$. The numerical values of $f(\varepsilon)$, calculated from

$$f(\varepsilon) = \frac{1}{e^{(\varepsilon-\mu)/kT} + 1} \tag{4.15-4}$$

for various values of ε show that $f(\varepsilon)$ is greater than 0.993 for all values of ε from zero up to $0.950\varepsilon_F$. Since $kT = 0.01\varepsilon_F$, this corresponds to an energy of $5\,kT$ below ε_F. When $\varepsilon = 0.990\varepsilon_F$, $f(\varepsilon)$ is down to 0.729. These results show that the occupancy of the single-particle states only drops by an appreciable amount in an energy range of the order of $kT = 0.01\varepsilon_F$ immediately below the Fermi energy. In other words, at a temperature $T << T_F$, it is mainly electrons near the Fermi energy which are found to be excited above ε_F. As a first approximation, let it be assumed that all the electrons within a range of kT from ε_F are affected. Now, the number of particle states (orbitals) which have energy eigenvalues in the range ε to $\varepsilon + d\varepsilon$ is

$$g(\varepsilon)\,d\varepsilon = \frac{gV}{4\pi^2}\left(\frac{2m}{\hbar^2}\right)^{3/2}\varepsilon^{1/2}\,d\varepsilon \tag{4.15-5}$$

Using this, with $\Delta\varepsilon$ equal to kT, gives the number of electrons N_{exc} in excited states at temperature T:

$$N_{exc} \simeq \frac{gV}{4\pi^2}\left(\frac{2m}{\hbar^2}\right)^{3/2}\varepsilon_F^{1/2}\,kT \tag{4.15-6}$$

Also, the Fermi energy

$$\varepsilon_F = kT_F = \frac{\hbar^2}{2m}\left(\frac{6\pi^2 N}{gV}\right)^{2/3} \tag{4.15-7}$$

can be rearranged as

$$\frac{gV}{\pi^2}\left(\frac{2m}{\hbar^2}\right)^{3/2} = \frac{6N}{\varepsilon_F^{3/2}} \tag{4.15-8}$$

Substituting in Eq. (4.5-6) then gives

$$N_{\text{exc}} \simeq \frac{3}{2}N\left(\frac{\hbar T}{\varepsilon_F}\right) = \frac{3}{2}N\left(\frac{T}{T_F}\right) \tag{4.15-9}$$

The energy gained by each of these solutions is of the order of kT. Hence, at temperature T, the internal energy of the electron gas (compared to that at $T = 0$) is

$$U_{\text{elec}} \simeq \frac{3}{2}N\left(\frac{T}{T_F}\right)kT = \frac{3\ NkT^2}{2T_F} \tag{4.15-10}$$

and the heat capacity is

$$C_{\text{elec}} = \left(\frac{\partial U_{\text{elec}}}{\partial T}\right) = 3\ Nk\left(\frac{T}{T_F}\right) \tag{4.15-11}$$

This result confirms that C_{elec} is proportional to T, though the numerical factor is 3, not the more correct factor 4.93 in Eq. (4.15-2).

The total heat capacity of a metal is the sum of the contributions of the free electrons and the lattice vibrations. Since T is approximately $0.01T_F$ for a typical metal at room temperatures, Eq. (4.15-2) gives the heat capacity of the free electrons at room temperatures to be around 0.05 Nk, where N is the number of free electrons. Now, the contribution of the lattice vibrations to the heat capacity of a solid at room temperatures is generally of the order of 3 Nk, where N is the number of atoms (generally equal to the number of free electrons, to within a factor of two). Hence at room temperatures, the contribution of the free electrons to the specific heat of a metal is swamped by the contribution of the lattice vibrations. The statement that the heat capacity of a classical monatomic solid is 3 Nk is the *law of Dulong and Petit.* It agrees fairly well with experimental results for most elemental solids at high temperatures, but at low temperatures, the specific is much lower than 3 Nk, being proportional to T^3 and approaching zero with $T \to 0$ as shown in Fig. 4.15-2(b).

■ EXAMPLE 4.15-1

Show the Fermi–Dirac distribution and the location of the chemical potential for a system of eight identical fermions at three different temperatures. How does the chemical potential vary with temperature for a fermion gas?

Solution. The distribution of particles among the energy states and the location of the chemical potential is shown in Fig. 4.15-3. (a) At $T = 0$, all the single-particle states up to the Fermi energy ε_F are filled. All the orbits above ε_F are empty. The chemical potential $\mu(0)$ is equal to ε_F. (b) For $T \neq 0$ but $T << T_F$, some of the fermions which at $T = 0$ would be just below ε_F, are now just above ε_F. In this case, the chemical potential $\mu(T)$ is just below ε_F. (c) When $T >> T_F$, $f(\varepsilon)$ is less than 0.5, and the chemical potential becomes negative.

For $T << T_F$, the chemical potential of a fermion gas is given, to a good approximately, by

$$\mu \simeq \varepsilon_F \left[1 - \frac{\pi^2}{12}\left(\frac{T}{T_F}\right)^2\right] \tag{4.15-12}$$

According to this equation, if $T = 0.01T_F$, μ is equal to $0.999\varepsilon_F$, the chemical potential is just fractionally less than the Fermi energy ε_F. In the high-temperature limit when $T >> T_F$, the fermion gas approximates to the classical monatomic ideal gas

$$e^{-\mu/kT} = \left(\frac{gV}{N}\right)\left(\frac{mkT}{2\pi\hbar^2}\right)^{3/2} \tag{4.15-13}$$

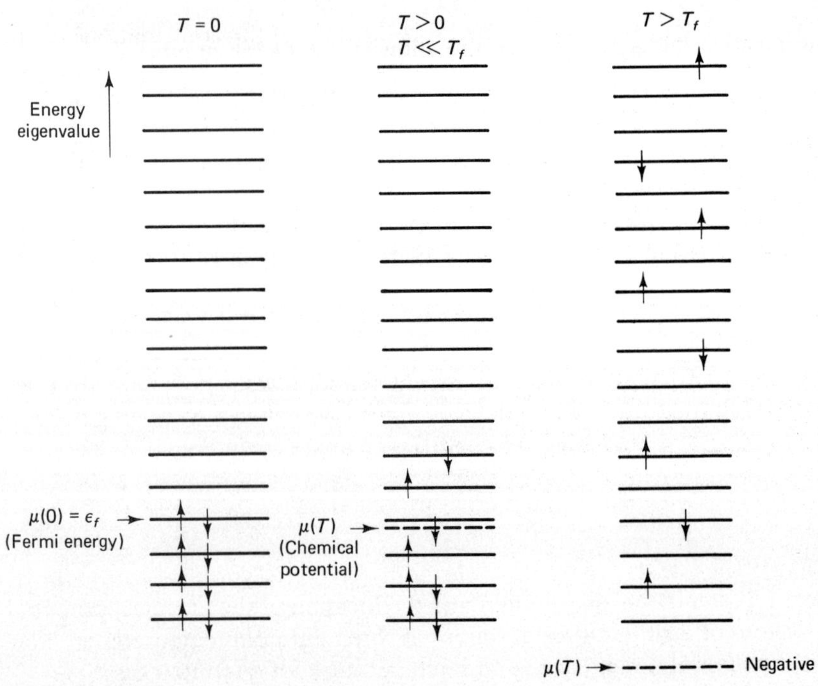

Figure 4.15-3 Distribution of eight identical fermions of spin $\frac{1}{2}$ among energy states.

where g is the spin degeneracy (having a value of 2 for fermions). As an example of a fermion gas in the classical limit, ${}^3_2$He atoms at STP give $\exp(-\mu/kT) = 3.4 \times 10^5$, and $\mu/kT - -12.7$. Thus, in the high-temperature limit, the chemical potential μ is negative. At $T = 0$, the fermion gas is said to be *completely degenerate*; at $T << T_F$, it is *degenerate*; and at $T >> T_F$, it approximates to the *ideal classical gas*. ■

■ EXAMPLE 4.15-2

(a) Develop an expression for the number of orbitals with the magnitude n of the quantum number $\mathbf{n} = n_x, n_y, n_z$ in the range Δn about n.

(b) Find the number of orbitals with energy between ε and $\varepsilon + d\varepsilon$ (the density of states).

(c) Plot the Fermi–Dirac distribution function $f(\varepsilon)$ versus ε/μ for zero temperature and for a temperature $kT = \frac{1}{5}\mu$.

(d) Plot the density of orbitals as a function of energy, for a free-electron gas in three dimensions. Show the distribution of occupied orbitals $g(\varepsilon)f(\varepsilon)$.

Solution

(a) The enumeration of orbitals is of central importance to many applications of the *FD* and *BE* distribution functions. Apart from spin, the number of orbitals with magnitude n of the quantum number is equal to the volume of an octant of a sphere of radius n in the space of n_x, n_y, n_z, as in Fig. 4.15-4(a). The limitation to an octant occurs because the independent orbitals are described by the positive integers alone. In the allowed octant the number of orbitals with some radius n is

$$g \cdot \frac{1}{8} \cdot \frac{4\pi}{3} n^3 = \frac{1}{6} g \pi n^3 \tag{4.15-14}$$

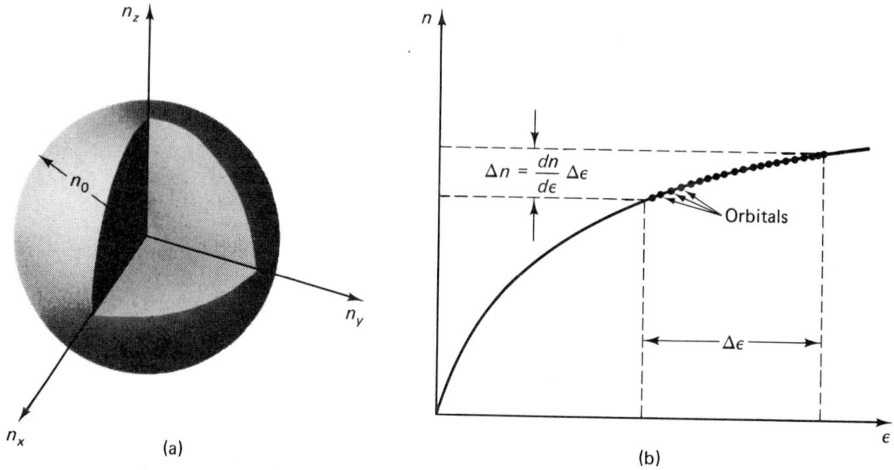

Figure 4.15-4 (a) Orbitals in quantum space. (b) Density of states function.

where g is the spin degeneracy. For electrons, $g = 2$. The number of orbitals with the magnitude n of the quantum number in the range Δn about n is

$$\left[\frac{d}{dn}\left(\frac{1}{6}g\pi n^3\right)\right]\Delta n = \frac{1}{2}g\pi n^2\,\Delta n \tag{4.15-12}$$

(b) The number of orbitals with energy between ε and $\varepsilon + d\varepsilon$ is equal to

$$g(\varepsilon)\,d\varepsilon = \frac{1}{2}g\pi n^2 \frac{dn}{d\varepsilon}\,d\varepsilon \tag{4.15-13}$$

For free particles of mass m the energy is related to the quantum number by

$$\varepsilon = \frac{\hbar^2}{2m}\left(\frac{\pi n}{L}\right)^2 \qquad n = \left(\frac{2m\varepsilon L^2}{\hbar^2\pi^2}\right)^{1/2}$$

or

$$\frac{dn}{d\varepsilon} = \left(\frac{mL^2}{2\hbar^2\pi^2\varepsilon}\right)^{1/2} \qquad n^2\frac{dn}{d\varepsilon} = \frac{L^3(2m)^{3/2}}{2\pi^3\hbar^3}\,\varepsilon^{1/2} \tag{4.15-14}$$

These relations, when substituted into (4.15-13), give

$$g(\varepsilon) = \frac{gV}{4\pi^2}\left(\frac{2m}{\hbar^2}\right)^{3/2}\varepsilon^{1/2} \tag{4.15-15}$$

This is the density-of-orbitals function for free particles in the volume V.

(c) The Fermi–Dirac distribution function

$$f(\varepsilon) = \frac{1}{e^{(\varepsilon-\mu)/kT}+1} \tag{4.15-16}$$

is plotted versus ε/μ for zero temperature and for a temperature $kT = \frac{1}{5}\mu$ in Fig. 4.15-5(a). When the system is heated from absolute zero, electrons are transferred from the shaded region at $\varepsilon/\mu < 1$ to the shaded region at $\varepsilon/\mu > 1$. For a metal, μ is around 50,000 K.

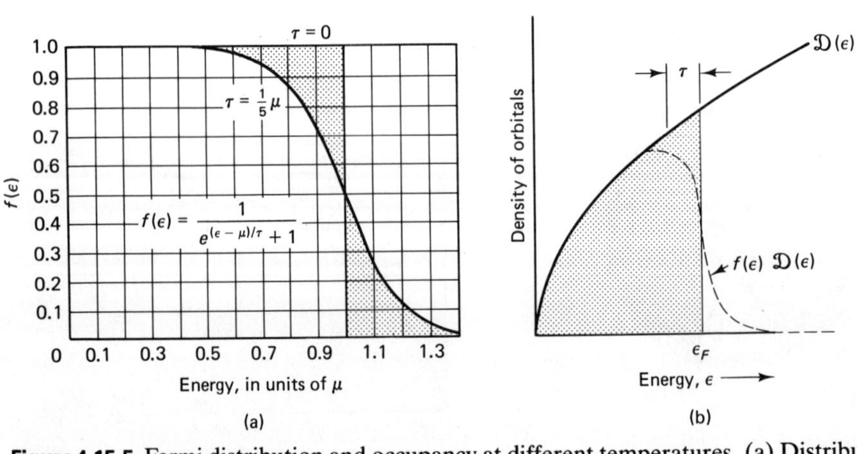

Figure 4.15-5 Fermi distribution and occupancy at different temperatures. (a) Distribution. (b) Occupation.

(d) The density of orbitals as a function of energy is shown in Fig. 4.15-5(b). The dashed curve represents the density $g(\varepsilon)f(\varepsilon)$ of occupied orbitals at a finite temperature, but such that kT is small compared to ε_F. The shaded area represents the occupied orbitals at absolute zero. ■

■ EXAMPLE 4.15-3

For the free-electron gas, the product of the density of states function and the Fermi–Dirac distribution function gives

$$n(\varepsilon) = g(\varepsilon)f(\varepsilon) = \frac{\hbar V}{2}\left(\frac{8m}{h^2}\right)^{3/2} \frac{\varepsilon^{1/2}}{e^{(\varepsilon-\varepsilon_F)/kT}+1}$$

$$= \frac{3N}{2\varepsilon_F^{3/2}} \cdot \frac{\varepsilon^{1/2}}{e^{(\varepsilon-\varepsilon_F)/kT}+1} \qquad (4.15\text{-}17)$$

(a) Sketch the graph of $n(\varepsilon)$.

(b) What fraction of the free electrons in Cu have kinetic energies between 3.95 eV and 4.05 eV at room temperature?

(c) Derive an expression for the low-temperature heat capacity of a free-electron gas and show that it is directly proportional to the absolute temperature.

Solution

(a) Equation (4.15-17) shows that $n(\varepsilon)$ varies as $\varepsilon^{1/2}$. The sketch of $n(\varepsilon)$ is shown in Fig. 4.15-6(a). At $T=0$, $(\varepsilon-\varepsilon_F)/kT = -\infty$ for $\varepsilon<\varepsilon_F$, and $(\varepsilon-\varepsilon_F)/kT = +\infty$ for $\varepsilon>\varepsilon_F$. The integral $\int_0^\infty n(\varepsilon)\,d\varepsilon$ gives the total number of particles, as can be seen from

$$\int_0^\infty n(\varepsilon)\,d\varepsilon = \frac{3N}{2\varepsilon_F^{3/2}}\left[\int_0^{\varepsilon_F} \varepsilon^{1/2}\,d\varepsilon + \int_{\varepsilon_F}^{\infty} \varepsilon^{1/2}\,d\varepsilon\right]$$

$$= \frac{3N}{2\varepsilon_F^{3/2}} \cdot \frac{2\varepsilon_F^{3/2}}{2} = N \qquad (4.15\text{-}18)$$

using $1/(e^{-\infty}+1)=1$ and $1/(e^{\infty}+1)=0$. Figure (4.15-6a) shows the graph drawn for two different temperatures, $T=0$ and $T>>0$. For most metals, the $T>>0$ curve as drawn would correspond to a temperature nearer the melting point than to room temperature. Note that the Fermi energy ε_F changes very little with temperature.

(b) Cu has a Fermi energy $\varepsilon_F = 7.0$ eV which is the same at $T=0$ as at room temperature of 20°C = 293 K. The fraction of free electrons having kinetic energies between 3.95 eV and 4.05 eV is given by

$$\frac{1}{N}\int_{3.95\ \text{eV}}^{4.05\ \text{eV}} n(\varepsilon)\,d\varepsilon$$

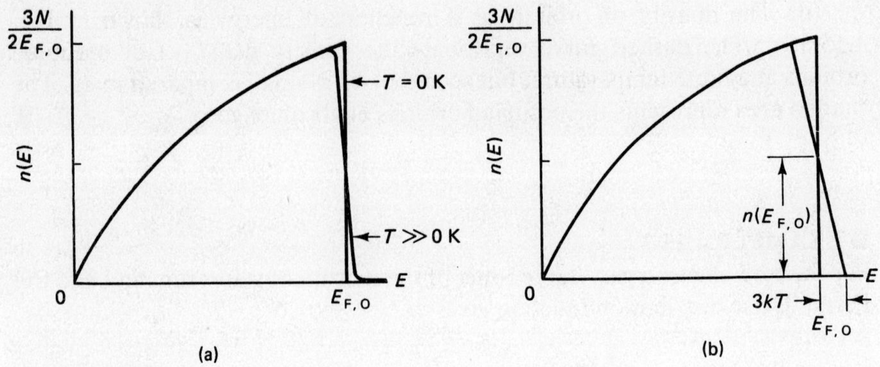

Figure 4.15-6 Electron distribution with energy. (a) Occupation at $T = 0$ and $T >> 0$. (b) Population change at $T > 0$.

Since $\Delta\varepsilon = 4.05 - 3.95 = 0.10$ eV is reasonably small, the integral can be approximated by

$$\frac{1}{N} n(\bar{\varepsilon})\,\Delta\varepsilon \tag{4.15-19}$$

with $\bar{\varepsilon}$ being the average energy, or 4.00 eV. Thus, from (4.15-17), with $kT = (1/40)$ eV at room temperature,

$$\frac{n(\bar{\varepsilon})\,\Delta\varepsilon}{N} = \left[\frac{3}{2(7.0)^{3/2}}\right]\left[\frac{(4.0)^{1/2}}{e^{(4.0-7.0)(1/40)}+1}\right](0.10)$$

$$= \frac{0.0162}{e^{-120}+1} = 0.016$$

(e) The (shaded) area under the $T > 0$ curve in Fig. 4.15-6(a) equals the total number of free electron, that is,

$$\int_0^\infty n(\varepsilon)\,d\varepsilon = N \tag{4.15-20}$$

as seen in part (a). The total energy of the free-electron gas is given by $\int_0^\infty \varepsilon n(\varepsilon)\,d\varepsilon$ and can be evaluated exactly only at $T = 0$. The result is

$$\int_0^\infty \varepsilon n(\varepsilon)\,d\varepsilon = \frac{3}{5} N\varepsilon_F \tag{4.15-21}$$

The shaded area under the curve in Fig. 4.15-6(b) equals the number of electrons promoted above the Fermi energy when the gas is warmed to temperature T. This area is a triangle (approximately) with base kT and height equal to the value of $n(\varepsilon)$ at ε_F, that is from (4.15-17),

$$n(\varepsilon_F) = \frac{3N}{2\varepsilon_F}\cdot\frac{\varepsilon_F^{1/2}}{e^0+1} = \frac{3N}{4\varepsilon_F} \tag{4.15-22}$$

The number of electrons promoted above ε_F is therefore (area of a triangle = one-half base times height)

$$\frac{1}{2} \cdot \frac{3N}{4\varepsilon_F} \cdot kT = \frac{3}{8} \cdot \frac{NkT}{\varepsilon_F} \tag{4.15-23}$$

Now, an electron being promoted from kT below ε_F to kT above ε_F must be given $2kT$ of heat. Therefore, the heat added is

$$\frac{3NkT}{8\ \varepsilon_F} \cdot 2kT = \frac{3}{4} \frac{Nk^2T^2}{\varepsilon_F} \tag{4.15-24}$$

and the total internal energy of the electron gas is

$$U = \frac{3}{5} N\varepsilon_F + \frac{3}{4} \frac{Nk^2T^2}{\varepsilon_F} = \frac{3}{4} N \left[\varepsilon_F + \frac{k^2T^2}{\varepsilon_F} \right] \tag{4.15-25}$$

The electronic heat capacity of the gas is then

$$C_{\text{elec}} = \frac{dU}{dT} = \frac{3}{2} \frac{Nk^2T}{\varepsilon_F} \tag{4.15-26}$$

It can be seen that the low-temperature heat capacity of the free-electron gas is proportional to the absolute temperature, approaching zero as T approaches zero. The heat capacity of a classical free-electron gas is a constant at $\frac{3}{2}Nk$. ■

■ EXAMPLE 4.15-4

What is the mean-free path, Fermi speed, and collision time of electrons in Cu at room temperature? The electrical conductivity of Cu at room temperature is $5.9 \times 10^7\ (\Omega\ \text{m})^{-1}$ and the electron concentration is $8.5 \times 10^{28}\ \text{m}^{-3}$ for its free-electron gas. The Fermi energy, ε_F, for Cu is 7.0 eV.

Solution. The electrical conductivity, σ, of a free-electron gas is

$$\sigma = \frac{N}{V} \frac{e^2 t}{m} \tag{4.15-27}$$

where N/V is the concentration, $m = 9.11 \times 10^{-31}$ kg, $e = 1.60 \times 10^{-19}$ C, and t = collision time (seconds). Thus,

$$t = \frac{\sigma m}{e^2(N/V)} = \frac{[5.9 \times 10^7\ (\Omega \cdot \text{m})^{-1}][9.11 \times 10^{-31}\ \text{kg}]}{(1.60 \times 10^{-19}\ \text{C})^2(8.5 \times 10^{28}\ \text{m}^{-3})} = 2.5 \times 10^{-14}\ \text{s}$$

To find the mean-free path (average distance between collisions), l, the Fermi speed, v_F, is first determined from

$$\frac{1}{2} m v_F^2 = \varepsilon_F \tag{4.15-28}$$

With $\varepsilon_F = (7.0\ \text{eV})(1.60 \times 10^{-19}\ \text{J/eV}) = 1.12 \times 10^{-18}$ J, this gives

$$v_F = \left[\frac{2(1.12 \times 10^{-18}\ \text{J})}{9.11 \times 10^{-31}\ \text{kg}} \right]^{1/2} = 1.57 \times 10^6\ \text{m/s}$$

The mean-free path, l, of an electron in Cu is

$$l = v_F t = (1.57 \times 10^6 \text{ m/s})(2.5 \times 10^{-14} \text{ s})$$
$$= 3.9 \times 10^{-8} \text{ m} = 39 \text{ nm}$$

It is seen that the electrons in Cu at room temperature get past an average of about 200 atoms before being scattered. At low temperatures, this value may be 10^5 times larger for very pure and defect-free single crystals of Cu. ■

■ EXAMPLE 4.15-5

Estimate the Fermi energy of the nuclear gas in nuclear matter.

Solution. Nuclear matter is composed of protons and neutrons, but no electrons. The number of protons is specified by the letter Z, and the number of neutrons by the letter N, the sum of the two being called the mass number A. Thus, for a given atom,

$$A = Z + N \tag{4.15-29}$$

The term nucleon is applied collectively to both protons and neutrons, when there is no necessity to distinguish between them. In other words, there are A nucleons in the nucleus. Both protons and neutrons are fermions of spin $s = \frac{1}{2}$. There are very strong attractive forces between the nucleons inside the nucleus. The mean binding energy of a nucleon in a nucleus is typically about 8 MeV, whereas the binding energy of an electron in an atom is only in the eV to keV range. Due to the strong interactions between nucleons, the protons and neutrons in an atomic nucleus do not constitute two noninteracting ideal fermion gases. However, it is appropriate to use the ideal fermion gas model to determine the energies for the neutrons and the protons inside an atomic nucleus, assuming that they are two independent ideal fermion gases.

Experiments on the scattering of high-energy atomic particles show that the radius R of a nucleus of atomic weight A is given approximately by $R = 1.3 \times 10^{-15} A^{1/3}$ m. Also, the concentration of neutrons is

$$\frac{N}{V} = \frac{A - Z}{V} = \frac{A - Z}{\frac{4}{3}\pi R^3} = 1.09 \times 10^{44} \left(\frac{A - Z}{A}\right) \text{m}^{-3} \tag{4.15-30}$$

Using the values $m = 1.67 \times 10^{-27}$ kg for the mass of a neutron, $g = (2s + 1) = 2$, and $\hbar = 1.05 \times 10^{-34}$ J s in the equation

$$\varepsilon_F = kT_F = \frac{\hbar^2}{2m}\left(\frac{6\pi^2 N}{gV}\right)^{2/3} \tag{4.15-31}$$

and assuming that $(A - Z)/A = N/A$ is equal to 0.5 for a light nucleus, it is found that $(N/V)_{\text{protons}} = (N/V)_{\text{neutrons}} = 0.05 \times 10^{39}$ cm^{-3} and that the Fermi energy of an ideal gas of neutrons in a volume equal to the volume of the atomic nucleus is $\varepsilon_F = 4.5 \times 10^{-12}$ J $\equiv 28$ MeV, as illustrated in Fig. 4.15-7(a). According to Eq. (4.15-18), the corresponding Fermi temperature is 3.3×10^{11} K, showing that the ideal neutron gas is highly degenerate.

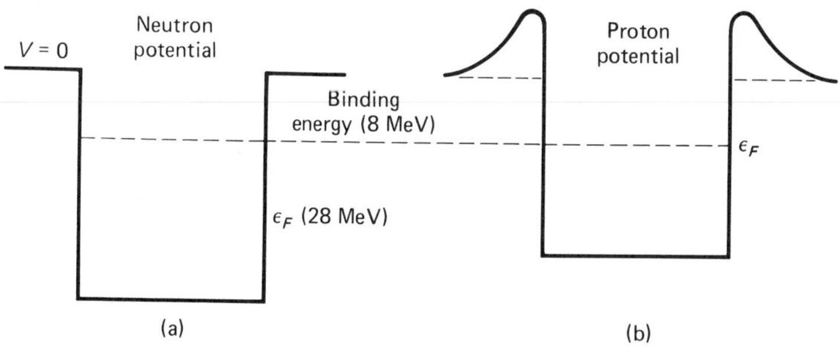

Figure 4.15-7 Potential wall for neutrons and protons in an atomic nucleus.

The average kinetic energy of a particle in a degenerate Fermi gas is $\frac{3}{5}$ of the Fermi energy, so the average kinetic energy in nuclear matter is 16 MeV per nucleon. The binding energy of each neutron in an atomic nucleus is about 8 MeV. On the ideal fermion gas model, the total depth of the potential well for the neutrons inside the nucleus should be about 36 MeV, as shown in Fig. 4.15-7(a). On the ideal fermion gas model the kinetic energies of the neutron should range up to about 28 MeV. These rough orders of magnitude are confirmed by scattering experiments using high-energy nucleons.

For atomic masses $A > 100$, there are significantly more neutrons than protons in a nucleus, so that for a heavy nucleus the concentration and hence the Fermi energy is less for the protons than for the neutrons. The total depth of the potential well is less for protons than for neutrons, as shown in Fig. 4.15-7(b). When a proton approaches the nucleus from outside the nucleus, there is an electric force of repulsion due to the positive electric charges of the approaching proton and the protons inside the atomic nucleus. The coulomb potential barrier outside the nucleus is due to the electric charges of the protons in the nucleus. ■

■ EXAMPLE 4.15-6

What is the state of matter in white dwarf and neutron stars? Consider Sirius B to be a typical white dwarf star.

Solution. White dwarf stars have masses comparable to that of the Sun. Sirius B is a typical white dwarf star. Its estimated mass (determined by measurements on orbits) is 2.09×10^{30} kg, which is 1.05 times the mass of the Sun. The estimated radius of Sirius B is 5.57×10^3 km and its volume is 7.23×10^{20} m^3. For comparison, the radii of the Earth and the Sun are 6.4×10^3 km and 7.0×10^5 km, respectively. A typical white dwarf star has about the mass of the Sun and about the size of the Earth. The matter of which the white dwarf is composed is, indeed, very tightly packed.

The main source of energy in main sequence stars is thermonuclear reactions near the centers of the stars, which convert hydrogen into helium. It is assumed that the helium is completely ionized. Each helium nucleus (α particle) consists of four nucleons, namely, two protons and two neutrons, each of mass M_P approximately equal to 1.67×10^{-27} kg. Each helium atom loses two electrons, so that, on the average, there are 0.5 free electrons per nucleon. The total number N of nucleons in the star is equal to

$$N = \frac{2.09 \times 10^{30} \text{ kg}}{1.67 \times 10^{-27} \text{ kg}} = 1.25 \times 10^{57} \tag{4.15-32}$$

Let N_e denote the total number of free electrons in the white dwarf star. Let x stand for the ratio of N_e/N. For completely ionized helium, $x = 0.5$. The nonrelativistic equation for the Fermi energy gives

$$\varepsilon_F = \frac{\hbar^2}{2m}\left(\frac{6\pi^2 N_e}{gV}\right)^{2/3} = \frac{\hbar^2}{2m}\left(\frac{3\pi^2 xN}{V}\right)^{2/3} \tag{4.15-33}$$

where m is the mass of an electron and $g = (2s + 1) = 2$ for an electron of spin $\frac{1}{2}$. For Sirius B, with $x = 0.5$, $N = 1.25 \times 10^{57}$, $V = 7.23 \times 10^{20}$ m^3, $m = 9.1 \times 10^{-31}$ kg, and $\hbar = 1.055 \times 10^{-34}$ J · s, this gives

$$\varepsilon_F = 5.3 \times 10^{-14} \text{ J} = 0.33 \text{ MeV} \tag{4.15-34}$$

This value of the Fermi energy of the free electron is getting close to the rest mass of an electron, which is $mc^2 = 0.5$ MeV. For larger values of (N/V), the full relativistic theory should be used. The Fermi temperature for Sirius B is

$$T_F = \frac{\varepsilon_F}{k} = 3.8 \times 10^9 \text{ K} \tag{4.15-35}$$

The estimated temperature at the center of Sirius B is 2×10^7 K, so that $T << T_f$ and the electron gas is *degenerate*. As a rough approximation, using the completely degenerate, nonrelativisic, ideal fermion gas at $T = 0$, the pressure for Sirius B is

$$\begin{aligned} p_0 &= \frac{2}{5}\left(\frac{N_e}{V}\right)\varepsilon_F = \frac{2}{5}\left(x\frac{N}{V}\right)\varepsilon_F \\ &= 1.8 \times 10^{22}\ N/\text{m}^2 \end{aligned} \tag{4.15-36}$$

This is approximately 1.8×10^{17} atmospheres and is sufficient to prevent the further gravitational collapse of a white dwarf star such as Sirius B. ■

4.16 BOSON GAS: LIQUID HELIUM

Bosons are particles of integral spin. There is no limit to the number of bosons that can be in any single-particle state (orbital). Because of this, the properties of bosons are quite different from those of fermions.

The mean number of bosons in a single-particle state, which has energy eigenvalue ε, is given by the Bose–Einstein distribution function

$$f_{BE}(\varepsilon) = \frac{1}{\exp[(\varepsilon - \mu)/kT] - 1} \tag{4.16-1}$$

This is the probability of occupation of a given state. When (4.16-1) is multiplied by the density of states available, $g(\varepsilon)$, the population distribution, $n(\varepsilon)$, is obtained:

$$n(\varepsilon) = g(\varepsilon) f_{BE}(\varepsilon) \tag{4.16-2}$$

The first step is then to find an expression for $g(\varepsilon)\,d\varepsilon$, the number of states that have energy values between ε and $(\varepsilon + d\varepsilon)$. Recalling that for a particle in a box V,

$$\varepsilon = \frac{\hbar^2\pi^2}{2mV^{2/3}}(n_x^2 + n_y^2 + n_y^2) = \frac{\hbar^2\pi^2}{2mV^{2/3}} n^2$$

or

$$n_x^2 + n_y^2 + n_z^2 = n^2 = \frac{2mV^{2/3}\,\varepsilon}{\hbar^2\pi^2}, \tag{4.16-3}$$

the possible combinations of n_x, n_y, and n_z, which give the same value of ε, lie on the surface of a sphere of radius $(n_x^2 + n_y^2 + n_z^2)^{1/2}$ in number space. Since n_x, n_y, and n_z must be positive, only one octant need be considered. The number of states with energy eigenvalues less than or equal to ε, is equal to the number of points in the octant

$$\begin{aligned} \frac{1}{8}\left(\frac{4\pi}{3}\right)(n_x^2 + n_y^2 + n_z^2)^{3/2} &= \frac{\pi}{6}\left(\frac{2mV^{2/3}\,\varepsilon}{\hbar^2\hbar^2}\right)^{3/2} \\ &= \frac{V}{6\pi^2}\left(\frac{2m}{\hbar^2}\right)^{3/2}\varepsilon^{3/2} \end{aligned} \tag{4.16-4}$$

Differentiating this gives the number of states having energy values between ε and $\varepsilon + d\varepsilon$,

$$g(\varepsilon)\,d\varepsilon = \frac{gV}{4\pi^2}\left(\frac{2m}{\hbar^2}\right)^{3/2}\varepsilon^{1/2}\,d\varepsilon \tag{4.16-5}$$

where g is the spin degeneracy. Equation (4.15-5) is the expression for the density of states. Substitution into (4.16-2) gives

$$n(\varepsilon)\,d\varepsilon = \frac{gV}{4\pi^2}\left(\frac{2m}{\hbar^2}\right)^{3/2}\frac{\varepsilon^{1/2}\,d\varepsilon}{\exp[(\varepsilon-\mu)/kT]-1} \tag{4.16-6}$$

This energy distribution differs from the Fermi–Dirac distribution by virtue of a negative (-1) rather than a positive $(+1)$ in the denominator.

Integrating Eq. (4.16-6) for an ideal boson gas consisting of N bosons gives

$$N = \frac{gV}{4\pi^2}\left(\frac{2m}{\hbar^2}\right)^{3/2}\int_0^\infty \frac{\varepsilon^{1/2}\,d\varepsilon}{\exp[(\varepsilon-\mu)/kT]-1} \tag{4.16-7}$$

This is an integral equation which gives the chemical potential μ in terms of T, V, and N. The variation of the chemical potential μ with temperature may at first glance be obtained by noting that the denominator in Eq. (4.16-7) must always be positive (because the numbers of bosons in any single-particle state cannot be negative). Thus, $\exp[(\varepsilon-\mu)/kT]$ must always be greater than unity. This implies that $\varepsilon > \mu$. In other words, the chemical potential of an ideal boson gas must always be less than the energy of the lowest single-particle state. If the energy of this state is chosen to be zero on the energy scale, the chemical potential μ of a boson gas will always be negative. It is less negative as T decreases, becoming zero when the temperature reaches a value T_B called the Bose temperature, and stays zero as T goes to zero. The Bose temperature will be shown to be

$$T_B = \frac{2\pi\hbar}{mk}\left(\frac{N}{2.612\,gV}\right)^{2/3} = \frac{3.31\,\hbar^2}{mk}\left(\frac{N}{gV}\right)^{2/3} \tag{4.16-8}$$

For a boson gas composed of 6.02×10^{23} bosons ($^4_2$He atoms), each of mass 6.65×10^{-27} kg and spin zero [meaning $g=(2s+1)=1$], in a box of volume 22.4×10^{-3} m^3, the Bose temperature is 0.036 K. If there were no interactions between the bosons, the ideal boson gas would not liquify. In practice, due to interactions between the bosons, all real boson gases liquify before their temperatures reach their Bose temperatures. In the case of helium, the gas liquifies at a temperature of 4.21 K at atmospheric pressure. [If liquid helium were treated as an ideal boson gas, the Bose temperature given by Eq. (4.16-8) would be 3.1 K.]

There is no restriction on the number of bosons which can be in any single-particle state. At the absolute zero of temperature, all the bosons tend to be in the single-particle ground state. If this state is chosen as the zero of the energy scale, the $\varepsilon^{1/2}$ term in Eq. (4.16-7) is not included in the integration of the equation, and the result gives only N_{exc}, the number of bosons in excited states. If N is the total number of bosons, the number of bosons in the single-particle ground state is

$$N_0 = N - N_{\text{exc}} \tag{4.16-9}$$

Near absolute zero of temperature, with $\varepsilon=0$ in Eq. (4.16-7), in the limit of $T\to0$, the population of the ground state becomes large (close to N), and $\exp(-\mu/kT)$ equals unity. Making the change of variable $x=\varepsilon/kT$, Eq. (4.16-7) then becomes

$$N = \frac{gV}{4\pi^2}\left(\frac{2mkT}{\hbar^2}\right)^{3/2}\int_0^\infty \frac{x^{1/2}\,\alpha x}{e^x-1} \tag{4.16-10}$$

The definite integral is equal to the product of the gamma function $\Gamma(\frac{3}{2})=\sqrt{\pi}/2$ and the Riemann zeta function $\zeta(\frac{3}{2})=2.612$. Hence, at absolute zero, the number of bosons in a state is

$$N = 2.612\,gV\left(\frac{mkT}{2\pi\hbar^2}\right)^{3/2} \tag{4.16-11}$$

The *Bose temperature* (also known as the *Bose–Einstein condensation* temperature), is that temperature at which, according to Eq. (4.16-11), all the bosons should become excited. Letting N_{exc} equal to N and T equal to T_B, Eq. (4.16-11) gives

$$N = 2.612 gV \left(\frac{mkT_B}{2\pi\hbar^2}\right)^{3/2} \tag{4.16-12}$$

or, upon rearrangement,

$$T_B = \frac{2\pi\hbar^2}{mk}\left(\frac{N}{2.612 gV}\right)^{2/3} = \frac{3.31\hbar^2}{mk}\left(\frac{N}{gV}\right)^{2/3} \tag{4.16-13}$$

which is Eq. (4.16-8).

The chemical potential of an ideal boson gas is *zero* from absolute zero temperature up to the Bose temperature, then *negative* as the temperature increases. The assumption that $\exp(-\mu/kT)$ is equal to unity at very low temperatures is valid up to the Bose temperature. Also, to a good approximation, when $T < T_B$,

$$N_{\text{exc}} = N\left(\frac{T}{T_B}\right)^{3/2} \qquad N_0 = N\left[1 - \left(\frac{T}{T_b}\right)^{3/2}\right] \tag{4.16-14}$$

For large N, at temperatures close to the absolute zero,

$$-\frac{\mu}{kT} \simeq \frac{1}{N} \tag{4.16-15}$$

According to Eq. (4.16-15)), the chemical potential of an ideal boson gas is zero at $T = 0$. For a small macroscopic system, such as $N = 10^{18}$ bosons, the value of $(-\mu/kT)$ is equal to 10^{-18}, which is extremely small.

The observed properties of liquid ^{4}He are, qualitatively, in keeping with the predictions. When liquid helium is cooled from its boiling point at atmospheric pressure, it undergoes a phase transition at 2.19 K, to a species called *liquid helium II*. At low temperatures the gas particles will be preferentially distributed over the lowest energy states. As previously seen, the particles of a Fermi–Dirac gas cannot all collect in the lowest energy states as $T \rightarrow 0$ but are forced to form a "Fermi sphere" distribution because of the exclusion principles. Since this principle does not apply for Bose–Einstein particles, it is clear that they will tend to collect in the lowest energy state ($\varepsilon = 0$) as $T \rightarrow 0$. Since this state is one of essentially zero momentum, it is also clear that the pressure of the gas will tend to zero (as compared with the large Fermi–Dirac pressure at $T = 0$ K).

The fact is that for a gas of noninteracting bosons at a temperature below a certain transition temperature, a substantial fraction of the total number of particles in the system occupies the single orbital of lowest energy, which is called the *ground orbital.* Any other single orbital, including the orbital of second lowest energy, will at the same temperature be occupied by a relatively negligible number of particles. The total occupancy of all orbitals equals the specified number of particles in the system. The effect is called the *Einstein condensation* and is reminiscent of the condensation of a vapor.

Helium gas consists mainly of the isotope $^4_2$He, which has a resultant spin of zero and is a *boson*. The isotope $^3_2$He has a resultant spin of $\frac{1}{2}$ and is a *fermion*. Since $^3_2$He (a fermion) forms only 1.3 parts per million of naturally occurring helium, the properties of liquid helium are dominated by the properties of the isotope $^4_2$He. The critical temperature of helium is 5.2 K, so that helium cannot exist as a liquid above 5.2 K, whatever the external pressure. At normal atmospheric pressure (760 mm Hg), the boiling point of liquid $^4_2$He is 4.21 K. The interatomic (van der Waals) forces are too weak for liquid $^4_2$He to solidify at atmospheric pressure, even at the lowest temperatures that have been attained. Figure 4.16-1 shows the melting curve of liquid and solid helium, and the transition curve between the two forms of liquid helium, He I and He II. The liquid He II exhibits superfluid properties as a consequence of the condensation of atoms into the ground orbital of the system. At absolute zero and pressures below 25 atm, helium is a liquid. The liquid–vapor curve can follow from the critical point of 5.2 K down to absolute zero without any appearance of the solid. A temperature called the λ point is the triple point at which liquid He I, liquid He II, and vapor coexist.

When the temperature of liquid $^4_2$He is progressively reduced, there is an anomaly in the heat capacity at 2.17 K, as shown in Fig. 4.16-2. The sharp

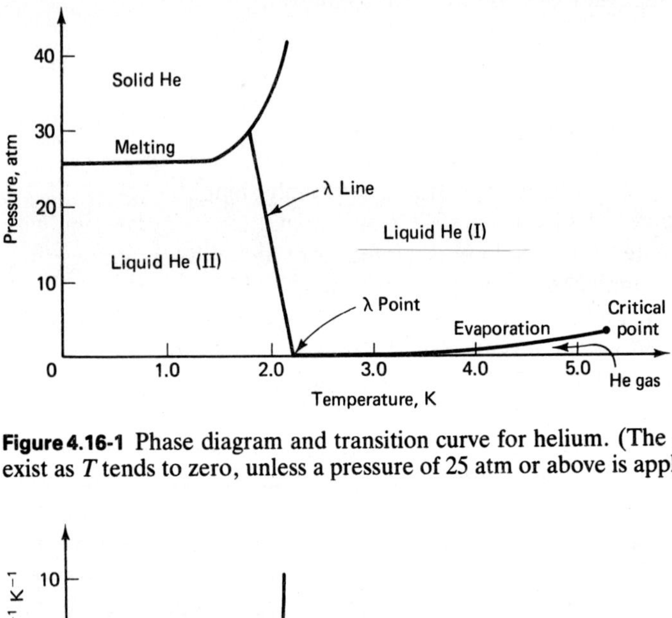

Figure 4.16-1 Phase diagram and transition curve for helium. (The solid phase cannot exist as T tends to zero, unless a pressure of 25 atm or above is applied.)

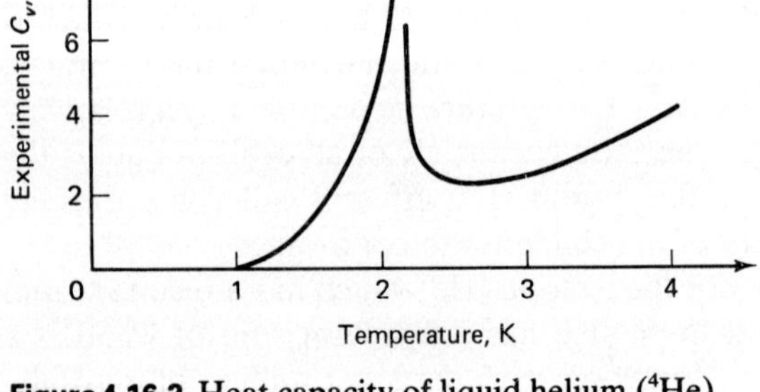

Figure 4.16-2 Heat capacity of liquid helium (^{4}He).

peak near 2.17 K is evidence of an important transition in the nature of the liquid. The viscosity of the liquid above the transition temperature is typical of normal liquids, whereas the viscosity below the transition (as determined by rate of flow through narrow slits) is vanishingly small, at least 10^6 times smaller than the viscosity above the transition. The transition is called a *lambda transition* because of the shape of the graph.

Liquid ^4_2He at temperatures above the λ point is liquid He I. Below the λ point it is He II. Apart from its very low mass density, liquid He I is a fairly normal liquid. On the other hand, liquid He II has some remarkable properties. For example, the viscosity of liquid He II, measured by the rate of flow of the liquid through a capillary tube, is effectively zero. Also, the thermal conductivity is very high. (Note that helium is a liquid at absolute zero at pressures below 25 atm.)

In the two-fluid theory of liquid He II, it is assumed that, at temperatures below the λ point, liquid He II is a mixture of two fluids, namely, a normal fluid, which behaves like a normal liquid having viscosity, and a superfluid, which has no viscosity and which gives rise to the superfluid properties of liquid He II. It is assumed that the proportion of the superfluid component increases from zero at the λ point to unity at the absolute zero of temperature.

The theoretical heat capacity curve, calculated from the theory is shown in Fig. 4.16-3. It exhibits a singularity, at T_0. The shape of the experimental C_V versus T curve must, of course, be influenced by the presence of intermolecular forces, which have been neglected in the treatment. Nevertheless, the transition at 2.19 K is undoubtedly closely related to the quantum effects described.

To summarize, ordinary helium gas is composed almost wholly of neutral atoms of the isotope ^4He. The spin angular momentum of such an atom is zero so that the Bose distribution is used to treat the behavior of this gas. At normal atmospheric pressure, helium gas condenses to a liquid at 4.18 K. It remains as

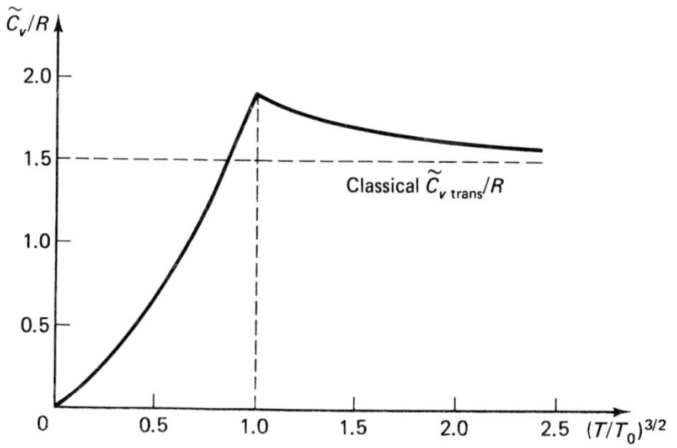

Figure 4.16-3 Theoretical heat capacity of a Bose–Einstein gas.

a liquid, that is, it does not freeze into a solid, down to the absolute zero of temperature if it is cooled at a pressure equal to its own vapor pressure. (To obtain solid helium it is necessary to pressurize the liquid, about 26 atm of pressure being needed near absolute zero.) If, by pumping off the vapor, the temperature of liquid helium is reduced to 2.18 K, a drastic change in its properties is observed. The λ point is at 2.18 K. Liquid helium is He I when it is above this temperature and He II when below. He I is essentially a classical fluid, its behavior not being unusual, but He II contains a superfluid component which causes it to show large-scale quantum effects, including the following.

1. As the temperature of liquid helium is lowered by evaporation and the vapor is pumped away, the liquid boils in the usual manner. But as the λ point is reached and passed, the boiling suddenly stops throughout the liquid. Though evaporation continues, and the temperature and vapor pressure fall, the liquid is completely calm. This is explained by the fact that heat can be conducted out of the liquid with practically no resistance, the heat conductivity being increased by a factor of about one million below the λ point.
2. The most spectacular and unusual property of liquid helium below the λ point is its ability to creep as a thin film along the walls of its container, as shown in Fig. 4.16-4. The speed of this ordered mass motion may be 30 cm or more per second. The effect involves helium first absorbing on the entire surface of the cold container to form a thin film. The film then acts like a siphon through which the liquid flows with almost no viscosity. The absence of viscosity means that, when flowing, He II does not develop the small-scale turbulences that cause the frictional energy loss responsible for the viscosity of ordinary fluids. This high degree of order is due to the $(1 + n)$ enhancement factor associated with the low-energy behavior of a system of bosons. All the helium atoms tend to condense into the same lowest-energy quantum state (the Bose condensation). The superfluid component, whose concentration rapidly approaches 100% as the temperature decreases below the λ point, is comprised of those atoms in that quantum

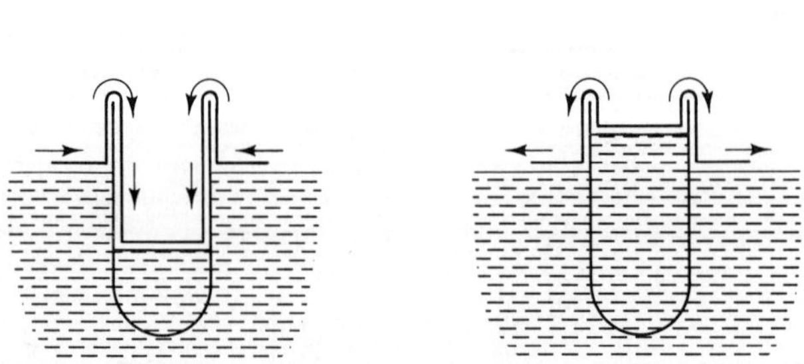

Figure 4.16-4 Superfluidity behavior of liquid He(II).

state. To the extent that all the atoms do get into the same microscopic state, it becomes the state of the entire macroscopic system and the system behaves in a completely ordered way. (The characteristic liquid flow through a siphon, is shown schematically for liquid levels in the container (a) below and (b) above the level of the liquid helium reservoir. He II creeps up along the inside wall, over the rim, and down along the outside wall as a thin film, collecting as a drop on the bottom. When this drop falls another will form, and so on, until the vessel is empty.)

■ EXAMPLE 4.16-1

Consider a system of six identical bosons of spin zero [meaning $s = 0$, $g = (2s + 1) = 1$]. Sketch the population distribution at $T = 0$, $0 < T < T_B$, and $T > T_B$, where T_B is the Bose temperature. What is the value of the chemical potential μ in each case?

Solution. At $T = 0$, all the bosons are in the single-particle ground state. When $T > 0$ but $T < T_B$, some of the bosons are in excited single-particle states, but a significant fraction of the bosons are still in the single-particle ground state. When $T > T_B$, there are very few, if any, bosons in the single-particle ground state. The case when the occupation number $b(\varepsilon) << 1$ for all the single-particle states corresponds to the classical limit.

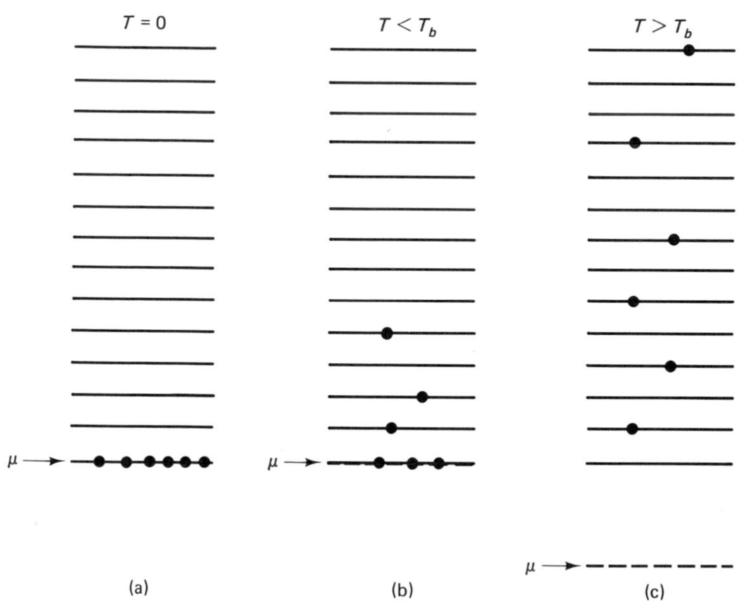

Figure 4.16-5 Distribution among single-particle states of a system of six bosons of spin zero.

The value of the chemical potential μ is as shown. At $T=0$, μ is equal to the energy of the lowest-energy single-particle state. (This energy has been chosen as the zero of the energy scale, in which case $\mu=0$ at $T=0$.) When $T>0$ but $T<T_B$, μ is only fractionally less than the energy of the lowest-energy single-particle state. As T increases above T_B, μ gets more and more negative. ■

■ EXAMPLE 4.16-2

Consider a gas composed of a fixed number, N, of bosons in a container of volume V, and let the Bose–Einstein distribution function be written

$$f_{BE}(\varepsilon)=\frac{1}{e^{\alpha}e^{\varepsilon/kT}-1} \tag{4.16-16}$$

(a) Show that α is a strictly increasing function of the temperature T.
(b) Evaluate the total energy of the boson gas.

Solution

(a) The density-of-states function is

$$g(\varepsilon)=\frac{2\pi V(2m)^{3/2}}{h^3}\,\varepsilon^{1/2} \tag{4.16-17}$$

Therefore, the number of bosons having an energy between ε and $\varepsilon+d\varepsilon$ is

$$\begin{aligned} n(\varepsilon)\,d\varepsilon &= g(\varepsilon)f_{BE}(\varepsilon)\,d\varepsilon \\ &= \frac{2\pi V(2m)^{3/2}\,\varepsilon^{1/2}}{h^3}\,d\varepsilon \end{aligned} \tag{4.16-18}$$

Normalization to N gives

$$N=\int n(\varepsilon)\,d\varepsilon=\frac{2\pi V(2m)^{3/2}}{h^3}\int_0^{\infty}\frac{\varepsilon^{1/2}\,d\varepsilon}{e^{\alpha}e^{\varepsilon/kT}-1}$$

which can be rewritten this in terms of the variable $x=\varepsilon/kT$.

$$N=\frac{2\pi V(2mkT)^{3/2}}{h^3}\int_0^{\infty}\frac{x^{1/2}\,dx}{e^{\alpha}e^{x}-1} \tag{4.16-19}$$

As T increases, the expression ahead of the integral increases. Therefore, since N is fixed, the integral decreases, which implies that α increases (as a function of T). Also, because N is finite, the integral must converge, so α has to be positive.

(b) The total energy of the boson gas is obtained from

$$E=\int \varepsilon\, n(\varepsilon)\,d\varepsilon=\frac{2\pi V(2m)^{3/2}}{h^2}\int_0^{\infty}\frac{\varepsilon^{3/2}\,d\varepsilon}{e^{\alpha}e^{\varepsilon/kT}-1} \tag{4.16-20}$$

Making the change of variable $x = \varepsilon/kT$, this gives

$$E = \frac{2\pi V(2m)^{3/2}(kT)^{5/2}}{h^3}\int_0^\infty \frac{x^{3/2}\,dx}{e^{\alpha}e^{x}-1}$$
$$= \frac{3V(2\pi mkT)^{3/2}kT}{2h^3}\sum_{n=1}^{\infty}\frac{e^{-n\alpha}}{n^{5/2}}$$
$$= \frac{3V(2\pi mkT)^{3/2}kT}{2h^3}\sum_{n=1}^{\infty}\frac{\varphi^{n}}{n^{5/2}} \qquad (4.16\text{-}21)$$

where $\varphi = e^{-\alpha}$. ■

4.17 BOSE–EINSTEIN CONDENSATION

The key to the Einstein condensation is the behavior of the chemical potential of a boson system at low temperatures. The chemical potential is responsible for the establishment of a large population of particles in the ground orbital.

Consider an ideal boson gas consisting of N identical bosons, each of mass m in a box of volume V at an absolute temperature T, immersed in a heat bath and particle reservoir. It is assumed that there are no interactions between the bosons and that the bosons occupy single-particle states (orbitals).

To bring out some results analytically, let the energy of the ground orbital chosen to be zero, and let the Bose–Einstein distribution function be written to express the number of bosons at energy ε and temperature T:

$$n(\varepsilon, T) = \frac{1}{e^{(\varepsilon-\mu)/kT}-1} \qquad (4.17\text{-}1)$$

Since there is no limit to the number of bosons that can be in a single-particle state, at the absolute zero of temperature, all the bosons will be in the single-particle ground state. However, it will be shown that a substantial fraction of bosons remain in the ground state at finite temperature.

When the temperature of the heat reservoir is raised above the absolute zero, some of the bosons occupy excited single-particle states, though at sufficiently low temperatures a significant proportion of the bosons may still be in the single-particle ground state. At high temperatures, the average number of bosons in any single-particle state is much less than unity, which is the classical limit.

At a temperature $T > 0$, the occupancy of the ground orbital at $\varepsilon = 0$ is

$$n(0, T) = \frac{1}{e^{-\mu/kT}-1}. \qquad (4.17\text{-}2)$$

When $T = 0$ the occupancy of the ground orbital is equal to the total number of particles in the system, so that

$$n(0,0) = N = \lim_{T\to 0}\frac{1}{e^{-\mu/kT}-1} = \lim_{T\to 0}\frac{1}{1-\dfrac{\mu}{kT}-1} \qquad (4.17\text{-}3)$$

where use has been made of the series expansion $e^{-x} = 1 - x + \cdots$. It is seen that x, which is μ/kT, must be small in comparison with unity, for otherwise the total number of particles N could not be large. From (4.17-3),

$$N = \frac{kT}{\mu}; \qquad \mu = -\frac{kT}{N} \tag{4.17-4}$$

as $T \to 0$. This result for μ is confirmed by forming the limit:

$$\lim_{T\to 0} n(0, T) = \frac{1}{e^{1/N} - 1} \cong \frac{1}{\left(1 + \frac{1}{N} + \cdots\right) - 1} \cong N \tag{4.17-5}$$

which is valid to high accuracy when $N >> 1$.

Consider now the energy spectrum of the ideal boson gas in a box. This is obtained by multiplying the density of states

$$g(\varepsilon)\, d\varepsilon = \frac{gV}{4\pi^2}\left(\frac{2m}{\hbar^2}\right)^{3/2} \varepsilon^{1/2}\, d\varepsilon \tag{4.17-6}$$

by the mean number of bosons in each of these states:

$$n(\varepsilon)\, d\varepsilon = \frac{gV}{4\pi^2}\left(\frac{2m}{\hbar^2}\right)^{3/2} \frac{\varepsilon^{1/2}\, d\varepsilon}{\{\exp[(\varepsilon - \mu)/kT] - 1\}} \tag{4.17-7}$$

This energy distribution differs from the Fermi–Dirac energy distribution, given by virtue of a negative rather than a postive sign inside the bracket in the denominator.

Integrating (4.17-7) for an ideal boson gas consisting of N bosons in a box gives

$$N = \frac{gV}{4\pi^2}\left(\frac{2m}{\hbar^2}\right)^{3/2} \int_0^\infty \frac{\varepsilon^{1/2}\, d\varepsilon}{\{\exp[(\varepsilon - \mu)/kT] - 1\}}. \tag{4.17-8}$$

This is an integral equation which can be solved for $\mu(T)$ by numerical methods.

Since the number of bosons in any single-particle state cannot be negative, the denominator in Eq. (4.17-8) must always be positive, so that, for an ideal boson gas, $\exp[(\varepsilon - \mu)/kT]$ must always be greater than unity. This means $\varepsilon > \mu$. The chemical potential of an ideal boson gas is thus always less than the energy of the lowest-energy single-particle state. Choosing the energy of the lowest-energy single-particle state (the single-particle ground state) as the zero of the energy scale, the chemical potential μ of the ideal boson gas is then as shown in Fig. 4.17-1.

At low temperature, when $T < T_B$, the chemical potential μ has a fairly constant numerical value just below zero. When T increases above T_B, the chemical potential becomes more and more negative. Changes in V and N of the system at a fixed value of T lead to a change in the value of μ through their effect on the value of T_B in Fig. 4.17-1.

Returning to the integration of Eq. (4.17-8), it should be pointed out that it holds only for $T > T_B$, but for $T < T_B$, it must be modified. As the tem-

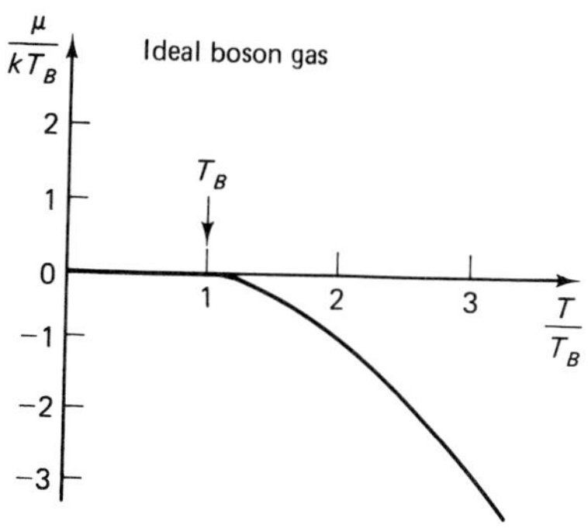

Figure 4.17-1 Chemical potential of ideal boson gas.

perature of the gas is lowered sufficiently, particles will begin to crowd into the single-particle ground with energy $\varepsilon = 0$. This state is completely ignored in Eq. (4.17-8). Because of the factor $\varepsilon^{1/2}$ in the integrand, it is given zero weight. At higher temperatures this introduces no error, but at very low temperatures, this state must not simply be omitted since it contains an appreciable number of particles. To explicitly retain the first term, which gives the number of particles in the ground state, and integrate only the remainder, Eq. (4.17-8) will be written as

$$N = \frac{1}{e^{-\mu/kT} - 1} + V\frac{2\pi(2m)^{3/2}}{h^3}\int_0^\infty \frac{\varepsilon^{1/2}\, d\varepsilon}{e^{(\varepsilon-\mu)/kT} - 1} \tag{4.17-9}$$

In this equation,

$$N_0 = \frac{1}{e^{-\mu/kT} - 1} \tag{4.17-10}$$

is the number of particles in the ground state with zero energy and zero momentum, while

$$N_{\varepsilon>0} = V\frac{2\pi(2m)^{3/2}}{h^3}\int_0^\infty \frac{\varepsilon^{1/2}\, d\varepsilon}{e^{(\varepsilon-\mu)kT} - 1} \tag{4.17-11}$$

is the number of particles whose energy and momentum is nonzero.

Above the temperature T_B, the number of particles in the ground state is negligible, the first term in Eq. (4.17-9) can be omitted, and the chemical potential is given by the original Eq. (4.17-8). Below the temperature T_B, the chemical potential stays extremely close to zero, and the number of particles with nonzero energy is given by Eq. (4.17-11) with $\mu = 0$. When the integral in this equation is evaluated by introducing a new variable $x = \varepsilon/kT$, the result is

$$N_{\varepsilon>0} = N\left(\frac{T}{T_c}\right)^{3/2} \tag{4.17-12}$$

This gives the fraction of particles $N_{\varepsilon>0}/N$ in states with energy $\varepsilon > 0$. The remaining particles, namely, those in the ground state with zero energy and zero momentum, are given by the fraction

$$\frac{N_0}{N} = 1 - \left(\frac{T}{T_B}\right)^{3/2} \tag{4.17-13}$$

It is seen from Eq. (4.17-13) that the number of atoms in excited orbitals varies as $T^{3/2}$ at temperatures below T_B. For $T > T_B$, the number of particles in the ground state is negligible. As T decreases below T_B, the number of such particles increases rapidly. These particles possess zero energy and zero momentum. Because of the latter, they contribute neither to the pressure nor do they possess viscosity (since viscosity is related to transport of momentum). Figure 4.17-2 shows all this.

The process of concentrating particles into the zero-energy ground state constitutes the *Bose–Einstein condensation.* It differs from the condensation of a vapor into a liquid in that no spatial separation into phases with different properties occurs. Nevertheless, there are similar features. For example, the pressure of a boson gas at $T < T_B$, just like the pressure of a saturated vapor, depends only on its temperature but not on its volume. A boson gas at temperatures below T_B is known as *degenerate*.

The speed distribution of a boson gas is shown in Fig. 4.17-3. When $T < T_B$, there is a thick line just to the right of the ordinate. This corresponds

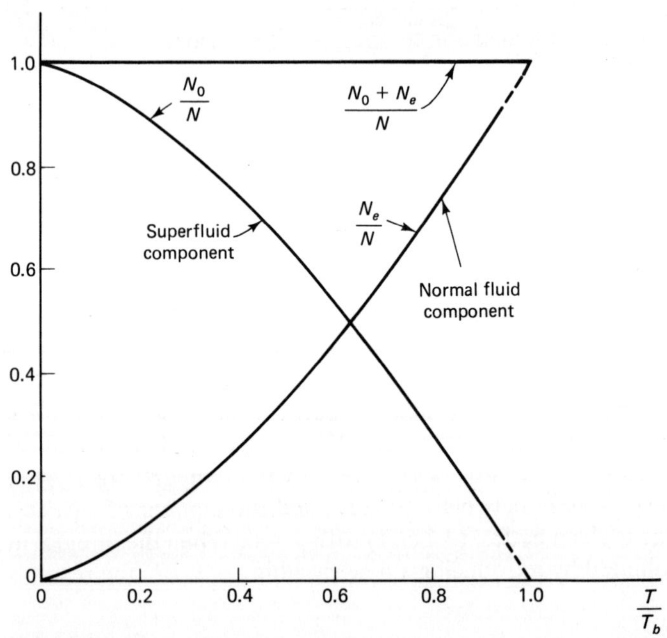

Figure 4.17-2 Condensed boson gas: fraction of particles in the zero-energy ground state.

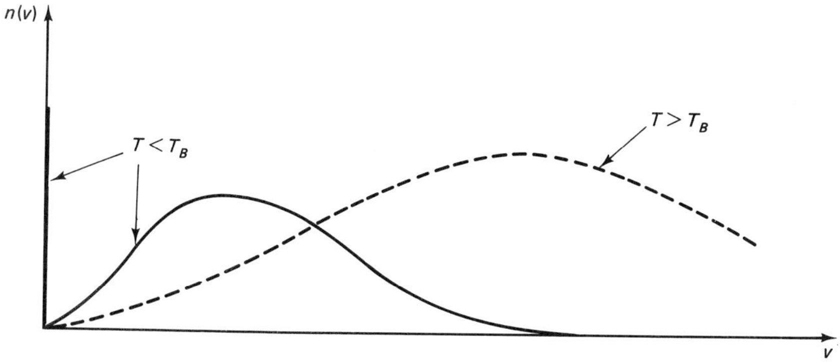

Figure 4.17-3 Speed distribution of boson gas at different temperatures.

to the bosons in the single-particle ground state. (According to the uncertainty principle, however, the speed of the bosons in the single-particle ground state cannot be zero.) The case when $T > T_B$ corresponds to the classical Maxwell–Boltzmann speed distribution.

■ EXAMPLE 4.17-1

Discuss the values of the Fermi temperature for metals and the Bose temperature for helium 4. What is the large difference in values due to?

Solution. The Einstein condensation temperature for bosons

$$T_B = \frac{2\pi\hbar^2}{mk}\left(\frac{N}{2.612\,gV}\right)^{2/3}$$

is compared to the Fermi temperature for metals

$$T_F = \frac{\hbar^2}{2mk}\left(\frac{6\pi^2 N}{gV}\right)^{2/3}$$

If the particle masses and concentrations are equal,

$$\frac{T_F}{T_B} = \frac{[3\pi^2(2.612)]^{2/3}}{4\pi} \simeq 1.45$$

However, the value of T_F for electrons in metals is around 5×10^5 K, while the calculated value of T_B for atoms ^{4}He is around 3 K. The large difference in values here is due to the large difference in masses. The behavior of a system below T_B or T_F is quite different for bosons and fermions, because of the differences in occupancies. The particles in the ground orbital below T_B form the condensed phase or superfluid phase. In liquid ^{4}He the experimental value of the transition temperature between the low-temperature (He II) phase which shows superfluid characteristics and the high-temperature phase (He I) which behaves as a normal fluid is 2.17 K. ■

4.18 ENERGY AND SPECIFIC HEAT OF BOSON GAS

To determine the thermodynamic properties of the ideal boson gas, consider N noninteracting identical bosons inside a box.

The expression for the grand potential Ω follows from relations developed earlier:

$$pV = kT \ln \Xi = -\Omega \tag{4.18-1}$$

$$p = \frac{2}{3}\frac{U}{V} \tag{4.18-2}$$

$$U = \int_0^\infty \varepsilon\, n(\varepsilon)\, d\varepsilon = \frac{gV}{4\pi^2}\left(\frac{2m}{\hbar^2}\right)^{3/2} \int_0^\infty \frac{\varepsilon^{3/2}\, d\varepsilon}{e^{(\varepsilon-\mu)/kT} - 1} \tag{4.18-3}$$

For $T < T_B$ and $\mu = 0$, Eq. (4.18-3) can be integrated by changing the variable of integration to $x = \varepsilon/kT$. Evaluating the integral and substituting for V, the result is

$$U = 0.77\, Nk \frac{T^{5/2}}{T_B^{3/2}} \qquad T < T_B \tag{4.18-4}$$

whence the variation of U with T is

$$C_V = 1.93\, R \left(\frac{T}{T_B}\right)^{3/2} \qquad T < T_B \tag{4.18-5}$$

The relation of U and T, at constant V, is shown in Fig. 4.18-1(a). According to Chapter 3, U is equal to $3NkT/2$ for an ideal monatomic gas in the classical limit. This relation is shown by the dotted line. It can be seen that the total energy of an ideal boson gas is always less than the energy an ideal monatomic classical gas would have at the same values of V, T, and N. Also, in

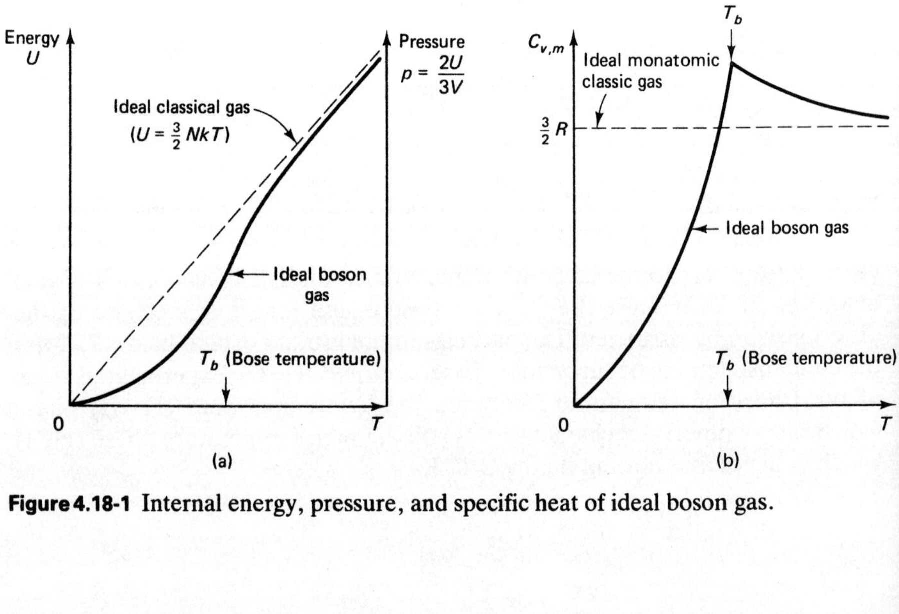

Figure 4.18-1 Internal energy, pressure, and specific heat of ideal boson gas.

the high-temperature limit, the energy of the ideal boson gas approaches that of the ideal monatomic classical gas.

When $T = 0$, all the bosons are in the single-particle ground state, which is chosen as the zero of the energy scale. Hence, for an ideal boson gas, $U = 0$ at $T = 0$, as shown in Fig. 4.18-1(a). Contrast this with the behavior of the ideal fermion gas illustrated in Fig. 4.13-1(a). At $T = 0$ the fermions fill all the single-particle states up to the Fermi energy, and U is not zero at $T = 0$.

According to Eq. (4.13-9), the pressure p of the ideal boson gas is equal to $2U/3V$, so that the variation of p with T at constant volume has the same form as the variation of U and T. It can be seen from Fig. 4.18-1(a) that the pressure of an ideal boson gas is always less than the pressure of an ideal monatomic classical gas at the same values of V, T, and N.

The specific heat exceeds the classical value $3R/2$ at the condensation temperature. At high temperatures the classical result of course holds. The variation of the molar specific heat, $C_{V,m}$, with T is shown in Fig. 4.18-1(b). The curve has a cusp at the Bose temperature. The experimental curve was previously shown in Fig. 4.16-2. The difference between the two curves is ascribed to the effect of the interactions between the atoms.

The internal energy, the heat capacity, and the entropy of a Boson gas are given by

$$U = 0.770\, NkT \left(\frac{T}{T_B}\right)^{3/2} \qquad (T < T_B)$$

$$C_V = \left(\frac{\partial U}{\partial T}\right)_V = 1.93\, Nk \left(\frac{T}{T_B}\right)^{3/2} \qquad (T < T_B)$$

$$S = \left(\frac{\partial \Omega}{\partial T}\right)_{\mu, V} = 1.28\, Nk \left(\frac{T}{T_B}\right)^{3/2} \qquad (T < T_B)$$

where T_B is the Bose temperature.

The chemical potential μ of a boson gas is always negative. It is virtually zero up to the Bose temperature. As T increases above T_B, μ gets more and more negative.

If the energy eigenvalue of the single-particle ground state is chosen as the zero of the energy scale, the chemical potential μ of an ideal boson gas is very close to zero for all temperatures below the Bose temperature. In this temperature range, putting $\mu = 0$ and making the substitution $x = \varepsilon/kT$, the expression for the total energy U of the ideal boson gas,

$$U = \int_0^\infty \varepsilon n(\varepsilon)\, d\varepsilon = \frac{gV}{4\pi^2}\left(\frac{2m}{\hbar^2}\right)^{3/2} \int_0^\infty \frac{\varepsilon^{3/2}\, d\varepsilon}{\exp[(\varepsilon - \mu)/kT] - 1} \tag{4.18-6}$$

approximates to

$$U = \frac{gV}{4\pi^2}\left(\frac{2m}{\hbar^2}\right)^{3/2} (kT)^{5/2} \int_0^\infty \frac{x^{3/2}\, dx}{e^x - 1} \tag{4.18-7}$$

The definite integral in (4.18-7) is equal to the product of the gamma function

$\Gamma(\frac{5}{2}) = 3\pi^{1/2}/4$ and the Riemann zeta function $\zeta(\frac{5}{2}) = 1.34$. Using the expression for the Bose temperature,

$$T_B = \frac{2\pi\hbar^2}{mk}\left(\frac{N}{2.612\,gV}\right)^{2/3} = \frac{3.31\hbar^2}{mk}\left(\frac{N}{gV}\right)^{2/3} \tag{4.18-8}$$

it is finally found that

$$U = 0.770\,NkT\left(\frac{T}{T_B}\right)^{3/2} \tag{4.18-9}$$

The heat capacity for $T < T_B$ is

$$C_V = \left(\frac{\partial U}{\partial T}\right)_V = 1.93\,Nk\left(\frac{T}{T_B}\right)^{3/2} \qquad (T < T_B) \tag{4.18-10}$$

Notice that C_V is proportional to $T^{3/2}$.

As for the entropy,

$$S = -\left(\frac{\partial \Omega}{\partial T}\right)_{\mu, V} \tag{4.18-11}$$

where the grand potential Ω is defined in terms of the grand partition function Ξ by the equation

$$\Omega = -kT \ln \Xi \tag{4.18-12}$$

Furthermore,

$$pV = kT \log \Xi = -\Omega \tag{4.18-13}$$

so that for both an ideal fermion and boson gas,

$$pV = \tfrac{2}{3}\, U \tag{4.18-14}$$

Equations (4.18-9) and (4.18-13) give, for $T < T_B$,

$$pV = \frac{2}{3}\, U = 0.513\,NkT\left(\frac{T}{T_B}\right)^{3/2} = -\Omega \tag{4.18-15}$$

Thus, using (4.18-11) and (4.18-15),

$$S = -\left(\frac{\partial \Omega}{\partial T}\right)_{\mu, V} = 1.28\,Nk\left(\frac{T}{T_B}\right)^{3/2} \tag{4.18-16}$$

Notice that the entropy S is zero when $T = 0$, in agreement with the third law of thermodynamics.

4.19 PHOTON GAS

The term photon denotes a quantum of energy of the electromagnetic field. What is now viewed as an application of the boson distribution function

$$\langle n \rangle = \frac{1}{e^{(\varepsilon - \mu)/kT} - 1} \tag{4.19-1}$$

for the occupancy of an orbital of energy ε was originally made by Planck in 1900 to explain the experimentally obtained spectrum of blackbody radiation. This is the first work in which the concept of a quantum of energy was ever used.

Planck was concerned with the problem of thermal radiation, namely, the distribution in frequency of electromagnetic energy in thermal equilibrium with the walls of a cavity. In agreement with experiment, he found that

$$\langle n \rangle = \frac{1}{e^{\hbar\omega/kT} - 1} \tag{4.19-2}$$

where $\langle n \rangle$ is the number of photons in a cavity mode of angular frequency ω. ($\hbar\omega = \varepsilon$ is the energy of a mode or orbital when occupied by a single light quantum or photon.) This result follows from (4.19-1) if the chemical potential is set equal to zero. Equation (4.19-2) will be derived from first principles, as a special case of the ideal boson gas.

Thermal radiation is a simple and interesting example of a perfect quantal gas (photons do not interact with each other). All bodies emit electromagnetic radiation by virtue of their temperature but usually this radiation is not in thermal equilibrium. However, within an opaque enclosure whose walls are maintained at a uniform temperature T, radiation and walls reach thermal equilibrium, and in this state the radiation has quite definite properties.

To study this equilibrium radiation, a small hole is cut in the walls of the enclosure. Such a hole, if sufficiently small, will not disturb the equilibrium in the cavity and the emitted radiation will have the same properties as the cavity radiation (the radiation emitted by a perfectly blackbody at the same temperature T as the enclosure).

Historically, there are two distinct statistical approaches to cavity radiation, giving identical answers. The first method, due to Planck, considers the normal modes of electromagnetic waves in a cavity (a problem similar to that of the harmonics of a vibrating string). Planck modified the expression, given by the classical equipartition theorem, for the mean energy of each normal mode, and this led him to postulate the existence of energy quanta. The second approach treats radiation as a gas of radiation quanta, that is, of photons, obeying BE statistics, and it is this approach that is followed in statistical mechanics.

Consider the radiation inside a cavity, which is inside the heat reservoir shown in Fig. 4.19-1. Each normal mode (standing wave solution) of angular frequency ω is treated as a harmonic oscillator which can have the energy eigenvalues $n\hbar\omega$ where $n = 0, 1, 2, \ldots$.

A photon gas has three important characteristics: First, photons behave in many respects like particles of spin 1; hence, photons obey BE statistics. Second, photons do not interact with one another. (This follows, for example, from the linearity of Maxwell's equations.) Third, photons are continually absorbed and emitted by the walls of the cavity inside the heat reservoir, so that the total number of photons in the photon gas is not constant.

Consider as the system, a single particle with the properties of a harmonic

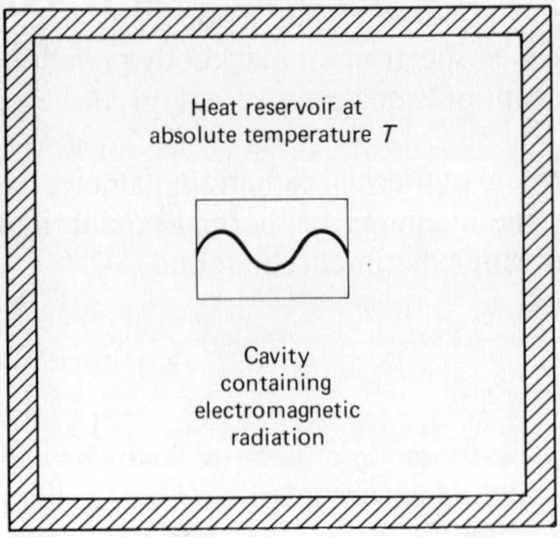

Figure 4.19-1 Radiation inside a cavity in equilibrium with a heat reservoir.

oscillator. Such an oscillator in quantum theory has orbitals of energy 0ε, ε, $2\varepsilon, \ldots, n\varepsilon, \ldots$. The oscillator is treated as a system in thermal contact with a reservoir, but the oscillator itself is not allowed to diffuse into the reservoir. An oscillator is equivalent to a mode of the electromagnetic field in the cavity. When the oscillator is in the orbital of energy $n\varepsilon = n\hbar\omega$, this is equivalent to n photons in the mode.

In the quantum theory of a harmonic oscillator it is shown that the ground-state energy is $\frac{1}{2}\hbar\omega$ above the energy of a classical harmonic oscillator at rest. The energy of the nth orbital of a quantum harmonic oscillator is $(n + \frac{1}{2})\,\varepsilon$, where $\frac{1}{2}\varepsilon$ is the zero-point energy of the oscillator. However, the zero-point energy will be omitted from the discussion. To do so will simplify the appearance of the equations without altering their thermodynamic consequences, for the entropy of a system of harmonic oscillators at a given temperature is not affected by the omission or inclusion of the zero-point energy.

When the oscillator is excited to occupy the orbital of energy $n\varepsilon_1$, the system is equivalent in every respect to a system with n photons in the state 1. A change in the orbital that is occupied in the oscillator is equivalent to a change in the number of photons in the real system. But the oscillator system is made up of one and only one oscillator. The oscillator is conserved, that is, the system is not allowed to exchange oscillators with the reservoir, but only to exchange energy. By this conservation device the chemical potential does not appear in the problem, and the partition function for the system of one oscillator is simply

$$Z = \sum_{n=0}^{\infty} e^{-n\varepsilon/kT} \tag{4.19-3}$$

where n is the number of photons having energy ε. The thermal average value of the occupancy is

$$\langle n \rangle = \frac{\Sigma n e^{-n\varepsilon/kT}}{\Sigma e^{-n\varepsilon/kT}} = \frac{\Sigma n x^n}{\Sigma x^n} \tag{4.19-4}$$

where $x \equiv e^{-\varepsilon/kT}$. Now, from the binomial theorem,

$$\Sigma x^n = (1 - x)^{-1} \tag{4.19-5}$$

Also,

$$x \left(\frac{d}{dx}\right) \Sigma x^n = x \Sigma n x^{(n-1)} = \Sigma n x^n \tag{4.19-6}$$

Hence,

$$\Sigma n x^n = x \left(\frac{d}{dx}\right)\left(\frac{1}{1-x}\right) = \frac{x}{(1-x)^2} \tag{4.19-7}$$

Using (4.19-5) and (4.19-7) Eq. (4.19-4) becomes

$$\begin{aligned} \langle n \rangle &= \frac{x}{1-x} = \frac{e^{-\varepsilon/kT}}{e^{\hbar\omega/kT} - 1} \\ &= \frac{1}{e^{\varepsilon/kT} - 1} = \frac{1}{e^{\hbar\omega/kT} - 1} \end{aligned} \tag{4.19-8}$$

where $\hbar\omega = \varepsilon$.

Equation (4.19-8) gives the thermal average number of photons in a mode of frequency ω. It is the *Planck distribution function,* and is plotted in Fig. 4.19-2 as a function of temperature. Note that the Bose–Einstein distribution (4.19-1) reduces to the Planck distribution on setting the chemical potential equal to zero. Note also that there are many more photons in a cavity when the walls are at a high temperature than when the walls are at a low temperature. In Fig. 4.19-2, $\langle n(\omega) \rangle$ is the thermal average of the number of photons in the mode of frequency ω. A plot of $\langle n(\omega) \rangle + \frac{1}{2}$ is also given, where $\frac{1}{2}$ is the effective zero-point occupancy of the mode. The dashed line is the classical limit.

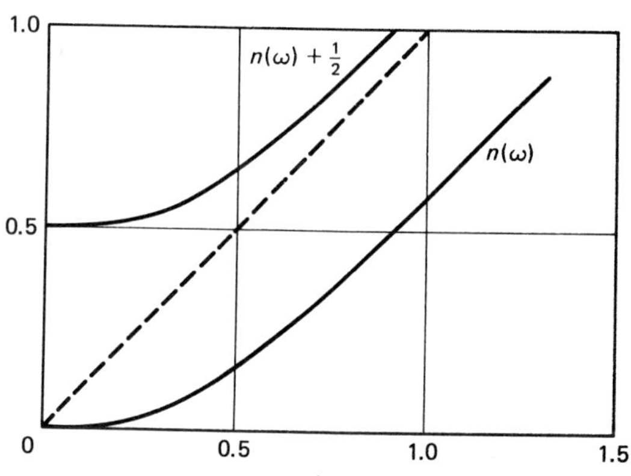

Figure 4.19-2 Planck distribution as function of reduced temperature.

■ EXAMPLE 4.19-1

Show that the density of photon states in a frequency interval $d\nu$ is

$$g(\nu) = \frac{8\pi V}{c^3} \nu^2$$

where V is the volume of the cavity.

Solution. Consider, for simplicity, a cubic cavity of side L whose edges define a set of axes and whose walls are maintained at a temperature T. Maxwell's equations for electromagnetic waves show that the rectangular components of the wave vector **k** must satisfy the boundary conditions

$$\frac{k_x L}{\pi} = n_x; \qquad \frac{k_y L}{\pi} = n_y; \qquad \frac{k_z L}{\pi} = n_z \tag{4.19-9}$$

where n_x, n_y, and n_z are positive integers. Each triplet (n_x, n_y, n_z) represents an electromagnetic mode of oscillation for the cavity. These modes are the photon states. Each allowed frequency ν corresponds to a certain mode (n_x, n_y, n_z) and can be written as

$$\nu = \frac{c}{\lambda} = \frac{ck}{2\pi} = \frac{c}{2L}(n_x^2 + n_y^2 + n_z^2)^{1/2} = \frac{cn}{2L} \tag{4.19-10}$$

where $n = (n_x^2 + n_y^2 + n_z^2)^{1/2}$.

The number of points in a spherical shell of radius N and thickness dN in the first octant of (n_x, n_y, n_z)-space is the volume

$$\frac{1}{8}(4\pi n^2\, dn) = \frac{\pi n^2}{2}\, dn \tag{4.19-11}$$

This is the number of states between n and $n + dn$, as well as the number of states between ν and $d\nu$. From $\nu = cn/2L$,

$$dn = \frac{2L}{c}\, d\nu \tag{4.19-12}$$

Thus, the number of states between ν and $\nu + d\nu$ is

$$\frac{\pi n^2}{2} = \left(\frac{\pi}{2}\right)\left(\frac{2L\nu}{c}\right)^2\left(\frac{2L}{c}\, d\nu\right) = \frac{4\pi L^3}{c^3} \nu^2\, d\nu$$

$$= \frac{4\pi V}{c^3} \nu^3\, d\nu \tag{4.19-13}$$

where $V = L^3$ is the volume of the cavity. Because there are two possible independent polarization directions for each mode, this result should be multiplied by two to obtain the number of photon states:

$$2\left(\frac{4\pi V}{c^3} \nu^2\right) d\nu = \frac{8\pi V}{c^3} \nu^2\, d\nu = g(\nu)\, d\nu \tag{4.19-14}$$

where

$$g(\nu) = \frac{8\pi V}{c^3} \nu^2 \tag{4.19-15}$$

is the density of states function for photons in the interval $d\nu$.

4.20 RADIATION PROPERTIES

The energy distribution for blackbody radiation is derived by treating the radiation as a photon gas.

A photon of energy ε has a (circular) frequency ω given by

$$\varepsilon = \hbar\omega \tag{4.20-1}$$

The momentum of a photon with this energy has magnitude

$$p = \frac{\hbar\omega}{c} \tag{4.20-2}$$

since photons have zero rest mass.

The total energy per unit volume of all the photons having energies between $\hbar\omega$ and $\hbar(\omega + d\omega)$ is equal to the number of single-particle states (normal modes) per unit volume in the angular frequency range ω to $(\omega + d\omega)$, times the mean number of photons in each single-particle state, times $\hbar\omega$, the energy of each photon. For a particle in an enclosure of volume V, the number of states in which the molecule has a momentum in the interval p to $(p + dp)$ is given by

$$f(p)\,dp = \frac{V 4\pi p^2\,dp}{h^3} = \frac{V 4\pi(\hbar\omega/c)^2\,dp}{h^3} \tag{4.20-3}$$

However, a photon of given momentum p can still exist in two different states of polarization (right- and left-circular polarizations, for example). The net effect is that the expression (4.20-3) for the density of states is multiplied by a factor 2. This gives for the number of photon states in which the photon has a frequency in the range ω to $\omega + d\omega$,

$$f(\omega)\,d\omega = 2\,\frac{V 4\pi(\hbar\omega/c)^2\hbar\,d\omega/c}{h^3}$$

$$= \frac{V\omega^2\,d\omega}{\pi^2 c^3} \tag{4.20-4}$$

Multiplying (4.20-4) by the average number of photons $\langle n \rangle$ in each single-particle state yields the number of photons in the frequency range ω to $\omega + d\omega$

$$dN(\omega) = \frac{V}{\pi^2 c^3} \frac{\omega^2\,d\omega}{\exp(\hbar\omega/kT) - 1} \tag{4.20-5}$$

Hence, for the energy of the radiation in this frequency range

$$\hbar\omega \, dN(\omega) = \frac{V\hbar}{\pi^2 c^3} \frac{\omega^3 \, d\omega}{\exp(\hbar\omega/kT) - 1} \tag{4.20-6}$$

so that the photon density (number of photons per unit volume) in the frequency range ω to $\omega + d\omega$ is

$$u(\omega, T) \, d\omega = \frac{\hbar\omega^3 \, d\omega}{\pi^2 c^3 [\exp(\hbar\omega/kT) - 1]} \tag{4.20-7}$$

This is *Planck's radiation law:* it gives the distribution of energy density as a function of frequency for radiation in thermal equilibrium. It is a function of only the temperature T of the system. Planck's law gives a complete description of blackbody radiation, in excellent agreement with observations. It can be expressed in alternative forms, at a given temperature, as

$$u(\nu) \, d\nu = \frac{8\pi h \nu^3 \, d\nu}{c^3[\exp(h\nu/kT) - 1]} \tag{4.20-8}$$

where $u(\nu) \, d\nu$ is the radiant energy per unit volume in the frequency range ν to $(\nu + d\nu)$, and

$$u(\lambda) \, d\lambda = \frac{8\pi \, d\lambda}{\lambda^5} \left\{ \frac{hc}{[\exp(hc/\lambda kT) - 1]} \right\} \tag{4.20-9}$$

where $u(\lambda) \, d\lambda$ is the radiant energy per unit volume in the wavelength range λ to $(\lambda + d\lambda)$. The variation of u with λ is shown in Fig. 4.20-1(a), and the variation of u with ν is shown in Fig. 4.20-1(b).

The $u(\lambda)$ curves show that it tends to zero as λ tends to zero and as λ tends to infinity. The area under the curves is proportional to T^4. The values of the wavelength λ_{max} at which u is a maximum lie on the rectangular hyperbola $\lambda_{max} T = \text{const}$. At 6000 K, the maximum of the distribution lies at the edge of the visible part of the spectrum. Six-thousand degrees Kelvin is the temperature of the surface of the sun (which approximates a blackbody fairly closely).

It is seen from Fig. 4.20-1(b) that the maximum of the distribution shifts to higher frequencies with increasing temperature. This maximum is obtained from Eq. (4.20-7) by applying the condition

$$\left[\frac{\partial u(\omega, T)}{\partial \omega} \right]_T = 0 \tag{4.20-10}$$

This differentiation leads to the transcendental equation

$$(3 - x) \, e^x = 3 \tag{4.20-11}$$

where $x \equiv \hbar\omega/kT_{max} \equiv h/kT\nu_{max}$. The solution of Eq. (4.20-11), which must be found graphically or numerically, is

$$\frac{\hbar\omega_{max}}{kT} = \frac{h\nu_{max}}{kT} = 2.822 \tag{4.20-12}$$

[It can be seen that Eq. (4.20-11) is nearly satisfied by $x = 3$.] Equation (4.10-12), known as *Wien's displacement law,* is well confirmed by experiment.

Interestingly, two separate distribution laws (prior to Planck's) can be recovered: In the long-wavelength region, as λ tends to infinity, corresponding to the extreme infrared, $(hc/\lambda kT)$ tends to zero and

$$\exp\left(\frac{hc}{\lambda kT}\right) - 1 \simeq 1 + \frac{hc}{\lambda kT} - 1 \simeq \frac{hc}{\lambda kT} \tag{4.20-13}$$

Hence, as λ tends to infinity, Eq. (4.20-9) becomes

$$u_\lambda \, d\lambda \simeq \left(8\pi \frac{d\lambda}{\lambda^4}\right) kT \tag{4.20-14}$$

This is the *Rayleigh–Jeans equation.* In its original theory it was assumed that the mean energy of each of the harmonic oscillators was equal to the classical value of kT, given by the equipartition theorem.

In the short-wavelength region, as λ tends to zero, corresponding to the extreme ultraviolet, $(hc/\lambda kT)$ tends to infinity, $\exp(hc/\lambda kT) >> 1$, and Eq. (4.20-9) becomes

$$u(\lambda)\, d\lambda \simeq \frac{8\pi hc}{\lambda^5} \exp\left(\frac{-hc}{\lambda kT}\right) d\lambda \tag{4.20-15}$$

This is *Wien's law.* As λ tends to zero, the exponential term dominates the right-hand side of (4.20-15) and $u(\lambda)$ tends to zero.

The total energy per unit volume in the cavity can be obtained by integrating Eq. (4.20-8) over all frequencies:

$$u = \int_0^\infty u(\nu)\, d\nu = \frac{8\pi h}{c^3} \int_0^\infty \frac{\nu^3 \, d\nu}{\exp(h\nu/kT) - 1} \tag{4.20-16}$$

Setting

$$x = \frac{h\nu}{kT} \tag{4.20-17}$$

so that

$$\nu^3 \, d\nu = \left(\frac{kT}{h}\right)^4 x^3 \, dx \tag{4.20-18}$$

this gives, at an absolute temperature T,

$$u = \frac{8\pi k^4 T^4}{h^3 c^3} \int_0^\infty \frac{x^3 \, dx}{\exp(x) - 1} \tag{4.20-19}$$

Since the definite integral in Eq. (4.20-19) is a number, it can be seen immediately that the total energy u per unit volume is proportional to T^4. The definite integral is a standard integral, which has the numerical value of $\pi^4/15$. Thus,

$$u = \left(\frac{8\pi^5 k^4}{15 h^3 c^3}\right) T^4 = aT^4 \tag{4.20-20}$$

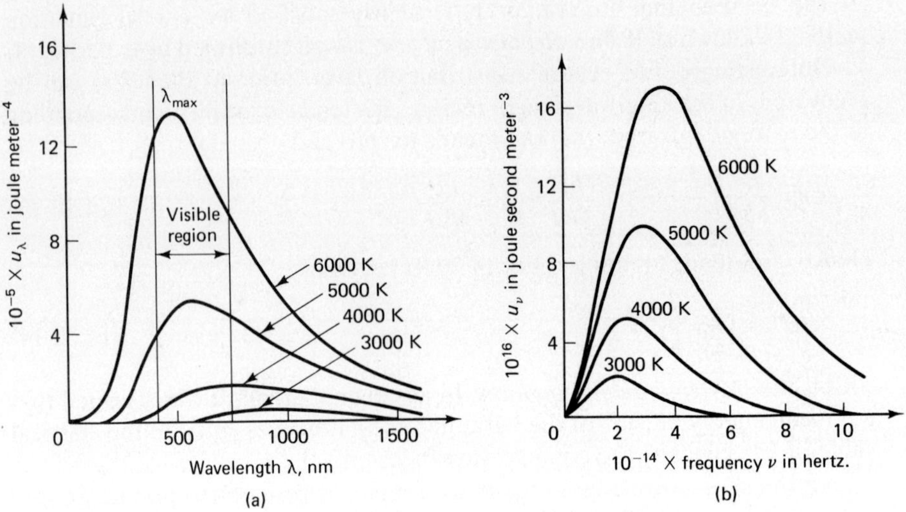

Figure 4.20-1 Planck's radiation law: (a) u_λ versus λ, (b) u_ν versus ν.

This is the *Stefan–Boltzmann law.* Substituting numerical values for k, h, and c, it is found that

$$u = 7.56 \times 10^{-16} T^4 \text{ J} \cdot \text{m}^{-3} \tag{4.20-21}$$

Equation (4.20-21) gives the total radiant energy per unit volume inside the cavity, which is surrounded by a heat reservoir of constant absolute temperature T. The total radiant energy is equal to the area under the curve of u against λ in Fig. 4.20-1(a). The higher the temperature of the enclosure the larger the area under the curve.

It is possible to relate the energy of radiation per unit volume, Eq. (4.20-20), which is always moving at the speed of light c, to the energy emitted per unit area per unit time. The result is *Stefan's law:*

$$e = (2\pi^5 k^4/15h^3c^2)\, T^4 = \sigma T^4 \tag{4.20-22}$$

Substituting numerical values for k, h, and c, it is found that

$$\sigma = 5.67 \times 10^{-8} \text{ W m}^{-2} \text{ K}^{-4} \tag{4.20-23}$$

This is in agreement with the experimental results. Also, substitution for h, c, and k in Wien's displacement law gives

$$\lambda_{max} T = 2.9 \times 10^{-3} \text{ m K} \tag{4.20-24}$$

■ EXAMPLE 4.20-1

(a) How many photons per second does a 100-W yellow lamp generate at $\lambda = 560$ nm (5.4×10^{14} Hz)?

(b) A small lamp bulb has an interior cavity of volume 1.0 cm^3 which is heated to 1500 K. Calculate the energy inside the cavity due to radiation in the wavelength range 550–575 nm.

Solution

(a) In one second, a 100-W yellow lamp (100 W = 100 J s^{-1}) generates

$$\frac{E}{h\nu} = \frac{(1s)(100 \text{ J s}^{-1})}{(6.626 \times 10^{-34} \text{ J s})(5.4 \times 10^{14} \text{ Hz})} = 2.8 \times 10^{20} \text{ photons}$$

(b) The energy density in the range λ to $\lambda + h\lambda$ is

$$\rho(\lambda) = \frac{8\pi hc}{\lambda^5}\left[\frac{c^{-hc/\lambda kT}}{1 - e^{-hc/\lambda kT}}\right]$$

calculated at the middle of the wavelength range (562.5 nm) times the range $\Delta\lambda = 25$ nm. Here,

$$\frac{8\pi hc}{\lambda^5} = \frac{8\pi(6.626 \times 10^{-34} \text{ J s})(2.998 \times 10^8 \text{ m s}^{-1})}{(5.625 \times 10^{-7} \text{ m})^5} = 8.866 \times 10^7 \text{ J m}^{-4}$$

$$c^{-hc/\lambda kT} = \exp\left[\frac{-(6.626 \times 10^{-34} \text{ J s})(2.998 \times 10^8 \text{ m s}^{-1})}{(5.625 \times 10^{-7} \text{ m})(1.381 \times 10^{-23} \text{ J K}^{-1})(1500 \text{ K})}\right]$$

$$= \exp(-17.05)$$

$$= 3.94 \times 10^{-8}$$

Thus,

$$\rho(562.5 \text{ nm}) = (8.866 \times 10^7 \text{ J m}^{-4})\left(\frac{3.94 \times 10^{-8}}{1 - 3.94 \times 10^{-8}}\right)$$

$$= 3.49 \text{ J m}^{-4}$$

Multiplying this by the range $\Delta\lambda = 25$ nm gives

$$(3.49 \text{ J m}^{-4})(25 \times 10^{-9} \text{ m}) = 8.7 \times 10^{-8} \text{ J m}^{-3}$$

The total energy due to the radiation in the range 550–575 nm in the 1.0 cm^3 cavity is therefore

$$(8.7 \times 10^{-8} \text{ J m}^{-3})(1 \times 10^{-6} \text{ m}^3) = 8.7 \times 10^{-14} \text{ J}$$

Note that each photon has energy hc/λ, and that there are about 2.5×10^5 yellow photons in the cavity. ■

■ **EXAMPLE 4.20-2**

The Sun can be treated as a blackbody of absolute temperature 5800 K. Its radius is 7.00×10^8 m and its distance from the Earth is 1.5×10^{11} m. Using Stefan's constant $\sigma = 5.67 \times 10^{-8}$ W m^{-2} K^{-4}, show that the total radiant energy emitted by the Sun is 3.95×10^{26} J/s and the energy reaching the top of the Earth's atmosphere in 1.40 kW/m^2. This is known as the *solar constant*.

Solution. Stefan's law says that the energy emitted per unit area per unit time is

$$c = \left(\frac{2\pi^5 k^4}{15\hbar c^2}\right) T^4 = \sigma T^4$$

where $\sigma = 5.67 \times 10^{-8}$ W m^{-2} K^{-4}. This is in agreement with experimental results. The surface area of the sun is $4\pi(7.00 \times 10^8 \text{ m})^2 = 615.4 \times 10^{16}$ m^2. Its temperature is 5800 K and so $T^4 = 1131.6 \times 10^{12}$ K^4. The total radiant energy emitted by the sun is

$$5.67 \times 10^{-8} \text{ W}(615.4 \times 10^{16})(11.31.6 \times 10^{12}) = 3.95 \times 10^{26} \text{ W}$$
$$= 3.95 \times 10^{26} \text{ J/s}$$

This radiation is spread over an area of $4\pi(1.5 \times 10^{11} \text{ m})^2 = 28.26 \times 10^{22}$ m^2. The rate at which energy is reaching the top of the Earth's atmosphere, or *solar constant,* is

$$\frac{3.95 \times 10^{26} \text{ W}}{28.26 \times 10^{22} \text{ m}^2} = \frac{1.4 \text{ kW}}{\text{m}^2}$$

■

4.21 THERMODYNAMICS OF BLACKBODY RADIATION

All thermodynamic properties of blackbody radiation follow from the partition function, but some properties follow more directly from the description of a perfect photon gas. For example, the equilibrium radiation in a cavity exerts a pressure on the walls of the enclosure given by

$$p = \frac{1}{3}u(T) \tag{4.21-1}$$

where

$$u(T) = \int_0^\infty u(\omega, T)\, d\omega = aT^4 \tag{4.21-2}$$

is the total energy density. This result follows directly from Maxwell's electromagnetic theory and from the description of a perfect photon gas. For a gas of particle of mass m, kinetic theory gives the pressure as

$$p = \frac{1}{3}\rho\langle v^2\rangle \tag{4.21-3}$$

where ρ is the density of the gas and $\langle v^2\rangle^{1/2}$ is the rms-speed of the molecules. For the photon gas, all the photons have the same velocity c, the velocity of light, so that ρc^2 can be interpreted, from Einstein's mass–energy relation, as the energy density $u(T)$. Hence, for the photon gas, Eq. (4.21-3) reduces to Eq. (4.21-1).

The partition function of a photon gas is

$$Z = \Sigma \exp[-(n_1\hbar\omega_1 + n_2\hbar\omega_2 + \cdots)/kT] \tag{4.21-4}$$

where $n_1, n_2, \ldots,$ are the numbers of photons in single-particle states having energy eigenvalues $\hbar\omega_1, \hbar\omega_2, \ldots,$ respectively. All the numbers $n_1, n_2, \ldots,$ in Eq. (4.21-4) can vary independently from 0 to ∞. Thus, the partition function can be written

$$Z = \left(\sum_{n_1=0}^{\infty} e^{-n_1\hbar\omega_1/kT} \right) \left(\sum_{n_c=0}^{\infty} e^{-n_2\hbar\omega_2/kT} \right) \cdots \tag{4.21-5}$$

Letting $x = \exp(-\hbar\omega_i/kT)$ and using the binomial theorem $\Sigma x^n = (1-x)^{-1}$, it is seen that

$$\sum_{n_i=0}^{\infty} \exp(-n_i\,\hbar\omega_i\,/kT) = [1 - \exp(-\hbar\omega_i/kT)]^{-1} \tag{4.21-6}$$

Substituting into Eq. (4.21-5) and taking logarithm of both sides gives

$$\ln Z = -\Sigma \ln [1 - \exp(-\hbar\omega_i/kT)] \tag{4.21-7}$$

Now, the density of states, or number of photon states in a volume V having angular frequencies in the range ω to $\omega + d\omega$ is

$$g(\omega)\,d\omega = \frac{V\omega^2\,d\omega}{\pi^2 c^3} \tag{4.21-8}$$

Multiplying the right-hand side of Eq. (4.21-7) by Eq. (4.21-8) and replacing the sum by an integral gives

$$\ln Z = \frac{V}{\pi^2 c^3} \int_0^{\infty} \omega^2 \ln[1 - \exp(-\hbar\omega/kT)]\,d\omega \tag{4.21-9}$$

Integrating by parts by letting du be equal to $\omega^2 d\omega$ and v be equal to ln $[1 - \exp(-\hbar\omega/kT)]$ in the equation $\int v\,du = uv - \int u\,dv$, and using the standard integral

$$\int_0^{\infty} \frac{x^3\,dx}{e^x - 1} = \frac{\pi^4}{15}$$

Eq. (4.21-9) becomes

$$\ln Z = \frac{\pi^2}{45}\left(\frac{kT}{\hbar c}\right)^3 V = \frac{8\pi^5}{45}\left(\frac{kT}{hc}\right)^3 V \tag{4.21-10}$$

Knowing the partition function, all the thermodynamic functions fall into place. For example, the chemical potential

$$\mu = -kT\left(\frac{\partial \ln Z}{\partial N}\right)_{V,T} = 0 \tag{4.21-11}$$

turns out to be zero for a photon gas, since Z in Eq. (4.21-10) is independent of N. The internal energy,

$$U = kT^2\left(\frac{\partial \ln Z}{\partial T}\right)_{V,N} \tag{4.21-12}$$

for a photon gas, using Eq. (4.21-10), is

$$U = \frac{8\pi^5 k^4 T^4 V}{15h^3c^3} \tag{4.21-13}$$

Since U/V is the energy density u, Eq. (4.21-13) is equal to the previously listed equation for the energy density

$$u = \left(\frac{8\pi^5 k^4}{15h^3c^3}\right) T^4 = aT^4 \tag{4.21-14}$$

As for the pressure, it is

$$p = kT \left(\frac{\partial \ln Z}{\partial V}\right)_{T,N} = \frac{8\pi^5 k^4 T^4}{45h^3c^3} \tag{4.21-15}$$

Combining Eqs. (4.21-13) and (4.21-15) gives

$$pV = \frac{1}{3} V \tag{4.21-16a}$$

or

$$p = \frac{1}{3} u \tag{4.21-16b}$$

thereby reproducing earlier results.

■ EXAMPLE 4.21-1

Show that for an isentropic expansion of a photon gas, the relations between T and V, and between u and V are, respectively,

$$T^3 V = \text{const} \qquad uV^{4/3} = \text{const} \tag{4.21-17}$$

Solution. The entropy in terms of the partition function is

$$s = k \ln Z + \frac{V}{T} \tag{4.21-18}$$

Using Eqs. (4.21-10) and (4.21-13), this gives, for the entropy of a photon gas,

$$S = \frac{32\pi^5 k^4 T^3 V}{45h^3c^3} \tag{4.21-19}$$

For an isentropic expansion, s remains constant, so that

$$T^3 V = \text{const} \tag{4.21-20}$$

Using Eq. (4.21-13) to eliminate T from Eq. (4.21-20) results in

$$uV^{4/3} = \text{const}$$

where $u = U/V$ is the energy density of the photon gas. ■

4.22 PHONONS IN SOLIDS

The quantum of energy in an elastic wave is called a *phonon*. Just as the energy of an electromagnetic wave in a cavity is quantized, so the energy of an elastic wave in a solid is quantized. Phonons, like photons obey BE statistics, but they are not conserved, that is, they can be emitted or absorbed. Most of the previous discussions have considered systems where the molecular spacing is very large compared with the size of the molecules. Because of this wide spacing, it was reasonable to assume that the molecules were free to move as independent entities, with their energies essentially unaffected by the force fields of other molecules.

The solid state represents a different circumstance. In this instance the atoms are so closely spaced that their force fields essentially lock them in place, and they are only free to oscillate about some equilibrium position. One of the problems of a statistical analysis is to describe satisfactorily the contribution of this oscillatory motion to the internal energy and specific heat of the solid. Classical statistical mechanics cannot predict many of the observed values of specific heats of gases, particularly at low temperatures. A similar difficulty arises in predicting the heat capacities of solids. Not only do the heat capacities differ from the classical equipartition results, even at fairly high temperatures (200 K), but the fact that the electrons fail to contribute their share to the heat capacities also is inexplicable in classical theory.

One model of a solid is to view it as a collection of $3N$ harmonic oscillators, or waves (if there are N atoms). The energy of each oscillator is, according to quantum mechanics, $\varepsilon_i = (n_i + \frac{1}{2})\, h\nu_i$, where $n_i = 0, 1, 2, \ldots$, and ν_i characterizes the oscillator. The total energy of the solid is then

$$E = \sum_{i=1}^{3N} \left(n_i + \frac{1}{2}\right) h\nu_i \tag{4.21-1}$$

and the partition function for this system is

$$Z = \sum_{\text{states}} e^{-\beta E} = \sum e^{-\beta(n_i\varepsilon_i + n_2\varepsilon_2 + \cdots)} \tag{4.22-2}$$

where $n_1, n_2, \ldots$, are the numbers of photons in single-particle states which have energy eigenvalues $\varepsilon_1, \varepsilon_2, \ldots$, respectively, and β rather than $1/kT$ has been used to shorten the writing somewhat. All the numbers $n_1, n_2, \ldots$, in Eq. (4.21-2) can vary independently from 0 to ∞. Hence, the partition function can be factorized:

$$\begin{aligned} Z &= \sum_{n_i=0}^{\infty} \varepsilon^{-\beta n_1 \varepsilon_1} \sum_{n_2=0}^{\infty} \varepsilon^{-\beta n_2 \varepsilon_2 \cdots} \\ &= \sum_{n_1=0}^{\infty} \exp\left[-\beta\left(n_1 + \frac{1}{2}\right)\right] h\nu_1 \sum_{n_2=0}^{\infty} \exp\left[-\beta\left(n_2 + \frac{1}{2}\right)\right] h\nu_2 \cdots \\ &\left\{\exp\left[-\beta\left(0 + \frac{1}{2}\right)\right] h\nu_1 + \exp\left[-\beta\left(1 + \frac{1}{2}\right)\right] h\nu_1 + \cdots\right\} \\ &\left\{\exp\left[-\beta\left(0 + \frac{1}{2}\right)\right] h\nu_2 + \exp\left[-\beta\left(1 + \frac{1}{2}\right)\right] h\nu_2 + \cdots\right\} \end{aligned} \tag{4.22-3}$$

Putting $\chi = \varepsilon^{-\beta h\nu}$ and using the fact that

$$\sum_{n=0}^{\infty} \chi^n = \frac{1}{1-\chi}, \tag{4.22-4}$$

Equation (4.22-3) gives

$$Z = \left(\frac{\varepsilon^{-\beta h\nu_1/2}}{1-\varepsilon^{-\beta h\nu_1}}\right)\left(\frac{\varepsilon^{-\beta h\nu_2/2}}{1-\varepsilon^{-\beta h\nu_2}}\right)\cdots \equiv \prod_{i=1}^{3N}\left(\frac{\varepsilon^{-\beta h\nu_i/2}}{1-\varepsilon^{-\beta h\nu i}}\right) \tag{4.22-5}$$

The logarithm of Z is then

$$\ln Z = \sum_{i=1}^{3N} -\frac{\beta h\nu i}{2} - \ln(1-\varepsilon^{-\beta h\nu i}) \tag{4.22-6}$$

It remains to assign values to the frequencies ν_i. There are two basic ways of doing this. In the *Einstein model* (1907), all ν_i are taken equal to a common value ν (the value of which depends on the solid in question). In this case

$$\ln Z = \frac{3N\beta h\nu}{2} - 3N\ln(1-\varepsilon^{-1\beta h\nu}) \qquad \text{(Einstein model)} \tag{4.22-7}$$

Accordingly, the internal energy is

$$\begin{aligned} U &= -\frac{\partial \ln Z}{\partial\beta} = \frac{3}{2}Nh\nu + \frac{3Nh\nu}{\varepsilon^{\beta h\nu}-1} \\ &= \frac{3}{2}NkT\frac{\Theta_E}{T} + 3NkT\frac{\Theta_E/T}{\varepsilon^{\Theta E/T}-1} \end{aligned} \tag{4.22-8}$$

where

$$\theta_E = \frac{h\nu}{k} \tag{4.22-9}$$

is known as the *Einstein temperature* of the solid.

The first term in (4.22-8) is just a constant (independent of T) and represents the contribution of the zero-point energy of all the oscillators. It has no physical (i.e., measurable) significance. The molal constant value specific heat is, thus,

$$\overline{C}_V = 3R\left(\frac{\Theta_E}{T}\right)^2 \frac{\varepsilon^{\Theta_E/T}}{(\varepsilon^{\theta_E/T}-1)^2} \tag{4.21-10}$$

This relation for specific heat is plotted in Fig. 4.22-1 as a function of T/θ_E. At very high temperatures the specific heat approaches a constant value of $3R$. This is exactly what would be expected from a simple equipartition of energy appraisal of the problem. (The atoms have three directions of motion, and there are two vibrational energy modes for each coordinate, so that a total of six degrees of freedom are expected. Each degree of freedom contributes $\frac{1}{2}R$ to the specific heat; thus the total of $3R$ is quite reasonable.) The shape of the curve is in general agreement with experimental findings for specific heats of solids but predicts values for $\langle C_V \rangle$ which are too low at low temperatures.

The main drawback in the Einstein analysis is that all atoms in a solid

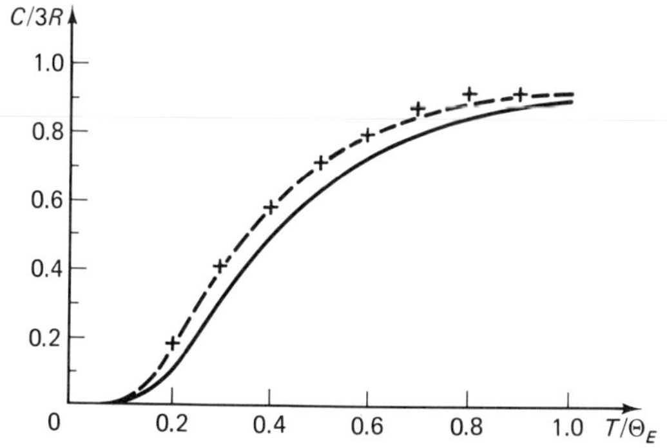

Figure 4.22-1 Specific heat of a solid according to Einstein theory. Comparison of the observed molar heat capacity of diamond, + experimental points, $\theta_E = 1325\ K$ with Einstein's model (full line). (After Einstein's original 1907 paper.)

simply do not oscillate with the same frequency as is assumed for the calculation. A more accurate model was proposed by *Debye (1912),* who assumed a continuous distribution of frequencies which could be computed from elastic wave theory for isotropic media. In this analysis, a crystalline solid is viewed as being composed of regularly spaced atoms in a three-dimensional lattice. If an atom is displaced from its equilibrium position, it will experience a restoring force due to the surrounding atoms, the atoms behaving like a collection of coupled oscillators. Any disturbance will be transmitted to the surrounding atoms, resulting in a wave propagating through the solid.

Waves propagating in the lattice can be either transverse or longitudinal, with velocities v_t and v_l, respectively. A transverse wave has 2 degrees of vibrational freedom, while a longitudinal wave has only 1 degree of freedom. Each vibrational degree of freedom (mode of vibration) of the crystal corresponds to a state of the system, and the phonons are distributed among these states according to the Bose–Einstein law. There is a maximum frequency of vibration, called the *Debye frequency,* ν_D. The maximum frequency exists because a system of N molecules possesses only $3N$ modes of vibration (each molecule having 3 degrees of independent vibrational freedom). The specific heat of a crystalline solid is obtained by first finding the Debye frequency and the density of states for the crystal, and then using this information to find the vibrational kinetic energy and molar specific heat.

To find the number of vibrational states with frequencies in the interval between ν and $(\nu + d\nu)$, consider the sound waves traveling in a solid to be exactly analogous to the blackbody oscillations of the phonons in a cavity. By this reasoning, the number dM modes of vibration with frequencies in the interval between ν and $(\nu + d\nu)$ is

$$dM = \frac{4\pi V}{c^3}\nu^2\, d\nu \tag{4.22-11}$$

for each type of vibration, with c being the velocity of the wave propagating in the medium. Taking into account the 2 degrees of vibrational freedom for a transverse wave and the single degree of vibrational freedom for a longitudinal wave, the number of vibrational states, dS, with frequencies in the interval between ν and $(\nu + d\nu)$ is

$$dM = 4\pi V\left(\frac{2}{v_t^3} + \frac{1}{v_l^3}\right)\nu^2\,d\nu \tag{4.22-12}$$

Thus, the density-of-states function in the frequency interval $d\nu$ is

$$g(\nu) = 4\pi V\left(\frac{2}{v_t^3} + \frac{1}{v_l^3}\right)\nu^2 \tag{4.22-13}$$

This will be rewritten in terms of the Debye frequency ν_D, which is the maximum possible vibrational frequency in a solid composed of N molecules. Since the number of possible states is finite and equal to $3N$,

$$3N = \int dS = \int_0^{\nu_d} g(\nu)\,d\nu = 4\pi V\left(\frac{2}{v_t^3} + \frac{1}{v_l^3}\right)\int_0^{\nu_d}\nu^2\,d\nu = 4\pi V\left(\frac{2}{v_t^3} + \frac{1}{v_l^3}\right)\frac{\nu_D^3}{3} \tag{4.22-14}$$

Thus, the density of states in the frequency interval $d\nu$ is

$$g(\nu) = 4\pi V\left(\frac{2}{v_t^3} + \frac{1}{v_l^3}\right)\nu^2 = \frac{9N}{\nu_D^3}\nu^2 \tag{4.22-15}$$

Now, assuming that acoustical energy is transferred through a crystal lattice in quantized amounts of $h\nu$ by the phonons, which are bosons whose total number is not fixed, an expression for the total kinetic energy (vibrational energy) of the crystalline solid can be obtained.

The number of phonons $dn(\nu)$ with frequencies in the range between ν and $(\nu + d\nu)$ is

$$dn(\nu) = F_{BE}\,g(\nu)\,d\nu = \frac{1}{e^{E/kT} - 1}\,\frac{9N}{\nu_d^3}\,\nu^2\,d\nu \tag{4.22-16}$$

where $g(\nu)$ is given by (4.22-15). Since each phonon has an energy $E = h\nu$, the energy dE possessed by these $dn(\nu)$ phonons is

$$dE = h\nu\,dn(\nu) = \frac{9Nh}{\nu_d^3}\,\frac{\nu^3}{e^{h\nu/kT} - 1}\,d\nu \tag{4.22-17}$$

The total energy E_T of the solid will be the sum of all the phonon energies:

$$E_T = \int dE = \frac{9Nh}{\nu_D^3}\int_0^{\nu_D}\frac{\nu^3}{e^{h\nu/kT} - 1}\,d\nu = 9NkT\left(\frac{T}{T_D}\right)^3\int_0^{q_D}\frac{q^3\,dq}{e^q - 1} \tag{4.22-18}$$

where $q_D = h\nu_D/kT$ and

$$T_D = \frac{h\nu_D}{k} \tag{4.22-19}$$

is called the *Debye temperature*. The integral in (4.22-18) cannot be evaluated in terms of the more common functions. The most important features are its limiting values.

An expression for C_V, the molar specific heat at constant volume, in the limit $T << T_D$, can be obtained from (4.21-18), where $q_D = h\nu_D/kT = T_D/T$. For $T << T_D$, $q_D \to \infty$, and the value of the integral approaches $\pi^4/15$. This comes from the fact that

$$\int_0^\infty \frac{q^p}{e^q - 1}\, dq = \Gamma(p+1)\,\zeta(p+1) \tag{4.22-20}$$

where $\Gamma(x)$ is the gamma function, obeying $\Gamma(x+1) = x\Gamma(x)$ and having particular values of

$$\Gamma(1) = 1; \qquad \Gamma\left(\frac{1}{2}\right) = \sqrt{\pi}; \qquad \Gamma(n+1) = n! \tag{4.22-21}$$

and $\zeta(x)$ is the Riemann zeta function

$$\zeta(x) = \frac{1}{1^x} + \frac{1}{2^x} + \frac{1}{3^x} + \cdots \tag{4.22-22}$$

having particular values of

$$\zeta(2) = \frac{\pi^2}{6}; \qquad \zeta(4) = \frac{\pi^4}{90}; \qquad \zeta(6) = \frac{\pi^6}{945} \tag{4.22-23}$$

Thus, with $p = 3$, Eq. (4.22-18) gives

$$E_T = 9NkT\left(\frac{T}{T_D}\right)^3\left(\frac{\pi^4}{15}\right) \tag{4.22-24}$$

and, from the definition of the specific heat,

$$C_V = \frac{1}{n}\left(\frac{\partial E_T}{\partial T}\right)_V = \frac{36kN}{n}\frac{\pi^4}{15}\left(\frac{T}{T_D}\right)^3 = \frac{12\pi^4 R}{5}\left(\frac{T}{T_D}\right)^3 \tag{4.22-25}$$

where n is the number of moles, and R is the gas constant per mole. Experimentally, it is found that at low temperature, C_V does vary as T^3, as shown in Fig. 4.22-2.

On the other hand, for $T >> T_D$, $q_D = T_D/T \to 0$, so that in (4.22-18),

$$e^q \approx 1 + q \tag{4.22-26}$$

over the entire range of integration. Thus,

$$E_T \approx 9NkT\left(\frac{T}{T_D}\right)^3 \int_0^{T_D/T} q^2\, dq = 3NkT \tag{4.22-27}$$

and, from the definition of C_V,

$$C_V = \frac{1}{N}\left(\frac{\partial E_T}{\partial T}\right)_V = 3\frac{N}{n}k = 3R \tag{4.22-28}$$

This value, which holds only at high temperatures, is the *Dulong–Petit law*. The same value for C_V is also predicted by the Einstein model, Eq. (4.22-10).

As would be expected, the quantum mechanical models differ from the classical results only at "low" temperatures. Both the Einstein and Debye models predict that the heat capacity of a solid should go to zero as T goes to

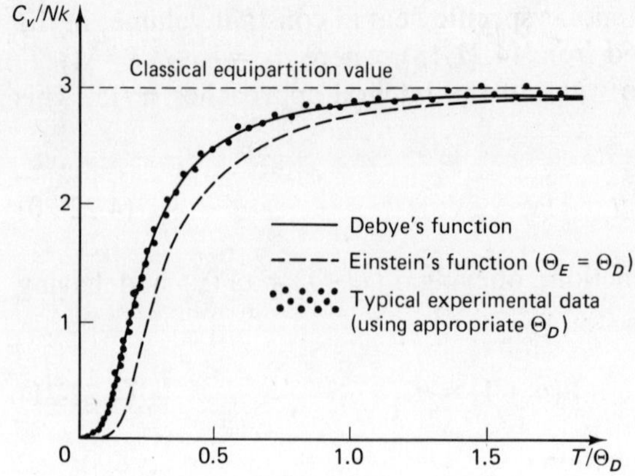

Figure 4.22-2 Specific heat curves of solids for Einstein and Debye models.

zero. The limiting forms for C_V are entirely different functions of the temperature. The Einstein model predicts much lower values of C_V as T goes to zero than does the Debye model. The more realistic Debye model predicts that C_V should go to zero proportional to T^3—the so-called Debye T^3 law. This is illustrated in Fig. 4.22-2, where the points represent typical experimental values of C_V. The points can be fitted to the Debye curve by making a suitable selection for the value of Θ_D. Since the scattering of the experimental data is greater at higher temperatures, Θ_D is usually picked to yield agreement with the experiments in the low-temperature region, where Debye's T^3 law applies. The important thing to note is that these points cannot be fitted to the Einstein model or any value of Θ_E. By way of comparison, the Einstein curve, Eq. (4.22-10), is also shown in Fig. 4.22-2, where Θ_E has been set equal to Θ_D.

It should be mentioned that, in addition to the transfer of energy by atomic vibrations, if the crystalline solid is a conductor, there will also be energy transfer by the free conduction electrons. The total specific heat will then be determined by the sum of the electronic specific heat and the lattice specific heat. However, the electronic contribution to energy transfer becomes important only at very low temperatures.

■ EXAMPLE 4.22-1

Silver has a Debye temperature of 210 K and a Fermi energy $E_F = 5.5$ eV. (a) What is the specific heat due to lattice vibrations and the electronic specific heat at room temperature (300 K)? (b) At what temperature will the electronic molar specific, $C_{V,el}$, and the lattice molar specific heat, C_V, be equal?

Solution

(a) At $T = 300$ K the Debye theory predicts a specific heat due to lattice vibrations of about $3R \approx 25$ J mol^{-1} K^{-1} (the Dulong–Petit limit). As for the electronic molar specific heat, it is given by the first-order relation

$$C_{V,el} = \frac{R\pi^2 kT}{2E_{Fo}}$$

$$= \frac{(8.31 \text{ J mol}^{-1} \text{ K}^{-1})\,\pi^2(8.62 \times 10^{-5} \text{ eV K}^{-1})(300 \text{ K})}{2(5.5 \text{ eV})}$$

$$= 0.19 \text{ J mol}^{-1} \text{ K}^{-1}$$

so at this temperature, the electron specific heat can be neglected.

(b) At low temperatures, the Debye theory result can be taken as

$$C_V = \frac{12\pi^4 R}{5}\left(\frac{T}{T_D}\right)^3.$$

Equating C_V and $C_{V,el}$ gives

$$\frac{12\pi^4 R}{5}\left(\frac{T}{T_D}\right)^3 = \frac{R\pi^2 kT}{2E_F}$$

$$T = \left(\frac{5kT_d^3}{24\pi^2 E_F}\right)^{1/2} = \left[\frac{5(8.62 \times 10^{-5} \text{ eV K}^{-1})(210 \text{ K})^3}{24\pi^2(5.5 \text{ eV})}\right]^{1/2} = 1.75 \text{ K}$$

This result shows that the electronic heat conduction will be noticeable only at very low temperatures. ■

■ EXAMPLE 4.22-2

At very low temperatures, the experimentally determined variation of the specific heat of a metal is given, to a good approximation, by

$$C_V = AT^3 + \gamma T \tag{4.22-29}$$

where A and γ are constants. The γT term represents the contribution of the conduction (free) electrons to the specific heat of the metal. This contribution is to be added to the contribution of the lattice vibrations, given by the Debye T^3 law. Evaluate C_V as $T \to \infty$ and $T \to 0$, respectively.

Solution. In 1907 Einstein proposed a model in which each atom vibrates with a single frequency ν, so that all phonons have the same frequency and therefore the same energy, $h\nu$. Increasing the temperature increases the vibrational energy and therefore the number of phonons. This number is obtained from the Bose–Einstein distribution function

$$f_{BE} = \frac{1}{e^{\alpha} e^{h\nu/kT} - 1} \tag{4.22-30}$$

with $\alpha = 0$. In three dimensions, the total phonon energy is

$$E = \frac{3Nh\nu}{e^{h\nu/kT} - 1} \tag{4.22-31}$$

whence

$$c_V = \frac{dE}{dT} = \frac{3Nk(h\nu/kT)^2}{(e^{h\nu/kT} - 1)^2} \tag{4.22-32}$$

The Einstein model agrees with the Dulong and Petit law ($c_V = 3Nk$) at high temperatures, but gives $c_V \to 0$ as $T \to 0$. At high temperatures, evaluating Eq. (4.22-32) as $T \to \infty$ gives

$$C_V = \frac{3Nk\,0}{0} = \frac{0}{0}$$

so, let $h\nu/kT = x$ and write

$$C_V = \lim_{x\to 0} \frac{3Nkx^2}{e^{2x} - 2e^x - 1}$$

Using l'Hopital's rule gives

$$C_V = \lim_{x\to 0} \frac{6Nkx}{2e^{2x} - 2e^x}$$

Using it again yields

$$C_V = \lim_{x\to 0} \frac{6Nk}{4e^{2x} - 2e^x} = 3Nk$$

which is the law of Dulong and Petit. However, at low temperatures, as $T \to 0$,

$$C_V \to \frac{3Nk\,\infty}{\infty}$$

Therefore, using l'Hopital's rule, this time with $x \to \infty$, gives

$$C_V = \frac{6Nk}{\infty} = 0$$

Einstein's model doesn't yield $c_V \propto T^3$ as the Debye model does. The latter is based on the assumption that the frequency varies (in such a way that the wave speed is constant up to a cutoff frequency). ■

4.23 FLUCTUATIONS IN OPEN SYSTEMS

Fluctuations in closed and classical systems have been discussed in Chapter 2 in terms of the canonical ensemble. In this section, the dispersion of the number of particles in a state l of an open system will be determined.

Consider the sum of Gibbs factors taken over all states of the system for all numbers of particles. This is the grand canonical partition function. It is the normalizing factor that converts relative probabilities to absolute probabilities:

$$\Xi(\mu, T) = \sum_{N=0}^{\infty} \sum_{l} e^{[N(\mu) - \varepsilon_l(N)]/kT} \tag{4.23-1}$$

$$P(N_l, \varepsilon_l) = \frac{e^{[N_l(\mu) - \varepsilon^l(N_l)]/kT}}{\Xi} \tag{4.23-2}$$

The sums are to be carried out over all states of the system for all numbers of particles. ε_l is being written as $\varepsilon_l(N)$ to emphasize the dependence of the state on the number of particles N. If $A(N, l)$ is the value of A, a physical quantity, when the system has N particles and is in the quantum state l, then the ensemble average of A is

$$\langle A \rangle = \sum_N \sum_l A(N, l) P(N, \varepsilon_l) = \frac{\sum_N \sum_l A(N, l)\, e^{(N_\mu - \varepsilon_l)/kT}}{\Xi} \tag{4.23-3}$$

Now, the number of particles in the system can vary because the system is in diffusive contact with a reservoir. The ensemble average of the number of particles in the system is thus

$$\langle N \rangle = \frac{\sum_N \sum_l N e^{(N_\mu - \varepsilon_l)/kT}}{z} \tag{4.23-4}$$

where, in the numerator, each Gibbs factor in the grand sum has been multiplied by N, and z is the canonical partition function.

Equation (4.23-4) for $\langle N \rangle$ can be written in a form more convenient for calculation by noting that

$$\frac{\partial \Xi}{\partial \mu} = \frac{1}{kT} \sum_N \sum_l N e^{(N_\mu - \varepsilon_l)/kT} \tag{4.23-5}$$

whence the general relation

$$\langle N \rangle = kT \frac{\partial z / \partial \mu}{z} = kT \frac{\partial \ln z}{\partial \mu} \tag{4.23-6}$$

Similarly, $\langle N^2 \rangle$ is given by

$$\langle N^2 \rangle = \frac{1}{\Xi} \sum_N \sum_l N^2 e^{(N_\mu - \varepsilon_l)/kT} \tag{4.23-7}$$

$$= \frac{k^2 T^2}{\Xi} \frac{\partial^2 \Xi}{\partial \mu^2} \tag{4.23-8}$$

The mean square deviation $\langle (\Delta N)^2 \rangle$ of N from $\langle N \rangle$ will now be calculated. By definition, this is

$$\begin{aligned} \langle (\Delta N)^2 \rangle &= \langle (N - \langle N \rangle)^2 \rangle = \langle (N^2 - 2N\langle N \rangle + \langle N \rangle^2) \rangle \\ &= \langle N^2 \rangle - 2\langle N \rangle \langle N \rangle + \langle N \rangle^2 = \langle N^2 \rangle - \langle N \rangle^2 \end{aligned} \tag{4.23-9}$$

or, by (4.23-8) and (4.23-6),

$$\langle (\Delta N)^2 \rangle = T^2 \left[\frac{1}{z} \frac{\partial^2 z}{\partial \mu^2} - \frac{1}{z^2} \left(\frac{\partial z}{\partial \mu} \right)^2 \right] \tag{4.23-10}$$

This may be rewritten by noting that

$$\langle N \rangle = \frac{\tau}{\Xi} \frac{\partial \Xi}{\partial \mu} \tag{4.23-11}$$

$$\frac{\partial \langle N \rangle}{\partial \mu} = \frac{\tau}{\Xi} \frac{\partial^2 \Xi}{\partial \mu^2} - \frac{\tau}{\Xi^2} \left(\frac{\partial \Xi}{\partial \mu} \right)^2 \tag{4.23-12}$$

where $\tau = kT$. Equation (4.23-10) thus becomes

$$\langle(\Delta N)^2\rangle = \tau \frac{\partial \langle N \rangle}{\partial \mu} \tag{4.23-13}$$

This result will now be applied to a *perfect gas,* a *Fermi gas,* and a *Bose gas,* respectively.

For a *perfect gas,* the number of particles is related to the chemical potential by

$$\varepsilon^{\mu/\varepsilon} = \frac{\langle N \rangle}{v} \left(\frac{2\pi\hbar^2}{M\tau}\right)^{3/2}$$

or

$$\langle N \rangle = e^{\mu/\varepsilon} \frac{v}{V_Q} \tag{4.23-14}$$

where

$$V_Q = \left(\frac{2\pi\hbar^2}{M\tau}\right)^{3/2} \tag{4.23-15}$$

is the quantum volume. Using (4.23-14) in (4.23-13), the mean square fluctuation is then

$$\begin{aligned} \langle(\Delta N)^2\rangle &= \varepsilon \frac{\partial \langle N \rangle}{\partial \mu} = \frac{\tau}{\tau} e^{\mu/\varepsilon} \frac{v}{V_Q} \\ &= \langle N \rangle \end{aligned} \tag{4.23-16}$$

Upon division by $\langle N \rangle^2$, this gives the mean square fractional fluctuation as

$$\frac{\langle(\Delta N)^2\rangle}{\langle N \rangle^2} = \frac{1}{\langle N \rangle} \tag{4.23-17}$$

The fractional fluctuations are extremely small if there is a macroscopic number of particles in the volume under consideration. If $\langle N \rangle$ is on the order of 10^{20} atoms, the root-mean-square fractional fluctuation is

$$\left[\frac{\langle(\Delta N)^2\rangle}{\langle N \rangle^2}\right]^{1/2} = 10^{-10} \tag{4.23-18}$$

This shows the accuracy with which a system in diffusive contact simulates a system with fixed total number of particles. In such a system the number of particles is well defined even though it is not held exactly constant.

For a *Fermi gas,* the distribution function or average fermion occupancy for a single orbital is

$$f(\varepsilon) = \langle n(\varepsilon) \rangle = \frac{1}{e^{(\varepsilon-\mu)/\tau} + 1} \tag{4.23-19}$$

Applying (4.23-13), this gives

$$\langle(\Delta n)^2\rangle = \tau \frac{\partial \langle n \rangle}{\partial \mu} = \frac{\tau}{\tau} \frac{e^{(\varepsilon-\mu)/\tau}}{[e^{(\varepsilon-\mu)/\tau} + 1]^2} \tag{4.23-20}$$

Adding and subtracting 1 to the numerator of (4.23-20) yields

$$\begin{aligned}\langle(\Delta n)^2\rangle &= \frac{1}{e^{(\varepsilon-\mu)/\tau}+1} - \frac{1}{[e^{(\varepsilon-\mu)/\tau}+1]^2} \\ &= \langle n\rangle - \langle n\rangle^2\end{aligned} \tag{4.23-21}$$

or

$$\langle(\Delta n)^2\rangle = \langle n\rangle(1-\langle n\rangle) \tag{4.23-22}$$

For an extremely degenerate FD gas in which the energetically lowest-lying single-particle states are occupied, $\langle n\rangle \approx 1$, and the fluctuation vanishes, that is,

$$\frac{\langle(\Delta n)^2\rangle}{\langle n\rangle^2} = \frac{1}{\langle n\rangle} - 1 = 0 \quad \text{as} \quad \langle n\rangle \to 1 \tag{4.23-23}$$

For a *Bose gas,* the distribution function or occupancy of an orbital is given by

$$b(\varepsilon) = \langle n(\varepsilon)\rangle = \frac{1}{e^{(\varepsilon-\mu)/\tau}-1} \tag{4.23-24}$$

Applying (4.23-13) gives

$$\langle(\Delta n)^2\rangle = \tau\frac{\partial\langle n\rangle}{\partial\mu} = \frac{\tau}{\tau}\frac{e^{(\varepsilon-\mu)/\tau}}{(e^{(\varepsilon-\mu)/\tau}-1)^2} \tag{4.23-25}$$

Subtracting and adding 1 to the numerator of (4.23-25) yields

$$\begin{aligned}\langle(\Delta n)^2\rangle &= \frac{1}{e^{(\varepsilon-\mu)/\tau}-1} + \frac{1}{(e^{(\varepsilon-\mu)/\tau}-1)^2} \\ &= \langle n\rangle + \langle n\rangle^2\end{aligned} \tag{4.23-26}$$

or

$$\langle(\Delta n)^2\rangle = \langle n\rangle(1+\langle n\rangle) \tag{4.23-27}$$

This shows that if the occupancy is large, with $\langle n\rangle \approx 1$, then the fractional fluctuations are of the order of unity:

$$\frac{\langle(\Delta n)^2\rangle}{\langle n\rangle^2} = \frac{1}{\langle n\rangle} + 1 = 1 \tag{4.23-28}$$

Thus, enormous fluctuations in numbers of particles in an orbital can occur for bosons and photons. In the latter case, the fractional fluctuations in the number of photons are not smoothed out as the average number of photons increases. Photons like to travel in packs.

■ EXAMPLE 4.23-1

Give some physical reasons for the Boltzmann, Bose–Einstein, and Fermi–Dirac distribution laws. Particles that come under these statistics are known as boltzons, bosons, and fermions, respectively.

Solution. Of course, all particles are in reality either bosons or fermions, and there are no boltzons. However, when the average population per state is much less than 1, that is, $\langle n_r \rangle << 1$, the BE and FD distribution laws reduce to the Boltzmann distribution law, and bosons and fermions behave like boltzons. To show when the assumption $\langle n_r \rangle << 1$ is justified, consider an ideal gas. The molecular energy is the sum of translational, rotational, vibrational, and electronic energies. For simplicity, only the translational states are counted; inclusion of rotational, vibrational, and electronic states would not only strengthen the results but complicate the problem.

For a cubic box of volume V, quantum mechanics give the translational energy of a molecule as

$$E_{\text{trans}} = \left(\frac{h^2}{8mV^{2/3}}\right)(n_x^2 + n_y^2 + n_z^2) \tag{4.23-29}$$

It is required to calculate the number of translational states whose energy is less than some maximum value $E_{\max}$. For $E_{\text{trans}} \leq E_{\max}$, the translational quantum numbers must satisfy the inequality

$$n_x^2 + n_y^2 + n_z^2 \leq \frac{8mV^{2/3}}{h^2} E_{\max} \tag{4.23-30}$$

Now, the average translational energy per molecule is $\frac{3}{2}kT$, and most molecules have E_{trans} within two or three times this average value. Therefore, $E_{\max} = 3kT$ is taken as the maximum translational energy for a typical molecule. In the n-space whose coordinates are the quantum numbers n_x, n_y, and n_z, the number of quantum states with $E_{\text{trans}} \leq E_{\max}$ equals the number of dots in one-eighth of a sphere whose radius $r_{\max}$ is

$$r_{\max} = (8mV^{2/3}h^{-2}E_{\max})^{1/2} = (24mV^{2/3}h^{-2}kT)^{1/2} \tag{4.23-31}$$

Thus, the number of molecular states with $E_{\text{trans}} \leq E_{\max}$ equals one-eighth the volume of a sphere of radius $r_{\max}$, and the number of available molecular translational states is

$$\frac{1}{8}\frac{4\pi r_{\max}^3}{3} = \frac{\pi}{6}(24mV^{2/3}h^{-2}kT)^{3/2} \approx 60\left(\frac{mkT}{h^2}\right)^{3/2}\frac{N}{V} << 1 \tag{4.23-32}$$

For $\langle N_r \rangle << 1$ to hold requires that $60(mkT/h^2)^{3/2}\,V >> N$, where N is the number of molecules. This inequality reads

$$\frac{1}{60}\left(\frac{h^2}{mkT}\right)^{3/2}\frac{N}{V} << 1 \tag{4.23-33}$$

A typical system is 1 mole of 0_2 at 0°C and 1 atm, for which $m = [32/(6 \times 10^{23})]\,g$, $N = 6 \times 10^{23}$, and $V = 22{,}400\ \text{cm}^3$. This gives 4×10^{-8} for the left side of (4.9-13). Hence, $\langle N_r \rangle << 1$ is well satisfied. The translational levels are spaced extremely close together, so there are far more translational states available than there are molecules in the gas.

The three distributions, Maxwell–Boltzmann, Bose–Einstein, and Fermi–Dirac, under guise of a single formula, can be written

$$\frac{n}{g} = \frac{1}{e^{(\varepsilon - \mu)/kT} - 1,\ +0,\ +1} \tag{4.23-34}$$

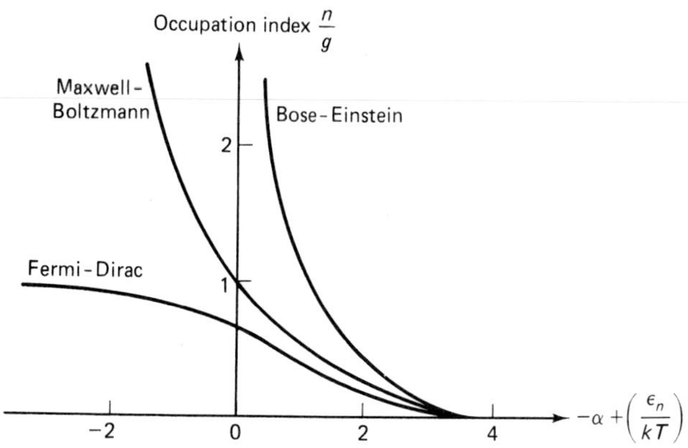

Figure 4.23-1 Occupation index under MB, BE, and FD statistics.

where n/g = the occupation index (average number of particles in an energy state), $\beta = 1/kT$ as was previously shown, and the minus sign is for BE, the plus sign is for FD, and the zero is for MB. It is seen that if for both the BE and FD statistics, $e^{(\varepsilon - \mu)/kT}$ is large enough, that is,

$$e^{(\varepsilon - \mu)/kT} >> 1 \tag{4.23-35}$$

then the unity term can be ignored in the denominator of (4.23-34), and the MB distribution is obtained. This is the *classical limit* of the BE and FD laws. Figure 4.23-1 shows the occupation index n/g for the three statistics as a function of $(\varepsilon - \mu/kT)$; the following characteristics are noteworthy:

(a) The occupation index for FD statistics can never exceed unity. This is in accord with the Pauli exclusion principle.

(b) For a given value of μ,

$$\left(\frac{n}{g}\right)_{FD} < \left(\frac{n}{g}\right)_{MB} < \left(\frac{n}{g}\right)_{BE} \tag{4.23-36}$$

It can be seen from the figure that for $(\varepsilon - \mu/kT) > 2$, the three occupation indices are the same, meaning that there is no essential difference in the high-energy state occupancy among the three statistics. ■

■ EXAMPLE 4.23-2

The chemical potential μ is related to the fractional change of the accessible states with a change in the number of particles. Its knowledge is a key to the understanding of the behavior of systems. What are the values of μ for the different types of systems?

Solution. The chemical potential for fermions and bosons is found from the total number of particles

$$N = \frac{2\pi Vg}{h^3}(2m)^{3/2}\int_0^\infty \varepsilon^{1/2}\left[\exp\left(\frac{\varepsilon - \mu}{kT}\right) \pm 1\right]^{-1} d\varepsilon \tag{4.23-37}$$

The extrication of $\mu(T)$ can be done by a variety of numerical methods. The result for fermions and bosons, plotted in terms of a common characteristic temperature,

$$T_0 = \frac{h^2}{2mk}\left(\frac{N}{2\pi Vg}\right)^{2/3} \tag{4.23-38}$$

is shown in Fig. 4.23-2. At high temperatures, the distinction between fermions and bosons disappear, for they are both in the classical limit. In that case, the ± 1 in Eq. (4.23-37) can be neglected in comparison with the exponential, and

$$N = \frac{2\pi Vg}{h^3}(2m)^{3/2}\exp\left(\frac{\mu}{kT}\right)\int_0^\infty \varepsilon^{1/2}\exp\left(-\frac{\varepsilon}{kT}\right)d\varepsilon \tag{4.23-39}$$

The integral can be performed explicitly, so that after rearrangement,

$$\mu = -kT\ln\left[\left(\frac{2\pi mkT}{h^2}\right)^{3/2}\frac{gV}{N}\right] \tag{4.23-40}$$

In the classical regime, the value of the chemical potential is negative and less than $-kT$.

The low-temperature limiting form of μ can also be obtained. As $T \to 0$, let $\mu \to \mu_0$. Then, for those states for which $\varepsilon > \mu_0$, $\exp[(\varepsilon - \mu_0)/kT] \to \exp[\infty]$; these states do not contribute to the integral in Eq. (4.23-37). For states with $\varepsilon < \mu_0$, $\exp[(\varepsilon - \mu_0)/kT] \to \exp[-\infty] = 0$. The upper limit of integration may be taken as μ_0 in Eq. (4.23-37) which, for a fermion gas, reduces to

$$N = \frac{2\pi Vg}{h^3}(2m)^{3/2}\int_0^{\mu_0} \varepsilon^{1/2}\,d\varepsilon \tag{4.23-41}$$

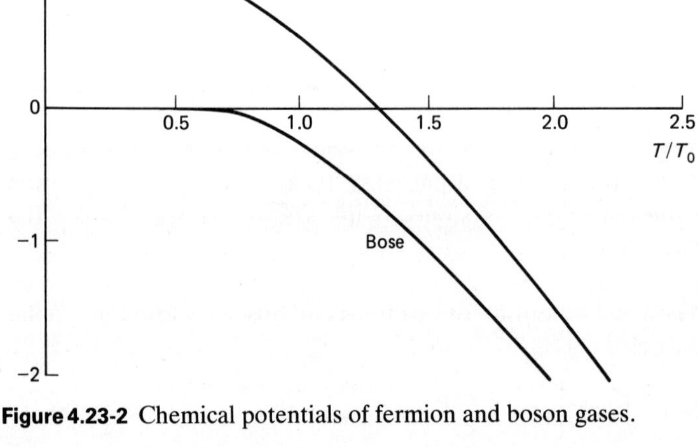

Figure 4.23-2 Chemical potentials of fermion and boson gases.

This gives

$$\mu_0 = \frac{h^2}{8m}\left(\frac{6N}{\pi Vg}\right)^{3/2} \tag{4.23-42}$$

which is the Fermi energy, ε_F, and a corresponding Fermi temperature, defined as $T_F = \mu_0/k$.

As for the boson gas, when $T \to 0$, μ becomes less negative, and at some temperature T_B, must be identically zero. Equation (4.23-37) then becomes

$$N = \frac{2\pi gV}{h^3}(2m)^{3/2}\int_0^\infty \varepsilon^{1/2}\left[\exp(\varepsilon/kT_B) - 1\right]^{-1} d\varepsilon \tag{4.23-43}$$

$$= \frac{2\pi gV}{h^3}(2m)^{3/2}(kT_B)^{3/2}\int_0^\infty x^{1/2}\left[\exp(x) - 1\right]^{-3/2} d\varepsilon \tag{4.23-44}$$

where the change of variable $x = \varepsilon/kT_B$ has been made. The definite integral is the Riemann zeta function $\zeta(\frac{3}{2})$. In general,

$$\zeta(n) = \frac{1}{\Gamma(n)}\int_0^\infty \frac{x^{n-1}}{e^x - 1}\,dx \tag{4.23-45}$$

$\zeta(\frac{3}{2})$ has the numerical value 2.612. Thus, after rearrangement, Eq. (4.23-44) gives

$$T_B = \frac{h^2}{2\pi mk}\left(\frac{N}{2.612\,gV}\right)^{2/3} \tag{4.23-46}$$

■

■ EXAMPLE 4.23-3

(a) Show that when a small macroscopic system is in thermal equilibrium with a heat reservoir, the fluctuations in the energy of the small system from the ensemble average energy are extremely small. What are the implications in regard to the statistical mechanics viewpoint and the classical thermodynamics approach to calculating properties?

(b) Consider a container of saturated vapor. Show that the probability of finding a droplet of radius R is proportional to $e^{-4\pi R^2\sigma/kT}$, where σ is the surface tension of the liquid.

Solution

(a) The relative fluctuation of the energy of a system in a heat bath is inversely proportional to the square root of its size: $\Delta E/\langle E\rangle \sim 1/\sqrt{N}$. For a macroscopic system where $N \sim 10^{23}$, $\Delta E/\langle E\rangle \sim 10^{-11}$, the fluctuations are extremely small. The energy of a macroscopic body in a heat bath is, for all practical purposes, completely determined. The fact that $\Delta E/\langle E\rangle$ is as small for macroscopic systems means that the probability distribution $g(E)e^{-E/kT}/Z$ has an extremely sharp maximum at energy $\langle E\rangle$. For a macroscopic system the density of states $g(E)$ is a very rapidly increasing function of E; namely, $g(E) \propto E^{1/2}$. The Boltzmann factor $\exp(-E/kT)$, on the other hand, being an exponential, is rapidly decreasing with energy. The product of these two factors yields a very sharp maximum.

The situation is quite different for a small system consisting of a few molecules or only one molecule! Also, there are cases where large relative fluctuations occur in macroscopic systems as for example a liquid in contact with its saturated vapor. The energy of the system is not well determined: it can vary widely between the extremes "only liquid present" and "only vapor present." There are also large fluctuations in the volume of the system corresponding to these two extremes. Another example of large fluctuations is that of density near the critical point. The approach in statistical mechanics is to calculate the properties of a system completely from first principles. If this is not possible in cases of systems which are too complex for a statistical treatment, directly measurable quantities must be used.

(b) Suppose that the quantum states of the system are divided into two categories, 1 and 2. Let P_1 be the probability that the system is in state 1. Then

$$P_1 = Z_1/Z \tag{4.23-47}$$

where Z is the partition function of the total system and Z_1 is the partition function counting only states with the property characteristic 1. Now, the free energy is

$$A_1 = -kT \ln Z \tag{4.23-48}$$

and

$$\Delta A = A_1 - A \tag{4.23-49}$$

Thus, (4.23-47) may be written

$$P_1 = e^{-\Delta A/kT} \tag{4.23-50}$$

Applying this to the probability of finding a droplet of radius R in the saturated vapor, the free energy in the liquid phase is equal to that in the vapor, but the droplet contains an additional term $\Delta A = 4\pi R^2\sigma$ because of the surface tension σ of the liquid. Thus, from (4.23-50), the probability of finding a droplet of radius will be proportional to

$$e^{-4\pi R^2\sigma/kT} \tag{4.23-51}$$

■

4.24 SUMMARY

The statistical treatment of quantum-mechanical systems is the same as that of classical statistical mechanics. Given a system with a large number N of constituent particles, it is possible in principle to follow the evolution of the system by solving Schrödinger's equation

$$H(p_1, q_1, \ldots, p_n, q_n, t)\psi(q_1, q_2, \ldots, q_n, t) = ih\frac{\partial\psi}{\partial t} \tag{4.24-1}$$

given the initial state $\psi(q_1, q_2, \ldots, q_n, 0)$. This procedure, however, may not be practical even when the number of particles N is small, since the initial conditions are not available with sufficient accuracy. In cases of large N it is simply impractical to attempt to compute ψ. It is more profitable to consider the statistical properties of a large ensemble of similar systems and assume that the ensemble statistics are a reliable predictor of the expected behavior of a single system.

A single quantum-mechanical system is taken to consist of N weakly interacting "particles" constrained to some volume V. These particles may be atoms, electrons, nuclei, protons, harmonic oscillators, electromagnetic radiation modes, normalized lattice vibrations, photons, and mixtures thereof. It is assumed that the particle density is low enough so that, to first order, the energy of the system can be considered as the sum of the individual particle energies. A given microscopic energy eigenstate of the system is constructed by assigning to each particle s a quantum state $\psi_i^{(s)}$, where

$$\hat{H}(p_s, q_s)\psi_i^{(s)} = \epsilon_i \psi_i^{(s)} \tag{4.24-2}$$

where $\hat{H}(p_s, q_s)$ is the Hamiltonian of particle s when occupying the volume V all by itself, so that ψ_i is the single-particle energy eigenfunction of the particle s in the ith quantum state with energy eigen value ϵ_i. The total energy of the system is then

$$E = \sum_{i=1}^{\infty} n_i \epsilon_i \tag{4.24-3}$$

where n_i is the number of particles with energy ϵ_i. There are many ways of distributing the N particles among the energy levels ϵ_i so as to result in a given total system energy of E. (More precisely, $E \pm \delta E$, where δE is the resolution of the energy-measuring apparatus.) Each individual arrangement that is describable, in principle, by a unique system eigenfunction, is called a distinguishable microscopic arrangement. It is a fundamental postulate of quantum statistics that in equilibrium different (distinguishable) microscopic arrangements of the system that correspond (within δE) to the same total energy are equally likely to occur.

The aim of statistical mechanics is to determine the most probable distribution $(n_1, n_2, \ldots, n_s, \ldots)$ of finding n_1 particles with energy ϵ_1, n_2 with energy ϵ_2, and so on, such that the total energy of the system as given by (4.24-3) is a constant E. Since, in general, there is more than one microscopically distinguishable arrangement corresponding to a given distribution $(n_1, n_2, \ldots, n_s, \ldots)$, it follows from the basic postulate of quantum statistics that the probability of finding $(n_1, n_2, \ldots, n_s, \ldots)$ is proportional to the number of such microscopic arrangements, each resulting in a total energy E. The most probable distribution $(n_1, n_2, \ldots, n_s, \ldots)$ is the one associated with the largest number of microscopically distinguishable arrangements. The plan of attack is to derive an expression for $P(n_1, n_2, \ldots, n_s, \ldots)$, the total number of microscopically distinct arrangements corresponding to a given arbitrary sequence $(n_1, n_2, \ldots, n_s, \ldots)$ and then find the particular sequence that maximizes P

subject to the conditions that the total number of particles and the total energy be constant. The particles of concern fall into one of three categories:

(a) Identical but distinguishable particles. Two examples of such particles are: (1) very heavy point masses that are confined inside a large box, and (2) a collection of harmonic oscillators.

(b) Identical indistinguishable particles of half-odd-integral spin—fermions (total spin angular momentum of $h/2$, $3h/2$, $5h/2$, etc.). Examples are electrons and protons. These particles obey the Pauli exclusion principle, according to which no two particles can occupy the same quantum state, or alternatively, that the total N particle eigenfunction be antisymmetric (change sign) upon the permutation of any two particles. This constraint has profound effect on the particle statistics.

(c) Identical indistinguishable particles of integral spin—bosons. Examples are photons, phonons, nuclei with an even number of elementary particles. The Pauli exclusion principle does not apply to systems made up of bosons, so that more than one particle in the system can possess the same set of quantum numbers. The eigenfunction is symmetric upon permutation of two particles.

The task is to solve for the most probable distribution of the different classes of particles, using a model system of N identical "particles," each with a spectrum of allowed energy levels (derived by solving the time-independent Schrödinger equation for a single particle).

Case (a): Identical but Distinguishable Particles (Boltzons)

To determine the number of microscopically distinguishable arrangements, consider N particles and s bins, and, let n_1 particles be placed in bin 1, ignoring for the moment the manner in which these particles are distributed between the quantum states of the bin. The first particle can be chosen from among a total of N particles, the second from the remaining $N-1$, and so on, so that the total number of distinguishable choices is

$$N(N-1)(N-2)\cdots(N-n_1+1)=\frac{N!}{(N-n_1)!} \tag{4.24-5}$$

In arriving at this result, all permutations among the group n_1 of particles are counted as distinct. The sequence of particles $(1,4,7)$ and $(7,4,1)$, for example, is counted as two distinct arrangements. Therefore (4.24-5) needs to be divided by $n_1!$, the number of permutations of n_1 particles among themselves. The result is

$$\frac{N!}{n_1!(N-n_1)!} \tag{4.24-6}$$

Now, for the internal arrangement of the n_1 particles in bin 1, let the number of quantum states combined in bin 1 be g_1. Particle 1 can be placed into any of the g_1 states of the bin. The second particle can, likewise, also be put into any of the g_1 states. The total number of distinguishable microscopic arrangements of the n_1 particles is thus $g_1^{n_1}$. This result, multiplicd by (4.24-6), gives

$$P_1 = \frac{N!\, g_1^{n_1}}{n_1!(N - n_1)!} \tag{4.24-7}$$

for the total number of microscopic distinguishable ways of putting n_1 particles out of a total N into the g_1 states of bin 1. The number of distinct microscopic arrangements of putting n_2 of the remaining $(N - n_1)$ particles into bin 2 is, in analogy with (4.24-7),

$$P_2 = \frac{(N - n_1)!\, g_2^{n_2}}{n_2!(N - n_1 - n_2)!} \tag{4.24-8}$$

The total number of distinct microscopic arrangements in which n_1 particles are in bin 1, n_2 in bin 2, and so on, is thus

$$\begin{aligned} P(n_s) &\equiv P(n_1, n_2, \ldots, n_s, \ldots) \\ &= P_1 P_2, \ldots, P_s, \ldots \\ &= N! \prod_{s=1}^{\infty} \frac{g_s^{n_s}}{n_s!} \qquad \text{case (a)} \end{aligned} \tag{4.24-9}$$

Case (b): Identical Indistinguishable Particles of Half-Odd-Integral Spin (Fermions)

Since the particles are deemed indistinguishable, it is not possible to tell which particle occupies a particular state. All that can be determined is whether a given state in a bin is occupied or not. Since each state is associated with a one-particle energy eigenfunction, the Pauli exclusion principle (which applies to particles with half-odd-integral spin) prevents the placing of more than one particle in any one state. Consider, for a moment, that the particles are distinguishable. The task of fitting n_s particles into g_s states is begun by putting one particle into any one state in bin s. There are g_s ways of doing so. Since no two particles can occupy the same state, the second particle can be placed into any of the remaining $(g_s - 1)$ states, so that altogether there are

$$g_s(g_s - 1) \cdots (g_s - n_s + 1) \tag{4.24-10}$$

configurations (because of the Pauli exclusion principle $g_s \geq n_s$). This last expression includes all possible permutations of the $n_s!$. Since the particles are indistinguishable, these permutations do not lead to distinguishable arrangements (they correspond to the same eigenfunction) so that the total number of distinct microscopic arrangements with n_s particles in bin s is

$$\begin{aligned} P_s &= \frac{g_s(g_s - 1) \cdots (g_s - n_s + 1)}{n_s!} \\ &= \frac{g_s!}{(g_s - n_s)!\, n_s!} \end{aligned} \tag{4.24-11}$$

and the total number of configurations with n_1 particles in bin 1, n_2 in bin 2, ..., is

$$P(n_s) = P(n_1, n_2, \ldots, n_s, \ldots) = \prod_{s=1}^{\infty} P_s$$
$$= \prod_{s=1}^{\infty} \frac{g_s!}{(g_s - n_s)!n_s!} \qquad \text{case (b)} \tag{4.24-12}$$

Case (c): Identical, Indistinguishable Particles of Integral Spin (Bosons)

In this case, as in (b), it is not possible to state which particle is placed in any particular bin or, within a given bin, which particle occupies any given quantum state. There is, however, no restriction on the number of particles occupying a given state. In other words, more than one particle can have the same single-particle eigenfunction. The number of distinct arrangements of n_s particles in the g_s states of bin s can be determined by means of Fig. 4.24-1.

Each space between two adjacent partitions, as well as between the two extreme partitions and the walls, represents one state. The $(g_s - 1)$ partitions thus account for the g_s states in bin s. The number of white dots is n_s and is equal to the number of particles in the bin. In the present example, $g_s = 8$ and $n_s = 14$. If the n_s white dots and the $(g_s - 1)$ black lines were all distinguishable, the number of distinct arrangements of white dots and black lines would be equal to the number of permutations of $(n_s + g_s - 1)$ objects, which is $(n_s + g_s - 1)!$. Since the particles and partitions are in fact indistinguishable, the $(g_s - 1)!$ permutations of the black lines among themselves or the $n_s!$ permutations of the white dots do not lead to distinguishable arrangements. The number of distinguishable arrangements is thus reduced to

$$P_s = \frac{(n_s + g_s - 1)!}{n_s!(g_s - 1)!} \tag{4.24-13}$$

and the total number of distinct arrangements involving $n_1, n_2 \ldots, n_s$ is thus

$$P(n_s) = P(n_1, n_2, \ldots, n_s, \ldots)$$
$$= \prod_{s=1}^{\infty} \frac{(n_s + g_s - 1)!}{n_s!(g_s - 1)!} \qquad \text{case (c)} \tag{4.24-14}$$

Having obtained $P(n_s) = P(n_1, n_2, \ldots, n_s, \ldots)$ for boltzons, fermions, and

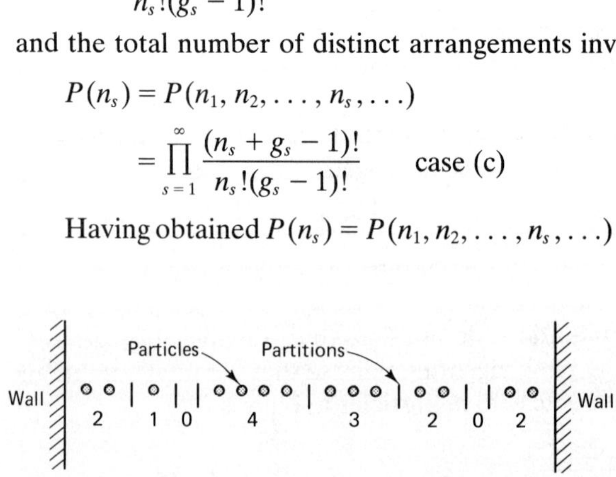

Figure 4.24-1 Distinct ways of arranging n_s indistinguishable particles (white dots) in the g_s states (each state is bounded by two partitions or by a partition and a wall). (In the figure $n_s = 14$, $g_s = 8$.)

bosons, the next task is to find the distribution $n_1, n_2, \ldots, n_s$ for which $P(n_s)$ is a maximum, subject to the auxiliary conditions

$$\sum_{s=1}^{\infty} n_s = N = \text{const} \qquad \text{(constancy of total number of particles)} \quad (4.24\text{-}15)$$

$$\sum_{s=1}^{\infty} \epsilon_s n_s = E = \text{const} \qquad \text{(constancy of total energy)} \quad (4.24\text{-}16)$$

It is more convenient to maximize $\ln P$, subject to conditions (4.24-15) and (4.24-16), rather than P itself. The problem of maximizing a general function $G(n_1, n_2, \ldots)$ subject to auxiliary conditions $f(n_1, n_2, \ldots) = 0$ and $g(n_1, n_2, \ldots) = 0$ is treated by Lagrange's method of undetermined multipliers. In applying this method, let

$$F(n_1, n_2, \ldots, \alpha, \beta) = G(n_1, n_2, \ldots) - \alpha f(n_1, n_2, \ldots) - \beta g(n_1, n_2, \ldots) \quad (4.24\text{-}17)$$

where α and β are multipliers and

$$G = \ln P(n_s) \quad (4.24\text{-}18)$$

$$f = \left(\sum_{s=1}^{\infty} n_s\right) - N \quad (4.24\text{-}19)$$

$$g = \left(\sum_{s=1}^{\infty} \epsilon_s n_s\right) - E \quad (4.24\text{-}20)$$

The values of $n_1, n_2, \ldots,$ are solved for maximum G by requiring that

$$\frac{\partial F}{\partial n_s} = 0 \qquad \text{for all } s \quad (4.24\text{-}21)$$

$$\frac{\partial F}{\partial \alpha} = 0 \qquad \frac{\partial F}{\partial \beta} = 0 \quad (4.24\text{-}22)$$

This yields

$$F = \ln P - \alpha\left[\left(\sum_{s=1}^{\infty} n_s\right) - N\right] - \beta\left[\left(\sum_{s=1}^{\infty} \epsilon_s n_s\right) - E\right] \quad (4.24\text{-}23)$$

which is now applied to cases (a), (b), and (c).

Case (a): Boltzons

Here, P is given by Eq. (4.24-9):

$$P(n_s) = P(n_1, n_2, \ldots, n_s \ldots)$$

$$= N! \prod_s^{\infty} \frac{g_s^{n_s}}{n_s!} \quad (4.24\text{-}24)$$

and (4.24-23) becomes

$$F = \ln\left(N! \prod_{s=1}^{\infty} \frac{g_s^{n_s}}{n_s!}\right) - \alpha\left[\left(\sum_{s=1}^{\infty} n_s\right) - N\right) - \beta\left[\left(\sum_{s=1}^{\infty} \epsilon_s n_s\right) - E\right]$$

$$= \ln N! + \prod_{s=1}^{\infty} (n_s \ln g_s - \ln n_s!) - \alpha\left[\left(\sum_{s=1}^{\infty} n_s\right) - N\right] - \beta\left[\left(\sum_{s=1}^{\infty} \epsilon_s n_s\right) - E\right] \tag{4.24-25}$$

For $n_s >> 1$ and using Stirling's approximation ($\ln n! = n \ln n - n$), this gives

$$\frac{\partial F}{\partial n_s} = \ln g_s - \ln n_s - \alpha - \beta\epsilon_s = 0 \qquad \text{(for all } s) \tag{4.24-26}$$

Case (b): Fermions

Equation (4.24-22) gives for $P(n_s)$

$$P(n_s) = P(n_1, n_2, \ldots, n_s, \ldots)$$

$$= \prod_{s}^{\infty} \frac{g_s!}{(g_s - n_s)!\, n_s!}$$

or

$$\ln P = \prod_{s=1}^{\infty} (\ln g_s! - [\ln n_s! + \ln (g_s - n_s)!]) \tag{4.24-27}$$

Using Stirling's formula, this last result can be approximated by

$$\ln P = \prod_{s=1}^{\infty} \{\ln g_s! - (n_s \ln n_s - n_s) - [(g_s - n_s) \ln (g_s - n_s) - (g_s - n_s)]\}$$

$$= \prod_{s=1}^{\infty} \{\ln g_s! - n_s \ln n_s - (g_s - n_s) \ln (g_s - n_s) + g_s\} \tag{4.24-28}$$

Then, from (4.24-23),

$$F = \ln P - \alpha\left[\left(\sum_{s=1}^{\infty}\right) - N\right] - \beta\left[\left(\sum_{s=1}^{\infty} \epsilon_s n_s\right) - E\right]$$

or

$$\frac{\partial F}{\partial n_s} = -\ln n_s + \ln (g_s - n_s) - \alpha - \beta\epsilon_s = 0 \tag{4.24-29}$$

for all s. It has been assumed that $g_s - n_s >> 1$.

Case (c): Bosons

Here, $P(n_s)$ is given by Eq. (4.24-14)

$$P(n_s) = P(n_1, n_2, \ldots, n_s, \ldots)$$
$$= \prod_s^{\infty} \frac{(n_s + g_s - 1)!}{n_s!\,(g_s - 1)!}$$

and, repeating the same procedure that lead to (4.24-29),

$$\ln(n_s + g_s) - \ln n_s - \alpha - \beta\epsilon_s = 0 \tag{4.24-30}$$

Finally, from Eqs. (4.24-26), (4.24-29), and (4.24-30), the distribution laws for thermal equilibrium are obtained:

$$\text{(a)}\ n_s = \frac{g_s}{e^{\alpha + \beta\epsilon_s}} \qquad \text{(Maxwell–Boltzmann)} \tag{4.24-31}$$

$$\text{(b)}\ n_s = \frac{g_s}{e^{\alpha + \beta\epsilon_s} + 1} \qquad \text{(Fermi–Dirac)} \tag{4.24-32}$$

$$\text{(c)}\ n_s = \frac{g_s}{e^{\alpha + \beta\epsilon_s} - 1} \qquad \text{(Bose–Einstein)} \tag{4.24-33}$$

In order to use the distribution laws (4.24-31), (4.24-32) and (4.24-33), it is necessary to determine the values of the constants α snd β, which appear in these laws. This is done by enforcing the auxiliary conditions (4.24-15) and (4.24-16), which cause (4.24-22) to be satisfied. The parameter β is of a more fundamental nature than α, since a single value of β is used in the distribution laws, even when the system contains more than one type of particle. The α parameter is different for different types of particles. Therefore, the parameter β will be evaluated first. To determine β, consider a gas made up of N massive atoms of zero spin. The large mass of these particles, which renders them relatively immobile, and the large volume V in which they are enclosed, make it possible to consider them as distinguishable so their distribution is governed by the Maxwell–Boltzmann law (4.24-31),

$$n_s = \frac{g_s}{e^{\alpha_1 + \beta\epsilon_s}} \tag{4.24-34}$$

where α_1 refers to a species of particles, and g_s is the density of states, that is, the number of energy eigenstates whose energies fall within a small interval $\Delta\epsilon_s$ centered on ϵ_s. This number is

$$g_s = \frac{4\pi\, Vm(2m)^{1/2}}{h^2}\, \epsilon_s^{1/2}\, \Delta\epsilon_s \tag{4.24-35}$$

where V is the volume of the physical space to which the particles are confined and m is their mass. It can be derived as follows: The energy eigenfunctions u of a free particle of mass m in a rectangular box with impenetrable walls satisfy the Schrödinger eigenvalue equation

$$-\frac{h^2}{2m}\left(\frac{\partial^2}{\partial x_2} + \frac{\partial^2}{\partial y^2} + \frac{\partial^2}{\partial z_2}\right) u = \epsilon u \tag{4.24-36}$$

subject to the requirement that u vanish at the walls. The appropriate solutions are of the form

$$u_{lmn} \propto \sin l\frac{\pi x}{a} \sin m\frac{\pi y}{b} \sin n\frac{\pi z}{c} = \sin(k_x x)\sin(k_y y)\sin(k_z z) \qquad (4.24\text{-}37)$$

$$k_x = \frac{l\pi}{a}, \qquad k_y = \frac{m\pi}{b}, \qquad k_z = \frac{n\pi}{c} \qquad (4.24\text{-}38)$$

where (a, b, c) are the sides of the box and l, m, n are any positive integers. A vector may be associated with each state

$$\mathbf{k} = \mathbf{i}k_x + \mathbf{j}k_y + \mathbf{k}k_z \ (\mathbf{x}, \mathbf{y}, \mathbf{z} = \text{unit Cartesian vectors})$$

The state energy is

$$\epsilon_{l,m,n} = \frac{\hbar^2}{2m}(k_x^2 + k_y^2 + k_z^2)_{l,m,n} \qquad (4.24\text{-}39)$$

$$= \frac{\hbar^2 k_{l,m,n}^2}{2m} \qquad (4.24\text{-}40)$$

Since a change of any of l or m or n by unity results in a new eigenstate u_{lmn}, a volume

$$d^3\mathbf{k} = \left(\frac{\pi}{a}\right)\left(\frac{\pi}{b}\right)\left(\frac{\pi}{c}\right) = \frac{\pi^3}{V} \qquad (4.24\text{-}41)$$

in k_x, k_y, k_z space may be associated with each eigenstate. The total number of eigenstates whose k parameter ($k^2 = k_x^2 + k_y^2 + k_z^2$) falls between zero and k is thus

$$N(k) = \frac{(\frac{1}{8})(4\pi k^3/3)}{\pi^3/V} = \frac{k^3 V}{6\pi^2} \qquad (4.24\text{-}42)$$

where $4\pi k^3/3$ is the volume in k_x, k_y, k_z space occupied by the states' $\mathbf{k}$ vectors. The $(\frac{1}{8})$ factor accounts for the fact that only positive values of k_x, k_y and k_z are counted. π^3/V is the volume per eigenstate.

The number of states with k values between k and $k + dk$ is then

$$\frac{dN(k)}{dk}\,dk = \frac{3k^2 V}{6\pi^2}\,dk \qquad (4.24\text{-}43)$$

and using $\epsilon_k = \hbar^2 k^2/2m$, $dk = m\,d\epsilon_k/\hbar^2 k$, gives

$$g_s = \frac{4\pi V m(2m)^{1/2}}{h^3}\epsilon_s^{1/2}\,\Delta\epsilon_s \qquad (4.24\text{-}44)$$

for the number of states in $\Delta\epsilon_s$. This is the desired result. The total number of particles is

$$\sum_{s=1}^{\infty} n_s = \sum_{s=1}^{\infty} \frac{g_s}{e^{\alpha_1 + \beta\epsilon_s}} = N \qquad (4.24\text{-}45)$$

If the density of particles is large, the summation can be replaced by integration

$$N = \frac{4\pi V m (2m)^{1/2}}{h^3} e^{-\alpha_1} \int_0^\infty \epsilon^{1/2} e^{-\beta\epsilon} \, d\epsilon \tag{4.24-46}$$

With the change of variable $u = \beta\epsilon$, Eq. (4.24-46) can be rewritten as

$$\begin{aligned} N &= \frac{4\pi V m (2m)^{1/2}}{h^3} e^{-\alpha_1} \int_0^\infty \frac{1}{\beta} \epsilon^{1/2} e^{-\beta\epsilon} \, d(\beta\epsilon) \\ &= \frac{4\pi V m (2m)^{1/2}}{h^3} e^{-\alpha_1} \beta^{-3/2} \int_0^\infty (\beta\epsilon)^{1/2} e^{-\beta\epsilon} \, d(\beta\epsilon) \\ &= \frac{4\pi V m (2m)^{1/2}}{h^3} e^{-\alpha_1} \beta^{-3/2} \int_0^\infty u^{1/2} e^{-u} \, du \\ &= \frac{4\pi V m (2m)^{1/2}}{h^3} e^{-\alpha_1} \beta^{-3/2} \Gamma\left(\frac{3}{2}\right) \end{aligned}$$

where the definition of the gamma function

$$\Gamma(x) = \int_0^\infty e^{-u_u x - 1} \, du \tag{4.24-48}$$

has been used. Similarly, the total energy of the N particles is given by

$$\begin{aligned} E &= \sum_{s=1}^{\infty} n_s \epsilon_s \\ &= \frac{4\pi V m (2m)^{1/2}}{h^3} e^{-\alpha_1} \int_0^\infty \epsilon^{3/2} e^{-\beta\epsilon} \, d\epsilon \\ &= \frac{4\pi V m (2m)^{1/2}}{h^3} e^{-\alpha_1} \beta^{-5/2} \Gamma\left(\frac{5}{2}\right) \end{aligned} \tag{4.24-49}$$

Dividing (4.24-49) by (4.24-47) gives

$$\frac{N}{E} = \beta \frac{\Gamma(\frac{3}{2})}{\Gamma(\frac{5}{2})} = \beta \frac{\Gamma(\frac{3}{2})}{\frac{3}{2}\Gamma(\frac{3}{2})} = \frac{2}{3}\beta \tag{4.24-50}$$

where the properties of gamma functions, such as $\Gamma(1) = 1$, $\Gamma(\frac{1}{2}) = \sqrt{\pi}$, $\Gamma(\frac{3}{2}) = \sqrt{\pi}/2$, and $\Gamma(x+1) = x\Gamma(x)$ have been used. Since the N particles have been chosen sufficiently heavy as to make them distinguishable, the results of classical statistical mechanics apply, namely, that the average energy per particle at thermal equilibrium is

$$\frac{E}{N} = \frac{3}{2} kT \tag{4.24-51}$$

where k is the Boltzmann constant and T is the temperature. Using this in (4.24-50) finally gives

$$\beta = \frac{1}{kT} \tag{4.24-52}$$

The value of β in the above derivation has been based on taking boltzons as an example. However, since β is a common characteristic of systems in thermal equilibrium, it will have the same value for fermions and bosons. To complete the derivation of the quantum-statistical laws, however, the parameter α will have to be determined. This requires more specifics about the system of particles. Some important examples are considered below.

Case (a): Boltzons

For particles obeying Maxwell–Boltzmann statistics, α is obtained directly from (4.24-47) after using $\beta = (kT)^{-1}$, $\Gamma(\frac{3}{2}) = \sqrt{\pi}/2$:

$$\alpha = \ln\left[\frac{2\pi^{3/2} V m (2m)^{1/2}}{h^2 N} (kT)^{3/2}\right] \tag{4.24-53}$$

where N is the number of particles, m is their mass, and V is the volume to which they are confined.

The fraction of the number of particles with energy ϵ_i is then obtained from (31):

$$f_i = \frac{n_i}{N} = \frac{g_i e^{-\alpha - \epsilon_i/kT}}{\sum_s g_s e^{-\alpha - \epsilon_s/kT}} = \frac{g_i e^{-\alpha - \epsilon_i/kT}}{\sum_s g_s e^{-\epsilon_s/kT}} \tag{4.24-54}$$

and is independent of α. The ratio of the populations of any two levels, say i and j, is thus

$$\frac{f_i}{f_j} = \frac{g_i}{g_j} e^{-(\epsilon_i - \epsilon_j)/kT} \tag{4.24-55}$$

and is known as the Boltzmann ratio.

Case (b): Fermions

The distribution function of identical, indistinguishable particles of half-odd-integral spin (Fermions) is given in (4.24-32) as

$$n_s = \frac{g_s}{e^{\alpha + \beta\epsilon_s} + 1} \tag{4.24-56}$$

Using $\beta = (kT)^{-1}$ and defining

$$\mu \equiv -\alpha kT \tag{4.24-57}$$

(4.24-56) may be rewritten as

$$n_s = \frac{g_s}{e^{(\epsilon_s - \mu)/kT} + 1} \tag{4.24-58}$$

Dividing (4.24-58) by g_s gives the occupation n_s/g_s (number of particles per state) as

$$n(\epsilon) \equiv f(\epsilon) = \frac{1}{e^{[\epsilon - \mu(T)]/kT} + 1} \tag{4.24-59}$$

The quantity μ is called the chemical potential of the system of particles. The function $f(\epsilon)$, the Fermi–Dirac distribution law, gives the probability that the quantum state ϵ is occupied. It follows that $f(\epsilon) \leq 1$. A plot of $f(\epsilon)$ at $T = 0$ and also at some finite temperature $T > 0$ is as shown in Fig. 4.24-2.

At $T = 0, f(\epsilon) = 1$ for $\epsilon < \mu$ and $f(\epsilon) = 0$ for $\epsilon > \mu$. In other words, at zero temperature the states whose energies lie below the chemical potential $\mu(0)$ are occupied, while those above it are empty. At finite temperatures the transition from occupied states is "fuzzed" over a region of width $\sim kT$, as shown in the figure. At energies ϵ such that $(\epsilon - \mu) >> kT, f(\epsilon) \sim e^{-[e - \mu(T)/kT]}$. This exponential dependence of the occupation probability is the same as that of the Boltzmann law (4.24-55) so that in this limit the statistical behavior of fermions and of classical particles is similar.

Case (c): Bosons and Photons

The general form of the Bose distribution is

$$n(\epsilon) = \frac{1}{e^{\alpha} e^{\varepsilon/kT} - 1} \tag{4.24-60}$$

Bosons are particles of integral spin (0, 1, 2, . . .) which obey Bose–Einstein statistics. There is no limit to the number of bosons that can be in any single-particle state. these single-particle states (orbitals) are given by solving Schrödinger's time-independent equation for a particle in a cubical box. The density of states, that is, the number of states lying between ϵ and $\epsilon + d\epsilon$ is

$$N(\varepsilon)\, d\varepsilon = \frac{4\pi V}{h^3} (2m^3)^{1/2}\, \varepsilon^{1/2}\, d\varepsilon \tag{4.24-61}$$

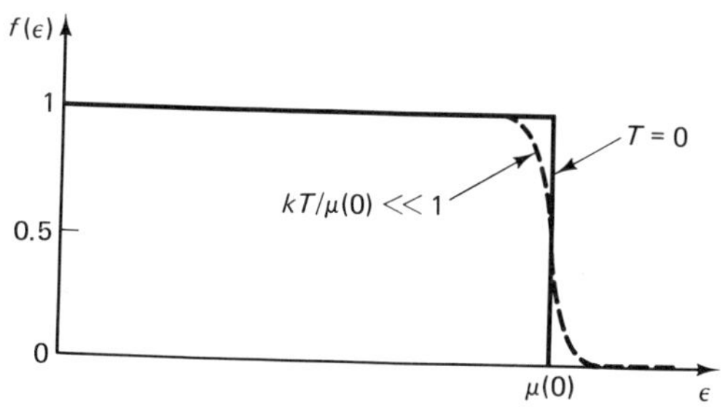

Figure 4.24-2 Fermi–Dirac distribution function at absolute zero and at a low temperature. The width of the transition region between $f(\epsilon) \simeq 1$ and $f(\epsilon) \simeq 0$ is of the order of kT.

where $V = a^3$, the volume of the box. To apply (4.24-60) to bosons whose total number N of particles remains fixed, like helium, the parameter α must first be determined. This is done by setting

$$N = \int_0^\infty n(\varepsilon)\, N(\varepsilon)\, d\varepsilon \tag{4.24-62}$$

where $N(\varepsilon)\, d\varepsilon$ is the number of quantum states of the system in an energy interval ε to $(\varepsilon + d\varepsilon)$, and $n(\varepsilon)$ is the number of bosons per quantum state, so that the integral is just the total number. Using Eqs. (4.24-60), (4.24-61), and carrying out the integration gives

$$\begin{aligned} N &= \int_0^\infty \frac{N(\varepsilon)\, d\varepsilon}{e^\alpha e^{\varepsilon/kT} - 1} \\ &= \frac{(2\pi mkT)^{3/2} V}{h^3} e^{-\alpha} \left(1 + \frac{1}{2^{3/2}} e^{-\alpha} + \frac{1}{3^{3/2}} e^{-2\alpha} + \cdots \right) \end{aligned} \tag{4.24-63}$$

To simplify the appearance of this equation, let $e^{-\alpha} = A$ so that

$$N = \frac{(2\pi mkT)^{3/2} V}{h^3} A \left(1 + \frac{1}{2^{3/2}} A + \frac{1}{3^{3/2}} A^2 + \cdots \right) \tag{4.24-64}$$

For large mass m and high temperature T, A must be very small since it is fixed. In these circumstances, terms beyond the first power in A can be dropped. But large m and high T should be the classical region. Indeed, it is found that the first term gives the classical Boltzmann result

$$N = \frac{(2\pi mkT)^{3/2} V}{h^3} A \tag{4.24-65}$$

or

$$A = \frac{Nh^3}{(2\pi mkT)^{3/2} V} = e^{-\alpha} \tag{4.24-66}$$

Note that $A = e^{-\alpha}$ is proportional to N, as in the Boltzmann result for a system of classical oscillators. Also note that since N is fixed α must be very large (as A is very small), in contrast to the fact that α is zero for a system of bosons in which N varies.

The total energy E of the ideal gas is

$$\begin{aligned} E &= \int_0^\infty \varepsilon\, n(\varepsilon)\, N(\varepsilon)\, d\varepsilon \\ &= \frac{(2\pi mkT)^{3/2} V}{h^3} V \left(\frac{3}{2} kT\right) A \left(1 + \frac{1}{2^{5/2}} A + \frac{1}{3^{5/2}} A^2 + \cdots \right) \end{aligned} \tag{4.24-67}$$

Once again the classical result follows for very small values of A. Neglecting terms beyond the first power in A, and using (4.24-66), this corresponds to an average energy per particle E/N equal to $(3/2)\, kT$, which is the classical equipartition of energy result for three-dimensional translation motion.

The general Bose result for the average energy per particle is obtained by dividing (4.24-67) by (4.24-64):

$$\bar{E} = \frac{E}{N} = \frac{3}{2}kT\left[1 - \frac{1}{2^{5/2}}\frac{Nh^3}{V(2\pi mkT)^{3/2}}\right] \tag{4.24-68}$$

The term beyond 1 in the bracketed expression represents the deviation of the Bose gas from the classical gas. This is sometimes called the *degeneracy effect.* Note that the degeneracy term is negative so that the average particle energy is less for a Bose gas than for a classical gas.

The chemical energy of an ideal boson gas is always less than the energy of the lowest-energy single-particle state. Choosing the energy of the lowest-energy single-particle state (the ground state) as the zero of the energy scale, it follows that the chemical potential μ of an ideal boson gas is always negative. Its variation with temperature is shown in Fig. 4.24-3, being always negative at temperatures above the Bose temperature T_B given by

$$T_B = \frac{3.31\hbar^2}{mk}\left(\frac{N}{gV}\right)^{2/3} \tag{4.24-69}$$

where m is the mass of a boson and g is the spin degeneracy. The chemical potential is virtually zero up to the Bose temperature as shown in Fig. 4.24-3.

The variation of the chemical potential μ of an ideal fermion gas with temperature is shown in Fig. 4.24-4. At $T = 0$, the chemical potential $\mu(0)$ is equal to the Fermi energy ϵ_F. When T is very much less than the Fermi temperature given by

$$kT_F = \frac{\hbar^2}{2m}\left(\frac{6\pi^2 N}{gV}\right)^{2/3}$$

the chemical potential is only just below the Fermi energy. As the temperature increases further, the chemical potential decreases significantly, and is zero at a temperature just below the Fermi temperature T_F. Then, as T increases above T_F, the chemical potential gets more and more negative.

The Bose distribution with $\alpha = 0$ applies to a photon and a phonon gas. Just as the wave particle duality for electromagnetic radiation was used to

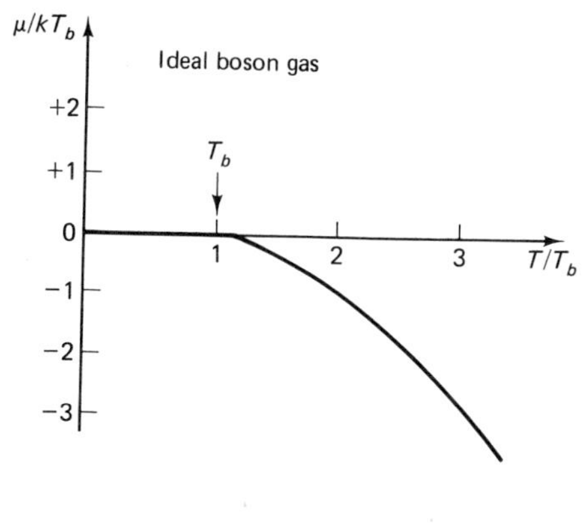

Figure 4.24-3 Variation of the chemical potential of an ideal boson gas with temperature.

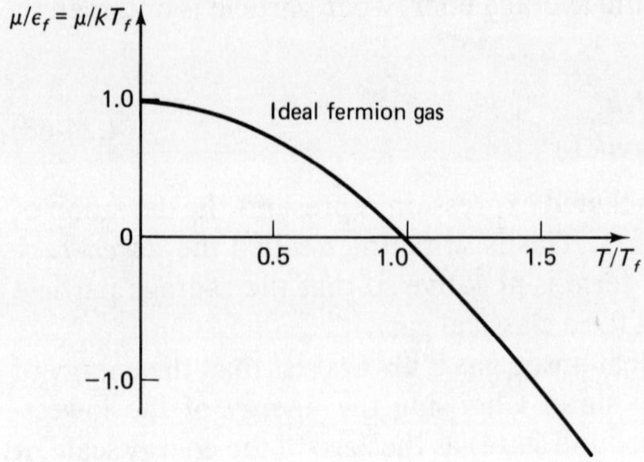

Figure 4.24-4 Variation of the chemical potential of an ideal fermion gas with temperature.

derive the thermally excited distribution of radiation in a cavity either on a wave picture or a particle picture, so the thermally excited distribution of elastic vibrations in a solid can be deduced by applying a wave particle duality for acoustic radiation. Just as photons are the quanta of electromagnetic radiation, so phonons are the quanta of acoustic radiation. Photons are emitted and absorbed by vibrations of the atoms in a cavity wall; phonons are emitted and absorbed by vibrating atoms at the lattice points in the solid. The sources of each type of radiation are quantized, and the energy gain or loss is discrete (the discrete energy transfer having a value $h\nu$, where ν is the frequency of the acoustic vibration for phonons and of the electromagnetic vibration for photons). Just as the number of photons is not fixed or conserved, so the number of phonons is not fixed or conserved. Thus, the Bose distribution with $\alpha = 0$ applies to phonon and to photon. (There are differences, of course, between photons and phonons. The photon propagates through vacuum, whereas the phonon propagates through a crystal lattice.)

SYMBOLS IN CHAPTER 4

A, a	Helmholtz function
C, c	Specific heat
$D(\varepsilon)$	Density of states
E, ε	Energy
F, f	Distribution function
G, g	Gibbs function, degeneracy
H, h	Enthalpy, Planck constant
k	Boltzmann constant
l	Mean free path
m	Mass
N, n	Number of particles
$\hat{o}$	Operator
P, p	Probability, pressure, momentum

Q, q	Heat transfer
R	Gas constant
S, s	Entropy
T	Temperature
U, u	Internal energy
V, v	Volume
W	Thermodynamic probability, work
Z, z	Partition function
α, β, γ	Lagrangian multipliers
λ	Absolute activity
μ	Chemical potential
ν	Frequency, Hz
Ξ	Grand canonical partition function
ψ	Wave function
Ω, ω	Grand potential, circular frequency

PROBLEMS

4.1. A deuteron is a bound system composed of a proton and a neutron. It is the simplest nucleus having more than one nucleon.

(a) What is its spin S?

(b) What is its binding energy?

(c) What is the energy of a proton (γ ray) emitted when a photon and neutron come together to form a deuteron?

4.2. For a cavity of volume V which contains blackbody radiation, show that the number of modes of the radiation field with frequencies between ν and $(\nu + d\nu)$ is given by

$$dN = \frac{8\pi\nu^2\,d\nu}{c^3}\,V$$

where c is the velocity of light.

4.3. Considering a cavity at temperature T to be a container of photons, which are spin-1 bosons, determine the spectral distribution (the amount of energy per frequency interval) of the blackbody radiation coming from a small hole in the container. Note that because photons are continuously being emitted and absorbed by the walls of the cavity, their number is not conserved. The density of states in a frequency interval $d\nu$ is

$$g(\nu) = \frac{8\pi V}{c^3}\,\nu^2$$

where V is the volume of the cavity.

4.4. Let the spectral distribution for photons, $F(\nu)$, be given by Planck's equation

$$dE_\nu = h\nu\,dn_\nu = \frac{8\pi Vh}{c^3}\,\frac{\nu}{e^{h\nu}/kT - 1}\,d\nu = F(\nu)\,d\nu$$

where dE_ν is the energy in the frequency range between ν and $(\nu + d\nu)$. Show that the total electromagnetic energy inside a cavity whose walls are maintained at a temperature T is proportional to T^4 (Stefan–Boltzmann law). Evaluate the proportionality factor. What is the emitting power of a blackbody?

4.5. A plot of the energy density $\rho(\lambda)$ of the radiation in equilibrium with a blackbody versus λ shows a maximum at $\lambda = \lambda_{\text{max}}$. It is seen that the peak of the spectral distribution curve shifts upward in frequency as the temperature increases. The Wien displacement law states that

$$\lambda_{\text{max}} T = \text{const}$$

Show that Wien's displacement law follows from Planck's radiation law.

4.6. The spectral distribution for a container of photons is given by

$$dE_\nu = h\nu \, dn_\nu = \frac{8\pi V h}{c^3} \frac{\nu^3}{e^{h\nu/kT} - 1} d\nu = F(\nu) \, d\nu$$

What is the ratio of the energy emitted from a 2000 K blackbody in wavelength bands of width 100 Å centered on 5000 Å (visible) and 50,000 Å (infrared)?

4.7. In the free electron theory of metals, the electrons in a metal are treated like a gas composed of noninteracting spin $\frac{1}{2}$ fermions confined to a three-dimensional box. For a three-dimensional well of side l, the allowed energies for noninteracting particles of mass m are

$$E = (n_x^2 + n_y^2 + n_z^2) E_0 ; \qquad E_0 = \frac{h^2}{8ml^2}$$

where n_x, n_y, and n_z are positive integers. Determine the number of states (n_x, n_y, n_z) with energies in the interval E and $E + dE$.

4.8. Draw a graph of the Fermi–Dirac distribution at $T \simeq 0$ K and $T > 0$ K. Obtain an expression for $E_{F,0}$, the Fermi energy at $T = 0$ K for an electron gas in a metal.

4.9. Metallic potassium has a density of 0.86×10^3 kg/m^3 and an atomic weight of 39. What is the Fermi energy for the electrons in the metal if each potassium atom donates one electron to the electron gas?

4.10. Show that the average kinetic energy per particle for a Fermi gas of N particles at $T = 0$ K is

$$E_{\text{av}} = \tfrac{3}{5} E_{F,0}$$

where $E_{F,0}$ is the Fermi energy at 0 K.

4.11. A series in p which is useful in statistical mechanics is

$$\int_0^\infty E^p F_{FD} \, dE = \int_0^\infty \frac{E^p}{e^{(E - E_F)/kT} + 1} dE$$

$$= \frac{1}{p+1} \left[E_f^{p+1} + \sum_{n=1}^{\infty} 2(kT)^{2n} \left(1 - \frac{1}{2^{2n-1}}\right) \zeta(2n) \frac{d^{2n}}{dE_f^{2n}} (E_f^{p+1}) \right] \tag{1}$$

where E is the energy, F_{FD} is the Fermi–Dirac distribution function, $p > -1$, $kT << E_F$, and $\zeta(x)$ is the Riemann zeta function,

$$\zeta(x) = \frac{1}{1^x} + \frac{1}{2^x} + \frac{1}{3^x} + \cdots$$

which is extensively tabulated. Some values are

$$\zeta(2) = \frac{\pi^2}{6} = 1.645 \qquad \zeta(4) = \frac{\pi^4}{90} = 1.082 \qquad \zeta(6) = \frac{\pi^6}{945} = 1.017$$

Using (1), and keeping only the first term of the series, obtain, for low temperatures, an expression for the Fermi energy E_F of an electron gas in terms of $E_{F,0}$, the Fermi energy at $T = 0$ K.

4.12. Silver has a Fermi energy of 5.5 eV at $T = 0$ K. What is the Fermi energy at $T = 300$ K and the size of the first-order correction?

4.13. The work function, ϕ, is the minimum energy required to remove an electron from a material. At 0 K, silver has a Fermi energy of 5.5 eV and a work function of 4.6 eV. What is the average electrostatic potential energy seen by the free electrons in silver? What is the prospect of finding a metal that will be an efficient emitter of electrons under illumination by infrared light ($\simeq$7000 Å)?

4.14. Consider the normal to a metal surface to be the x axis. The rate of electron emission is obtained by multiplyig the density of electrons with sufficient energy to pass into vacuum by the velocity component in the x direction, and then integrating over all velocities subject to the restriction that the energy $\frac{1}{2}mv_x^2$ must be larger than $(E_F + q\phi)$ in order for the electron to have sufficient energy to be emitted into the vacuum. For a density of occupied states in the metal given by

$$n(E)\,dE = f(E)\,N_v(E)\,dE = \frac{8\pi m^3}{h^3}\,\frac{v^2\,dv}{\exp[(E - E_F)]/kT] + 1}$$

where use has been made of the relationships, $\frac{1}{2}mv^2 = E$, and $dE = mv\,dv$, obtain the current density J_x (Richardson's equation).

4.15. Evaluate the average energy, $\langle\epsilon\rangle$, of a system of oscillators, **(a)** according to classical mechanics, **(b)** according to quantum mechanics, in which case the energy of the oscillator can only take on values of $\epsilon_i = ih\nu$. Each quantum $h\nu$ of acoustical energy is a phonon.

4.16. In a photoelectric effect experiment, photon of energy $h\nu$ strikes the surface of a metal. If the energy of the photon is greater than the *work function* $e\phi$, the electron may be ejected with a maximum kinetic energy of $\frac{1}{2}mv^2$. A "repeller" potential can be used to measure $\frac{1}{2}mv^2$ by finding the minimum voltage V_0 necessary to stop the electrons. Show how Planck's constant may be measured from a plot of stopping voltage versus ν. Determine the threshold frequency.

4.17. An Einstein crystal is defined to be a collection of N three-dimensional oscillators in thermal equilibrium at temperature T. From the equation for the average energy of an oscillator,

$$\langle\epsilon\rangle = \frac{h\nu}{e^{h\nu/kT} - 1}$$

derive an equation for the heat capacity C_V of such a crystal. Determine the high and low limits of C_V and plot C_V versus $kT/h\nu$.

4.18. In the Debye theory a crystalline solid is viewed as being composed of regularly spaced atoms in a three-dimensional lattice. If an atom is displaced from its equilibrium position, it will experience a restoring force due to the surrounding atoms, the atoms behaving like a collection of coupled oscillators. Any disturbance will be transmitted to the surrounding atoms, resulting in a wave propagating through the solid. The amount of energy transferred from one atom to its neighbor is quantized in amounts of $h\nu$, with ν being the classical frequency at which the atom vibrates about its equilibrium position. Each quantum $h\nu$ of acoustical energy is a phonon, in analogy to the photons of electromagnetic radiation.

In a crystal it is found that there is a maximum frequency of vibration, called the Debye frequency, ν_D. The maximum frequency exists because a system of N molecules possesses only $3N$ modes of vibration (each molecule having three degrees of independent vibrational freedom). Determine the Debye frequency and the density of states for the crystal.

4.19. Acoustical energy is transferred through a crystal lattice in quantized amounts of $h\nu$ by phonons, which are bosons whose total number, like that of photons, is not fixed. Obtain an expression for the total kinetic energy (vibrational energy) of the crystalline solid, if the density of states for phonons is

$$g(\nu) = \frac{9N}{\nu_D^3}\,\nu^2$$

where ν_D is the Debye frequency.

4.20. The total energy E of a solid is the sum of all the phonon energies:

$$E = \int dE = \frac{9Nh}{\nu_D^3}\int_0^{\nu_D} \frac{\nu^3}{e^{h\nu/kT}-1}\,d\nu = 9NkT\left(\frac{T}{T_D}\right)^3 \int_0^{q_D} \frac{q^3\,dq}{e^q - 1}$$

where $q_D = h\nu_D/kT$ and $T_D = h\nu_D/k$ (called the Debye temperature). Obtain an expression for C_V, the molar specific heat at constant volume, in the limit $T \ll T_D$.

4.21. The total energy of a solid is

$$E = 9NkT\left(\frac{T}{T_D}\right)^3 \int_0^{q_D} \frac{q^3\,dq}{e^q - 1}$$

where $q_D = h\nu_D/kT = T_D/T$. Show that $C_V = 3R$ (Dulong–Petit law) follows from this in the limit $T \gg T_D$.

4.22. The total energy E of a solid is the sum of all the phonon energies:

$$E = \int dE = \frac{9Nh}{\nu_D^3}\int_0^{\nu_D} \frac{\nu^3}{e^{h\nu/kT}-1}\,d\nu = 9NkT\left(\frac{T}{T_D}\right)^3 \int_0^{q_D} \frac{q^3\,dq}{e^q - 1}$$

where $q_D = h\nu_D/kT$ and $T_D = h\nu_D/k$ (called the Debye temperature). (The integral must be evaluated numerically.) Obtain an expression for C_V at an arbitrary temperature.

4.23. The total energy of an electron gas is given by

$$E_T = \int_0^{\infty} E F_{FD}\, g(E)\, dE$$

where

$$g(E) = \frac{2\pi V(2m)^{3/2}}{h^3} E^{1/2} = CE^{1/2} \quad \text{and} \quad F_{FD} = \frac{1}{e^{(E-E_F)/kT}+1}.$$

Find the total energy of an electron gas at low temperature.

4.24. In the free-electron theory of metals it is assumed that the weakly bound valence electrons of the atoms composing the metal are not bound to particular atoms but move throughout the entire solid. For such an electron gas, the total energy is given by the first-order approximation

$$E_T = \frac{3}{5} NE_{F,0}\left[1 + \frac{5\pi^2}{12}\left(\frac{kT}{E_{F,0}}\right)^2\right]$$

where $E_{F,0}$ is the Fermi energy at absolute zero.

(a) Find, to the first order, the molar electronic specific heat at constant volume.

(b) Calculate the electronic molar specific heat, $C_{V,el}$, for silver at room temperature ($T = 300$ K). Silver has valence 1, and a Fermi energy $E_{F,0} = 5.5$ eV.

4.25. The lattice molar specific heat of a solid at low temperature is given by

$$C_V = \frac{12\pi^4 R}{5}\left(\frac{T}{T_D}\right)^3$$

The electronic molar specific heat, on the other hand, is given by

$$C_{V,l} = \frac{R\pi^2 kT}{E_{F,0}}$$

where $E_{F,0}$ is the Fermi energy at absolute zero. For silver find the temperature at which the electronic molar specific heat, $C_{V,el}$, and the lattice molar specific heat, C_V, are equal. The Debye temperature for silver is 210 K.

4.26. An ideal gas is one in which the gas molecules are noninteracting. However, since an ideal gas will be composed of either fermions or bosons, it should be analyzed from a quantum-statistical viewpoint. The number of particles dn_E having energies in the interval E and $(E + dE)$ is

$$\mathrm{d}n_{\mathrm{E}} = \begin{cases} F_{FD}\, g(E)\, dE \\ F_{BE}\, g(E)\, dE \end{cases}$$

where $g(E)$ is the density of states, and

$$F_{FD} = \frac{1}{e^{\alpha} e^{E/kT} + 1}$$

$$F_{BE} = \frac{1}{e^{\alpha} e^{E/kT} - 1}$$

for fermions and bosons, respectively. Consider a gas composed of a fixed number, N, of bosons in a container of volume V. Show that α is a strictly increasing function of the temperature T.

4.27. The number of bosons in a container of volume V is given by

$$N = \frac{2\pi V(2mkT)^{3/2}}{h^3}\int_0^{\infty}\frac{q^{1/2}\,dq}{e^{\alpha}e^{q} - 1}$$

where $q = E/kT$. For a given particle density, N/V, find the lowest possible temperature of the boson gas consistent with Bose–Einstein statistics.

4.28. The number, N, of bosons in a container of volume V is given by

$$N = \int dn_E = \int_0^{\infty} EF_{BE}\, g(E)\, dE$$

where

$$g(E) = \frac{2\pi V(2m)^{3/2}E^{1/2}}{h^3}$$

is the density of states, and

$$F_{BE} = \frac{1}{e^{\alpha}e^{E/kT} - 1}$$

is the Bose–Einstein distribution function. Evaluate the total energy of the Bose gas.

4.29. The normalization condition for a gas composed of N spin $\frac{1}{2}$ fermions in a container of volume V is

$$N = \int dn_E = 2\int_0^\infty F_{FD}\, g(E)\, dE = \frac{4\pi V(2m)^{3/2}}{h^3}\int_0^\infty \frac{E^{1/2}\, dE}{e^\alpha e^{E/kT}+1}$$

$$= \frac{4\pi V(2m)^{3/2}}{h^3}\int_0^\infty \frac{q^{1/2}\, dq}{e^\alpha e^q + 1}$$

where

$$g(E) = \frac{4\pi V(2m)^{3/2}\, E^{1/2}}{h^3}$$

is the density of states,

$$F_{FD} = \frac{1}{e^\alpha e^{E/kT}+1}$$

is the Fermi–Dirac distribution, and $q = E/kT$. Show that α is a strictly increasing function of temperature. What is the effect of α being monotonically increasing with the mass of the fermions?

4.30. The total energy of a gas of N spin $\frac{1}{2}$ fermions in a container of volume V is given by

$$E_T = \int E\, dn_E = 2\int_0^\infty E F_{FD}\, g(E)\, dE$$

where

$$F_{FD} = \frac{1}{e^\alpha e^{E/kT}+1}$$

and

$$g(E) = \frac{2\pi V(2mkT)^{3/2}\, E^{1/2}}{h^3}$$

is the density of states. Evaluate the total energy of the fermion system assuming that $\alpha \geq 0$.

4.31. List and discuss examples of quantum systems which show marked differences from what is expected classically at or near absolute zero.

4.32. Bosons and fermions have fixed total numbers of particles given by

$$N_{BE} = \int_0^\infty F_{\mathrm{BE}}\, g(E)\, dE = \frac{2\pi V(2m)^{3/2}}{h^3}\int_0^\infty \frac{E^{1/2}\, dE}{e^\alpha e^{E/kT}-1}$$

$$N_{FD} = \int_0^\infty F_{\mathrm{FD}}\, g(E)\, dE = \frac{4\pi V(2m)^{3/2}}{h^3}\int_0^\infty \frac{E^{1/2}\, dE}{e^\alpha e^{E/kT}+1}$$

Find the kinetic energy per particle in the high-temperature limit $T >> 0$ K for the boson and fermion gases and show that the average energy for both gases is asymptotically equal to the classical value of $\frac{3}{2}kT$.

4.33. Classical thermodynamics gives the pressure of a gas as

$$p = \left(\frac{\partial E}{\partial V}\right)_s$$

Obtain an expression for the pressure of a quantum-mechanical gas of non-interacting particles in a three-dimensional cubic box of side L whose energy is

$$E_i = \frac{h^2}{8mL^2}(n_x^2 + n_y^2 + n_z^2) = \frac{h^2}{8mV^{2/3}}(n_x^2 + n_y^2 + n_z^2)$$

4.34. Classical thermodynamics gives the ideal gas law as

$$pV = NkT$$

What is the first-order correction at high temperatures when quantum-mechanical properties of the gas are taken into consideration?

4.35. The total energy of an electron gas at low temperature is given by

$$E \simeq \frac{3}{5} NE_{F,0}\left[1 + \frac{5\pi^2}{12}\left(\frac{kT}{E_{F,0}}\right)^2\right]$$

where $E_{F,0}$ is the fermion energy at $T = 0$ K. Obtain an expression for the pressure of an ideal spin $\frac{1}{2}$ fermion gas at low temperatures.

4.36. For a given energy level E_i in a system of fermions there will be a certain number g_i of states that will have this energy (i.e., E_i will have a degeneracy of order g_i). The maximum number of fermions that can occupy this level will therefore be g_i. According to the Pauli exclusion principle, no more than one particle can be found in each state. The number of different ways that N identical, indistinguishable fermions can be distributed among energy levels $E_1, E_2, \ldots, E_i, \ldots$, such that the ith energy level will have $n_i \leq g_i$ filled states $(n_1 + n_2 + \cdots = N)$, is given by

$$P(n_1, n_2, \ldots, n_s, \ldots) - \frac{g_1!}{n_2!(g_2 - n_2)!} \frac{g_2!}{n_2!(g_2 - n_2)} \cdots = \prod_{s=1}^{\infty} \frac{g_s!}{(g_s - n_s)!\, ns!}$$

Find the most probable distribution, subject to the conditions that the number of particles and total energy are held fixed.

4.37. Bosons are integer spin particles which do not obey the Pauli exclusion principle, so any number of bosons may be found in a given state. Suppose at the energy E_i there are g_i states. Find the number of different ways that N identical, indistinguishable bosons can be distributed among energy levels $E_1, E_2, \ldots, E_i, \ldots$, with a number n_i of particles in the ith level.

4.38. The total number of ways $x(n_1, n_2, \ldots, n_s, \ldots)$ to arrange $n_1, n_2, \ldots, n_s, \ldots$ bosons in the energy levels $E_1, E_2, \ldots, E_s, \ldots$ if there are $g_1, g_2, \ldots, g_s \ldots$ states in each level, is

$$x(n_s) = x(n_1, n_2, \ldots, n_s, \ldots)$$
$$= \prod_{s=1}^{\infty} \frac{(n_s + g_s - 1)!}{n_s!(g_s - 1)!}$$

Find the most probable distribution of bosons if the number of particles and total energy are held constant.

4.39. The Bose–Einstein distribution function is

$$F_{BE} = \frac{1}{e^{\alpha} e^{\beta E_i} - 1}$$

Show that a system of bosons which does not have a fixed number of particles (e.g., a system of photons in a blackbody cavity) has a Bose–Einstein distribution with $\alpha = 0$.

4.40. Statistical mechanics treats the photon as a boson of spin 1 and the photon gas as a special case of the ideal boson gas. Quantum theory of radiation, however, was first introduced by Planck in 1900 to explain the spectrum of blackbody radiation. Give a derivation for the radiant energy for unit volume in the angular frequency range $d\omega$ at ω (Planck's radiation law) based on the wave model.

4.41. To understand what does or does not constitute a microscopically distinguishable arrangement, consider a system of two identical particles with two allowed eigenstates, and sketch the possible microscopic arrangements if **(a)** the particles are distinguishable, **(b)** the particles are indistinguishable of half-odd-integral spin, and **(c)** the particles are indistinguishable of integral spin.

4.42. Determine the most probable distribution for a mixture of two species of particles A and B. Let A consist of particles obeying Maxwell–Boltzmann statistics and B consist of fermions.

4.43. The value of $\beta = 1/kT$ is usually determined by selecting a system of particles sufficiently heavy so as to make them distinguishable. Determine the value of β in the case of fermions and bosons.

4.44 **(a)** Evaluate the chemical potential $\mu(0)$, of a free-electron gas that is confined to a volume v at $T = 0$.

(b) What is the chemical potential (Fermi energy) of a metal with 10^{29} electrons/m^3?

(c) What is the velocity v_F of electrons in a metal near the top of the occupied states?

(d) How does the energy of electrons compare with those of bosons at zero temperature?

4.45. Explain what is meant by a degenerate and a nondegenerate Fermi gas, and give examples of applications.

4.46. Show that the chemical potential $\mu(T)$ corresponds to the increase of the total energy of a system when one particle is added to it. What happens when two materials with different chemical potentials are brought into contact with each other?

4.47. Viewed as a classical gas, the heat capacity per unit volume of an electron gas is

$$C = \frac{\partial E}{\partial T} = \frac{3}{2}\left(\frac{N}{V}\right) k$$

This result is bigger by some two orders of magnitude than the measured values at room temperature. Explain the nature of this discrepancy in light of the Pauli exclusion principle and the density of electrons $g(\epsilon) f(\epsilon)$.

4.48. The average energy per particle of a Bose gas is given by

$$\langle E \rangle = \frac{3}{2} kT \left[1 - \frac{1}{2^{5/2}} \frac{Nh^3}{V(2\pi mkT)^{3/2}} \right]$$

where the term beyond 1 in the bracketed expression represents the deviation of the Bose gas from the classical gas. Show that whenever the mean interparticle distance is comparable to or smaller than the DeBroglie wavelength assigned to particles on the basis of their temperature, quantum effects in the system of particles should be expected.

4.49. **(a)** What is the quantum theory of radiation?

(b) What maximum wavelength of photon will separate a molecule whose binding energy is 15 eV?

(c) What is the rate of emission of quanta from a radio station operating at a frequency of 103.7 MHz with a power output of 200 kW?

4.50. Solid-state physics uses a particle-type model for the excitation energy of a crystal lattice. The phonon is thought of as a quantum of energy of the lattice vibrations.

(a) Show that the density of phonons, that is, the number of phonon orbitals per unit frequency range is

$$d(\omega) = \frac{3V}{2\pi^2 v_s^3}\,\omega^2$$

where V is the volume of the solid and v_s is the average elastic wave velocity.

(b) Show that the Debye frequency is

$$\omega_D = \left(\frac{6\pi^2 N v_s^3}{V}\right)^{1/3}$$

(c) Obtain the Debye T^3 law for the heat capacity of a solid below the Debye temperature.

Chapter 5

Thermodynamics: Classical and Statistical

In this chapter the methods of statistical mechanics will be applied to determine the thermodynamic properties of systems in terms of their molecular properties. Apart from mass, the requisite molecular parameters are generally obtained from spectroscopy, statistical calculation of thermodynamic functions being frequently a more accurate method than direct thermal measurement.

All the key thermodynamic properties can be obtained in terms of the molecular partition function and the temperature for an assembly of independent systems. Molecular energy levels, as determined by spectroscopy, are directly applicable to the evaluation of partition functions. The most convenient line of development is therefore based on the canonical ensemble. Once the partition function $Z(V, N, T)$ has been evaluated, all the thermodynamic functions follow easily from the results of Chapter 2. This is the essential achievement of statistical thermodynamics, but on the way it provides much insight and information into molecular behavior.

5.1 THE LAWS OF THERMODYNAMICS

Thermodynamics (classical) is an experimental science based on a small number of principles that are generalizations made from experience. It is concerned only with macroscopic or large-scale properties of matter and makes no hypotheses about the structure of matter. The number of atoms in a macroscopic amount of matter is of the order of 10^{20} or more, and is so large that the behavior of the individual atoms or molecules is not explicitly considered; indeed, from the point of view of thermodynamics, the atomic constitution of matter is of little interest. General relations between such quantities as coefficients of expansion, compressibilities, heat capacities, heats of transformation, and magnetic and dielectric coefficients may be derived, especially in relation to temperature. The principles of thermodynamics indicate which few of these relations must be determined experimentally in order to completely specify all the properties of a system. Classical thermodynamics is based upon four laws of experience to which no exceptions have been found. These laws make no mention of the molecular concept. They may be stated as follows:

Zeroth Law If two bodies are independently in thermal equilibrium with a third, they are in thermal equilibrium with each other. Alternatively: the degree of hotness (or temperature) is a measurable property of any body internally at equilibrium.

First Law The internal energy of a completely isolated assembly is invariant with respect to time.

The first law of thermodynamics is the law of conservation of energy, that is, that energy is neither created nor destroyed. Its initial formulation resulted from the demonstration that the expenditure of a given amount of mechanical work resulted not in the loss of that energy but in the production of an equivalent amount of heat energy. The modern statement includes all forms of energy, electrical, gravitational, and electromagnetic energies, and matter itself, for which the equivalence relation is $E = mc^2$.

Let E equal the internal energy of the system. Then the mathematical statement of the first law is

$$dE = dQ - dW \tag{5.1-1}$$

where dE, the change in internal energy of the system, is equal to the heat dQ absorbed by the system less dW, the work done by the system. If the system goes from state 1 by any path to state 2, then to other states, eventually returning to state 1, the first law requires that

$$dE = 0 \tag{5.1-2}$$

If the right-hand side of Eq. (5.1-2) is not zero, then energy is not conserved and perpetual motion machines become possible.

Mathematically, Eq. (5.1-2) shows that dE is an exact differential. The internal energy E is thus a state function, and the change in internal energy of a system in going from state 1 to state 2 depends only on these states and not on the path between them. In contrast, dQ and dW are not exact, and the changes in these quantities do depend on the path followed in going from state 1 to state 2.

Second Law There are several ways of introducing the second law of thermodynamics, based on the experimental obervations that certain processes, allowed by the first law of thermodynamics, do not take place in practice. For example, let a copper block at an empirical temperature of 350 K be placed in a beaker of water, which is at an empirical temperature of 300 K. It is consistent with the first law of thermodynamics for the copper block to gain energy from the water, and for the water to lose an equal amount of energy. In practice, this process is never observed, but the copper block and water will reach thermal equilibrium at a temperature somewhere between 300 and 350 K. This result is summarized in the Clausius statement of the second law of thermodynamics:

> No process is possible whose sole result is the transfer of heat from a colder to a hotter body.

The term *sole result* is important, because it is possible to transfer heat from a cold system to a hotter system using a refrigerator, but in this case external work must be done on the working substance.

Another approach to the second law of thermodynamics is based on the experimental result that there is a strict upper limit to the efficiency of a heat engine. Though it would be consistent with the first law of thermodynamics to do so, it is not possible in practice to convert all the heat taken from a heat reservoir into work during a cyclic process which brings the working substance back to its original thermodynamic state. This is illustrated in terms of the Carnot cycle (see Fig. 5.1-1).

This type of experimental result is summarized in the Kelvin–Planck statement of the second law of thermodynamics:

> No process is possible whose sole result is the absorption of heat from a reservoir and the conversion of all of this heat into work.

It is possible to add a quantity of heat Q to a system and allow the system to do external work equal to, or even greater than Q, but then the working substance is not back in its initial state. The Kelvin–Planck statement of the second law shows that there must be a fundamental difference between the effects of heat and work on a system.

The traditional way of developing the concepts of thermodynamic temperature and entropy from the second law of thermodynamics is based on the

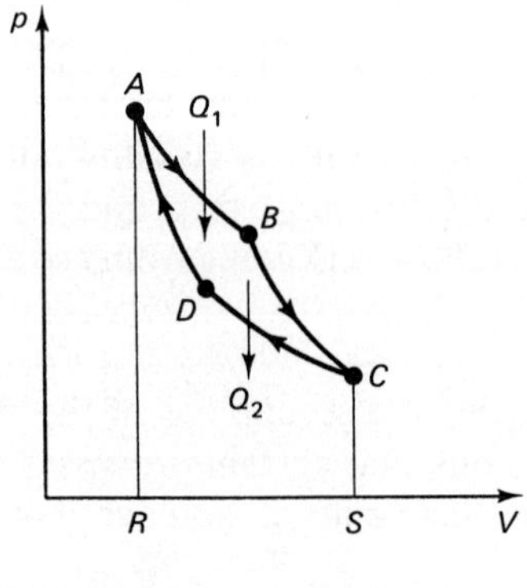

Figure 5.1-1 Carnot cycle indicator diagram.

application of these statements to heat engines. The second law is used to derive Carnot's theorems:

(a) No engine can be more efficient than a reversible engine working between the same temperature limits.
(b) All reversible engines working between the same two limits of temperature have the same efficiency.

Thus, the efficiency of a Carnot engine is independent of the working substance, and depends only on the temperatures of the hot and cold reservoirs. Kelvin defined the thermodynamic scale of temperature using the relation

$$\frac{Q_1}{Q_2} = \frac{T_1}{T_2} \tag{5.1-3}$$

where Q_1 is the heat taken in by the Carnot engine from the hot reservoir and Q_2 is the heat given up by the Carnot engine to the cold reservoir. According to (5.1-3), the ratio of the two temperatures T_1 and T_2 on the thermodynamic scale is equal to the ratio of the heats absorbed and rejected by a Carnot engine working between reservoirs at these temperatures.

The internationally agreed fixed point on the Kelvin thermodynamic scale is the temperature of the triple point of water, which by definition is exactly 273.16 K. The Kelvin thermodynamic scale is an absolute scale since, according to Carnot's theorem (a), the ratio Q_1/Q_2 is independent of the working substance. If a different working substance is used, the values of Q_1 and Q_2 would be different, but for all reversible engines the ratio of Q_1 to Q_2 would be the same, provided the temperatures of the reservoir and sink are unchanged.

The net work done by the gas per cycle is

$$W_1 - W_2 = Q_1 - Q_2 \tag{5.1-4}$$

From this, the efficiency of the Carnot engine is

$$\eta = \frac{W_1 - W_2}{Q_1} = \frac{Q_1 - Q_2}{Q_1} \tag{5.1-5}$$

Substitution of (5.1-3) into (5.1-5) gives

$$\eta = \frac{T_1 - T_2}{T_1} \tag{5.1-6}$$

Equation (5.1-6) is an important restriction on the maximum possible efficiency of a heat engine. For example, if T_1 is 400 K and T_2 is 300 K, the maximum possible efficiency of the Carnot engine is 25%. No heat engine can be more efficient than this, if it is working between the same two limits of temperature.

If the convention is adopted that the heat given to a system is positive and the heat given up by the system is negative, then for a reversible Carnot cycle, Eq. (5.1-3) becomes

$$\frac{Q_1}{T_1} + \frac{Q_2}{T_2} = 0 \tag{5.1-7}$$

where T_1 and T_2 are the thermodynamic temperatures of the hot and cold reservoirs, respectively. Equation (5.1-7) can be extended to any arbitrary reversible cycle to give

$$\oint \frac{dQ}{T} = 0 \tag{5.1-8}$$

This is Clausius' theorem. According to it, if a system is taken around a reversible cycle, and if the heat dQ_{rev} added reversibly to the system at each point is divided by the thermodynamic temperature of the system at that point, the sum of all such quotients is zero. This means that dQ_{rev}/T is an exact differential:

$$dS = \frac{dQ}{T} \tag{5.1-9}$$

According to (5.1-9), dS is the change in the entropy S of the system, when a quantity of heat dQ is added to the system in a reversible change from one equilibrium state to another equilibrium state. Thus, while dQ is not an exact differential, dQ/T is exact for a reversible heat engine operating through a Carnot cycle. Furthermore, any reversible cyclic process can be represented by a number of Carnot cycles, which leads to the important result that (5.1-8) applies to all reversible cyclic processes. Hence, dQ/T is the differential of a property of the system that, analogous to the internal energy, depends only on initial and final states and thus is a state function. This property is the *entropy*, and substitution of $dS = dQ/T$ into (5.1-1) gives

$$dE = T\,dS - dW = T\,ds - p\,dV$$

or

$$T\,dS = dE + p\,dV \tag{5.1-10}$$

Equation (5.1-10) is a combined statement of the first and second laws of thermodynamics. For a reversible process, $\Delta S = Q/T$; for a reversible process, $\Delta S = 0$; and for irreversible processes, $\Delta S > 0$. These considerations apply to macroscopic systems.

Third Law Integration of Eq. (5.1-9) from the absolute zero to a thermodynamic temperature T gives

$$S = \int \frac{dQ}{T} + S_0 \tag{5.1-11}$$

where S_0 is the entropy of the system at absolute zero of temperature. The value of S_0 cannot be determined using only the second law of thermodynamics. Nernst (1906) proposed that:

> The entropy change in any reaction between perfect crystals at a thermodynamics temperature of zero is zero.

Historically the third law arose from the heat theorem (1906) of Nernst. It states that, in an isothermal reaction involving condensed phases only, the entropy change approaches zero as the absolute temperature approaches zero, that is,

$$\lim_{T \to 0} \Delta S = 0 \tag{5.1-12}$$

As a consequence of Eq. (5.1-12), all condensed substances would have the same entropy as the absolute zero. Planck reasoned that this common entropy must be equal to zero on the basis that translational motion ceases at the absolute zero. Later work, a large part of which was done by G. N. Lewis and colleagues, showed that only perfect crystalline solids have zero entropy at 0 K. Statistically the entropy at absolute zero is best determined by the degeneracy of the lowest state.

The above laws are completely macroscopic in content. The general connection with statistical mechanism is

$$S = k \ln W \tag{5.1-13}$$

which allows thermodynamic properties to be related to the microscopic characteristics of a system. A large body of conclusions in thermodynamics follow from purely macroscopic considerations. This approach is fruitful and historically the oldest. However, the use of microscopic information and statistical methods to calculate W not only enhances the powers of prediction, but provides a more inclusive discipline. This is statistical thermodynamics, as contrasted to classical thermodynamics.

■ EXAMPLE 5.1-1

Temperature may be established as a function of a property of substances. Any system that changes in a regular, observable manner when its temperature changes may serve as a thermometer. Give examples of physical properties that may be used as a basis for a thermometer, and indicate how such a thermometer may be calibrated (determination of a base point and choice of a graduated scale).

Solution. Examples of physical properties that may be used as a basis for thermometers are the height of a liquid in a tube (alcohol or mercury), the volume of a gas (argon), and the electrical resistance of a wire (platinum). For example, extensive investigations have shown that, as the ice point pressure is reduced, the values of the ratio p_s/p_i for various gases approach each other, and in the limit, become equal at zero pressure:

$$\frac{T_s}{T_i} = \lim_{p_i \to 0} \frac{p_s}{p_i} \tag{5.1-14}$$

where T_s and T_i are the temperatures at the steam and ice points. Careful experimental determination shows that the right-hand side of Eq. (5.1-14) has a value of 1.36609 ± 0.00004. In other words, when the pressure in a gas thermometer at the ice-point temperature is reduced and approaches zero,

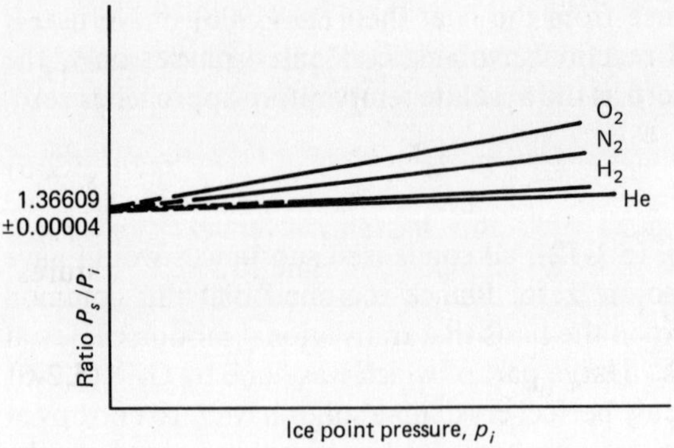

Figure 5.1-2 Convergence of p_s/p_i for various gases.

then a temperature may be measured that is independent of the gas. For the Celsius scale of temperature, $T_s - T_i = 100$, and $T_s/T_i = 1.36609$ gives

$$T_i = 273.15 \tag{5.1-15}$$

This temperature of 273.15 degrees is above the lowest conceivable temperature. Hence it is known as an absolute temperature. Absolute temperatures are expressed in degrees Kelvin in the Celsius system. Thus, Eq. (5.1-14) becomes

$$T_i = 273.15 \text{ K} \tag{5.1-15}$$

■

5.2 THERMODYNAMIC FUNCTIONS

The laws of thermodynamics give rise to a number of general equations which are of great practical value. According to the first law the change in internal energy when an infinitesimal reversible change takes place in a system of fixed material content is given by

$$dU = dE = dQ - dW \tag{5.2-1}$$

The heat absorbed in a reversible process is given by the second law as

$$dQ = T\,dS \tag{5.2-2}$$

and the work done by the system via a wall that can transmit pressure is

$$dW = p\,dV \tag{5.2-3}$$

where V is the volume. Reversibility in regard to work implies that the internal and external pressures differ only by infinitesimal amounts. Equation (5.2-1) thus becomes

$$dU = T\,dS - p\,dV \tag{5.2-4}$$

The restriction to fixed material content does not, of course, rule out chemical or phase changes within the assembly because these do not alter quantities of elements present. Thus, for a reversible process at constant E and V,

$$(dS)_{E,V} = 0 \quad \text{(reversible)} \tag{5.2-5}$$

Now, a heat engine working in a cycle that has an irreversible step is less efficient than a Carnot engine working between the same two temperatures. Hence,

$$(dS)_{E,V} > 0 \quad \text{(irreversible)} \tag{5.2-6}$$

Since spontaneous irreversible processes are moving toward equilibrium, Eqs. (5.2-5) and (5.2-6) give the criterion of equilibrium, that for a closed system at constant internal energy and volume, the entropy is a maximum at equilibrium.

While Eq. (5.2-4) embodies much of the first and second laws, it is convenient in practice to define three other thermodynamic quantities, the enthalpy H, the Helmholtz free energy A, and the Gibbs free energy G:

$$H \equiv U + pV \tag{5.2-7}$$

$$A \equiv U - TS \tag{5.2-8}$$

$$G \equiv H - TS \equiv U + pV - TS \tag{5.2-9}$$

Differentiation and substitution of (5.2-4) then yields

$$dH = T\,dS + V\,dp \tag{5.2-10}$$

$$dA = S\,dT - p\,dV \tag{5.2-11}$$

$$dG = -S\,dT + V\,dp \tag{5.2-12}$$

From Eqs. (5.2-11) and (5.2-12) and the knowledge of entropy, two additional criteria for equilibrium are obtained. In a closed system at constant temperature and volume the Helmholtz free energy is a minimum, and at constant temperature and pressure the Gibbs' free energy is a minimum. Also, two alternate definitions of entropy are possible:

$$-S = \left(\frac{\partial A}{\partial T}\right)_V = \left(\frac{\partial G}{\partial T}\right)_p \tag{5.2-13}$$

Incorporating these into (5.2-11) and (5.2-12) yields the Gibbs–Helmholtz equations

$$\left\{\frac{\partial(A/T)}{\partial T}\right\}_V = -\frac{U}{T^2} \tag{5.2-14}$$

$$\left\{\frac{\partial(G/T)}{\partial T}\right\}_p = -\frac{H}{T^2} \tag{5.2-15}$$

When a reversible process is carried out at constant temperature or constant temperature and pressure, (5.2-11) and (5.2-12) give

$$dA_T = -p\,dV = dW_{\text{rev}} \tag{5.2-16}$$

$$dG_{p,T} = 0 \tag{5.2-17}$$

The second of these expressions is widely used in thermodynamics to derive practical equations for chemical and phase equilibria.

■ EXAMPLE 5.2-1

A Carnot cycle using 1 mol of an ideal diatomic gas is operating between 500°C and 0°C. The volumes V_1 and V_2 at the beginning and end of the heat addition process are 0.0100 m^3 and 0.1000 m^3, respectively. Represent the cycle on $p-V$ and $T-S$ diagrams and calculate pertinent properties at all states of the cycle as well as the overall heat transfer and work done.

Solution. A Carnot engine is an idealized heat engine that consists of four processes: two isothermals and two reversible adiabatics. This results in the maximum net work that can be obtained from a quantity of heat taken in at the high temperature.

Denoting the temperatures T_1 and T_2 as T_h (high) and T_3 and T_4 as T_l (low), application of the first law ($dQ = dE + p\,dV$) to steps 2 and 4 yields

$$p\,dV = -dE = -nc_v\,dT \tag{5.2-18}$$

where n is the number of moles of the gas. Using the ideal gas law

$$pV = nRT \tag{5.2-19}$$

Eqs. (5.2-18) and (5.2-19) combine to give

$$\frac{nRT\,dV}{V} = -nc_v\,dT$$
$$\frac{dV}{V} = -\frac{c_v}{R}\frac{dT}{T} \tag{5.2-20}$$

This can be integrated for step 4 between the limits V_4, $T_4 = T_l$, and V_1, $T_1 = T_h$:

$$\int_{V_4}^{V_1} \frac{dV}{V} = -\frac{c_v}{R}\int_{T_l}^{T_h} \frac{dT}{T} \tag{5.2-21}$$

or

$$\ln\left(\frac{V_1}{V_4}\right) = -\frac{c_v}{R}\ln\left(\frac{T_h}{T_l}\right) \tag{5.2-22}$$

Taking exponents and rearranging:

$$V_1 T_h^{c_v/R} = V_4 T_l^{c_v/R} \tag{5.2-23}$$

and, similarly for step 2 between states 2 and 3,

$$V_2 T_h^{c_v/R} = V_3 T_l^{c_v/R} \tag{5.2-24}$$

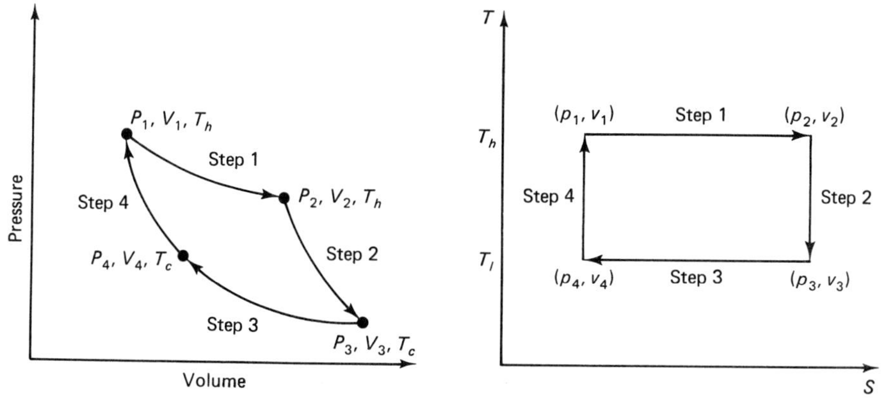

Figure 5.2-1 *p-v* and *T-S* diagram for Carnot cycle.

For a diatomic ideal gas, $c_v = \frac{5}{2}R$, so Eq. (5.2-24) gives

$$V_3 = V_2\left(\frac{T_h}{T_l}\right)^{5/2} = (0.1000\ \text{m}^3)\left(\frac{773\ \text{K}}{273\ \text{K}}\right)^{5/2}$$
$$= 1.349\ \text{m}^3$$

Similarly, dividing Eq. (5.2-23) by Eq. (5.2-24) gives

$$V_4 = V_3\left(\frac{V_1}{V_2}\right) = (1.349\ \text{m}^3)\left(\frac{0.0100}{0.1000}\right) = 0.1349\ \text{m}^3$$

Applying the first law ($Q = \Delta E + W$) to step 1 and noting that $\Delta E = 0$ in an isothermal process for a perfect gas,

$$\Delta E = 0 = Q - W = Q - \int_{V_1}^{V_2} p\, dV$$

or

$$Q = \int_{V_1}^{V_2} p\, dV = \int_{V_1}^{V_2} \frac{nRT\, dV}{V} = nRT_h \ln\frac{V_2}{V_1}$$
$$= (1\ \text{mol})\left(\frac{8.314\ \text{J}}{\text{mol K}}\right)\left(\frac{1\ \text{kJ}}{1000\ \text{J}}\right)(773\ \text{K}) \ln\left(\frac{0.1000}{0.0100}\right)$$
$$= 14.80\ \text{kJ} \tag{5.2-25}$$

For step 2, reversible adiabatic expansion from T_h to T_l, $Q = 0$, and so

$$W = -\Delta E = -nc_v(T_l - T_h) = n\frac{5R}{2}(T_h - T_l)$$
$$= (1\ \text{mol})\left(\frac{5}{2}\right)\left(\frac{8.314\ \text{J}}{\text{mol K}}\right)\left(\frac{1\ \text{kJ}}{1000\ \text{J}}\right)(773\ \text{K} - 273\ \text{K})$$
$$= 10.39\ \text{kJ} \tag{5.2-26}$$

For step 3, isothermal compression at T_l from V_3 to V_4, $\Delta E = 0$, and

$$Q = W = \int_{V_3}^{V_4} p\,dV = nRT_l \ln\frac{V_4}{V_3}$$

$$= (1 \text{ mol})\left(\frac{8.314 \text{ J}}{\text{mol K}}\right)\left(\frac{1 \text{ kJ}}{1000 \text{ J}}\right)(273 \text{ K}) \ln\left(\frac{0.1345}{1.345}\right)$$

$$= -5.23 \text{ kJ} \tag{5.2-27}$$

For step 4, reversible adiabatic compression from T_l to T_h, $Q = 0$, and

$$W = -\Delta E = -nc_v(T_h - T_l) = -n\frac{5R}{2}(T_h - T_l)$$
$$= -10.39 \text{ kJ} \tag{5.2-28}$$

For the entire cycle,

$$Q_{\text{overall}} = 14.80 + 0 - 5.23 + 0 = 9.57 \text{ kJ} = \oint dQ$$

$$\Delta E_{\text{overall}} = 0 - 10.39 + 0 + 10.39 = 0 = \oint dE$$

$$\text{Net work output} = 9.57 \text{ kJ} = \oint dW$$

$$\text{Heat input} = {}_1Q_2 = 14.80 \text{ kJ}$$

$$\text{Thermal efficiency} = \frac{W_{\text{net}}}{Q_{\text{in}}} = \frac{9.57}{14.80} = 0.647$$

The thermal efficiency for a Carnot engine can also be obtained from Carnot's formula of

$$\frac{T_h - T_l}{T_h} = \frac{773 - 273}{773} = 0.647$$

■

5.3 MAXWELL'S RELATIONS

In the preceding sections, the laws of thermodynamics were expressed in terms of the extensive properties, internal energy, entropy, and volume, and the intensive properties, temperature and pressure. The introduction of the functions H, A, and G will now be shown to result in some useful changes of independent variables. Consider

$$dE = T\,dS - p\,dV \tag{5.3-1}$$

$$dH = dE + p\,dV + V\,dp = T\,dS + V\,dp \tag{5.3-2}$$

$$dA = dE - T\,dS - S\,dT = -p\,dV - S\,dT \tag{5.3-3}$$

$$dG = dE + p\,dV + V\,dp - T\,dS - S\,dT = V\,dp - S\,dT \tag{5.3-4}$$

It is seen that

$$T = \left(\frac{\partial E}{\partial S}\right)_V = \left(\frac{\partial H}{\partial S}\right)_p \tag{5.3-5}$$

$$-p = \left(\frac{\partial E}{\partial V}\right)_S = \left(\frac{\partial A}{\partial V}\right)_T \tag{5.3-6}$$

$$V = \left(\frac{\partial H}{\partial p}\right)_S = \left(\frac{\partial G}{\partial p}\right)_T \tag{5.3-7}$$

$$-S = \left(\frac{\partial A}{\partial T}\right)_V = \left(\frac{\partial G}{\partial T}\right)_p \tag{5.3-8}$$

Since the differentials are exact, the Euler reciprocity relationship, namely that, if

$$dx = u\,dy + v\,dz \tag{5.3-9}$$

then

$$\left(\frac{\partial u}{\partial z}\right)_y = \left(\frac{\partial v}{\partial y}\right)_z = \frac{\partial^2 x}{\partial z\,\partial y} \tag{5.3-10}$$

can be applied to the four preceding equations, whence

$$\left(\frac{\partial T}{\partial V}\right)_S = -\left(\frac{\partial p}{\partial S}\right)_V \tag{5.3-11}$$

$$\left(\frac{\partial T}{\partial p}\right)_S = \left(\frac{\partial V}{\partial S}\right)_p \tag{5.3-12}$$

$$\left(\frac{\partial p}{\partial T}\right)_V = \left(\frac{\partial S}{\partial V}\right)_T \tag{5.3-13}$$

$$\left(\frac{\partial V}{\partial T}\right)_p = -\left(\frac{\partial S}{\partial p}\right)_T \tag{5.3-14}$$

which are known as the *Maxwell relations*. An immediate application is that the fundamental equation $T\,dS = dE + p\,dV$, written as

$$\begin{aligned} dE &= T\left[\left(\frac{\partial S}{\partial T}\right)_V dT + \left(\frac{\partial S}{\partial V}\right)_T dV\right] - p\,dV \\ &= T\left(\frac{\partial S}{\partial T}\right)_V dT + \left[T\left(\frac{\partial S}{\partial V}\right)_T - p\right] dV \end{aligned} \tag{5.3-15}$$

leads to

$$\left(\frac{\partial E}{\partial V}\right)_T = T\left(\frac{\partial S}{\partial V}\right)_T - p = T\left(\frac{\partial p}{\partial T}\right)_V - p = 0 \tag{5.3-16}$$

when $pV = RT$. In other words, for a perfect gas the internal pressure is zero and the internal energy is a function of temperature only. Other applications of Maxwell's equations may be found in classical thermodynamics.

Because of the large number of possible mathematical relations between the thermodynamic properties and their derivatives, it is desirable to have a simple method for deriving any desired partial derivative. The following illustrates how this can be done. From the combined statements of the first and second laws (the fundamental equation),

$$dE = T\,dS - p\,dV \tag{5.3-17}$$

the definition of H, A, and G,

$$H = E + pV \tag{5.3-18}$$

$$A = E - TS \tag{5.3-19}$$

$$G = E + pV - TS \tag{5.3-20}$$

their differentials,

$$dH = T\,dS + V\,dp \tag{5.3-21}$$

$$dA = -p\,dV - S\,dT \tag{5.3-22}$$

$$dG = V\,dp - S\,dT \tag{5.3-23}$$

and two of Maxwell's relations,

$$\left(\frac{\partial p}{\partial T}\right)_V = \left(\frac{\partial S}{\partial V}\right)_T \tag{5.3-24}$$

$$\left(\frac{\partial V}{\partial T}\right)_p = \left(\frac{\partial S}{\partial p}\right)_T \tag{5.3-25}$$

along with the expression for C_V and C_p,

$$C_V = \left(\frac{\partial E}{\partial T}\right)_V \tag{5.3-26}$$

$$C_p = \left(\frac{\partial H}{\partial T}\right)_p \tag{5.3-27}$$

any desired derivative can be obtained. For example, suppose that it is desirable to determine $(\partial G/\partial S)_H$, which is indicated as X. Then

$$dG = X\,dS + Y\,dH \tag{5.3-28}$$

Choosing the frequently used pair T and p for independent variables, the differentials of (5.3-28) can be expressed in terms of dT and dp using the 11 given relations (5.3-17) to (5.3-27). In other words, from (5.3-23) and (5.3-28),

$$\begin{aligned} dG &= -S\,dT + V\,dp = X\,dS + Y\,dH \\ &= X\,dS + Y(T\,dS + V\,dp) \end{aligned} \tag{5.3-29}$$

where (5.3-21) has been used for dH.

Now, dS can be written as

$$dS = \left(\frac{\partial S}{\partial T}\right)_p dT + \left(\frac{\partial S}{\partial p}\right)_T dp$$
$$= \left(\frac{C_p}{T}\right) dT - \left(\frac{\partial V}{\partial T}\right)_p dp \qquad (5.3\text{-}30)$$

So, Eq. (5.3-29) becomes

$$-S\,dT + V\,dp = X\left[\frac{C_p\,dT}{T} - \left(\frac{\partial V}{\partial T}\right)_p dp\right] + Y\left\{C_p\,dT + \left[V - T\left(\frac{\partial V}{\partial T}\right)_p\right] dp\right\} \qquad (5.3\text{-}31)$$

Since Eq. (5.3-31) must hold both for $dT = 0$ and for $dp = 0$, the coefficients of dT and dp can be equated to obtain two simultaneous equations in X and Y,

$$-S = \left(\frac{C_p}{T}\right) X + C_p Y \qquad (5.3\text{-}32)$$

$$V = -\left(\frac{\partial V}{\partial T}\right)_p X + \left[V - T\left(\frac{\partial V}{\partial T}\right)_p\right] Y \qquad (5.3\text{-}33)$$

Equation (5.3-33) is then readily solved for

$$X = \left(\frac{\partial G}{\partial S}\right)_H = \frac{-V(S + C_p) + ST(\partial V/\partial T)_p}{C_p V/T} \qquad (5.3\text{-}34)$$

The quantity $(1/V)(\partial V/\partial T)_p$ is the coefficient of volume expansion.

5.4 ENTHALPY, INTERNAL ENERGY, AND SPECIFIC HEAT RELATIONS

Let the internal energy of a pure substance be expressed as a function of the temperature and volume. Based on a unit mass, for convenience,

$$u = u(T, v) \qquad (5.4\text{-}1)$$

whereupon

$$du = \left(\frac{\partial u}{\partial T}\right)_v dT + \left(\frac{\partial u}{\partial v}\right)_T dv \qquad (5.4\text{-}2)$$

Recalling the definition

$$c_v = \left(\frac{\partial u}{\partial T}\right)_v \qquad (5.4\text{-}3)$$

(5.4-2) becomes

$$du = c_v\,dT + \left(\frac{\partial u}{\partial v}\right)_T dv \qquad (5.4\text{-}4)$$

From the fundamental relation $T\,ds = du + p\,dv$,

$$ds = \frac{du}{T} + \frac{p}{T}dv \tag{5.4-5}$$

$$= c_v\frac{dT}{T} + \left[\frac{1}{T}\left(\frac{\partial u}{\partial v}\right)_T + \frac{p}{T}\right]dv \tag{5.4-6}$$

But the entropy may be written as a function of the independent variables T and v so that

$$ds = \left(\frac{\partial s}{\partial T}\right)_v dT + \left(\frac{\partial s}{\partial v}\right)_T dv \tag{5.4-7}$$

Equating the coefficients of dT and dv in (5.4-6) and (5.4-7) gives

$$\frac{c_v}{T} = \left(\frac{\partial s}{\partial T}\right)_v \tag{5.4-8}$$

$$\left(\frac{\partial s}{\partial v}\right)_T = \frac{1}{T}\left[\left(\frac{\partial u}{\partial v}\right)_T + p\right] \tag{5.4-9}$$

Now, the Maxwell relation

$$\left(\frac{\partial s}{\partial v}\right)_T = \left(\frac{\partial p}{\partial T}\right)_v \tag{5.4-10}$$

enables (5.4-9) to be written as

$$\left(\frac{\partial u}{\partial T}\right)_T = T\left(\frac{\partial p}{\partial T}\right)_v - p \tag{5.4-11}$$

and the relation for the internal energy, (5.4.2), becomes

$$du = c_v\,dT + \left[T\left(\frac{\partial p}{\partial T}\right)_v - p\right]dv \tag{5.4-12}$$

Similarly, for the entropy,

$$ds = c_v\frac{dT}{T} + \left(\frac{\partial p}{\partial T}\right)_v dv \tag{5.4-13}$$

and for the enthalpy,

$$dh = c_p\,dT + \left[v - T\left(\frac{\partial v}{\partial T}\right)_p\right]dp \tag{5.4-14}$$

where the definition of the specific heat at constant pressure is recalled as

$$c_p = \left(\frac{\partial h}{\partial T}\right)_p \tag{5.4-15}$$

Also, taking the entropy as a function of pressure and temperature, and considering these independent variables instead of volume and temperature, results in

$$ds = c_p \frac{dT}{T} - \left(\frac{\partial v}{\partial T}\right)_p dp \tag{5.4-16}$$

The equations for the internal energy, (5.4-12), the entropy, (5.4-13), and the enthalpy, (5.4-14), involve the five quantities: p, v, T, c_v, and c_p. Similar relationships between the specific heats may be developed in the following way. The two expressions for the entropy, (5.4-13) and (5.4-16), may be equated to give

$$c_v \frac{dT}{T} + \left(\frac{\partial p}{\partial T}\right)_v dv = c_p \frac{dT}{T} - \left(\frac{\partial v}{\partial T}\right)_p dp \tag{5.4-17}$$

Rearranging,

$$c_p - c_v = T\left[\left(\frac{\partial p}{\partial T}\right)_v \frac{dv}{dT} + \left(\frac{\partial v}{\partial T}\right)_p \frac{dp}{dT}\right] \tag{5.4-18}$$

For a variation at constant volume, $dv = 0$ and

$$(c_p - c_v)_v = T\left(\frac{\partial v}{\partial T}\right)_p \left(\frac{\partial p}{\partial T}\right)_v \tag{5.4-19}$$

or for a variation at constant pressure, $dp = 0$ and

$$(c_p - c_v)_p = T\left(\frac{\partial p}{\partial T}\right)_v \left(\frac{\partial v}{\partial T}\right)_p \tag{5.4-20}$$

These relations are identical. Using the cyclic relation

$$\left(\frac{\partial x}{\partial y}\right)_z \left(\frac{\partial v}{\partial z}\right)_x \left(\frac{\partial z}{\partial x}\right)_y = -1 \tag{5.4-21}$$

$(\partial p/\partial T)_v$ may be eliminated from Eq. (5.4-20).

$$\left(\frac{\partial p}{\partial T}\right)_v = -\left(\frac{\partial v}{\partial T}\right)_p \left(\frac{\partial p}{\partial v}\right)_T \tag{5.4-22}$$

Thus, the final relation is

$$c_p - c_v = -T\frac{(\partial v/\partial T)_p^2}{(\partial v/\partial p)T} \tag{5.4-23}$$

With (5.4-23), the determination of u, h, and s has been reduced to the specification of only four properties: p, v, T, and either c_v or c_p. Since the Gibbs and Helmholtz functions are expressed in terms of u, h, and s, this also specifies the information necessary for their determination.

The foregoing is important because it establishes the experimental measurements that must be performed to describe fully all the thermodynamic properties of a pure substance. For a single phase of a substance (i.e., liquid or vapor phase), it is presumed that there exists an equation of state relating pressure, volume, and temperature:

$$v = v(T, p) \tag{5.4-24}$$

The p, v, and T measurements may be employed to obtain an empirical relationship of the form of (5.4-24). This relation may then be combined with specific heat data to calculate other thermodynamic properties.

Additional useful expressions relating specific heats may be developed. Equation (5.4-16) is an exact differential equation of the form

$$dz = M\,dx + N\,dy \tag{5.4-25}$$

Equating the second mixed partial derivatives yields

$$\left[\frac{\partial(c_p/T)}{\partial p}\right]_T = -\left[\frac{\partial}{\partial T}\left(\frac{\partial v}{\partial T}\right)_p\right]_p \tag{5.4-26}$$

$$\left(\frac{\partial c_p}{\partial p}\right)_T = -T\left(\frac{\partial^2 v}{\partial T^2}\right)_p \tag{5.4-27}$$

Similarly, using (5.4-13) as starting point, an expression for the variation of c_v with volume can be obtained:

$$\left[\frac{\partial(c_v/T)}{\partial v}\right]_T = \left[\frac{\partial}{\partial T}\left(\frac{\partial p}{\partial T}\right)_v\right]_v \tag{5.4-28}$$

or

$$\left(\frac{\partial c_v}{\partial v}\right)_T = T\left(\frac{\partial^2 p}{\partial T^2}\right)_v \tag{5.4-29}$$

Equations (5.4-11), (5.4-27), and (5.4-29) provide very useful results when applied to an ideal gas. In that case,

$$v = \frac{RT}{p} \qquad p = \frac{RT}{v} \tag{5.4-30}$$

Differentiating and inserting these expressions in (5.4-11), (5.4-27), and (5.4-29), gives

$$\begin{aligned} \left(\frac{\partial c_p}{\partial p}\right)_T &= -T(0) = 0 \\ \left(\frac{\partial c_v}{\partial v}\right)_T &= T(0) = 0 \end{aligned} \tag{5.4-31}$$

$$\left(\frac{\partial u}{\partial v}\right)_T = T\left(\frac{R}{v}\right) - p = 0 \tag{5.4-32}$$

That is, the specific heats and internal energy of an ideal gas are functions of temperatures only.

■ EXAMPLE 5.4-1

Show, from microscopic arguments, that the molar specific heat at constant volume of a monatomic ideal gas is $c_v = 3/2\,R$, and that $c_p = 5/2\,R$, $\gamma = 5/3 = 1.667$.

Solution. Using microscopic information, many interesting statements can be made. The case of a monatomic ideal gas is particularly appropriate: The number of states of such a gas in some small energy range, namely, the density-of-states function, is given by

$$g(E, V) = BV^N E^{3N/2} \tag{5.4-33}$$

where N is the number of molecules in the gas and B is some constant independent of E and V. Hence

$$\ln g = \ln B + N \ln V + \frac{3N}{2} \ln E \tag{5.4-34}$$

The temperature parameter $\beta = (kT)^{-1}$ is then given by

$$\beta = \frac{\partial \ln g}{\partial E} = \frac{3N}{2}\frac{1}{E} \tag{5.4-35}$$

Thus

$$E = \frac{3N}{2\beta} = \frac{3N}{2} kT \tag{5.4-36}$$

This expresses directly the relation between the internal energy and the absolute temperature of the gas. If N is taken to be Avogadro's number N_A, (5.4-36) can be written

$$E = \frac{3}{2} RT \tag{5.4-37}$$

where R is the gas constant.

The molar specific heat at constant volume of a monatomic ideal gas is then

$$c_v = \left(\frac{\partial E}{\partial T}\right)_v = \frac{3}{2} R \tag{5.4-38}$$

Furthermore, from $c_p = c_v + R$,

$$c_p = \frac{3}{2} R + R = \frac{5}{2} R \tag{5.4-39}$$

and

$$\gamma \equiv \frac{c_p}{c_v} = \frac{5}{3} = 1.667 \tag{5.4-40}$$

These simple microscopic arguments lead to very definite quantitative predictions. The experimental values for the monatomic gases helium and argon show very satisfactory agreement with the theoretical values. ■

■ EXAMPLE 5.4-2

From purely microscopic considerations, it can be shown that an ideal gas is described by

$$pV = NRT \tag{5.4-41a}$$

$$E = E(T) \tag{5.4-41b}$$

Show that the latter property is actually a direct consequence of the equation of state, which can be considered to be a purely empirical relation. Also, describe Joule's experimental verification of $E = E(T)$ using "free expansion."

Solution. Quite generally, the internal energy E of N mol of any gas can be considered a function of T and V:

$$E = E(T, V) \tag{5.4-42}$$

and therefore,

$$dE = \left(\frac{\partial E}{\partial T}\right)_V dT + \left(\frac{\partial E}{\partial V}\right)_T dV \tag{5.4-43}$$

But the fundamental thermodynamic relation for a quasistatic change of parameters can be written as

$$T\,dS = \frac{1}{T}dE + \frac{NR}{V}dV \tag{5.4-44}$$

Using (5.4-41a) to express p in terms of V and T, (5.4-44) becomes

$$dS = \frac{1}{T}dE + \frac{NR}{V}dV$$

or by (5.4-44),

$$dS = \frac{1}{T}\left(\frac{\partial E}{\partial T}\right)_V dT + \left[\frac{1}{T}\left(\frac{\partial E}{\partial V}\right)_T + \frac{NR}{V}\right] dV \tag{5.4-45}$$

The mere fact that dS on the left side is the exact differential of a well-defined function allows some important conclusions to be drawn. Consider S dependent on T and V; then

$$dS = \left(\frac{\partial S}{\partial T}\right)_V dT + \left(\frac{\partial S}{\partial V}\right)_T dV \tag{5.4-46}$$

Since this expression must be true for all values of dT and dV, comparison with (5.4-43) shows immediately that

$$\left(\frac{\partial S}{\partial T}\right)_V = \frac{1}{T}\left(\frac{\partial E}{\partial T}\right)_V \tag{5.4-47a}$$

$$\left(\frac{\partial S}{\partial V}\right)_T = \frac{1}{T}\left(\frac{\partial E}{\partial V}\right)_T + \frac{NR}{V} \tag{5.4-47b}$$

But the equality of the second derivatives, irrespective of order of differentiation,

$$\frac{\partial^2 S}{\partial V\,\partial T} = \frac{\partial^2 S}{\partial T\,\partial V} \tag{5.4-48}$$

implies a definite connection between the expressions on the right side of (5.4-47a) and (5.4-47b):

$$\frac{1}{T}\left(\frac{\partial^2 E}{\partial V\,\partial T}\right)=\left[-\frac{1}{T^2}\left(\frac{\partial E}{\partial V}\right)_T+\frac{1}{T}\left(\frac{\partial^2 E}{\partial T\,\partial V}\right)\right]+0 \tag{5.4-49}$$

Since the second derivatives of E are again equal irrespective of the order of differentiation, this last relation shows immediately that

$$\left(\frac{\partial E}{\partial V}\right)_T=0 \tag{5.4-50}$$

Thus E is independent of V and is solely dependent on T. The fact that the internal energy E of a gas does not depend on its volume (if the gas is sufficiently dilute that it can be considered ideal) was verified in a classical experiment by Joule. He made use of the "free expansion"of an ideal gas as illustrated in Fig. 5.4-1. A container consisting of two compartments separated by a valve is immersed in water. Initially, the valve is closed and one compartment is filled with the gas under investigation, while the other compartment is evacuated. Suppose that the valve is now opened so that the gas is free to expand and fill both compartments. In this process no work gets done by the system consisting of the gas and container. (The container walls are rigid and nothing moves.) Hence, by the first law, the heat Q absorbed by this system equals its increase in internal energy:

$$Q=\Delta E \tag{5.4-51}$$

Assuming that the internal energy change of the (thin-walled) container is negligibly small, ΔE measures simply the energy change of the gas. Joule found that the temperature of the water did not change in this experiment. The water absorbed no heat from the gas; consequently, the heat Q absorbed by the gas also vanishes. All that happens in the experiment is that the temperature of the gas remains unchanged while its volume changes from its

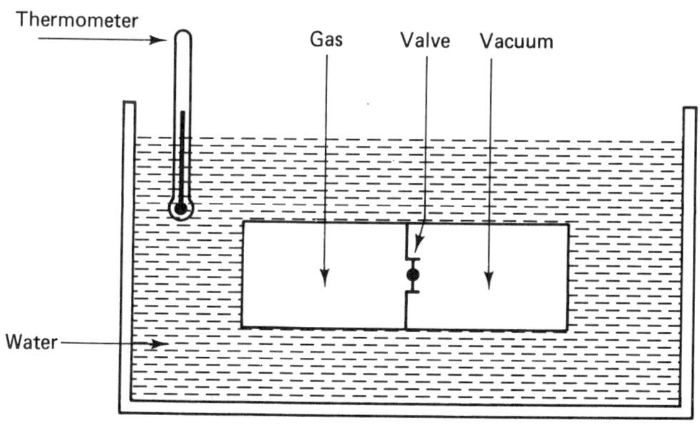

Figure 5.4-1 Experimental setup for studying the free expansion of a gas.

initial value V_i to its final value V_f. Since $Q = 0$, Joule's experiment leads by virtue of (5.4-51) to the conclusion

$$E(T, V_f) - E(T, V_i) = 0 \tag{5.4-52}$$

which verifies that $E(T, V)$ is independent of the volume V. ■

5.5 OPEN SYSTEMS AND THE CHEMICAL POTENTIAL

If the mass of a system is varied, the extensive thermodynamic properties must necessarily change. Accordingly, for an open system of one or more components each present in amounts n_i, the combined statement of the first and the second law becomes

$$dE = T\,dS - p\,dV + \sum_i \mu_i\,dn_i \tag{5.5-1}$$

where the summation is over all chemical species in the phase.

Equation (5.5-1) serves to define the chemical potential μ_i of the ith component, because, by the rule of partial differentiation,

$$\mu_i \equiv \left(\frac{\partial E}{\partial n_i}\right)_{S, V, n_j \neq i} \tag{5.5-2}$$

The chemical potentials, as shown by Eq. (5.5-1) do not depend on the total amounts but rather on the relative amounts or composition of the system. Hence, the μ_i's are intensive properties.

If dn_i mol of component 1, dn_2 mol of component 2, and so forth are added reversibly to an assembly, changes in internal energy, enthalpy, Helmholtz free energy, and Gibbs free energy result from the extra material added. They are given by

$$dU = T\,dS - p\,dV + \Sigma\mu_i\,dn_i \tag{5.5-3}$$

$$dH = T\,dS + V\,dp + \Sigma\mu_i\,dn_i \tag{5.5-4}$$

$$dF = -S\,dT - p\,dV + \Sigma\mu_i\,dn_i \tag{5.5-5}$$

$$dG = -S\,dT + V\,dp + \Sigma\mu_i\,dn_i \tag{5.5-6}$$

From these, the following additional expressions for the chemical potential can be obtained:

$$\mu_i = \left(\frac{\partial H}{\partial n_i}\right)_{S, p, n_j \neq i} = \left(\frac{\partial A}{\partial n_i}\right)_{V, T, n_j \neq i} = \left(\frac{\partial G}{\partial n_i}\right)_{p, T, n_j \neq i} \tag{5.5-7}$$

From the equalities of Eq. (5.5-7), it is noted that the chemical potential is equal to the partial molal free energy.

Equation (5.5-6) gives, for an open system,

$$dG = V\,dp - S\,dT + \Sigma\mu_i\,dn_i \tag{5.5-8}$$

The meaning of the chemical potential (and indeed, by similar arguments, of any partial molar quantity) may be clarified if the system is imagined to be increased in size at constant temperature and pressure by adding to it reversibly small amounts of each constituent in proportions that keep the overall composition constant. The change in G is then

$$dG_{T,p} = \Sigma\mu_i\,dn_i \tag{5.5-9}$$

If the assembly initially contained n_i mol of component 1, n_2 mol of 2 and so on, then the dn_i are by definition in the same ratio as the n_i, and $dn_1 = n_1\,d\chi$, where $d\chi$ is now the fractional increase in the size of the system resulting from the small additions. Thus,

$$dG_{T,p} = d\chi\,\Sigma\mu_i n_i \tag{5.5-10}$$

This equation may be integrated from $\chi = 0$ to $\chi = 1$, giving

$$G = \Sigma\mu_i n_i \tag{5.5-11}$$

The total free energy of an open system is therefore the sum over all components of the product of the chemical potential and the number of moles. The same result may be obtained from use of Euler's theorem on homogeneous functions (as seen in the example below).

Differentiation of (5.5-11) gives

$$dG = \sum_i \mu_i\,dn_i + \sum_i n_i\,d\mu_i \tag{5.5-12}$$

Subtracting this result from (5.5-8) gives the extended Gibb–Duhem equation,

$$V\,dp - S\,dT = \sum_i n_i\,d\mu_i \tag{5.5-13}$$

At constant T and p, the more familiar form of the Gibbs–Duhem equation is

$$\sum_i n_i\,d\mu_i = 0 \tag{5.5-14}$$

For a system consisting of a single phase and C components in terms of the $C + 2$ intensive properties T, p, and $\mu_i, \ldots, \mu_c$, from (5.5-13) it is seen that only $C + 1$ of these quantities are independently variable. Moreover, if the system consists of several phases, the Gibbs–Duhem relation applies to each phase, so that the number of independent variables is

$$f = C + 2 - P \tag{5.5-15}$$

where P is the number of phases. This is the *Gibbs phase rule* relating the number of degrees of freedom of a system, the number of phases, and the number of components.

■ **EXAMPLE 5.5.1**

From Eq. (5.5-1) and the definition of G, show, by means of Euler's theorem on homogeneous functions, that $G = \sum_i \mu_i n_i$.

Solution. Euler's theorem is that if f is a homogeneous function of order n of a set of variables y_i, then

$$nf = \sum_i y_i \frac{\partial f}{\partial y_i} \tag{5.5-16}$$

Observing that E in Eq. (5.5-1) is homogeneous and first order in the extensive properties S, V, and the n_i's, this gives

$$E = \left(\frac{\partial E}{\partial S}\right) S + \left(\frac{\partial E}{\partial V}\right) V + \sum_i \left(\frac{\partial E}{\partial n_i}\right) n_i$$

$$= TS - pV + \sum_i \mu_i n_i \tag{5.5-17}$$

and therefore

$$G = E - TS + pV = \sum_i \mu_i n_i \tag{5.5-18}$$

which is Eq. (5.5-11). ■

5.6 PHASE EQUILIBRIA

The introduction of the chemical potential in the preceding section calls for a discussion of systems consisting of two or more phases in equilibrium. A phase is a homogeneous part of a system bounded by surfaces across which the properties change discontinuously. In the most general situation each phase will contain several components, that is, it will contain several different species of molecules or ions. For example, a gaseous phase might consist of a mixture of gases; a liquid phase might be a solution.

The properties of a phase are specified by the concentrations of the various components in it. There is the possibility of transfer of matter between different phases and also of chemical reactions (which change the amounts of different components). The latter case, known as multicomponent systems, will be discussed in Section 5.8. The present analysis is restricted to the discussion of systems containing one component only. A one-component system can, of course, still exist in different phases, corresponding to gaseous, liquid, and different crystalline forms of matter. For different phases to be in equilibrium certain conditions must hold so that no mass transfer occurs. These conditions are additional to those for temperature and pressure equilibrium.

To derive the criteria for two phases to coexist in equilibrium, consider the system shown in Fig. 5.6-1. The two phases could be solid and liquid (e.g., ice

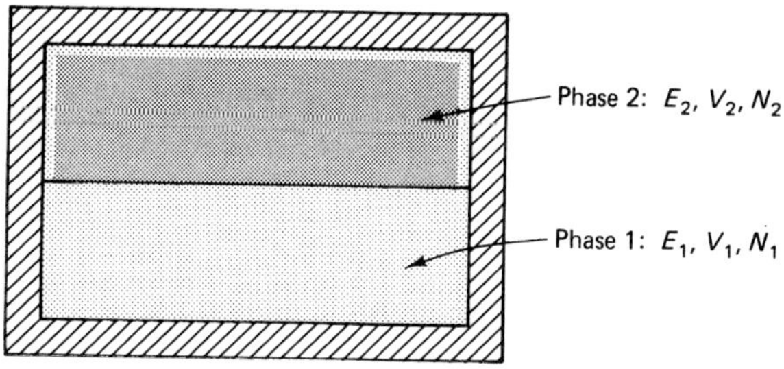

Figure 5.6-1 Equilibrium of two phases of a one-component system.

and water), liquid and vapor, solid and vapor, or other possibilities, such as different crystalline forms of the same substance (e.g., grey and white tin).

Together, the two phases form an isolated system, so that

$$E_1 + E_2 = E \tag{5.6-1}$$

$$V_1 + V_2 = V \tag{5.6-2}$$

$$N_1 + N_2 = N \tag{5.6-3}$$

express the division of constant energy E, volume V, and particle number N between the two phases. The entropy of the system is

$$S(E, V, N, E_1, V_1, N_1) = S_1(E_1, V_1, N_1) + S_2(E_2, V_2, N_2) \tag{5.6-4}$$

where subscripts 1 and 2 label quantities in the two phases 1 and 2, respectively. For equilibrium of the system, the entropy must be a maximum, that is, treating E_1, V_1, and N_1 as the independent variables and eliminating E_2, V_2, and N_2 by means of Eqs. (5.6-1), (5.6-2), and (5.6-3),

$$\begin{aligned} dS = &\left[\left(\frac{\partial S_1}{\partial E_1}\right)_{V_1, N_1} - \left(\frac{\partial S_2}{\partial E_2}\right)_{V_2, N_1}\right] dE_1 \\ &+ \left[\left(\frac{\partial S_1}{\partial V_1}\right)_{E_1, N_1} - \left(\frac{\partial S_2}{\partial V_2}\right)_{E_2, N_2}\right] dV_1 \\ &+ \left[\left(\frac{\partial S_1}{\partial N_1}\right)_{E_1, V_1} - \left(\frac{\partial S_2}{\partial N_2}\right)_{E_2, V_2}\right] dV_1 = 0 \end{aligned} \tag{5.6-5}$$

Since E_1, V_1, and N_1 are independently variable, each of the square brackets in (5.6-5) must vanish for the system to be in equilibrium. The first two of these conditions (coefficient of dE_1 and dV_1 zero) are the conditions for equal temperatures and equal pressures for the two phases. The third condition

$$\left(\frac{\partial S_1}{\partial N_1}\right)_{E_1, V_1} - \left(\frac{\partial S_2}{\partial N_2}\right)_{E_2, V_2} = 0 \tag{5.6-6}$$

is the condition for particle equilibrium, that is, no transfer of molecules between the two phases. Now, the chemical potential μ_i for each phase is

$$\mu_i = -T_i\left(\frac{\partial S_i}{\partial N_i}\right)_{E_i V_i} \qquad i = 1,2 \tag{5.6-7}$$

so condition (5.6-6) becomes

$$\mu_1 = \mu_2 \tag{5.6-8}$$

that is, in the equilibrium, the chemical potentials of the two phases must be the same.

For an open system consisting of only a single phase, $S = S(E, V, N)$, so

$$\begin{aligned} dS &= \left(\frac{\partial S}{\partial E}\right)_{V,N} dE + \left(\frac{\partial S}{\partial V}\right)_{E,N} dV + \left(\frac{\partial S}{\partial N}\right)_{E,V} dN \\ &= \frac{1}{T}\, dE + \frac{1}{T}\, dV - \frac{1}{T}\mu\, dN \end{aligned} \tag{5.6-9}$$

where it is recalled that

$$\left(\frac{\partial S}{\partial E}\right)_{V,N} = \frac{1}{T}, \qquad p = T\left(\frac{\partial S}{\partial V}\right)_{E,N}, \qquad \mu = -T\left(\frac{\partial S}{\partial N}\right)_{E,V}$$

or equivalently,

$$dE = T\, ds - p\, dv + \mu\, dN \tag{5.6-10}$$

Equation (5.6-10) is the generalization of the fundamental thermodynamic relation for a one-component system whose size, that is, particle number, is not kept fixed.

From (5.6-10), the chemical potential can be expressed in various ways depending on which of the variables E, S, *and* V are held constant. Also, different independent variables can be chosen, so that, expressed in terms of the Helmholtz and Gibbs free energies, (5.6-10) becomes

$$dF = -S\, dT - p\, dV + \mu\, dN \tag{5.6-11}$$

$$dG = -S\, dT + V\, dp + \mu\, dN \tag{5.6-12}$$

whence

$$\mu = \left(\frac{\partial F}{\partial N}\right)_{T,V} = \left(\frac{\partial G}{\partial N}\right)_{T,p} \tag{5.6-13}$$

It is seen that the chemical potential is always the derivative with respect to the particle number N. For a single one-component phase, the Gibbs free energy $G(T, p, N)$ is an extensive variable, that is, it is proportional to the particle number N:

$$G(T, p, N) = Ng(T, p, 1) \equiv NG(T, p) \tag{5.6-14}$$

where

$$g(T, p) = \frac{1}{N} G(T, p, N) \tag{5.6-15}$$

is the corresponding Gibbs free energy per particle. It follows that

$$\mu = g(T,p) \tag{5.6-16}$$

where the chemical potential μ for a homogeneous one-component system is simply the Gibbs free energy per particle.

Reverting back to a system consisting of one component in two phases, each phase by itself is, of course, a homogeneous one-component system so that (5.6-16) applies to it. Hence, the equilibrium condition (5.6-8) is written in the form

$$g_1(T,p) = g_2(T,p) \tag{5.6-17}$$

The Gibbs free energies per particle for the two phases must be the same.

So far one of the equilibrium conditions has always been that the system be at a uniform pressure. This condition frequently applies, but it depends on the neglect of surface effects. In cases, for example, of drops of liquid suspended in their vapor, surface tension effects will be important if the drops are small, and the pressures within and outside them are no longer the same.

5.7 CLAUSIUS–CLAPEYRON EQUATION

The equilibrium condition that the Gibbs free energies for two phases must be the same, namely,

$$g_1(T,p) = g_2(T,p) \tag{5.7-1}$$

defines a curve in the (T,p) plane, as shown in Fig. 5.7-1. At a point on the curve the chemical potentials of the two phases are the same; at a point off the curve they differ. The Clausius–Clapeyron equation which will now be derived is the differential equation for a phase equilibrium curve. The equation specifies the slope dp/dT at each point of the curve. It thus gives information about the dependence on pressure and temperature of the phase equilibrium.

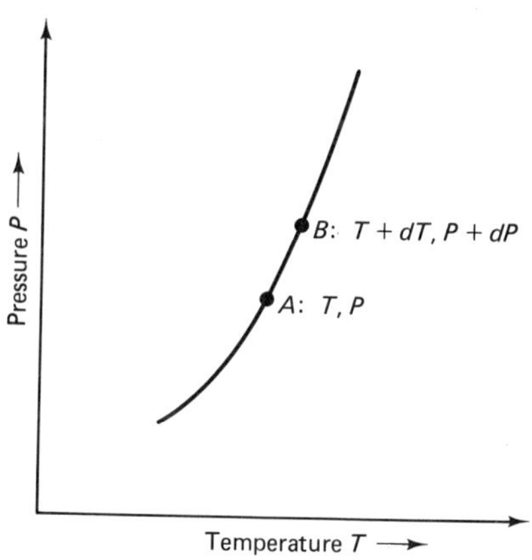

Figure 5.7-1 The phase equilibrium curve $g_1(T,P) = g_2(T,P)$.

For two points A and B on the equilibrium curve, the Gibbs free energy or chemical potential of the two phases are equal:

$$g_1(T, p) = g_2(T, p) \tag{5.7-2}$$

$$g_1(T + dT, p + dp) = g_2(T + dT, p + dp) \tag{5.7-3}$$

whence by subtraction

$$\left(\frac{\partial g_1}{\partial T}\right)_P dT + \left(\frac{\partial g_1}{\partial P}\right)_T dp = \left(\frac{\partial g_2}{\partial T}\right)_P dT + \left(\frac{\partial g_2}{\partial P}\right)_T dp \tag{5.7-4}$$

so that

$$\frac{dp}{dT} = \frac{\Delta(\partial g/\partial T)_p}{\Delta(\partial g/\partial p)_T} \tag{5.7-5}$$

Now, for each phase separately,

$$dG_i = S_i\, dT + V_i\, dp + \mu_i\, dN_i \tag{5.7-6}$$

For a one-component phase, $\mu_i = g_i$ and $G_i = N_i g_i$. Hence, Eq. (5.7-6) becomes

$$N_i\, dg_i = S_i\, dT + V_i\, dp \tag{5.7-7}$$

so that

$$\left(\frac{\partial g_i}{\partial T}\right)_P = -\frac{S_i}{N_i}, \qquad \left(\frac{\partial g_i}{\partial p}\right)_T = \frac{V_i}{N_i} \tag{5.7-8}$$

Substituting (5.7-8) into (5.7-5) gives

$$\frac{dp}{dT} = \frac{\Delta(S/N)}{\Delta(V/N)} = \frac{\Delta S}{\Delta V} \tag{5.7-9}$$

where in the last expression on the right-hand side ΔS and ΔV must refer to the same quantity of substance (it could be one molecule or 1 mol). The value ΔS is the change in entropy, ΔV the change in volume of this same quantity of substance as it undergoes a phase transformation from phase 1 to phase 2. Equation (5.7-9) is the *Clausius–Clapeyron equation*. It relates changes occurring at a temperature T and pressure p corresponding to a definite point on the equilibrium curve to the slope of the curve at that point.

A phase change, which is accompanied by a change in entropy ΔS, involves the absorption or emission of heat. If L_{12} is the latent heat of transformation when a given quantity of the substance is transformed from phase 1 to phase 2, at a temperature T, then the entropy change is

$$\Delta S = S_2 - S_1 = \frac{L_{12}}{T} \tag{5.7-10}$$

Combining (5.7-9) and (5.7-10) results in an alternate form of the *Clausius–Clapeyron equation:*

$$\frac{dp}{dT} = \frac{L_{12}}{T\,\Delta V} \qquad (5.7\text{-}11)$$

Phase changes are characterized by discontinuities in the numerator and denominator of Eq. (5.7-5) when going from one phase to another. These discontinuities imply differences in entropy and molar volume (i.e., density) between the phases. The entropy difference implies that the phase change involves a latent heat. This type of phase change is called a *first-order transition.* Solid–liquid–vapor phase changes and many allotropic transitions, for example, grey to white tin, are of this kind. There are phase changes in which the numerator and denominator of Eq. (5.7-5) are continuous, that is, entropy and density are continuous (so that no latent heat is involved) but some higher-order derivatives of g_i and other thermodynamic quantities, such as the specific heat, are discontinuous or may become infinite. These kinds of phase changes will not be considered here.

For the processes of melting, evaporation, and sublimation, the entropy changes ΔS (and hence the corresponding latent heats) are positive. That ΔS is positive for these phase changes is a consequence of the fact that these are changes from more- to less- ordered states of the system. The atoms in a crystal are arranged regularly at the lattice sites, so there is long-range order over the whole size of the crystal. Except near the boiling point, liquids show a considerable degree of short-range order, that is, liquids possess localized structure, as in a crystal, except that there are many vacant sites, about one in ten. In a gas there is almost no structure left. These different degrees of order show up on X-ray diffraction experiments.

For a transformation from solid or liquid to gas the density decreases so that ΔV is positive (i.e., the molar volume increases). It follows from the Clausius–Clapeyron equation that the slope dp/dT of the vapor–pressure and sublimation curve is positive, as drawn in Fig. 5.7-1.

For the process of melting ΔV may be positive or negative depending on whether the solid expands or contracts on melting. Most substances expand upon melting. This is a consequence of the large number of vacant sites in a liquid, so that the density of the liquid is smaller than that of the solid. (These vacancies also allow the molecules in the liquid to migrate easily and this accounts for their property of fluidity.) For a substance that expands on melting, the slope dp/dT of the melting curve is positive, as seen from Eq. (5.7-11), with $\Delta V > 0$. The behavior of ice ($\Delta V < 0$) is an exception, due mainly to its very open tetrahedral crystal structure.

■ EXAMPLE 5.7-1

The boiling point of water at atmospheric pressure is 100°C. What is the change in the boiling point per Pa change in the pressure? The enthalpy of vaporization is 40.69 kJ mol^{-1}, the molar volume of liquid water is

0.019×10^{-3} m^3 mol^{-1}, and the molar volume of steam is 30.199×10^{-3} m^3 mol^{-1}, all at 100°C and 1.01325 bar.

Solution. From Eq. (5.7-11)

$$\frac{dp}{dT} = \frac{\Delta H_{\text{vap}}}{T(V_{\text{vap}} - V_{\text{liq}})} = \frac{40{,}690 \text{ J mol}^{-1}}{(373.15 \text{ K})(30.199 - 0.019) \times 10^{-3} \text{ m}^3 \text{ mol}^{-1}}$$

$$= 3613 \text{ Pa K}^{-1}$$

Thus $dT/dp = 2.768 \times 10^{-4}$ KPa^{-1}. ■

■ EXAMPLE 5.7-2

What is the change in pressure required to change the freezing point of water 1°C? At 0°C the heat of fusion of ice is 333.5 J g^{-1}, the density of water is 0.9998 g cm^{-3}, and the density of ice is 0.9168 g cm^{-3}.

Solution. The reciprocals of the densities, 1.0002 and 1.0908, are the volumes in cubic centimeters of 1 g. The volume change upon freezing $(V_l - V_s)$ is therefore -9.06×10^{-8} m^3 g^{-1}. For small changes, ΔH_{fus}, T and $(V_l - V_s)$ are virtually constant, so that

$$\frac{\Delta p}{\Delta T} = \frac{\Delta H_{\text{fus}}}{T(V_l - V_s)} = \frac{333.5 \text{ J g}^{-1}}{(273.15 \text{ K})(-9.06 \times 10^{-8} \text{ m}^3 \text{ g}^{-1})}$$

$$= -1.348 \times 10^7 \text{ Pa K}^{-1}$$

The change in freezing point of water per bar pressure is

$$\frac{\Delta T}{\Delta p} = \frac{10^5 \text{ Pa bar}^{-1}}{-1.348 \times 10^7 \text{ Pa K}^{-1}} = -0.0075 \text{ K bar}^{-1}$$

The negative sign indicates that an increase in pressure of 1 bar lowers the freezing point 0.0075 K. The lowering of the melting point of ice under pressure is responsible for the motion of glaciers. The deeper parts of a glacier melt under the weight of ice on top of them, allowing them to flow. They freeze again when the pressure decreases. This is the way a glacier can flow around an obstructing rock. Ice skaters similarly obtain lubrication between the melting ice and the skate. ■

5.8 MULTICOMPONENT SYSTEMS: LAW OF MASS ACTION

So far it has been assumed that the particles in the system are not conserved because the system can exchange particles with a particle reservoir. Chemical reactions represent another mechanism which allows the numbers of particles of different species to change. So, consider a system consisting of several components, that is, different substances. For example, the system might contain molecular oxygen (O_2), molecular nitrogen (N_2), and nitric oxide (NO), and undergo the chemical reaction

$$N_2 + O_2 \leftrightarrow 2NO \tag{5.8-1}$$

In general, a multicomponent system will undergo several reactions. For the above system another reaction would be

$$N_2 + 2O_2 \leftrightarrow 2NO_2 \tag{5.8-2}$$

Although the arguments are easily generalized, it will be assumed for simplicity that only one reaction occurs. (In practice one reaction frequently dominates under given conditions of temperature and pressure.)

Equation (5.8-1) and similar equations represent two reactions proceeding in opposite directions at the same rates. The situation is quite analogous to the equilibrium between liquid water and water vapor where water molecules in the liquid are continually escaping from the liquid while others in the vapor, striking the surface of the liquid, are recaptured by the liquid.

Using a thermodynamic approach, the general chemical reaction is written as

$$\nu_1 A_1 + \nu_2 A_2 + \cdots + \nu_s A_s = 0 \tag{5.8-3}$$

where $A_1, A_2, \ldots, A_s$, stands for the chemical formulas of the different components, s of them in all, and $\nu_1, \nu_2, \ldots, \nu_s$, are the relative number of molecules that take part in the reaction. For example, in the reaction (5.8-1),

$$A_1 = N_2, \qquad A_2 = O_2, \qquad A_3 = NO$$

$$\nu_1 = 1, \qquad \nu_2 = 1, \qquad \nu_3 = -2 \tag{5.8-3a}$$

To study the reaction (5.8-3), assume that the system is kept at a constant temperature T and a constant pressure p. Its equilibrium is then characterized by the fact that the Gibbs free energy G of the system is a minimum. The Gibbs free energy for a given state of the system is now a function of T, p, and $N_1, N_2, \ldots, N_s$, the number of molecules of the various components $A_1, A_2, \ldots, A_s$:

$$G = G(T, p, N_1, N_2, \ldots, N_s) \tag{5.8-4}$$

The thermodynamic expressions for open systems apply, as for example,

$$dG = -S\,dT + V\,dP + \sum_i \mu_i\,dN_i \tag{5.8-5}$$

(summation with respect to i being over all components $i = 1, 2, \ldots, s$) where μ_i is the chemical potential of the ith component in the system:

$$\mu_i \equiv \mu_i(T, P, N_1, N_2, \ldots, N_s) = \left(\frac{\partial G}{\partial N_i}\right)_{T, P, N_j (j \neq i)} \tag{5.8-6}$$

For equilibrium at constant temperature and constant pressure, G is a minimum. Hence, from (5.8-5),

$$dG = \sum_i \mu_i \, dN_i = 0 \tag{5.8-7}$$

The particle numbers $N_1, N_2, \ldots, N_i$, of the different components cannot change in an arbitrary way, independently of one another. The changes are related to one another through the chemical reaction (5.8-3). This requires that

$$dN_i \propto \nu_i \qquad i = 1, 2, \ldots, s \tag{5.8-8}$$

and (5.8-7) becomes

$$\sum_i \nu_i \mu_i = 0 \tag{5.8-9}$$

Equation (5.8-9) is the basic equilibrium condition for chemical reactions from which the properties of such a system follow. This equation has a simple interpretation: it shows that the chemical potentials of the different components must be added in the same proportions in which the components occur in a reaction. For reaction (5.8-1), for example, Eq. (5.8-9) gives

$$\mu_{\mathrm{NO}} = \frac{1}{2}(\mu_{\mathrm{N_2}} + \mu_{\mathrm{O_2}}) \tag{5.8-10}$$

Thus, the chemical potential of a compound is made up additively from the chemical potentials of its constituents.

Applying (5.8-9) to a reaction in a gaseous phase, treating the gases as perfect gases so that the chemical potentials can be easily obtained from the Helmholtz free energy, gives

$$F \equiv F(T, V, N_1, N_2, \ldots, N_s) = \sum_i F_i(T, V, N_i) \tag{5.8-11}$$

where F_i refers to the ith component, and F refers to the system as a whole.

Now, the partition function for a perfect gas of N particles, applied to the ith component, is

$$Z_i(T, V, N_i) = \frac{1}{N_i!}[Z_i(T, V, 1)]^{N_i} \tag{5.8-12}$$

Hence, from the relationship between free energy and partition function,

$$\begin{aligned} F_i(T, V, N_i) &= -kT \ln Z_i(T, V, N_i) \\ &= -kT[N_i \ln Z_i(T, V, 1) - N_i \ln (N_i - 1)] \end{aligned} \tag{5.8-13}$$

The chemical potential μ_i of the ith is

$$\mu_i = \left(\frac{\partial F}{\partial N_i}\right)_{T, V, N_j (j \neq 1)} \tag{5.8-14}$$

Equations (5.8-11), (5.8-13), and (5.8-14) give

$$\mu_i = -kT[\ln Z_i(T, V, 1) - \ln N_i] \tag{5.8-15}$$

Substitution of (5.8-15) into the equilibrium condition (5.8-9) leads to

$$\sum_i \nu_i \ln N_i = \sum_i \nu_i \ln Z_i(T, V, 1) \tag{5.8-16}$$

and taking exponentials:

$$\prod_i (N_i)^{\nu_i} = K_N(T, V) \equiv \prod_i [Z_i(T, V, 1)]^{\nu_i} \tag{5.8-17}$$

This equation is known as the *law of mass action*. It states how the particle numbers $N_1, N_2, \ldots, N_i$, of the different components depend on the conditions of the system. $K_N(T, V)$ is called the *equilibrium constant* for the reaction (5.8-3).

It is convenient to express the law of mass action in terms of the particle concentrations

$$c_i \equiv \frac{N_i}{V} \tag{5.8-18}$$

Equation (5.8-17) then becomes

$$\prod_i (c_i)^{\nu_i} = K_c(T) \equiv \prod_i \left[\frac{1}{V} Z_i(T, V, 1)\right]^{\nu_i} \tag{5.8-19}$$

In this form of the law of mass action the equilibrium constant $K_c(T)$ depends on the temperature only, as indicated by the notation. For a perfect gas,

$$\frac{1}{V} Z_i(T, V, 1) = \left(\frac{2\pi m_i kT}{h^2}\right)^{3/2} Z_i^{\text{int}} \tag{5.8-20}$$

where m_i is the mass and Z_i^{int} is the internal partition function of a molecule of the ith component. It is seen from Eqs. (5.8-19) and (5.8-20) that for a reaction involving only perfect gases, the equilibrium constant can be calculated from knowledge of the internal partition functions of the gas molecules of the system. In principle these internal partition functions can be calculated from the rotational and vibrational properties of the molecules. In practice, spectroscopic data are used to deduce the energy levels of the molecules and hence the partition functions.

■ EXAMPLE 5.8-1

The equilibrium constant K_p at a certain temperature for the reaction

$$CO + \frac{1}{2}O_2 \rightarrow CO_2$$

is 6.44. Determine the composition and partial pressures of the equilibrium mixture at this temperature if the total pressure is (a) 1 atm and (b) 10 atm.

Solution. Let x be the fraction of 1 mol of CO_2 that dissociates. In other words, consider the reaction to have formed 1 mol of CO_2; then, of this mole of CO_2, x amount dissociates back into x moles of CO and $x/2$ moles of O_2. The result of this sequence of events is

$$CO + \frac{1}{2}O_2 \rightarrow (1-x)CO_2 + xCO + \frac{x}{2}O_2$$

The total number of moles at equilibrium is thus $1 - x + x + (x/2) = 1 + (x/2)$, and the partial pressures of the constituents are

$$p_{CO_2} = \frac{1-x}{1+x/2}, \qquad p_{CO} = \frac{x}{1+x/2}, \qquad p_{O_2} = \frac{x/2}{1+x/2} = \frac{x}{2+x}$$

From Eq. (5.8-19), the equilibrium constant, in terms of pressures, is

$$K_p = \frac{(1-x)/(1+x/2)}{[x/(1+x/2)][(x/2)/(1+x/2)]^{1/2}} = \frac{(1-x)(2+x)^{1/2}}{x^{3/2}} = 6.44$$

or

$$(1-x)(2+x)^{1/2} = (6.44)^{3/2}$$

Squaring both sides and solving for x gives

$$x = 0.3$$

The composition of the equilibrium mixture, in mole fractions, is then

$$x_{CO_2} = \frac{1-0.3}{1+0.15} = 0.609 \quad x_{CO} = \frac{0.3}{1+0.15} = 0.261 \quad x_{O_2} = \frac{0.15}{1+0.15} = 0.130$$

and the partial pressures for CO_2, CO, and O_2 are 0.609 atm, 0.261 atm, and 0.130 atm, respectively.

(b) When the total pressure is increased to 10 atm, the partial pressures become

$$p_{CO_2} = \frac{10-10x}{1+x/2}, \qquad p_{CO} = \frac{10x}{1+x/2}, \qquad p_{O_2} = \frac{10x/2}{1+x/2}$$

and similar calculations as above give $x = 0.16$, that is, the reaction is more complete (since a smaller fraction of CO_2 dissociates back to CO and O_2). Thus, increasing the total pressure causes the reaction

$$CO + \frac{1}{2}O_2 \rightarrow CO_2$$

to proceed more to the right, because the volume of the products is less than the volume of the reactants, and an increase in pressure tends to drive the reaction in the direction of smaller volume. ■

5.9 STATISTICAL THERMODYNAMICS OF PERFECT GASES

It has been shown that the thermodynamic properties of a system at fixed T, V, and composition can be found by averaging over the systems of a hypothetical ensemble, each system in the ensemble having the same T, V, and composition, but not necessarily the same quantum state. Using the postulate

postulate that quantum states of equal energy have equal probability of occurring in the ensemble, and using various thermodynamic relations, the average energy and pressure of the ensemble systems can be found.

Equating these averages to the thermodynamic U and the pressure p, the statistical mechanical formulas for U and p in terms of the canonical partition function Z can be found. The expressions for U and p are then used in $dU = T\,dS - p\,dV$ to find the statistical mechanical expression for $S = U/T + k \ln Z$. The key statistical mechanical formulas for calculating thermodynamic properties are

$$Z \equiv \sum_j e^{-E_j/kT} \tag{5.9-1}$$

$$P = kT\left(\frac{\partial \ln Z}{\partial V}\right)_{T,N_B} \tag{5.9-2}$$

$$U = kT^2\left(\frac{\partial \ln Z}{\partial T}\right)_{V,N_B} \tag{5.9-3}$$

$$S = \frac{U}{T} + k \ln Z \tag{5.9-4}$$

$$A = -kT \ln Z \tag{5.9-5}$$

$$\mu_B = -RT\left(\frac{\partial \ln Z}{\partial N_B}\right)_{T,V,N_{C \neq B}} \tag{5.9-6}$$

where E_j is the energy of quantum state j of the thermodynamic system, and the sum in (5.9-1) goes over all possible quantum states of the thermodynamic system. If species B is electrically charged, μ_B in (5.9-6) is to be replaced by the electrochemical potential; this is true of all equations containing μ_B in this chapter. Equations (5.9-1) to (5.9-6) are valid for gases, liquids, and solids, and for solutions as well as pure substances.

The procedure in statistical mechanics is therefore as follows:

1. The Schrödinger equation for the entire thermodynamic system is solved to obtain the quantum-mechanical energies E_i of the system's possible quantum states.
2. The canonical partition function $Z \equiv \Sigma e^{-E_i/kT}$ is evaluated.
3. Z is used to calculate the system's thermodynamic properties from Eqs. (5.9-2) to (5.9-6).

To a good approximation, the energy of a molecule can be partitioned as follows:

$$\epsilon = \epsilon_{\text{trans}} + \epsilon_{\text{rot}} + \epsilon_{\text{vib}} + \epsilon_{\text{elec}} \tag{5.9-7}$$

where the nuclear mode of energy storage has been neglected. (Generally, temperatures are not sufficiently high to cause excitation of nuclear energy levels.)

To the extent that the energies associated with different degrees of freedom in a molecule are independent and additive, the molecular partition function is correspondingly factorable. The latter is given by

$$z=\sum_i e^{-\epsilon_i/kT}=\sum_\epsilon g(\epsilon)e^{-\epsilon/kT} \tag{5.9-8}$$

Making use of (5.9-7) and the compounding of multiplicities, this gives

$$z=\sum_{\epsilon_t}\sum_{\epsilon_r}\sum_{\epsilon_v}\sum_{\epsilon_e} g_r g_t g_v g_\epsilon e^{-(\epsilon_t+\epsilon_r+\epsilon_v+\epsilon_e)/kT} \tag{5.9-9}$$

which can be written

$$z=z_{\text{trans}}\, z_{\text{rot}}\, z_{\text{vib}}\, z_{\text{elec}} \tag{5.9-10}$$

where each factor is a sum analogous to (5.9-8) over the allowed energies of the appropriate degree of freedom.

The contribution a degree of freedom makes to the thermodynamic functions depends on the relative magnitude of kT and $\Delta\epsilon$, the typical energy level spacing associated with that degree of freedom. The latter parameter can alternatively be expressed in terms of a characteristic temperature

$$\Theta=\frac{\Delta\epsilon}{k} \tag{5.9-11}$$

Its units are degrees Kelvin.

Three cases are to be distinguished:

(a) $kT \ll \Delta\epsilon$ or $T \ll \Theta$ (unexcited) (5.9-12)

(b) $kT \sim \Delta\epsilon$ or $T \sim \Theta$ (active) (5.9-13)

(c) $kT \gg \Delta\epsilon$ or $T \gg \Theta$ (fully excited) (5.9-14)

When (5.9-12) applies, only the lowest energy level is significantly populated and the partition function becomes

$$z=\sum_j g_j e^{-\epsilon_j/kT}\approx g_0 e^{-\epsilon_0/kT} \tag{5.9-15}$$

or, if ϵ_0 is chosen as the energy zero,

$$z\approx g_0 \qquad (\epsilon_0\equiv 0) \tag{5.9-16}$$

The corresponding degree of freedom is said to be unexcited (or inactive), since there is effectively no excitation of systems to higher energy states.

If (5.9-13) holds, T and Θ are of comparable magnitude, and several energy levels are significantly populated. The degree of freedom, termed active, can then make sizable contributions to the thermodynamic functions. As in case (a), the system must be treated by quantum mechanics.

Condition (5.9-14) is identical to the energy spacing criterion associated with the classical limit of quantum statistics. In this instance a large number of energy levels are significantly populated. The degree of freedom is termed fully excited or classical.

TABLE 5.9-1 THERMODYNAMIC PROPERTIES AND THE PARTITION FUNCTION Z

$Z = \sum_i \exp\left(\frac{-E_i}{kT}\right)$	$c_p = \left(\frac{\partial H}{\partial T}\right)_{p,N}$
$U = kT^2\left(\frac{\partial \ln Z}{\partial T}\right)_{V,N}$	$S = k \ln Z + \frac{U}{T}$
$p = kT\left(\frac{\partial \ln Z}{\partial V}\right)_{T,N}$	$A = kT \ln Z$
$H = kT^2\left(\frac{\partial \ln Z}{\partial T}\right)_{V,N} + kTV\left(\frac{\partial \ln Z}{\partial V}\right)_{T,H}$	$G = -kT \ln Z + kTV\left(\frac{\partial \ln Z}{\partial V}\right)_{T,N}$
$c_v = \left(\frac{\partial U}{\partial T}\right)_{V,N}$	$\mu = -kT\left(\frac{\partial \ln Z}{\partial N}\right)_{T,V}$

The characteristic temperatures encountered in statistical thermodynamics are roughly of the following orders of magnitude:

$$\Theta_{\text{trans}} \sim 10^{-16} \text{ to } 10^{-8} \text{ K} \tag{5.9-17}$$

$$\Theta_{\text{rot}} \sim 10 \text{ K} \tag{5.9-18}$$

$$\Theta_{\text{vib}} \sim 1000 \text{ K} \tag{5.9-19}$$

$$\Theta_{\text{elec}} \sim 10^6 \text{ K} \tag{5.9-20}$$

Thus, at ordinary temperatures, say 300 K, translations and rotations are always fully excited, and vibrations are generally active. Electronic degrees of freedom are, however, inactive except in a few cases where there is a low-lying excited state. A listing of thermodynamic properties in terms of the canonical partition function, as obtained in Chapter 2, is shown in Table 5.9-1.

5.10 TRANSLATIONAL PARTITION FUNCTION

A molecule of a perfect gas behaves as a free particle in translational motion within a containing vessel. The quantum-mechanical particle in a box, treated in Section 2.6, is the appropriate model. In a rectangular container of dimensions $a \times b \times c$, the allowed energy levels are given by

$$\epsilon_{n_1,n_2,n_3} = \frac{h^2}{8m}\left(\frac{n_1^2}{a^2} + \frac{n_2^2}{b^2} + \frac{n_3^2}{c^2}\right) \qquad n_1, n_2, n_3 = 1, 2, \ldots \tag{5.10-1}$$

The molecular translational partition function

$$z_{\text{trans}} = \sum_{\epsilon_i} g_t e^{-\epsilon_i/kT} \tag{5.10-2}$$

is, in this instance, a triple sum over the quantum numbers n_1, n_2, and n_3:

$$z_{\text{trans}} = \sum_{n_1=1}^{\infty} \sum_{n_2=1}^{\infty} \sum_{n_3=1}^{\infty} \exp\left[-\frac{h^2}{8mkT}\left(\frac{n_1^2}{a^2}+\frac{n_2^2}{b^2}+\frac{n_3^2}{c^2}\right)\right]$$

$$= \sum_{n=1}^{\infty} e^{-h^2n^2/8ma^2kT} \sum_{n=1}^{\infty} e^{-h^2n^2/8mb^2kT} \sum_{n=1}^{\infty} e^{-h^2n^2/8mc^2kT} \tag{5.10-3}$$

Before replacing the sums in (5.10-3) by integrals, it should be ascertained that this is permissible, namely, that the translational levels are effectively continuous, and the number of available translational states far exceeds the number of molecules in the gas. Otherwise stated, the average occupation number $\langle N_i \rangle$ is much less than unity. This is shown in Example 5.10-1. Therefore, to a good approximation,

$$\sum_{n=1}^{\infty} e^{-h^2n^2/8ma^2kT} = \int_0^{\infty} e^{-h^2x^2/8ma^2kT}\,dx$$

$$= \frac{1}{2}\pi^{1/2}\left(\frac{8ma^2kT}{h^2}\right)^{1/2} = (2\pi mkT)^{1/2}\frac{a}{h} \tag{5.10-4}$$

where

$$\int_0^{\infty} e^{-ax^2}\,dx = \frac{1}{2}\left(\frac{\pi}{a}\right)^{1/2}$$

has been used in the integration. Proceeding analogously with the second and third sums, and noting that $abc = V$, the volume of the containing vessel,

$$z_{\text{trans}} = \frac{(2\pi mkT)^{3/2}V}{h^3} = \left(\frac{2\pi mkT}{h^2}\right)^{3/2} V \tag{5.10-5}$$

Equation (5.10-5) is the final result. However, the partition function can alternatively be evaluated in the classical manner from the phase integral

$$\int\cdots\int dq_1 \cdots dp_3\, e^{-(p_1^2+p_2^2+p_3^2)/2mkT}$$

$$= \left(\int_{-\infty}^{\infty} e^{-p^2/2mkT}\,dp\right)^3 V = (2\pi mkT)^{3/2}V \tag{5.10-6}$$

where the threefold integration over the spatial coordinates gives V. Dividing by h^3 for the three translational degrees of freedom retrieves Eq. (5.10-5).

The molar partition function in the Boltzmann limit for an assembly of independent indistinguishable particles is applicable:

$$Z = \frac{z^N}{N!} \tag{5.10-7}$$

$$= \frac{(2\pi mKT)^{3N/2}V^N}{h^{3N}N!} \tag{5.10-8}$$

This can also be obtained classically from the corresponding phase integral.

Once the canonical ensemble partition function is known, all the thermodynamic functions for a perfect monatomic gas can be derived by the methods of Chapter 3. For example, the internal energy is

$$\langle U\rangle_{\text{trans}} = NkT^2\left(\frac{\partial \ln z_{\text{trans}}}{\partial T}\right)_V$$

$$= NkT^2\left[\frac{\partial}{\partial T}\ln\left(\frac{2\pi mkT}{h^2}\right)^{3/2} V\right]_V$$

$$= NkT^2\frac{d}{dT}\left(\frac{3}{2}T + \text{terms independent of } T\right)$$

$$= \frac{3}{2}NkT \tag{5.10-9}$$

The specific heat is then

$$C_V = \left(\frac{\partial U}{\partial T}\right)_V = \frac{3}{2}Nk \tag{5.10-10}$$

in agreement with the equipartition-of-energy theorem.

Before taking up other properties, it is desirable to note that

$$\ln(N!) = N\ln N - N = N\ln\left(\frac{N}{e}\right) \tag{5.10-11}$$

whereupon from (5.10-8),

$$\ln Z = N\ln\left(\frac{q^e}{N}\right) = N\ln\left[\frac{(2\pi mkT)^{3/2}Ve}{h^2 N}\right] \tag{5.10-12}$$

The Helmholtz free energy is (cf. Table 5.9-1)

$$A = kT\ln Z = -NkT\ln\left[\frac{(2\pi mkT)^{3/2}Ve}{h^3 N}\right] \tag{5.10-13}$$

The pressure is

$$p = kT\left(\frac{\partial \ln Z}{\partial V}\right)_{N,T}$$

$$= NkT\frac{d}{dV}(\ln V + \text{terms independent of } V) = \frac{NkT}{V} \tag{5.10-14}$$

which reproduces the ideal gas law. This particular result applies as well to a perfect polyatomic gas since the rotational, vibrational, and electronic factors in the molecular partition function are independent of V. In other words, the only contribution to the pressure of a gas comes from its translational motion.

The Gibbs free energy is

$$G = A + PV = -NkT\ln\left[\frac{(2\pi mkT)^{3/2}V}{h^3 N}\right] \tag{5.10-15}$$

noting that $pV = NkT = NkT\ln e$.

The chemical potential per molecule is

$$\mu = -kT\left(\frac{\partial \ln Z}{\partial N}\right)_{V,T} = -kT \ln\left[\frac{(2\pi mkT)^{3/2} V}{h^3 N}\right] \tag{5.10-16}$$

The same result also follows from using $\mu = G/N$.

The entropy of a perfect monatomic gas is given by

$$S = \left(\frac{\partial kT \ln Z}{\partial T}\right)_{V,N} = \frac{U}{T} + Nk \ln z \tag{5.10-17}$$

$$= \frac{3}{2}Nk + Nk \ln\left[\frac{(2\pi mkT)^{3/2} Ve}{h^3 N}\right] \tag{5.10-18}$$

Noting that $\frac{3}{2}$ can be replaced by $\ln(e^{3/2})$,

$$S = Nk \ln\left[\frac{(2\pi mkT)^{3/2} Ve^{5/2}}{h^3 N}\right] \tag{5.10-19}$$

This equation was first obtained (using different arguments) by Sackur (1911) and by Tetrode (1912). It is known as the *Sackur–Tetrode equation*. Through it, the translational entropy of an ideal gas is immediately obtainable from a knowledge of the molecular weight, temperature, and concentration or pressure. No other parameters are required. A correlation of translational spectroscopic entropies of monatomic gases with entropies determined calorimetrically using the third law is shown in Table 5.10-1. It is an impressive affirmation of statistical thermodynamics.

TABLE 5.10-1 SPECTROSCOPIC AND CALORIMETRIC ENTROPIES OF MONATOMIC GASES AT 1 ATM PRESSURE

Gas	Molecular weight, M (amu)	Boiling point, T/(K)	S_{calor} ($J\ K^{-1}\ mol^{-1}$)	S_{spec} ($J\ K^{-1}\ mol^{-1}$)
Argon	39.948	87.29	129.1 ± 0.4	129.20
Krypton	83.80	119.93	144.9 ± 0.4	145.04
Xenon	131.30	165.1	157.6 ± 0.4	157.29

■ EXAMPLE 5.10-1

A mole of perfect gas of 3000 K and 1 bar pressure would be contained in a volume of 0.2494 m^3. What is the molecular partition function for translational motion of a hydrogen atom at 3000 K in this volume? What is the de Broglie wavelength?

Solution. The molecular translational partition function is proportional to the volume in which the molecule moves:

$$z_{\text{trans}} = \left(\frac{2\pi mkT}{h^2}\right)^{3/2} V \tag{5.10-20}$$

With $m = 1.0080 \times 10^{-3}$ kg mol^{-1}, $k = 1.3806 \times 10^{-23}$ J K^{-1}, $T = 3000$ K, $h = 6.626 \times 10^{-34}$ J s, $V = 0.2494$ m^3, this gives

$$z_{\text{trans}} = \left[\frac{2\pi(1.0080 \times 10^{-3})(1.3806 \times 10^{-23})(3000)}{6.626 \times 10^{-34}}\right]^{3/2} (0.2494)$$
$$= 7.791 \times 10^{30}$$

This is the number of energy levels accessible to the single hydrogen atom at 3000 K in a volume of 0.2494 m^3.

The quantity $(h^2/2\pi mkT)^{1/2}$ is known as the *thermal de Broglie wavelength*, Λ. It is related to the molecular translational partition function and the volume by

$$z_{\text{tr}} = \frac{V}{\Lambda^3} \tag{5.10-21}$$

The condition for the application of Boltzmann statistics is that the thermal de Broglie wavelength be small compared with the dimensions of the container. Here,

$$\Lambda = \left(\frac{V}{z_{\text{tr}}}\right)^{1/3} = \left(\frac{0.2494 \text{ m}^3}{7.791 \times 10^{30}}\right)^{1/3} = 3.175 \times 10^{-11} \text{ m}$$ ■

■ EXAMPLE 5.10-2

(a) Show that the condition for the average population at an energy level, $\langle N_r \rangle \ll 1$, is equivalent to the inequality

$$\frac{1}{60}\left(\frac{h^2}{mkT}\right)^{3/2} \frac{N}{V} \ll 1 \tag{5.10-22}$$

(b) Show that, for a typical system of 1 mol of O_2 at 0°C and 1 atm, the left-hand side is on the order of 4×10^{-8}.

Solution. (a) To show that the assumption $\langle N_r \rangle \ll 1$ is justified, consider an ideal gas. The molecular energy is the sum of translational, rotational, vibrational, and electronic energies. Only the translational states will be counted; inclusion of rotational, vibrational, and electronic states would merely strengthen the result. For a cubic box of volume V, Section 2.6 gives the translational energy of a molecule as

$$\epsilon_{\text{trans}} = \left(\frac{h^2}{8mV^{2/3}}\right)(n_x^2 + n_y^2 + n_z^2) \tag{5.10-23}$$

Let the number of translational states whose energy is less than some maximum values $\epsilon_{\max}$ be calculated. For $\epsilon_{\text{trans}} \le \epsilon_{\max}$, the translational quantum numbers must satisfy the inequality

$$n_x^2 + n_y^2 + n_z^2 \le 8mV^{2/3}h^{-2}\epsilon_{\max} \tag{5.10-24}$$

It is known that the average translational energy per molecule is $\frac{3}{2}kT$, and that most molecules have ϵ_{trans} within two or three times this average value. Therefore, take $\epsilon_{max} = 3kT$ as the maximum ϵ_{trans} for a typical molecule. Figure 5.10-1(a) shows axes whose coordinates are the quantum numbers n_x, n_y, and n_z. In Fig. 5.10-1(b), a dot is shared by four cubes at the same elevation and four more cubes immediately above.

Imagine a dot placed at each point whose coordinates are integers. The distance of a dot from the origin is $r = (n_x^2 + n_y^2 + n_z^2)^{1/2}$. The number of quantum states with $\epsilon_{trans} \leq \epsilon_{max}$ equals the number of dots in one-eighth of a sphere whose radius r_{max} is, from Eq. (5.10-24),

$$r_{max} = (8mV^{2/3}h^{-2}\epsilon_{max})^{1/2} = (24mV^{2/3}h^{-2}kT)^{1/2} \tag{5.10-25}$$

Consider only the first octant, because n_x, n_y, and n_z must all be positive. Figure 5.10-1(a) shows eight dots that form a cube whose edge length is 1. Imagine many such cubes to fill up the first octant out to r_{max}. The cube shown has eight dots, but each dot is shared by eight cubes, as shown in Fig. 5.10-1(b): four at the same elevation, and four more either immediately above or below. The density of translational states is thus 8/8 = 1 per unit volume of space. The number of molecular states with $\epsilon_{trans} \leq \epsilon_{max}$ equals one-eighth the volume of a sphere of radius r_{max}. The number of available molecular translational states is then

$$\frac{1}{8}\frac{4\pi r_{max}^3}{3} = \frac{\pi}{6}(24mV^{2/3}h^{-2}kT)^{3/2} \approx 60\left(\frac{mkT}{h^2}\right)^{3/2} V \tag{5.10-26}$$

The condition $\langle N_r \rangle \ll 1$ to hold requires that $60(mkT/h^2)^{3/2}V \gg N$, where N is the number of molecules. This inequality reads

$$\frac{1}{60}\left(\frac{h^2}{mkT}\right)^{3/2}\frac{N}{V} \ll 1 \tag{5.10-27}$$

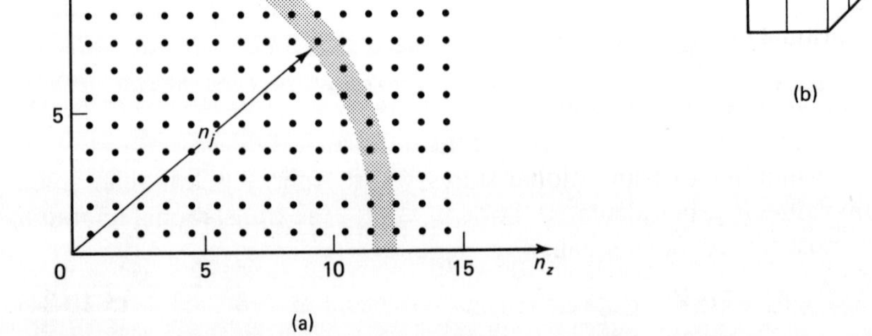

Figure 5.10-1 Translational density of states with energy less than ϵ_{max}. (a) $n_z = 0$ plane. (b) Dot shared by four cubes.

(b) A typical system of 1 mol of O_2 at 0°C and 1 atm, for which $m = [32/(6 \times 10^{23})]$ g, $N = 6 \times 10^{23}$, and $V = 22,400\ \text{cm}^3$, yields 4×10^{-8} for the left side of (5.10-27). Hence, $\langle N_r \rangle \ll 1$ is well satisfied. The translational levels are spaced extremely close together, so there are far more translational states available than there are molecules in the gas. This is the reason why summations in (5.10-3) can be replaced by integrations in (5.10-4). This condition does not hold for (a) extremely low temperatures, (b) extremely high densities, or (c) extremely small particle masses. Examples of these are (a) liquid helium at 2 or 3 K, (b) white dwarf stars and neutron stars (pulsars), and (c) the conduction electrons in metals. ■

■ EXAMPLE 5.10-3

(a) What are the translational contributions to C_p°, H°, S°, and G° for $O_2(g)$ at 298.15 K?

(b) What are the translational contributions to C_p°, H°, S°, and $0(g)$ at 298.15 K?

Solution. (a) The molar thermodynamic properties are obtained by letting N equal Avogadro's number N_A in Nk, and replacing the quantity V/N_A by kT/p°, where the standard pressure is 1 bar ($p^\circ = 10^5$ Pa $= 10^5$ N m^{-2}). This gives, for one mole, along with $pV = RT$,

$$U^\circ_{\text{trans}} = \frac{3}{2}RT \qquad C^\circ_{V,\text{trans}} = \frac{3}{2}R$$

$$H^\circ_{\text{trans}} = \frac{5}{2}RT \qquad C^\circ_{p,\text{trans}} = \frac{5}{2}R$$

$$S^\circ_{\text{trans}} = R\left\{\frac{5}{2} + \ln\left[\left(\frac{2\pi mkT}{h^2}\right)^{3/2}\frac{kT}{p^\circ}\right]\right\}$$

$$A^\circ_{\text{trans}} = -RT\left\{1 + \ln\left[\left(\frac{2\pi mkT}{h^2}\right)^{3/2}\frac{kT}{p^\circ}\right]\right\}$$

$$G^\circ_{\text{trans}} = RT\ \ln\left[\left(\frac{2\pi mkT}{h^2}\right)^{3/2}\frac{kT}{p^\circ}\right]$$

Applying these to the present case,

$$C^\circ_{p,\text{trans}} = \frac{5}{2}(8.31441\ \text{J K}^{-1}\ \text{mol}^{-1}) = 20.786\ \text{J K}^{-1}\ \text{mol}^{-1}$$

$$H^\circ_{\text{trans}} = \frac{5}{2}(8.31441\ \text{J K}^{-1}\ \text{mol}^{-1})(298.15\ \text{K}) = 6.197\ \text{kJ mol}^{-1}$$

$$m = \frac{M}{N_A} = \frac{31.9988 \times 10^{-3}\ \text{kg mol}^{-1}}{6.022045 \times 10^{23}\ \text{mol}^{-1}} = 5.31361 \times 10^{-26}\ \text{kg}$$

$$\left(\frac{2\pi mkT}{h^2}\right)^{3/2}\frac{kT}{p^\circ} = \left[\frac{2\pi(5.31361\times10^{-26})(1.380662\times10^{-23})(298.15)}{(6.626176\times10^{-34})^2}\right]^{3/2}$$

$$\times\frac{(1.380662\times10^{-23})(298.15)}{10^5} = 7.20893\times10^6$$

$$S^\circ_{\text{trans}} = (8.31441\text{ J K}^{-1}\text{ mol}^{-1})(2.5+\ln 7.20893\times10^6) = 152.077\text{ J K}^{-1}\text{ mol}^{-1}$$

$$G^\circ_{\text{trans}} = -(8.31441\text{ J K}^{-1}\text{ mol}^{-1})(298.15\text{ K})\ln(7.20893\times10^6)$$
$$= -39.145\text{ kJ mol}^{-1}$$

(b) The translational contributions to $C^\circ_{p,\text{trans}}$ and H°_{trans} are the same as for $O_2(g)$, or any other gas. However, because the mass of the atom is one-half that of the diatomic molecule, the translational entropy for the atomic gas is smaller by $-\frac{3}{2}R\ln 2 = -8.645\text{ J K}^{-1}\text{ mol}^{-1}$, and the translational Gibbs energy is more positive by $\frac{3}{2}RT\ln 2 = 2.577\text{ kJ mol}^{-1}$. Thus, at 298.15 K,

$$C^\circ_{p,\text{trans}} = 20.786\text{ J K}^{-1}\text{ mol}^{-1}$$

$$S^\circ_{\text{trans}} = 143.432\text{ J K}^{-1}\text{ mol}^{-1}$$

$$H^\circ_{\text{trans}} = 6.197\text{ kJ mol}^{-1}$$

$$G^\circ_{\text{trans}} = -36.568\text{ kJ mol}^{-1}$$

■

5.11 PERFECT GAS MIXTURES

For a multicomponent ideal gas, such as a homogeneous binary gas mixture, the molar partition function is given by

$$Z = \frac{z_1^{N_1} z_2^{N_2}}{N_1!N_2!} \tag{5.11-1}$$

Using the expressions in Section 5.10 for z_1 and z_2, involving m_1 and m_2, Eq. (5.11-1) becomes

$$\ln Z = N_1\ln\left[\frac{(2\pi m_1 kT)^{3/2}Ve}{h^3N_1}\right] + N_2\ln\left[\frac{(2\pi m_2 kT)^{3/2}Ve}{h^3N_2}\right] \tag{5.11-2a}$$

It is of particular interest to evaluate the pressure and entropy of the mixture. The pressure, which reproduces the ideal gas law, is

$$p = (N_1+N_2)\frac{kT}{V} \tag{5.11-2b}$$

Equation (5.11-2b) is equivalent to Dalton's law for ideal mixtures.

In analogy with (5.10-14),

$$S(V,N_1,N_2,T) = N_1k\ln\left[\frac{(2\pi m_1 kT)^{3/2}Ve^{5/2}}{h^3N_1}\right]$$
$$= N_2k\ln\left[\frac{(2\pi m_1 kT)^{3/2}Ve^{5/2}}{h^3N_2}\right] \tag{5.11-3}$$

This is compared to the entropy of the unmixed gas at the same pressure and temperature; N_1 molecules in volume

$$V_1 = \frac{N_1}{N_1 + N_2} V$$

N_2 molecules in volume

$$V_2 = \frac{N_2}{N_1 + N_2} V$$

so that

$$S_1(V_1, N_1, T) + S_2(V_2, N_2, T) = N_1 kT \ln\left[\frac{(2\pi m_1 kT)^{3/2} V e^{5/2}}{h^3(N_1 + N_2)}\right] + N_2 kT \ln\left[\frac{(2\pi m_2 kT)^{3/2} V e^{5/2}}{h^3(N_1 + N_2)}\right] \qquad (5.11\text{-}4)$$

Subtracting (5.11-4) from (5.11-3) gives the entropy of mixing,

$$\Delta S_{\text{mix}} = -N_1 k \ln\frac{N_1}{N_1 + N_2} - N_2 k \ln\frac{N_2}{N_1 + N_2} \qquad (5.11\text{-}5)$$

which is identical with the classical thermodynamic result,

$$\overline{\Delta S}_{\text{max}} = \sum_{i=1}^{2} N_i k \ln\frac{N}{N_i} = \sum_{i=1}^{2} N_i k \ln\frac{V}{V_i}$$

$$= -R \sum_i x_i \ln x_i \qquad \text{(per mole of mixture)} \qquad (5.11\text{-}6)$$

However, Eq. (5.11-6) carries an inconsistency. Consider two equal volumes V, both of which contain N atoms of the same ideal monatomic gas at temperature T, and which are separated, initially, by a solid partition. When the partition is removed, there are $2N$ atoms in a volume $2V$, and Eq. (5.11-6) gives an entropy increase of $2Nk \ln 2$, which is certainly incorrect. This was the *Gibbs paradox* of the nineteenth century. It was resolved by the realization that the atoms of a monatomic gas are indistinguishable (by the inclusion of the factor $N!$ in the molar partition function).

The Gibbs paradox arises unless the factor $N!$ is included, as in (5.10-7). Setting $m_1 = m_2 = m$ and $N_1 + N_2 = N$ in (5.11-4), so that it represents the entropy of a partitioned single-component gas, an entropy equal to (5.10-19) correctly shows that $\Delta S_{\text{mix}} = 0$ when the partition is removed. If, however, $Z = z^N$ and $Z = z_1^{N_1} z_2^{N_2}$ were used in place of (5.10-7) and (5.11-1), respectively, a spurious entropy of mixing equal to (5.11-5) would show up.

5.12 POLYATOMIC GASES

In Chapter 3, it was shown that a monatomic gas should have an atomic heat capacity of $\frac{3}{2}k$ at constant volume, due to the kinetic energy of translation of the atoms. This result agrees with experiment. For a simple diatomic gas such as carbon monoxide the molecular heat capacity at constant volume is normally about $\frac{5}{2}k$, but rises quite sharply around 1000 K and approaches $\frac{7}{2}k$ before it finally dissociates. This difference is due to the energy of internal motions, which will now be discussed.

It is first noted that even the monatomic gas may have internal motions. In a monatomic gas, however, internal motions are motions of the electrons. If the atom is "rotating," the electrons are rotating around the nucleus, and if the atom is "vibrating," the electrons are in excited states. It turns out that the energy spacing between the lowest electron levels in an atom is almost always large compared with kT. (Typically the spacing is about 10^{-19} J, while kT at room temperature is about 4×10^{-21} J.) Thus, the Boltzmann probability for any excited electron state is totally negligible at all ordinary temperatures, as is the electronic excitation when calculating the heat capacity. The electronic heat capacity is quenched at all ordinary temperatures.

In a molecule the orbiting electrons move much faster than the nuclei. Therefore, the motion of the nuclei may be ignored in evaluating the electronic state of the molecule; this is the *Born–Oppenheimer* approximation. The electrons remain in their ground state at normal temperatures, as in a monatomic gas. For a diatomic molecule such as carbon monoxide, the molecule has six independent modes of motion, corresponding to its six spatial degrees of freedom, as seen in Fig. 5.12-1. It takes six numbers, in three dimensions, to specify where the two nuclei are. These six independent modes are three modes corresponding to translation of the center of mass in the x, y, and z directions; one mode of vibration parallel to the line of centers; and two modes of rotation about the center of mass. (Note that rotation around the line of centers is an electronic excitation which is quenched at ordinary temperatures.)

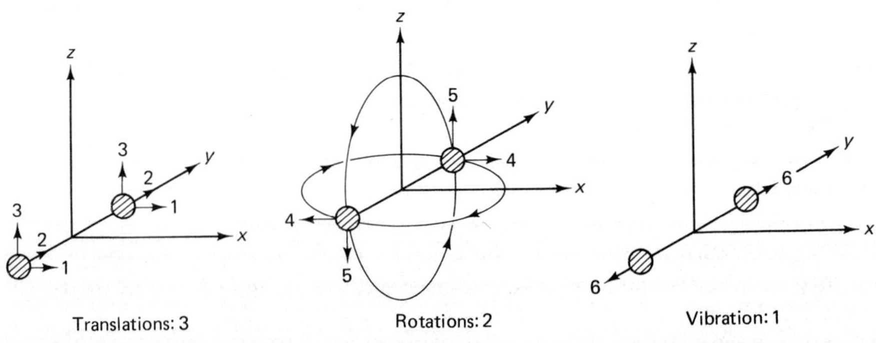

Figure 5.12-1 The six modes of motion of the nuclei of a diatomic molecule.

In general, for polyatomic molecules there are rotational, vibrational, and electronic contributions to the partition function in addition to the translational contribution already computed. The Boltzmann limit is applicable and is, in fact, further enhanced, since internal degrees of freedom serve to increase the number of quantum states available to a given number of molecules. Thus, the molecular partition function becomes

$$Z = \frac{1}{N!}(z_{\text{trans}}\, z_{\text{rot}}\, z_{\text{vib}}\, z_{\text{elec}})^N \tag{5.12-1}$$

This can be expressed as

$$Z = Z_{\text{trans}}\, Z_{\text{rot}}\, Z_{\text{vib}}\, Z_{\text{elec}} \tag{5.12-2}$$

with

$$Z_{\text{trans}} = \frac{z_{\text{trans}}^N}{N!} \tag{5.12-3}$$

and

$$Z_{\text{rot}} = z_{\text{rot}}^N \tag{5.12-4}$$

$$Z_{\text{vib}} = z_{\text{vib}}^N \tag{5.12-5}$$

$$Z_{\text{elec}} = z_{\text{elec}}^N \tag{5.12-6}$$

Note that the indistinguishability factor $N!$ is generally included with the translational contribution so that the above monatomic formulas can be used. Now,

$$\ln Z = \ln Z_{\text{trans}} + \ln Z_{\text{rot}} + \ln Z_{\text{vib}} + \ln Z_{\text{elec}} \tag{5.12-7}$$

where, using (5.10-12),

$$\ln Z_{\text{trans}} = N \ln\left(\frac{z_{\text{trans}}\, e}{N}\right) \tag{5.12-8}$$

and

$$\ln Z_{\text{rot}} = N \ln z_{\text{rot}} \tag{5.12-9}$$

$$\ln Z_{\text{vib}} = N \ln z_{\text{vib}} \tag{5.12-10}$$

$$\ln Z_{\text{elec}} = N \ln z_{\text{elec}} \tag{5.12-11}$$

Since all the thermodynamic functions are derived from $\ln Z$ by multiplications and/or differentiations, the individual molecular degrees of freedom contribute additively to the thermodynamics functions of a perfect gas. The formulas of Chapter 2 are applicable to each contribution separately. For example, the Helmholtz function is

$$A = -kT \ln Z = A_{\text{trans}} + A_{\text{rot}} + A_{\text{vib}} + A_{\text{elec}} \tag{5.12-12}$$

where

$$A_{\text{trans}} = -kT \ln Z_{\text{trans}} = -NkT \ln\left(\frac{z_{\text{trans}}\, e}{N}\right) \tag{5.12-13}$$

and

$$A_{\text{rot}} = -kT \ln Z_{\text{rot}} = -NkT \ln z_{\text{rot}} \tag{5.12-14}$$

$$A_{\text{vib}} = -kT \ln Z_{\text{vib}} = -NkT \ln z_{\text{vib}} \tag{5.12-15}$$

$$A_{\text{elec}} = -kT \ln Z_{\text{elec}} = -NkT \ln z_{\text{elec}} \tag{5.12-16}$$

Translational contributions in polyatomic gases are identical to those derived for the monatomic gas. Contributions from the internal degrees of freedom are considered in the following sections.

5.13 ROTATIONAL PARTITION FUNCTION

The rotational motion of a diatomic molecule can be considered as the planar rotation of a dumbbell about its center of mass, as in Fig. 5.13-1. Initially, to avoid complications that result from molecular symmetry, unsymmetrical diatomic molecules such as HCl, HD, CO, will be considered. These are known as *heteronuclear diatomic molecules*.

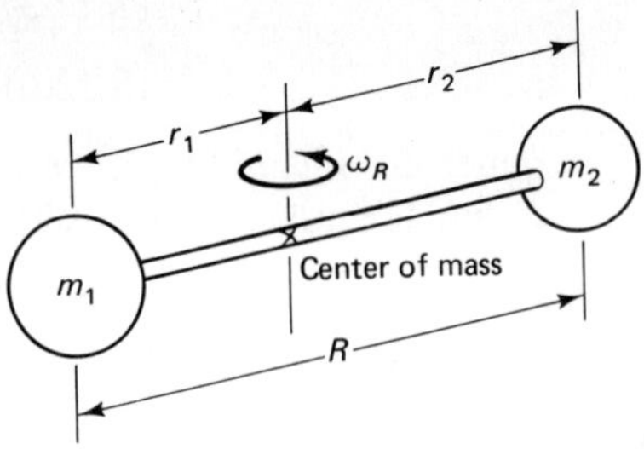

Figure 5.13-1 Rotational motion of a diatomic molecule.

This is the simplest model for a diatomic or linear polyatomic molecule. Its motion is governed by the rotational Hamiltonian

$$\hat{h}_{\text{rot}} = \frac{\hat{L}^2}{2I} \tag{5.13-1}$$

where $\hat{L}$ represents the angular momentum operator and I the moment of inertia. For a diatomic molecule,

$$I = \mu R^2 \tag{5.13-2}$$

μ is the reduced mass and R the internuclear separation. The allowed rotational energies are given by

$$\epsilon_J = \frac{J(J+1)\hbar^2}{2I} \qquad J = 0, 1, 2, \ldots \tag{5.13-3}$$

The Jth rotational level is $(2J + 1)$-fold degenerate, namely,

$$g_J = 2J + 1 \tag{5.13-4}$$

The partition function is then given by

$$z_{\text{rot}} = \sum_{\substack{\text{rotational}\\ \text{states, } r}} e^{-\epsilon_{\text{rot}(r)}/kT} \equiv \sum_{\substack{\text{rotational}\\ \text{levels, } j}} g_{\text{rot}(J)}\, e^{-\epsilon_{\text{rot}(J)}/kT}$$

$$= \sum_{J=0}^{\infty} (2J+1) e^{-J(J+1)h^2/8\pi^2 IkT} \tag{5.13-5}$$

It is found by calculation that, for almost every case of practical interest (the exceptions are the hydrogen isotopes at low temperatures), $h^2/8\pi^2 IkT$ is small, and successive terms in the summation (5.13-5) have values sufficiently close together for the summation to be approximated by integration. Thus,

$$z_{rot} = \int_0^{\infty} (2J + 1)e^{-J(J+1)C}\, dJ \tag{5.13-6}$$

where

$$C = \frac{h^2}{8\pi^2 IkT} \tag{5.13-7}$$

Let

$$x = J^2 + J \tag{5.13-8}$$

Then

$$dx = (2J + 1)\, dJ$$

and

$$z_{rot} = \int_0^{\infty} e^{-Cx}\, dx \tag{5.13-9}$$

$$= -\frac{1}{C}(e^{-\infty} - e^0) = \frac{1}{C}$$

Thus

$$z_{rot} = \frac{8\pi^2 IkT}{h^2} \tag{5.13-10}$$

The quantity $h^2/8\pi^2 Ik$ has the dimension of temperature, and is called the characteristic temperature for rotation, denoted by Θ_{rot}; that is,

$$\Theta_{rot} = \frac{h^2}{8\pi^2 Ik} \tag{5.13-11}$$

and

$$z_{rot} = \frac{T}{\Theta_{rot}} \qquad \text{heteronuclear} \tag{5.13-12}$$

Some values of Θ_{rot} are given in Table 5.13-1.

Equation (5.13-12) is valid only for *heteronuclear* diatomic molecules. For a *homonuclear* diatomic molecule, such as H_2, which is composed of two identical atoms, the nuclei of the two atoms in the molecule are indistinguishable. The Pauli principle requires that ψ_N (the wave function for nuclear motion) be symmetric with respect to interchange of the identical nuclei if the nuclei are bosons or antisymmetric if the nuclei are fermions. This restriction on ψ_N cuts the number of available quantum states in half compared with a heteronuclear diatomic, for which there is no symmetry restriction. In order to have a single formula for both homonuclear and heteronuclear diatomics, a symmetry

TABLE 5.13-1 CHARACTERISTIC TEMPERATURES FOR ROTATION

Molecule	Θ_{rot} (K)	z_{rot} (300 K)	z_{rot} (1000 K)
H_2	85·38	1·76	5·86
HD	64·27	4·67	15·56
D_2	43·03	3·49	11·62
N_2	2·863	52·41	174·7
O_2	2·069	72·51	241·7
CO	2·766	108·5	361·5
$^{35}Cl_2$	0·3495	429·3	1431
I_2	0·0537	2793	9310

number $\sigma(=2$ for a homonuclear diatomic and 1 for a heteronuclear diatomic) is defined so that a factor $1/\sigma$ is included in z_{rot}. Thus,

$$z_{rot} \approx \frac{T}{\sigma\Theta_{rot}} = \frac{2IkT}{\sigma\hbar^2} \qquad \text{for} \quad T >> \Theta_{rot} \tag{5.13-13}$$

Equation (5.13-13) is a good approximation, except for H_2 and its isotopic species D_2 and HD. The light molecule H_2 has a small moment of inertia and hence a relatively high Φ_{rot}. A more rigorous derivation (Example 5.13-1) shows that

$$z_{rot} \simeq \frac{T}{\Theta_{rot}} + \frac{1}{3} + \frac{1}{15}\left(\frac{\Theta_{rot}}{T}\right) + \cdots \tag{5.13-14}$$

This is known as *Mulholland's expansion.*

It is worthwhile to note that z_{trans} and z_{rot} are obtained by using the fact that the translational and rotational energy spacings are very small compared with kT, so that the sums over discrete energy levels can be replaced by integrals over a continuous range of energy. Since continuous values for energy correspond to classical mechanics, classical results for $C_{v,\text{trans}}$ and $C_{v,\text{rot}}$ are obtained. But this applies only at temperatures for which $T >> \Theta_{rot}$. At low temperatures, the series (5.13-14) for z_{rot} should be used, and U_{rot} and $C_{v,\text{rot}}$ calculated from it. At very low temperatures, (5.13-14) fails, and z_{rot} must be calculated by direct term-by-term summation. The plot of $C_{v,\text{rot}}$ versus T is shown in Fig. 5.13-2. The classical value R is reached at $T \approx 1.5\Theta_{rot}$.

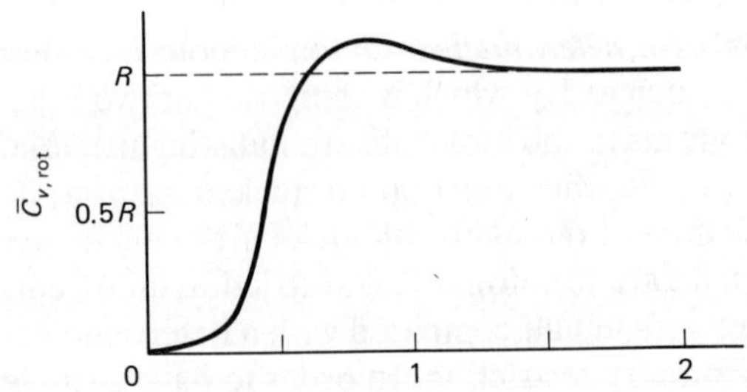

Figure 5.13-2 Rotational contribution to C_v of a diatomic gas.

■ EXAMPLE 5.13-1

The Euler–Maclaurin summation formula is

$$\sum_{x=x_0}^{x_n} F(x) = \int_{x_0}^{x_n} F(x)\,dx + \frac{1}{2}[F(x_0) + F(x_n)] + \phi \tag{5.13-15}$$

where

$$\phi = -\frac{1}{12}[F^{(1)}(x_0) - F^{(1)}(x_n)] + \frac{1}{720}[F^{(2)}(x_0) - F^{(2)}(x_n)]$$
$$- \frac{1}{30240}[F^{(3)}(x_0) - F^{(3)}(x_n)] + \cdots \tag{5.13-16}$$

with

$$F^{(k)}(x_0) = \left[\frac{\partial^k F(x)}{\partial x^k}\right]_{x=x_0}, \text{ etc.} \tag{5.13-17}$$

Applying this to the terms of the rotational partition function

$$F(x) = F(J) = (2J + 1)e^{-J(J+1)\Theta_{rot}/T} \tag{5.13-18}$$

derive Mulholland's equation (5.13-14).

Solution. For $x_0 = J_0 = 0$, $F(x_0) = e^0 = 1$. Also, $F(\infty) = 0$. Now,

$$F^{(1)}(x_0) = 2 - \frac{\Theta_{rot}}{T} \tag{5.13-19}$$

$$F^{(2)}(x_0) = -12\frac{\Theta_{rot}}{T} + 12\left(\frac{\Theta_{rot}}{T}\right)^2 - \left(\frac{\Theta_{rot}}{T}\right)^3 \tag{5.13-20}$$

Higher derivatives contain only powers of Θ_{rot}/T greater than 1. Therefore, neglecting powers of Θ_{rot}/T higher than the first, and noting that all the derivatives which contain the factor $\exp[-J(J+1)\Theta_{rot}/T]$ vanish at $J = \infty$,

$$\phi = -\frac{1}{12}\left(2 - \frac{\Theta_{rot}}{T}\right) - \frac{12}{720}\frac{\Theta_{rot}}{T}$$
$$= -\frac{1}{6} + \frac{1}{15}\left(\frac{\Theta_{rot}}{T}\right) \tag{5.13-21}$$

Also,

$$\int_0^\infty F(x)\,dx = \frac{T}{\Theta_{rot}} \tag{5.13-22}$$

Substituting these into (5.13-15) results in Mulholland's equation (5.13-14). ■

■ EXAMPLE 5.13-2

Discuss how the symmetry number is obtained and how it is used in the partition function.

Solution

For any type of molecule, the symmetry number is defined as the number of equivalent ways in which a given orientation of the molecule can be obtained by rotation, when like atoms are regarded as indistinguishable. Thus, for a homonuclear diatomic molecule,

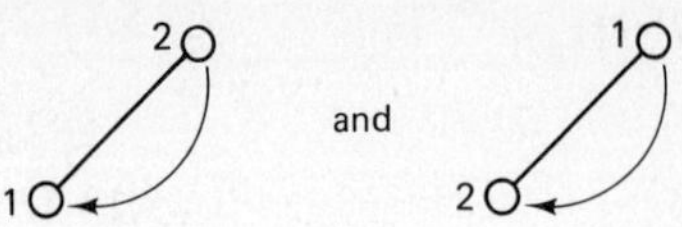

Figure 5.13-3a Orientations of a diatomic molecule.

There are two ways of obtaining an indistinguishable orientation ($\sigma = 2$).

For the square-planar molecule XeF_4, there are eight indistinguishable orientations obtainable from one another by rotations, so $\sigma = 8$. The second, third, and fourth orientations are obtained by 90°, 180°, and 270° rotations about the C_4 axis; the fifth orientation is obtained from the first by a 180° rotation about the C_2 axis passing through atoms 1 and 3; the sixth, seventh, and eighth orientations are obtained by 90°, 180°, and 270° rotations applied to the fifth orientation. (See Figure 5.13-3b.)

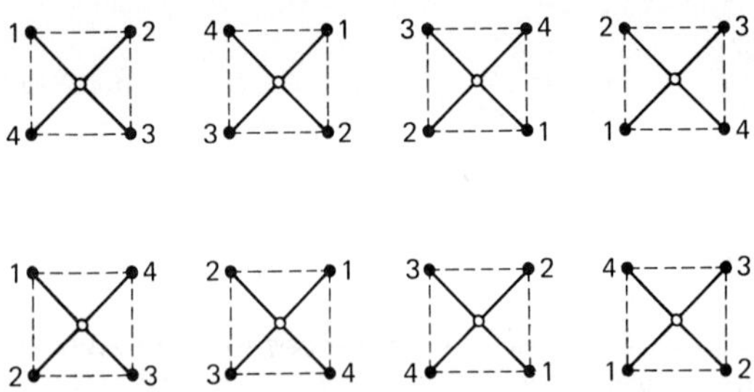

Figure 5.13-3b Orientations of a square-planar molecule.

For a nonlinear polyatomic molecule, the rotational partition function is given by

$$z_{\text{rot}} = \frac{\pi^{1/2}}{\sigma}\left(\frac{2kT}{\hbar^2}\right)^{3/2}(I_a I_b I_c)^{1/2} \tag{5.13-23}$$

where I_a, I_b, and I_c are the molecule's principal moments of inertia. Equation (5.13-23) holds provided T is not extremely low. ■

■ EXAMPLE 5.13-3

Show that when $T >> \Theta_r$, the mean energy of rotation of a diatomic molecule is kT.

Solution. The rotational partition function is

$$z_r = \frac{T}{\Theta_r} = T\left(\frac{2Ik}{\hbar^2}\right) = \frac{2I}{\beta\hbar^2} \tag{5.13-24}$$

where β has been used for $1/kT$ to shorten the writing. Taking the logarithm of (5.13-23),

$$\ln z_r = -\ln \beta + \ln\left(\frac{2I}{\hbar^2}\right) \tag{5.13-25}$$

Using $u = -(\partial \ln z/\partial\beta)_{V,N}$, when $T >> \Theta_r$, gives

$$u = \left(\frac{\partial \ln \beta}{\partial\beta}\right)_{V,N} = \frac{1}{\beta} = kT \tag{5.13-26}$$

■

■ EXAMPLE 5.13-4

The moment of inertia for $O_2(g)$ is $I = 1.9373 \times 10^{-46}$ kg m². What is the characteristic rotational temperature and the molar rotational contributions to C_p°, S°, H°, and G° for $O_2(g)$ at 298.15 K?

Solution. The characteristic rotational temperature is

$$\Theta_r = \frac{h^2}{8\pi^2 Ik}$$

where $h = 6.62618 \times 10^{-34}$ J s, $I = 1.9373 \times 10^{-46}$ kg m², and $k = 1.38066 \times 10^{-23}$ J K^{-1}. Thus,

$$\Theta_r = \frac{(6.62618 \times 10^{-34})^2}{8\pi^2(1.9373 \times 10^{-46})(1.38066 \times 10^{-23})} = 2.079 \text{ K}$$

From $Z_r = z_r^N$, the following expressions and results for the molar rotational contributions at 298.15 K and the standard-state pressure of $p^\circ = 1$ bar are obtained:

$$C_{p,r}^\circ = R = 8.314 \text{ J K}^{-1} \text{ mol}^{-1}$$

$$H_r^\circ = RT = (8.314 \text{ J K}^{-1} \text{ mol}^{-1})(298.15 \text{ K}) = 2.479 \text{ kJ mol}^{-1} = U_r^\circ$$

$$S_r^\circ = R \ln\left(\frac{eT}{\sigma\Theta_r}\right) = (8.314 \text{ J K}^{-1} \text{ mol}^{-1}) \ln\left[\frac{(2.718)(298.15 \text{ K})}{2(2.079 \text{ K})}\right]$$
$$= 43.838 \text{ J K}^{-1} \text{ mol}^{-1}$$

$$G_r^\circ = -RT \ln\left(\frac{T}{\sigma\Theta_r}\right) = -(8.314 \text{ J K}^{-1} \text{ mol}^{-1})(298.15 \text{ K}) \ln\left(\frac{298.15 \text{ K}}{2 \times 2.079 \text{ K}}\right)$$
$$= -10.591 \text{ kJ mol}^{-1}$$

Note that $H_r^\circ = U_r^\circ$ because Z_r is independent of V, and there is no rotational contribution to pressure.

■

5.14 VIBRATIONAL PARTITION FUNCTION

To a first approximation, the vibrational motion of the two atoms in a diatomic molecule along the line joining them can be treated as simple harmonic motion.

The harmonic oscillator approximation gives the vibrational energies of a diatomic molecule as $(n + \frac{1}{2})h\nu$, where ν goes from 0 to ∞ and there is no degeneracy [Section (2.9)]. The choice of zero level of energy is arbitrary, and it is customary in molecular statistical mechanics to take the energy zero at the lowest available energy level of the molecule. This level has zero rotational energy and $\frac{1}{2}h\nu$ vibrational energy. Therefore,

$$\epsilon_{\text{vib}} = (n + \tfrac{1}{2})h\nu - \tfrac{1}{2}h\nu = nh\nu \tag{5.14-1}$$

where ϵ_{vib} is measured from the $n = 0$ level, that is, $n = 0, 1, 2, \ldots$. The expression for the thermodynamic internal energy that is obtained from the partition function will then be U relative to the lowest molecular energy level. (To be fully consistent, the zero-point translational energy should be subtracted from ϵ_{trans}, but this is so small as to be utterly negligible.) The partition function is

$$z_{\text{vib}} = \sum_{n=0}^{\infty} e^{-nh\nu/kT} = \sum_{n=0}^{\infty} e^{-n\Theta_{\text{vib}}/T} \tag{5.14-2}$$

where

$$\Theta_{\text{vib}} \equiv \frac{h\nu}{k} \tag{5.14-3}$$

is the *characteristic vibrational temperature*. Some values of Θ_{vib} for ground electronic states are

Molecule	H_2	$H^{35}Cl$	N_2	O_2	$^{35}Cl_2$	I_2
Θ_{vib}/K	5990	4151	3352	2239	798	307

The generally high values of Θ_{vib} compared with ordinary temperatures show that the vibrational levels are not closely spaced compared with kT. (For most diatomic molecules there is very little occupancy of excited vibrational levels at room temperature.) Hence, the vibrational sum cannot be replaced by an integral. This is no cause for alarm, however, because the sum is easily evaluated exactly. Equation (5.14-2) is recognized as the sum of a geometric series:

$$1 + x + x^2 + x^3 + \cdots = \sum_{n=0}^{\infty} x^n = \frac{1}{1-x} \qquad \text{for } |x| < 1 \tag{5.14-4}$$

The sum in (5.14-2) corresponds to (5.14-4) with $x = e^{-h\nu/kT} < 1$. Therefore,

$$z_{\text{vib}} = \frac{1}{1 - e^{-h\nu/kT}} \tag{5.14-5}$$

In terms of Θ_{vib},

$$z_{vib} = (1 - e^{-\Theta_{vib}/T})^{-1} \tag{5.14-6}$$

and, just as $Z_{rot} = z_{rot}^N$,

$$\ln Z_{vib} = -N \ln(1 - e^{-\Theta_{vib}/T}) \tag{5.14-7}$$

Vibrational contributions to thermodynamic functions can now be calculated. First,

$$\frac{\partial \ln Z_{vib}}{\partial T} = \frac{N(\Theta_{vib}/T^2)e^{-\Theta_{vib}/T}}{1 - e^{-\Theta_{vib}/T}} \tag{5.14-8}$$

The internal energy is, therefore,

$$E_{vib} = H_{vib} = kT^2 \frac{\partial \ln Z_{vib}}{\partial T} = \frac{Nk\Theta_{vib}}{e^{\Theta_{vib}/T} - 1} \tag{5.14-9}$$

The vibrational heat capacity is then given by

$$C_V^{vib} = C_P^{vib} = \frac{\partial E_{vib}}{\partial T} = \frac{Nk(\Theta_{vib}/T)^2 e^{\Theta_{vib}/T}}{e^{\Theta_{vib}/T} - 1} \tag{5.14-10}$$

The vibrational contribution to molar heat capacity, as a function of temperature, is represented in Fig. 5.14-1. Since 300 K is very much below Θ_{vib} for most of the lighter diatomic molecules, it is seen that the vibrational degree of freedom is effectively unexcited and C_V^{vib} is close to zero at room temperature. For the heavier diatomic molecules, however, vibration is active at room temperature and contributes significantly to heat capacity.

At sufficiently high temperature,

$$e^{\Theta_{vib}/T} \approx 1 + \frac{\Theta_{vib}}{T} \tag{5.14-11}$$

so that

$$\lim_{T \to \infty} C_V^{vib} = R \tag{5.14-12}$$

in agreement with the equipartition result for two quadratic contributions.

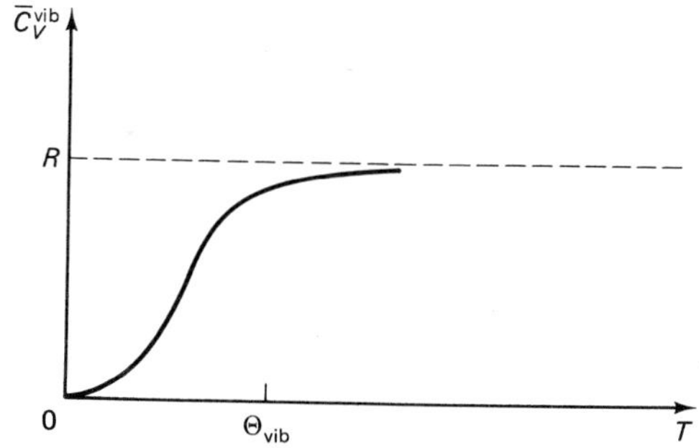

Figure 5.14-1 Vibrational contribution to molar specific heat C_v.

Returning to (5.14-7), the vibrational contributions to the other thermodynamic functions can be evaluated:

$$A_{\text{vib}} = G_{\text{vib}} = NkT \ln(1 - e^{-\Theta_{\text{vib}}/T}) \tag{5.14-13}$$

$$\mu_{\text{vib}} = \mu^{\circ}_{\text{vib}} = RT \ln(1 - e^{-\Theta_{\text{vib}}/T}) \qquad \text{(per mole)} \tag{5.14-14}$$

and

$$S_{\text{vib}} = \frac{Nk\Theta_{\text{vib}}/T}{e^{\Theta_{\text{vib}}/T} - 1} - Nk \ln(1 - e^{-\Theta_{\text{vib}}/T}) \tag{5.14-15}$$

Vibrational contributions in the equipartition (high-temperature) limit can be obtained by taking limits of the above expressions as $T \to \infty$. The results are

$$E_{\text{vib}} = NkT, \qquad C_V^{\text{vib}} = Nk \tag{5.14-16}$$

$$A_{\text{vib}} = NkT \ln \frac{\Theta_{\text{vib}}}{T} \tag{5.14-17}$$

$$S_{\text{vib}} = -Nk \ln \frac{\Theta_{\text{vib}}}{eT} \tag{5.14-18}$$

The harmonic oscillator approximation can be extended to polyatomic molecules as well. If forces between atoms in a polyatomic molecule are assumed to be harmonic (proportional to interatomic displacements), then all vibrational motions of the molecule can be represented as a superposition of dynamically independent motions, known as the normal modes of vibration.

In general, an N atomic molecule has $3N - 6$ normal modes ($3N - 5$ for linear molecules). The concerted atomic motions associated with each normal mode take place with a characteristic frequency ν_i. The vibrational energy of a polyatomic molecule is the sum of vibrational energies associated with the $3N - 6$ (or $3N - 5$) normal modes of vibration. The vibrational quantum numbers vary independently of one another. When the energy is the sum of independent energies, the partition function is the product of partition functions, one for each kind of energy:

$$Z_{\text{vib}} = \prod_{s=1}^{3N-6} Z_{\text{vib},s} = \prod_{s=1}^{3N-6} \frac{1}{1 - e^{-h\nu_s/kT}} \tag{5.14-19}$$

where ν_s is the frequency of the sth normal mode and (5.14-5) is used for the partition function of a single vibrational mode. (If the molecule is linear, $3N - 6$ is changed to $3N - 5$.)

■ EXAMPLE 5.14-1

What are the vibrational contributions to $C^{\circ}_{p,v}$, S°_{v}, H°_{v}, and G°_{v} for $O_2(g)$ at 298.15 K? The spectroscopic parameter or vibrational frequency is $\tilde{\nu} = 1580.246\ \text{cm}^{-1}$.

Solution. The characteristic vibrational temperature in terms of the vibrational frequency $\tilde{\nu}$ (cm^{-1}) is

$$\Theta_v = \frac{hc}{k}\tilde{\nu}$$

where $h = 6.26176 \times 10^{-24}$ J s, $c = 2.997925 \times 10^8$ m s^{-1}, $\tilde{\nu} = 1.580246 \times 10^{-5}$ m^{-1}, and $k = 1.380662 \times 10^{-23}$ J K^{-1}. Thus,

$$\Theta_v = \frac{(6.26176 \times 10^{-24})(2.997925 \times 10^8)(1.580246 \times 10^5)}{1.380662 \times 10^{-23}}$$

$$= 2273.64 \text{ K}$$

With $Z_v = z_v^N$, the vibrational contributions to the various thermodynamic quantities for diatomic molecules are

$$C^\circ_{p,v} = R\left(\frac{\Theta_v}{T}\right)^2 \frac{e^{\Theta_v/T}}{(e^{\Theta_v/T}-1)^2} = (8.31441 \text{ J K}^{-1}\text{ mol}^{-1})$$

$$\cdot\left(\frac{2273.64}{298.15}\right)\frac{e^{2273.64/298.15}}{(e^{2273.64/298.15}-1)^2}$$

$$= 0.236 \text{ J K}^{-1}\text{ mol}^{-1}$$

$$S^\circ_v = R\left[\frac{\Theta_v/T}{e^{\Theta_v/T}-1} - \ln(1-e^{-\Theta_v/T})\right]$$

$$= (8.31441)\left[\frac{(2273.64)/(298.15)}{e^{2273.64/298.15}-1} - \ln(1-e^{-2273.64/298.15})\right]$$

$$= 0.035 \text{ J K}^{-1}\text{ mol}^{-1}$$

$$H^\circ_v = RT\frac{\Theta_v/T}{e^{\Theta_v/T}-1} = (8.31441)(298.15)\left[\frac{(2273.64)/(298.15)}{e^{2273.64/298.15}-1}\right]$$

$$= 0.009 \text{ kJ mol}^{-1}$$

$$G^\circ_v = RT\ln(1-e^{-\Theta_v/T}) = (8.31441)(298.15)\ln(1-e^{-2273.64/298.15})$$

$$= -0.001 \text{ kJ mol}^{-1}$$

■

5.15 ELECTRONIC AND NUCLEAR PARTITION FUNCTIONS

Let the electronic energy levels be numbered $0, 1, 2, \ldots$, in order of increasing energy, and let their degeneracies be $g_{\text{elec},0}, g_{\text{elec},1}, g_{\text{elec},2}, \ldots$. Take the zero level of energy at the lowest available level, the energy of the ground electronic level is zero: $\epsilon_{\text{elec},0} = 0$. The energy of each excited electronic level is then measured relative to the ground electronic level. With $\epsilon_{\text{elec},0} = 0$, this gives

$$Z_{\text{elec}} = g_{\text{elec},0} + g_{\text{elec},1}\, e^{-\beta\epsilon_{\text{elec},1}} + g_{\text{elec},2}\, e^{-\beta\epsilon_{\text{elec},2}} + \cdots$$

$$= \sum_{\text{levels}} g_{\text{elec}}\, e^{-\beta\epsilon_{\text{elec}}} \qquad (5.15\text{-}1)$$

Since there is no general formula for the ϵ_{elec}'s, the series is added term by term using spectroscopically observed electronic energies.

For most diatomic molecules, the characteristic temperature Θ_{elec} for electronic excitation is very much higher than room temperature, so that the higher-energy electronic states are not excited to any significant extent at room temperatures, and all the $\exp(-\beta\epsilon_e)$ terms in (5.15-1) are negligible compared with $\exp(-\beta\epsilon_0)$. Hence, at room temperatures when $T \ll \Theta_e$,

$$Z_{\text{elec}} = g_0 \exp(-\beta\epsilon_0) \qquad (5.15\text{-}2)$$

where g_0 is the degeneracy (statistical weight) of the lowest energy level (ground state). With ϵ_0, the energy of the lowest level, chosen as zero, Eq. (5.15-2) further reduces to

$$Z_{\text{elec}} = g_0 \qquad (5.15\text{-}3)$$

Since the higher electronic states are not excited significantly at room temperatures, they do not contribute significantly to the heat capacity of a diatomic molecule at room temperatures.

As for the nuclear partition function, the characteristic temperature Θ_{nuc} for nuclear excitation is very much higher than the characteristic temperature Θ_{elec} for electronic excitation. The higher-energy nuclear states are therefore not excited significantly at room temperatures, and the atomic nuclei do not contribute to any significant extent to the heat capacity of a diatomic molecule.

Summarizing, the partition function of a diatomic molecule is given by

$$z = z_t z_r z_v z_e z_n \qquad (5.15\text{-}4)$$

where the subscripts t, r, v, e, and n refer to translation, rotational, vibrational, electronic, and nuclear, respectively. It is convenient to use the symbol z_{int}, which is defined as

$$z_{\text{int}} = z_r z_v z_e z_n \qquad (5.15\text{-}5)$$

to represent the contributions of all the internal degrees of freedom to the partition function. Since z_r, z_v, z_e, and z_n are all independent of the volume, z_{int} is a function of temperature only. Using β for $1/kT$ to simplify the writing and the fact that $z_t = (m/2\beta\pi\hbar^2)^{3/2}\, V$, Eq. (5.15-4) can be rewritten in the form

$$z_1 = \left(\frac{m}{2\beta\pi\hbar^2}\right)^{3/2} V z_{\text{int}} \qquad (5.15\text{-}6)$$

where the subscript 1 (really superfluous) refers to a single molecule. For a system of N indistinguishable diatomic molecules in a box, the partition function is (if there are no interaction between the molecules)

$$z = \frac{z_1^N}{N!} = \frac{1}{N!}\left(\frac{m}{2\beta\pi\hbar^2}\right)^{3N/2} V^N z_{\text{int}}^N \qquad (5.15\text{-}7)$$

Taking the logarithm of both sides and applying Stirling's approximation gives

$$\ln z(\beta, V, N) = N\left[\ln V - \frac{3}{2}\ln\beta + \frac{3}{2}\ln\left(\frac{m}{2\pi\hbar^2}\right) + \ln z_{\text{int}} - \ln(N+1)\right] \tag{5.15-8}$$

here z_{int} is a function of β only. (For an ideal monatomic gas, z_{int} reduces to $g_0 g_n$, the total degeneracy associated with the internal degrees of freedom.) Applying the relation for the pressure in terms of the partition function yields

$$p = \frac{1}{\beta}\left(\frac{\partial \ln z}{\partial v}\right)_{\beta, N} = \frac{N}{\beta v} = \frac{NkT}{v} \tag{5.15-9}$$

This is the equation of state of an ideal diatomic gas.

For the specific heat of a diatomic gas, consider a mole of gas, consisting of N_A diatomic molecules. At temperatures $T \ll \Theta_r$, only the higher-energy translational states are excited to any significant extent. In this limit, the mean translational energy of each molecule is approximately $3kT/2$, the total energy of the N_A diatomic molecules is approximately $3N_A kT/2$, and the molar heat capacity at constant volume is

$$C_{V,M} = \left(\frac{\partial U}{\partial T}\right)_V \simeq \frac{3}{2} N_A k = \frac{3}{2} R \qquad (T \ll \Theta_r) \tag{5.15-10}$$

Figure 5.15-1 shows a sketch of the variation of the molar heat capacity of an ideal diatomic gas with temperature. The temperature scale is not linear. At room temperature, the molar heat capacity is, generally, equal to $\frac{5}{2}R$.

When the temperature is of the order of Θ_r, which for H_2 is 85.5 K, the higher-energy rotational states begin to be excited significantly. When $T \gg \Theta_r$, according to (5.13-26), the mean rotational energy per molecule is kT, and the contribution to the total energy U due to the rotation of the molecules is $N_A kT$. In this temperature range, provided $T \ll \Theta_v$, the molar heat capacity is

$$C_{V,M} \simeq N_A\left(\frac{\partial}{\partial T}\right)\left(\frac{3}{2}kT + kT\right) = \frac{5}{2} N_A k = \frac{5}{2} R \tag{5.15-11}$$

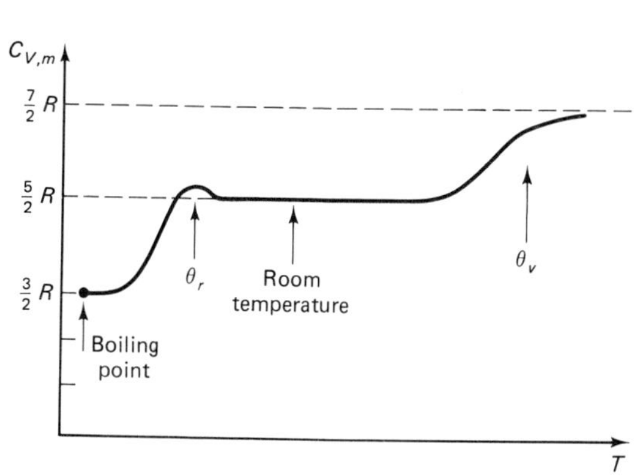

Figure 5.15-1 Molar constant volume specific heat of ideal diatomic gas.

In other words, in the vicinity of $T = \Theta_r$, $C_{V,M}$ increases from $3R/2$ to $5R/2$ as shown in Fig. 5.15-1. The molar constant volume, specific heat of a diatomic gas at room temperature (generally between Θ_r and Θ_v) is $5R/2$. At higher temperatures ($T \simeq \Theta_v$), higher-energy vibrational states are excited when $T \simeq \Theta_v$. For hydrogen, $\Theta_v = 6140$ K. When $T >> \Theta_v$, the mean vibrational energy per molecule is

$$\langle \epsilon_v \rangle = \frac{1}{2} k\Theta_v + \frac{k\Theta_v}{e^{\Theta_v/T} - 1} = kT \qquad \text{for} \quad T >> \Theta_v \tag{5.15-12}$$

Thus, when $T >> \Theta_v$, $C_{V,M}$ is given to a good approximation by

$$C_{V,M} = N_A\left(\frac{\partial}{\partial T}\right)\left(\frac{3}{2} kT + kT + kT\right) = \frac{7}{2} N_A k = \frac{7}{2} R \tag{5.15-13}$$

so that in the vicinity of $T = \Theta_v$, $C_{V,M}$ increases from $5R/2$ to $7R/2$, as shown in Fig. 5.15-1.

As for the chemical potential,

$$\mu = -kT\left(\frac{\partial \ln z}{\partial N}\right)_{V,T} \tag{5.15-14}$$

it becomes, using (5.15-8),

$$\mu = -kT \ln\left[\frac{V}{N}\left(\frac{mkT}{2\pi\hbar^2}\right)^{3/2} Z_{\text{int}}\right] \tag{5.15-15}$$

With Eq. (5.15-6), this can be rewritten as

$$\mu = -kT \ln\left(\frac{z_1}{N}\right) \tag{5.15-16}$$

where z_1 is the single-particle partition function. Just as the absolute temperature is an indicator that shows the direction of heat flow during approach to thermodynamic equilibrium, so the chemical potential is an indicator of the direction of particle flow during approach to thermodynamic equilibrium. Particles tend to go from regions of high chemical potential to regions of lower chemical potential, which, for an ideal diatomic gas, is from regions of high concentration to regions of lower concentration.

■ EXAMPLE 5.15-1

Determine the electronic contributions to $S°$ and $G°$ for 0(g) at 298.15 K. The ground state (zero energy) has a degeneracy of 5, the first excited state has an energy of 158.2 cm^{-1} and a degeneracy of 3, and the second excited state has an energy of 226.5 cm^{-1} with no degeneracy.

Solution. The characteristic electronic temperatures are obtained using

$$\Theta_{\text{elec}} = \frac{hc}{k}\tilde{\nu}$$

where $h = 6.626176 \times 10^{-34}$ J s, $c = 2.997925 \times 10^{8}$ m s^{-1}, $\tilde{\nu}_1 = 1.582 \times 10^4$ m^{-1}, $\tilde{\nu}_2 = 2.265 \times 10^4$ m^{-1}, and $k = 1.380662 \times 10^{-23}$ J K^{-1}. Thus,

$$\Theta_{\text{elec},1} = \frac{(6.626176 \times 10^{-34})(2.997925 \times 10^{8})(1.582 \times 10^{4})}{1.380662 \times 10^{-23}} = 228.1 \text{ K}$$

$$\Theta_{\text{elec},2} = (228.1)\left(\frac{226.5}{158.2}\right) = 325.9 \text{ K}$$

$$z_{\text{elec}} = g_0 + g_1 e^{-\epsilon_1/kT} + g_2 e^{-\epsilon_2/kT} + \dots$$
$$= 5 + 3e^{-228.1/298.15} + e^{-325.9/298.15} = 6.7311$$

$$S^\circ_{\text{elec}} = R \ln z_e = (8.31441 \text{ J K}^{-1}\text{ mol}^{-1}) \ln 6.7311 = 15.853 \text{ J K}^{-1}\text{ mol}^{-1}$$

$$G^\circ_{\text{elec}} = -TS^\circ_{\text{elec}} = -(298.15 \text{ K})(15.853 \text{ J K}^{-1}\text{ mol}^{-1}) = -4.726 \text{ kJ mol}^{-1}$$

5.16 IDEAL GAS PROPERTIES; REACTION CONSTANTS

Having found the molecular partition function $z = z_{\text{trans}}\, z_{\text{rot}}\, z_{\text{vib}}\, z_{\text{elec}}$, it is now possible to calculate the thermodynamic properties of a diatomic or monatomic ideal gas in terms of molecular properties. Starting with the equation for pressure,

$$p = kT\left(\frac{\partial \ln Z}{\partial V}\right)_{T,N} \tag{5.16-1}$$

and the logarithm of Z being given by

$$\ln Z = N \ln z_{\text{trans}} + N \ln z_{\text{rot}} + N \ln z_{\text{vib}} + N \ln z_{\text{elec}} - N(\ln N - 1) \tag{5.16-2}$$

use is made of the fact that the molecular vibrational, rotational, and electronic energies all depend on properties of the gas molecules but are independent of V. (In contrast, ϵ_{trans} does depend on V.) Therefore only z_{trans} is a function of V, and (5.16-2) becomes

$$p = kT\left(\frac{\partial \ln Z}{\partial V}\right)_{T,N} = kT\left[\frac{\partial(N \ln z_{\text{trans}})}{\partial V}\right]_{T,N} = NkT\left(\frac{\partial \ln z_{\text{trans}}}{\partial V}\right)_T \tag{5.16-3}$$

Now,

$$z_{\text{trans}} = \left(\frac{2\pi mkT}{h^2}\right)^{3/2} V \tag{5.16-4}$$

or

$$\ln z_{\text{trans}} = \frac{3}{2}\ln\left(\frac{2\pi mk}{h^2}\right) + \frac{3}{2}\ln T + \ln V \tag{5.16-5}$$

Hence,

$$p = NkT\left(\frac{\partial \ln z_{\text{trans}}}{\partial V}\right)_T = \frac{NkT}{V} \tag{5.16-6a}$$

or

$$pV = NkT \tag{5.16-6b}$$

Equation (5.16-6) is the equation of state for an ideal monatomic, diatomic, or polyatomic gas.

As for the internal energy of an ideal diatomic gas, consider the zero of energy to be at the lowest available molecular energy level to find $U - U_0$, where U_0 is the thermodynamic internal energy of a hypothetical ideal gas in which every molecule is in the lowest translational, rotational, vibrational, and electronic state. This would be the internal energy at absolute zero if the gas did not condense and if the molecules were not fermions. In other words,

$$U - U_0 = U_{\text{trans}} + U_{\text{rot}} + U_{\text{vib}} + U_{\text{elec}} \tag{5.16-7}$$

where

$$U_{\text{trans}} = \frac{3}{2}nRT \tag{5.16-8}$$

$$U_{\text{rot}} = nRT \tag{5.16-9}$$

$$U_{\text{vib}} = nR\frac{h\nu}{k}\frac{1}{e^{h\nu/kT} - 1} = nR\Theta_{\text{vib}}\frac{1}{e^{\Theta_{\text{vib}}/T} - 1} \tag{5.16-10}$$

$$U_{\text{elec}} = 0 \tag{5.16-11}$$

For a monatomic gas there are no rotational or vibrational degrees of freedom and $\overline{U} - \overline{U}_0 = \overline{U}_{\text{trans}} = 3RT/2$, in agreement with classical kinetic theory. For a polyatomic gas,

$$U_{\text{trans}} = \tfrac{3}{2}nRT; \qquad U_{\text{elec}} = 0 \tag{5.16-12}$$

$$U_{\text{rot}} = \begin{cases} \tfrac{3}{2}nRT & \text{(nonlinear)} \\ nRT & \text{(linear)} \end{cases} \tag{5.16-13}$$

$$U_{\text{vib}} = nR\sum_{s=1}^{3N-6}\frac{\Theta_{\text{vib},s}}{e^{\Theta_{\text{vib},s}/T} - 1} \qquad \Theta_{\text{vib},s} = \frac{h\nu_s}{k} \tag{5.16-14}$$

where ν_s is the frequency of the sth normal mode. The sum of the four energies is $U - U_0$.

The specific heat at constant volume is

$$C_V = C_{V,\text{trans}} + C_{V,\text{rot}} + C_{V,\text{vib}} + C_{V,\text{elec}} \tag{5.16-15}$$

where $C_{V,\text{trans}} \equiv (\partial U_{\text{trans}}/\partial T)_{V,n}$, and so on. Differentiation of (5.16-8) to (5.16-11) gives for an ideal gas of diatomic molecules at moderate temperatures

$$C_{V,\text{trans}} = \frac{3}{2}nR \tag{5.16-16}$$

$$C_{V,\text{rot}} = nR \tag{5.16-17}$$

$$C_{V,\text{vib}} = nR\left(\frac{\Theta_{\text{vib}}}{T}\right)^2\frac{e^{\Theta_{\text{vib}}/T}}{(e^{\Theta_{\text{vib}}/T} - 1)^2} \tag{5.16-18}$$

$$C_{V,\text{elec}} = 0 \tag{5.16-19}$$

Note that U and C_V of the ideal gas are functions of T only, and that Eqs. (5.16-16) and (5.16-17) for $\overline{C}_{V,\text{trans}}$ and $\overline{C}_{V,\text{rot}}$ agree with the classical–statistical–mechanical equipartition theorem of $\frac{1}{2}R$ for each quadratic term in the energy, but $\overline{C}_{V,\text{vib}}$ does not agree with the equipartition theorem. For a polyatomic gas, differentiation of the U's in (5.16-12), (5.16-13), and (5.16-14) gives

$$C_{V,\text{trans}} = \frac{3}{2}nR; \qquad U_{\text{el}} = 0 \tag{5.16-20}$$

$$C_{V,\text{rot}} = \begin{cases} \frac{3}{2}nR & \text{(nonlinear)} \\ nR & \text{(linear)} \end{cases} \tag{5.16-21}$$

$$C_{V,\text{vib}} = \sum_{s=1}^{3N-6} C_{V,\text{vib},s} \tag{5.16-22}$$

Large polyatomic molecules often have low-frequency vibrations that contribute to C_V at room temperature. The high-temperature (classical) limit of $C_{V,\text{vib},s}$ is R as seen in Fig. 5.14-1.

The entropy, $S = U/T + k \ln Z$, of an ideal gas is obtained from the proper substitution of the U's and Z's. The result, for a diatomic gas, is

$$S_{\text{trans}} = \frac{5}{2}nR + nR \ln\left[\frac{(2\pi m)^{3/2}}{h^3}\frac{(kT)^{5/2}}{p}\right] \tag{5.16-23}$$

$$S_{\text{rot}} = nR + nR \ln \frac{T}{\sigma\Theta_{\text{rot}}} \tag{5.16-24}$$

$$S_{\text{vib}} = nR\frac{\Theta_{\text{vib}}}{T}\frac{1}{e^{\Theta_{\text{vib}}/T} - 1} - nR \ln(1 - e^{-\Theta_{\text{vib}}/T}) \tag{5.16-25}$$

$$S_{\text{elec}} = nR \ln g_{\text{elec},0} \tag{5.16-26}$$

where Θ_{rot} and Θ_{vib} are given by (5.13-11) and (5.14-3). S is the sum of (5.16-23) to (5.16-26) (provided T is not extremely high or low). For a monatomic gas with a nondegenerate ground electronic state, $S_{\text{rot}} = S_{\text{vib}} = S_{\text{elec}} = 0$, and $S = S_{\text{trans}}$. For a polyatomic gas, S_{trans} is given by (5.16-23), and S_{rot} and S_{vib} by

$$S_{\text{rot}} = \frac{3}{2}nR + nR \ln z_{\text{rot}} \qquad \text{(nonlinear)} \tag{5.16-27}$$

$$S_{\text{vib}} = \sum_{s=1}^{3N-6} S_{\text{vib},0}, \qquad S_{\text{elec}} = nR \ln g_{\text{elec},0} \tag{5.16-28}$$

where

$$z_{\text{rot}} = \frac{\pi^{1/2}}{\sigma}\left(\frac{2kT}{\hbar^2}\right)^{3/2}(I_a I_b I_c)^{1/2} \tag{5.16-29}$$

as mentioned in Section 5.13.

To discuss heats of reactions, enthalpies of formation, and equilibrium constants, it is first noted that $U - U_0$, S, and C_v of an ideal gas are readily calculated using statistical mechanics. The molecular properties needed for these calculations are: (a) the molecular weight (which occurs in z_{trans}), (b) the molecular geometry (which is needed to calculate the moments of inertia in z_{rot}), (c) the molecular vibration frequencies (which occur in z_{vib}), and (d) the degeneracy of the ground electronic level and the energies and degeneracies of any low-lying excited electronic levels (which occur in z_{elec}). This information is obtained spectroscopically. From observations on a molecule's spectrum, its thermodynamic properties can be calculated. For molecules of simple structure, the gas phase thermodynamic properties calculated by statistical mechanics are usually more accurate than the values determined from calorimetric measurements. For liquids and solids, the intermolecular forces make the statistical mechanical values less accurate than the calorimetric values.

To examine the relation between the tabulated gas phase thermodynamic quantities ΔH_f°, S°, and ΔG_f° and the calculated statistical mechanical quantities $U - U_0$ and S°, consider the gas phase reaction $\Sigma_B \nu_B B = 0$, where ν_B is the stoichiometric coefficient for species B. The reaction's standard change in internal energy at temperature T is $\Delta U_T^\circ \equiv \Sigma_B \nu_B \overline{U}_{B,T}^\circ$, where $\overline{U}_{B,T}^\circ$ is the standard molar internal energy of B at T. To relate ΔU_T° to the statistical mechanical quantities $\overline{U}_{B,T} - \overline{U}_{B,0}$, consider the following two paths for transforming standard-state stoichiometric numbers of moles of reactants to products at T (see Fig. 5.16-1).

The ΔU for step (a) is $\Delta U_a = \Delta U_T^\circ$, the reaction's standard change in internal energy at temperature T. Steps (c) and (d) convert reactants to products at 0 K, so $\Delta U_c + \Delta U_d = \Delta U_0^\circ$, where ΔU_0° is the reaction's standard change in U at 0 K. Steps (b) and (e) involve only the internal energy differences between temperatures T and 0 for each species, so $\Delta U_b + \Delta U_e = \Sigma_B \nu_B(\overline{U}_{B,T}^\circ - \overline{U}_{B,0}^\circ)$; note that ν_b is negative for reactants.

Since path (a) and path (b) + (c) + (d) + (e) start and end at the same states, $\Delta U_a = \Delta U_b + \Delta U_c + \Delta U_d + \Delta U_e$. Substitution of the expressions for these ΔU's gives

$$\Delta U_T^\circ = \sum_B \nu B(\overline{U}_{B,T}^\circ - \overline{U}_{B,0}^\circ) + \Delta U_0^\circ \qquad (5.16\text{-}30)$$

Each $\overline{U}_{B,T}^\circ - \overline{U}_{B,0}^\circ$ is calculated by statistical mechanics. For ΔU_0° at $T = 0$, all the molecules of each species are in their ground electronic, vibrational, and rotational states. Step (c) involves dissociation of each reactant in its ground state into atoms. The energy required to dissociate one molecule of ground state B into atoms is the spectroscopic dissociation energy $D_{0,B}$.

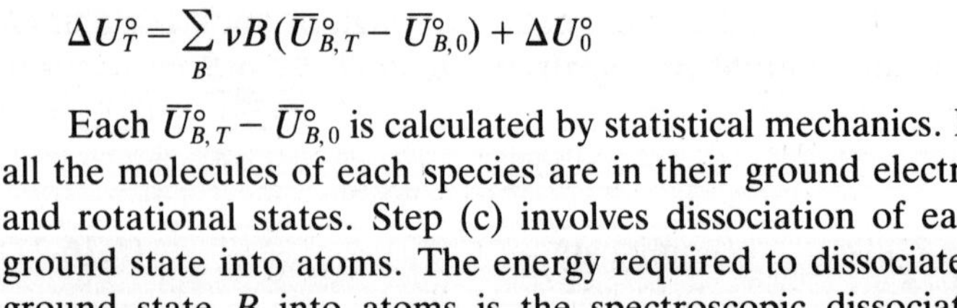

Figure 5.16-1 Alternate paths for transforming reactants to products.

The energy neeced to dissociate $|\nu_B|$ moles of B is $-\nu_B N_0 D_{0,B}$, where N_0 is the Avogadro constant and ν_B is negative for a reactant. For the H_2 ground electronic state, D_0 is 4.48 eV. Step (d) is the reverse of dissociation of the products, and the energy change for each product species is $-\nu_B N_0 D_{0,B}$. Therefore, $\Delta U_0^\circ = \Delta U_c + \Delta U_d = -N_0 \Sigma_B \nu_B D_{0,B}$ and

$$\Delta U_0^\circ = -N_0 \Delta D_0 \tag{5.16-31}$$

where

$$\Delta D_0 \equiv \Sigma_B \nu_B D_{0,B} \tag{5.16-32}$$

For (5.16-31) to be used to calculate ΔU_T° for a gas phase reaction from statistical mechanics, the dissociation energies D_0 of the species are needed.

The standard enthalpy of formation, $\Delta H_{f,T}^\circ$, of a gaseous compound is readily calculated from its definition (formed from the corresponding separated elements, each element being in its standard state at T), using

$$\Delta H_{f,T}^\circ = \Delta U_{f,T}^\circ + \Delta|\nu_g|RT \tag{5.16-33}$$

where $\Delta|\nu_g|$ is the absolute value of the change in number of moles of gases in the reaction. If the elements forming the compound are all gaseous at T and 1 atm, then $\overline{U}_T^\circ - U_0^\circ$ for each element and for the compound can be calculated from statistical mechanics; $\Delta U_{f,0}^\circ$ is calculated from (5.16-32) and then $\Delta U_{f,T}^\circ$ is found from (5.16-31). If one or more of the elements are not gaseous, additional data (calorimetric) will be needed to find $\Delta U_{f,T}^\circ$.

The standard Gibbs energy of formation is given by

$$\Delta G_{f,T}^\circ = \Delta H_{f,T}^\circ - T\,\Delta S_{f,T}^\circ \tag{5.16-34}$$

Calculation of $\Delta H_{f,T}^\circ$ has been discussed above. $\Delta S_{f,T}^\circ$ is found from the entropies of the compound and of its elements. If these are all gases, their entropies can be calculated by statistical mechanics. For solid and liquid elements, the experimental S must be used.

Equilibrium constants were discussed in Section 5.8. For the ideal gas reaction $\Sigma_B \nu_B B = 0$, the equilibrium constant is calculated from $\Delta G^\circ = -RT \ln K_p^\circ$. To calculate K_p° using statistical mechanics, the $\Delta G_{f,T}^\circ$ of each gas is calculated as above, ΔG_T° is calculated as $\Sigma_B \nu_B \Delta G_{f,B,T}^\circ$, and then K_p° is calculated from ΔG_T°. For theoretical reasons, it is also useful to express the equilibrium constant in terms of the partition functions of the reactants and products (Section 5.8).

■ EXAMPLE 5.16-1

(a) What are the values of C_p°, S°, H°, and G° for $O_2(g)$ at 298.15 K?
(b) What are the values of C_p°, S°, H°, and G° for $0(g)$ at 298.15 K?

Solution. (a) Adding the translational, rotational, vibrational, and electronic contributions obtained in Examples 5.10-3, 5.13-4, 5.14-1, and 5.15-1 gives

$$C_p^\circ = (20.786 + 8.314 + 0.236 + 0)\ \mathrm{J\ K^{-1}\ mol^{-1}} = 29.336\ \mathrm{J\ K^{-1}\ mol^{-1}}$$

$$S^\circ = (152.077 + 43.838 + 0.035 + 9.134)\ \mathrm{J\ K^{-1}\ mol^{-1}}$$

$$= 205.085\ \mathrm{J\ K^{-1}\ mol^{-1}}$$

$$H^\circ = (6.197 + 2.479 + 0.009 - 491.888)\ \mathrm{kJ\ mol^{-1}} = -483.203\ \mathrm{kJ\ mol^{-1}}$$

$$G^\circ = (-39.145 - 10.591 - 0.001 - 494.611)\ \mathrm{kJ\ mol^{-1}}$$

$$= -544.349\ \mathrm{kJ\ mol^{-1}}$$

(b) Atoms have only translational and electronic contributions, and, therefore, from Examples 5.10-3 and 5.15-1,

$$C_p^\circ = 20.786\ \mathrm{J\ K^{-1}\ mol^{-1}} \qquad C_{p,\mathrm{elec}}^\circ \ \text{taken as zero}$$

$$S^\circ = (143.432 + 15.853)\ \mathrm{J\ K^{-1}\ mol^{-1}} = 159.286\ \mathrm{J\ K^{-1}\ mol^{-1}}$$

$$H^\circ = 6.197\ \mathrm{kJ\ mol^{-1}} \qquad H_{\mathrm{elec}}^\circ \ \text{taken as zero}$$

$$G^\circ = (-36.568 - 4.727)\ \mathrm{kJ\ mol^{-1}} = -41.294\ \mathrm{kJ\ mol^{-1}}$$ ■

5.17 STATISTICAL THERMODYNAMICS OF IMPERFECT GASES

As the density of a gas of N molecules at temperature T in a volume V is increased, the forces between the molecules become more important. Because of these intermolecular forces, the perfect gas equation of state

$$pV = NkT \tag{5.17-1}$$

no longer accurately describes the state of the gas. In this section, the methods of statistical mechanics will be applied to obtain a first correction to Eq. (5.17-1) which is produced by intermolecular forces. An approximate evaluation of the partition function will be obtained for the case when the interaction energy is not zero.

For an imperfect gas, the energy of a system of N molecules is written in the form

$$E = \sum_{i=1}^{N} \epsilon_i + \sum_{i>j}^{N} \Theta(r_{ij}) \tag{5.17-2}$$

The first summation in (5.17-2) represents the energy of the molecules when there are no forces between the molecules. Only this term is present in the case of a perfect gas. The second summation in (5.17-2) represents the interaction potential energy *between the molecules (not between the atoms in one molecule,* which is contained in the ϵ's). Thus, $\Theta(r_{ij})$ is the potential energy between the molecules i and j, which are separated by a distance $r_{ij} = |r_i - r_j|$. The summation is for $i > j$ so that the interaction energy is not counted twice, otherwise it would be written $\frac{1}{2}\sum_{i \neq j}^{N} \Theta(r_{ij})$

Now, the partition function can be written as

$$Z = \int_V d^3 r_1 \cdots d^3 r_N \int e^{-\beta E} \text{ (differential of remaining variables)} \tag{5.17-3}$$

where the first group of integrals are over the volume of the system, for the position of the center of mass r_i of each molecule, while the second group of integrals contains the translational and "internal" variables (describing the rotational and vibrational motion). When (5.17-2) is substituted into (5.17-3),

$$Z = \int_V d^3 r_1 \cdots d^3 r_N e^{-\beta \Sigma \epsilon_i} \text{ (differential of remaining variables)} \tag{5.17-4}$$

and since only the first group of integrals in (5.17-4) depend on V,

$$\ln Z = \ln\left(\int_V d^3 r_1 \cdots d^3 r_N e^{-\beta \Sigma \Theta(r_{ij})}\right) + \text{(terms independent of } V) \tag{5.17-5}$$

Now, the pressure is given by

$$p = kT\left(\frac{\partial \ln Z}{\partial V}\right) \tag{5.17-6}$$

so that substitution of (5.71-5) into (5.17-6) gives

$$p = kT\frac{\partial(\ln Z_{\text{conf}})}{\partial V} \tag{5.17-7}$$

where

$$Z_{\text{conf}} = \int_V d^3 r_1 \cdots d^3 r_N e^{-\beta \Sigma \Theta(r_{ij})} \tag{5.17-8}$$

is known as the *configuration integral*. If $\Theta(r_{ij}) = 0$ (no intermolecular forces), then

$$Z_{\text{conf}} = V^N \tag{5.17-9}$$

and (5.17-7) immediately yields the perfect gas equation of state (5.17-1). The molar partition function for an imperfect gas (5.17-4) can thus be written, with the aid of (5.10-8) for a perfect gas, as

$$Z(V, N, T) = \frac{(2\pi mkT)^{3N/2}}{h^{3N} N!} Z_{\text{conf}} \tag{5.17-10}$$

Gas imperfections result in a correction term that is to be added to the translational, rotational, vibrational, and electronic contributions.

The whole problem is how to evaluate (5.17-8) when $\Theta(r)$ is not zero. This is a complicated integral, because it involves many variables and the function $\Theta(r)$ is not simple. To keep the discussion from being too laborious, it is assumed that when a gas is not too dense, most of the interactions between the molecules are due to interaction between *pairs of molecules* (and not among three or four molecules at a time). The integrand (5.17-8) of Z_{conf} is written in the following form:

$$\exp\left[-\beta\sum_{i>j}\Theta(r_{ij})\right]=\prod_{i>j}e^{-\beta\Theta(r_{ij})} \tag{5.17-11}$$

$$=\prod_{i>j}(1+u_{ij})=(1+u_{21})(1+u_{31})(1+u_{32})\cdots \tag{5.17-12}$$

$$=1+(u_{21}+u_{31}+u_{32}+\cdots) \tag{5.17-13}$$

$$+\text{ terms containing higher powers of } u \tag{5.17-14}$$

where the factors

$$1+u_{ij}\equiv e^{-\beta\Theta(r_{ij})} \tag{5.17-15}$$

represent the deviation from a perfect gas due to the interaction between pairs of molecules. Typical forms of the functions $u(r_{ij})$ and $f(r_{ij})$ are sketched in Fig. 5.17-1. The terms $u_{21}+u_{31}+\cdots=\sum_{i>j}u_{ij}$ in (5.17-14) represent the interactions between pairs of molecules, whereas the higher powers of u (such as $u_{21}u_{31}$) vanish unless three or more molecules are in interaction

If just the terms that represent the interaction between one pair of molecules at a time are kept, then (5.17-8) becomes

$$Z_{\text{conf}}=\int_V d^3r_1\cdots d^3r_N\left(\sum_{i>j}^{N}u_{ij}\right)=V^N+\int_V d^3r_1\cdots d^3r_N\sum_{i>j}^{N}u_{ij} \tag{5.17-16}$$

The last integral can also be written as

$$\sum_{i>j}^{N}\int d^3r_1\cdots d^3r_N u_{ij}=V^{N-2}\sum_{i>j}^{N}\int_V d^3r_i d^3r_j u_{ij} \tag{5.17-17}$$

retaining the two variables r_i and r_j while integrating out the rest. Each integral in the summation has the same value [they differ only by the labels (i,j) on the integration variables]. Moreover, there are $\frac{1}{2}N(N-1)$ terms in this sum [i can be any of N numbers, and j any of the $(N-1)$ remaining numbers; also, division by 2 because $i>j$]. Therefore,

$$Z_{\text{conf}}\simeq V^N+V^{N-2}\frac{1}{2}N(N-1)\int_V d^3r_1 d^3r_2 u_{12} \tag{5.17-18}$$

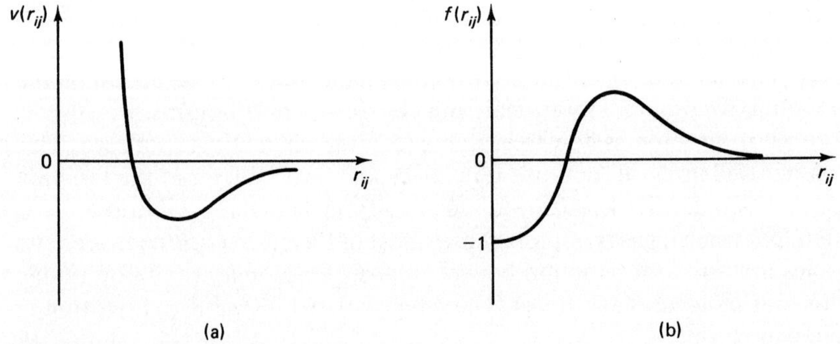

Figure 5.17-1 Typical $v(r_{ij})$ and $f(r_{ij})$.

The notation can be further simplified by introducing the variable $r = r_1 - r_2$, and since $d^3 r_1\, d^3 r_2 = d^3 r_1\, d^3 r$,

$$Z_{\text{conf}} = V^N + V^{N-2}\frac{1}{2}N(N-1)\int_V d^3 r_1 \int_V d^3 r u(r) \tag{5.17-19}$$

The first integral gives another factor of V, and for large N, there follows $N(N-1) \simeq N^2$, so (5.17-19) gives

$$Z_{\text{conf}} \simeq V^N - V^{N-1} N^2 B(T) \tag{5.17-20}$$

where

$$B(T) = \frac{1}{2}\int d^3 r (1 - e^{-\beta\Theta(r)}) \tag{5.17-21}$$

is known as the *second virial coefficient.* Since $\Theta(r)$ goes to zero if r is larger than a few angstroms (say 5×10^{-8} cm), the integrand in (5.17-21) goes to zero unless r is very small. Because of this, $B(T)$ is independent of the volume of the system.

The expression (5.17-20) for Z_{conf} contains the first correction due to the intermolecular forces. Taking the logarithm yields

$$\ln Z_{\text{conf}} \simeq \ln V^N[1 - V^{-1}N^2 B(T)] = \ln V^N + \ln[1 - V^{-1}N^2 B(T)]$$

or, using $\ln(1-x) \simeq -x$ for small x, this reduces to

$$\ln Z_{\text{conf}} \simeq \ln V^N - V^{-1}N^2 B(T) \tag{5.17-22}$$

The equation of state can now be obtained by substituting (5.17-22) into (5.17-7):

$$p = kT\frac{\partial}{\partial V}[N \ln V - V^{-1}N^2 B(T)] = kT\left[\frac{N}{V} + \frac{N^2}{V^2}B(T)\right]$$

or

$$pV = NkT\left[1 + \frac{NB(T)}{V}\right] \tag{5.17-23}$$

This is the first correction in the so-called virial series

$$pV = NkT\left[1 + \frac{NB(T)}{V} + \frac{N^2 C(T)}{V^2} + \cdots\right] \tag{5.17-24}$$

where $B(T)$, $C(T)$, and so on are called the *second, third, . . . , virial coefficients.* Equation (5.17-24) is known as the *virial equation of state* (Kammerlingh–Onnes, 1901). It is a power series in (N/V), which is useful if (N/V) is not too large.

5.18 INTERMOLECULAR POTENTIALS

To estimate the second virial coefficient $B(T)$, an expression for the interaction potential $V(r)$ must be used. The nature of $V(r)$ is of the form shown in Fig. 5.18-1. For small values of r, $V(r)$ drops rapidly as r increases (corresponding to a strong repulsive force). For larger values of r, $V(r)$ slowly increases (corresponding to a weak attractive force).

Perhaps the most widely used intermolecular potential is the Lennard–Jones 6–12 potential (1924), shown in Fig. 5.18-1(a):

$$V(r) = 4\epsilon\left[\left(\frac{\sigma}{r}\right)^{12} - \left(\frac{\sigma}{r}\right)^{6}\right] \tag{5.18-1}$$

where ϵ is the depth of the minimum in the potential and σ is the intermolecular distance at which $V = 0$. For $r < \sigma$, the $1/r^{12}$ term dominates, and V decreases as r decreases. The parameter σ is an approximation to the sum of the average radii of the two molecules. Although it is possible to evaluate $B(T)$ using this relation, the analysis is cumbersome and involves numerical evaluations of an infinite series. To avoid such problems, consider instead, the potential shown in Fig. 5.18-1(c), described by

$$V(r) = \begin{cases} \infty, & r < \sigma \\ 0, & r > \sigma \end{cases} \tag{5.18-2}$$

This is the *hard sphere* potential. According to this model, molecules behave as impenetrable spheres of diameter σ which interact only upon direct collision. Substituting (5.18-2) into (5.17-21) and writing the latter in the form

$$B(T) = 2\pi N_0 \int_0^\infty (1 - e^{-V(r)/kT}) r^2\, dr \tag{5.18-3}$$

gives

$$\begin{aligned} B(T) &= 2\pi N_A \int_0^\sigma r^2\, dr + 2\pi N_A \int_0^\infty (1-1) r^2\, dr \\ &= 2\pi N_A \left(\frac{r^3}{3}\right)_0^\sigma = \frac{2\pi N_A \sigma^3}{3} = b \end{aligned} \tag{5.18-4}$$

where N_A is the Avogadro number. Since the radius of the hard sphere molecule is half the distance σ of closest approach, and the volume of a hard sphere molecule is $\frac{4}{3}\pi(\sigma/2)^3 = \frac{1}{6}\pi\sigma^3$, Eq. (5.18-4) says that $B(T)$ or the quantity b is equal to four times the molecular volume per mole.

Another model, which takes account, in a rudimentary fashion, of attractive intermolecular forces is the *square-well potential* [Fig. 5.18-1(b)]:

$$V(r) = \begin{cases} \infty & r < \sigma \\ -\varepsilon & \sigma < r < R\sigma \\ 0 & r > R\sigma \end{cases} \tag{5.18-5}$$

According to this model, a hard core of diameter σ is surrounded by an attractive shell of depth ϵ extending to a diameter $R\sigma$. Using (5.18-5) in (5.18-3) gives

$$B(T) = 2\pi N_A \left[\int_0^\sigma r^2\, dr + \int_\sigma^{R\sigma} r^2 (1 - e^{\epsilon/kT})\, dr\right] \tag{5.18-6}$$

Identifying the first integral with the hard sphere result (5.18-4), Eq. (5.18-6) yields

$$B(T) = b[1 - (R^3 - 1)(e^{\epsilon/kT} - 1)] \tag{5.18-7}$$

With $\epsilon/kT \ll 1$, and consequently, $(e^{\epsilon/kT} - 1) \approx \epsilon/kT$, this can be written as

$$B(T) = b - \frac{a}{RT} \tag{5.18-8}$$

where

$$a = N_0 b(R^3 - 1)\epsilon \tag{5.18-9}$$

(5.18-8) results in van der Waals equation in virial form.

As for the Lennard–Jones potential, quantum-mechanical calculations by London (1930) show that, for sufficiently large distance, the potential energy of interaction between two nonpolar molecules varies as the inverse sixth power of distance, that is,

$$V_{\text{attr}}(r) = -\text{const}\, r^{-6} \tag{5.18-10}$$

On the other hand, the short-range repulsive intermolecular interaction is approximated by the inverse power relation

$$V_{\text{rep}}(r) \approx \text{const}\, r^{-n} \qquad ((5.18\text{-}11)$$

where n must be greater than 6 for this term to dominate at small internuclear separations. For convenience, the choice $n = 12$ (twice 6) is most often made. The complete intermolecular potential is thus obtained by adding (5.18-11) (with $n = 12$) to give

$$V(r) = c_1 r^{-12} - c_2 r^{-6} \tag{5.18-12}$$

which is the well-known *Lennard–Jones 6–12 potential.* It is extensively employed in evaluations of the second and third virial coefficients and of transport properties in fluids.

■ EXAMPLE 5.18-1

Express the Lennard–Jones potential in terms of the empirical parameters σ and ϵ, the collision diameter, and potential well depth, respectively, represented in Figure 5.18-1(a).

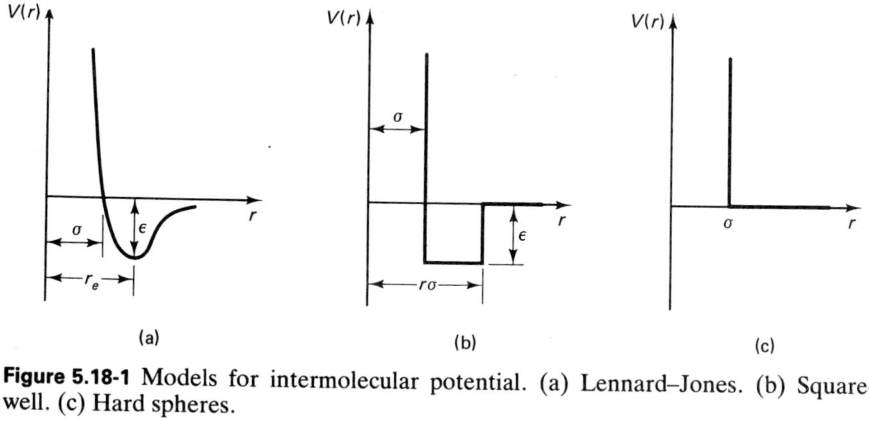

Figure 5.18-1 Models for intermolecular potential. (a) Lennard–Jones. (b) Square-well. (c) Hard spheres.

Solution. The collision diameter σ is the point at which the potential curve crosses the axis, that is,

$$c_1\sigma^{-12} - c_2\sigma^{-6} = 0 \tag{5.18-13}$$

where σ represents an effective hard-core diameter. The parameter ϵ represents the maximum energy of attraction, that is,

$$V(r_e) = -\epsilon \tag{5.18-14}$$

At the equilibrium internuclear separation r_e, the derivative of $V(r)$, that is, the force, vanishes:

$$V'(r_e) = 12c_1 r_e^{-13} + 6c_3 r_e^{-7} = 0 \tag{5.18-15}$$

Solving for r_e,

$$r_e = \left(\frac{2c_1}{c_2}\right)^{1/6} \tag{5.18-16}$$

and, from (5.18-12),

$$\epsilon = \frac{c_2^2}{4c_1} \tag{5.18-17}$$

Solving (5.18-13) and (5.18-17) for c_1 and c_2 and substituting into (5.18-12) gives

$$V(r) = 4\epsilon\left[\left(\frac{\sigma}{r}\right)^{12} - \left(\frac{\sigma}{r}\right)^{6}\right] \tag{5.18-18}$$

This is the standard form of the Lennard–Jones potential. ■

■ EXAMPLE 5.18-2

Weak intermolecular bonds between molecules are called secondary bonds. Consider the secondary bonding described by the following "6–12" potential:

$$E = -\frac{K_A}{a^6} + \frac{K_B}{a^{12}}$$

where K_A and K_B are constants for attraction and repulsion, respectively. Given $K_A = 10.37 \times 10^{-78}$ J m^6 and $K_B = 16.16 \times 10^{-135}$ J m^{12}, calculate the bond energy and bond length for argon.

Solution. The (equilibrium) bond length occurs at $dE/da = 0$:

$$\left(\frac{dE}{da}\right)_{a=a_0} = \frac{6K_A}{a_0^7} - \frac{12K_B}{a_0^{13}} = 0$$

Rearranging and solving for a_0 gives

$$a_0 = \left(2\frac{K_B}{K_A}\right)^{1/6} = \left(2 \times \frac{16.16 \times 10^{-135}}{10.37 \times 10^{-78}}\right)^{1/6} \text{ m}$$

$$= 0.382 \times 10^{-9} \text{ m} = 0.382 \text{ nm}$$

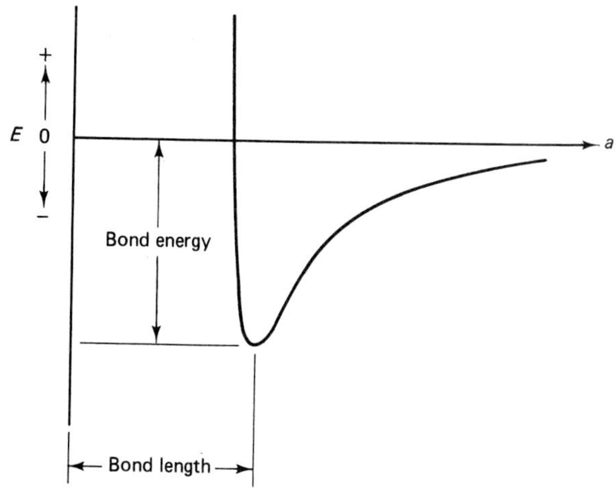

Figure 5.18-2 Bond energy curve.

The bond energy $E(a_0)$ is

$$E(0.382\text{ nm}) = \frac{K_A}{a_0^6} + \frac{K_B}{a_0^{12}}$$

$$= -\frac{(10.37 \times 10^{-78}\text{ J m}^6)}{(0.382 \times 10^{-9}\text{ m})^6} + \frac{(16.6 \times 10^{-135}\text{ J m}^{12})}{(0.382 \times 10^{-9}\text{ m})^{12}}$$

$$= -1.66 \times 10^{-21}\text{ J}$$

For 1 mol of Ar,

$$E_{\text{bond}} = (-1.66 \times 10^{-21}\text{ J/bond}))(6.02 \times 10^{23}\text{ bonds/mole})$$
$$= -0.999 \times 10^3\text{ J mol}^{-1}$$

This bond energy is less than 1% of the magnitude of a primary (covalent) bond (Chapter 6). The bond energy is negative (indicating an attractive force between the molecules of argon). ■

5.19 SUMMARY

Classical thermodynamics rests on an experimental foundation of great scope. Fundamental relations are inductively derived and regarded as a set of laws or postulates. Statistical thermodynamics develops these from first principles. The partition function characterizing a canonical ensemble (V, N, T specified) is of central importance. Once $Z(V, N, T)$ is known, all the thermodynamic functions of the system can be determined.

The laws of thermodynamics are the zeroth law, first law, second law, and third law. Classically, they are completely macroscopic. Statistical-mechanically, the connection between macroscopic and microscopic concepts is through Boltzmann's relation $S = k \ln W$. Thermodynamic functions for internal energy, enthalpy, specific heats, compressibility, and coefficient of volume expansion are established. Maxwell's relations further enable useful exchange of variables.

An open system is one in which the mass varies. The fundamental equation of thermodynamics $dU = T\,dS - p\,dV$ now takes the form $dU = T\,dS - p\,dV + \Sigma\mu_i\,dn_i$, where $\mu_i = (\partial G/\partial n_i)_{p,T}$ is the chemical potential. The equilibrium condition for chemical reaction or phase change is characterized by the same value of the Gibbs function in both phases. From this, useful relations such as the Clausius–Clapeyron equation and the equilibrium constant is obtained.

The postulate that quantum states of equal energy have equal probability of occurrence in an ensemble enables the average energy and pressure of an ensemble system to be found. Equating the averages to the thermodynamic U and the pressure p, the statistical-mechanical relations for U and p in terms of the canonical partition function Z can be found. The expressions for U and p are then used in $dU = T\,dS - p\,dV$ to find the expression for $S = U/T + k \ln Z$. Key statistical-mechanical relations for calculating thermodynamic properties are those for p, U, S, A, and μ. The calculation of Z is more conveniently achieved by introducing characteristic temperatures for translation, rotation, vibration, and electronic excitation. Contribution of these modes to the partition function and thermodynamic properties of monatomic and diatomic ideal gases are obtained. Imperfect gases are treated by introducing the configuration integral and calculating the first and second virial coefficients. Intermolecular potentials are introduced, among these, the hard-sphere, square-well, and Lennard-Jones 6–12.

SYMBOLS IN CHAPTER 5

A, a	Helmholtz-free energy
B	Second virial coefficient
C_p, C_v	Specific heat at constant pressure, at constant volume
D_e, D_0	Equilibrium dissociation energy, spectroscopic dissociation energy; $D_e = D_0 + \frac{1}{2}\hbar\omega_0$
E, ε	Energy
G, g	Gibbs-free energy, degeneracy
H, h	Enthalpy
I	Moment of inertia
J	Rotational quantum number
k	Boltzmann constant
L	Latent heat
M	Molecular mass
N, n	Number of particles
p	Pressure
Q, q	Heat transfer

R, r	Gas constant, radial distance
S, s	Entropy
T	Tempcrature
U, u	Internal energy
V, v	Volume; potential energy
W, w	Work
Z, z	Partition function
α	Coefficient of thermal expansivity
β	Lagrangian multiplier $1/kT$
ε	Energy
θ_r, θ_v	Characteristic temperature of rotation, of vibration
K	Coefficient of compressibility
Λ	De Broglie wavelength
ν	Frequency (Hz), stoichiometric coefficient

PROBLEMS

5.1. Using the first law of thermodynamics and other relevant definitions, show that the specific heat at constant volume is

$$C_V = -\left(\frac{\partial E}{\partial V}\right)_T \left(\frac{\partial V}{\partial T}\right)_E$$

and hence E, the internal energy, is a function of both the volume and the temperature.

5.2. Show that the absolute temperature T has the general property of indicating the direction of heat flow between two systems in thermal contact, that is, positive heat is always absorbed by the system with higher β (lower T) and given off by the system with lower β (higher T).

5.3. Show that the specific heat at constant volume, C_V, for an ideal diatomic gas is given by

$$C_V = \frac{5}{2}R + \frac{Rx^2 e^x}{(e^x - 1)^2}$$

where $x = h\nu/kT$ is the vibrational energy.

5.4. Calculate the heat capacity at constant pressure, C_p, for each of the substances in the reaction at 25°C:

$$C(s) + 2H_2O(g) = CO_2(g) + 2H_2(g)$$

where $C(s)$ obeys Dulong and Petit's law, $CO_2(g)$ is a linear molecule, and $H_2O(g)$ has 20% contribution to its C_p from the vibrational mode.

5.5. Discuss the assertion that, to excellent approximation, the entropy becomes vanishingly small as the system approaches its ground-state energy, that is,

$$\text{as } E \to E_0 \qquad S \to 0$$

5.6. A certain triatomic gas ABA has a ratio of specific heats $C_p/C_V = 1.38$. What can be said about the structure of the gas if the literature gives $C_V = 2.5\,R$ for a linear triatomic gas excluding the vibrational contribution and $C_V = 3.0\,R$ for a nonlinear triatomic gas (excluding vibrational contribution)?

5.7. The ratio $C_p/C_V = \gamma$ for an ideal ozone gas is 1.15. Using the fact that the heat capacity C_V is made up of contributions due to translational, rotational, and vibrational motions, determine whether ozone is linear or nonlinear.

5.8. A mole of ideal gas at 300 K is expanded adiabatically and reversibly from 20 to 1 atm. What is the final temperature of the gas if $C_V = \frac{3}{2}R$ per mole? 90.5K

5.9. Enthalpy is a thermodynamic property of a substance, and at a given temperature and pressure, it is the same regardless of how the substance is produced. This makes it possible to calculate enthalpy changes for reactions that cannot be studied directly. Consider the reaction:

$$C(\text{graphite}) + 2H_2(g) = CH_4(g)$$

It is not practical to measure directly the heat formed. However, the following reactions can be arranged, along with their given ΔH° values:

$$CH_4(g) + 2O_2(g) = CO_2(g) + 2H_2O(l) \qquad \Delta H^\circ = -212.80 \text{ kcal mol}^{-1}$$

$$C(\text{graphite}) + O_2(g) = CO_2(g) \qquad \Delta H^\circ = -94.05 \text{ kcal mol}^{-1}$$

$$H_2(g) + \frac{1}{2}O_2(g) = H_2O(l) \qquad \Delta H^\circ = -68.32 \text{ kcal mol}^{-1}$$

Determine the heat of formation of $CH_4(g)$.

5.10. From the following $T\,ds$ equations:

$$T\,ds = dh - v\,dp$$

$$T\,ds = du + p\,dv$$

determine, on the $T-s$ diagram the slope of (a) a constant-pressure line and (b) a constant-volume line. Consider a perfect gas.

5.11. Statistical mechanics calculations are usually more easily performed on a fixed volume basis, while experimental measurements are more readily carried out under constant (say atmospheric) pressure. It is therefore desirable to have a relation between the specific heats at constant volume and constant pressure. Show that

$$C_p - C_V = VT\frac{\alpha^2}{K}$$

where α is the coefficient of volume expansion, and K is the isothermal compressibility.

5.12. From the relation for specific heats

$$C_p - C_V = VT\frac{\alpha^2}{K}$$

show that, for a perfect gas,

$$C_p - C_V = R$$

5.13. The third law of thermodynamics asserts that, as the temperature $T \to 0$, the entropy S of a system approaches smoothly some limiting constant value S_0 independent of all parameters of the system. Show that as $T \to 0$, $S \to S_0$, $C_V \to 0$, $C_p \to 0$, $\alpha \to 0$, but the compressibility K remains constant.

5.14. **(a)** Calculate the work done at 25°C when 100 g of nitrogen at a pressure of 260 T is expanded reversibly and isothermally to a pressure of 100 T.

(b) The valve on a cylinder containing initially 10 liters of an ideal gas at 25 atm and 25°C is opened to the atmosphere, where the pressure is 260 T and the temperature is 25°C. Assuming the process to be isothermal, how much work is done on the atmosphere by the expanding gas?

5.15. Show that the entropy of a van der Waals gas

$$\left(p + \frac{a}{v^2}\right)(v - b) - RT$$

is given by

$$ds = \frac{C_v(T)}{T}\,dT + \frac{R}{v - b}\,dv$$

5.16. Free expansion is a complicated irreversible process involving turbulence and gross nonuniformities of pressure and temperature (to the extent that these quantities can be defined at all for such a marked nonequilibrium situation). Equilibrium conditions, however, prevail in the initial and final situations. Show that to predict the outcome of the process, the only knowledge required is that of the energy function E characteristic of equilibrium macrostates of the system.

5.17. Calculate the temperature change in a free expansion of a van der Waals gas. Show that $T_2 < T_1$. What practical limitation is placed on using the free expansion to cool a gas to low temperatures?

5.18. Show that in a throttling (Joule-Thomson) process, the enthalpy remains constant, and the Joule-Thomson coefficient, $\mu \equiv (\partial T/\partial p)_h$, is given by

$$\mu = \frac{v}{c_p}(T\alpha - 1)$$

where $\alpha \equiv 1/v\,(\partial v/\partial T)_p$ is the coefficient of volume expansion.

5.19. Neither free-expansion nor throttling processes result in a temperature change in the case of an ideal gas. However, both of these processes become interesting if the gas is not ideal, that is, when the mutual interaction between molecules is of importance. What effect does molecular attraction or repulsion have on the Joule–Thomson coefficient μ?

5.20. For a system at constant temperature and pressure, that is, in contact with a heat reservoir at T_0 and p_0, the stable equilibrium situation is characterized by the minimum value of $E - T_0 S + p_0 V$, which is the availability function. Show that this leads to

$$\left(\frac{\partial E}{\partial T}\right)_v \geq 0 \quad \text{or} \quad C_v \geq 0$$

for stability (Le Châtelier's principle).

5.21. Show that, for a system, whose external parameters are fixed, but is in thermal contact with a heat reservoir, the stable equilibrium situation is characterized by the condition that

$$A_0 = \text{minimum}$$

where $A_0 = E - T_0 S$ is the Helmholtz function (with the temperature of the system equal to that of the reservoir T_0).

5.22. Show that, for a single-component system consisting of two phases, say solid and liquid, or liquid, or liquid and gas, application of the equilibrium condition (Gibbs free energy $G = E + pV - TS = \text{minimum}$) leads to the Clausius–Clapeyron equation.

$$\frac{dp}{dT} = \frac{L_{12}}{T\Delta V}$$

where L is the latent heat and V is the volume.

5.23. From the Clausius–Clapeyron equation

$$\frac{dp}{dT} = \frac{L}{T\Delta V}$$

where L is the latent heat per mole and V is the molar volume, show that the pressure of a vapor in equilibrium with the liquid (or gas) at temperature T is given by

$$p = p_0 e^{-L/RT}$$

where p_0 is a constant.

5.24. For a homogeneous system of energy E and volume V consisting of m different kinds of molecules, show that the chemical potential per molecule of the jth species,

$$\mu_j = -T\left(\frac{\partial S}{\partial N_j}\right)^{E,V,N}$$

is given by

$$\mu_j = \left(\frac{\partial E}{\partial N_j}\right)_{S,V,N} = \left(\frac{\partial A}{\partial N_j}\right)_{T,V,N} = \left(\frac{\partial G}{\partial N_j}\right)_{T,p,R}$$

where $A = E - TS$ and $G = E + pV - TS$ are the Helmholtz and Gibbs free energies, respectively.

5.25. The total energy of a homogeneous system of different kinds of molecules is

$$E = E(S, V, N_1, N_2, \ldots, N_m).$$

From the requirement that extensive quantities must scale properly, show that

$$G = E + pV - TS = \Sigma \mu_i N_i$$

and

$$S\,dT - V\,dp + \sum_i N_i\,d\,\mu_i = 0$$

The latter is known as the "Gibbs–Duhem relation."

5.26. A chemical reaction such as

$$2H_2 + O_2 \rightarrow 2H_2O$$

can be written in the general form

$$\sum_{i=1}^{m} b_i B_i = 0$$

where b_i is the number of molecules B_i in the chemical transformation. Suppose that the molecules form gases that can be considered ideal and that they are confined in a container of volume V at an absolute temperature T. Show the relationship between the mean numbers of the reacting molecules resulting in the equilibrium constant and the law of mass action.

5.27. The molecular partition function

$$z_i = \sum_s e^{-\beta\epsilon(s)}$$

has the simplifying property that it is proportional to the volume V. Show that the temperature dependence of the equilibrium constant is given by

$$\frac{d \ln K_n}{dT} = \frac{\Delta E}{kT^2}$$

where

$$K_n(T) = z_1^{b_1} z_2^{b_2} \cdots z_m^{b_m} \qquad z_i = z_i(T)$$

differs from

$$K_N(T, V) = z_1^{b_1} z_2^{b_2} \cdots z_m^{b_m} \qquad z_i = z_i(T, V)$$

only by a factor involving the volume.

5.28. The ideal gas law may be written

$$pV = \frac{m}{M} RT$$

where M is the molecular weight. Show that in a gravitational field, the pressure at height h is given by

$$p = p_0 e^{-gMh/RT}$$

What is the pressure in the earth's atmosphere at an altitude of 10 km if the average molar mass of the air is 0.29 kg mol^{-1} and the temperature is 0°C?

5.29. **(a)** Show that in a reversible adiabatic expansion of a perfect gas, the relations between pressure and volume, pressure and temperature, and volume and temperature, are given by

$$\frac{p_1}{p_2} = \left(\frac{v_2}{v_1}\right)^{\gamma} \qquad \frac{T_1}{T_2} = \left(\frac{p_1}{p_2}\right)^{(\gamma-1)/\gamma} \qquad \frac{T_1}{T_2} = \left(\frac{v_2}{v_1}\right)^{(\gamma-1)}$$

where $\gamma = c_p/c_v$.

(b) What is the pressure, temperature, and work done when a perfect monatomic gas is allowed to expand adiabatically from 22.7 L mol^{-1} at 1 bar and 0°C to a volume of 45.4 L mol^{-1}?

5.30. The heat evolved or absorbed by a chemical reaction is measured with a calorimeter. There are two types of calorimetric experiments: constant volume and constant pressure. When a reaction is carried out in a sealed bomb, no pV work is done, and the first law may be written $\Delta U = (Q)_v$. Thus the change in internal energy for the reaction is obtained. When the reaction is carried out at constant pressure, the first law may be written $\Delta H = (Q)_p$. Chemists are usually more interested in ΔH than ΔU because chemical reactions are generally carried out at constant pressure. Show that the relation between ΔH and ΔU for perfect gases is

$$\Delta H = \Delta U + \Delta nRT$$

The combustion of $C_2H_5OH(l)$ in a constant-volume calorimeter produces 1364.34 kJ mol^{-1} at 25°C. What is the value of ΔH° for the combustion reaction?

5.31. Lavoisier and Laplace recognized in 1780 that the heat absorbed in decomposing a compound must be equal to the heat evolved in its formation under the same conditions. Hess pointed out in 1840 that the overall heat of a chemical reaction at constant pressure is the same, regardless of the intermediate steps involved. These findings stem from the fact that the enthalpy is a state function. Show how this makes it possible to calculate the enthalpy changes for reactions that cannot be studied directly, as, for example, when carbon burns to carbon monoxide in a limited amount of oxygen.

5.32. In the chemical reaction involving r substances A_i,

$$\sum_{i=1}^{r} \nu_i A_i = 0$$

the change in enthalpy is

$$\Delta H = \sum_{i=1}^{r} \nu_i H_i$$

For reactants and products in their standard states, this becomes

$$\Delta H^\circ = \sum_{i=1}^{r} \nu_i H_i^\circ$$

Since absolute enthalpies are not known, enthalpies relative to a defined reference state are used instead. These "relative" enthalpies of substances are called enthalpies of formation and are represented by ΔH_f°. Since the same reference state is used for reactants and products, the same ΔH° is obtained as would be obtained with absolute enthalpies. Thus standard enthalpy changes for reactions may be calculated using enthalpies of formation as

$$\Delta H^\circ = \sum_{i=1}^{r} \nu_i \Delta H_{f,i}^\circ$$

where the enthalpy of formation of a substance is the change in enthalpy for the reaction in which the substance is formed from its elements. Write out the above equations for the reaction

$$H_2(g) = \frac{1}{2} O_2(g) = H_2O(l)$$

The reaction of heated coal (approximated by graphite) with superheated steam absorbs heat. This heat is usually provided by burning some of the coal. Calculate ΔH° (500 K) for both reactions. How many moles of graphite are required to produce one mole of hydrogen?

5.33. **(a)** Show, from classical as well as statistical considerations, that the entropy of mixing C moles of perfect gas is

$$\Delta S_{\text{mix}} = -R \sum_{i=1}^{C} x_i \ln x_i$$

and that this is always positive.

(b) What is the entropy of mixing of $\frac{1}{2}$ mol of oxygen with $\frac{1}{2}$ mol of nitrogen at 25°C, assuming that they are perfect gases?

5.34. **(a)** Show that the entropy, as a function of temperature and pressure, is given by

$$\Delta S = \frac{C_p}{T} dT - V\alpha \, dp$$

where $\alpha = (1/V)(\partial V/\partial T)_p$ is the coefficient of thermal expansivity.

(b) What is the change in entropy of liquid water at 90°C when the pressure is raised from 1 bar to 100 bar? For water at 90°C, $\alpha = 8.3 \times 10^{-4}\ \text{K}^{-1}$.

(c) What is the change in entropy of 1 mol of helium when it is heated from 298 K to 1000 K at 1 bar?

5.35. **(a)** An equation of state is a relation between thermodynamic variables that define the state of a material. From the fundamental equation for a closed system, $dV = T\,dS - p\,dV$, show that

$$\left(\frac{\partial U}{\partial V}\right)_T = \frac{\alpha T - \kappa P}{\kappa}$$

$$\left(\frac{\partial H}{\partial P}\right)_T = V(1 - \alpha T)$$

where α is the coefficient of thermal expansivity and κ is the coefficient of compressibility.

(b) The pressure on liquid benzene at 25°C is raised from 1 to 11 bar. What is the change in enthalpy? For liquid benzene under these conditions $\alpha = 1.237 \times 10^{-3}\ \text{K}^{-1}$ and the density is $0.879\ \text{g cm}^{-3}$.

(c) What is $(\partial U/\partial V)_T$ and $(\partial H/\partial p)_T$ for a perfect gas?

5.36. For an open system containing k different substances the Gibbs energy may be considered to be a function of T, p, n_1, n_2, . . . , n_k, where n_i is the amount of substance i in the system. The total differential of G is

$$dG = \left(\frac{\partial G}{\partial T}\right)_{P,n_i} dT + \left(\frac{\partial G}{\partial P}\right)_{T,n_i} dp + \sum_{i=1}^{k} \left(\frac{\partial G}{\partial n_i}\right)_{T,P,n_j} dn_i$$

where the subscript n_i means that the amounts of all substances are held constant, except for the one being varied (i.e., $j \neq i$).

(a) Show that

$$G = \sum_{i=1}^{k} n_i \mu_i$$

and show how this can be used to obtain a simple expression for chemical equilibrium in terms of the chemical potential $\mu_i = (\partial G/\partial n_i)_{T,p,n_j}$ of the components.

(b) Obtain the Gibbs–Duhem equation

$$S\,dT - V\,dp + \sum_{i=1}^{k} n_i\,d\mu_i = 0$$

and illustrate its application to a two-component system.

5.37. (a) In a one-component system, two phases (α and β) may be in equilibrium. The pressure, temperature, and chemical potential must then be the same in the two phases. When the temperature is changed at constant pressure, or the pressure is changed at constant temperature, one of the phases will disappear. However, if the temperature and pressure are both changed in such a way as to keep the two chemical potentials equal to each other, the two phases will continue to coexist. The plot of pressure versus temperature along the two phases is referred to as the coexistence curve. What is the necessary relation for dp/dT along the coexistence curve?

(b) What is the change in the boiling point of water at 100°C per Pa change in atmospheric pressure? The enthalpy of vaporization is $40.69\ \text{kJ mol}^{-1}$, the molar volume of liquid water is $0.019 \times 10^{-3}\ \text{m}^3\ \text{mol}^{-1}$, and the molar volume of steam is $30.199 \times 10^{-3}\ \text{m}^3\ \text{mol}^{-1}$, all at 100°C and 1.01325 bar.

(c) Calculate the change in pressure required to change the freezing point of water 1°C. At 0°C the heat of the fusion of ice is $333.5\ \text{J g}^{-1}$, the density of water is $0.9998\ \text{g cm}^{-3}$, and the density of ice is $0.9168\ \text{g cm}^{-3}$.

5.38. The relationship between the standard Gibbs energy of reaction and the equilibrium constants K_p, K_c, is given by

$$\Delta G° = -RT \ln K_p \qquad K_p = \prod_i \left(\frac{p_i}{p°}\right)^{\nu_i}$$

(a) What is the relation between K_p and K_c for a mixture of perfect gases?

(b) Tables give $\Delta G_f° = 4.833\ \text{kJ mol}^{-1}$ for $NH_3(g)$ at 500 K. Determine the value K_p for the ammonia synthesis reaction

$$\frac{1}{2} N_2(g) + \frac{3}{2} H_2(g) = NH_3(g)$$

at 500 K. What is the value of K_c? What is the value of $\Delta G°$ calculated from K_c?

5.39. **(a)** Calculate the extent of dissociation of $H_2(g)$ at 3000 K and 1 bar: $H_2(g) = 2H(g)$. From tables, $\Delta G_f^\circ = 46.006$ kJ K^{-1} mol^{-1} for H(g) at 3000 K.

(b) If 1.588 g of nitrogen tetroxide gives a total pressure of 1.0133 bar when partially dissociated in a 500-cm^3 glass vessel at 25°C, what is the extent of reaction? What is the value of K_p? What is the extent of reaction at a total pressure of 0.5 bar? The reaction is $N_2O_4(g) = 2NO_2(g)$.

(c) Develop a relation between the equilibrium constant K_p and the extent of reaction when two different product molecules are formed as in $PCl_5(g) = PCl_3(g) + Cl_2(g)$.

5.40. **(a)** Show that the molecular translational partition function is

$$z_{tr} = \frac{V}{\Lambda^3}$$

where $\Lambda = (h^2/2\pi mkT)^{1/2}$ is the thermal de Broglie wavelength.

(b) What is the molecular partition function for translational motion of a hydrogen atom at 3000 K in a volume of 0.2494 m^3? (This is the volume that would contain 1 mol of perfect gas at this temperature and 1 bar pressure.) What is the de Broglie wavelength?

5.41. **(a)** The molecular translational partition function is

$$z_t = \left(\frac{2\pi mkT}{h^2}\right)^{3/2} V$$

Substituting this in the canonical ensemble partition function

$$Z_t = \frac{z_t^N}{N!},$$

obtain the following molar thermodynamic quantities at a standard-state pressure of p°:

$$p = \frac{RT}{V} \qquad S_t^\circ = R\left\{\frac{5}{2} + \ln\left[\left(\frac{2\pi mkT}{h^2}\right)^{3/2}\frac{kT}{p^\circ}\right]\right\}$$

$$U_t^\circ = \frac{3}{2}RT \qquad A_t^\circ = -RT\left\{1 + \ln\left[\left(\frac{2\pi mkT}{h^2}\right)^{3/2}\frac{kT}{p^\circ}\right]\right\}$$

$$(C_V^\circ)_t = \frac{3}{2}R \qquad G_t^\circ = -RT\ln\left[\left(\frac{2\pi mkT}{h^2}\right)^{3/2}\frac{kT}{p^\circ}\right]$$

$$H_t^\circ = \frac{5}{2}RT \qquad (C_P^\circ)_t = \frac{5}{2}R$$

(b) What are the translational contributions to C_p°, H°, S°, and G° for O(g) at 298.15 K?

(c) What are the translational contributions to C_p°, S°, H° and G° for O(g) at 298.15 K?

5.42. **(a)** The rotational energy of a rigid diatomic molecule is given by

$$\epsilon_r = \frac{J(J+1)h^2}{8\pi^2 I}$$

where J is the rotational quantum number, and I is the moment of inertia of the diatomic molecule. Show that the rotational energy levels are $(2J+1)$-fold degenerate and that the rotational partition function is given by

$$z_r = \sum_{J=0}^{\infty} (2J+1)e^{-J(J+1)h^2/8\pi^2 IkT}$$

(b) Show that for molecules with large moments of inertia,

$$z_r = \frac{T}{\Theta_r}$$

where

$$\Theta_r = \frac{h^2}{8\pi^2 Ik}$$

is the characteristic rotational temperature.

(c) What is the characteristic rotational temperature Θ_r for $H_2(g)$? What is the value of the molecular partition function for rotation at 3000 K? (The moment of inertia is 4.6054×10^{-47} kg m^2.)

(d) What are the rotational contributions to C_p°, S°, H°, and G° for $O_2(g)$ at 298.15 K? The moment of inertia for $O_2(g)$ is $I = 1.9373 \times 10^{-46}$ kg m^2.

5.43. (a) Show that, if vibrational energy is measured from the ground state ($n = 0$) rather than from the bottom of the potential energy curve, the vibrational partition function is

$$z_v = \frac{1}{1 - e^{-h\omega/kT}} = \frac{1}{1 - e^{-\Theta_v/T}}$$

where Θ_v is the characteristic vibrational temperature; $\Theta_v = h\omega/k = hc\overline{\omega}/k$.

(b) What are the vibrational contributions to C_p°, S°, H°, and G° for $O_2(g)$ at 298.15 K? The vibrational frequency is 1580.246 cm^{-1}.

5.44. (a) What is the electronic partition function when the energies of the excited states are large with respect to kT, and what is the electronic contribution to the canonical ensemble?

(b) What are the electronic contributions to S° and G° for $O(g)$ at 298.15 K? The degeneracy in the ground state is 5, the first excited state has an energy of 158.2 cm^{-1} with respect to the ground state and has a degeneracy of 3, and the second excited state has an energy of 226.5 cm^{-1} with no degeneracy.

5.45. What are the values of C_p°, S°, H°, and G° for $O_2(g)$ and $O(g)$ at 298.15 K obtained by summation of the translational, rotational, vibrational, and electronic contributions from Problems 5.41 through 5.44?

5.46. What are the values of ΔC_p°, ΔS°, ΔH°, and ΔG° for the reaction

$$\frac{1}{2} O_2(g) = O(g)$$

calculated statistically and what is the equilibrium constant at 298.15 K? The total of the translational, rotational, vibrational, and electronic contributions are as follows: For $O_2(g)$: $C_p^\circ = 29.336$ J K^{-1} mol^{-1}, $S^\circ = 205.085$ J K^{-1} mol^{-1}, $H^\circ = -483.203$ kJ mol^{-1}, $G^\circ = -544.349$ kJ mol^{-1}. For $O(g)$: $C_p^\circ = 20.786$ J K^{-1} mol^{-1}, $S^\circ = 159.286$ J K^{-1} mol^{-1}, $H^\circ = 6.197$ kJ mol^{-1}, $G^\circ = -41.294$ kJ mol^{-1}.

Chapter 6

Solids, Liquids, Conductors, Semiconductors, Transport Properties, Boltzmann Equation

The subject matter of the previous chapters is basic material in statistical mechanics and thermodynamics. The concepts and techniques, however, can be applied to many areas of study. From the atoms, the applications of molecular theory branch into two separate channels, one leading to systems larger than atoms (it goes from atoms to molecules, and to solids), the other from atoms to smaller systems (the nuclei and their constituents, the elementary particles). This chapter sets out to explore the first channel: the statistical mechanics of solids and liquid. The properties of condensed matter, especially crystalline solids, will be discussed.

Two or more atoms can combine to form a stable molecule, the atoms retaining their identity. Since a very large number of atoms may join together to make a solid, in much the same way as a few do to form a molecule, the phenomenon of molecular binding is relevant to the properties of solids. When a molecule is formed from two atoms, the inner-shell electrons of each atom remain tightly bound to the original nucleus and are barely disturbed at all. The outermost loosely bound electrons, known as the valence electrons, are influenced by all the particles (ions + electrons) of the system, and their wave functions are significantly modified when the atoms are brought together. Indeed, it is this very interaction that leads to binding, that is, to a lower total energy of the system.

Also discussed in this final chapter is the concept of transport processes. These are situations that lack equilibrium in character. They are concerned with the transfer of mass, momentum, and energy from one point to another through molecular motion and collision. Unless a uniform condition exists, that is, thermodynamics equilibrium prevails, transfers will always take place.

There are three categories of transfer processes: diffusion, radiation, and convection. In the diffusion category are such processes as mass diffusion, viscous transfer of momentum, and heat conduction. Diffusion occurs in dense matter. Slow neutrons or molecules in a gas at normal density are able to go but very short distances before colliding with matter. Diffusion is thus a short mean-free-path process. In the radiation category are such processes as heat radiation, neutron transport, and free molecule flow. Radiation occurs in low-density matter. Fast neutrons travel vast open spaces between atomic nuclei. Photons emitted by the sun are unimpeded on their way to earth or to a spacecraft. Molecules in a vacuum chamber traverse from one wall to another without a collision. Radiation is thus a long mean-free-path process compared to diffusion.

The term convection refers to transport by bulk motion of a moving fluid. A globule of the medium, by virtue of its mass and velocity, will transport momentum. Likewise, by virtue of its temperature and composition, it will transport energy and mass species. However, it is common practice to use the term more broadly to describe transfer processes involving moving fluids as convective, even though diffusion or radiation plays a dominant role in the transfer from a wall to the fluid. In this sense, convection is both convective and diffusive transport in a moving fluid.

6.1 NATURE OF SOLIDS, LIQUIDS, AND GASES

A solid is classified as crystalline or amorphous. A crystalline solid has a sharp melting point and shows crystals having well-developed faces and a characteristic shape. X-ray diffraction reveals a crystalline solid to have a regular, ordered structure composed of identical repeating units having the same orientation throughout the crystal; the repeating unit is a group of one or more atoms, molecules, or ions.

An amorphous solid on the other hand, does not have a characteristic crystal shape. When heated, it softens and melts over a wide temperature range. X-ray diffraction shows a disordered structure. Polymers often form amorphous solids; as the liquid polymer is cooled, the polymer chains become twisted and tangled together in a random, irregular way. (Some polymers do form crystalline solids; others are partly crystalline and partly amorphous.) The various kinds of glass contain chains and rings involving Si—O bonds; the structure is disordered and irregular. Amorphous materials like glass and certain plastics exist because their structure is not in equilibrium. Indeed, over a period of centuries, glass may crystallize out. If the glass is kept close to its softening temperature, the crystallization is much more rapid (a fact known to anyone with experience in glass blowing).

This chapter begins with a look at some of the possible ways in which atoms are arranged in crystalline solids and at the relationship between the form of a crystal and its internal energy. The survey of crystallography will necessarily be fairly superficial, for only those structures that are relevant to statistical mechanics will be discussed. In order to understand the central common features of the solid state, it is sufficient to know that a crystal is a periodic structure in space, with a basic repeating unit called the unit cell.

First, however, it should be pointed out why, from a microscopic viewpoint, the same substance can sometimes exist as a solid, sometimes as a liquid, and sometimes as a gas. The answer involves making comparisons between the kinetic energy of the substance's atoms and the pair dissociation energy ΔE. (For economy, the single word "atom" is used to mean either an atom or a molecule—whichever is appropriate—of the substance under consideration.)

Since each atom in a collection of atoms can interact with any other atom in the manner described in Section 5.18, one or more bound pairs of atoms is expected to form. To split up such a pair into its component atoms requires an energy ΔE, the dissociation energy of the pair. Consider, as in Fig. 6.1-1, what may happen when another atom with a kinetic energy $\frac{1}{2}mu^2$ collides with the bound pair.

If $\frac{1}{2}mu^2$ exceeds ΔE, there is enough energy available to dissociate the pair. Indeed if the average kinetic energy, written $\frac{1}{2}\overline{mu^2}$, greatly exceeds ΔE, virtually all collisions will lead to dissociation of the bound pairs. Under these conditions no bound pairs can persist; the system will consist mainly of individual atoms moving rapidly and with no apparent affinity for each other. In fact, the atoms will have to be forced to occupy any given volume. These are just the properties displayed by gases, so that for the gaseous phase,

$$\tfrac{1}{2}\overline{mu^2} >> \Delta E \tag{6.1-1}$$

At the other extreme, when the average kinetic energy $\frac{1}{2}\overline{mu^2}$ of a colliding atom is very much less than the dissociation energy of the bound pair, no dissociation will occur, and the result is that the colliding atom is easily captured by the pair. Before long, this process will be repeated many times over and a large conglomeration of atoms will form. Such a conglomeration—one that requires a significant amount of energy to dissociate it into a gas—is

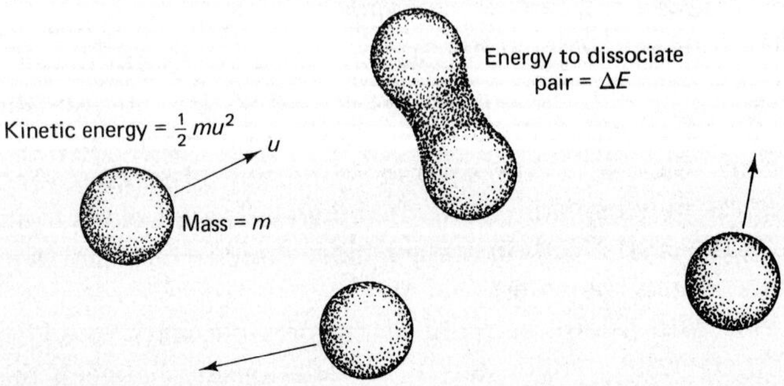

Figure 6.1-1 Competing roles of kinetic and potential energies in a system of many atoms.

characteristic of the solid phase. Therefore, on the microscopic scale, the solid phase occurs when

$$\tfrac{1}{2}\overline{mu^2} \ll \Delta E \tag{6.1-2}$$

Lastly, when $\tfrac{1}{2}\overline{mu^2}$ is the same order of magnitude as ΔE, some clustering is expected to occur, for within the group of colliding atoms possessing an average energy $\tfrac{1}{2}\overline{mu^2} \approx \Delta E$ there will be some with a kinetic energy greater than ΔE and these will tend to dissociate any clusters. The result is a system in which a limited amount of association is possible; a fluctuating system, part bound, part free. The system will not possess the rigidity of a solid nor the unrestrained freedom of a gas. The conditions characteristic of a liquid are, therefore,

$$\tfrac{1}{2}\overline{mu^2} \approx \Delta E \tag{6.1-3}$$

So by considering both the atoms' kinetic and potential energies, it is possible, in a qualitative manner, to account for the solid, liquid, and gaseous phases. There is also a fourth phase of matter—the plasma phase—which consists of separated nuclei and electrons. This arises when the kinetic energy of the colliding atoms is so great that they knock one another asunder. The plasma phase is by far and away the most common—stars are plasmas—but it is also the most difficult to discuss quantitatively.

6.2 ATOMIC BONDS

The preceding section has shown that the atoms in a crystalline solid are arranged in neatly ordered structures, but the nature of the forces that hold the atoms together has not yet been described. The various ways atoms are bound together will now be discussed. A crystal is classified as *ionic, covalent, metallic,* or *molecular,* according to the nature of the chemical bonding and the intermolecular forces in the crystal.

Consider, first, the general situation of two identical atoms in their ground states being brought together from an infinite separation. Initially, the energy of their interaction is zero. As the atoms approach, the attractive forces increase and the energy increases in a negative sense (the energy of attraction is negative since the atoms do the work, while that of repulsion is positive as work has to be done on the atoms to force them closer together). At a separation of a few atomic radii, repulsive forces begin to assert themselves, and the atoms reach an equilibrium separation r_0 at which the repulsive and attractive forces are equal and the mutual potential energy is a minimum. The situation is shown in Fig. 6.2-1.

The repulsive forces have a much shorter range than the attractive forces and are due partly to electrostatic repulsion of like charges (which obey the Pauli exclusion principle). As the atoms come closer together the Pauli principle requires that some of the electrons move to higher energy states; the

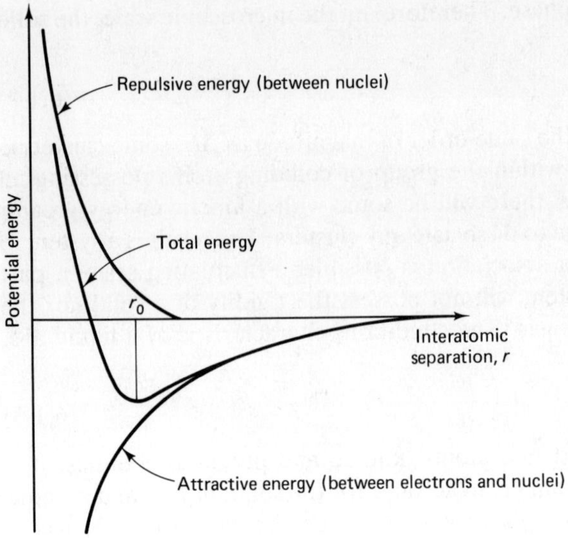

Figure 6.2-1 Interaction of two atoms as a function of their separation.

system therefore gains energy, a situation that is equivalent to the action of an interelectron repulsive force. Different charge distributions in the atoms give rise to different types of bonding. Roughly speaking, solids can be divided into four classes.

Ionic Crystals

This is the simplest of interatomic forces; it results from the mutual electrostatic attraction of opposite charges. Ionic crystals consist of an array of positive and negative ions and are held together by the Coulomb attraction between oppositely charged ions. Examples are NaCl, MgO, $CaCl_2$, and KNO_3.

If an atom gains or loses one or more electrons it is said to be ionized, and it becomes either negatively or positively charged. Atoms of many elements lose or gain electrons very easily because in so doing they acquire a completely filled outer electron shell, which is a configuration of great stability. In Fig. 6.2-2 sodium (Na), which has only one electron in its outer occupied shell (the 3*s* shell), will readily lose this electron to become a singly charged positive ion (Na^+). Similarly, chlorine (Cl), having an almost completely full outer shell, will readily accept an additional electron to fill its outer shell to become a negative ion. The result is the formation of a sodium chloride molecule.

In practice, Na^+ ions and Cl^- ions do not link up in pairs as shown. Rather, Na^+ ions will surround themselves with Cl^- ions, and Cl^- ions with Na^+ ions in the crystal in such a way that the attraction between neighboring unlike charges exceeds the repulsion due to like charges. The resulting sodium chloride structure has two interpenetrating face-centered cubic structures, one formed from sodium and the other from chlorine ions. In general, the ionic

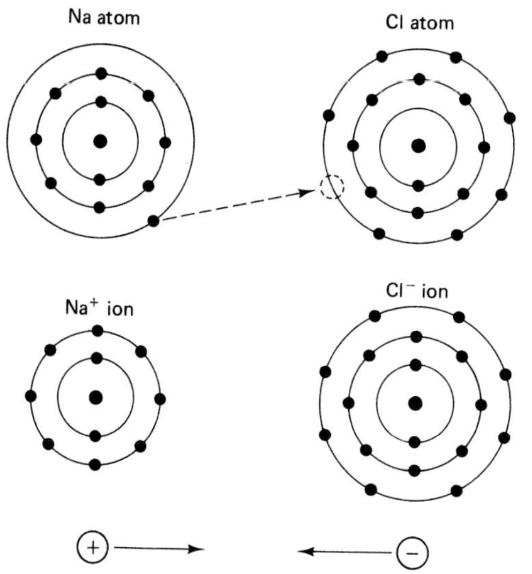

Figure 6.2-2 Ionic bonding in the sodium chloride molecule.

bond is quite strong, as evidenced by the fact that many ionic solids are hard and have quite high melting points (for example, lithium fluoride and sodium chloride, which have melting points of 845 and 801°C, respectively).

Covalent Crystals

Another way in which the outer electron shell of atoms can be filled to achieve a stable configuration is by the sharing of electrons. This defines the covalent bond. For example, chlorine requires one electron to complete the outer shell. This can be accomplished if two chlorine atoms join together and share their outer electrons. Figure 6.2-3 shows the covalent bonding that enables the chlorine molecule to be formed.

Covalent (or nonmetallic network) crystals consist of an infinite network of atoms held together by covalent bonds. There are no identifiable molecular

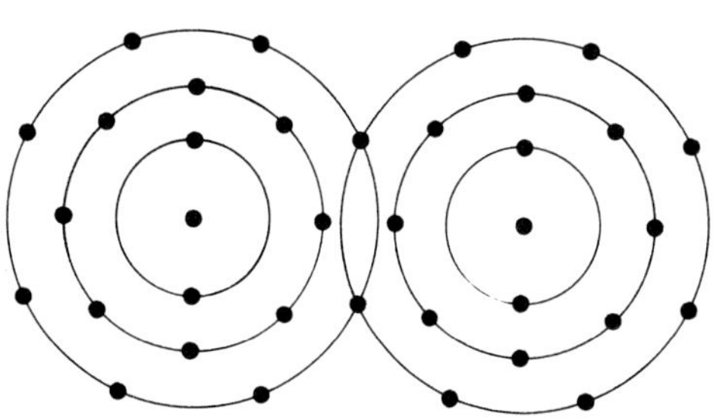

Figure 6.2-3 Sharing of electrons in a chlorine molecule.

subunits within the crystal. Such solids are best thought of as supermolecules in which atoms are joined to their neighbors by covalent bonds. Examples are carbon in the forms of diamond and graphite, Si, SiO_2, and SiC. In diamond each carbon is bonded to four others that surround it tetrahedrally, giving a three-dimensional network that extends throughout the crystal. Silicon has the same structure. SiO_2 has a three-dimensional network in which each Si is bonded to four O's at tetrahedral angles, and each O is bonded to two Si atoms. Covalent crystals are usually hard and have high melting points. Indeed diamond is the hardest material and has the highest known melting point at over 3550°C.

Metallic Crystals

These are bonded metal atoms. Here, positive metal ions are set in a sea of free (or nearly free) electrons. As with a covalent solid, a piece of metal is like a supermolecule. The binding can be thought of as a limiting case of covalent binding in which electrons are shared by all the ions in the crystal. Examples are Na, Cu, Fe, and various alloys. The free electrons contribute to the high electrical and thermal conductivities of metals together with their high ductility.

Molecular Crystals

Molecular crystals are composed of individual molecules. The atoms within each molecule are held together by covalent bonds. Relatively weak intermolecular forces hold the molecules together in the crystal. Molecular crystals are subdivided into van der Waals crystals, in which the intermolecular attractions are dipole–dipole, dipole–induced dipole, and dispersion forces, and hydrogen-bonded crystals, in which the main intermolecular attraction is due to hydrogen bonding. Examples are iodine and the inert gases.

It should be pointed out that the distinction between the various kinds of crystals is not always clear-cut. Also, in ionic, metallic, and three-dimensional covalent crystals, it is not possible to pick out individual molecules. The entire crystal is a single giant molecule. All told, chemical bonds, as determined by the electronic structure of the material, may be summarized as follows: the ionic bond between ions or positively and negatively charged atoms; opposite charged ions are attracted to each other. In the covalent bond, pairs of atoms share their outer electrons and thus fill their outer electron shell. In a metal all the outer-shell electrons are shared among atoms, moving through the lattice of atoms as conducting electrons. The van der Waals bond is a weak attraction between nearby neutral atoms or molecules. The weak hydrogen bond is mediated by a hydrogen atom that is shared by two molecules.

The strength of a bond can be measured by the energy required to break it, that is, the amount of heat that must be supplied to vaporize the solid and hence separate the constituent atoms. If the heat of vaporization of 1 kg mol is measured, this must be the energy used in breaking N_0 bonds ($N_0 =$

Avogadro's number). Values obtained in this way are shown in Table 6.2-1, together with the type of bonding for the solid concerned. As might be expected, the weakest arc the van der Waals bond and the hydrogen bond. Next comes the metallic bond, followed by the ionic and covalent bonds whose strengths are nearly comparable.

Melting points are also indicative of bond strengths for the following reason. The atoms in a solid are in constant vibration at normal temperatures. Heat that is stored in the material when its temperature is raised, is stored in the vibrational energy of the atoms (which is proportional to the absolute temperature). Thus, melting occurs when the thermal vibration becomes so great that the bonds are broken and the atoms become mobile. By comparing melting points it is possible to illustrate some interesting variations of bond strength. It is clear from the list of melting points of Group IV elements in Table 6.2-2 that bond strength decreases with increasing atomic number. From these and other data, it is concluded that electrons of higher principal quantum number form weaker bonds.

To evaluate the interaction among atoms, imagine the following experiment. Take one atom of an element, say X, and fix its position. Take another atom of X with a Newton balance attached to it, Fig. 6.2-4(a), and place this second atom at various values of r from the fixed atom. For each setting of r, decide whether the force $F(r)$ between the two atoms is attractive, repulsive, or zero. The general form of the interatomic force between the two atoms is

TABLE 6.2-1 HEATS OF VAPORIZATION

Bond type	Material	Heat of vaporization H (kcal mol^{-1})
van der Waals	He	0.02
	N_2	1.86
	CH_4	2.4
Hydrogen:		
O—H——O	Phenol	4.4
F—H——F	HF	6.8
Metallic	Na	26
	Fe	94
	Zn	28
Ionic	NaCl	153
	LiI	130
	NaI	121
	MgO	242
Covalent	C (Diamond)	170
	SiC	283
	SiO_2	405
	Si	85

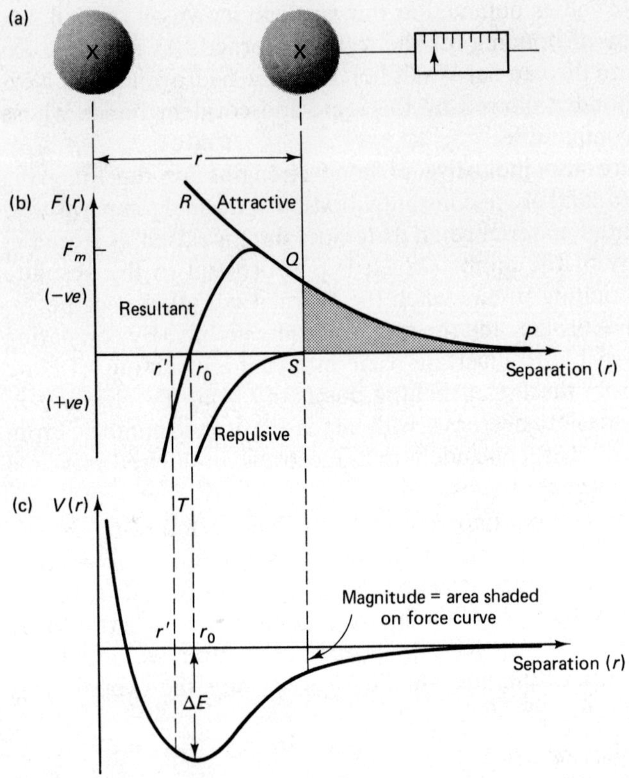

Figure 6.2-4 Interatomic force and potential energy between two atoms.

TABLE 6.2-2 MELTING POINTS

Material	Atomic number	Melting point (°C)
Carbon (diamond)	6	3,750
Silicon	14	1,421
Germanium	32	937
Tin	50	232

plotted in Fig. 6.2-4(b), and the corresponding variation in potential energy is shown in Fig. 6.2-4(c).

From a microscopic point of view, the following properties of matter are brought out by the experiment:

1. When the two lumps of matter are well separated there is no significant force between them. This says that $F(r) \to 0$ as $r \to \infty$.
2. The lumps will only "weld" together when they are in very close contact. This says that as r is reduced the force becomes attractive, as suggested by portion PQ of Fig. 6.2-4(b).

3. Were the curve PQ to continue ever upward along QR, the two atoms would coalesce into a single entity. To prevent this from happening, repulsive forces, as suggested by the curve ST, must come into play. The combined effect of the short-range attractive force followed, at smaller separations, by an even shorter range repulsive force is shown by the heavy line in Fig. 6.2-4(b).

Under appropriate conditions of temperature and pressure all elements can exist in the solid phase. Although the discussion concerns the interaction between two atoms of an element, the arguments are equally true of the interaction between two molecules of a molecular solid provided that the molecules do not interact "chemically" in coming together to form the material. Were this to happen, $F(r)$ would have different forms in the solid and gaseous phases.

The interaction between two atoms may be expressed in terms of potential energy. The energy $dV(r)$ required to displace the movable atom of Fig. 6.2-4(b) through a distance dr is $-F(r)\,dr$; the minus sign arises because the "person" holding the movable atom must provide a force equal in magnitude, but opposite in direction, to the interatomic force $F(r)$. Thus,

$$dV(r) = -F(r)\,dr \tag{6.2-1}$$

and

$$V(r) = -\int_{\infty}^{r} F(r)\,dr = \int_{r}^{\infty} F(r)\,dr \tag{6.2-2}$$

The potential energy of the pair of atoms at a separation r is the area under the $F(r)$ curve between r and infinity. This operation of measuring an area and plotting the result as a point is shown in Figs. 6.2-4(b) and (c). Note that the minimum in $V(r)$ occurs when $F(r) = 0$, the reason being that Eq. (6.2-1) can be rewritten as

$$F(r) = -\frac{dV(r)}{dr} \tag{6.2-3}$$

This says that $F(r) = 0$ at a minimum in $V(r)$ (where $dV(r)/dr = 0$). More informally, the reason is that as the movable atom is brought in from infinity to r_0 all the energy is transferred to the person holding the atom (he gets pulled in). In coming in from infinity to $r' < r_0$ part of the energy gained between infinity and r_0 must be fed back as the person pushes against the repulsive forces existing between r_0 and r'. Thus $V(r)$ is a minimum at $r = r_0$. Continuing in this vein, it is seen that at some separation smaller than r' all the energy acquired from the attractive force will be spent in acting against the repulsive force and here the potential energy will be zero. At still-smaller separations the energy gained from the attractive force will be insufficient to push X against the repulsive force; the person must provide the missing energy, and thus the potential energy will be positive at such separations.

It is often convenient to express the results in mathematical form. The objective is to find some mathematical expression that reproduces the main features of Fig. 6.2-4. Now, $F(r)$ and $V(r)$ are interrelated by Eq. (6.2-3). Since differentiation is simpler to perform than integration it makes more sense to model $V(r)$, knowing that $F(r)$ follows via (6.2-3) rather than to model $F(r)$, leaving $V(r)$ to be obtained by integration. The potential energy curve of Fig. 6.2-4 can be viewed as being the sum of the potential energy curve due to attractive forces (portion PQR) and the potential energy due to repulsive forces (portion ST). This is illustrated in Fig. 6.2-5 where the component potential energies are indicated by the light line and the total potential energy is indicated by the heavy line. The job is to find suitable analytic functions for the two component energies and then to add them.

Perhaps the simplest possible representation is that proposed by Mie in 1907. He represented the negative component by $-A/r^m$ and the positive component by B/r^n. Here A and B are positive constants and m and n are positive integers with $n > m$. (It is necessary to have $n > m$ so that the positive term will dominate as $r \to 0$ and will vanish more rapidly than the negative term as $r \to \infty$.) Adding together the two components gives

$$V(r) = -\frac{A}{r^m} + \frac{B}{r^n} \tag{6.2-4}$$

Another potential is the Lennard–Jones potential, which has the form

$$V(r) = 4\epsilon\left[-\left(\frac{\sigma}{r}\right)^p + \left(\frac{\sigma}{r}\right)^q\right] \tag{6.2-5}$$

where ϵ and σ are positive constants and p and q are positive integers. Like Mie's potential, Lennard–Jones' potential contains two continuously variable constants, ϵ and σ. Unlike Mie's constants A and B, the constant ϵ is solely associated with energy, the constant σ solely with the interatomic separation. The Lennard–Jones potential is employed almost exclusively in modeling van der Waals' interactions, where $p = 6$ and $q = 12$ is assumed. In such a case,

$$V(r) = 4\epsilon\left[-\left(\frac{\sigma}{r}\right)^6 + \left(\frac{\sigma}{r}\right)^{12}\right] \tag{6.2-6}$$

This is known as the Lennard–Jones 6–12 potential.

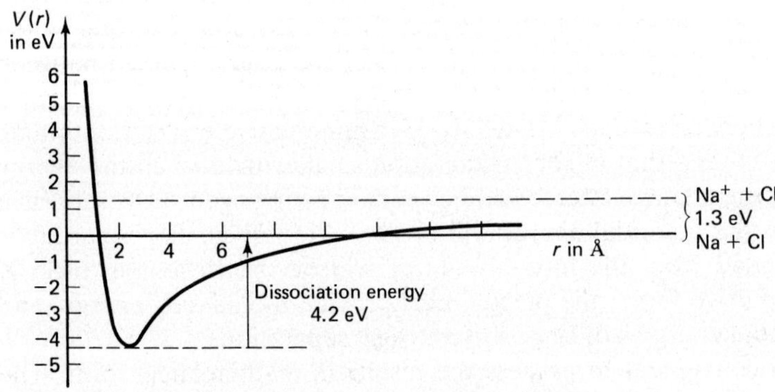

Figure 6.2-5 Mie potential and Lennard–Jones potential.

■ EXAMPLE 6.2-1

The forces that hold crystals together are mainly electrostatic. Ionic binding is caused by Coulomb's law of attraction between the plus and minus ions that make up the ionic crystal. What is the potential energy of two point charges $+e$ and $-e$ separated by 280 pm (as in Na^+Cl^-)?

Solution. The potential energy of two point charges equals $(1/4\pi\epsilon_0)(q_1 q_2/r)$, where $1/4\pi\epsilon_0)$, the Coulomb constant, has the value of 9.0×10^9 N m^2/C^2, and the quantum of charge is $e = 1.6 \times 10^{-19}$ C. Thus,

$$PE = 9.0 \times 10^9 \frac{\text{N}\cdot\text{m}^2}{\text{C}^2} \frac{(-)(1.6 \times 10^{-19}\ \text{C})^2}{280 \times 10^{-12}\ \text{m}} \left(\frac{1\ \text{eV}}{1.6 \times 10^{-19}\ \text{J}}\right)$$

$$= -5.13\ \text{eV}$$

This is a very strong bond. ■

■ EXAMPLE 6.2-2

In the ionic bond, the Coulombic attractive force follows the simple, well-known relationship

$$F_C = -\frac{K}{a^2} = \frac{k_0(Z_1 q)(Z_2 q)}{a^2}$$

where a is the separation distance between the ions, Z_i is the valence of the charge (+1 for Na^+ and −1 for Cl^-), q is the charge of an electron (1.6×10^{-19} C), and k_0 is a proportionality constant (9×10^9 V m/C). The repulsive force is a function of a having the exponential form

$$F_R = \lambda e^{-a/\rho}$$

where λ and ρ are experimentally determined constants for a given ion pair. F_C, F_R, and the net bonding force $F(= F_C + F_R)$ are plotted against a in Fig. 6.2-6(a) for the $Na^+{-}Cl^-$ pair having an equilibrium bond length of $a_0 =$ 0.28 nm. The bonding energy, related to the bonding force through

$$F = \frac{dE}{da}$$

is plotted in Fig. 6.2-6(b). Note that the equilibrium bond length, a_0, corresponding to $F = 0$, also corresponds to a minimum on the energy curve.

(a) Using the ionic radii $r_{Na} = 0.098$ nm, $r_{Cl^-} = 0.181$ nm, calculate the Coulombic attractive force between Na^+ and Cl^- in NaCl.

(b) What is the repulsive force?

Solution

(a) From $a_0 = r_{Na} + r_{Cl^-} = 0.098\ \text{nm} + 0.181\ \text{nm} = 0.278$ nm, the equilibrium bond length is obtained. The Coulombic force is

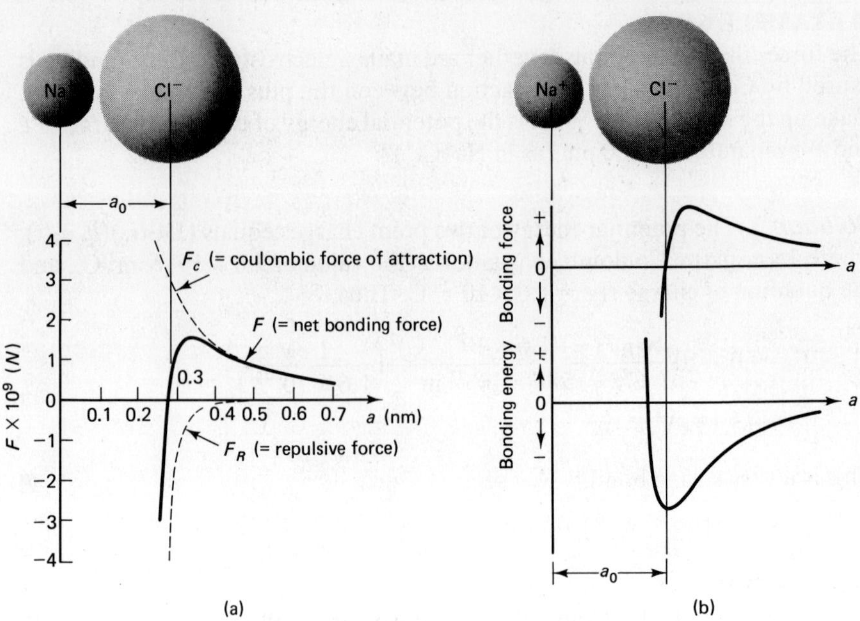

Figure 6.2-6 Bonding force curve and bonding energy curve for Na^+–Cl^- pair.

$$F_C = -\frac{k_0(Z_1 q)(Z_2 q)}{a_0^2}$$

$$= \frac{(9 \times 10^9 \text{ V m/C})(+1)(0.16 \times 10^{-18} \text{ C})(-1)(0.16 \times 10^{-18} \text{ C})}{(0.278 \times 10^{-9} \text{ m})^2}$$

$$= 2.98 \times 10^{-9} \text{ N}$$

(b) At the equilibrium distance, $F = F_C + F_R = 0$, so

$$F_R = -F_C = -2.98 \times 10^{-9} \text{ N}$$

■

■ EXAMPLE 6.2-3

Molecules such as H_2, Cl_2, and NO are bound by the mechanism of covalent bonding. The H_2 molecule can be visualized as two positive charges with two electrons moving in their electromagnetic fields. The most probable location for the electrons is between the hydrogen nuclei. Consider the two protons in the H_2 molecule to be separated by 0.74 Å. How much negative electric charge must be placed at a point midway between the two protons to give the system the observed binding energy of 4.5 eV?

Solution. The binding energy, the energy necessary to remove all particles to infinity, is equal in magnitude to the total Coulomb energy of the charges forming the H_2 molecule. The Coulomb energy consists of a positive contribution from the two protons,

$$E_+ = \frac{ke^2}{r_0} = \frac{14.40 \text{ eV}\cdot\text{Å}}{0.74 \text{ Å}} = 19.5 \text{ eV}$$

and a negative contribution from the unknown negative charge $\delta(-e)$ located midway between the two protons,

$$E_- = \delta(-e)V_{\text{mid}} = -\delta e\left(\frac{ke}{r_0/2} + \frac{ke}{r_0/2}\right) = -2\delta\frac{ke}{r_0/2}$$

$$= -2\delta\frac{14.40 \text{ eV Å}}{(0.74 \text{ Å})/2} = -(77.8 \text{ eV})\delta$$

where $ke^2 = 14.40$ eV · Å (from the Appendix) has been used. The sum gives the binding energy (negative):

$$-\text{BE} = E_+ + E_-$$

$$-4.5 \text{ eV} = 19.5 \text{ eV} - (77.8 \text{ eV})\delta$$

$$\delta = 0.308$$

This number is quite a bit smaller than that actually existing in the H_2 molecule. The two electrons are not located exactly at the mid-point, but can be found anywhere around the two protons. They spend most of their time, however, around the midpoint, thereby producing the attractive force that results in the covalent bonding of the H_2 molecule. ■

6.3 LATTICE STRUCTURE

A crystal may be defined as a three-dimensional pattern in which a structural motif is repeated in such a way that the environment of every motif is the same throughout the crystal. The rich variety of phenomena that present themselves in solids, the differences between solids, depend on the geometrical structure of the unit cell, and the nature and arrangement of the atoms within it. This section outlines the principles by which complex crystal structures are organized into categories with common geometrical properties.

A crystal contains a structural unit, called the basis (or motif), that is repeated in three dimensions to generate the crystal structure. The first step in organizing things is to mentally separate the crystal into unit cells, each with some arrangement of atoms within it. The usual nomenclature is to speak of a lattice, and a basis, or group of atoms that must be placed at each lattice point to form the crystal. The unit cell is the repeating structural unit of the lattice.

Figure 6.3-1 shows a two-dimensional lattice: an array of points (denoted by ●). Associated with each lattice point is a basis: a group of atoms. The whole—the lattice plus its associated basis—is referred to as a structure. The structure is formed by associating with each lattice point a basis consisting of an M atom and a W atom. The W atoms do not lie at lattice points. (Of course, the lattice points could have been chosen to coincide with the W atoms, in which case the M atoms would not lie at lattice points.) In fact, no atom need lie at a lattice point. The basis may, as in copper, for example, consist of a single atom or even, as in crystallized proteins, consist of some tens of thousands of atoms. (Remember that the lattice as such does not exist; it merely helps in the construction of the basis.) Repeating the unit cell throughout space reproduces the lattice.

In practice the lattice is normally specified by giving the form of the primitive cell. This is the smallest cell which, when repeated endlessly throughout space, will fill all space and reproduce the lattice. Figure 6.3-2 shows four possible primitive cells for the lattice. The construction of cells 1, 2, and 3 is self-evident. Cell 4—known as a Wigner–Seitz primitive cell—is constructed by drawing lines to connect a given lattice point to all neighboring lattice points and then bisecting each such line with a perpendicular line. Cell 5 is a nonprimitive cell; it is larger than the minimum-size primitive cell.

By changing the magnitude of the so-called primitive translation vectors **a** and **b** and the magnitude of the angle ϕ in Fig. 6.3-1, an unlimited number of lattices can be generated, all different from one another. However, some are unaltered on rotation through 90° about a lattice point, or on reflection across a plane passing through a lattice point and parallel to **a** or **b**. It turns out that in

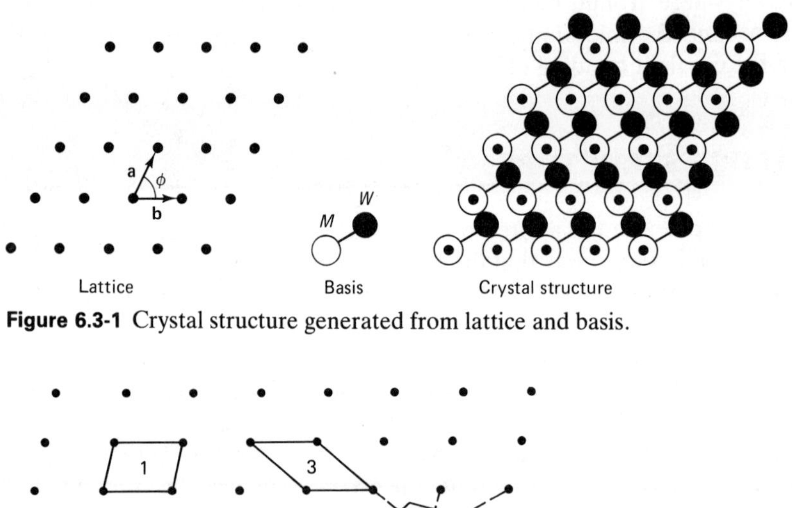

Figure 6.3-1 Crystal structure generated from lattice and basis.

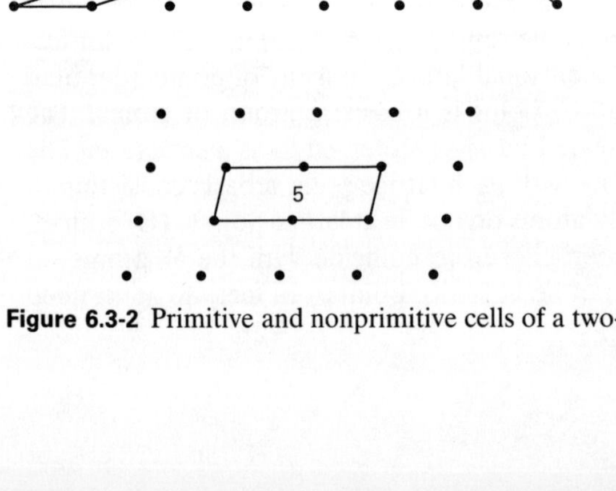

Figure 6.3-2 Primitive and nonprimitive cells of a two-dimensional lattice.

two dimensions there are only five fundamentally different types of lattice as shown in Fig. 6.3-3.

While in two dimensions a unit cell is a parallelogram with sides a and b and angle ϕ between them; in three dimensions a unit cell is a parallelepiped with sides a, b, c and angles α, β, γ, where α is the angle between edges b and c, and so on. In three dimensions, there are 14 different kinds of lattices (called *Bravais* lattices). These are illustrated by the cells, some of which are nonprimitive, in Fig. 6.3-4. As in two dimensions, endless repetition of these cells throughout space generates the lattice. Associating a basis with each lattice point yields the structure of the crystal.

To designate the location of any point in the unit cell, a coordinate system is set up with the origin at one corner of the unit cell and axes coinciding with the a, b, c edges of the cell. Note that these axes are not necessarily mutually perpendicular. The position of a point in the cell is specified by giving its coordinates as fractions of the unit cell lengths a, b, and c. Thus, the point at the origin is $(0, 0, 0)$; the interior lattice point in an I lattice is at $(\frac{1}{2}, \frac{1}{2}, \frac{1}{2})$; the point at the center of the face bounded by the b and c axes is $(0, \frac{1}{2}, \frac{1}{2})$.

The orientation of a crystal plane is described by its Miller indices (hkl), which are obtained by the following steps: (1) find the intercepts of the plane on the a, b, c axes in terms of multiples of the unit cell lengths a, b, c; (2) take the reciprocals of these numbers; (3) if fractions are obtained in step 2, multiply the three numbers by the smallest integer that will give whole numbers; (4) if an intercept is negative, indicate this with a bar over the corresponding Miller index. As an example, the shaded plane in Fig. 6.3-5(a) is determined by the points $(2, 0, 0)$, $(0, 3, 0)$ and $(0, 0, 1)$ and the plane ABC by $(5, 0, 0)$, $(0, 4, 0)$, and $(0, 0, 2)$, where the coordinates are in units of the lattice constants a, b, and c. Step 1 for the plane ABC gives $H = ha = 5$, $K = kb - 4$, and $L = lc - 2$. Step 2 gives $\frac{1}{5}$, $\frac{1}{4}$, and $\frac{1}{2}$. Step 3 then yields $(hkl) = (4\,5\,10)$ for the smallest three integers having the same ratio. These are the Miller indices of

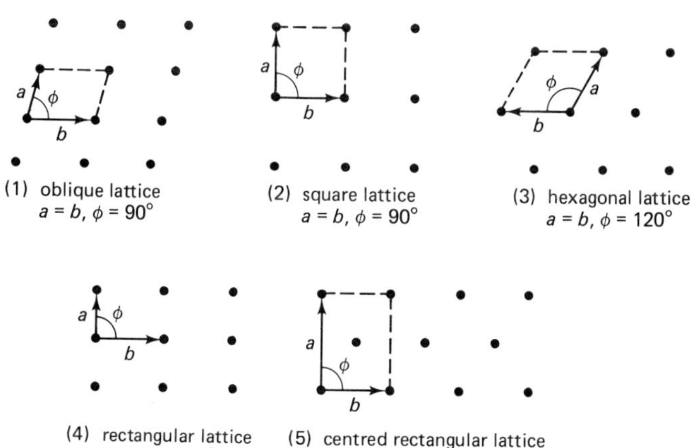

Figure 6.3-3 Fundamental two-dimensional lattices. (The cell shown in (5) is non-primitive.)

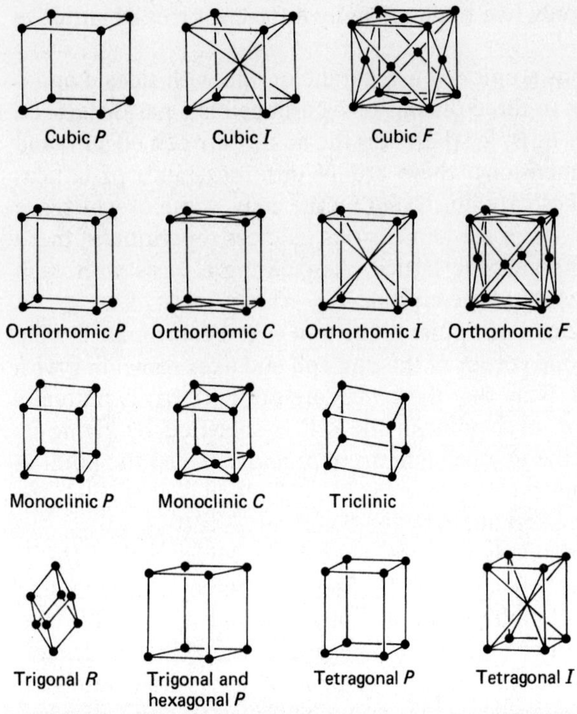

Figure 6.3-4 The fourteen fundamental (Bravais) lattices. (The cells are not always primitive. Letters *P*, *I*, *C*, and *F* are used to denote details of the cell.)

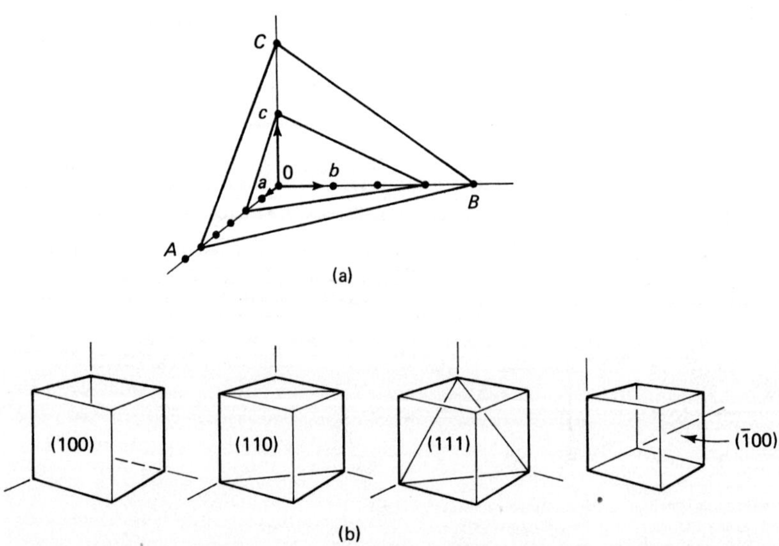

Figure 6.3-5 Miller indices for a plane *ABC* and for planes in a cubic lattice.

the plane ABC as well as parallel planes. Other planes, such as those in a cubic lattice, are shown in Fig. 6.3-5(b). Note that a Miller index of zero means the plane only intercepts a particular axis at infinity. Although the parallel planes $(\bar{1}00)$ and (100) in Fig. 6.3-5 can be described by either of these Miller indices to denote only the orientation of a particular plane, the two sets are required if it is desired to indicate the location of the planes as seen from a particular viewpoint.

■ EXAMPLE 6.3-1

Iron crystallizes in the cubic system. At room temperature, there is an iron atom at each corner of the unit cell, and another atom at the body center of the unit cell, as shown by the schematic view in Fig. 6.3-6(a).

The hard balls model in Fig. 6.3-6(b) shows that every iron atom in the body-centered cubic (bcc) metal structure is surrounded by eight adjacent iron atoms. There are two atoms per bcc unit cell. One atom is at the center of the cube, and eight octants are at the eight corners. Develop an expression for the *lattice constant*, a, and the *packing factor*, PF.

Solution. The materials with a bcc metal structure have atom contact along the body diagonal (b.d.) of the unit cell. Thus, from Fig. 6.3-6(c),

$$(\text{b.d.})_{\text{bcc}} = 4R = a_{\text{bcc}}\sqrt{3}$$

or

$$a_{\text{bcc}} = \frac{4R}{\sqrt{3}} \tag{6.3-1}$$

This is the lattice constant.

The packing factor of a bcc metal is obtained by calculating the volume fraction of the unit cell that is occupied by the atoms:

$$\text{PF} = \frac{\text{Volume of atoms}}{\text{Volume of unit cell}} = \frac{2(4\pi R^3/3)}{a^3} = \frac{2(4\pi R^3/3)}{(4R/\sqrt{3})^3}$$

$$= 0.68 \tag{6.3-2}$$

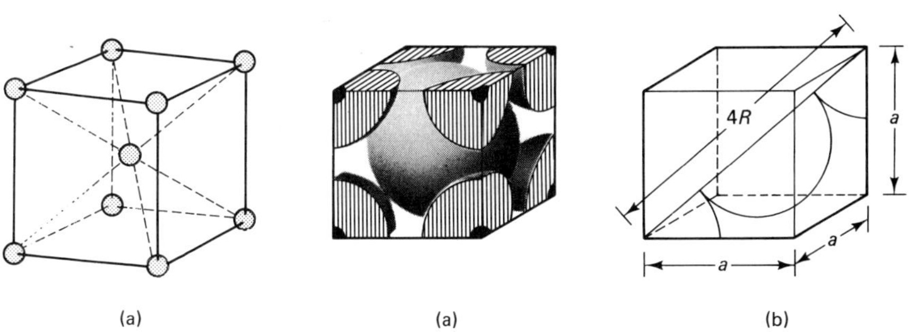

Figure 6.3-6 Body-centered cubic (bcc) metal structure. (a) Location of atom centers. (b) Hardball model. (c) Body diagonal.

Iron is the most common metal with this bcc structure. But it is not the only one. Chromium and tungsten, among other metals, also have bcc structures. The atomic arrangements in copper, however, are not the same as those in iron, although both are cubic. In addition to an atom at the corner of each unit cell of copper, there is one at the center of each face, but none at the center of the cube. This face-centered cubic (fcc) structure is characteristic of aluminum, copper, lead, silver, nickel, and iron (at elevated temperatures). For an fcc metal structure, the lattice constant is

$$a_{\text{fcc metal}} = \frac{4R}{\sqrt{2}} \tag{6.3-3}$$

■

■ EXAMPLE 6.3-2

Many properties of crystals are directional. For instance, the elastic modulus of bcc iron is greater parallel to the body diagonal than it is to the cube edge. Also, planes of atoms influence the properties and behavior of a material. It is therefore advantageous to have ways of identifying various planes and directions within a crystal. State and illustrate some of these ways.

Solution. A direction is labeled by the parallel line that passes through the origin, which is $(0, 0, 0)$. In Fig. 6.3-7(a), the [111] direction passes from $(0, 0, 0)$ through $(1, 1, 1)$. This direction also passes through $(\frac{1}{2}, \frac{1}{2}, \frac{1}{2})$ and $(2, 2, 2)$. Likewise, [112] passes through $(\frac{1}{2}, \frac{1}{2}, 1)$; but for simplicity's sake, the integer notation is used. The direction indices are enclosed in square brackets $[uvw]$, using the letters u, v, and w. A negative coefficient is designated with an overbar, such as $[11\bar{1}]$. There are no commas between the indices. Parallel directions always have the same indices.

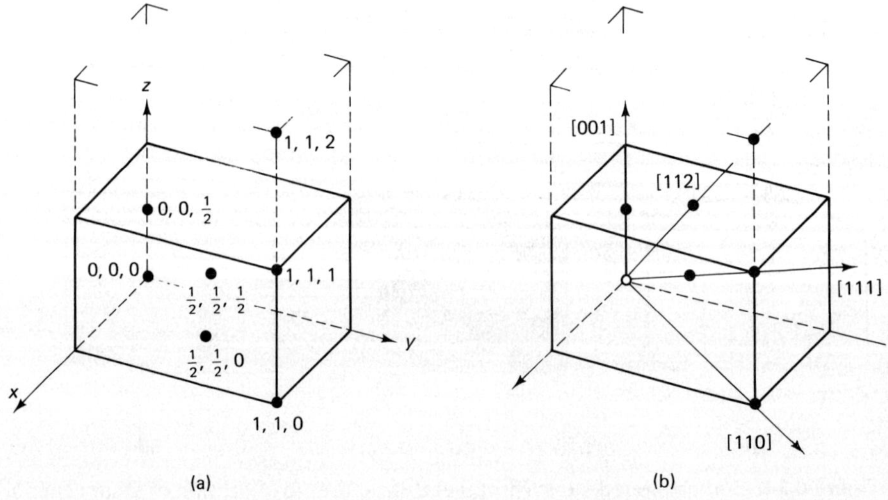

Figure 6.3-7 Orthorhombic unit cell. (a) Indices of points. (b) Indices of directions.

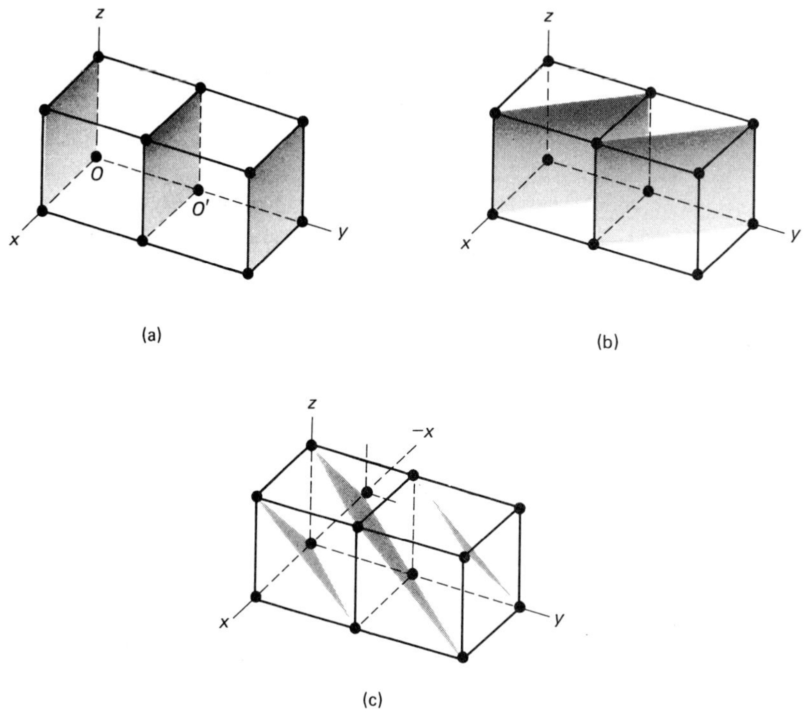

Figure 6.3-8 (010), (110), and (111) planes in cubic structures.

Lattice planes most readily visualized are those that outline the unit cell. However, the more important planes for practical purposes are those sketched in Fig. 6.3-8. These are labeled (010), (110), and $(\bar{1}11)$, respectively, where the numbers within the parentheses (hkl) are the Miller indices. More precisely, Miller indices are the reciprocals of the three intercepts that a plane makes with the axes, cleared of fractions and common multipliers. All parallel planes bear the same (hkl) notation. A perpendicular distance from the origin between adjacent planes measures the interplanar distance, d. For cubic crystals, the general rule for d spacings is

$$d_{hkl} = \frac{a}{\sqrt{h^2 + k^2 + l^2}} \tag{6.3-4}$$

where a is the lattice constant and h, k, and l are the indices of the planes. ■

■ EXAMPLE 6.3-3

Excellent experimental verification for the crystal structure of solids is available through X-ray diffraction. High-frequency electromagnetic waves with wavelengths slightly greater than the interplanar spacings of crystals are deflected according to definite physical laws. Derive Bragg's law, and show how it can be used to decipher crystal structures.

Solution. Diffraction is the result of radiation being scattered by a regular array of scattering centers whose spacing is about the same as the wavelength of the radiation. For X rays, atoms are the scattering centers (the specific mechanism being the interaction of a photon of electromagnetic radiation with an orbital electron in the atom). Repeated stacking of crystal planes serve the same purpose as the parallel scratch lines on a diffraction grating. For a simple crystal lattice, the condition for diffraction is shown in Fig. 6.3-9.

Diffraction occurs when X-ray beams scattered off adjacent crystal planes are in phase. The precise geometry for constructive interference is that the difference in path between adjacent beams is some integral number (n) of radiation wavelengths (λ). The waves may be reflected from the atom of H or H' and remain in phase at K. Also, X rays are reflected not only from the surface plane, but from adjacent subsurface planes. If these reflections are to remain in phase and be coherent, the distance $MH''P$ must be equal to one or more integer wavelengths of the rays. The value of n in the equation

$$n\lambda = \lambda d \sin\theta \tag{6.3-5}$$

is the integer number of waves that occur in the distance $MH''P$. This condition is known as *Bragg's equation*; d is the spacing between adjacent crystal planes, and 2θ is referred to as the *diffraction angle*.

Most experiments in X-ray diffraction use very fine powder of the material to be analyzed. This powder is mixed with a plastic cement and is formed into a very thin filament that is placed at the center of a circular camera. A collimated beam of X rays is directed at the powder as shown in Fig. 6.3-10, and an exposure of diffraction patterns taken.

The angle 2θ from the straight-through exit port is fixed by the lattice spacing d and the wavelength as given by Eq. (6.3-5). Every cone of reflection is recorded in two places on the strip of film. Crystal structure and lattice constants may thus be calculated from patterns such as those shown in Fig. 6.3-10. For example, what is the distance between planes if the first-order diffraction

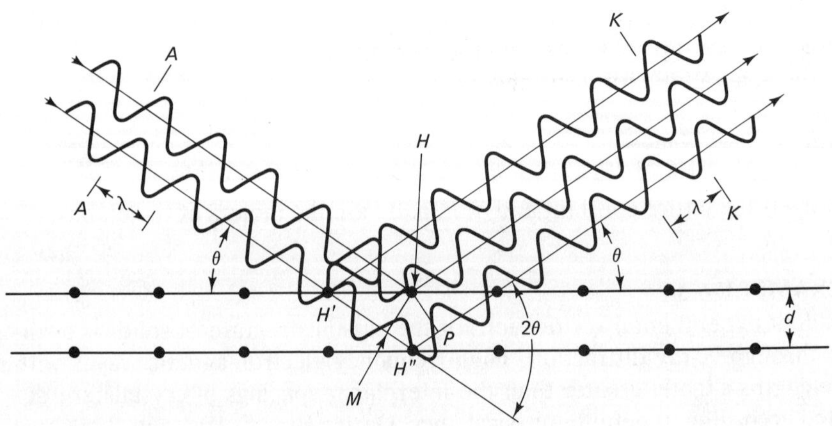

Figure 6.3-9 Geometry for Bragg diffraction.

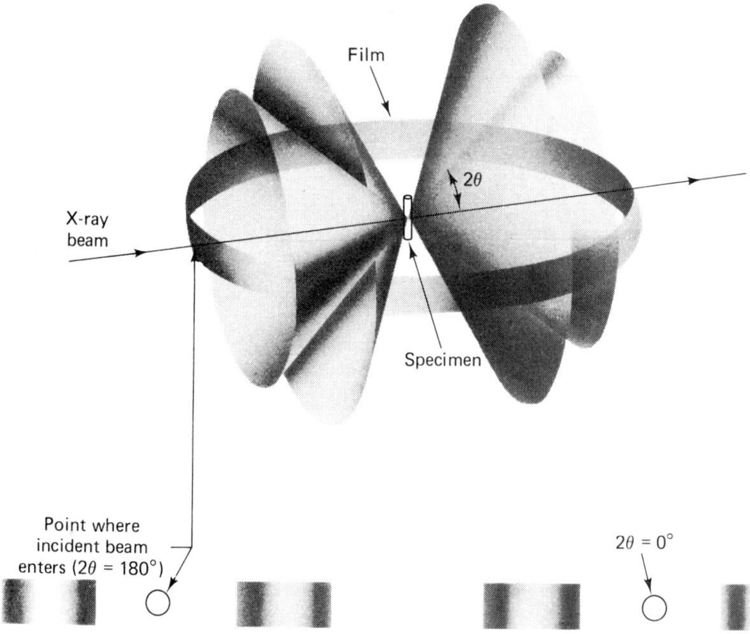

Figure 6.3-10 Exposure of X-ray diffraction patterns. (Each cone of reflection is recorded in two places on the strip of film.)

maximum occurs at $\theta = 15°$ for 150-pm X rays? From Bragg's equation, $d = n\lambda/(2 \sin\theta)$. Now, $n = 1$ for the first order, $\lambda = 150$ pm, and $\sin 15° = 0.2588$. Thus, $d = 1(150 \text{ pm})/(2 \times 0.2588) = 290$ pm. ■

■ EXAMPLE 6.3-4

All physical properties of a crystal are related to the nature and arrangement of the atoms comprising the crystal. Once the symmetry of the crystal structure is known, the symmetry of its physical properties is also known. Take for instance, the elastic behavior of solids. Elasticity may be defined as the response of a material to an applied force. An elastic change specifically refers to a change that exists only during the application of the force. The equilibrium arrangement of atoms in a crystal corresponds to a balance of the repulsive and attractive electromagnetic forces on each atom exerted by its neighbors. When the crystal is subjected to an external force, the internal electromagnetic force balance is disturbed, the atoms change their positions in response, and a resulting dimensional change of the crystal occurs. If the displacements from equilibrium are small, the atoms should experience restoring forces proportional to their displacements.

Consider a uniform solid bar of cross-sectional area A and length L in its equilibrium state. If a force F is applied to the ends of the bar normal to a cross section, as shown in Fig. 6.3-11, the length changes by an amount ΔL; if ΔL is

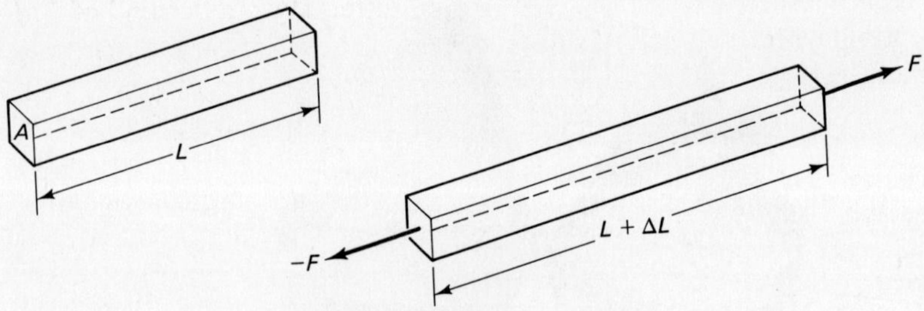

Figure 6.3-11 Strain in a metal bar under stress.

positive the bar is under tension, and if ΔL is negative the bar is under compression. The stress σ in the bar is defined as the force per unit area; that is,

$$\sigma = \frac{F}{A} \tag{6.3-6}$$

The strain ϵ resulting from the application of stress σ is defined as the relative change in length of the bar; that is,

$$\epsilon = \frac{\Delta L}{L} \tag{6.3-7}$$

Experiments show that for sufficiently small stresses the strain is proportional to the stress. Therefore,

$$\sigma = Y\epsilon \tag{6.3-8}$$

where the constant Y is called Young's modulus. This equation is known as Hooke's law. Substituting for the stress and strain gives

$$\frac{F}{A} = Y\frac{\Delta L}{L} \tag{6.3-9}$$

(a) Show that Young's modulus for the bar is related to the force constant for the constituent atoms by

$$Y = \frac{k_A}{d}$$

where k_A is the force constant of the atom, and d is the spacing of the atoms.

(b) Given that the speed of sound c in a copper bar is 3.5×10^3 m s^{-1} and that the density ρ of copper is 9.0×10^3 kg m^{-3}, estimate the atomic force constant of the copper atoms.

Solution

(a) Suppose that the bar is isotropic and composed of atoms spaced a distance d apart when no force is applied. Since the length of the bar is initially L, there are L/d atoms along the length of the bar. Also, the number of atoms in a cross-sectional area is A/d^2. Therefore, when a force of magnitude F is applied uniformly over an area A at each end of the bar, each atom on the end surface feels a force Fd^2/A, which is then transmitted along a "chain" of L/d atoms, each of which is displaced a distance x from equilibrium. The change in length must be

$$\Delta L = \left(\frac{L}{d}\right)x \tag{6.3-10}$$

Assuming the force on each atom to be proportional to the displacement,

$$\frac{Fd^2}{A} = k_A x = k_A \, \Delta L \frac{d}{L} \tag{6.3-11}$$

where k_A is the force constant of the atom. Therefore,

$$\begin{aligned} k_A &= \left(\frac{F}{A}\right)\left(\frac{L}{\Delta L}\right) d \\ &= Yd \end{aligned}$$

so that

$$Y = \frac{k_A}{d} \tag{6.3-12}$$

(b) The speed of sound in a solid bar is related to Young's modulus for the material by

$$c = \left(\frac{Y}{\rho}\right)^{1/2} \tag{6.3-13}$$

Young's modulus for the copper bar is then

$$\begin{aligned} Y &= \rho c^2 \\ &= 9.0 \times 10^3 \times (3.5)^2 \times 10^6 \\ &= 1.1 \times 10^{11} \text{ N m}^{-2} \end{aligned} \tag{6.3-14}$$

Now, to estimate the distance between the copper atoms, recall that there are 6.0×10^{26} atoms (Avogadro's number) in every kilogram-mole of the solid. (That is, for copper there are $N = 6.0 \times 10^{26}$ atoms in every 64 kg.) The volume V occupied by the amount of copper is

$$V = \frac{64}{9.0 \times 10^3} \text{ m}^3$$

An estimate of the distance between the atoms is

$$\begin{aligned} d &= \left(\frac{V}{N}\right)^{1/3} \\ &= \left(\frac{64}{9.0 \times 10^3} \times \frac{1}{6.0 \times 10^{26}}\right)^{1/3} \\ &= 2.3 \times 10^{-10} \text{ m} \end{aligned} \tag{6.3-15}$$

The atomic force constant of the copper atoms is thus

$$\begin{aligned} k_A &= Yd \\ &= 1.1 \times 10^{11} \times 2.3 \times 10^{-10} \\ &= 25 \text{ N m}^{-1} \end{aligned}$$

■

6.4 STATISTICAL MECHANICS OF CRYSTALLINE SOLIDS

The solid state, like the gaseous state, lends itself to quantitative application of statistical thermodynamics, because its structure can be fairly accurately represented by an idealized model. Indeed, some of the results have already been presented in Chapter 4. In this and subsequent sections, they will be reviewed and extended to include insulators, conductors, and semiconductors.

A crystalline solid can be regarded as a single giant molecule with a set of allowed quantum-mechanical states having energies E_j. This description is a natural one for metallic, covalent, and ionic crystals. Also, a molecular crystal is taken as a single giant molecule containing weak bonds due to van der Waals forces and strong bonds due to ordinary chemical bonding. All the thermodynamic properties of the crystal are derivable from the crystal's canonical partition function $Z = \Sigma e^{-E_j/kT}$.

Consider first a covalent or metallic crystal or an ionic crystal composed of monatomic ions. Let there be N atoms or ions present. (For 1 mol of diamond, N is Avogadro's number. For 1 mol of NaCl, N is twice Avogadro's number.) The giant molecule has $3N - 6$ vibrational modes of motion (degrees of freedom). There are also three translational and three rotational degrees of freedom of the crystal as a rigid body; these modes of motion can be activated by moving the crystal around the laboratory. Since energy of bulk motion is not included in the thermodynamic internal energy U, there is no need to consider these six modes of motion.

In addition to the $3N - 6$ vibrational mode energies, there is electronic energy. Most solids are insulators or semiconductors and have a significant energy gap between the highest occupied and lowest vacant electronic energy levels. Provided T is not extremely high, the electronic energy for these solids can be ignored. For metals, excitation of electrons to higher levels occurs at all temperatures and must be allowed for. For now, the contribution of electronic energy to E_j for metals will be ignored; this contribution will be added on at the end of the calculation. Thus, only vibrational energy is considered in finding the energies E_j in the crystal's canonical partition function $Z = \Sigma e^{-E_j/kT}$.

An oversimplified picture of a crystal at temperature T was taken by Einstein (1907) who assumed that all $3N$ lattice vibrations have the same fundamental frequency ν_0. For each oscillator, the partition function is

$$z = \sum_{n=0}^{\infty} e^{-\epsilon_n/kT} \tag{6.4-1}$$

where, according to quantum mechanicals,

$$\epsilon_n = (n + \tfrac{1}{2}) h\nu_0 \qquad n = 0, 1, 2, \ldots \tag{6.4-2}$$

In this case, the energy origin is chosen at the bottom of the harmonic potential well. The oscillator partition function (6.4-1) has previously been evaluated, the result being

$$z = e^{-h\nu_0/2kT}(1 - e^{-h\nu_0/kT})^{-1} \tag{6.4-3}$$

The molar partition function for lattice vibrations is given by a product of contributions from each mode. In this case there are $3N$ equal factors, so that

$$Z = z^{3N} = e^{-3Nh\nu_0/2kT}(1 - e^{-h\nu_0/kT})^{-3N} \tag{6.4-4}$$

The internal energy of the crystal is evaluated, using

$$E = kT^2 \frac{\partial \ln Z}{\partial T} \tag{6.4-5}$$

First,

$$\ln Z = -\frac{3Nh\nu_0}{2kT} - 3N \ln(1 - e^{h\nu_0/kT}) \tag{6.4-6}$$

and

$$-\frac{\partial \ln Z}{\partial T} = \frac{3Nh\nu_0}{2kT^2} - \frac{3N(-e^{h\nu_0/kT})(h\nu_0/kT^2)}{1 - e^{-h\nu_0/kT}} \tag{6.4-7}$$

Thus,

$$E = \frac{3Nh\nu_0}{2} + \frac{3Nh\nu_0 e^{-h\nu_0/kT}}{1 - e^{-h\nu_0/kT}} \tag{6.4-8}$$

or

$$E = E_0 + \frac{3Nh\nu_0}{e^{h\nu_0 kT} - 1} \tag{6.4-9}$$

where the first term represents the "zero-point energy" of the crystal, a sum of $3N$ equal contributions $\frac{1}{2}h\nu_0$. E_0 is independent of temperature and consequently does not contribute to C_V. Now

$$C_V = \frac{\partial E}{\partial T} = \frac{3Nh\nu_0 e^{h\nu_0/kT}(h\nu_0/kT^2)}{(e^{h\nu_0/kT} - 1)^2} \tag{6.4-10}$$

In terms of the parameter

$$\Theta_E \equiv \frac{h\nu_0}{k} \tag{6.4-11}$$

known as the Einstein characteristic temperature, an "equation of state" for heat capacity can be expressed in the following dimensionless form:

$$\frac{C_V}{Nk} = \frac{3(\Theta_E/T)^2 e^{\Theta_E/T}}{(e^{\Theta_E/T} - 1)^2} \tag{6.4-12}$$

This is in line with the fact that for a majority of crystalline solids made up of atoms or monatomic ions, the temperature dependence of the gram-atomic heat capacity takes the form

$$\frac{\overline{C}_V}{R} = f\left(\frac{T}{\Theta_D}\right) \tag{6.4-13}$$

where $f(T/\Theta_D)$ represents a universal function for all monatomic crystals. The parameter Θ_D, the Debye characteristic temperature, must be appropriately chosen for each substance. In accordance with (6.4-13), different crystals at corresponding temperatures—the same multiple of their respective Θ_D—have the same heat capacity per gram-atom. For most solids at 1 atm, the heat capacities C_V and C_P are practically the same.

The Einstein theory thus leads to a law of corresponding states for heat capacity, in agreement with (6.4-13). Figure 6.4-1 shows the Einstein and Debye (essentially experimental) heat capacity curves for equal values of the characteristic temperatures, $\Theta_E = \Theta_D$. The functional form of $f(T/\Theta)$ is not accurately predicted by the Einstein model. In particular, the limiting dependence near absolute zero is incorrect, the calculated heat capacity falling off more sharply than it should.

The Einstein theory does predict correctly the $T = 0$ and $T = \infty$ limits of C_V. As $T \to 0$, $\Theta_E/T \to \infty$ and $(e^{\Theta_E/T} - 1)^2 \to e^{2\Theta_E/T}$. Thus, $\overline{C}_V/T \to 3(\Theta_E/T)^2 \times e^{\Theta_E/T}$, which goes to zero in the limit $T \to 0$. As $T \to \infty$, $\Theta_E/T \to 0$, $e^{\Theta_E/T} \to 1$, while $(E^{\Theta_E/t} - 1) \to \Theta_E/T$. Then, $\overline{C}_V/R \to 3$, in agreement with the equipartition theorem.

The entropy of a perfect crystal according to the Einstein model is found from

$$S = \frac{E}{T} + k \ln Z = 3Nk\left[\frac{h\nu_0/kT}{e^{h\nu_0/kT} - 1} - \ln(1 - e^{-h\nu_0/kT})\right] \tag{6.4-14}$$

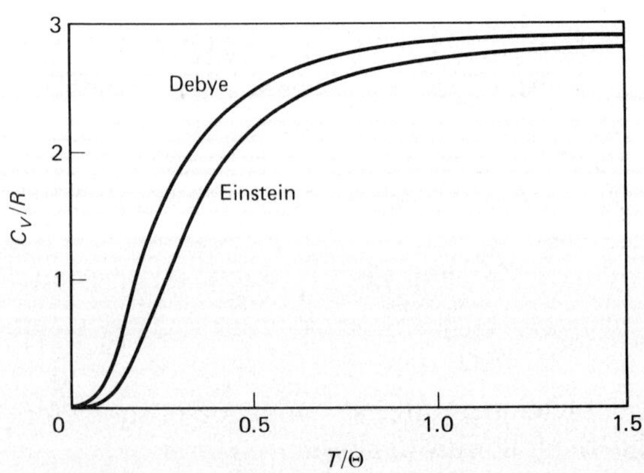

Figure 6.4-1 Einstein and Debye specific heat curves for $\theta_E = \theta_D$.

or, in terms of Θ_E,

$$S = 3Nk\left[\frac{\Theta_E/T}{e^{\Theta_e/T} - 1} - \ln(1 - e^{-\Theta_E/T})\right] \tag{6.4-15}$$

As $T \to 0$, $S \to 0$, which shows that the lattice modes of perfect crystals conform to Planck's statement of the third law of thermodynamics.

To apply the Einstein equation, a single measured value of $\overline{C}_V$ is used to evaluate the Einstein vibrational frequency ν_E (and the Einstein characteristic temperature Θ_E). Equation (6.4-12) then shows fairly good overall agreement with experiment. However, at very low T (below about 40 K), the agreement is poor. This is because of the very crude assumption that all normal mode vibrations of the crystal have the same frequency. Θ_E values are 200 to 300 K for most elements, so most elements have nearly reached their high-temperature C_V values at room temperature. Some values of Θ_E are 67 K for Pb, 240 K for Al, and 1220 K for diamond.

In 1912, Debye presented a treatment that gives much better agreement with C_V values at low temperatures than the Einstein theory. The large number of vibrational modes ($\approx 10^{24}$) makes it possible to treat the crystal's vibrational frequencies as continuously distributed over a range from a very low frequency (which can be taken as essentially zero) to a maximum frequency ν_m. Recall that originally, the crystal's canonical partition function is a product of normal modes:

$$Z = e^{-U_0/kT} \prod_{i=1}^{3N} \frac{1}{1 - e^{-h\nu_i/kT}} \tag{6.4-16}$$

or

$$\ln Z = \frac{U_0}{kT} \sum_{i=1}^{3N} \ln(1 - e^{-h\nu_i/kT}) \tag{6.4-17}$$

From ln Z, all thermodynamic properties can then be readily calculated.

Let $g(\nu)\,d\nu$ be the number of vibrational frequencies in the range between ν and $\nu + d\nu$. Changing the summation in (6.4-17) to a sum over vibrational frequencies (rather than over normal modes), gives

$$\ln Z = -\frac{U_0}{kT} - \sum_{\nu} \ln(1 - e^{-h\nu/kT}) g(\nu)\,d\nu \tag{6.4-18}$$

or, going over to an integral,

$$\ln Z = -\frac{U_0}{kT} \int_0^{\nu_m} \ln(1 - e^{-h\nu/kT}) g(\nu)\,d\nu \tag{6.4-19}$$

Debye assumed that for frequencies in the range 0 to ν_m the distribution function $g(\nu)$ is the same as the distribution function of the elastic vibrations of a continuous solid calculated ignoring the solid's atomic structure. This latter quantity is

$$g(\nu) = 12\pi V \nu_s^{-3} \nu^2 \tag{6.4-20}$$

where v_s is essentially the average speed of sound waves in the solid. The total number of vibrational frequencies is $3N$, so

$$\int_0^{\nu_m} g(\nu)\,d\nu = 3N \tag{6.4-21}$$

Together, Eqs. (6.4-20) and (6.4-21) give

$$\nu_m = \left(\frac{3N v_s^3}{4\pi V}\right)^{1/3} \tag{6.4-22}$$

Using this to eliminate v_s from (6.4-20) yields

$$g(\nu) = 9N\nu_m^{-3}\nu^2 \qquad 0 \le \nu \le \nu_m \tag{6.4-23}$$

For $\nu > \nu_m$, the Debye $g(\nu)$ is 0.

Substitution of (6.4-23) in (6.4-19) gives the Debye partition function Z. The use of $U = kT^2(\partial \ln z/\partial T)_{V,N}$ and $C_V = (\partial U/\partial T)_V$ then gives

$$C_V = 9Nk\left(\frac{T}{\Theta_D}\right)^3 \int_0^{\Theta_D/T} \frac{x^4 e^x}{(e^x-1)^2}\,dx \tag{6.4-24}$$

where x is a dummy variable, and Θ_D is the Debye characteristic temperature, defined as

$$\Theta_D \equiv \frac{h\nu_m}{k} \tag{6.4-25}$$

The integral in (6.4-24) cannot be expressed in terms of elementary functions but can be evaluated numerically by expanding the exponentials in infinite series and then integrating. At high T, the upper limit of the integral in (6.4-24) is close to zero, so the integration variable x is small. Hence, $e^x \approx 1$, and $e^x - 1 = 1 + x + \cdots - 1 \approx x$. The integrand becomes x^2, and integration gives $C_V = 3Nk$, as found by the Einstein theory. At low T, the upper limit in (6.4-24) can be taken as infinity. A table of definite integrals gives

$$\int_0^\infty \frac{x^4 e^x}{(e^x-1)^2\,dx} = \frac{4\pi^4}{15}$$

so the low T limit is

$$C_V = \frac{12\pi^4 Nk}{5}\left(\frac{T}{\Theta_D}\right)^3 \propto T^3; \qquad \text{low } T \tag{6.4-26}$$

This is the *Debye T^3 law*. It agrees well with experimental data, and is used to extrapolate heat capacity data to absolute zero to evaluate entropies.

The Debye theory works quite well at all temperatures but is far from perfect. The value of Θ_D for a given solid found by fitting C_V at one temperature differs somewhat from the value obtained using C_V at another temperature. Some Θ_D values (found from low-temperature C_V data) are 160 K for Na, 343 K for Cu, and 2230 K for diamond. The Debye quadratic expression (6.4-23) with a cutoff at ν_m turns out to be a rather crude approximation to the actual $g(\nu)$ as seen in Fig. 6.4-2. At low frequencies, the Debye $g(\nu)$

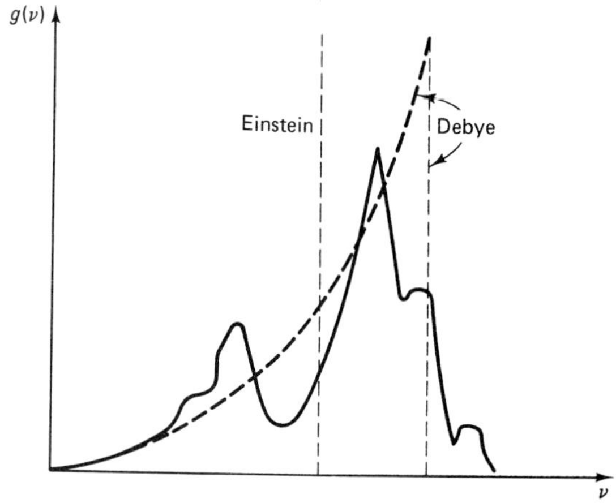

Figure 6.4-2 Einstein and Debye models for the spectral density function. (Typical vibrational distribution function is shown by the solid curve.)

coincides with the true $g(\nu)$ curve. At T near zero, it is the low-frequency modes that are excited and that determine C_V, so the Debye theory gives the correct low T behavior of C_V.

6.5 BAND THEORY OF SOLIDS

Most properties of metals can be explained with the aid of the free-electron model, but when it comes to insulators and semiconductors, the theory fails. This is not surprising, because the term "free electron," by definition, means an electron free to roam around and conduct electricity. On the other hand, the main job of an insulator is *not* to conduct electricity, so it is only sufficient to imagine that the valence electrons cling to their respective lattice ions and are unwilling to move away. Basic theory is therefore successful at the two extremes: free electrons denote high conductivity; tightly bound electrons mean no conductivity. Now, what about semiconductors? They are neither good conductors, nor insulators, so neither model really applies. The theory that has gained wide popularity is the one based on the concept of energy bands.

A crystalline solid can be regarded as a single giant molecule. Before discussing the effect of putting many atoms together in a solid, it is first recalled that quantum mechanics enables the energy levels in atoms to be predicted and calculated. Moreover, if the angular dependences of probability distribution functions in three dimensions are plotted, a fairly realistic representation of $|\psi|^2\,dx\,dy\,dz$ for the hydrogen atom can be obtained as a shown in Fig. 6.5-1. The ground state, a so-called s orbital, is spherically symmetric, whereas the shown excited state (p orbitals) has three equivalent distributions for each coordinate direction.

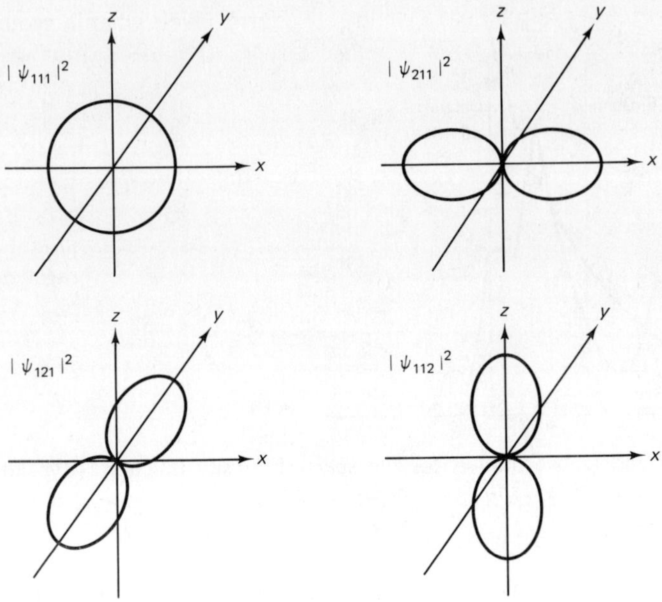

Figure 6.5-1 Electron probability distributions for the ground state (1s) and excited state (2p) for the hydrogen atom.

The principal predictions resulting from the wave-mechanical treatment of electrons bound n atoms can be summarized as follows: (1) Electrons can only exist in certain specific states, termed energy levels, corresponding to specific energy values, each governed by a certain set of quantum numbers. (2) It is possible for a given energy level to be degenerate (that is, more than one electron probability distribution may be associated with the same energy). (3) The actual distribution of electron charge in space is found by plotting the square of the amplitude of the wave function in three dimensions.

As was seen in the discussion of the three-dimensional box problem in Section 2.5, most energy levels are degenerate except for some specific cases. In fact, the introduction of the fourth quantum number s means that every level must be at least twofold degenerate, since s can have the two values $\pm\frac{1}{2}$. All atomic energy levels, therefore, consist of two or more overlapping energy states, and some of the higher levels (with large values of n) must be highly degenerate.

Now, it is possible to "remove" the degeneracy of an energy level by splitting it into a set of closely separated sublevels. This is achieved by applying some external agency to the atom, such as a magnetic or an electric field, which has the effect of depressing some and raising other degenerate levels within a given electronic state. (Examples are the Zeeman effect, 1896, and the Stark effect, 1913.) The removal of degeneracy as a result of the application of external agencies is the key to the understanding of energy bands in solids.

Consider the changes occurring in electronic energy levels when a group of identical atoms with initially wide physical separations are gradually brought closer and closer together to form a solid (that is, when condensation from the gaseous to the solid phase takes place). At the outset the energy levels of any given atom are identical to those in the free state, but as the spacing is gradually reduced a stage is reached at which the outer valence electrons begin to interact with one another. The external fields produced by the electrons of neighboring atoms cause the valence levels to split in a manner similar to that observed in the Zeeman or Stark effects. As the interatomic spacing is still further reduced, the inner electronic levels also start to split. An energy level diagram as a function of separation is then as shown in Fig. 6.5-2. In (a), the characteristic spacing, d, of atoms in a crystal, is such that there are bands of energies allowed to electrons separated by bands of forbidden energy. In (b), d is such that the uppermost bands overlap.

The lower-lying energy levels are spread less than those that lie higher, because the electrons in lower levels are electrons in inner subshells of the atoms, which are not significantly influenced by the presence of nearby atoms. These electrons are localized in particular atoms, even when the separation distance r is small, because the potential barriers between the atoms are for them relatively high and wide. The valence electrons, on the other hand, are not localized at all for small r, but they become part of the whole system. The overlapping of their wave functions results in a spreading of their energy levels.

It is worthwhile to point out that the 1s level in an individual atom becomes a band of N levels, as does the 2s level, when counting in such a way that each of these can accommodate two electrons of opposite spin. But the 2p

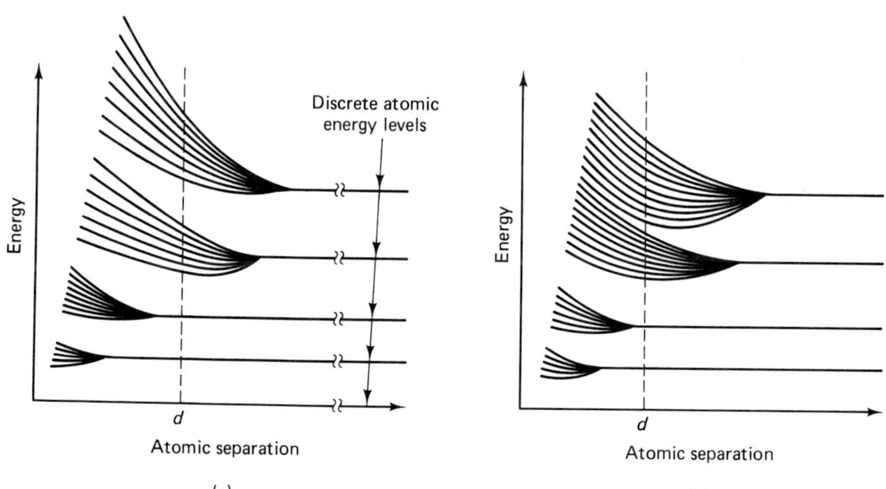

Figure 6.5-2 Splitting of discrete energy levels into bands of allowed energies in a crystal.

level is triply degenerate in the space quantum number m_l in the isolated atom, since m_l can assume any of the values $-1, 0, 1$. Thus, the $2p$ level in the atom leads to $3N$ levels in the solid.

Consider Na(c) as an example. The electron configuration of an Na atom is $1s^2 2s^2 2p^6 3s$. Let the crystal contain N atoms, where N is on the order of 10^{26}. The crystal or molecular orbitals (MOs) are constructed as linear combinations of the atomic orbitals (AOs) of the N sodium atoms in the manner shown in Fig. 6.5-3(a). In the Na(c) solids, which is being regarded as a single Na_N molecule, the N $1s$ AOs of the isolated atoms form N crystal orbitals with energies identical to that of a sodium $1s$ AO. These N $1s$ crystal orbitals have a capacity of $2N$ electrons and hold the $2N$ $1s$ electrons of the atoms. Similarly, the inner-shell $2s$ and $2p$ AOs form crystal orbitals with energies from the AO energies, and these orbitals are filled in Na_N, and so on.

The $3s$ band contains N orbitals and so has a capacity of $2N$ electrons. Since there are N $3s$ electrons in the N sodium atoms, the $3s$ band is only half-filled. Hence, there are many vacant electronic energy levels in Na(c) that lie a negligible distance above the highest occupied electronic level. This makes Na(c) an excellent conductor of electricity, since electrons are easily excited to higher energy levels by an applied electric field.

In Fig. 6.5-3(b), the Mg atom has the electron configuration $1s^2 2s^2 2p^6 3s^2$. The $3s$ band in Mg(c) is exactly filled by the $2N$ electrons of the N magnesium atoms. The $3p$ band (formed by unoccupied $3p$ AOs of the Mg atoms) overlaps the $3s$ band in Mg(c). There is no energy gap between the highest occupied and lowest vacant electronic energy levels, and Mg is an excellent conductor of electricity.

Consider now Ne(c). The Ne electron configuration is $1s^2 2s^2 sp^6$. The $1s$ and $2s$ electrons give bands of narrow width. The interaction between the $2p$ AOs is not great (van der Waals forces are rather weak), and the $2p$ band is narrow. Above the $2p$ band of Fig. 6.5-3(c) lies a fairly narrow $3s$ band. Because the $2p$ and $3s$ bands are narrow, and because the $2p$ and $3s$ levels in an Ne atom are very widely separated, a substantial energy gap (several

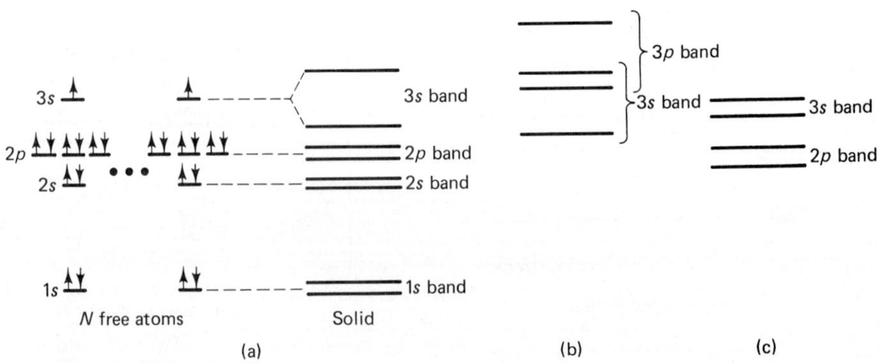

Figure 6.5-3 (a) Construction of crystal orbitals in sodium. (b) Energy bands in magnesium. (c) In Ne.

electron-volts) occurs between the top of the filled $2p$ band in Ne(c) and the bottom of the vacant $3s$ band. It takes a great deal of energy to excite electrons from the $2p$ band to the $3s$ band, and Ne(c) is therefore a poor conductor of electricity (an insulator).

A rough estimate of the average level separation within a band may be obtained as follows. A crystal weighing 1 kg has approximately 10^{26} atoms and bands typically 1.0 eV wide. There are thus 10^{26} levels spread over 1.0 eV, so the spacing is less than 10^{-26} eV. The energy levels within a band are therefore discrete, but of such small separation that the band can be taken as continuous to the first approximation.

A question arises of how the electrons fit into the levels available within a band. As with isolated atoms, the Pauli principle, which states that no two electrons can have the same set of four quantum numbers, can be applied. In a band, the energy states are not coincident, but discrete and separate, available and one electron can be assigned to each of them. The lowest level is filled first, then the next highest, and so on until all the available electrons are used. The Pauli principle therefore is still adhered to within a band.

Summarizing, the band theory of solids predicts the existence of a series of energy bands of finite width E_b, separated from one another by certain energy gaps. Of major importance is the interval between the top of the valence and the bottom of the conduction bands, which is called the forbidden energy gap and denoted by E_g. Neglecting variations in band width and shape and the lower-lying states, the relationship between the free atom and the atom in a solid is represented as in Fig. 6.5-4.

The general classification of solids according to their particular energy band arrangement is given in Fig. 6.5-5, from which it can be seen that two broad divisions exist: solids in which the valence band is full and separated from the conduction band by a definite energy gap, and those in which the valence band is only partially filled or overlapping a partially filled conduction band. To the former group belong insulators and semiconductors, and to the latter group metals. The feature distinguishing insulators and semiconductors is the width of the energy gap. In insulators E_g is on the order of 7.0 eV and in semiconductors it is typically 1.0 eV, so in semiconductors there is possibility of excitation of electrons into the conduction band even at room temperature.

The development of the band theory of solids has been influenced by the

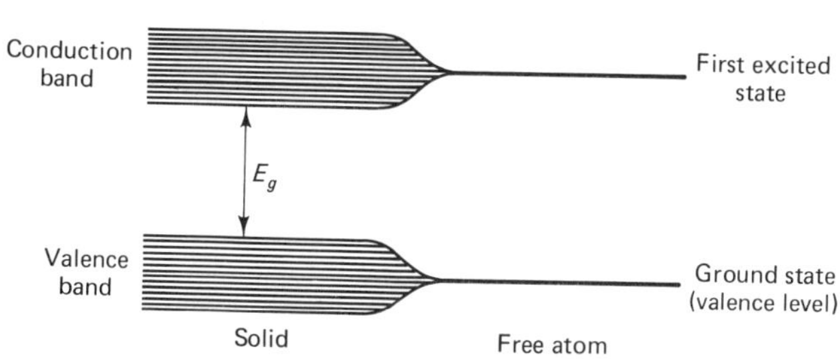

Figure 6.5-4 Energy levels for the free atom and energy bands for the solid.

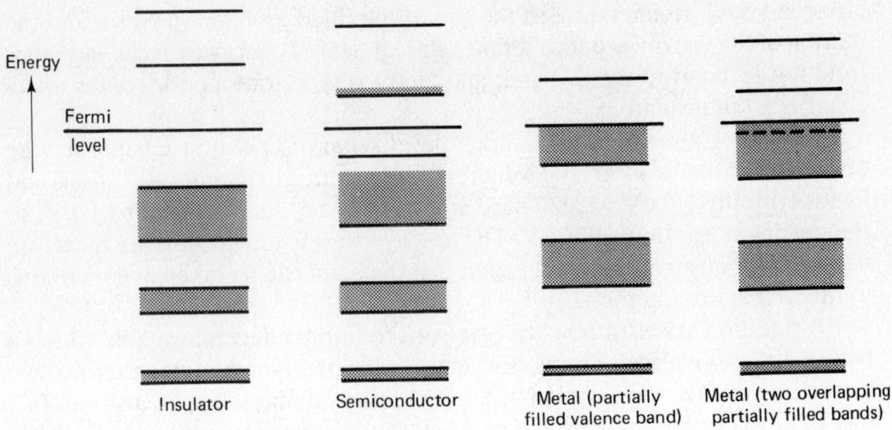

Figure 6.5-5 Energy band for different classes of materials.

severe limitations of the free-electron theory. First, the electrons in a real metal are not under the perfectly uniform potential of a *particular* electron in a box, but rather the potential of a regular assembly of ion cores and other electrons. Second, the free-electron theory gives no clue about the relationship between metals and other solids. Third, the theory gives a totally inadequate picture of all the properties of metals except those that depend on the excited electrons and corresponding holes that exist near the surface of the Fermi sea at temperatures above 0 K. In the next section the band theory will be developed to make the model consistent with some of the ideas about the structure of solids and the motion of electrons in a periodic lattice.

■ EXAMPLE 6.5-1

Discuss the limits of the free-electron model and the situations where the application of Fermi statistics has proved successful to the study of monovalent metals.

Solution. The hypothesis of a free-electron gas was first used by Drude in 1900. Later, Sommerfeld proposed that the properties of the electron gas be determined by quantum-mechanical principles. Reducing the task to a single-electron problem, the theory is based on the following assumptions: (1) the valence electrons move independently of one another, (2) the potential in the interior of the metal is constant, (3) a wave function is attributed to every electron which has been obtained as a solution of the time-independent Schrödinger equation, (4) the electron energy distribution is given by Fermi–Dirac statistics. This latter assumption contains Pauli's exclusion principle according to which each eigenstate can be occupied by no more than two electrons with opposite spins. From it, the results in Chapter 4 were obtained and are now summarized:

The occupation probabilities of the state of energy ϵ is given by the Fermi–Dirac distribution function

$$f_{FD} = \frac{1}{e^{(\epsilon-\mu)/kT} + 1} \tag{6.5-1}$$

where μ is the chemical potential or Fermi energy. Multiplying this by the density of states

$$g(\epsilon)d\epsilon = 2\left(\frac{V}{4\pi^2}\right)\left(\frac{2m}{\hbar^2}\right)^{3/2} \epsilon^{1/2}\, d\epsilon \tag{6.5-2}$$

gives the population distribution or number $n(\epsilon)d\epsilon$ of electrons with energies between ϵ and $\epsilon + d\epsilon$:

$$n(\epsilon)d\epsilon = \frac{V}{2\pi^2}\left(\frac{2m}{\hbar^2}\right)^{3/2} \frac{\epsilon^{1/2}\, d\epsilon}{e^{(\epsilon-\mu)/kT} + 1} \tag{6.5-3}$$

The probability function $f(\epsilon)$, density of states function $g(\epsilon)$, and population distribution function $n(\epsilon)$ are plotted in Fig. 6.5-6 for $T = 0$ K and $T > 0$ K.

At $T = 0$ K, all energy states are filled up to $\epsilon_{F,0}$, the Fermi energy, with probability equal to unity; for higher energies, the occupation probability is equal to zero. For $T > 0$ K, the occupation probability for states of energy

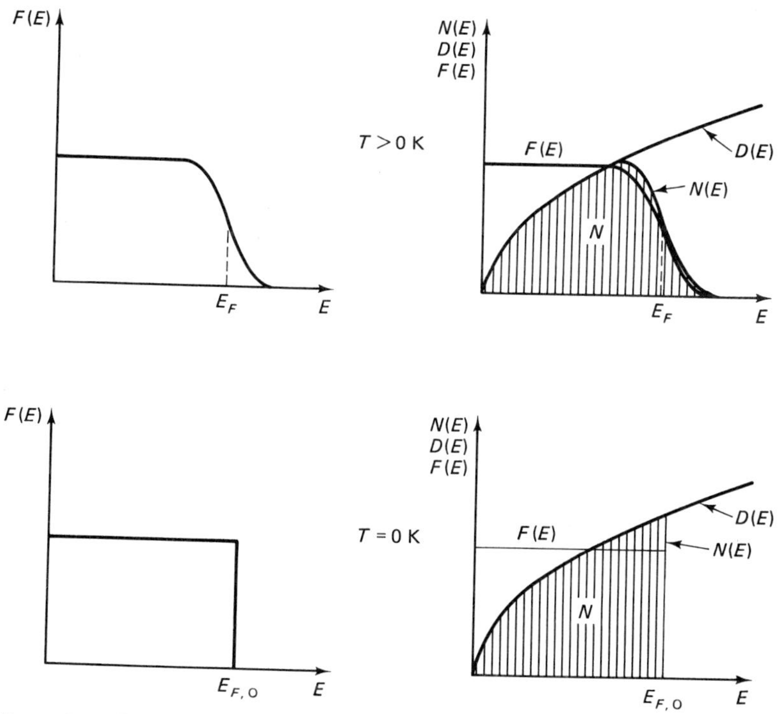

Figure 6.5-6 Fermi–Dirac distribution function $f(\varepsilon)$, density of state $g(\varepsilon)$, and population $n(\varepsilon)$ at $T = 0$ K and $T < 0$ K.

$\epsilon = \epsilon_F$ is equal to $\frac{1}{2}$. The Fermi energy is just slightly temperature dependent and hence is often written ϵ_F instead of $\epsilon_{F,0}$.

Properties of the electron gas at T = 0 K. At absolute zero, all states are filled up to the Fermi energy $\epsilon_{F,0}$. The value of the limiting energy depends on the total number N of valence electrons contained in the volume V. The equation determining $\epsilon_{F,0}$ is

$$N = \int_0^{\epsilon_{F,0}} n(\epsilon)\,d\epsilon = \frac{V}{2\pi^2}\left(\frac{2m}{\hbar^2}\right)^{3/2} \int_0^{\epsilon_{F,0}} \frac{\epsilon^{1/2}\,d\epsilon}{e^{(\epsilon-\epsilon_{F,0})/kT}+1} \tag{6.5-4}$$

Noting that for $T = 0$ K, $f(\epsilon) = 1$ up to $\epsilon_{F,0}$ and is zero thereafter, the integration gives

$$\epsilon_{F,0} = \frac{\hbar^2}{2m}\left(3\pi^2 \frac{N}{V}\right)^{2/3} = \frac{\hbar^2}{2m}(3\pi^2 n_e)^{1/3} \tag{6.5-5}$$

where $n_e = N/V$ is the electron concentration. (The electron concentration in metals is of the order of the concentration of atoms. For monovalent metals, $n_e \simeq n_{\text{atoms}}$.) Thus, $\epsilon_{F,0}$ has very high values, of the order of several electron-volts.

The total energy of all valence electrons in a volume V at absolute zero is given by

$$U_0 = \int_0^{\epsilon_{F,0}} \epsilon\, n(\epsilon)\,d\epsilon \tag{6.5-6}$$

Upon substitution of Eq. (6.5-3), this yields

$$U_0 = \frac{V}{5\pi^2}\left(\frac{2m}{\hbar^2}\right)^{3/2} \epsilon_{F,0}^{5/2} \tag{6.5-7}$$

and, with Eq. (6.5-5), a mean energy per electron

$$\langle u_0\rangle = \frac{U_0}{N} = \frac{3}{5}\epsilon_{F,0} \tag{6.5-8}$$

In contrast to the classical Maxwell–Boltzmann gas whose particles have no energy at absolute zero, the electron gas has considerable zero-point energy. It corresponds to a temperature of the order of 10^4 K for a classical gas. This can be easily seen from the relations

$$\langle u_0\rangle = \frac{3}{5}\epsilon_{F,0} = \frac{3}{2}kT_F \tag{6.5-9}$$

$$T_F = \frac{2}{5}\frac{\epsilon_{F,0}}{k} \tag{6.5-10}$$

where T_F is the degeneracy temperature. Also, in contrast to the ideal gas, the electron gas at absolute zero has a high zero-point pressure

$$p_0 = \frac{2}{3}n_e\langle u_0\rangle = \frac{2}{5}n_e\epsilon_{F,0} \tag{6.5-11}$$

which is of the order of 10^5 atm.

Properties of the electron gas at $T > 0$ K. If the temperature is raised, the mean energy of the electron gas is also raised via the interaction with the lattice vibrations (phonons). The excitation energies transferred from the phonons to the electrons amount to about kT. As all states with energies $\epsilon < \epsilon_{F,0}$ are occupied, only the fastest electrons of energies $\epsilon \simeq \epsilon_{F,0}$ can be excited to higher levels. Thus, the curve of the Fermi–Dirac function is changed only near the Fermi level, in an energy interval of the order of kT. The value of the Fermi energy ϵ_F is only slightly temperature dependent and is obtained from

$$N = \int_0^\infty n(\epsilon)\, d\epsilon = \frac{V}{2\pi^2}\left(\frac{2m}{\hbar^2}\right)^{3/2} \int_0^\infty \frac{\epsilon^{1/2}\, d\epsilon}{e^{(\epsilon - \epsilon_F)/kT} + 1} \tag{6.5-12}$$

Using the substitution $x = \epsilon/kT$ and $\alpha = \epsilon_F/kT$, this can be rewritten in the form

$$N = \frac{V}{2\pi^2}\left(\frac{2mkT}{\hbar^2}\right)^{3/2} \int_0^\infty \frac{x^{1/2}\, dx}{e^{x-\alpha} + 1} \tag{6.5-13}$$

or, upon division by V,

$$n_e = \frac{2}{\sqrt{\pi}}\, n_0\, F_{1/2}(\alpha) \tag{6.5-14}$$

where

$$n_0 = \frac{1}{4}\left(\frac{2mkT}{\pi\hbar^2}\right)^{3/2} \tag{6.5-15}$$

is the "effective density of states" or critical electron concentration, and

$$F_{1/2}(\alpha) = \int_0^\infty \frac{x^{1/2}\, dx}{e^{x-\alpha} + 1} \tag{6.5-16}$$

is a Fermi integral. In Fermi statistics, integrals of the form

$$F(\alpha) = \int_0^\infty \frac{f(x)\, dx}{e^{x-\alpha} + 1} \tag{6.5-17}$$

are often encountered. They cannot be solved analytically, but can be approximated by series expansion. Equation (6.5-16) will be evaluated for the two special cases, $n_e \gg n_0$ and $n_e \ll n_0$.

1. For metals, the condition $n_e \gg n_0$ always prevails, and thus $F_{1/2}(\alpha) \gg 1$. Accordingly also $\alpha \gg 1$ and for this range of α values,

$$F_{1/2}(\alpha) = \frac{2}{3}\alpha^{3/2}\left(1 + \frac{\pi^2}{8\alpha^2} + \cdots\right) \tag{6.5-18}$$

Substituting in Eq. (6.5-14) gives

$$n_e = \frac{1}{3\pi^2}\left(\frac{2m}{\hbar^2}\right)^{3/2} \epsilon_F^{3/2}\left[1 + \frac{\pi^2}{8}\left(\frac{kT}{\epsilon_F}\right)^2 + \cdots\right] \tag{6.5-19}$$

As $\alpha = \epsilon_F/kT \gg 1$, the term ϵ_F can be replaced by $\epsilon_{F,0}$, and use of Eq. (6.5-5) gives

$$\epsilon_F(T) = \epsilon_{F,0}\left[1 - \frac{\pi^2}{12}\left(\frac{kT}{\epsilon_{F,0}}\right)^2\right] \tag{6.5-20}$$

for the temperature dependence of the Fermi energy. It is seen that it decreases slightly with increasing temperature; the relative drop of ϵ_F between 0 and 300 K being 5×10^{-3} (for a value of $\epsilon_{F,0} = 5$ eV).

2. For semiconductors, $n_e \ll n_0$ is of importance. The condition $F_{1/2}(\alpha) \ll 1$, $\alpha \ll 0$ introduces the concept of negative ϵ_F values and means that the Fermi-Dirac distribution function goes over to the Maxwell–Boltzmann function. Equation (6.5-3) then becomes

$$\begin{aligned} n(\epsilon)\,d\epsilon &= \frac{V}{2\pi^2}\left(\frac{2m}{\hbar^2}\right)^{3/2} \epsilon^{1/2} \exp\left(-\frac{|\epsilon_F|}{kT}\right) \exp\left(-\frac{\epsilon}{kT}\right) d\epsilon \\ &= \text{const}\,\epsilon^{1/2} \exp\left(-\frac{\epsilon}{kT}\right) d\epsilon \end{aligned} \tag{6.5-21}$$

For $\alpha \ll 0$, the Fermi integral $F_{1/2}(\alpha)$ can be solved analytically, yielding

$$\begin{aligned} F_{1/2}(\alpha) &= \exp(-|\alpha|) \int_0^\infty x^{1/2} \exp(-x)\,dx \\ &= \frac{\sqrt{\pi}}{2} \exp(-|\alpha|) \end{aligned} \tag{6.5-22}$$

With this, Eq. (6.5-14) yields an expression for determining ϵ_F:

$$n_e = n_0 \exp\left(-\frac{|\epsilon_F|}{kT}\right) \tag{6.5-23}$$

or

$$\frac{|\epsilon_F|}{kT} = \ln\frac{n_0}{n_e} = \ln\frac{1}{4n_e}\left(\frac{2mkT}{\pi\hbar^2}\right)^{3/2} \tag{6.5-24}$$

Thus, in the case of sufficiently low electron concentration n_e and high temperatures T, the electron gas behaves like a classical gas (characterized by a Maxwell–Boltzmann distribution). The electron gas is then said to be "nondegenerate." Nondegeneracy occurs when $n_e \ll n_0$, or $\epsilon_F/kT > 0$ and $|\epsilon_F|/kT \gg 1$.

Generally, an electron gas is highly degenerate as long as its temperature T is considerably below the Fermi or degeneracy temperature $T_F = (\frac{2}{5})(\epsilon_F/k)$, that is, when $\epsilon_F \simeq \epsilon_{F,0} \gg kT$. Degeneracy is also defined in terms of a critical electron concentration by

$$n_e \gg n_{\text{crit}} = \frac{1}{3\pi^2}\left(\frac{2mkT}{\hbar^2}\right)^{3/2} \tag{6.5-25}$$

At $T = 300$ K, the critical density is $n_{\text{crit}} \simeq 10^{19}/\text{cm}^3$, but the electron concentration in monovalent metals is $n_e \simeq 10^{22}/\text{cm}^3$. Thus, the electron gas of metals is highly degenerate. This degeneracy would disappear only at temperatures considerably above the melting point of the metal.

Limitations of Sommerfeld's free-electron model. Despite the neglect of the electrostatic forces between the positive ions and the electrons, and the fact that the theory takes into account only two characteristics of a metal, the electron concentration and the work function, the free-electron model explains satisfactorily the following properties:

1. The specific heat of an electron gas. Only a small fraction of electrons (the fastest) can take up thermal energy. For this reason, the specific heat of the degenerate electron gas is much smaller than that of a chemical gas. From

$$U(T) = \int_0^\infty \epsilon n(\epsilon)\, d\epsilon$$

with the proper substitutions and $x = \epsilon/kT$, $\alpha = \epsilon_F/kT$,

$$U(T) = \frac{V}{2\pi^2}\left(\frac{2m}{\hbar^2}\right)^{3/2}(kT)^{5/2}F_{3/2}(\alpha) \tag{6.5-26}$$

where

$$F_{3/2}(\alpha) = \int_0^\infty \frac{x^{3/2}\,dx}{e^{x-a}+1} \tag{6.5-27}$$

is a Fermi integral which in the case of degeneracy ($\alpha = \epsilon_F/kT >> 1$) expands into

$$F_{3/2}(\alpha) = \frac{2}{5}\alpha^{5/2}\left[1 + \frac{5}{8}\left(\frac{\pi}{\alpha}\right)^2 + \cdots\right] \tag{6.5-28}$$

giving, for the electron gas,

$$\begin{aligned} U(T) &= \frac{V}{5\pi^2}\left(\frac{2m}{\hbar^2}\right)^{3/2}\epsilon_F^{5/2}\left[1 + \frac{5\pi^2}{8}\left(\frac{kT}{\epsilon_F}\right)^2 + \cdots\right] \\ &= \frac{3}{5}N\epsilon_{F,0}\left[1 + \frac{5\pi^2}{12}\left(\frac{kT}{\epsilon_{F,0}}\right)^2 + \cdots\right] \end{aligned} \tag{6.5-29}$$

or a mean energy per electron

$$\langle u(T)\rangle = \frac{U(T)}{N} = \frac{3}{5}\epsilon_{F,0}\left[1 + \frac{5\pi^2}{12}\left(\frac{kT}{\epsilon_{F,0}}\right)^2 + \cdots\right] \tag{6.5-30}$$

From this, the specific heat of the electron gas is

$$c_V = \left(\frac{\partial U}{\partial T}\right)_V = \frac{\pi^2}{2}\cdot\frac{Nk^2}{\epsilon_{F,0}}T = \gamma T \tag{6.5-31}$$

where

$$\gamma = \frac{\pi^2}{2}\cdot\frac{Nk^2}{\epsilon_{F,0}} \tag{6.5-32}$$

Unlike the classical gas, the electron gas has a temperature-dependent specific heat that is directly proportional to the temperature. For temperatures above the Debye temperature (of the order of 100 K), the value of c_V is only about 1% of the Dulong-Petit value $3Nk$ of the lattice-specific heat.

2. Thermionic emission due to high temperatures.

3. Field emission due to high electric field strengths. These two effects are limiting cases (electric field $\mathscr{E} \to 0$ and temperature $T \to 0$, respectively) of the so-called thermionic field (TF) emission that can be described satisfactorily within the framework of the Sommerfeld free-electron model. The emission of electrons at high temperatures (thermionic emission) will be discussed first.

As recalled in Chapter 4, an electron needs at least $\epsilon_F + \phi$ energy to escape from the metal. ϕ is the work function. The available energy is to be in the form of kinetic energy. This is indeed the case in the free-electron model, because the potential energy is zero and the electrons do not interact. The relationship between energy and momentum is thus simply

$$\epsilon = \frac{1}{2m}(p_x^2 + p_y^2 + p_z^2) \tag{6.5-33}$$

Furthermore, the electron besides having the right amount of energy, must go in the right direction. Taking x as the coordinate perpendicular to the surface of the metal, the electron momentum must satisfy the inequality

$$\frac{p_x^2}{2m} > \frac{p_{x,0}^2}{2m} = \epsilon_{F,0} + \phi \tag{6.5-34}$$

According to the rules of quantum mechanics, the electron may be reflected. The probability of escape is then $1 - r(p_x)$, where $r(p_x)$ is the reflection coefficient. If the number of electrons having momentum between p_x and $p_x + dp_x$ is $n(p_x)\,dp_x$, then the number of electrons arriving at the surface per second per unit area is

$$\frac{p_x}{m} n(p_x)\,dp_x \tag{6.5-35}$$

and the number of those escaping is

$$[1 - r(p_x)]\frac{p_x}{m} n(p_x)\,dp_x \tag{6.5-36}$$

Multiplying this by the quantum of charge and summing the contributions from all electrons that have momenta in excess of $p_{x,0}$ yields the emission current density

$$J = \frac{e}{m}\int_{p_{x,0}}^{\infty} [1 - r(p_x)]p_x n(p_x)\,dp_x \tag{6.5-37}$$

Now, the density of states in momentum space, that is, the number of states between p_x and $p_x + dp_x$, p_y and $p_y + dp_y$, p_z and $p_z + dp_z$ is

$$\frac{2}{h^3}\,dp_x\,dp_y\,dp_z \tag{6.5-38}$$

where the spin of the electrons has been included. Multiplying this by the distribution function gives the number of electrons

$$n(p_x, p_y, p_z)\,dp_x\,dp_y\,dp_z = \frac{2}{h^3}\frac{dp_x\,dp_y\,dp_z}{\exp\{[(1/2m)(p_x^2+p_y^2+p_z^2)-\epsilon_F]/kT\}+1} \tag{6.5-39}$$

To get the number of electrons in the momentum range dp_x at p_x, Eq. (6.5-39) is integrated for all values of p_y and p_z. Thus,

$$n(p_x)\,dp_x = \frac{2}{h^3}dp_x \int_{-\infty}^{\infty}\int_{-\infty}^{\infty} \frac{dp_y\,dp_z}{\exp\,\{[(1/2m)(p_x^2+p_y^2+p_z^2)-\epsilon_F]/kT\}+1} \tag{6.5-40}$$

This complicated-looking integral can be simplified by considering only those electrons exceeding the threshold $\phi(>> kT)$, in which case the unity term in the denominator can be neglected. What is left then are Gaussian functions whose integrals between $-\infty$ and $+\infty$ have been listed in Chapter 1. This leads to

$$n(p_x)\,dp_x = \frac{4\pi mkT}{h^3}\,e^{\epsilon_F/kT}\,e^{-p_x^2/2mkT}\,dp_x \tag{6.5-41}$$

Substituting Eq. (6.5-41) into (6.5-37) and taking $r(p_x) = r$ to be independent of p_x gives the current density

$$J = \frac{4\pi emk^2}{h^3}(1-r)T^2 e^{-\phi/kT} = A_0(1-r)T^2 e^{-\phi/kT} \tag{6.5-42}$$

where

$$A_0 = \frac{4\pi emk^2}{h^3} = 1.2\times 10^6\ \mathrm{A\ m^{-2}\ K^{-2}} \tag{6.5-43}$$

Equation (6.5-42) is the *Richardson–Dushman equation,* first derived independently by Sommerfeld and Nordheim. The important thing is the factor $e^{-\phi/kT}$, which is strongly dependent on both temperature and work function (e.g., for tungsten, $\phi \simeq 4.5$ eV). Figure 6.5-7 illustrates the thermionic process. Note that at any finite temperature, there is some emission of electrons from a metal surface because there is some occupancy of higher-energy states in the metal above the vacuum level, corresponding to the high-energy tail of the Fermi distribution.

Field emission. Electrons can be extracted from metals, even at low temperatures, if a sufficiently high electric field is applied to the metal surface. The presence of an electric field increases the emission current, because more electrons can escape over the reduced barrier. If the electric field is increased further (towards 10^9 V/m), then a new escape route opens up. Instead of going over the potential barrier, the electrons *tunnel* across it. This is called *field emission,* and it is practically independent of temperature. The rate of emission depends primarily on the product of the density of occupied states at a

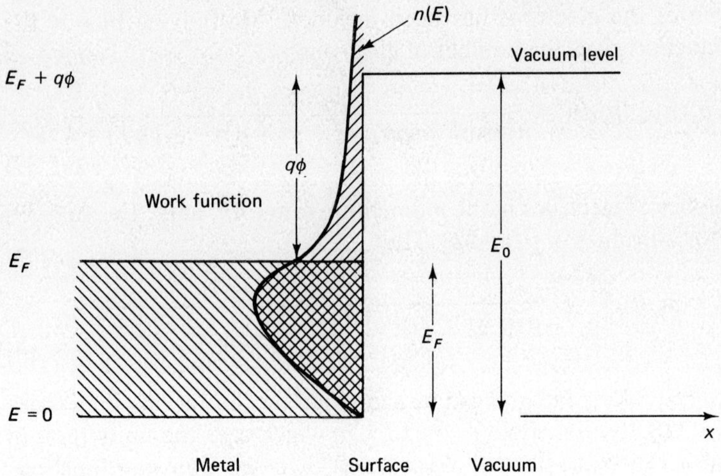

Figure 6.5-7 Free-electron model for thermionic emission.

given energy and the tunneling transmission coefficient corresponding to that energy. The total emission then corresponds to an integration of this product over all energies. Most of the contribution comes from states close to the Fermi energy.

Before presenting an analysis of the problem, it should be mentioned that the forces on an electron in front of an infinitely conducting sheet are correctly given by replacing the sheet by the mirror charge (a positively charged particle the same distance behind the sheet). This results in the *Schottky effect*, by which the value of the work function is lowered to ϕ_{eff}. Consider the case when the electric field is constant: $V(x) = -e\mathscr{E}x$. Figure 6.5-8 shows the reduction in height of the potential barrier due to the Schottky effect. The main contribution to the tunnelling current comes from electrons situated around the Fermi level. For them the width of the barrier is calculated from $-\phi = e\mathscr{E}x_F$ (corresponding to the Fermi energy), and the height of the potential barrier they face is ϕ_{eff}. The tunnelling current varies exponentially with barrier width:

$$J \sim \exp\left[-\frac{2m\phi_{\text{eff}})^{1/2}}{\hbar}x_F\right] \tag{6.5-44}$$

$$\sim \exp\left[-\frac{(2m\phi_{\text{eff}})^{1/2}}{\hbar}\frac{\phi}{\mathscr{E}}\right]. \tag{6.5-45}$$

The exponential factor is fairly well confirmed by experiment. A more exact solution is too long to quote. It may be noted that the role of temperature in the above equations has been taken over by the electric field. The free-electron model of a metal surface suitable for discussion of the field emission process is shown in Fig. 6.5-9.

The properties of metals discussed above, namely, the specific heat of

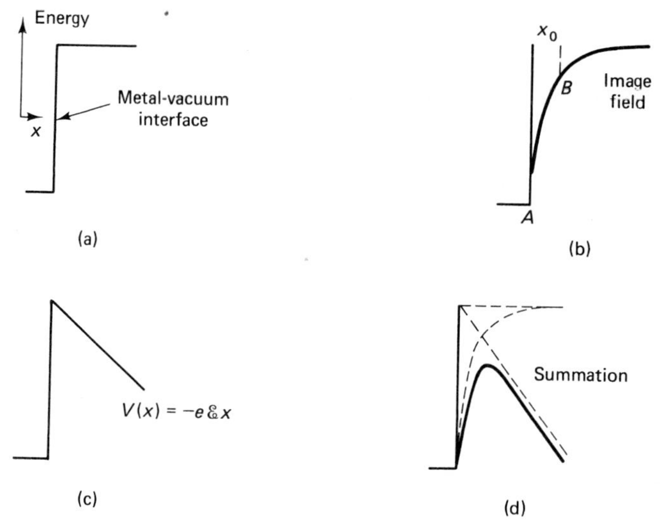

Figure 6.5-8 Schottky effect. (a) Potential at metal-vacuum interface. (b) Potential by image charge field. (c) Potential due to applied anode voltage. (d) Reduction in height of the potential barrier.

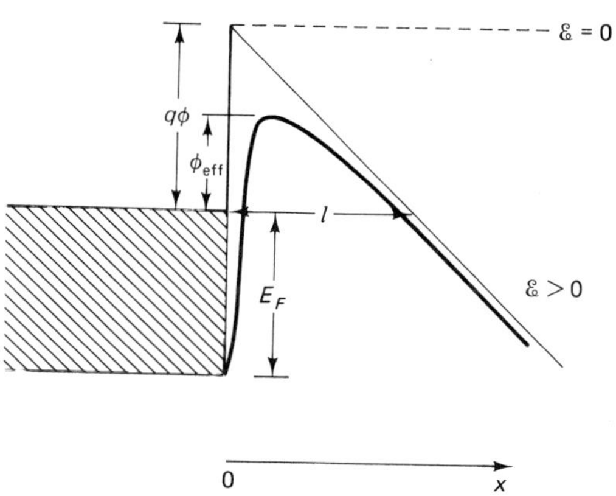

Figure 6.5-9 Free-electron model for field emission.

an electron gas, the thermionic emission, the field emission, along with the Wiedemann–Franz law, and the dia- and paramagnetism of the electron gas, are problems that have been solved adequately by Sommerfeld's free-electron model. The band theory of solids, however, bases the discussion on a more accurate model. It automatically answers the question why some substances are metals or insulators, semiconductors or semimetals, and it no longer needs to restrict which model applies, for instance, only to metals, or only to ionic crystals. ■

■ EXAMPLE 6.5-2

The free-electron theory of metals, as discussed in Chapter 4, does not explain why some materials are good conductors of electricity, while others, insulators, are poor conductors; and why some materials, called semiconductors, have conducting properties somewhere in between the other two. The band theory of solids is a more successful theory. It explains why the ratios of resistivities of solids may be as large as 10^{30}. Show the splitting of energy levels and the formation of energy bands when two or more atoms are brought together. Discuss the way electrons occupy the available bands as governed by the Pauli exclusion principle, and compare the energy required to excite a $3s$, $2s$, or $1s$ electron to a higher-energy ($4s$) unoccupied level.

Solution. A solid can be thought of as being formed by bringing together isolated single atoms. Any single atom will possess a large number of discrete energy levels that can be occupied by the electrons of the atom. Consider first the combination of two atoms. If there were no interaction between the two atoms, the value of each energy level would be the same as for each isolated atom [Fig. 6.5-10(a)], with the number of levels at a particular energy simply being doubled, as shown in Fig. 6.5-10(b). Because of interactions, however, each previously single energy level is split into two levels, as in Fig. 6.5-10(c). In a similar fashion, if more atoms were brought together there would be a larger number of splittings of each energy level, one split for each additional atom. Figure 6.5-10(d) shows the splitting of energy levels when five atoms are brought together. Because there are of the order of 10^{23} atoms cm^{-3} in solids, each previously single energy level of an isolated atom will be split into an enormous number of parts. Since the values of the energy levels remain approximately the same, the net effect of assembling a large number of interacting atoms is to form bands of practically continuous energy levels, separated by gaps where no electron states exist, as illustrated in Fig. 6.5-10(e).

The Pauli exclusion principle governs the question of what quantum states electrons can populate in any particular atom, that is, how many electrons can be put into any particular n, l quantum set of levels. Clearly the number of

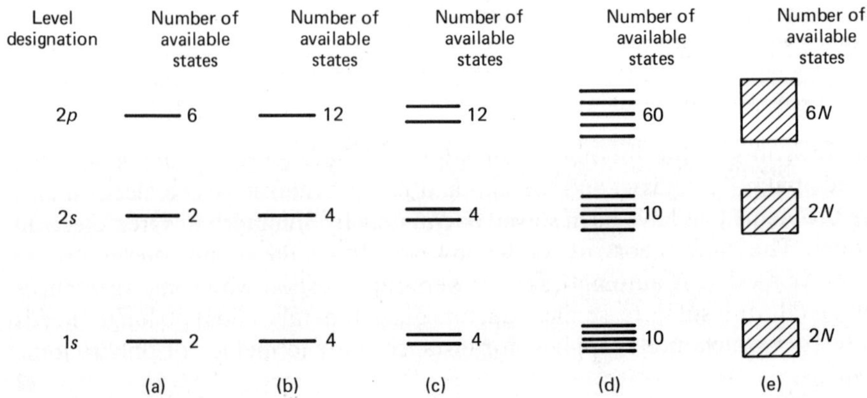

Figure 6.5-10 Interacting atoms forming energy bands.

electrons that can exist together in any n, l quantum level is the total number of different l "orientations" and s "orientations" that are possible. This number is $(2l + 1)(2s + 1) = 2(2l + 1)$ for a particular l value, followed by $m_s = \pm\frac{1}{2}$ according to the Pauli principle. (Table 6.5-1).

In this manner, the bands are filled with electrons for the case of many-electron atoms. As an example, $_{11}$Na has all its energy levels filled up to the 3s level, which has one electron, and its electron configuration is $1s^2 2s^2 2p^6 3s^1$. Since the 3s level can accommodate two electrons, it is only half-filled. The bands in solids may be filled, partialy filled, or empty. The highest energy band occupied by the valence electrons and the unoccupied band directly above it determine the conduction properties of a crystalline solid. If the band containing the valence electrons is filled, it is referred to as the valance band, and the next higher band is referred to as the conduction band; if the band containing the valance electrons is not filled, it is called the conduction band.

An atom can be "active" only in terms of any interaction involving the Coulomb force (e.g., chemically) if there is an unfilled shell. Consequently inner shells (ones with lower values of n than the maximum value occupied) can only contribute if one or more electrons are removed from that inner shell, either through ionization or excitation to higher shells. Under most circumstances this means that only electrons in unfilled shells, or at the very most the outermost filled shell, are likely to take part in Coulomb interactions or be excited to higher levels since these take by far the least energy to excite. An example is sodium. It has all levels filled to the $n = 2$ shell and one electron in the 3s level when in its ground state. The energy required to excite a 3s, 2s, or 1s electron to a higher-energy unoccupied level (for example, the 4s level) can be evaluated. For the transition $3s \rightarrow 4s$, the effective nuclear charge that the 3s electron "sees" is approximately 1. Therefore,

$$\Delta E_{3s\rightarrow 4s} \simeq -Z_{\text{eff}}^2 E_0\left(\frac{1}{4^2} - \frac{1}{3^2}\right) = -E_0\left(\frac{1}{16} - \frac{1}{9}\right)$$

$$= 0.05E_0 = 0.05(13.58) = 0.7 \text{ eV}$$

TABLE 6.5-1

	n	l	m_l	s	m_s
$n = 1,\ l = 0$	1	0	0	$\frac{1}{2}$	$+\frac{1}{2}$
1s	1	0	0	$\frac{1}{2}$	$-\frac{1}{2}$
$n = 1,\ l = 0$	2	0	0	$\frac{1}{2}$	$+\frac{1}{2}$
2s	2	0	0	$\frac{1}{2}$	$-\frac{1}{2}$
$n = 2,\ l = 1$	2	1	0	$\frac{1}{2}$	$+\frac{1}{2}$
2p	2	1	0	$\frac{1}{2}$	$-\frac{1}{2}$
	2	1	1	$\frac{1}{2}$	$+\frac{1}{2}$
	2	1	1	$\frac{1}{2}$	$-\frac{1}{2}$
	2	1	-1	$\frac{1}{2}$	$+\frac{1}{2}$
	2	1	-1	$\frac{1}{2}$	$-\frac{1}{2}$

An electron from the 2*s* level, however, sees the full nucleus $Z = 11$ shielded essentially by the two 1*s* electrons. (Energies are evaluated by ignoring all electrons that are further from the nucleus than itself.) Therefore,

$$\Delta E_{1s\to 4s} \simeq \frac{(11-2)^2 E_0}{2^2} - \frac{(1)^2 E_0}{4^2} \simeq 20E_0 = 275 \text{ eV}$$

The energy required has increased by two orders of magnitude. If the 1*s* electron is to be excited, it will require an energy of the order

$$\Delta E_{1s\to 4s} \simeq (11)^2 E_0 - \frac{1^2 E_0}{4} = 120E_0 = 1.64 \text{ keV}$$

which is even greater. Clearly all chemical and optical properties are most likely to be determined only by the outermost, least tightly bound electrons. This is the statement made by chemists that the chemistry of an atom is determined by its outermost "valence" electrons. In fact, the whole periodic table of elements can be thought of correctly as a listing of elements in order of increasing nuclear charge, grouped according to the orbital angular momentum of their outermost electrons. ■

■ **EXAMPLE 6.5-3**

(a) Calculate the number of conduction electrons per unit volume in lithium. The Fermi energy, ϵ_F, for lithium is 4.72 eV at $T = 0$.

(b) Estimate the relative number of conduction electrons in a metal that are thermally excited to higher-energy states at room temperature.

Solution

(a) The Fermi energy, ϵ_F, is related to the electron concentration, N/V, by

$$\epsilon_F = \frac{h^2}{8m}\left(\frac{3N}{\pi V}\right)^{2/3} \tag{6.5-46}$$

Solving for N/V, the number of free electrons per unit volume, gives

$$n_e = \frac{N}{V} = \frac{\pi}{3}\left(\frac{8m}{h^2}\epsilon_F\right)^{3/2} \tag{6.5-47}$$

With the mass of the electron $m = 9.11 \times 10^{-31}$ kg, the Fermi energy $\epsilon_F = 4.72$ eV $= 4.72 \times 1.60 \times 10^{-19}$ J, and $h = 6.63 \times 10^{-34}$ J s:

$$n_e = \frac{\pi}{3}\left[\frac{8 \times 9.11 \times 10^{-31} \times 4.72 \times 1.60 \times 10^{-19}}{(6.63 \times 10^{-34})^2}\right]^{3/2} = 4.64 \times 10^{28}/\text{m}^3$$

$$= 4.64 \times 10^{22} \text{ electrons/cm}^3$$

This is equivalent to one free electron per lithium atom, since the number of lithium atoms per unit volume, in solid lithium of density 0.534 g/cm³, is

$$\left(0.534 \frac{\text{g}}{\text{cm}^3}\right)\left(\frac{\text{mole}}{6.94 \text{ g}}\right)\left(6.02 \times 10^{23} \frac{\text{atoms}}{\text{mole}}\right) = 4.64 \times 10^{22} \text{ atoms/cm}^3$$

(b) Most of the excited electrons are in a range $\Delta\epsilon \simeq 2kT$ around the Fermi energy. Taking into account that the probability of occupation at the Fermi energy is $\frac{1}{2}$, the number of excited electrons is

$$\frac{1}{2} n(\epsilon_F)\Delta\epsilon = \frac{1}{2} n(\epsilon_F)(2kT) = n(\epsilon_F)kT \tag{6.5-48}$$

where, from previous derivations on fermions (Chapter 4), the number of electrons per unit volume having energy ϵ_F is

$$n(\epsilon_F) = \frac{\pi}{2}\left(\frac{8m}{h^2}\right)^{3/2} \epsilon_F^{1/2} \tag{6.5-49}$$

Now, the electron concentration, as related to the Fermi energy, is

$$n_e = \frac{\pi}{3}\left(\frac{8m}{h^2}\right)^{3/2} \epsilon_F^{3/2} \tag{6.5-50}$$

The relative number of conduction electrons that are thermally excited to higher energies is then

$$\begin{aligned} \frac{n(\epsilon_F)}{n_e} &= \left[\frac{\pi}{2}\left(\frac{8m}{h^2}\right)^{3/2} \epsilon_F^{1/2} kT\right] \Big/ \left[\frac{\pi}{3}\left(\frac{8m}{h^2}\right)^{3/2} \epsilon_F^{3/2}\right] \\ &= \frac{3}{2}\frac{kT}{\epsilon_F} \simeq \frac{kT}{\epsilon_F} \end{aligned} \tag{6.5-51}$$

which is small. At room temperature, $kT \simeq 0.025$ eV, and, typically for metals, $\epsilon_F \simeq 4$ eV. Hence the fraction of thermally excited conduction electrons is 1/160. However, the absolute number of excited electrons is large, because the number of conduction electrons is so large. ■

6.6 ELECTRONS IN CRYSTALS

A crystal is a collection of atoms arranged in a regular array, the spacing between atoms being of the same order of magnitude as the dimensions of the atoms. Each atom is more or less anchored to one point—called its *site* in the lattice—by the electrostatic forces produced by all the other atoms.

To understand the effect of putting a great many atoms close together, consider first two atoms only that are initially far apart. As explained previously, when the atoms are brought close enough together, the wave functions of the individual atoms overlap, and a given energy level of the system is split into two distinct levels, the splitting increasing as the separation of the atoms decreases. (An example of this is found in the ground-state energy level of a system containing two hydrogen atoms.)

With three isolated atoms, there would be a threefold exchange degeneracy of the energy levels. If the atoms are brought together in a uniform linear lattice, each of the levels splits into three distinct levels. For a system containing N atoms of a given species, each level of one of these atoms leads to an N-fold degenerate level of the system when the atoms are separated. With decreasing separation, each of these splits into a set of N levels. The spread in energy between the lowest and highest level of a particular set depends on the separation distance R, but not significantly on the number of atoms in the system if the same separation distance is maintained. Thus, as more and more atoms are added to the system, each set of split levels contains more and more levels spread over about the same energy range at a particular R. At the values of R found in a solid, a few angstroms, the energy spread is of the order of a few electron volts. For a solid containing something like 10^{23} atoms per mole, it is seen that the levels of each set in a solid are so closely spaced in energy that they effectively form a continuous band. The process just described is illustrated in Fig. 6.6-1.

The free-electron model can be used to evaluate the width in energy of a band for the case of a one-dimensional metal. The eigenfunctions for an electron in the deep square well that represents the ion cores are essentially standing waves like

$$\Psi \propto \cos\frac{2\pi x}{\lambda} = \cos kx \quad \text{and} \quad \Psi \propto \sin\frac{2\pi x}{\lambda} = \sin kx \tag{6.6-1}$$

where λ is the wavelength and $k = 2\pi/\lambda$ is the wave number. The eigenfunctions have nodes at each end of the well. These boundary conditions lead immediately to the requirement that

$$\frac{n\lambda}{2} = L \tag{6.6-2}$$

where L is the length of the well. Each value of the integer $n = 1, 2, 3, \ldots$ corresponds to a different eigenfunction, or energy level if two electrons of opposite spin are allowed per level. Since for free electrons the energy is

$$\epsilon = \frac{p^2}{2m} = \frac{h^2}{2m\lambda^2} = \frac{h^2 n^2}{8mL^2} \tag{6.6-3}$$

the minimum value of n corresponds to the level of essentially zero energy at the bottom of the band, and the maximum value of n corresponds to the level

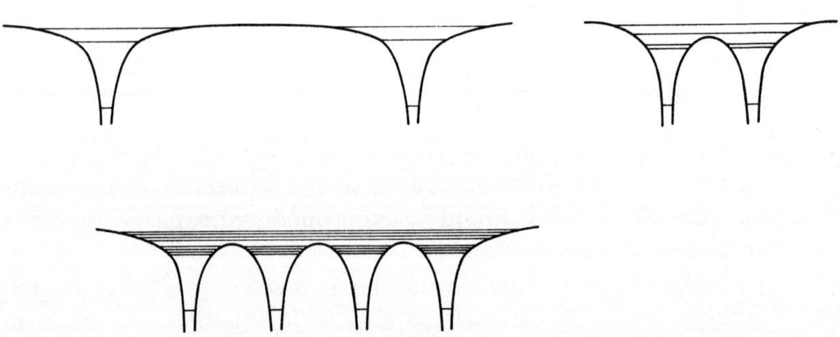

Figure 6.6-1 Energy level schemes for (a) two isolated atoms, (b) two atoms in a diatomic molecule, and (c) four atoms in a one-dimensional crystal.

of maximum energy at the top, the width of the band being roughly equal to that maximum. If there are N ions each separated by distance a in the one-dimensional metal of length L, then $N = L/a$. Now, the number of levels in the band is just equal to N, so the maximum value of n will also be equal to N. Thus, the maximum energy, or width of the band, is

$$\epsilon_{\max} = \frac{h^2 N^2}{8mL^2} = \frac{h^2 L^2}{8mL^2 a^2} = \frac{\hbar^2 \pi^2}{2ma^2} \tag{6.6-4}$$

This result, which depends on a but is independent of N, confirms what was said earlier that the width of a band depends on the separation of the ions and not on the number of ions in the lattice.

Consider a one-dimensional crystal that is so long that the boundary conditions can be ignored at its end. The most convenient eigenfunctions for a free electron are sinusoidal traveling waves of the form

$$\Psi \propto e^{ikx} \quad \text{and} \quad \Psi \propto e^{-ikx} \tag{6.6-5}$$

describing an electron moving in the positive or negative direction, respectively, of the x axis. It is convenient to take only the form $\Psi \propto e^{ikx}$, and let k be either positive or negative. Writing the energy ϵ of the free electron in terms of its wave number $k = p/\hbar$, where p is the momentum,

$$\epsilon = \frac{p^2}{2m} = \frac{\hbar^2 k^2}{2m} \tag{6.6-6}$$

Equation (6.6-6) is plotted in Fig. 6.6-2 over a range of k values, both positive and negative. A positive value of k corresponds to an electron moving in the positive x direction, and a negative value of k corresponds to motion in the opposite direction. The energy depends on k^2, so the curve is symmetric about $k = 0$.

It can be seen by comparing (6.6-4) and (6.6-6) that

$$-\frac{\pi}{a} \le k \le \frac{\pi}{a} \tag{6.6-7}$$

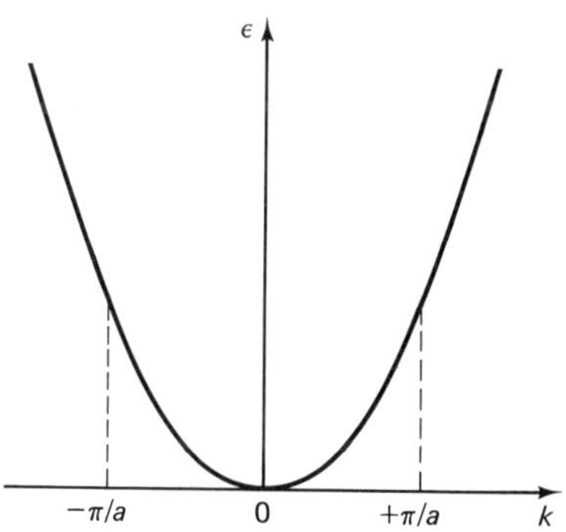

Figure 6.6-2 Energy of free electron as function of wave number.

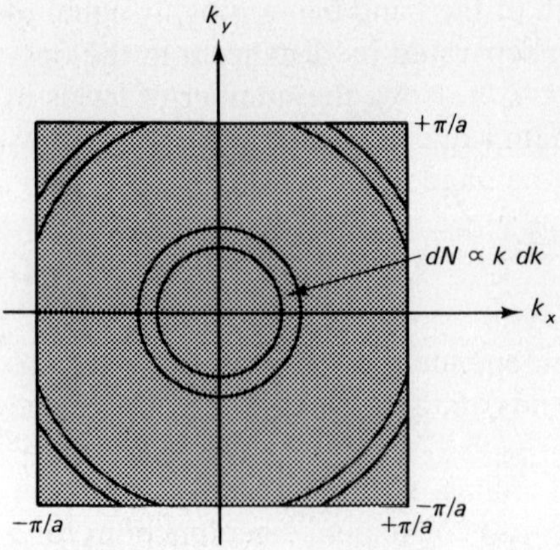

Figure 6.6-3 Allowed values of k_x and k_y for a free electron in the first band of a two-dimensional square lattice with ion separation a.

That is, the energy for the first band of the crystal falls within the range $-\pi/a \leq k \leq \pi/a$, where a is the ion separation of the one-dimensional lattice. The maximum value of ϵ is found at $-\pi/a$ and π/a, while the minimum value $\epsilon = 0$ is found at $k = 0$. Since $k \propto 1/\lambda \propto n$ and $n = 1, 2, 3, \ldots$ the values of k allowed are evenly spread throughout this range. Each of them is associated with a different quantum state the electron.

For a two-dimensional metal with ions spaced by the distance a in both the x and y directions, the allowed values of the component wave numbers k_x and k_y are uniformly distributed within a square over ranges extending from $-\pi/a$ to π/a as shown in Fig. 6.6-3. Each pair of k_x and k_y values defines a point that specifies a quantum state for a free electron of the metal. A circle surrounding the origin of radius k, where $k^2 = k_x^2 + k_y^2$, represents states having the same energy since. In two dimensions, the energy is

$$\epsilon = \frac{\hbar^2}{2m}(k_x^2 + k_y^2) = \frac{\hbar^2}{2m}k^2 \tag{6.6-8}$$

The number of states dN, for values of k ranging from k to $k + dk$, is equal to the number of points contained within the area limited by k and $k + dk$. As long as $k < \pi/a$, dN increases with increasing k; specifically $dN = 2\pi k\,dk$.

6.7 KRONIG-PENNEY MODEL

This model, historically the first (1931), is concerned with the solution of Schrödinger's equation, using an array of rectangular potential wells and barriers. The essential feature is its periodicity. In contrast to the free-electron model with its uniform potential inside the solid, the Kronig–Penney model goes one step further by taking into account the variation of potential due to the presence of immobile lattice ions.

Consider the periodic potential seen by an electron in a one-dimensional lattice as in Fig. 6.7-1. The ions are a distance a apart, and the highest potential is halfway between the ions. The actual potential is replaced by an array of rectangular wells of height V_0 and barriers of width b. It will be shown that as a consequence of periodicity, a series of allowed and forbidden energy bands arises.

Schrödinger's equation will be solved for regions where $V = V_0$ and $V = 0$ separately. Within the wells, the SWE equation is

$$-\frac{\hbar^2}{2m}\Psi_V'' + V\Psi_V = E\Psi_V \tag{6.7-1}$$

Between wells, it is

$$-\frac{\hbar^2}{2m}\Psi_0'' = E\Psi_0 \tag{6.7-2}$$

The latter has the solution

$$\Psi_0 = A\ \exp(i\beta x) + B\ \exp(-i\beta x) \tag{6.7-3}$$

where $\beta = (2mE)^{1/2}/\hbar$. The former has the solution

$$\Psi_V = C\ \exp(\alpha x) + D\ \exp(-\alpha x) \tag{6.7-4}$$

where $\alpha = [2m(V_0 - E)]^{1/2}/\hbar$.

For continuity at $x = 0$,

$$\Psi_0(0) = \Psi_V(0) \tag{6.7-5}$$

$$\left.\frac{\partial \Psi_0}{\partial x}\right|_{x=0} = \left.\frac{\partial \Psi_V}{\partial x}\right|_{x=0} \tag{6.7-6}$$

According to Block's theorem, periodicity is obtained if a solution of the form $\Psi_k(x + a) = \Psi(x)\exp(ika)$ is used. Therefore, at the barrier boundaries $x = -b$ and $x = a - b$,

$$\Psi_V(-b) = \Psi_0(a - b)\exp(-ika) \tag{6.7-7}$$

$$\left.\frac{\partial \Psi_V}{\partial x}\right|_{x=-b} = \exp(-ika)\left.\frac{\partial \Psi_0}{\partial x}\right|_{x=a-b} \tag{6.7-8}$$

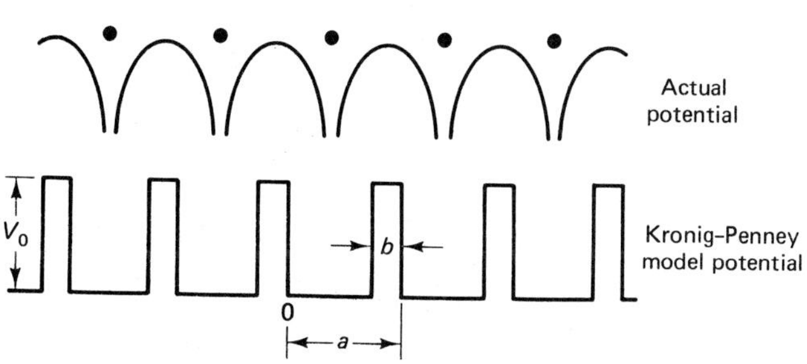

Figure 6.7-1 Potential wells and barriers for a one-dimensional crystals.

Applying the boundary conditions, four equations in four unknowns are obtained:

$$A + B = C + D \tag{6.7-9}$$

$$i\beta(A - B) = \alpha(C - D) \tag{6.7-10}$$

$$C \exp(-\alpha b) + D \exp(\alpha b) = \exp(-ika)\{A \exp[i\beta(a - b)] + B \exp[-i\beta(a - b)]\}$$

$$\alpha C \exp(-\alpha b) - \alpha D \exp(\alpha b) = \exp(-ika) i\beta\{A \exp[i\beta(a - b)] - B \exp[-i\beta(a - b)]\}$$

These are solved simultaneously by requiring the determinant of the coefficients to vanish. The result is that

$$\cos ka = [(\alpha^2 - \beta^2)/2\alpha\beta] \sinh \alpha b \, \sin \beta(a - b) + \cosh ab \, \cos \beta(a - b) \tag{6.7-11}$$

Since k is a real constant, Eq. (6.7-11) requires that the right-hand side, being an oscillating function with increasing energy, must lie between 1 and -1. In other words, there will be certain ranges of E for which the value of the function lies between 1 and -1 (these are the allowed energy values), separated by ranges of E for which the function lies outside the range of 1 to -1 (these are the forbidden energy values). Equation (6.7-11) is simplified and solved by means of a device first introduced by Kronig and Penney, yielding

$$\cos ka = \left(\frac{P}{\beta a}\right) \sin \beta a + \cos \beta a \tag{6.7-12}$$

where

$$P = \left(\frac{ma}{xh^2}\right) b V_0 = \text{const} \tag{6.7-13}$$

The device is that the width of the barrier b is allowed to decrease to zero, while the height of the barrier V_0 is allowed to increase without limit while maintaining at all times the relationship given by Eq. (6.7-13). This is equivalent to maintaining a constant barrier area while proceeding to the limit in which the rectangular wells and barriers are replaced by delta functions. A plot of the right-hand side of Eq. (6.7-12) as a function of βa is shown in Fig. 6.7-2 for $P = 3\pi/2$.

It is seen that there is a solution for the shaded region and no solution outside the shaded region. Since α is related to E, this means that the electron may possess energies within certain bands but not outside them. Expressed in another way: There are allowed and forbidden bands of energy. This is a consequence of the periodicity of the potential. Also, some conclusions can be drawn at this stage.

1. If $V_0 b$ is large, that is, if P is large the function described by the right-hand side of Eq. (6.7-12) crosses the 1, -1 region at a steeper

angle, as shown in Fig. 6.7-3. Thus, the allowed bands are narrower and the forbidden bands wider. In the limit $P \to \infty$, the allowed band reduces to one single energy level, for it follows from Eq. (6.7-12) that

$$\sin \beta a = 0 \tag{6.7-14}$$

and the permissible values of energy are

$$E_n = \frac{\pi^2 \hbar^2}{2ma^2} n^2 \tag{6.7-15}$$

which may be recognized as the energy levels for a potential well of width a. Accordingly, all electrons are independent of one another and each one is confined to one atom by an infinite potential barrier. This is the case of the discrete energy spectrum for isolated atoms.

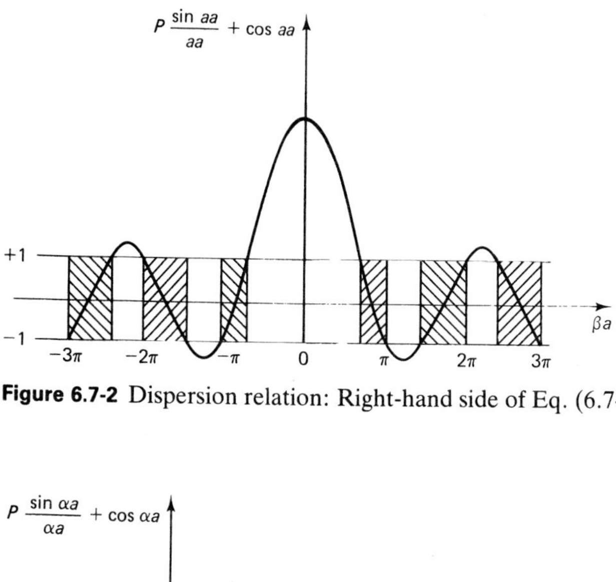

Figure 6.7-2 Dispersion relation: Right-hand side of Eq. (6.7-12) as function of βa.

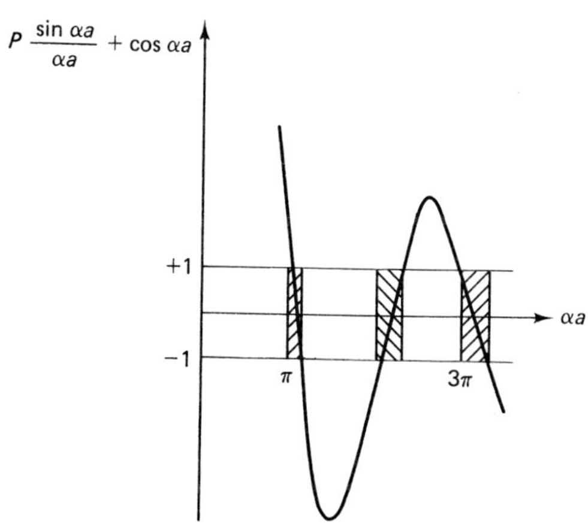

Figure 6.7-3 Right-hand side of Eq. (6.7-12) for $P = 6\pi$, larger than $P = 3\pi/2$ in Fig. 6.7-2.

2. In the other limit, $P \to 0$, Eq. (6.7-12) yields $\cos \beta a = \cos ka$, with,

$$E = \frac{\hbar^2 k^2}{2m} \tag{6.7-16}$$

This is the case of the bound electron. Thus, by ranging P from zero to infinity, the whole range from completely free electron to completely bound electron is covered.

3. At the boundary of an allowed band, $\cos ka = \pm 1$; that is,

$$k = \frac{n\pi}{a} \qquad n = 1, 2, 3, \ldots \tag{6.7-17}$$

But this is exactly Bragg's reflection rule ($n\lambda = 2d \sin\theta$) cast into its one-dimensional form ($n\lambda = 2d$). It is seen that the conditions described by Eq. (6.7-17) are the same as those set forth by the Bragg reflection rule. Accordingly, an allowed energy band in the Kronig–Penney model terminates at a condition corresponding to a Bragg reflection. Electron waves with $k = n\pi/a$ cannot propagate in the crystal and hence correspond physically to forbidden energies.

6.8 PERIODIC POTENTIAL AND FREE ELECTRONS

As a final approach toward the existence of energy bands in crystalline solids, consider what would happen if a small periodic potential were to be superposed on the free electrons, such as they would experience in a crystal. For free electrons,

$$E = \frac{\hbar^2 k^2}{2m} \tag{6.8-1}$$

and k can take on all values. Once a periodic potential is established, however, electron waves with particular values of k cannot propagate through the crystal because they undergo Bragg reflections. For those values of

$$k = \frac{n\pi}{a} \tag{6.8-2}$$

electron waves are represented by standing waves, not traveling waves. The effect of the periodic potential is to open up energy gaps in the previously continuous $E(k)$ variation whenever $k = n\pi/a$.

Consider the free electron as a propagating plane wave. Its wave function is then

$$\Psi_k = e^{ikx} \tag{6.8-3}$$

It may move across the crystal lattice and be reflected, so the wave function should contain a term representing a wave in the opposite direction:

$$\Psi_{-k} = e^{-ikx} \tag{6.8-4}$$

For each value of k two distinct wave functions Ψ_+ and Ψ_- can be constructed, with corresponding potential energies V_n and $-V_n$, respectively. Thus, the total energies are, for each n,

$$E_\pm = \frac{\hbar^2 k^2}{2m} \pm V_n \tag{6.8-5}$$

For $k = \pi/a$, the energy of the electron may be

$$\frac{\hbar^2 k^2}{2m} - V_1 \quad \text{or} \quad \frac{\hbar^2 k^2}{2m} + V_1 \tag{6.8-6}$$

but cannot be of any value in between. There is an energy gap. This is shown in Fig. 6.8-1 on the $E - k$ curve, which is constructed in the following manner. A parabola (dotted line) is drawn for the free electron and V_n added and subtracted to it at points $k = n\pi/a$.

A typical free-electron-like $E(k)$ plot for the Kronig–Penney model is shown in Fig. 6.8-2, with energy gaps at those values of k corresponding to Bragg reflection. Each separate segment of the curve can be considered to still exhibit free-electron-like behavior except for values of k near $n\pi/a$. The representation is often called an "extended representation" because $E(k)$ is plotted for all values of k. Because of the periodic nature of the solution, a displacement of k by $n2\pi/a$ does not alter the solution. It is therefore possible to present the same information by means of the "reduced representation" shown in Fig. 6.8-2(b). Here, all values of E are plotted for values of k between $-\pi/a$ and π/a, the individual segments of Fig. 6.8-2(a) each having been translated by multiples of $2\pi/a$ to bring them into this range of values. Each band in the reduced zone can be characterized by an index l. For example, $l = 1$ corresponds to an untranslated segment of the original curve, $l = 2$ and $l = 3$ each correspond to segments that have been translated by $2\pi/a$, etc. The shaded areas represent the allowed bands which are separated by forbidden gaps.

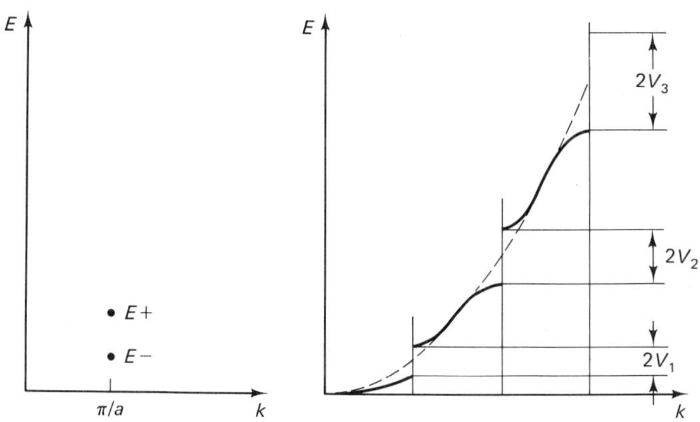

Figure 6.8-1 Construction of the $E - k$ curve from the free electron parabola.

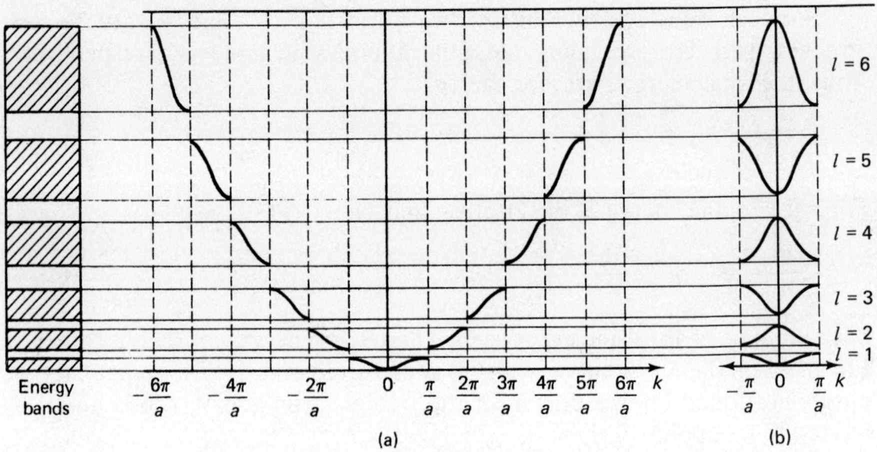

Figure 6.8-2 $E - k$ plot for Kronig–Penney model.

The reduced representation consisting of values of k between $-\pi/a$ and π/a is often called the *first Brillouin zone* for the one-dimensional lattice. Values of k between $-2\pi/a$ and $-\pi/a$, and between π/a and $2\pi/a$ constitute the *second Brillouin zone,* etc. in the extended representation, as detailed in Fig. 6.8-3.

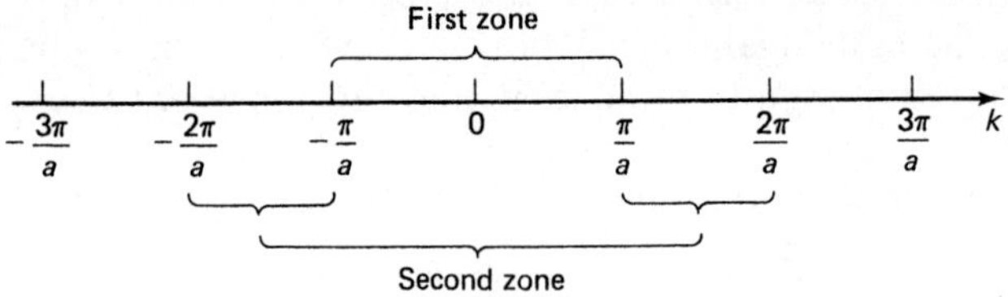

Figure 6.8-3 First two Brillouin zones for a one-dimensional lattice.

For two dimensions the wave number has components k_x and k_y. The first Brillouin zone is identified by the values $\pm\pi/a$ along the k_x axis and $\pm\pi/a$ along the k_y axis. The first Brillouin zone is thus a square passing through the points A, B, C, and D, as shown in Fig. 6.8-4. The second Brillouin zone passes through the points E, F, G, and H, identified by

$$\left.\begin{aligned} k_x + k_y &= \frac{2\pi}{a} \\ -k_x + k_y &= \frac{2\pi}{a} \\ -k_x - k_y &= \frac{2\pi}{a} \\ -k_x + k_y &= \frac{2\pi}{a} \end{aligned}\right\} \qquad (6.8\text{-}7)$$

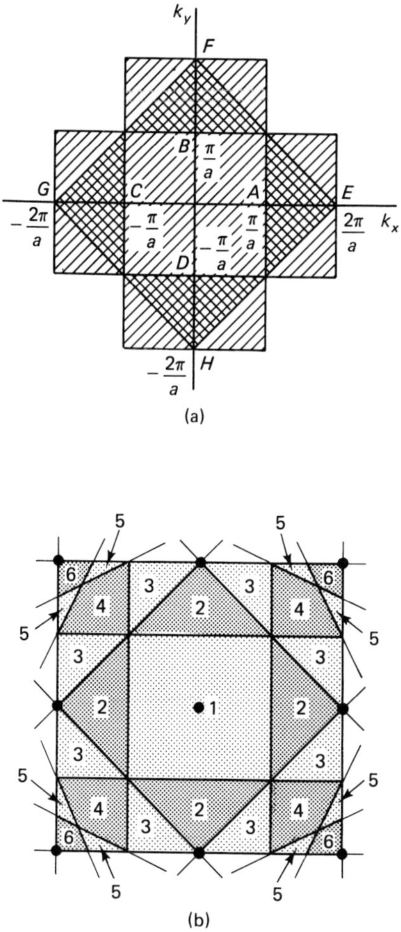

Figure 6.8-4 Two-dimensional Brillouin zones.

Generally, in two dimensions,

$$n_1 k_x + n_2 k_y = (n_1^2 + n_2^2)\,\frac{\pi}{a} \tag{6.8-8}$$

These four equations describe a set of four lines at 45° to the k_x and k_y axes passing through E, F, G, and H. The second Brillouin zone is thus the region between the squares $ABCD$ and $EFGH$. The third Brillouin zone is identified by the multipliers 0, ±1, ±2 for the k_x and k_y. Figure 6.8-4(a) shows the boundaries. Bragg planes (lines, in two dimensions) delineate zones for a two-dimensional square Bravais lattice in Fig. 6.8-4(b).

As for three dimensions, the Brillouin zones may be evaluated, using the generalized equation

$$n_1 k_x + n_2 k_y + n_3 k_z = (n_1^2 + n_2^2 + n_3^2)\,\frac{\pi}{a} \tag{6.8-9}$$

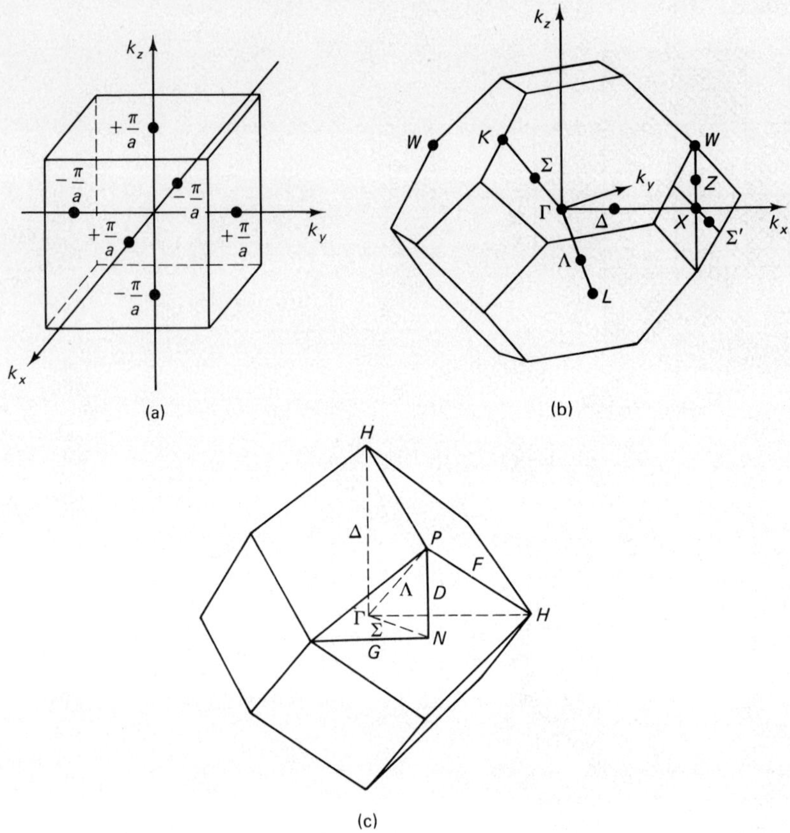

Figure 6.8-5 Three-dimensional Brillouin zones for crystal structures. (a) Simple cubic, (b) fare-centered cubic, and (c) body-centered cubic.

For a simple cubic lattice, the first Brillouin zone is a cube with equal extensions in k_x, k_y, and k_z between $-\pi/a$ and π/a. For other crystal structures, Brillouin zones can be determined by considering the crystal symmetry. Examples for simple cubic, face-centered cubic, and body-centered cubic are shown in Fig. 6.8-5.

■ EXAMPLE 6.8-1

Illustrate, in terms of the band picture, the basic electrical differences between solids (metals, semiconductors, and insulators). Also, give a summary of the different band representations.

Solution. Figure 6.8-6 shows the variation of $N(E)$ with E for metals, insulators, and semiconductors. If the highest-energy-allowed band to be

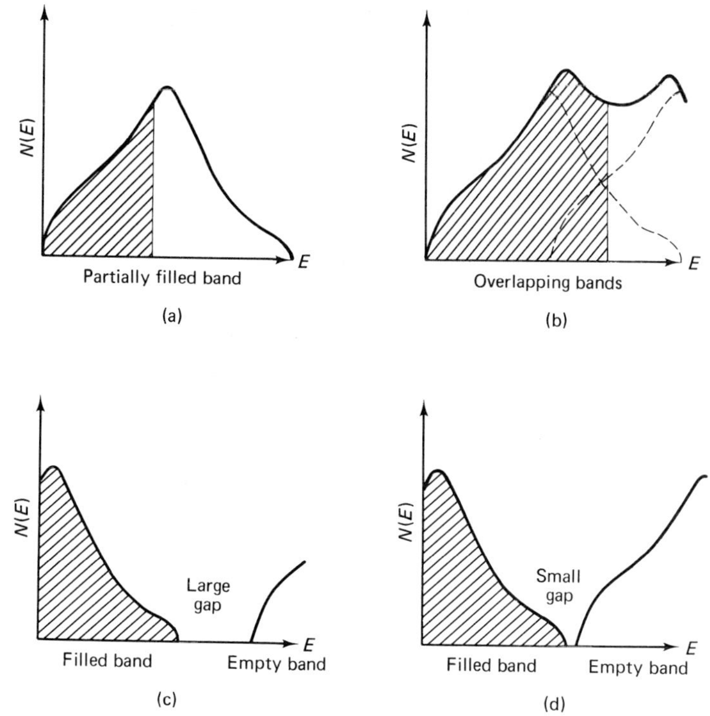

Figure 6.8-6 Band filling and bond spacing for metals, semiconductors, and insulators.

occupied by electrons is only partially occupied, the solid is a metal, and a drift of electrons in an electric field can be readily achieved. Even if the highest-energy-allowed band is completely filled, metallic behavior holds if the next higher-lying band overlaps the filled band. Metals are therefore depicted in Fig. 6.8-6(a) and (b). If the highest-energy-allowed band to be occupied by electrons is filled and the next-higher-lying band lies an appreciable distance above the top of the filled band, the solid is an insulator [Fig. 6.8-6(c)]. Thermal excitation across a large energy gap is required to raise electrons from the filled valence to the empty conduction band. Finally, if the gap between the top of the filled band and the higher-lying empty band is small, as in Fig. 6.8-6(d), intermediate conductivity occurs, and the material is classified as a semiconductor.

There are four different ways of describing the properties of an energy band in a crystalline solid: (1) energy as a function of position in the crystal: this is the familiar "flat band" E versus x plot shown in Fig. 6.8-7(a). It emphasizes the nonlocalized nature of the band states that extend throughout the crystal. (2) The E versus k curve of Fig. 6.8-7(b) is the dispersion relation for the Kronig–Penney model. (3) The $N(E)$ versus E curves in Fig. 6.8-7(c) give the variation of the density of states within a band, and (4) in Fig. 6.8-7(d), the

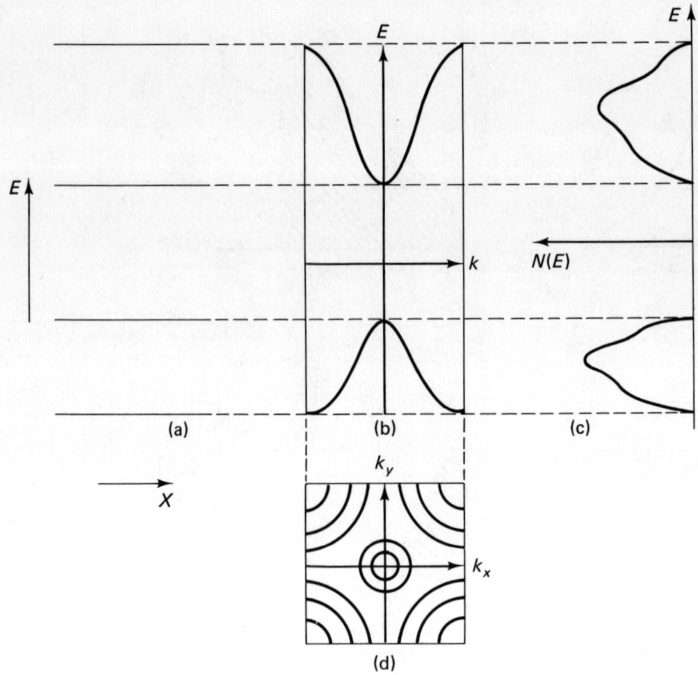

Figure 6.8-7 Description of band properties in solids.

equal-energy surfaces in k space that correlate with the shape of the E versus k waves. ■

■ EXAMPLE 6.8-2

The Kronig–Penney model shows that within each Brillouin zone the relationship between energy E and wave number k is given by the dispersion relation, Eq. (6.7-12):

$$\frac{P \sin \alpha a}{\alpha a} + \cos \alpha a = \cos ka$$

Discuss how constant-energy curves may be obtained for the cases of one- and two-dimensional lattices.

Solution. In one dimension, the relation between E and k is as given, and constant-energy curves, that is, curves relating energy and k values, are as shown in Fig. 6.8-7. In two dimensions, all points of equal k value, that is, $(k_x^2 + k_y^2)^{1/2}$, have the same energy, as the curve is a circle, except when it approaches a Bragg line. Figure 6.8-8 shows the loci of points with equal energy for the first Brillouin zone of a square lattice. Since points close to the

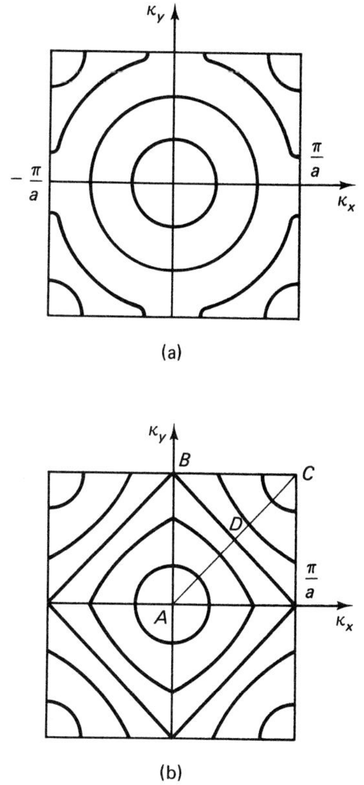

Figure 6.8-8 Equal energy contours in the first Brillouin zone of a two-dimensional square lattice.

origin correspond to electrons with small values of k (low momentum), which suffer the least perturbation by the periodic lattice, the loci of equal energy will be circles near the center of the zone. For k values that reach further out into the zone, the effect of the periodic perturbations becomes of greater importance and the curves of equal energy deviate markedly from circular form. The degree to which these curves deviate from circular depends, of course, on the strength of the electron binding, that is, on the value of P in the dispersion relation, the stronger the binding the closer to the origin will departure from the circular form be found, and vice versa.

Consider what happens at $T = 0$ as the available states are filled with electrons. Figure 6.8-9 shows contours of $E = 1.0$ and $E = 1.5$ in the first Brillouin zone. When all the states (starting with $E = 0$ at the origin) are filled as in Fig. 6.8-9(a), the next electrons will begin the $E = 1.5$ bounded area until it, too, is filled, and so on. One thing should be made clear: If an energy gap is large, all states in a lower band will be filled, and the material is an insulator; if the energy gap is small some states will remain unfilled in the lower band, and some states will be filled in the higher band. Both bands will then contribute to electrical conduction.

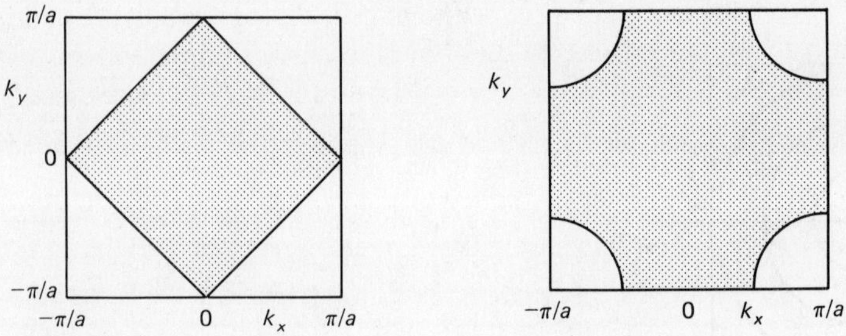

Figure 6.8-9 Filling of bands in a two-dimensional square lattice. ■

■ EXAMPLE 6.8-3

Consider two highest bands (valence and conduction bands) as in Fig. 6.8-10. At $T = 0$ K, the demarcation line between filled and unfilled states is sharp. Discuss what happens in a metal, semiconductor, and insulator at higher temperatures.

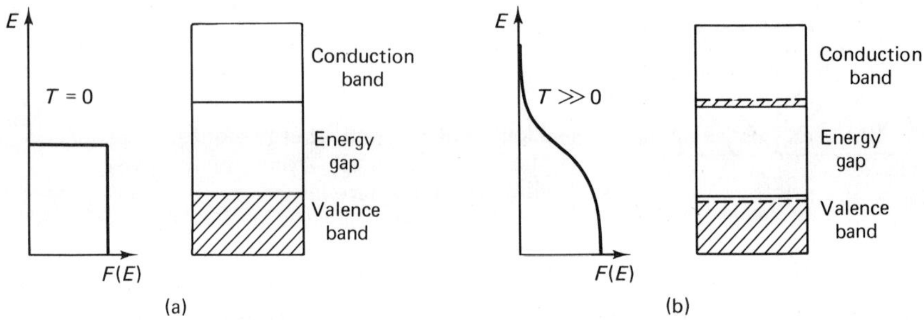

Figure 6.8-10 The two highest bands at $T = 0$ K and $T >> 0$ K. (Electrons at the bottom of the conduction band and holes at the top of the valence band.)

Solution. For temperatures above absolute zero (say, about room temperature, at which most electronic devices are supposed to work), it is no longer valid to assume that all states up to the Fermi energy are filled and all states above that are empty. The demarcation line between filled and unfilled states becomes less sharp.

In a metal, its highest-energy band is about half filled at absolute zero; at higher temperatures some of the electrons acquire somewhat higher energies in the band, but that is all. There is very little change in the effective number of electrons. A metal stays a metal at higher temperatures.

For the case of an insulator, Fig. 6.8-10(b) tells the story. Consider the two highest bands (the valence and conduction bands) and take the zero of energy at the top of the valence band. At finite temperatures (above zero), the Fermi function near the top of the valence band being less than 1, some states in the valence band will remain empty. By the same token, the probability function near the bottom of the conduction being greater than 0, some electrons will occupy states in the conduction band. Thus, at finite temperatures, an insulator is no longer an insulator. There is conduction by electrons in the conduction band and by holes in the valence band.

The same holds for semiconductors, except that the energy gap is much less. For example, diamond, an insulator, has an energy gap of 7 eV, whereas silicon and germanium, semiconductors, have energy gaps of 1.1 and 0.65 eV, respectively. There is no difference in principle: Insulators and semiconductors are distinguished only by the magnitudes of their respective energy gaps.

In a more general way, the Kronig–Penney model shows that for electrons in a one-dimensional solid, energy gaps appear at $k_x = \pm\frac{1}{2}N(2\pi/L_x)$, $\pm\frac{3}{2}N(2\pi/L_x), \ldots$, where N is the number of ions present and L_x/N is the nearest neighbor separation. The region between $k_x = -\frac{1}{2}N(2\pi/L_x)$ and $k_x = \frac{1}{2}N(2\pi/L_x)$ is referred to as the first Brillouin zone, the region between $k_x = -N(2\pi/L_x)$ and $k_x = N(2\pi/L_x)$ the second Brillouin zone, and so on. The first Brillouin zone contains $\frac{1}{2}N + \frac{1}{2}N = N$ allowed values of k_x. Since each such value of k_x can describe the translational momenta of two electrons of opposite spins it follows that the first Brillouin zone (and hence the first allowed energy band) can accommodate $2N$ electrons. The same is also true of all other Brillouin zones.

The final step is to see the effect of having different numbers of electrons present in a solid. Sodium, for example, has one conduction electron per atom. Thus (picturing it to be in the form of a one-dimensional crystal) its electrons will only fill the states from $k_x = -\frac{1}{4}N(2\pi/L_x)$ to $k_x = \frac{1}{4}N(2\pi/L_x)$, as shown in Fig. 6.8-11(a). It is seen that its highest-energy electrons are well removed from a band edge, where the electrostatic potential energy of an electron becomes significant. This explains the conduction characteristics of sodium. Extending the argument further, it shows that elements with an odd number of valence electrons per atom should be conductors, as is generally the case. Figure 6.8-11(b) illustrates the situation of three outer electrons per atom (a conductor). If there are an even number of valence electrons per atom, the solid is an insulator.

Diamond, for example, has four such electrons [Fig. 6.8-11(c)]. In the case of diamond the energy gap E_g which the top electrons have to cross before they can enter the next allowed energy band is around 7 eV. This is much larger than the thermal energy $\frac{3}{2}kT$(=0.04 eV at 300 K) possessed by these electrons. If, however, the material has a small energy gap, say around 1 eV, an appreciable number of electrons can be thermally excited across the gap. Under these circumstances, the material is an intrinsic semiconductor [Fig.6.8-11(d)]. Examples are silicon and germanium. The distinction between an intrinsic semiconductor and an insulator lies only in their values of E_g and how these values compare with $\frac{3}{2}kT$. In fact, all intrinsic semiconductors are insulators at $T = 0$.

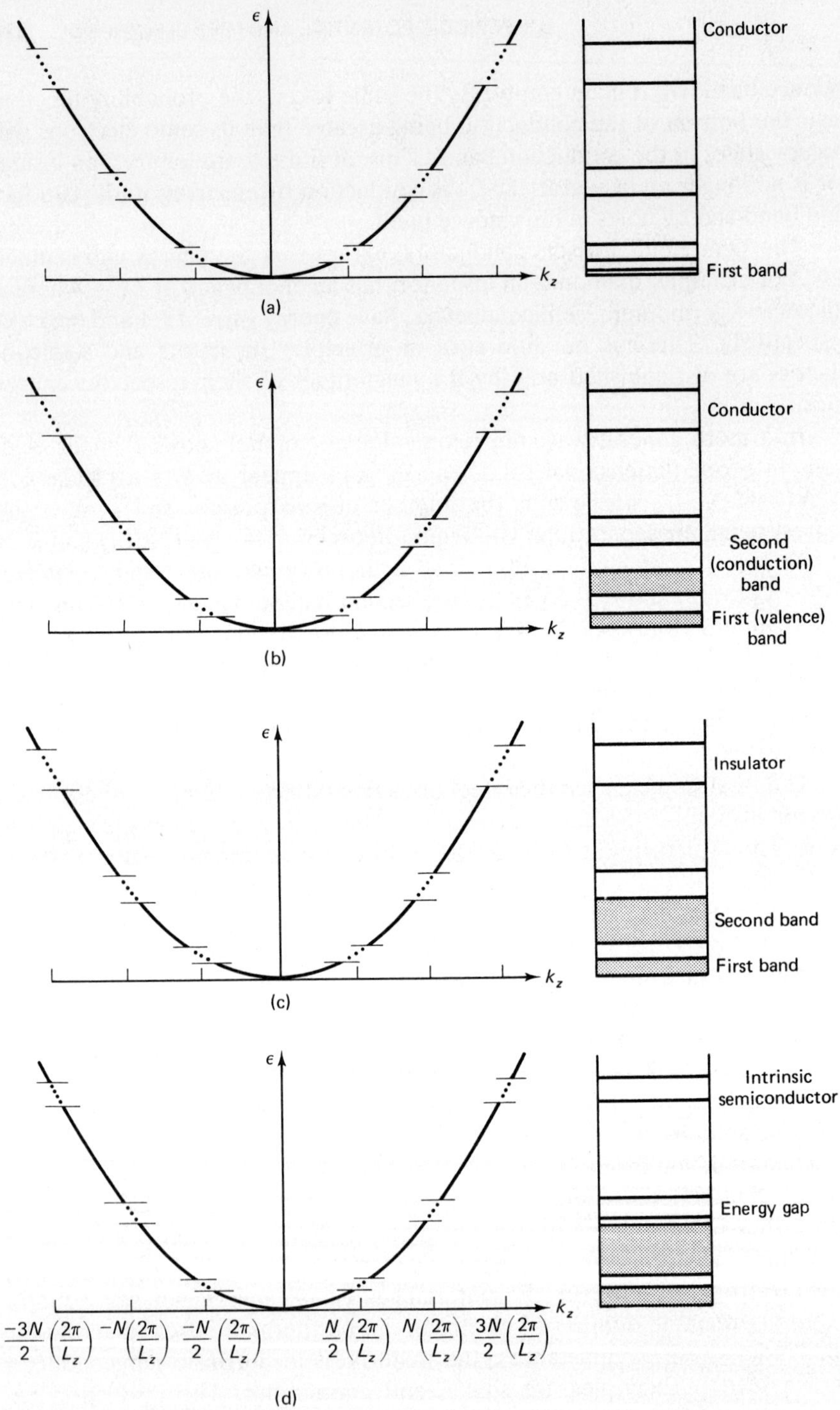

Figure 6.8-11 Energy bands for (a) a conductor with one valence electron per atom, (b) a conductor with three valence electrons per atom, (c) an insulator with four valence per atom, and (d) an intrinsic semiconductor with four valence electrons per atom. ■

■ EXAMPLE 6.8-4

(a) Draw and explain energy band diagrams for conductors, semiconductors, and insulators.

(b) The conductivity σ of conductors decreases as the temperature increases. How does σ change with temperature for intrinsic semiconductors and insulators?

Solution

(a) Solids may be conductors because of a partially filled conduction band or overlapping valence and conduction bands as shown in Fig. 6.8-12(a). In semiconductors and insulators, the valence and conduction bands don't overlap, but remain separated as in Fig. 6.8-12(b).

In Fig. 6.8-13, gray tin, germanium, silicon, and diamond have two s and two p outer electrons and the same structure. In gray Sn, the valence and conduction bands overlap, making it a metal. In Ge and Si, an energy gap of 1 eV between these bands makes the materials semiconductors. In diamond, the forbidden energy gap is greater than 5 eV, making diamond an insulator.

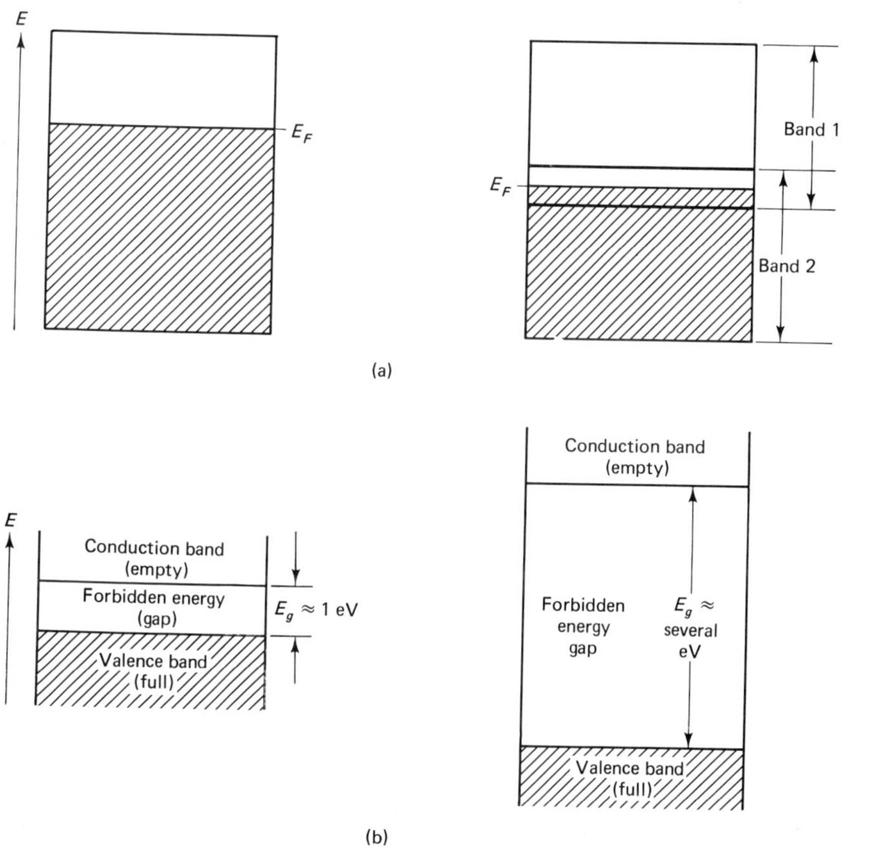

Figure 6.8-12 Energy band representation for conductors, semiconductors, and insulators.

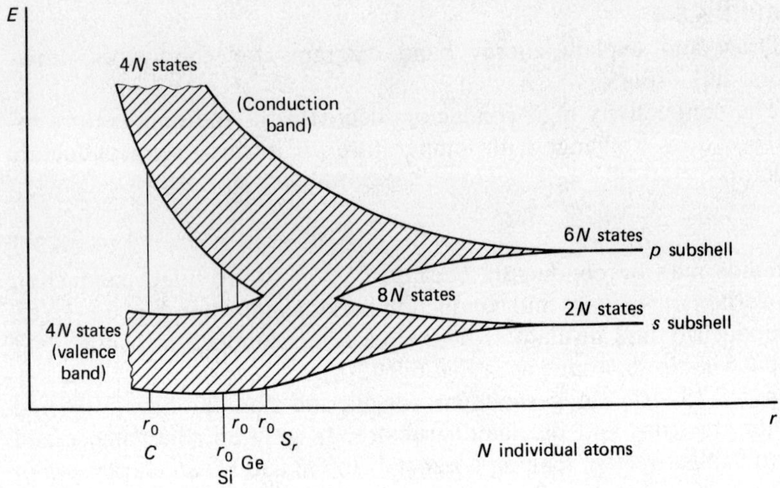

Figure 6.8-13 Band diagram for intinsic semiconductor and insulator.

(b) Because of an increasing number of electron–phonon interactions, the conductivity σ of conductors decreases as the temperature increases. For intrinsic semiconductors and insulators, at $T = 0$ K, all the electrons are in the full-valence band, and there is no conduction. Therefore, σ is zero at absolute zero. As T increases above absolute zero, electrons are thermally excited into the conduction band, and, σ increases with T. ■

6.9 EFFECTIVE MASS

When discussing the behavior of an electron in a periodic lattice under the influence of an external electric field, it is convenient to introduce the concept of an *effective mass*. It is known that an electron has a well-defined mass, and when accelerated by an electric field it obeys Newtonian mechanics. What happens when the electron to be accelerated is inside a crystal? The mass of an electron in a crystal is, in general, different from the free-electron mass.

From quantum mechanics, the velocity of an electron in a one-dimensional lattice is given by its group velocity

$$v_g = \frac{1}{\hbar}\frac{\partial E}{\partial k} \tag{6.9-1}$$

(Recall that information, energy, and particles move with the group velocity and that the maximum magnitude of v_g is c.) The acceleration is obtained by differentiating Eq. (6.9-1):

$$\frac{dv_g}{dt} = \frac{1}{\hbar}\frac{d}{dt}\frac{\partial E}{\partial k} = \frac{1}{\hbar}\frac{\partial^2 E}{\partial k^2}\frac{dk}{dt} \tag{6.9-2}$$

Classically, the work done by a particle traveling a distance $v_g\,dt$ under the influence of a force $e\mathscr{E}$ is

$$dE = e\mathscr{E} v_g\,dt = e\mathscr{E}\frac{1}{\hbar}\frac{\partial E}{\partial k}\,dt \qquad (6.9\text{-}2)$$

Expressing dk/dt from Eq. (6.9-2) and substituting it into Eq. (6.9-1a) gives

$$\frac{dv_g}{dt} = \frac{1}{\hbar^2}\frac{\partial^2 E}{\partial k^2}\,e\mathscr{E} \qquad (6.9\text{-}3)$$

Comparing this with the expression for a free classical particle,

$$\frac{m\,dv}{dt} = e\mathscr{E} \qquad (6.9\text{-}4)$$

it is seen that an effective mass m^* may be defined as

$$m^* = \hbar^2\left(\frac{\partial^2 E}{\partial k^2}\right)^{-1} \qquad (6.9\text{-}5)$$

Thus, an electron in a crystal lattice reacts to an electric field $\mathscr{E}$, but its mass is given by Eq. (6.9-5) in contrast to the mass of a free electron. Graphs of energy, group velocity, and effective mass are plotted as a function of k between $-\pi/a$ and π/a in Fig. 6.9-1 for two illustrative band shapes. It is seen that m^* may go to infinity and also take on negative values. When an electron,

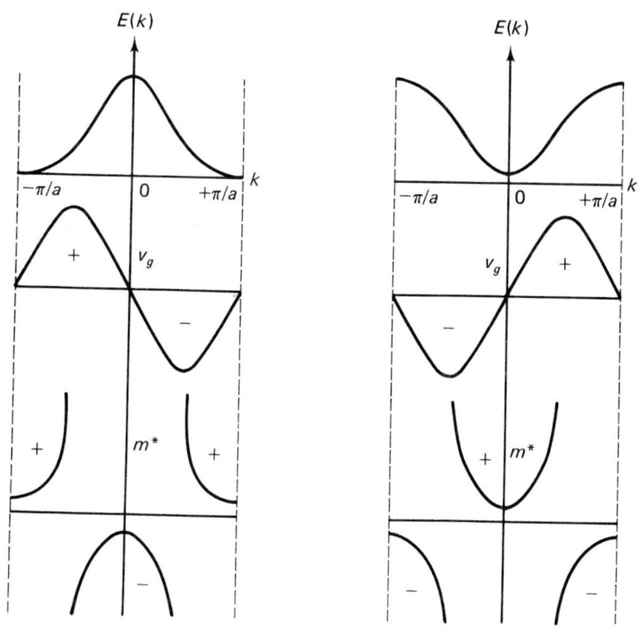

Figure 6.9-1 Energy, group velocity, and effective mass as function of k for two different band shapes.

initially at rest at $k = 0$, is accelerated by an electric field, it will move to higher values of k, and will become heavier and heavier, reaching infinity at $k = \pi/2a$. For even higher values of k, the effective mass becomes negative, presaging the advent of a new particle, the hole.

The definition of effective mass, as given by Eq. (6.9-5), is for a one-dimensional crystal. It can be generalized to three dimensions by noting that the energy is in terms of k_x, k_y, and k_z. For example, the effective mass in the x direction is

$$m_x^* = \hbar^2\left(\frac{\partial^2 E}{\partial k_x^2}\right)^{-1} \tag{6.9-6}$$

In the y direction it is

$$m_y^* = \hbar^2\left(\frac{\partial^2 E}{\partial k_y^2}\right)^{-1} \tag{6.9-7}$$

and similarly in the z direction. Oddly enough, the effective mass may be different in different directions. Physically, this means that the same elective field applied in different directions will cause varying amounts of acceleration.

■ EXAMPLE 6.9-1

The quantity $1/m^*$ is the reciprocal of the effective mass of the electron in a crystal lattice. Show, by means of the function $E(k)$, how the properties of the lattice determine $1/m^*$.

Solution. From the definition of m^* through Eq. (6.9-5),

$$\frac{1}{m^*} = \frac{1}{\hbar^2}\frac{d^2 E}{dk^2} \tag{6.9-8}$$

it is seen that the form of $E(k)$ and its derivatives determine $1/m^*$. Figure 6.9-2 shows the first, and part of the second, Brillouin zones of a one-dimensional crystal. The solid curve is $E(k)$, and the parabolic dashed curve in the free-electron relation

$$E = \frac{\hbar^2 k^2}{2m} \tag{6.9-9}$$

Near the center of the first zone, where Eq. (6.9-9) holds,

$$\frac{1}{m^*} = \frac{1}{\hbar^2}\left(\frac{d^2 E}{dk^2}\right) \simeq \left(\frac{2\hbar^2}{2m}\right)\hbar^2 = \frac{1}{m}$$

So in this region the lattice has very little effect on the electron, because its reciprocal effective mass is almost the same as its reciprocal true mass, and it responds to an applied electric field as if it were an essentially free electron.

The curvature of the function $E(k)$ changes significantly from the curvature of the parabola in proceeding in either direction from the center of the zone. Since d^2E/dk goes through zero, and then becomes negative and of

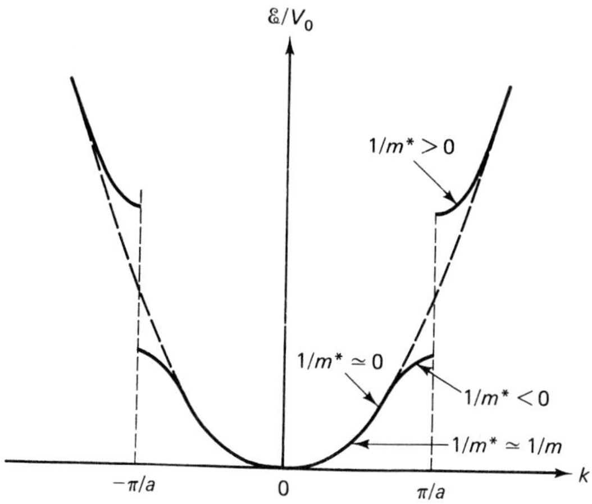

Figure 6.9-2 Reciprocal effective mass at different locations in the first and second Brillouin zones of a one-dimensional lattice.

large magnitude as k approaches either boundary of the first zone, $1/m^*$ does the same. Where $1/m^*$ is zero, a given applied force $e\mathscr{E}$ causes no acceleration of the electron. Where $1/m^*$ is negative the force causes an acceleration in the opposite direction of that experienced by a free electron. At the bottom of the energy band for the second Brillouin zone, $1/m^*$ is positive, but larger than $1/m$ for a free electron, so an applied force produces a relatively large acceleration of the electron in the lattice. ■

6.10 ELECTRICAL CONDUCTION IN SOLIDS

The electrical conductivity of materials varies over a wide range of values, from $10^9\ (\Omega\ \text{cm})^{-1}$ for metals to less than $10^{-15}\ (\Omega\ \text{cm})^{-1}$ for insulators. Semiconductors have a room temperature conductivity of $1\ (\Omega\ \text{cm})^{-1}$, although this depends on both the purity and temperature.

In simple discussion of Ohm's law, it is usual to relate I, the current flowing in a circuit, to the resistance R and the applied voltage V by

$$I = \frac{V}{R} \tag{6.10-1}$$

The current and voltage are the variables, and the resistance is the proportionality factor (constant independent of I or V, if Ohm's law holds). Equation (6.10-1) can also be written in an equivalent form,

$$J = \sigma\mathscr{E} \tag{6.10-2}$$

where J is the current density (current per unit area), σ is the electrical conductivity, and $\mathscr{E}$ is the electrical field. If A is the cross-sectional area of the material and l is the distance between electrodes with which the field is applied, the correlation between the two forms gives $J = I/A$, $\mathscr{E} = V/l$, and $\sigma = (1/R)(l/A)$. Again, σ must be independent of J or $\mathscr{E}$ for Ohm's law to hold.

The current density is

$$J = nqv_d \tag{6.10-3}$$

where n is the density of the free carriers, q is the quantum of charge, and v_d is the drift velocity induced by the electric field. The drift velocity is different from the thermal velocity $\langle v\rangle = (2kT/m^*)^{1/2}$ which has random direction. The drift speed is much less than the instantaneous speed of random motion. In copper for example, v_d is of the order of 10^{-2} cm/sec, whereas $\langle v\rangle$ is of the order of 10^8 cm/sec. The drift speed is proportional to the electric field:

$$v_d = \mu\mathscr{E} \tag{6.10-4}$$

μ is the *mobility* (velocity per unit field). Its units are (m/s) per (V/m), or m^2/V sec.

Combining Eqs. (6.10-2), (6.10-3), and (6.10-4) gives

$$\sigma = nq\mu \tag{6.10-5}$$

In general, since conduction involves holes as well as electrons,

$$\sigma = q(n\mu_n + p\mu_p) \tag{6.10-6}$$

where n and p are the densities of electrons and holes, and μ_n and μ_p are the mobilities of electrons and holes. Ohm's law requires that neither n nor μ be a function of the electric field. In a metal, the density of free carriers, n, is the density of valence electrons (a constant independent of temperature). It is of the order of $10^{22}/cm^3$. Combined with values of mobility μ of 10^2 to 10^3 cm^2/V sec this gives a conductivity of 10^5 to 10^6 $(\Omega\ cm)^{-1}$ that are observed in metals. For semiconductors, n is a rapidly increasing function of temperature (due to thermal excitation of electrons, either from imperfections or across the band gap).

Note that if conduction is by negative charge carriers, the charge q is negative, whereas if conduction is by positive carriers, q is positive. Since the sign of μ also depends on the sign of q, each term in the expression (6.10-6) for σ is always positive. The sign of the charge carrier can be determined from measurements of the *Hall effect*. This occurs when a voltage and magnetic field are applied to a rectangular piece of material as shown in Fig. 6.10-1. Current, by convention, flows from the positive side to the negative side, that is, in the direction of the negative z axis. But electrons flow in a direction opposite to the conventional current, that is, from left to right.

Let a magnetic field B be applied perpendicular to the current density J as shown. This charging will cause an electric field ε_H, which at equilibrium will produce a force to balance the magnetic force. The Hall electric field per current density per magnetic field, called the *Hall coefficient*, is

$$R_H = \frac{\varepsilon_H}{JB} \tag{6.10-7}$$

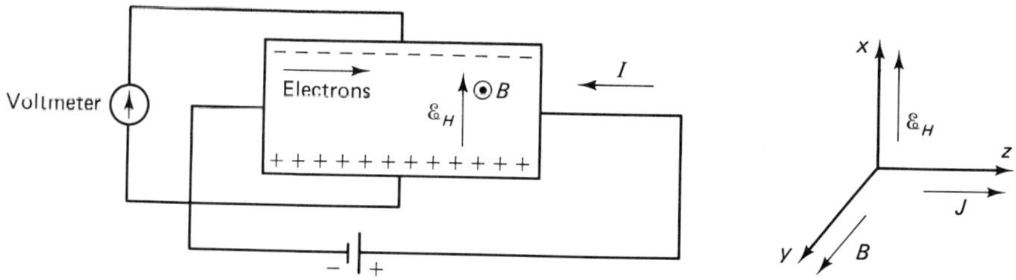

Figure 6.10-1 Measurement of the Hall effect.

R_H has the units of m^3/C. Since Eq. (6.10-3) gives $v_d = J/n(-e)$ for electrons and $v_d = J/p(+e)$ for holes, that is,

$$\varepsilon_H = \frac{JB}{n(-e)} \quad \text{for electrons,} \quad \varepsilon_H = \frac{JB}{p(+e)} \quad \text{for holes,} \tag{6.10-8}$$

and (6.10-7) results in

$$R_H = -\frac{1}{ne} \quad \text{for electrons} \quad R_H = \frac{1}{pe} \quad \text{for holes} \tag{6.10-9}$$

If both electrons and holes are present, a more complicated derivation gives

$$R_H = \frac{1}{e}\frac{(p\mu_p^2 - n\mu_n^2)}{(p\mu_p + n\mu_n)^2} \tag{6.10-10}$$

In an experiment, ε_H, J, and B are measurable; R_H, and with it the density of electrons or holes, can be determined. If only one type of carrier is present, the measurement gives immediately the sign and density of the carrier. If both carriers are simultaneously present, more algebra is needed to extract the information.

■ EXAMPLE 6.10-1

(a) The electron mobility and the hole mobility of Si at room temperature is 0.135 m^2/V s and 0.048 m^2/V s, respectively. The conductivity is 3.8×10^{-4} $(\Omega\ \text{m})^{-1}$. Calculate the free-electron and free-hole concentrations of intrinsic Si.

(b) Calculate the conductivity of n-type Si with 1.3×10^{19} free electrons/m^3 and 1.3×10^{13} free holes/m^3.

(c) A large field $\mathscr{E} = 2.0 \times 10^6$ V/m is applied in the x direction. What are the electron and hole drift velocities in intrinsic Si?

Solution

(a) In an intrinsic material, the number of electrons in the conduction band equals the number of holes in the valence band. That is, $n = p$ in Eq. (6.10-6):

$$\sigma = e(n\mu_n + p\mu_p) = en(\mu_n + \mu_p) \tag{6.10-11}$$

$$n = \frac{\sigma}{e(\mu_n + \mu_p)} = \frac{3.38 \times 10^{-4}\ (\Omega\ \text{m})^{-1}}{(1.60 \times 10^{-19}\ \text{C})(0.135 + 0.048)\ \text{m}^2/\text{V s}}$$

$$= 1.3 \times 10^{16}\ \text{m}^{-3} = p$$

(Note that $\Omega = V/A$ and $C = \text{A s}$.)

(b) Direct substitution into Eq. (6.10-6) gives

$$\sigma = q(n\mu_n + p\mu_p) \tag{6.10-12}$$

$$= (1.60 \times 10^{-19}\ \text{C})[(1.3 \times 10^{19}\ \text{m}^3)(0.135\ \text{m}^2/\text{V s})$$

$$+ (1.3 \times 10^{13}\ \text{m}^{-3})(0.048\ \text{m}^2/\text{V s})] = 0.28\ (\Omega\ \text{m})^{-1}$$

(c) From Eq. (6.10-4), $v_d = \mu\varepsilon$. For the electrons in Si,

$$v_d = (0.135\ \text{m}^2/\text{V s})(2.0 \times 10^6\ \text{V/m}) = 2.7 \times 10^5\ \text{m/s}$$

in the $-x$ direction (opposite the direction of ε). For the positive holes,

$$v_d = (0.048\ \text{m}^2/\text{V s})(2.0 \times 10^6\ \text{V/m}) = 9.6 \times 10^4\ \text{m/s}$$

in the x direction.

Note that there are about 4.99×10^{28} Si atoms/m^3, and $n = p = 1.\times 10^{16}/\text{m}^3$ in part (a). Therefore, the fraction of electrically active impurities must be less than $(1.3 \times 10^{16})/(4.99 \times 10^{28}) = 2.6 \times 10^{-13}$ for Si to be intrinsic at room temperature. This is less than three parts per 10 million, so that making such pure material is not an easy task. ■

■ EXAMPLE 6.10-2

Calculate the Hall coefficient for (a) intrinsic Si at room temperature having electron and hole concentration of $1.3 \times 10^{16}/\text{m}^3$ and electron and hole mobilities of 0.135 m^2/V s and 0.048 m^2/V s, respectively; (b) n-type Si having 1.3×10^{19} free electrons/m^3 and 1.3×10^{13} free holes/m^3.

Solution

(a) For an intrinsic semiconductor, $n = p$, and Eq. (6.10-10) simplifies to

$$R_H = \frac{1}{e}\frac{p\mu_p^2 - n\mu_n^2}{(p\mu_p + n\mu_n)^2} = \frac{1}{ne}\left(\frac{\mu_p - \mu_n}{\mu_p + \mu_n}\right) \tag{6.10-13}$$

$$= \frac{1}{(1.3 \times 10^{16}\ \text{m}^{-3})(1.6 \times 10^{-19}\ \text{C})}\left(\frac{0.048 - 0.135}{0.048 + 0.135}\right) = -2.3 \times 10^2\ \ \text{m}^3/\text{C}$$

(b) For n-type Si, $n >> p$, and $\mu_n > \mu_p$, so Eq. (6.10-10) reduces to (6.10-9):

$$R_H = -\frac{1}{ne} = -\frac{1}{(1.3 \times 10^{19}\ \text{m}^{-3})(1.60 \times 10^{-19}\ \text{C})} = -0.48\ \text{m}^3/\text{C}$$ ■

■ EXAMPLE 6.10-3

Calculate the Hall electric field for a current density of 5.0×10^3 A/m² and a magnetic field of 0.30 T in (a) intrinsic Si and (b) n-type Si of Example 6.10-2.

Solution. From Eq. (6.10-7),

$$\varepsilon_H = R_H J B \tag{6.10-14}$$

using the R_H values previously obtained,

(a) For intrinsic Si, $R_H = -2.3 \times 10^2$ m³/C. Therefore,

$$\varepsilon_H = -(2.3 \times 10^2 \text{ m}^3/\text{C})(5.0 \times 10^3 \text{ A/m}^2)(0.30 \text{ T}) = -3.4 \times 10^5 \text{ V/m}$$

(Note that $C = \text{A s}$ and $T = \text{V s/m}^2$.)

(b) For n-type Si, using $R_H = -0.48$ m³/C,

$$\varepsilon_H = -(0.48 \text{ m}^3/\text{C})(5.0 \times 10^3 \text{ A/m}^2)(0.30 \text{ T}) = -7.2 \times 10^2 \text{ V/m}$$ ■

■ EXAMPLE 6.10-4

The drift velocity can be evaluated in terms of the applied electric field $\mathscr{E}$, the instantaneous speed $\langle v \rangle$ of random motion, and the mean free path λ (average distance traveled by an electron between collisions with imperfections in the crystal lattice). Derive an expression for v_d in terms of these parameters and discuss the influence of the reciprocal effective mass $1/m^*$ on the electrical conductivity.

Solution. In the absence of an applied electric field, the directions in which the electrons move are random. The electrons frequently collide with imperfections in the crystal lattice, through thermal motion of the ions about their equilibrium positions or from the presence of impurity ions. In colliding with imperfections, the electrons suffer changes in speed and direction, and this makes their motion random. The frequency of electron–lattice imperfection collisions is described by a mean free path λ, which is the average distance that an electron travels between collisions. When an electric field is applied to a metal, the electrons modify their random motion in such a way that, on the average, they drift slowly in the direction opposite to that of the field with a speed v_d. This speed is much less than the instantaneous speed $\langle v \rangle$ of random motion. In copper, v_d is of the order of 10^2 cm/s, whereas $\langle v \rangle$ is of the order of 10^8 cm/s.

Consider a two-dimensional momentum space (p_x, p_y) in which the Fermi sphere (circle) at $t = 0$ with $\varepsilon = 0$ is at the origin. At absolute zero, all states are filled up to the momentum that corresponds to the Fermi energy. According to Newton's second law of motion, $F = dp/dt$ where $F = -e\mathscr{E}$ for electrons. Therefore, applying an electric field to the electrons for a time τ will change their momentum by $\Delta p = -e\mathscr{E}t$. *If the electric field is turned on at* $t = 0$, the electrons will begin to scatter off impurities, other defects, and phonons after the collision time τ. The average momentum of the electrons in the Fermi sphere then increases from 0 to Δp as shown in Fig. 6.10-2(b), where it remains in equilibrium.

Since electrons are negative, the direction of the drift velocity is opposite that of the electric field, with $v_d = \Delta p/m = -e\mathscr{E}\tau/m$. The current density J is $(N/V)qv_d$, so for electrons,

$$J = \left(\frac{N}{V}\right)(-e)\left(-\frac{e\mathscr{E}\tau}{m}\right) = \left(\frac{N}{V}\right)\left(\frac{e^2\mathscr{E}\tau}{m}\right) \tag{6.10-15}$$

Finally, the electrical conductivity σ, being defined by $J = \sigma\mathscr{E}$, is

$$\sigma = \frac{N}{V} \cdot \frac{e^2\tau}{m} \tag{6.10-16}$$

This may be equivalently expressed as

$$\sigma = \frac{1}{\rho} = \frac{ne^2\lambda}{m\langle v\rangle} \tag{6.10-17}$$

where ρ is the resistivity and n is the carrier density (number of electrons per unit volume).

The behavior of charge carriers under the influence of an applied electric field is summarized by Eq. (6.10-17), which predicts that the electrical conductivity σ of the material containing the carriers is proportional to the reciprocal of their masses. Taking into account the quantum behavior of charge-carrying electrons, this means that

$$\sigma \propto \frac{1}{m^*} \tag{6.10-18}$$

where the mass of the electron has been replaced by its effective mass m^*. For iron, the value of $1/m^*$ is about one-tenth that of $1/m$. Consequently, this metal is not a particularly good conductor. Copper, on the other hand, with a reciprocal effective mass $1/m^*$ roughly equal to $1/m$, is a good conductor. Using the reciprocal effective mass amounts to accounting for the influence of a perfect crystal lattice on the motion of an electron under an applied field.

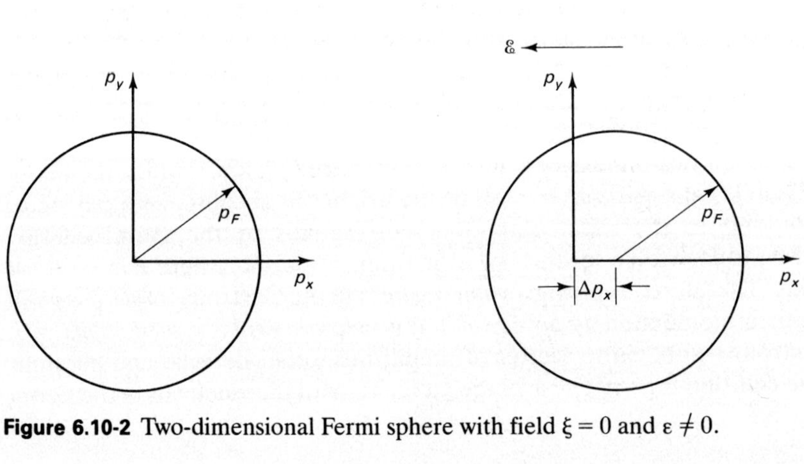

Figure 6.10-2 Two-dimensional Fermi sphere with field $\xi = 0$ and $\varepsilon \neq 0$.

Accelerated motion takes place between collisions of the electron with the imperfections in the lattice of a real material. These collisions tend to randomize the motion, causing the overall electron to drift with a velocity proportional to the strength of the applied field in contrast to an ever-increasing velocity. If there were no lattice imperfections, the electron current would increase in time with no theoretical limit. In such circumstances, the material would have zero resistance, and Ohm's law would not hold. So the presence of nonzero resistance or noninfinite conductivity is due primarily to the lattice imperfections. ■

■ EXAMPLE 6.10-5

Copper at room temperature has an electrical conductivity of 5.9×10^7 $(\Omega\ \text{m})^{-1}$, an electron concentration of $8.5 \times 10^{28}/\text{m}^3$, and a Fermi energy of $(7.0\ \text{eV})(1.60 \times 10^{-19}\ \text{J/eV}) = 1.12 \times 10^{-18}$ J. Find (a) the collision time, (b) the Fermi speed, and (c) the mean free path.

Solution

(a) From Eq. (6.10-12): $\sigma = (N/V)(e^2\tau/m)$. Solve for

$$\tau = \left(\frac{\sigma m}{e^2}\right)\Bigg/\left(\frac{N}{V}\right) \tag{6.10-19}$$

$$= \frac{(5.9 \times 10^7\ \Omega^{-1}\ \text{m}^{-1})(9.11 \times 10^{-3}\ \text{kg})}{1.60 \times 10^{-19}\ \text{C}}\Bigg/(8.5 \times 10^{28}\ \text{m}^{-3}) = 2.5 \times 10^{-14}\ \text{s}$$

(b) The Fermi speed is defined by $\frac{1}{2}mv_F^2 = E_F$, which gives

$$v_F = \sqrt{2E_F/m} \tag{6.10-20}$$

$$= \sqrt{\frac{2(1.12 \times 10^{-18}\ \text{J})}{9.11 \times 10^{-31}\ \text{kg}}} = 1.57 \times 10^6\ \text{m/s}$$

[Note that $J = \text{N m}$, $kg = (\text{N s}^2)/\text{m}$.]

(c) Most active electrons are just below the surface of the Fermi sphere. Using the Fermi speed, the mean free path is

$$\lambda = v_F\tau$$

or

$$\lambda = (1.57 \times 10^6\ \text{m/s})(2.5 \times 10^{-14}\ \text{s}) = 3.9 \times 10^{-8}\ \text{m} = 39\ \text{nm}$$

A few noteworthy comments: The Fermi speed (1.57×10^6 m/s) is considerably higher than the drift speed (which for a current density of $J = 1\ \text{A/mm}^2 = 10^6\ \text{A/m}^2$ would be $v_d = 10^{-4}$ m/s). With a mean free path of 39 nm, Cu electrons would get past an average of about 200 atoms before being scattered. At low temperatures and in ultra-high-purity single-crystal metals the mean free path is many orders of magnitude higher, indicating that there are very few impurities or defects to scatter the electrons. Also, at low temperatures the atoms have low-vibrational energies, which means that there are very few phonons to scatter the electrons. ■

■ EXAMPLE 6.10-6

An aluminum alloy wire, 1 m in length and 1 mm in diameter, is placed in an electrical circuit such as shown in Fig. 6.10-3 for measuring electrical conductivity. A voltage drop of 432 mV is observed across the length of the wire when it is carrying a 10-A current. Calculate the conductivity of the alloy.

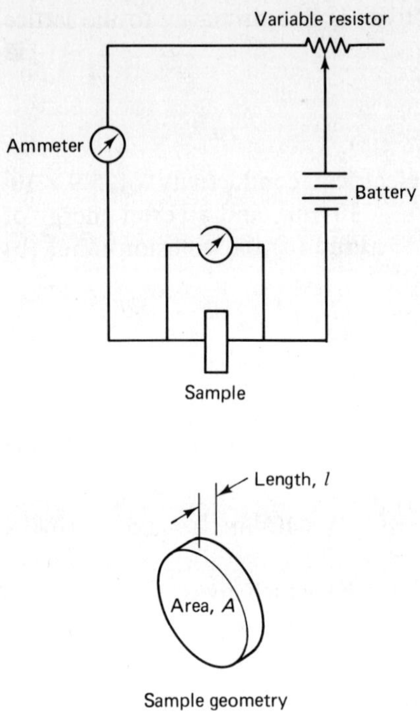

Figure 6.10-3 Measurement of electrical conductivity.

Solution. The resistance R (ohms) is the voltage divided by the current I (amperes);

$$R = \frac{V}{I} = \frac{432 \times 10^{-3}\ \text{V}}{10\ \text{A}} = 43.2 \times 10^{-3}\ \Omega \qquad (6.10\text{-}21)$$

The resistance R increases with sample length l, and decreases with sample area A:

$$R = \rho \frac{l}{A} \qquad (6.10\text{-}22)$$

where ρ is a constant of proportionality, the resistivity (Ω m). Thus,

$$\rho = \frac{RA}{l} = \frac{(43.2 \times 10^{-3}\ \Omega)[\pi(0.5 \times 10^{-3}\ \text{m})^2]}{1\ \text{m}}$$

$$= 33.9 \times 10^{-9}\ \Omega\ \text{m}$$

The conductivity is then

$$\sigma = \frac{1}{\rho} = \frac{1}{33.9 \times 10^{-9}\ \Omega\ \mathrm{m}} = 29.5 \times 10^{6}\ \Omega^{-1}\ \mathrm{m}^{-1}$$ ■

6.11 CONDUCTORS, INSULATORS, AND SEMICONDUCTORS

The band spectrum is not unique to the Kronig–Penney potential; the phenomenon occurs in two-dimensional and three-dimensional situations as well. In one form or another, it holds for all solids, whether they be insulators or conductors, ionic, molecular, covalent, or metallic. The differences among them are due to (1) the widths of the bands relative to the excitation energies of the separated atoms or molecules, and (2) the way the electrons fill the bands.

The high electrical conductivity of metals is due to the ease with which electrons in a metal can move under the influence of a static or low-frequency electric field. In the original Drude kinetic theory, it was assumed that each particle in the electron gas tended to move against the applied field while still obeying classical statistics. This implies that each electron is capable of taking on any value of energy within the range of the distribution. The electrons in a solid, however, cannot take on every possible energy value. A continuous range of energies is not available and a set of energy bands occurs, each band being separated from the next one by an energy gap. The top of this set of bands will be full, partially filled, or empty, and it is this band that is paramount in determining the nature of the solid. The behavior of the electrons in the valence and conduction bands will now be established, completely and partially filled bands being dealt with separately.

Consider first the free-electron model of a metal in which each valence electron is treated as a particle in a three-dimensional box the size of the metal crystal. In this simplified theory the potential in which the electron moves can be approximated by a rectangular well. This constant potential smooths out the actual periodic variation due to the ion cores and includes the average effect of all the remaining electrons. It is equivalent to treating the electrons as an ideal gas of fermions inside the solid. This approximation, which greatly reduces quantum-mechanical calculations, turns out to be surprisingly good in determining many of the observed properties of solids, such as contact potential and electronic specific heats. Chapter 4 gave the distribution with energy of free electrons in a metal as

$$n(E)N(E)\,dE = \frac{8\pi V(2m^3)^{1/2}}{h^3}\frac{E^{1/2}\,dE}{e^{(E-E_F)/kT}+1} \tag{6.11-1}$$

where $n(E)N(E)\,dE$ is the number of electrons with energy from E to $E + dE$ in a metal at temperature T. The justification is that the distribution of energy states in a band is nearly the same as that for free electrons if the Fermi energy E_F is not close to the top of the band. This condition applies to the alkali metals, for example, and accounts for the success that the free-electron model has in describing their electrical properties.

The distribution with energy is shown on the left side of Fig. 6.11-1 for the absolute zero temperature in a partly filled band, with energy being measured from the lowest energy in the band. The maximum energy allowed in the band is $E_{\max}$ and $E_F < E_{\max}$, as shown in that figure. At a temperature greater than zero, the uppermost electrons are excited to occupy nearby available higher states, and the distribution function takes the form shown on the right side of Fig. 6.11-1. The number of quantum states in an energy interval E to $E + dE$ is the factor $N(E)\,dE$ of (6.11-1), namely

$$N(E)\,dE = \frac{8\pi V(2m)^3)^{1/2}}{h^3} E^{1/2}\,dE \tag{6.11-2}$$

In Fig. 6.11-1, $N(E)$ is shown by a dashed curve and, for unit volume, is the density of states. The dash–dot curve is $n(E)$, the Fermi distribution for the number of electrons per state. The solid curve gives the product $n(E)N(E)$, the energy distribution of electrons, or number of electrons per unit energy interval.

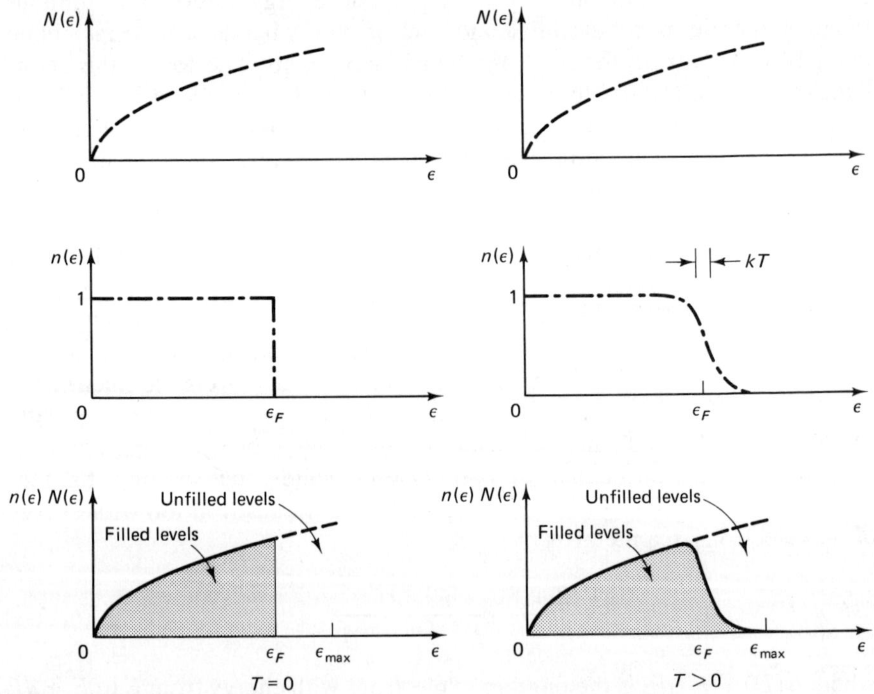

Figure 6.11-1 Distribution with energy of conduction electrons in an unfilled band of width $E_{\max}$ in a solid at $T = 0$ and at $T > 0$.

As electrons are added at 0 K, the energy levels are filled up to some energy E_F, the Fermi energy. The Fermi energy is the electrochemical potential of the electrons and determines their tendency to move at an interface, just like the chemical potential μ does for a substance. The quantity μ is a constant at any given temperature. The value of μ can be obtained by integrating (6.11-1) over all energies since this integration must yield the number of electrons per unit volume. Since $f(E, T) = \frac{1}{2}$ when $E = \mu$, the quantity μ is equal to the energy at which $f(E, T)$ has half its maximum value. At 0 K, $f(E,0) = 1$ for energies E less than the Fermi energy E_F, and $f(E,0) = 0$ for energies E greater than the Fermi energy E_F. As $T \rightarrow 0, \mu \rightarrow E_F$. It is a good approximation to take $\mu \approx E_F$ at other temperatures provided that $E_F >> kT$. The approximation that $\mu \approx E_F$ is a good one because $E_F \approx 5$ eV for most metals.

At room temperature the distribution of electron energies differs only slightly from that at 0 K for $E_F \approx 5$ eV. A small fraction of the electrons have energies greater than the Fermi energy E_F, and they leave behind holes at $E < E_F$. The excited electrons and holes both contribute to the electrical conductivity. Figure 6.11-1 provides the explanation for the very small contribution of electrons to the heat capacity for a metal. When the temperature of a metal is raised a small amount, only a small fraction of the electrons have their energies raised. Since the energies of most electrons are not affected by raising the temperature, the electronic heat capacity is negligible compared with the value $3R$ predicted by the Einstein and Debye models.

The free-electron theory of metals, however, is only an approximation. It would predict an infinite electrical conductivity because the excited electrons could in principle be accelerated to infinite velocities. In a real crystal friction is introduced by the scattering of electrons from vibrating atoms. According to the simplest theory the electrons in a metal have a characteristic electric mobility μ. Then the electric conductivity σ is given by

$$\sigma = Nq\mu \tag{6.11-3}$$

where N is the number of current electrons per unit volume and q is the charge carried by an electron. A more general theory is the band theory of solids.

To recall the linear combinations of wave functions of atoms in a solid, consider as in Fig. 6.11-2, the splitting that occurs when six hydrogen atoms are brought together in a linear array. The combination of six 1s wave functions produces six orbitals, three bonding and three antibonding. As electrons are fed into this energy level scheme, they first fill the lower energy bonding orbitals, two at a time. As the number of interacting atoms is increased the number of energy levels increases, and they become more and more closely spaced within a band, for the width of the band at a given internuclear separation does not increase substantially. In Fig. 6.11-2 there is the 1s band and the 2s band.

Thus, in contrast to an electron in free space which has an essentially continuous range of energy values, there are, in a solid, bands of energy levels containing very large numbers of discrete levels, with relatively large sepa-

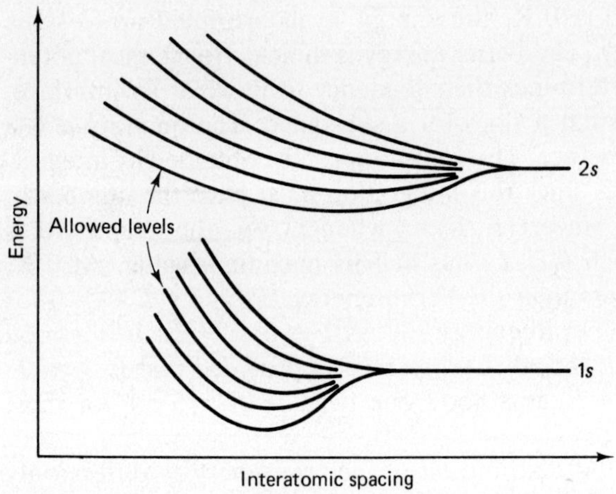

Figure 6.11-2 Energy levels for a linear array of six hydrogen atoms.

rations in energy between bands. The separations between allowed bands are the energy gaps. The different properties of conductors, insulators, and semiconductors may be understood in terms of these energy bands and band gaps. Metals have an electronic structure such that E_F lies within an allowed band. The cause of this may be that there are insufficient electrons available to fill an available band if the band structure is as shown in Fig. 6.11-3(a), or alternatively, that there are overlapping bands as in Fig. 6.11-3(b). In a metal the highest occupied band is only partially filled. This band is the conduction band. Since the energy levels within it are continuous, an electron at the top of the conduction band, where it is immobile, may be excited to a higher level with an infinitesimally higher energy. Because the levels are continuous a conduction electron is essentially free.

Insulators have one band fully occupied by electrons and a large energy gap between this band and the next highest band, which is devoid of electrons at low temperatures. The Fermi level E_F lies within the forbidden band. A band in which there are no electrons cannot make any contribution to current flow in a crystal. Neither can a completely full band. The electrons in them are immobile and do not contribute to electrical conductivity. In order for an electron to be able to move it must be excited to the next band. In an insulator the band gap to the first unfilled band is much larger than kT, and so the population of excited electrons is negligible.

A semiconductor is just an insulator with a narrow forbidden band gap. At low temperatures, it does not conduct. At higher temperatures, there is sufficient smearing of the Fermi–Dirac distribution function to ensure that some levels in the originally completely filled band (valence band) are now vacant and some in the next highest band (conduction band) are occupied. The electrons in the conduction band, with an abundance of unoccupied energy states in the vicinity, can contribute to current flow. Similarly, because there are now unoccupied levels in the valence band, an additional contribution also comes from electrons in this band.

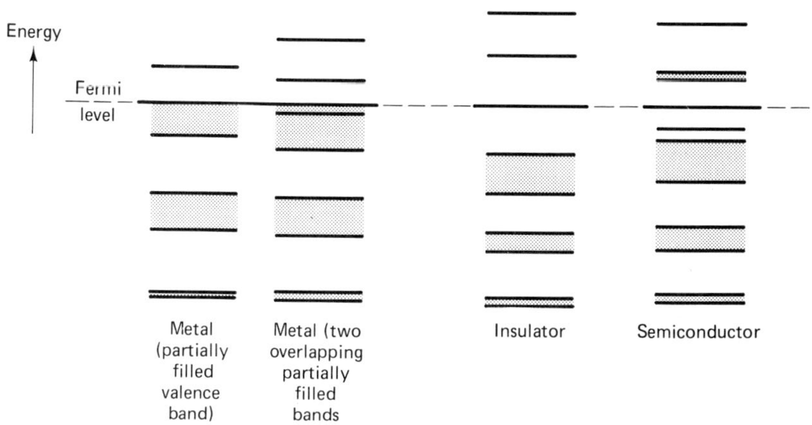

Figure 6.11-3 Electron occupation of allowed states in (a) a metal, (b) an insulator, (c) a semiconductor.

Semiconductors are covalent solids that may be regarded as "insulators" because the valence band is completely full and the conduction band is completely empty at the absolute zero of temperature, but they have an energy gap between the valence and conduction bands of no more than about 2 eV. (For silicon the energy gap is 1.14 eV and for germanium the gap is >0.67 eV.) Since $kT > 0.025$ eV at room temperature, the number of available states in the conduction band is high. Hence, a significant number of electrons is excited from the valence band into the conduction band, this number being the product of the number of electrons per quantum state and the number of quantum states per energy interval. Furthermore, the conductivity of a semiconductor increases rapidly with rising temperature, the number of excited electrons in silicon, for example, increasing by a factor of about one billion with a doubling of temperature from 300 to 600 K.

■ EXAMPLE 6.11-1

(a) Compare the Fermi distribution function with the energy band structure for an insulator and a semiconductor, respectively.

(b) What is the probability of an electron being thermally promoted to the conduction band in silicon ($E_g = 1.07$ eV) at room temperature (298 K) compared to the same happening in diamond ($E_g = 5.6$ eV)?

Solution

(a) The distribution function, with E_F falling in the center of the band gap (E_g), is as shown in Fig. 6.11-4(a) for an insulator. The important feature is that $f(E)$ is essentially equal to 1 throughout the valence band and equal to 0 throughout the conduction band. Virtually no electrons are promoted to the conduction band because of the magnitude of the band gap (>2 eV). Diamond, with an E_g of 5.6 eV is a poor conductor in this respect.

For silicon, element 14, residing just below carbon in the periodic table

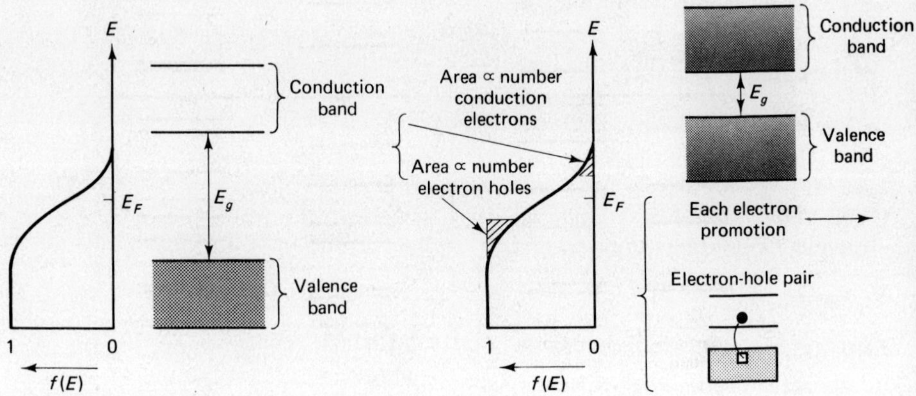

Figure 6.11-4 Fermi function and energy band structure for insulator and semiconductor.

and behaving chemically in a similar way, the energy band structure of silicon is as shown in Fig. 6.11-4(b). The main difference between the two figures is that silicon has a smaller band gap ($E_g = 1.107$ eV). The result is that, at room temperature (298 K), thermal energy promotes a small but significant number of electrons from the valence to the conduction band. Consequently, holes (positive charge carriers) are produced in the valence band as well as electrons (negative charge carriers) in the conduction band. Thus, silicon (a semiconductor) displays a moderate value of electrical conductivity, intermediate between that for metals and insulators.

(b) For diamond, the bottom of the conduction band corresponds to

$$E - E_F = \frac{5.6}{2}\ \text{eV} = 2.8\ \text{eV}$$

The Fermi distribution function

$$f(E) = \frac{1}{e^{(E - E_F)/kT} + 1} \qquad (6.11\text{-}4)$$

with $(E - E_F) = 2.8$ eV, $k = 86.2 \times 10^{-6}$ eV K^{-1}, and $T = 298$ K, is

$$f(E) = \frac{1}{e^{(2.8\ \text{eV})/(86.2 \times 10^{-6}\ \text{eV K}^{-1})(298\ \text{K})} + 1} = 4.58 \times 10^{-48}$$

For silicon,

$$E - E_F = \frac{1.107}{2}\ \text{eV} = 0.5535\ \text{eV}$$

$$f(E) = \frac{1}{e^{(0.5535\ \text{eV})/(86.2 \times 10^{-6}\ \text{eV K}^{-1})(298\ \text{K})} + 1} = 4.39 \times 10^{-10}$$

This number is small indeed; but is 38 orders of magnitude greater than the probability for diamond and is sufficient to create enough charge carriers (electron–hole pairs) to give silicon its semiconducting properties. ■

■ EXAMPLE 6.11-2

The conductivity is the product of the density of charge carriers, n, the charge of the carrier, q, and the mobility of the carrier, μ:

$$\sigma = nq\mu \tag{6.11-5}$$

For typical conductors, it is on the order of $10 \times 10^6\ \Omega^{-1}\ \mathrm{m}^{-1}$. The basis for this large value is that for a metal such as copper, the conductivity is entirely due to free electrons. Thus, for conductors,

$$\sigma = n_e q_e \mu_e \tag{6.11-6}$$

where the subscript e refers to purely electronic conduction. What effect does temperature and composition have on the conductivity of metals?

Solution. In the band model of metals, the electron mobility, μ_e, is dominant in determining the conductivity. In general, an increase in temperature above room temperature results in a drop in conductivity. This is due to a drop in electron mobility, μ_e, with increasing temperature. Electrons have a wavelike nature; their "wave packets" move through crystalline structure more effectively when that structure is nearly perfect. Irregularities produced by thermal vibrations diminish electron mobility. Conductivity, σ, and resistivity, ρ, are inversely related. For typical conductors, the resistivity is on the order of $0.1 \times 10^{-6}\ \Omega$ m. The resistivity increases as the temperature increases above room temperature. The relationship $\rho(T)$ is more often used than $\sigma(T)$ because the resistivity is found experimentally to increase linearly with temperature:

$$\rho(T) = \rho_0[1 + \alpha(T - T_0)] \tag{6.11-7}$$

where ρ_0 is the room temperature value of resistivity, α the temperature coefficient of resistivity, T the temperature, and T_0 the room temperature. This is plotted for several metals in Fig. 6.11-5(a).

Inspection of tables of properties reveals that ρ_0 is a function of compo-

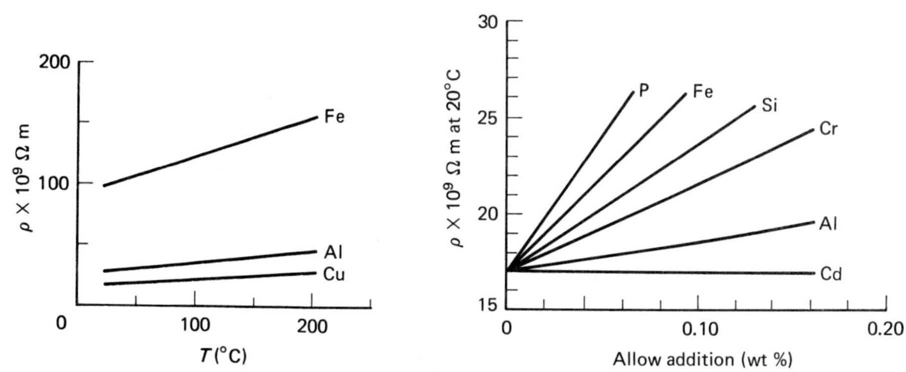

Figure 6.11-5 Variation of resistivity with temperature and composition.

sition. For example, $\rho_{0,\,\text{pure}\,F_e} < \rho_{0,\,\text{steel}}$. For small additions of impurity to a nearly pure metal, the increase in ρ with amount of impurity according to:

$$\rho(x) = \rho_{0,\,\text{pure}}(1 + \beta x) \tag{6.11-8}$$

where $\rho_{0,\,\text{pure}}$ is the resistivity of the pure metal, β a constant for a given metal-impurity system, and x the amount of impurity addition. A plot of this relation for various copper alloys with small levels of elemental additions is shown in Fig. 6.11-5(b). ■

■ EXAMPLE 6.11-3

(a) On a ρ–T diagram, show the difference between the temperature dependence of the electrical resistivity of a normal conductor and that of a superconductor.

(b) Describe the Meissner effect.

(c) Outline the BCS theory.

(d) What is the dc potential difference across a Josephson junction if a 6.0 GHz alternating current is produced?

Solution

(a) At extremely low temperatures, certain metals and alloys exhibit a characteristic discovered by Kamerlingh Onnes in 1911. He showed, for example, that mercury loses all its electrical resistance at a temperature below 4.2 K. Calculations indicate that current flow, once initiated, persists in a superconducting loop in the absence of applied fields for an indefinite time. The difference between the temperature dependence of the electrical resistivity of a normal conductor and that of a superconductor is shown in Fig. 6.11-6. The superconducting state can be extinguished by applying a sufficiently large magnetic field (the critical field) or by raising the temperature above a critical temperature in the absence of a magnetic field.

(b) When a material is totally in the superconducting state, the current is essentially on the surface. Under these conditions, neither electric nor magnetic fields can penetrate the interior. When the critical field is exceeded, the

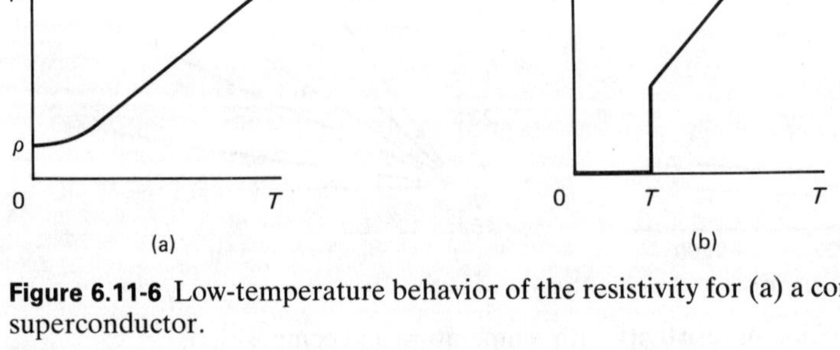

Figure 6.11-6 Low-temperature behavior of the resistivity for (a) a conductor and (b) a superconductor.

magnetic field totally penetrates the material, which is no longer superconducting. The expulsion of magnetic flux is called the *Meissner effect*.

(c) A quantum theory of superconductivity has been proposed by Bardeen, Cooper, and Schrieffer (the BCS theory) in which the electrons within kT of the Fermi energy are assumed to be assoicated in phonon-coupled pairs. At very low temperatures, mobile electrons attract lattice ions, moving them slightly from their equilibrium positions. This distorts the lattice, creating a phonon. If the lattice is sufficiently "soft" to have significant distorsion which persists for a short time after the electron has passed, there is a positive charge concentration left behind momentarily. This attracts a second electron to the site, most effectively one traveling in the opposite direction. These two electrons can be thought of as having undergone an attractive interaction through the medium of the lattice: They form a weakly bound pair state (a *Cooper pair*).

The interaction can only take place for electrons near the Fermi energy, since they both must be moving, and the result of the binding between them is that a small gap is produced in the level occupancy curve, centered at E_F (Fig. 6.11-7). The interaction energy is very small, on the order of 10^{-4} eV. If the effect of this pairing is to be significant, the temperature of the material must be very low, that is,

$$300 \text{ K} \times \frac{10^{-4} \text{ eV}}{2.5 \times 10^{-2} \text{ eV}} \simeq 1 \text{ K}$$

where kT at room temperature is taken to be $\frac{1}{40}$ eV. At these low temperatures where the interaction occurs, the paired electrons are unable to interact with the lattice and transfer energy to it, exciting both electrons across the gap, unless enough thermal energy is available to decouple the pair state. In this way, momentum can be imparted to the pair (by an external electric field, for example), which cannot be dissipated by the normal resistive process, so long as there is not enough thermal energy to break up the pair. Currents, once set in motion, will not stop until the superconducting state is altered. The situation is much like that of superfluid helium, described in Chapter 4. At sufficiently low temperatures all bosons enter the same state, the lowest available

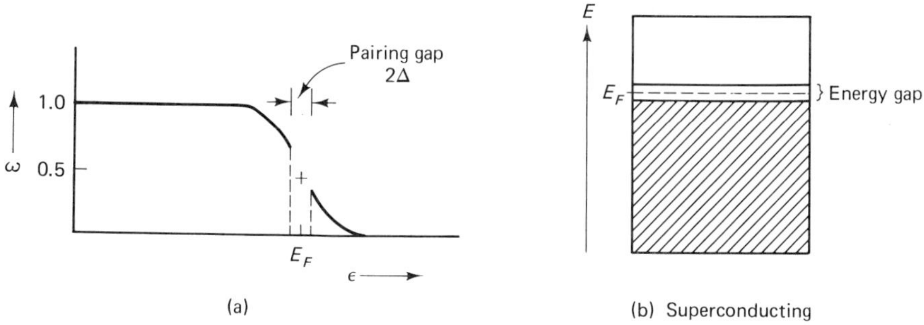

Figure 6.11-7 Pairing gap in Fermi–Dirac distribution for superconducting material.

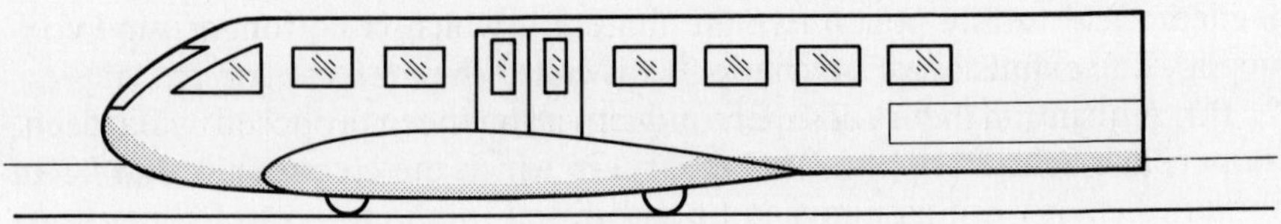

Figure 6.11-8 Magnetic and aerodynamic levitation train by J. E. Lay.

(ground) state. Cooper pairs are bosons having integral spin rather than the half-integral spin characteristic of single electrons and other fermions. Being all in the same state, Cooper pairs do not scatter. Once they start moving, they do not stop.

The complete absence of resistance in superconductors and consequent power savings make them ideal for applications where large currents are required, as, for example, in magnetic field generation or power transmission. Many ingenious ideas have been put forward for the use of superconductivity. One design, a magnetic and aerodynamic levitation train patented by the author, is shown in Fig. 6.11-8.

(d) A *Josephson junction* consists of two superconducting regions separated by an insulating tunnel barrier less than 2 nm thick. When a dc current below the threshold level i_c exists in the junction, it acts as a superconductor; that is, there is no ohmic resistance and hence no voltage drop across the device. Cooper pairs tunnel through the insulating barrier at currents below the threshold. Above the threshold, only single electrons tunnel across the barrier, and the device behaves like a normal conductor. When the dc current crosses the threshold, the device voltage rises from zero to a steady-state value in times on the order of 10 to 15 ps. Also, an alternating current is generated and has a frequency of $\nu = (2e/h)V$. The Josephson junction is a switching device with remarkably short switching time and very low power dissipation. In the present case, for a direct current greater than i_c,

$$V = \frac{\nu}{2e/h} = \frac{h\nu}{2e} = \frac{(4.136 \times 10^{-15}\text{ eV s})(6.00 \times 10^{9}\text{ Hz})}{2e} \tag{6.11-9}$$

$$= 12.41\ \mu\text{V}$$

■

6.12 GROUP IV SEMICONDUCTOR WITH GROUPS III AND V DOPANTS

Some of the more useful properties of semiconductors can be described in terms of those belonging to group IV of the periodic table of chemical elements. Although the following "bond model" does not apply to all semiconducting material, it does allow the effects of impurities upon the electronic properties of semiconductors to be discussed in a simple manner.

The characteristic lattice structure of a semiconductor from group IV of the periodic table is shown by the schematic two-dimensional representation of a silicon lattice in Fig. 6.12-1. Each silicon atom is bonded to four neighbors by covalent bonds. Each covalent bond requires two electrons. Silicon has four valence electrons, so each covalent bond shares an electron originating from the central atom and one originating from the neighboring atom. At low temperatures, for the case shown in Fig. 6.12-1(a), the semiconductor cannot conduct electricity, for all the bonds are complete at any instant. However, at higher temperatures, some electrons in the covalent bond can obtain enough energy to break free from the bond as shown in Fig. 6.12-1(b). In this case, the electrons released are free to move about throughout the crystal and can contribute to current flow. Furthermore, electrons in covalent bonds in the vicinity of the broken bond can move into the location left vacant in this bond, leaving another broken bond behind. This process also contributes to current flow. A broken bond can be identified as a hole in the valence band. The minimum energy required to release an electron from a covalent bond is then equal to the width of the forbidden band gap in the semiconductor.

It is apparent that there is a one-to-one correspondence between free electrons and holes; such is the case for intrinsic semiconductors. Even at quite high temperatures comparatively few bonds are broken as the energy required is approximately 1 eV and the conductivity is still quite low. The conductivity, however, can be increased by increasing the number of electrons in the conduction band or holes in the valence band. This is essentially what is done in impurity or extrinsic semiconductors. By adding minute amounts (a few parts per million) of suitable impurities to the melt during the growth of semiconductor crystals it is possible to increase their conductivities. For silicon (and germanium) two types of impurities are used. These are from groups III

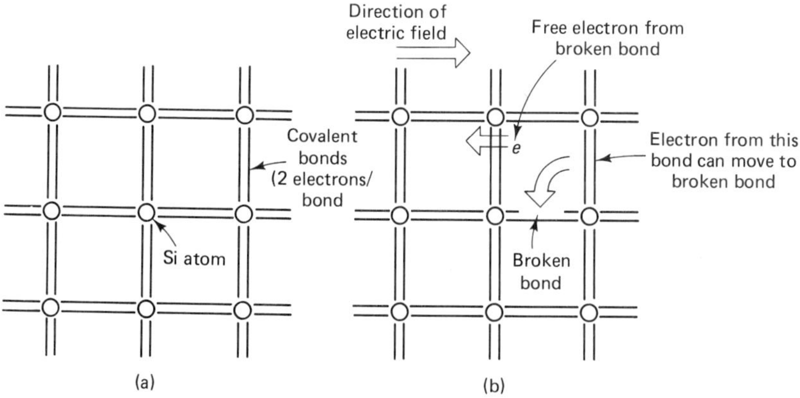

Figure 6.12-1 Intrinsic semiconductor lattice: (a) At absolute zero and (b) at temperature greater than zero.

and V of the periodic table, that is, trivalent elements such as boron and indium and pentavalent elements such as arsenic and phosphorus. The reason for this choice will now be given.

Consider first a portion of the lattice where a pentavalent group V impurity (e.g., phosphorus) has replaced a silicon atom as shown in Fig. 6.12-2(a). Four of the valence electrons are used up in the covalent bonds, but the fifth is in a different situation. It is not in a covalent bond, so it is not in the valence band. For the case shown, it is tied to the group V atom and is not free to move through the lattice. Hence, it is not in the conduction band either. But it takes only a small amount of energy to release this extra electron compared to that required to free electrons locked up in covalent bonds. Figure 6.12-2(b) shows the energy levels E_d of the impurity atoms just below the conduction band. They are termed donor levels since they donate electrons to the conduction band. At absolute zero the donor levels are all occupied, but even at moderately low temperatures most of the electrons are excited into the conduction band, as shown in Fig. 6.12-2(c), because E_d is so small ($\simeq 0.03$ eV), thereby increasing the free-electron concentration and hence the conductivity of the material. A semiconductor of this type is said to be n type since the conduction is predominantly due to negative charges (i.e., electrons). The electrons are then called *majority carriers,* and holes formed by the excitation of electrons from the valence band to the conduction band are *minority carriers.*

Consider, on the other hand, that silicon is doped with a group III trivalent impurity such as boron. Again it is found that the impurity atoms occupy sites normally occupied by silicon atoms. In this case there is one electron too few to complete the tetravalent covalent bonds and a vacancy is formed as shown in Fig. 6.12-3(a). This vacancy is not a hole as it is bound to the impurity atom but at any temperature above absolute zero an electron from a neighboring atom can fill the vacancy leaving a hole in the valence band.

For this reason the trivalent impurities are referred to as acceptors as they

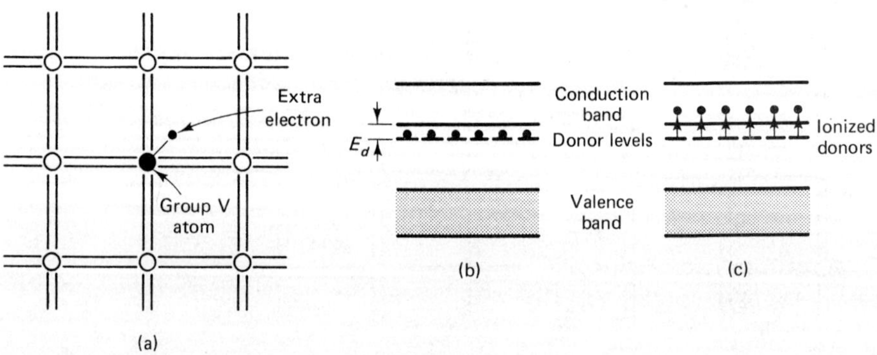

Figure 6.12-2 n-type semiconductor: (a) schematic lattice, (b) donor levels at absolute zero, and (c) donor levels at room temperature.

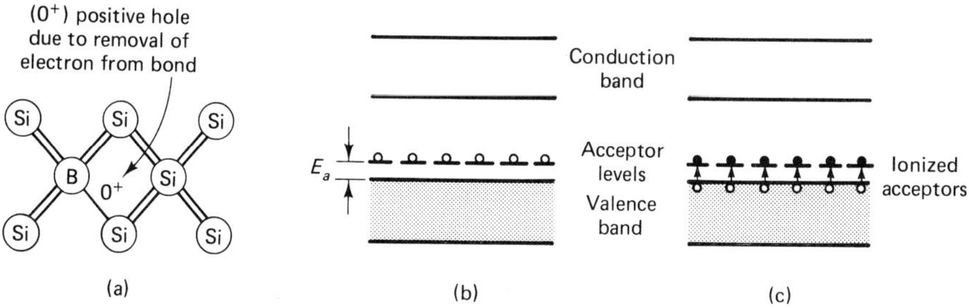

Figure 6.12-3 p-type semiconductor: (a) schematic lattice, (b) acceptor levels at absolute zero, and (c) acceptor levels at room temperature.

accept electrons excited from the valence band. It is often convenient to regard this situation as an acceptor atom with a positive hole orbiting around it. The energy E_a required to "free" the hole from its parent impurity atom can be estimated. (E_a is the energy required to excite an electron from the valence band to the energy levels associated with the acceptor impurity atoms.) It is found (by using the hole effective mass) that $E_a \simeq 0.036$ eV.

An extrinsic semiconductor with acceptor impurities is referred to as a p type since the conduction of current through it is predominantly by positive charges (i.e., holes). The holes are now the majority carriers and any electrons in the conduction band are minority carriers. The electronic properties of such extrinsic semiconductors as silicon or germanium have been described in terms of their energy band configurations plus certain interband energy levels introduced by the impurities that are rather sparingly distributed throughout the host material. The ionization of added atoms can be treated as a problem in chemical equilibrium. The donor D and acceptor A ionizations are

$$D = D^+ + e^- \tag{6.12-1}$$

$$A = A^- + h^+ \tag{6.12-2}$$

The concentrations of the species on the right-hand side increase with increasing temperature. The slope of the plot of the logarithm of the concentration of e^- or h^+ versus $1/T$ gives information about the ionization energy of the donor or acceptor.

It is of interest to summarize the above discussion by pointing out that the energy required to ionize impurity atoms to generate electrons in n-type germanium or holes in p-type germanium is much lower than that required to generate electron–hole pairs in intrinsic germanium. This is brought out in Fig. 6.12-4 which shows that the energy required to generate an electron–hole pair in pure germanium is about 0.7 eV, whereas the energy required to set an electron free from a donor atom or to replace a hole from an acceptor atom is about 0.01 eV.

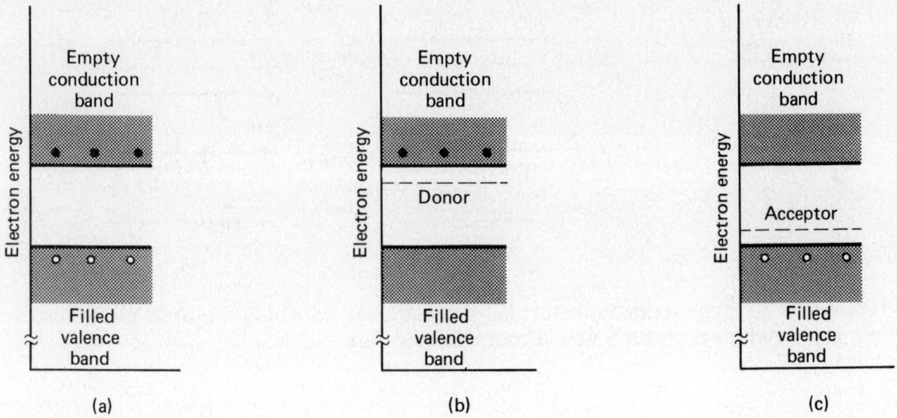

Figure 6.12-4 Energy band diagram for intrinsic germanium, *n*-type germanium, and *p*-type germanium.

■ EXAMPLE 6.12-1

Semiconductors have conductivities ranging from 10^{-4} to $10^4\ \Omega^{-1}\ \mathrm{m}^{-1}$. This range corresponds to band gaps of less than 2 eV. Both conduction electrons and electron holes are charge carriers in a simple semiconductor. For pure silicon, the number of conduction electrons is equal to the number of electron holes. Pure, elemental semiconductors of this type are called *intrinsic semiconductors*. The conductivity for intrinsic semiconductors is

$$\sigma = nq(\mu_e + \mu_h) \tag{6.12-3}$$

where n is the density of conduction electrons (equal to that of holes), q is the magnitude of electron charge (equal to that of hole charge), 0.16×10^{-8} C, μ_e is the mobility of a conduction electron, and μ_h is the mobility of a hole. Table 6.12-1 gives values of μ_e and μ_h together with E_g, at room temperature, for some semiconductors.

(a) Silicon (atomic mass = 28.09 amu) has a density of 2.33 g/cm^3 and a carrier density given by Table 6.12-1 of $n_e = 14 \times 10^{15}/\mathrm{m}^3$. What fraction of Si atoms provides a conduction electron at room temperature?

(b) Suppose that energy gaps of 0.2 eV, 1.0 eV, and 5.0 eV exist in intrinsic materials, with the Fermi level at the center of the gap. What are the probabilities of occupation of electron states at the bottom of the conduction band for these materials at room temperature (25°C)?

(c) Sketch the energy band diagram at $T = 0$ K, including the Fermi level, for pure Si, Si doped with Al, and Si doped with As.

Solution

(a) The atomic density for Si is first calculated:

TABLE 6.12-1 PROPERTIES OF COMMON SEMICONDUCTORS AT ROOM TEMPERATURE

Material	Energy gap E_g (eV)	Electron mobility μ_e (m^2/V s)	Hole mobility μ_h (m^2/V s)	Carrier density $n_e = n_h$ (m^{-3})
Si	1.107	0.140	0.038	14×10^{15}
Ge	0.66	0.364	0.190	23×10^{18}
Cd S	2.59	0.034	0.0018	
Ga As	1.47	0.720	0.020	1.4×10^{12}
In Sb	0.17	8.00	0.045	13.5×10^{21}

$$\rho_{\text{atomic}} = \left(2.33 \frac{\text{g}}{\text{cm}^3}\right)\left(\frac{10^6 \text{ cm}^3}{\text{m}^3}\right)\left(\frac{1 \text{ mol}}{28.09 \text{ g}}\right)\left(6.023 \times 10^{23} \frac{\text{atoms}}{\text{mol}}\right)$$
$$= 50.0 \times 10^{27} \text{ atoms/m}^3$$

Now, Table 6.12-1 gives

$$n_e = 14 \times 10^{15}/\text{m}^3$$

The fraction of atoms providing conduction electrons is thus

$$\frac{14 \times 10^{15}}{50 \times 10^{27}} = 2.8 \times 10^{-13}$$

that is, out of every 10^{13} atoms of pure Si, 2.8 atoms provide a conduction electron. (Compare this to the 1 : 1 ratio of conducting electrons to atoms in copper.)

(b) The probability of occupation of an electron state in a band is accurately given by the Fermi–Dirac distribution function

$$f(E) = \frac{1}{e^{(E - E_F)/kT} + 1} \tag{6.12-4}$$

where

$$kT = (8.62 \times 10^{-5} \text{ eV/K})(298 \text{ K}) = 0.025 \text{ eV}$$

Here, $E - E_F$ is the energy difference between the top of the energy gap and its middle, and so $E - E_F = E_g/2$. Substituting the 0.2 eV, 1.0 eV, and 5.0 eV values of E_g gives $(E - E_F)/kT$ values of 3.87, 19.3, and 96.7, respectively. The corresponding values of $f(E)$ are then 2.0×10^{-2}, 4.0×10^{-9}, and 1.0×10^{-42}.

(c) Pure Si is an intrinsic semiconductor. Si doped with Al is a p-type semiconductor. Si doped with As is an n-type semiconductor. The energy band diagram at $T = 0$ K for these semiconductors is shown in Fig. 6.12-5.

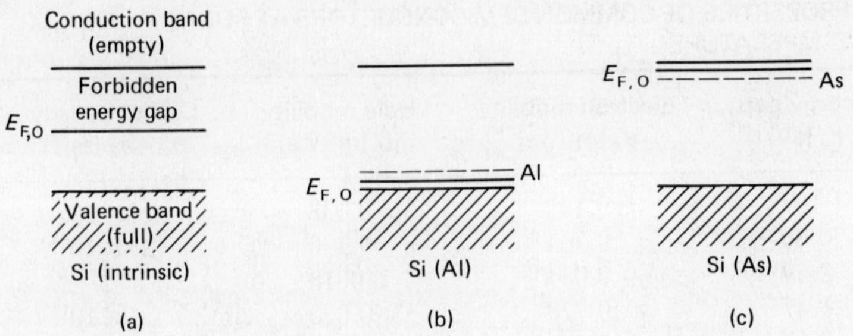

Figure 6.12-5 Energy band diagram at $T = 0$ K for intrinsic and extrinsic semiconductors. ■

■ EXAMPLE 6.12-2

(a) Silicon at room temperature has an electron mobility $\mu_e = 0.19\ \mathrm{m^2/V\ s}$, a hole mobility $\mu_h = 0.0425\ \mathrm{m^2/V\ s}$, and an intrinsic conductivity $\sigma = 5 \times 10^{-4}\ \Omega^{-1}\ \mathrm{m^{-1}}$. What fraction of the charge in intrinsic silicon is carried by electrons and by holes?

(b) Gallium arsenide has an electron mobility $\mu_e = 0.88\ \mathrm{m^2/V\ s}$, a hole mobility $\mu_h = 0.04\ \mathrm{m^2/V\ s}$, and an intrinsic conductivity of $10^{-6}\ \Omega^{-1}\ \mathrm{m^{-1}}$. How many carriers are there per $\mathrm{m^3}$?

Solution

(a) From $\sigma = \sigma_n + \sigma_p = (nq\mu)_n + (nq\mu)_p$ with $n = p$ for an intrinsic semiconductor,

$$\frac{\sigma_n}{\sigma} = \frac{\mu_e}{\mu_e + \mu_h} = \frac{0.19\ \mathrm{m^2/V\ s}}{(0.19 + 0.0425)\ \mathrm{m^2/V\ s}} = 0.82$$

$$\frac{\sigma_p}{\sigma} = 0.18$$

Typically, conductivity is higher from electrons in the conduction band than it is from the holes in the valence band, because $\mu_n > \mu_p$.

(b) Again, from $\sigma = (nq\mu)_n + (nq\mu)_p = nq(\mu_n + \mu_p)$,

$$n = \frac{\sigma}{q(\mu_n + \mu_p)}$$

$$= \frac{10^{-6}\ \Omega^{-1}\ \mathrm{m^{-1}}}{(0.16 \times 10^{-18}\ \mathrm{A\ s})(0.88 + 0.04)\ \mathrm{m^2/V\ s}} = 6.8 \times 10^{12}/\mathrm{m^3}$$

There will be $2 \times 6.8 \times 10^{12} = 1.36 \times 10^{13}$ carriers/m^3, since a hole remains in the valence band for each electron activated across the energy gap. ■

■ EXAMPLE 6.12-3

(a) What is the temperature dependence, $\sigma(T)$, of the conductivity of an intrinsic semiconductor?

(b) Calculate the conductivity of germanium at 200°C. Table 6.12-1 gives $n_e = 23 \times 10^{18}/\text{m}^3$, $E_g = 0.66$ eV, and electron and hole mobilities μ_e and μ_h of 0.364 m^2/V s and 0.190 m^2/V s, respectively.

(c) The resistivity of germanium at 20°C is 0.5 Ω m. What is its resistivity at 40°C? Germanium has an energy gap $E_g = 0.7$ eV.

Solution

(a) The conductivity $\sigma(T)$ is determined by the dependence of μ_e and μ_h as well as n. The latter increases exponentially with temperature, that is,

$$n \propto e^{-E_g/2kT} \tag{6.12-5}$$

where the factor 2 in the exponent arises from the fact that each thermal promotion of an electron produces two charge carriers (an electron–hole pair). This enables the temperature dependence of σ to be written as

$$\sigma = \sigma_0 e^{-E_g/2kT} \tag{6.12-6}$$

where σ_0 (not to be confused with σ at room temperature) is a constant that includes, among others, both the charge q and the electron mobility μ_e. Taking the logarithm of each side of Eq. (6.12-6) gives

$$\ln \sigma = \ln \sigma_0 - \frac{E_g}{2k} \cdot \frac{1}{T} \tag{6.12-7}$$

which indicates that a semilog plot of $\ln \sigma$ versus $1/T$ gives a straight line with a slope of $-E_g/2k$. Equation (6.12-7) is known as an *Arrhenius plot*.

(b) The conductivity, in terms of the carrier density and mobilities, is

$$\sigma = nq(\mu_e + \mu_h)$$

where $n_e = n_h = n$ for an intrinsic semiconductor. Using the data given,

$$\sigma_{300\text{ K}} = (23 \times 10^{18}\ \text{m}^{-3})(0.16 \times 10^{-18}\ \text{C})(0.364 + 0.190)\ \text{m}^2/\text{V s}$$
$$= 2.04\ \Omega^{-1}\ \text{m}^{-1}$$

This enables the constant σ_0 to be found:

$$\sigma = \sigma_0 e^{-E_g/2kT}$$

$$\sigma_0 = \sigma e^{E_g/2kT} = (2.04\ \Omega^{-1}\ \text{m}^{-1})e^{0.66\text{ eV}/2(86.2 \times 10^{-6}\text{ eV/K})(300\text{ K})}$$
$$= 7.11 \times 10^5\ \Omega^{-1}\ \text{m}^{-1}$$

Then

$$\sigma_{200°\text{C}} = (7.11 \times 10^5\ \Omega^{-1}\ \text{m}^{-1})e^{-0.66\text{ eV}/2(86.2 \times 10^{6}\text{ eV/K})(473\text{ K})}$$
$$= 217\ \Omega^{-1}\ \text{m}^{-1}$$

(c) Since resistivity ρ and conductivity σ are the inverse of each other,

$$\frac{\rho_1}{\rho_2} = \frac{\sigma_2}{\sigma_1} = \frac{\sigma_0 e^{-E_g/2kT_2}}{\sigma_0 e^{-E_g/2kT_1}}$$

$$\ln\frac{\rho_1}{\rho_2}=\frac{E_g}{2k}\left[\frac{1}{T_1}-\frac{1}{T_2}\right]$$

$$\ln\frac{\rho_{20^\circ\mathrm{C}}}{\rho_{40^\circ\mathrm{C}}}=\frac{0.7\ \mathrm{eV}}{2(86.1\times10^{-6}\ \mathrm{eV/K})}\left[\frac{1}{293\ \mathrm{K}}-\frac{1}{313\ \mathrm{K}}\right]=0.9$$

$$\ln\frac{0.5\ \Omega\ \mathrm{m}}{\rho_{40^\circ\mathrm{C}}}=0.9,\qquad \rho_{40^\circ\mathrm{C}}=0.2\ \Omega\ \mathrm{m}$$

■

■ EXAMPLE 6.12-4

Reproduce a small section of the periodic table of elements and show how small impurities alter the semiconductivity of materials by introducing excess electrons and holes. Compare the Fermi function and Arrhenius plot of an n-type semiconductor with that of an intrinsic semiconductor. Do likewise for a p-type semiconductor. Show exhaust and saturation regions.

Solution. A pure material has intrinsic properties (e.g., intrinsic conduction). Extrinsic semiconductivity results from impurities (known as dopants) and the process of adding these components is known as doping. There are two distinct types of extrinsic semiconductors: n-type (in which negative charge carriers dominate) and p-type (in which positive charge carriers dominate). The philosophy behind each type is illustrated in Fig. 6.12-6, which shows a small portion of the periodic table in the vicinity of silicon (group IV A). Intrinsic silicon has four valence (outer-shell) electrons. Phosphorus, an n-type material in group V A, has five valence electrons, the extra electron being independent of the electron pairs that serve as bonds between neighboring atoms. When supplied with a small amount of energy, this electron can be pulled away from the phosphorus atom and become a conduction electron (a negative charge carrier). The electron is located near the top of the energy gap, a position called the donor level, E_d. It can easily be donated to the conduction band [Fig. 6.12-7(a)]. The Fermi level is shifted upward.

On the other hand, aluminum (group III A) is a p-type dopant because it has three valence electrons. This deficiency of one electron compared to silicon's four valence electrons, can easily produce an electron hole (a positive charge carrier). A group III atom can accept an electron at the acceptor level

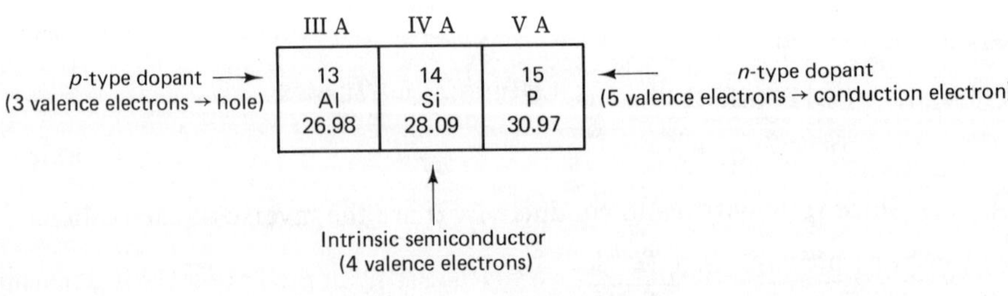

Figure 6.12-6 Small section of periodic table of elements.

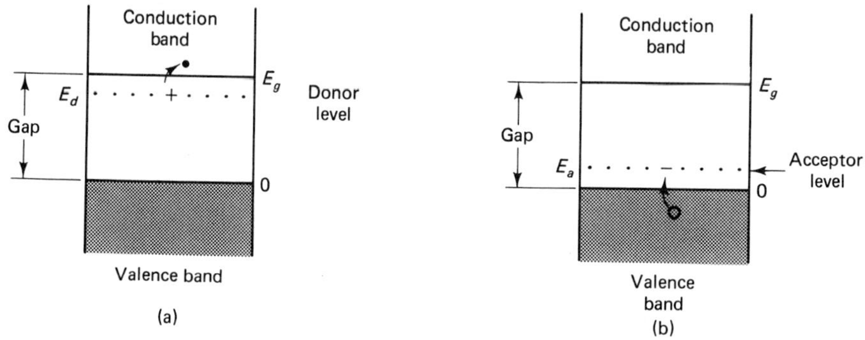

Figure 6.12-7 *n*-type and *p*-type extrinsic semiconductor.

E_a, leaving holes remaining in the valence band [Fig. 6.12-7(b)]. The Fermi function is shifted downward in a *p*-type material.

The Arrhenius behavior for *n*-type semiconductors is

$$\sigma = \sigma_0 e^{-(E_g - E_d)/kT} \tag{6.12-8}$$

where E_d is the donor level. There is no factor 2 in the exponent as there is in intrinsic semiconductors, because in extrinsic semiconductors, thermal activation produces a single charge carrier as opposed to two carriers produced in intrinsic semiconduction. The Arrhenius plot of electrical conductivity for an *n*-type semiconductor is shown in Fig. 6.12-8(a). At low temperatures (large $1/T$ values), extrinsic behavior prevails, until an upper limit is reached, corresponding to the temperature at which all possible extrinsic electrons have been promoted to the conduction band. The exhaustion range is a horizontal plateau in which the number of charge carriers is fixed and equal to the number of dopant atoms. As the temperature continues to rise (small $1/T$ values), intrin-

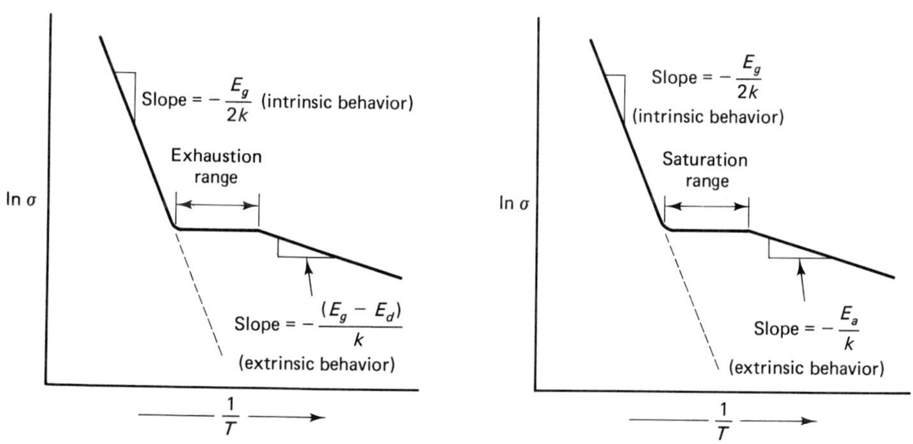

Figure 6.12-8 Arrhenius plot of electrical conductivity for *n*-type and *p*-type semiconductor.

sic conductivity of the pure silicon prevails. Similarly, for p-type semiconductors, the Arrhenius behavior is

$$\sigma = \sigma_0 e^{-E_a/kT} \tag{6.12-9}$$

and the plot of the conductivity over a wide temperature change is shown in Fig. 6.12-8(b). Again, there is a region between intrinsic and extrinsic behavior which is termed the saturation range, corresponding to all acceptor levels being saturated (occupied) with electrons. In terms of mobilities, the conductivity of extrinsic n-type and p-type semiconductors is given by $\sigma = nq\mu_e$ and $\sigma = nq\mu_h$, where n is the number of electrons and electron holes, respectively. Note that donor exhaustion and acceptor saturation are important to electrical engineers, because they provide a region of essentially constant conductivity, thereby minimizing the need for temperature compensation in electrical circuits. ■

■ EXAMPLE 6.12-5

(a) Silicon has a conductivity of $5 \times 10^{-4}\ \Omega^{-1}\ \text{m}^{-1}$ when pure. Specifications call for a conductivity of $200\ \Omega^{-1}\ \text{m}^{-1}$ when the silicon contains aluminum as an impurity. How many aluminum atoms are required per cubic meter of silicon?

(b) In a phosphorus-doped n-type silicon, the Fermi level E_F is shifted upward 0.1 eV. What is the probability of an electron being promoted to the conduction band ($E_g = 1.107$ eV) at room temperature (25°C)?

Solution

(a) The total conductivity, σ, is the sum $\sigma_{in} + \sigma_{ex}$. Tables give the intrinsic conductivity of silicon as $\sigma_n = 5 \times 10^{-4}\ \Omega^{-1}\ \text{m}^{-1}$. This is so small in comparison with $200\ \Omega^{-1}\ \text{m}^{-1}$ that it can be neglected. Therefore,

$$\sigma = n_p q \mu_p$$

or

$$n_p = \frac{\sigma}{q\mu_p} = \frac{200\ \Omega^{-1}\ \text{m}^{-1}}{(0.16 \times 10^{-18}\ \text{A s})(0.0425\ \text{m}^2/\text{V s})} = 3 \times 10^{22}/\text{m}^3$$

Since each aluminum atom contributes one acceptor site and hence one electron hole, the number of aluminum atoms required per cubic meter is 3×10^{22}. This may seem a large number, but it is only 0.6 ppm when compared with the number of silicon atoms per cubic meter.

(b) In intrinsic silicon, the Fermi level is midway between the band gap (1.107 eV), meaning that it is located at a distance 1.107/2 eV from the bottom of the conduction band. Since the Fermi level is shifted upward 0.1 eV with the addition of phosphorus, its distance from the bottom of the conduction band is now

$$E - E_F = \frac{1.107}{2}\ \text{eV} - 0.1\ \text{eV} = 0.4535\ \text{eV}$$

The Fermi distribution function

$$f(E) = \frac{1}{e^{(E - E_F)/kT} + 1}$$

then gives

$$f(E) = \frac{1}{e^{(0.4535 \text{ eV})/(86.2 \times 10^{-6} \text{ eV/K})(298 \text{ K})} + 1} = 2.20 \times 10^{-8}$$

While this is a small number, it is nevertheless two orders of magnitude greater than the value obtained for intrinsic silicon using $E - E_F = 1.107/2$ eV = 0.5535 eV:

$$f(E) = \frac{1}{e^{(0.5535 \text{ eV})/(86.2 \times 10^{-6} \text{ eV/K})(298 \text{ K})} + 1} = 4.39 \times 10^{-10}$$ ■

■ **EXAMPLE 6.12-6**

(a) A p-type semiconductor is produced with 10^{22} Al/m^3 in silicon which has intrinsic $\mu_e = 0.19$ m^2/V s, $\mu_h = 0.0425$ m^2/V s, and $E_g = 1.1$ eV at 20°C. At what temperature will the intrinsic conductivity of silicon be equal to the maximum extrinsic conductivity? Pure Si has a conductivity of 5×10^{-4} Ω^{-1} m^{-1} at 20°C.

(b) A hypothetical semiconductor with n-type doping and $E_g = 1$ eV, $E_d = 0.9$ eV has a conductivity of 100 Ω^{-1} m^{-1} at room temperture (25°C). What is the conductivity at 30°C?

(c) Tables give the following values for intrinsic Ga As at room temperature (300 K): $E_g = 1.47$ eV, $\mu_e = 0.720$ m^2/V s, $\mu_h = 0.020$ m^2/V s, $n_e = n_h = 1.4 \times 10^{12}$/m^3. Calculate the conductivity at room temperature and at 50°C.

Solution

(a) The general expression for conductivity in semiconductors is

$$\sigma = \sigma_{\text{intrinsic}} + (\sigma_n)_{\text{extrinsic}} + (\sigma_p)_{\text{extrinsic}}$$

Typically, only one of the three terms is significant at a time, and the others can be dropped. In this case, only $\sigma_{p,\text{ ex}}$ is significant at 20°C. Also, saturation corresponds to maximum extrinsic conductivity. Therefore, using μ_p for silicon,

$$\sigma_p = n_p q \mu_p$$

or

$$\sigma_p = (10^{22}/\text{m}^3)(0.16 \times 10^{-18} \text{ A s})(0.0425 \text{ m}^2/\text{V s}) = 68\ \Omega^{-1}\ \text{m}^{-1}$$

This is the maximum extrinsic conductivity. Now, the intrinsic conductivity of Si at 20°C is

$$\sigma_{20°\text{C}} = 5 \times 10^{-4}\ \Omega^{-1}\ \text{m}^{-1} = \sigma_0 e^{-1.1/2k(293 \text{ K})}$$

At temperature T (saturation), it is

$$\sigma_T = 68\ \Omega^{-1}\ \text{m}^{-1} = \sigma_0 e^{-1.1/2kT}$$

Therefore

$$\frac{\sigma_{20°C}}{\sigma_T} = \frac{5 \times 10^{-4}}{68} = \frac{e^{-1.1/2k(293\ K)}}{e^{-1.1/2kT}}$$

$$\ln\frac{5 \times 10^{-4}}{68} = -11.8 = \frac{-1.1\ eV}{2(86.1 \times 10^{-6}\ eV/K)}\left(\frac{1}{293\ K} - \frac{1}{T}\right)$$

This is solved for

$$T = 640\ K$$

(b) The temperature variation of the conductivity for an n-type semiconductor is

$$\sigma = \sigma_0 e^{-(E_g - E_d)/kT}$$

At 25°C, $\sigma = 100\ \Omega^{-1}\ m^{-1}$; this allows σ_0 to be found:

$$\sigma_0 = (100\ \Omega^{-1}\ m^{-1})e^{(1.0-0.9)\ eV/(86.2 \times 10^{-6}\ eV\ K^{-1})(298\ K)}$$
$$= 4.91 \times 10^3\ \Omega^{-1}\ m^{-1}$$

Then at 30°C:

$$\sigma = (4.91 \times 10^3\ \Omega^{-1}\ m^{-1})e^{-(1.0-0.9)\ eV/(86.2 \times 10^{-6}\ eV\ K^{-1})(303\ K)}$$
$$= 107\ \Omega^{-1}\ m^{-1}$$

(c) For intrinsic semiconduction,

$$\sigma = nq(\mu_e + \mu_h)$$

At 300 K, with the given data,

$$\sigma_{300\ K} = (1.4 \times 10^{12}\ m^{-3})(0.16 \times 10^{-18}\ C)(0.720\ m^2/V\ s + 0.020\ m^2/V\ s)$$
$$= 1.66 \times 10^{-7}\ \Omega^{-1}\ m^{-1}$$

From

$$\sigma = \sigma_0 e^{-E_g/2kT},$$
$$\sigma_0 = \sigma e^{E_g/2kT} = (1.66 \times 10^{-7}\ \Omega^{-1}\ m^{-1})e^{(1.47\ eV)/2(86.2 \times 10^{-6}\ eV\ K^{-1})(300\ K)}$$
$$= 3.66 \times 10^5\ \Omega^{-1}\ m^{-1}$$

At room temperature,

$$\sigma_{50°C} = (3.66 \times 10^5\ \Omega^{-1}\ m^{-1})e^{-(1.47\ eV)/2(86.2 \times 10^{-6}\ eV\ K^{-1})(323\ K)}$$
$$= 1.26 \times 10^{-6}\ \Omega^{-1}\ m^{-1}$$

■

■ EXAMPLE 6.12-7

The upper-temperature limit of extrinsic behavior for a phosphorus-doped germanium semiconductor is 100°C, with a conductivity of 60 $\Omega^{-1}\ m^{-1}$. The extrinsic conductivity at 27°C (300 K) is 54.8 $\Omega^{-1}\ m^{-1}$. The upper temperature

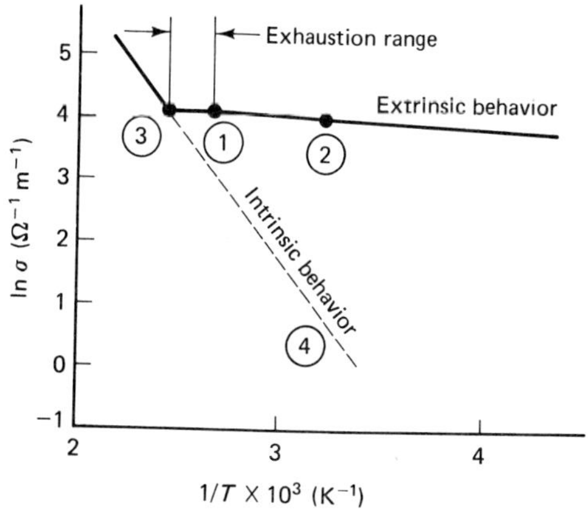

Figure 6.12-9 Arrhenius for phosphorus doped germanium semiconductor.

for the exhaustion range (point where the intrinsic conductivity equals the maximum extrinsic conductivity) is 135°C (408 K), with a conductivity of 60 Ω^{-1} m^{-1}. The intrinsic conductivity of germanium at 27°C (300 K) is 2.04 Ω^{-1} m^{-1}. Sketch the Arrhenius plot of electrical conductivity for the phosphorus-doped germanium semiconductor.

Solution. For the extrinsic semiconductor, let point 1 be 100°C (373 K) and $\sigma_{100°C} = 60\ \Omega^{-1}\ m^{-1}$. On the Arrhenius diagram, the coordinates of point 1 are then $\ln \sigma = 4.09\ \Omega^{-1}\ m^{-1}$ and $1/T = 2.68 \times 10^{-3}\ K^{-1}$. Let point 2 for the extrinsic semiconductor be 27°C (300 K) and $\sigma_{27°C} = 54.8\ \Omega^{-1}\ m^{-1}$. The coordinates of point 2 are then $\ln \sigma = 4.00\ \Omega^{-1}\ m^{-1}$ and $1/T = 3.33 \times 10^{-3}\ K^{-1}$.

For the intrinsic germanium semiconductor, let point 3 be 135°C (408 K) and $\sigma_{135°C} = 60\ \Omega^{-1}\ m^{-1}$. The coordintes on the Arrhenius diagram for this point are then $\ln \sigma = 4.09\ \Omega^{-1}\ m^{-1}$ and $1/T = 2.45 \times 10^{-3}\ K^{-1}$. Also, for intrinsic behavior, let point 4 be 27°C (300 K) and $\sigma_{27°C} = 2.04\ \Omega^{-1}\ m^{-1}$, with coordinates $\ln \sigma = 0.713\ \Omega^{-1}\ m^{-1}$ and $1/T = 3.33 \times 10^{-3}\ K^{-1}$.

The Arrhenius plot, on coordinates $\ln \sigma$ and $1/T$, is shown in Fig. 6.12-9. The extrinsic behavior (corresponding to low temperatures, high $1/T$) has slope $-(E_g - E_d)/k$, while the intrinsic behavior (corresponding to high temperatures, low $1/T$) has slope $-E_g/2k$. In between is the exhaustion range, in which all "extra electrons" have been promoted to the conduction band. ■

■ EXAMPLE 6.12-8

What is the combined effect of temperature and impurities on the Fermi energy? Plot the Fermi energy as a function of temperature for impurity semiconductors.

Solution. Consider an n-type semiconductor at absolute zero temperature. The donor levels are all occupied, but there are no electrons in the conduction

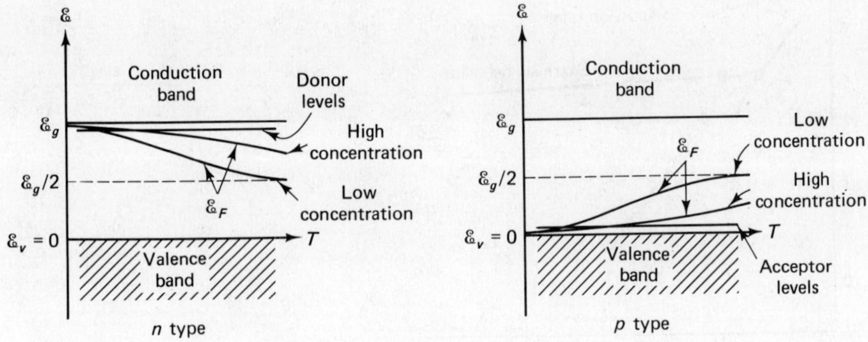

Figure 6.12-10 Fermi level as function of temperature for (a) n-type semiconductor and (b) p-type semiconductor.

band. The Fermi energy lies between the donor levels and the bottom of the conduction band. As the temperature is increased, electrons are raised from donor levels to the conduction band, and the Fermi energy begins to drop. When half the donor states are empty, the Fermi energy becomes the donor-level energy. With further increases in temperature, electrons in the valence band become excited, and the Fermi energy further drops. When the number of electrons from the valence band reaches a large fraction of those in the conduction band, the semiconductor acts as though it were intrinsic, and the Fermi energy drops to the center of the gap, as shown in Fig. 6.12-10(a). The same situation holds in Fig. 6.12-10(b) for a p-type semiconductor when the temperature is raised. The Fermi energy moves from between the top of the valence band and the acceptor level, at absolute zero, to the center of the gap at high temperatures. ■

6.13 ELECTROSTATICS OF *p-n* JUNCTIONS

When a p-type semiconductor is brought into contact with an n-type semiconductor, a double layer of charge is set up at the junction, as shown in Fig. 6.13-1. Such a junction is not usually made by joining two separate pieces of material, but by taking a single crystal of silicon and introducing donor and acceptor impurities to result in the double layer of charge seen in Fig. 6.13-1(a). On the p side of the junction there is an excess of free holes and a very low concentration of electrons. On the n side of the junction there is an excess of electrons relative to holes. The holes on the p side have a tendency to diffuse to the n side, and the electrons on the n side have a tendency to diffuse to the p side. This yields a depletion layer, which in practice is about 10^{-4} cm thick. Although there are few charge carriers in the depletion layer, the region is actually highly charged due to the presence of the ionized impurity atoms. Thus, the donor atoms in the n-type material are positively charged, while the

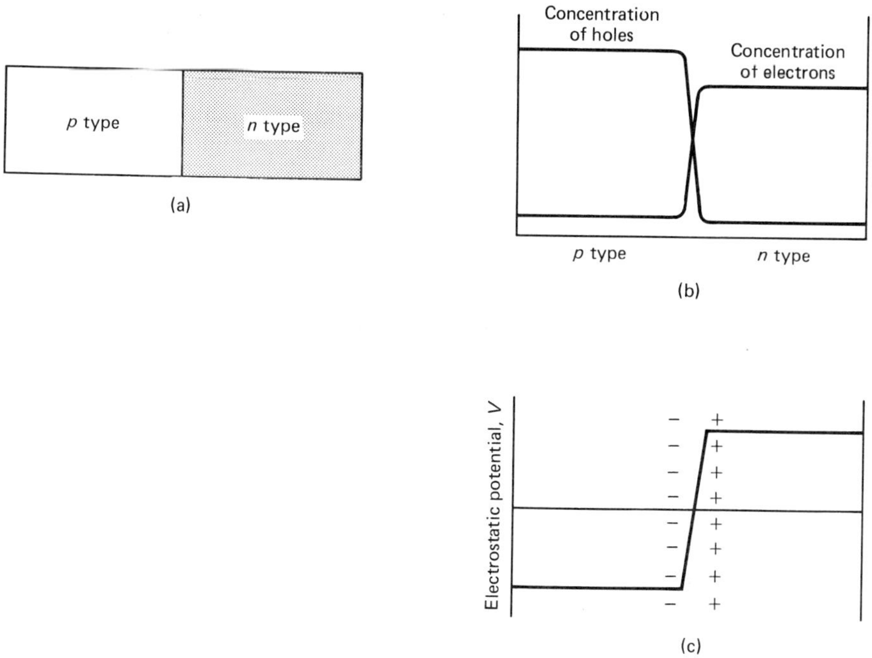

Figure 6.13-1 Depletion layer in a $p - n$ junction. (a) Single crystal modified into two separate regions. (b) Variation of hole and electron concentration across unbiased junction. (c) Electrostatic pøtential from acceptor (−) and donor (+) ions near junction.

acceptor atoms on the other side have negative charge. The double layer of charge formed is as shown in Fig. 6.13-1(b). The built-in electric field due to electrostatic potential changes at the junction is shown in Fig. 6.13-1(c).

To further discuss the electrical properties of semiconductor junctions, it is convenient to make use of the Fermi level. This is the energy below which all electron states are occupied at absolute zero. In a metal, the Fermi level coincides with the top level in a partially filled band. For intrinsic semiconductors, the Fermi level is situated at the center of the gap between the valence band and the conduction band. For n-type semiconductors the Fermi level is closer to the valence band. The comparison between the three cases is shown in Fig. 6.13-2.

Let two semiconductors be tightly brought together. An important property pertaining to the concept of the Fermi level is that when two solids are in contact at thermal equilibrium, the Fermi levels of the two solids come to alignment much as the water levels in connected tanks, so that the energy diagram is as shown in Fig. 6.13-3(b). Electrons slide down the potential hill from the n type to the p type material, and holes climb the hill from the p-type to the n-type material. As a result there are more electrons than normal in the p-type material and less electrons than normal in the n-type material. A

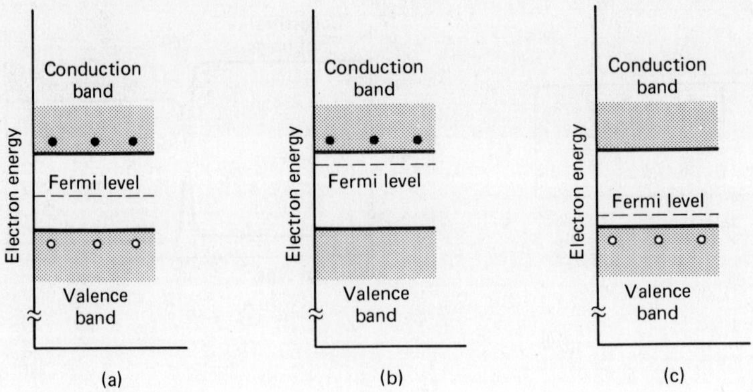

Figure 6.13-2 Fermi level for intrinsic semiconductor, *n*-type semiconductor, and *p*-type semiconductor.

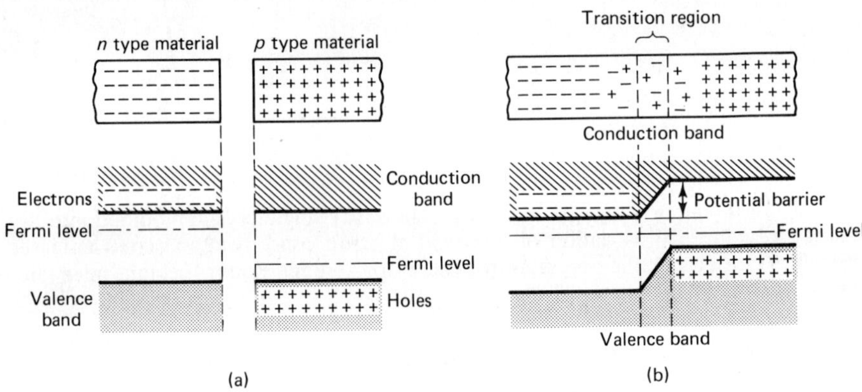

Figure 6.13-3 Energy levels in *n* and *p* materials before and after joining.

potential thus appears across the junction with the *n*-type region positive with respect to the *p*-type region.

By applying an external potential and making the *p*-type region positive with respect to the *n*-type region, more charge carriers are urged across the junction and current will flow from the *p*-type region to the *n*-type region. This situation is called *forward bias* and makes it easier for high-energy electrons and holes to be "emitted" from one material to another. It is shown in Fig. 6.13-4(a). With a *reverse bias,* as shown in Fig. 6.13-4(b), the height of the energy barrier is increased and only very few electrons and holes can pass across to carry a small current. The *n–p* diode then acts as a high resistance shunted by a capacitive component.

It will now be shown that diffusion potential or built-in voltage ψ_0 of a *p–n* junction is

$$\psi_0 = \frac{kT}{q} \ln\left(\frac{n_n p_p}{n_i^2}\right) \tag{6.13-1}$$

where q is the electronic charge, n_n is the electron density on the *n*-type side,

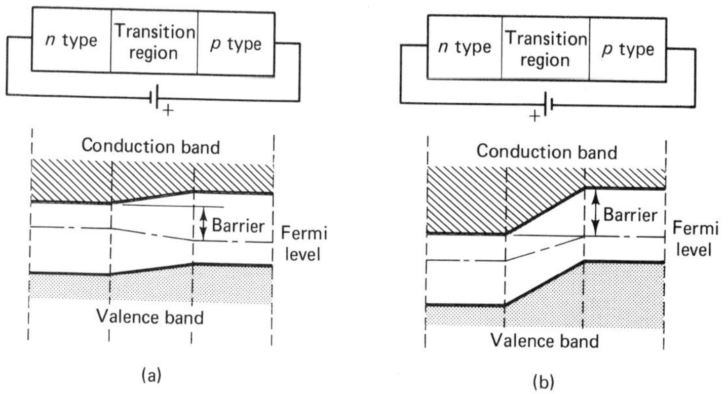

Figure 6.13-4 $p-n$ junction with forward and reverse bias.

p_p is the hole density on the p-type side, and n_i is the intrinsic carrier density of the semiconductor.

Since the density of electrons is greater in the n-type material, electrons diffuse across into the p-type region where they recombine with free holes. However, electrons leaving the n-type side will create a charge imbalance on this side by exposing ionized donors (positive charge). Similarly, holes leaving the p-type side will expose negative charge. These exposed charges will set up an electric field that will oppose the natural diffusion tendency of the electrons and holes and an equilibrium situation will be obtained. The characteristic of the equilibrium situation is that a system in thermal equilibrium can have only one Fermi level.

Far enough away from the metallurgical junction, conditions may be expected to remain unperturbed from those in isolated material. This means there must be a transition region near the junction in which a potential change, ψ_0 can be found from Fig. 6.13-5 by taking the difference of initial Fermi levels:

$$q\psi_0 = E_g - E_1 - E_2 \tag{6.13-2}$$

The expressions for E_1 and E_2 shown in Fig. 6.13-5 are

$$E_1 = kT \ln\left(\frac{N_V}{N_A}\right) \tag{6.13-3}$$

$$E_2 = kT \ln\left(\frac{N_C}{N_D}\right) \tag{6.13-4}$$

where N_V is the density of states in the valence band, N_C is the density of states in the conduction band, and N_D and N_A the densities of donors and acceptors, respectively. Thus, Eq. (6.13-2) gives

$$q\psi_0 = E_g - kT \ln\left(\frac{N_V}{N_A}\right) - kT \ln\left(\frac{N_C}{N_D}\right)$$

$$= E_g - kT \ln\left(\frac{N_C N_V}{N_A N_D}\right) \tag{6.13-5}$$

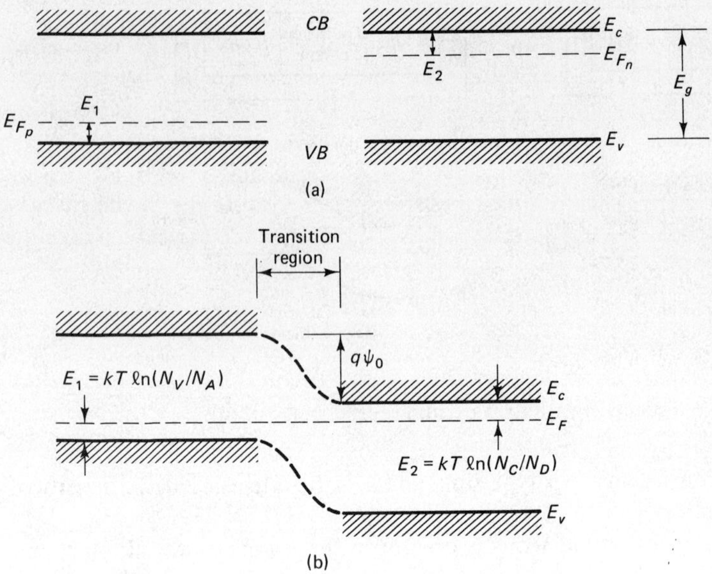

Figure 6.13-5 Energy-band diagram of a *p-n* homojunction. (a) Isolated pieces of *p*-type and *n*-type material. (b) Junction at thermal equilibrium.

Now, the number of electrons in the conduction band per unit volume of the crystal is

$$n = N_C e^{(E_F - E_C)/kT} \tag{6.13-6}$$

Similarly, the number of holes in the valence band per unit volume of the crystal is

$$p = N_V e^{(E_V - E_F)/kT} \tag{6.13-7}$$

For the idealized case of a pure and perfect semiconductor, n equals p because each electron in the conduction band leaves a vacancy or hole in the valence band. Hence,

$$n = p = n_i \tag{6.13-8}$$

$$np = n_i^2 = N_C N_V e^{(E_V - E_C)/kT} = N_C N_V e^{-E_g/kT} \tag{6.13-9}$$

where n_i is known as the "intrinsic concentration" and E_g is the width of the forbidden gap between the conduction and valence bands. Substitution of (6.13-9) into (6.13-5) yields

$$\psi_0 = \frac{kT}{q} \ln\left(\frac{N_A N_D}{n_i^2}\right) \tag{6.13-10}$$

which is the same as (6.13-1) and is the desired result. An applied voltage, V_a, will change the potential difference between the two sides of the diodes by V_a. Hence, the potential across the transition region will become $(\psi_0 - V_a)$.

■ EXAMPLE 6.13-1

The Fermi level is the energy level below which all electron states are occupied at absolute zero. In a metal the Fermi level coincides with the top occupied level in a partially filled band. When the temperature rises, electron transitions are most likely to take place from the highest occupied energy levels, that is, from near the Fermi level; thus the Fermi level acts as a reference level against which electron transitions can be measured. Calculate, in the cases of diamond ($E_g = 5.5$ eV) and germanium ($E_g = 0.67$ eV), the probabilities of the transitions.

Solution. An intrinsic semiconductor is defined as one that contains no electrically active impurities. Any electrical conduction must be due to the excitation of electrons from the valence band into the conduction band. Therefore, $n = p$. At $T = 0$ K in an intrinsic semiconductor the valence band is filled and no electrons are in the conduction band. For $T > 0$ K some electrons are excited to the conduction band, and their number equals the number of holes created in the valence band. This is illustrated in Fig. 6.13-6. The distribution function for the number of holes is the shaded area in Fig. 6.13-6(b). The number of holes with energies between E and $E + dE$ is given by

$$dn_h = (1 - F_{\mathrm{FD}})g(E)\,dE \tag{6.13-11}$$

where $g(E)$ is the density of states in the valence band. It will now be shown that E_F for an intrinsic semiconductor should be placed at the center of the energy gap E_g.

At temperature T above zero, a hole in the valence band is produced for each electron that is thermally excited into the conduction band. This implies that the probability of finding an electron at the energy value $E = E_g$ is equal to that of finding a hole at the energy value $E = 0$. Thus,

$$F(E_g) = 1 - F(0) \tag{6.13-12}$$

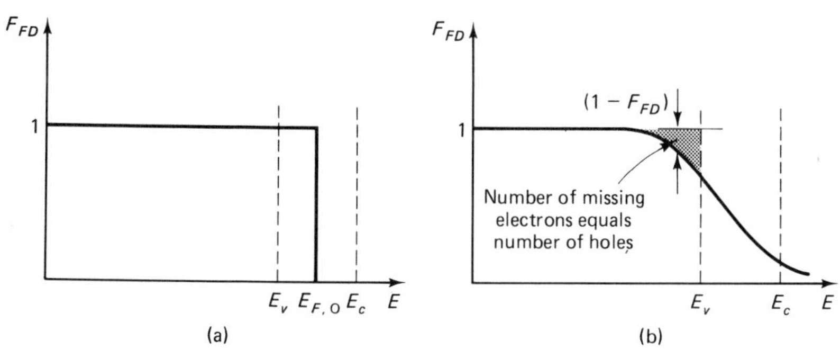

Figure 6.13-6 FD distribution for electrons and holes.

Inserting this into the energy distribution for fermions gives

$$\frac{1}{e^{(E_g - E_F)/kt} + 1} = 1 - \frac{1}{e^{-E_F/kT} + 1} = \frac{e^{-E_F/kT}}{e^{-E_F/kT} + 1}$$

$$e^{-E_F/kT} + 1 = e^{-E_F/kT}\left[\epsilon^{(E_g - E_F)/kT} + 1\right]$$

$$1 = e^{(E_g - 2E_F)/kT}$$

whereupon

$$E_F = \frac{E_g}{2} \qquad (6.13\text{-}13)$$

As an example, diamond has an energy gap E_g of approximately 5.5 eV. Therefore, at room temperature ($kT = 0.026$ eV) the occupation probability for the bottom of the conduction band is

$$\frac{1}{e^{E_g/2kT} + 1} = \frac{1}{e^{5.3/0.052} + 1} = e^{-5.5/0.052} \simeq e^{-100} \simeq 5 \times 10^{-45}$$

Less than one in every 10^{44} electrons in the valence band can have enough energy to be excited to the conduction band at room temperature. As there are only $\sim 10^{24}$ atoms in a gram-atomic weight of material it is not surprising that diamond is a good insulator.

For a typical semiconductor, say Ge ($E_g = 0.67$ eV), the probability is

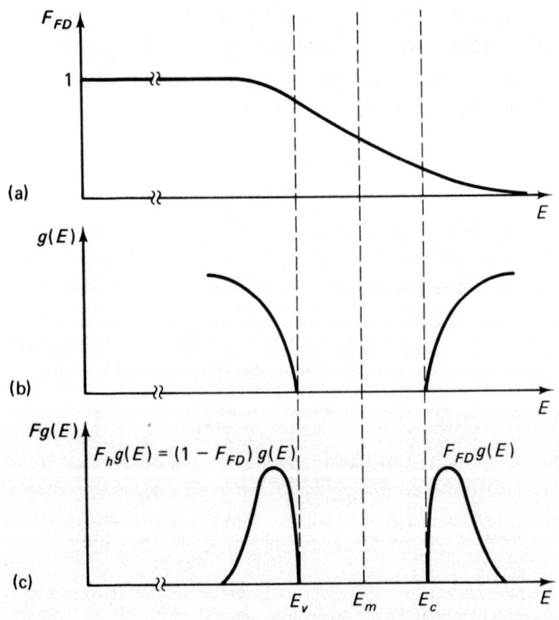

Figure 6.13-7 Density of states and probability distribution for electrons and holes.

$$P(E_c) = \frac{1}{e^{0.67/0.052} + 1} \simeq e^{-0.67/0.052} = e^{-12.8} = 2 \times 10^{-6}$$

Here, approximately one valence electron in a million can be excited to the bottom of the conduction band. It is certainly possible to find electrons in the conduction band for this material. The changes are much smaller than for a conductor (where P is on the order unity in the conduction band), but a reasonable number of electrons can still exist in the conduction band where they can contribute to electrical conduction properties. Notice the strong dependence of P for semiconductors on temperature. For an increase in temperature by 10 K, or 3%, at room temperature, the occupation probability would increase by approximately 30%. (This sensitivity to temperature increases as the gap energy of the semiconductor decreases.) For an intrinsic semiconductor, in the region immediately above the conduction band and below the valence band the density-of-states functions $g(E)$ are approximately symmetrical about the midpoint of the gap, $E_m = E_F$ as shown in Fig. 6.13-7(b). ■

■ EXAMPLE 6.13-2

Donors in an n-type silicon have energies that position the Fermi energy at 0.1 eV below the top of the energy gap at 20°C. Show that the negative carriers are the dominant charge carriers by calculating the ratio of the electron occupancy probability at the bottom edge of the conduction band to the probability of a hole being present at the top edge of the valence band. The energy gap for silicon is 1.1 eV.

Solution. The Fermi distribution function at $(E_F + 0.1)$, the bottom of the conduction band, is

$$f(E_F + 0.1) = \frac{1}{[e^{(0.1\ \text{eV})/(86.2 \times 10^{-6}\ \text{eV/K})(293\ \text{K})} + 1]} = 18 \times 10^{-3}$$

At the top of the valence band, which is a distance 1.0 eV from the Fermi energy, the probability of hole occupation is

$$f(E_F + 1.0) = \frac{1}{[e^{(1.0\ \text{eV})/(86.2 \times 10^{-6}\ \text{eV/K})(293\ \text{K})} + 1]} = 6 \times 10^{-18}$$

which is immeasurably smaller. For all practical purposes, the number of electron holes for this extrinsic silicon at 20°C can be ignored. The situation for donors and acceptors is illustrated in Fig. 6.13-8. In (a), the extra electron belonging to group V impurity is much more weakly bound to its parent atom than the electrons taking part in the covalent bond. This is equivalent to a donor level close to the conduction band. In (b), the group III impurity has one bonding electron missing; there is a "hole" that any valence electron with a little surplus energy can fall into. The acceptor level is just above the edge of the valence band. ■

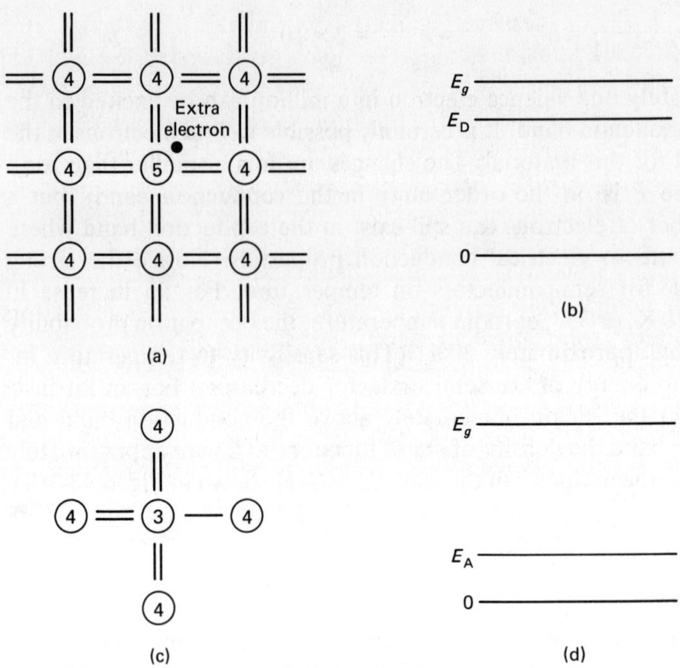

Figure 6.13-8 (a) Group V impurity and donor level. (b) Group III impurity and acceptor level.

6.14 SEMICONDUCTOR DEVICES

There are many semiconductor devices. Only a selected number will be discussed here.

Junction Devices (Diodes)

Consider a *p–n* junction (usually manufactured from a single piece of host crystal by varying the doping in different parts of it as the crystal is grown). Holes from the *p* side diffuse into the *n* side, giving the *n* side a positive charge. At the same time, electrons from the *n* side diffuse into the *p* side, giving the *p* side a negative charge. The results of these diffusions include the following: (1) An electric field, called the built-in field, is set up in the region of the junction. This field is directed from the side having a plus charge to the side having a minus charge (from the *n* side to the *p* side). A built-in voltage results from this built-in field. (2) The force on an electron acts opposite to the electric-field direction. Therefore, to move an electron from the *n* side to the *p* side requires external work. (3) The *p* side energy is raised and the *n* side energy is lowered until the Fermi energy is constant across the junction. (4) The recombination leaves the junction depleted of charge carriers compared to the remainder of the material. (5) There is a recombination current.

For electrons at zero external voltage, this current is $i_{r,n}(0)$. For holes, it is $i_{r,p}(0)$. (6) The electron-hole pairs constantly being generated in the junction region by thermal processes are separated and swept out of this region by the built-in field. The result is a generation current. For electrons at zero external voltage, this current is $i_{g,n}(0)$. For holes, it is $i_{g,p}(0)$. (7) At equilibrium, the recombination and generation currents are equal and opposite, that is, $i_{r,n}(0) = i_{g,n}(0)$ for electrons and $i_{r,p}(0) = i_{g,p}(0)$ for holes. These characteristics of the *p–n* junction are illustrated in Fig. 6.14-1.

A familiar application of the *p–n* junction is the *light-emitting diode (LED)*, which is found in the digital displays of many hand-held calculators and clocks. An LED operates on the principle shown in Fig. 6.14-2. When a forward bias is placed across the junction in the manner indicated, carriers of both types cross the junction; there they recombine emitting a photon, producing luminescence: $n + p \rightarrow$ photon.

Since $E = h\nu$, different frequencies of light can be obtained by varying the energy difference through which the electrons fall. This can be accomplished by using materials with different energy gaps. When GaAs is used, the pho-

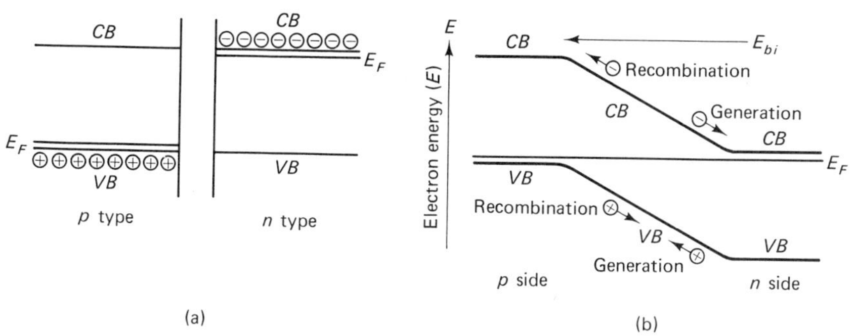

Figure 6.14-1 *p-n* junction, separated and together with no external voltage.

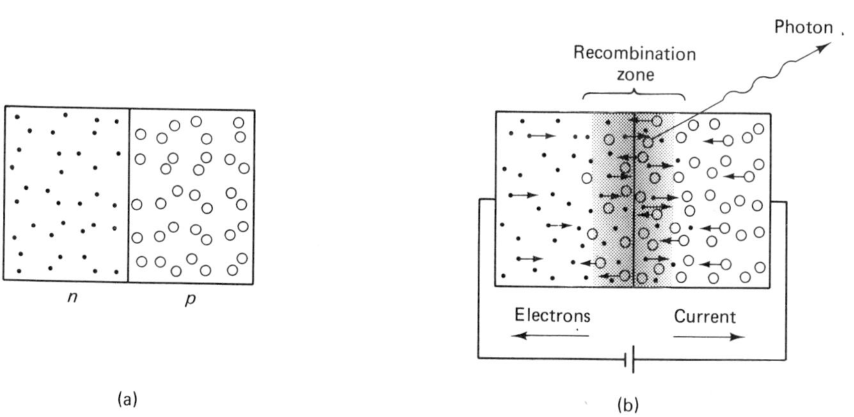

Figure 6.14-2 Light-emitting diode (LED).

tons emitted in the recombination zone are red; a phosphorus substitution, Ga (As, P), produces green photons. The *semiconductor laser* operates in much the same way as the LED, except that the transitions are stimulated rather than spontaneous. The sides of the junction are made into mirrors, thereby providing light that travels parallel to the *p–n* junction in coherent waves.

Semiconductor *optoelectronic devices* emit or absorb light as electrons change energies. Electrical energy is fed into light-emitting diodes and semiconductor lasers to produce light energy. In contrast, light energy is fed into solar cells to produce electrical energy. The *photovoltaic effect* is illustrated in Fig. 6.14-3. It occurs when photons are absorbed to create electron–hole pairs. If created close enough to the *p–n* junction region, the pairs can migrate to it. There, the built-in electric field sweeps the electrons to the *n* side and the holes to the *p* side of the junction. If the junction is connected to an external circuit, the resulting voltage and current can be used in that circuit. The device

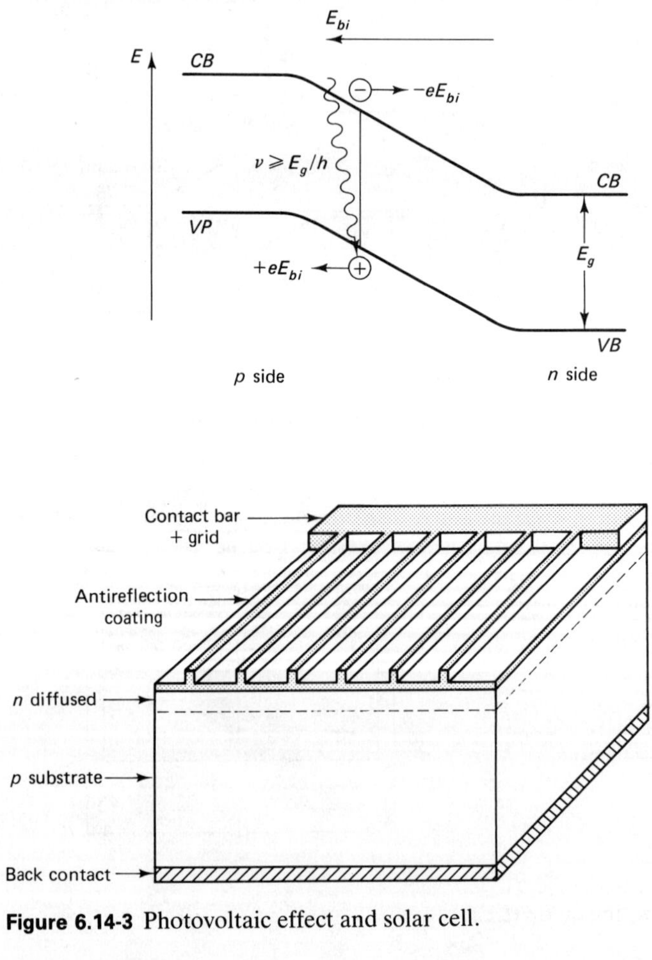

Figure 6.14-3 Photovoltaic effect and solar cell.

is called a *solar cell*, although sunlight isn't required. Any light source of sufficiently high frequency can provide the photons.

The *p–n* junction can also serve as a *rectifier*, that is, it can function as an electrical check valve that lets current pass one way but not the other. With the *forward bias* of Fig. 6.14-4(a), current can pass because carriers, both electrons and holes, can move through the junction. With the *reverse bias* of Fig. 6.14-4(b), the carriers are pulled away from each side of the junction to leave a carrier-depleted zone at the junction. Usable current passes with only the forward bias.

Figure 6.14-5(a) and (b) shows the current current flow as a function of voltage in an ideal rectifier (a) and in an actual rectifier (b). The ideal rectifier allows zero current to pass in reverse bias and has zero resistivity in forward bias. The actual rectifier has some small current in reverse bias (from minority carriers) and a small resistivity in forward bias.

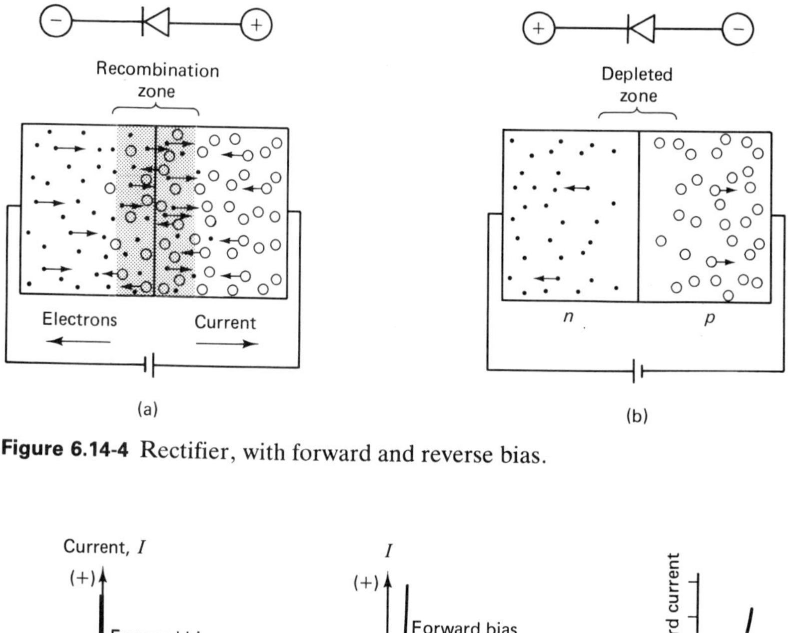

Figure 6.14-4 Rectifier, with forward and reverse bias.

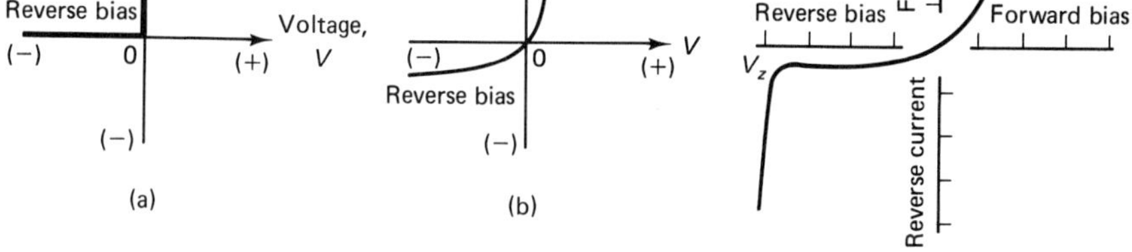

Figure 6.14-5 (a), (b) Ideal and actual rectifier. (c) Zener diode, when reverse voltage is sufficiently great, V_z electrons tunnel through the junction.

The small reverse current holds over a wide range of reverse voltages. There is, however, a point at which electrical breakdown in the depleted region occurs and a surge of current passes [Fig. 6.14-5(c)]. Specifically, those few carriers that are in the depleted zone are accelerated to high velocities by the steep potential drop. These energetic electrons can knock loose other electrons. An avalanche develops that leads to a high current. In effect, the device acts as a safety valve that opens at a definite voltage. Lightning arrestors operate that way. Figure 6.14-5(c) illustrates the principle of the *Zener diode,* which conducts current from a large reverse bias. Electrons and holes can tunnel across the junction to transport the current. Zener diodes are used for voltage regulation and overload protection.

Transistors

Perhaps the most heralded component in solid-state development has been the transistor, which is a junction device that amplifies a weak signal into a stronger, more usable output. It consists of a pair of nearby *p–n* junctions back-to-back in a single crystal of semiconductor, thereby forming a sandwich structure. Either the *p*-material or the *n*-material can constitute the filling. Thus there are two types of transistor: *p-n-p* or *n-p-n*, as shown in Fig. 6.14-6. The operation of junction transistors is based on the interaction of majority and minority carriers and as both carirer types are involved, such devices are classified as *bipolar*.

The three regions of a transistor are termed emitter, base, and collector. In Fig. 6.14-6(a), junction 1 (between emitter and base) is forward biased. As such, it looks identical to a rectifier. However, the function of the transistor involves behavior not considered in the description of the rectifier. Specifically, the recombination of electrons and holes in Fig. 6.14-6(a) does not occur immediately. In fact, many of the charge carriers move well beyond the junction. If the base (*n*-type) region is narrow enough, a large number of holes pass across junction 2. Once in the collector, the holes again move freely (as majority charge carriers). The extent of overshoot of holes beyond junction 1

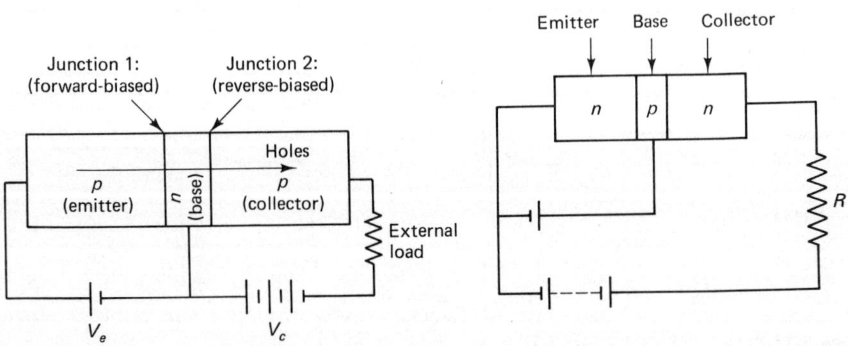

Figure 6.14-6 Schematic of $p + n - p$ and $n - p - n$ transistor.

is an exponential function of the emitter voltage, V_e. Because of this, the current in the collector, I_C, is an exponential function of V_e:

$$I_c = I_0 e^{V_e/B} \tag{6.14-1}$$

where I_0 and B are constants. The transistor is an amplifier, because slight increases in emitter voltage can produce large increases in collector current.

An *n-p-n* transistor works in a similar way; only the signs of the potential and currents are reversed, and electrons are the principal carriers. In Fig. 6.14-6(b), when the electrons reach the collector region from the base region conduction band, their energy is high enough for them to suffer many collisions and still cascade down the levels in the conduction band, before they are stopped. So almost all of them reach the collector terminal, or $I_c = I_e$. The emitter current I_e is controlled by the magnitude of the potential difference between the emitter and the base. This potential difference may be regarded as the input to the circuit; it may vary with time. The output of the circuit is the potential difference $I_c R$ across the resistor of Fig. 6.14-6(b). This can be much greater than the input, if R is large. (A power gain of 10^4 is not unusual.) The collector current is controlled by the emitter voltage according to Eq. (6.14-1).

Largely for constructional reasons, integrated circuits also use a different type of transistor, the *field-effect transistor* (FET). This device incorporates a channel between a source and drain (corresponding to the emitter and collector, respectively). There are various types of field-effect transistor; an example is the metal-oxide-silicon field-effect transistor (MOSFET), shown in Fig. 6.14-7. Assume that the substrate is an n-type semiconductor and the source and drain are made of p material. At zero gate bias, no drain current flows because one of the junctions is bound to be reverse bias. Suppose that the gate is made negative. Electrons will be repelled, and at sufficiently large negative gate voltage inversion will occur; that is, the material in the vicinity of the insulator SiO_2 will turn into a p-type semiconductor. Holes may then flow unimpeded from source to drain. The larger the gate voltage, the greater the carrier concentration in the inversion layer. Thus, a very small current to and from the gate can control a much larger current through the channel.

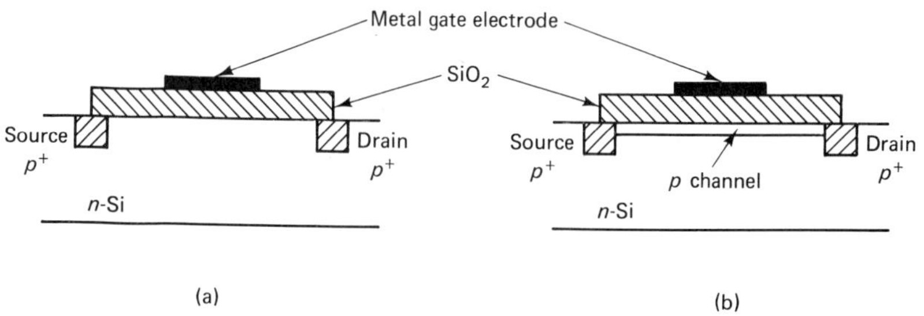

Figure 6.14-7 Schematic representation of a MOSFET. (a) Zero gate bias. (b) Forward bias introducing a p-channel.

■ **EXAMPLE 6.14-1**

Show, from thermodynamic considerations, that the Fermi level in a junction must be the same on both sides of the junction. Also, explain, on the basis of the Fermi distribution function, the operation of a *p–n* junction, a rectifying junction, and a Zener diode.

Solution. Consider two systems of particles (e.g., electrons) which are brought into contact at a temperature T. Before contact, the constant volumes V_1 and V_2 contain N_1 and N_2 particles, respectively. After contact, the particle numbers change in each volume until thermodynamic equilibrium has been established. The condition for this is that the free energy F of the whole system $F_1 + F_2$ reaches a minimum. Therefore,

$$\delta_F = \left(\frac{\partial F_1}{\partial N_1}\right)_{T,V_1} \delta N_1 + \left(\frac{\partial F_2}{\partial N_2}\right)_{T,V_2} \delta N_2 = 0 \tag{6.14-2}$$

with the additional constraint

$$\delta N_1 + \delta N_2 = 0 \tag{6.14-3}$$

yields (as in Chapter 2),

$$\left(\frac{\partial F_1}{\partial N_1}\right)_{T,V_1} = \left(\frac{\partial E_2}{\partial N_2}\right)_{T,V_2} \tag{6.14-4}$$

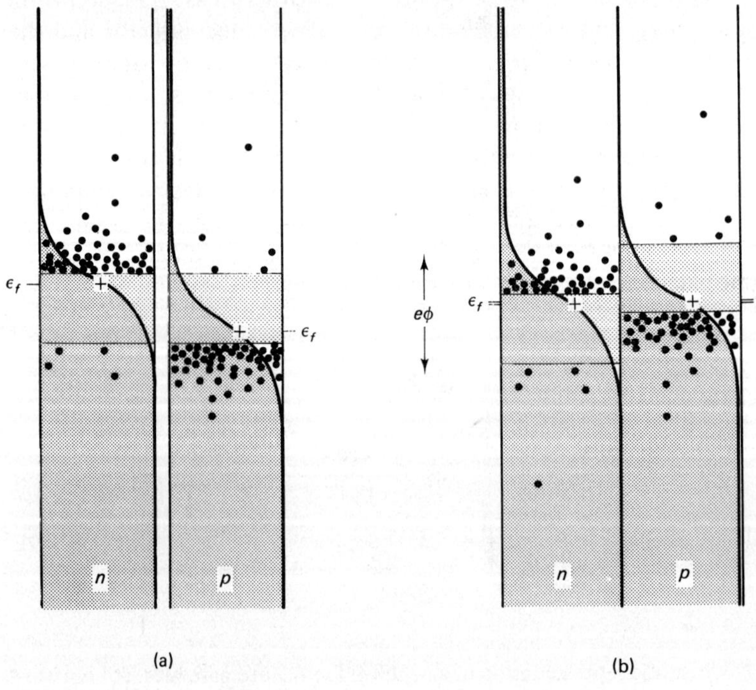

Figure 6.14-8 Fermi distribution for $n - p$ junction. (a) Separate n and p distributions. (b) Distributions in contact. $e\Phi$ equals initial ΔE_F.

Since the variation of free energy with particle number at constant temperature and volume is the chemical potential, Eq. 6.14-4 says that equilibrium is reached when the chemical potentials are equal. But the Fermi energy is also the chemical potential. Therefore, the condition of equilibrium is that Fermi energies on both sides of the junction be equal and line up on the same horizontal in the band representation of the contact. Figure 6.14-8(a) shows the n and p distributions when the two types are separated, with the energy gaps being equal. When the n–p junction is formed, the gaps shift so as to equalize the Fermi energy, as shown in Fig. 6.14-8(b). This introduces a contact potential, Φ, and requires the corresponding energy, $e\Phi$, equal to the initial ΔE_F.

When a rectifying junction is formed, as in Fig. 6.14-9(a), the conductivity through the junction is low, because few electrons are available for transfer across the conduction bands and few holes are available for transfer across the valence bands. If a forward bias is placed across the junction as in Fig. 6.14-9(b), the height of the potential hill is decreased, and the numbers of carriers in both bands increase exponentially with the applied voltage. On the other hand, a reverse bias [Fig. 6.14-9(c)] reduces the number of carriers drastically due to the increase in height of the energy hill. With appropriate circuitry, a rectifier can change ac current into dc current.

A reverse bias produces little reverse current initially. However, when the reverse voltage is sufficiently great (V_z), electrons and holes can tunnel through the junction, giving rise to a very large reverse-biased current, as in

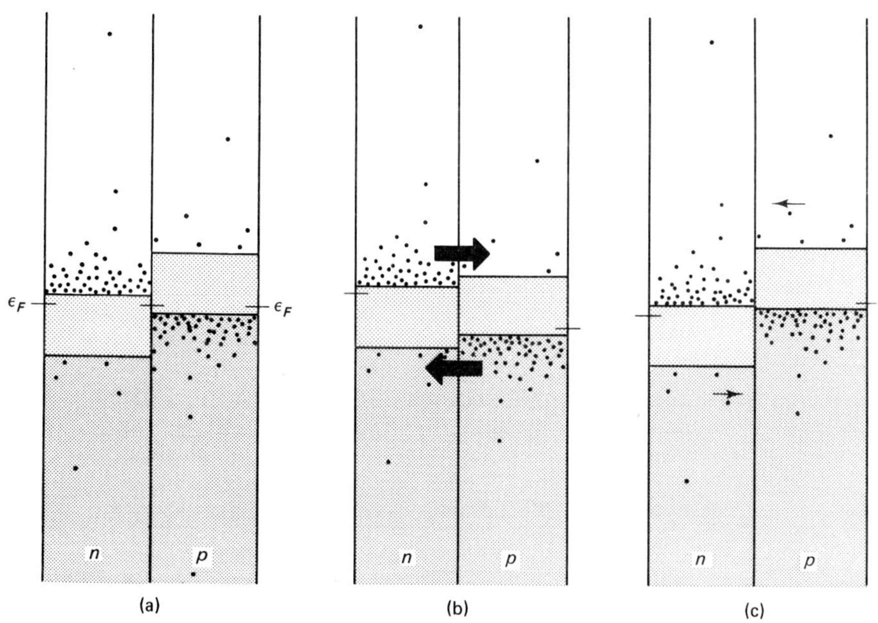

Figure 6.14-9 Rectifying junction. (a) No bias applied. (b) Forward bias applied. (c) Reverse bias applied.

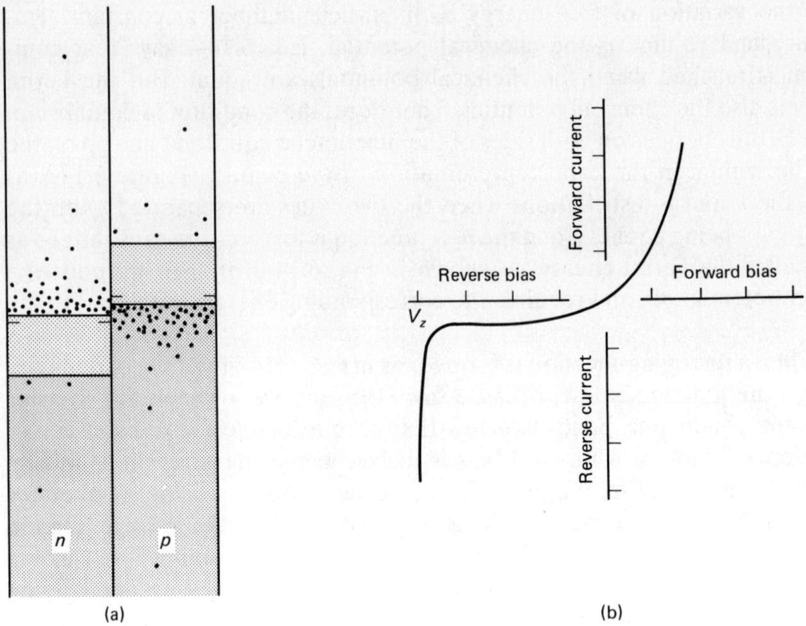

Figure 6.14-10 Zener diode. A reverse bias produces little reverse current initially. However, when the reverse voltage reaches V_z, electrons and holes can tunnel through the junction to produce a larger reverse current.

Fig. 6.14-10. The Zener diode, which conducts current from a large reverse bias, works on that principle. ■

■ EXAMPLE 6.14-2

Small forward resistance and large reverse resistance are typical of junction diodes. The saturation current (maximum theoretical reverse current) for a junction diode is 10 μA. What is the resistance at room temperature for forward and reverse biases of 0.20 V?

Solution. The total current through a p–n junction is

$$i = i_s(e^{eV/kT} - 1) \tag{6.14-5}$$

where i_s is the saturation current. Here, $eV = e(\pm 0.20 \text{ V})$ and $kT = \frac{1}{40}$ eV at room temperature, so that $eV/kT = \pm 8.0$. The forward and reverse bias resistances are

$$R = \frac{V}{i} = \frac{V}{i_s(e^{eV/kT} - 1)} = \frac{\pm 0.20 \text{ V}}{(10 \times 10^{-6} \text{ A})(e^{\pm 8.0} - 1)} \tag{6.14-6}$$

$$= 6.7\ \Omega \quad \text{and} \quad 20{,}000\ \Omega$$

An explanation can be given for why the forward resistance is so small and the reverse resistance so large: Forward bias drives holes in the p side and electrons in the n side toward the p–n junction where they can easily recombine, and so the forward resistance is low. However, reverse bias pulls the electrons and holes away from the junction, creating a large reverse resistance.

■ EXAMPLE 6.14.3

(a) Determine the photon wavelength (nanometers) necessary to promote an electron to the conduction band in intrinsic silicon.

(b) What photon wavelength (nm) is necessary to promote a donor electron to the conduction band in arsenic-doped silicon?

(c) At what frequencies will a GaAs solar cell at room temperature produce electrical energy? Relate this to a GaAs LED or laser.

Solution

(a) The energy of a photon of electromagnetic radiation is

$$E = \frac{hc}{\lambda} \tag{6.14-7}$$

where h is Planck's constant (0.663×10^{-33} J s), c is the speed of light (3.00×10^8 m /s), and λ is the wavelength. E here is the band gap of silicon, given in tables as $E_g = 1.107$ eV. Thus,

$$1.107 \text{ eV} = \frac{hc}{\lambda}$$

$$\lambda = \frac{hc}{1.107 \text{ eV}} = \frac{(0.663 \times 10^{-33} \text{ J s})(3.00 \times 10^8 \text{ m/s})}{(1.107 \text{ eV} \times 0.16 \times 10^{-18} \text{ J/eV})} \times 10^9 \text{ nm/m}$$

$$= 1120 \text{ nm}$$

(b) Electron promotion in an n-type extrinsic semiconductor requires only $E_g - E_d$ to be overcome. Tables give $E_g - E_d = 0.049$ eV for silicon. Therefore,

$$0.049 \text{ eV} = E = \frac{hc}{\lambda}$$

or

$$\lambda = \frac{(0.663 \times 10^{-33} \text{ J s})(3.00 \times 10^8 \text{ m/s})}{(0.049 \text{ eV} \times 0.16 \times 10^{-18} \text{ J/eV})} \times 10^9 \text{ nm/m}$$

$$= 25{,}400 \text{ nm}$$

(c) The energy of a photon may also be expressed as $E = h\nu$, where ν is the frequency (Hz). At room temperature GaAs has a band gap of 1.42 eV. Consequently, photons of energy greater than or equal to 1.42 eV are needed to produce electron–hole pairs:

$$h\nu \geq 1.42 \text{ eV}$$

$$\nu \geq \frac{1.42 \text{ eV}}{0.663 \times 10^{-33} \text{ J s} \times \text{eV}/1.602 \times 10^{-19} \text{ J}} = 3.43 \times 10^{14} \text{ Hz}$$

This is near the infrared region. GaAs solar cells will convert light energy into electrical energy for frequencies above 3.43×10^{14} Hz. Conversely, GaAs LEDs and lasers that operate by recombination between the bottom of the conduction band and the top of the valence band will emit 3.43×10^{14} Hz infrared radiation. ■

■ EXAMPLE 6.14-4

A *pnp* bipolar transistor is formed from two *p–n* junctions back to back, with a thin ($<10^{-1}$ mm) lightly doped, *n*-type region between two more heavily doped *p*-type regions. A battery is connected across the *p-n-p* transistor, as shown in Fig. 6.14-11(a). The base (center *n* portion) is lightly doped so that most (typically over 98 percent) of the holes from the heavily doped emitter (the left *p* portion) diffuse right across the base to the collector (the right *p* portion). In the particular *p-n-p* transistor, 99.0 percent of the holes diffuse from the emitter to the collector. What current from the base will allow a current of 2.0 A to flow to the collector?

Solution. For clarity, no donor or acceptor levels are shown in Fig. 6.14-11, and only the majority carriers are represented in each region. Although only 1 percent of the holes from the emitter recombine with electrons in the base, this is enough to quickly make the base positive, thereby setting up a strong electric field to repel any further hole current from the emitter. The hole current will be cut off unless the positive charge in the base portion of the *p-n-p* transistor is neutralized. This can be done by adding electrons from an external source to the base (a conventional electric current away from the base). In this way, a small current from the base allows a much larger emitter-collector current to flow.

In the present case, a 99.0 percent figure means that for every 100 holes flowing from the emitter, 99 get to the collector and 1 must be neutralized by adding an electron to the base. Therefore the current ratio is

$$\frac{99 \text{ charge carriers per time}}{1 \text{ charge carrier per time}} = \frac{2.0 \text{ A}}{i_b}$$

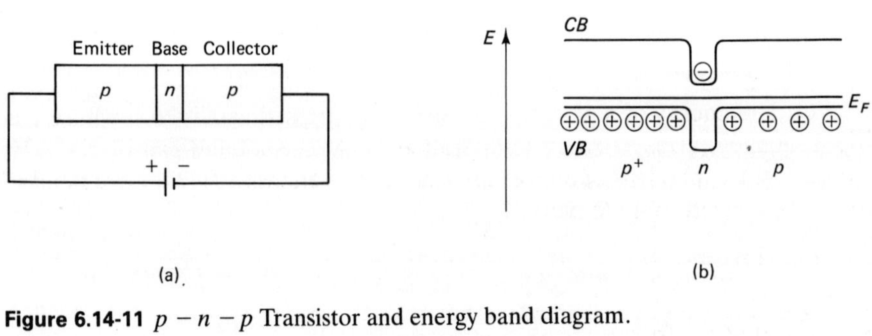

Figure 6.14-11 $p - n - p$ Transistor and energy band diagram.

or $i_b = (2.0\ \text{A})\left(\frac{1}{99}\right) = 0.020\ \text{A} = 20\ \text{mA}$

This small current of 20 mA turned on and off from the base controls the much larger current of 2.0 A to the collector. ■

■ EXAMPLE 6.14-5

(a) Consider the *p-n-p* transistor in Fig. 6.14-12(a) to be connected as shown. Explain how it can be used to amplify the ac source signals.

(b) How would the circuit in Fig. 6.14-12(a) be changed for an *n-p-n* bipolar junction transistor?

(c) A transistor has a collector current of 4.7 mA when the emitter voltage is 17 mV. Increasing the emitter voltage to 28 mV (a factor of 1.65) increases the collector current to 27.5 mA (a factor of 5.85). Determine the collector current when the emitter voltage is increased to 39 mV.

Solution

(a) The emitter–base *p–n* junction is forward biased and therefore has a small resistance. The base–collector *p–n* junction, however, is reverse biased and therefore has a large resistance. If the ac source has a small change in voltage, ΔV_{in}, the emitter current will change by Δi. Most of this Δi will pass through the base–collector junction, but since $\Delta V = R(\Delta i)$, the large R of this junction will give a large ΔV_{out}. Thus, a small voltage change at the input results in a large voltage change at the output, and the ac signals are amplified.

(b) In an *n-p-n* transistor, the emitter and collector would be *n* types and the base would be a *p* type. To maintain the emitter–base junction forward biased and the base–collector reverse biased, it is merely necessary to reverse the polarity of the batteries, as shown in Fig. 6.14-12(b).

(c) The current that moves through the collector is controlled by the

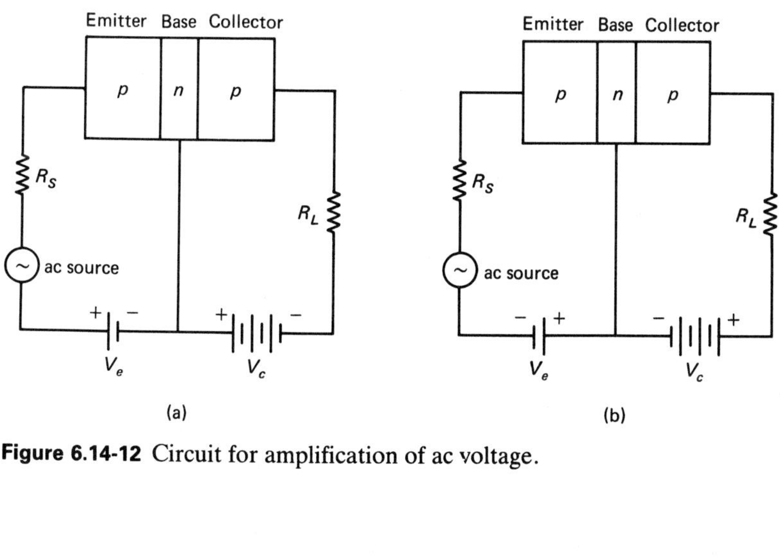

Figure 6.14-12 Circuit for amplification of ac voltage.

emitter voltage, V_e. As the emitter voltage fluctuates, the collector current, I_c, changes exponentially:

$$I_c = I_0 e^{V_e/B} \tag{6.14-8}$$

or

$$\ln I_c = \ln I_0 + V_e/B \tag{6.14-9}$$

where I_0 and B are constants for any given temperature. Here,

$$\ln\ 4.7 = 1.55 = \ln I_0 + 17/B$$

$$\ln 27.5 = 3.31 = \ln I_0 + 28/B$$

These two equations are solved simultaneously for

$$\ln I_0 = -1.17, \qquad B = 6.25\ \text{mV}$$

Thus, at $V_e = 39$ mV, Eq. (6.4-8) gives

$$\ln I_c = -1.17 + 39/6.25$$

$$I_c = 160\ \text{mA}$$

It is seen that the change in collector current is much greater than the change of the signal voltage. ■

6.15 LIQUIDS: RADIAL DISTRIBUTION FUNCTION

The preceding discussion has dealt primarily with the solid state. It is now time to proceed with an introduction to the liquid state. Of all the enterprises of physics, the study of the solid state is one of the most successful, whereas the study of the liquid state is one of the least successful. Liquids have neither the long-range structural order of solids nor the small intermolecular interaction energy of gases and so are much harder to deal with theoretically than solids or gases.

While there is no difficulty in defining a solid, there is some ambiguity in even stating what is meant by a liquid. It is possible, in the first instance, to distinguish between a liquid and a solid by the observation that liquids slosh and solids do not. To say the same thing in more sophisticated language: solids resist shear stress and support low-frequency transverse sound while liquids do not. That, however, still takes in amorphous solids and even glasses (which slosh but very, very slowly).

Along the line of argument of Section 6.1, it may be said that the liquid phase is characterized by the rough equality

$$\tfrac{1}{2}\overline{mu^2} \approx \Delta E \tag{6.15-1}$$

where $\tfrac{1}{2}\overline{mu^2}$ is, of course, an atom's mean kinetic energy, and ΔE is the pair dissociation energy. However, it is appropriate to review the p-V-T surface of Fig. 6.15-1 and pose the question of whether to label the shaded region a

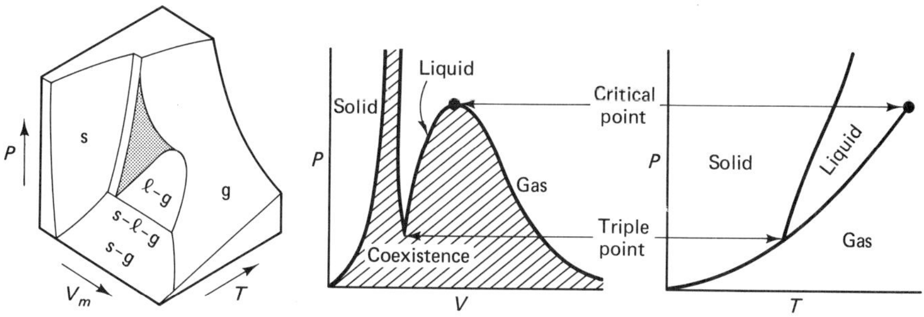

Figure 6.15-1 $p - V - T$ surface of a substance that expands on melting.

"liquid" or a "gas." It is seen that by manipulating p, V, and T, it is possible to go from a point in the liquid region to a point in the gas region without ever undergoing a phase transition—that is, without ever passing through a sharp differentiation between liquid and gas. It is for this reason that liquids and interacting gases are often treated as a fluid. By contrast, there is no way of entering the solid region without undergoing a phase transition.

There is just one instance in which there is a clear and unambiguous distinction between liquid and gas, and that is when they coexist with each other. When the two are seen together, with an interface between them, there is no trouble identifying which is the liquid. This is, of course, just the circumstance in which a phase transition separates the two.

If it is difficult to distinguish clearly between homogeneous liquid and homogeneous gas, it is not so difficult to distinguish between those states on the one hand and other possible states on the other. A crystalline solid is always separated from either type of fluid by a first-order phase transition. Moreover, a crystal can always be identified by its long-range order, operationally, by its X-ray diffraction pattern. A more subtle problem is to distinguish between a liquid and an amorphous or disordered solid, or a glass. The distinction here lies on the order of magnitude of the viscosity, since, for example, glasses do flow, very slowly, and may therefore be thought of as very viscous liquids. There is, however, a clear difference between these states and a liquid. The liquid is an equilibrium state of matter, whereas the others are not. (If an amorphous solid is heated and allowed to cool slowly, it will not return to precisely the same state; in fact, if the annealing is done slowly enough, it will crystallize.)

With the meaning of liquids (as interacting particles) established, discussion in the next section will turn to the question of how to study them. A number of approaches are used. One point of view is adopted that becomes increasingly accurate at low density, finally reducing to the ideal gas. Other approaches consider the liquid as a perturbation of the solid state—it is an imperfect solid, one that flows. The starting point is that of atoms in a regular lattice. Imperfections may be introduced into the lattice to give it more liquid-like properties. For example, there may be other defects as well. In lattice theory the atoms usually are not confined to lattice sites but are free to move around in a cell, with the cells being arranged in a lattice. Cell theories succeed in accounting for certain properties of liquids, especially near the melting curve.

The key idea in this line of reasoning is that of structure. An ideal gas has no structure at all, the average density being perfectly uniform. Liquids do have a short-range order that shows up in diffraction effects when X rays are scattered by liquids. The scattered X-ray intensity as a function of angle shows broad maxima, in contrast to the sharp maxima obtained from solids. Analysis of these X-ray diffraction patterns enables the radial distribution function (or pair correlation function) $g(r)$ for the liquid to be obtained; this function shows the variation in the average density of molecules with distance r from a given molecule. More precisely, $g(r) = \rho(r)/\rho$, where $\rho(=N/V)$ is the bulk molecular density, and $\rho(r)$ is the local molecular density in the thin spherical shell from r to $r + dr$ around a given molecule. For nonspherical molecules ρ/ρ_{bulk} depends on the directional angles θ and ϕ from the central molecule as well as on r; a radial pair correlation function can then be obtained by averaging over the angles. Theoretical $g(r)$ for a liquid of spherical molecules with a Lennard–Jones 6–12 molecular potential show maxima at $r = \sigma, 2\sigma, 3\sigma, \ldots$ indicating "shells" of molecules surrounding the central molecule. As r increases, the oscillations in $g(r)$ die out and $g(r)$ equals 1 for large r, reflecting the lack of long-range order. For $r < \sigma$, $g(r)$ goes rapidly to zero, as a result of the intermolecular Pauli repulsion. The experimentally observed $g(r)$ fairly well agrees with that calculated from the Lennard–Jones potential for gases.

Some of the most powerful methods of studying classical liquids involve a radial distribution function which expresses the probability that the positions of two molecules bear a certain relation to each other. Consider whatever structure does exist in a liquid to be represented by the snapshot in Fig. 6.15-2(a), showing the disposition of neighbors about an arbitrarily chosen central particle. (The picture is two-dimensional, but the liquid is actually three-dimensional.) Let $P(r)\,dr$ be the probability of finding a particle between r and $r + dr$, r being measured from one particle at the origin. For a

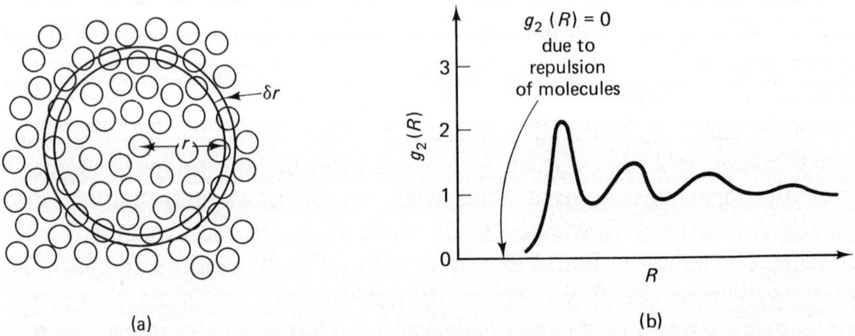

Figure 6.15-2 Conditions in a classical liquid.

uniform medium $P(r)$ would be proportional to $4\pi r^2$, and so it is convenient to introduce a pair correlation function $g_2(r)$ defined by

$$P(r) = 4\pi r^2 \rho g_2(r) \tag{6.15-2}$$

where ρ is the number density for the medium. If the medium were continuous and homogeneous, the pair correlation function $g_2(r)$ would be constant. As shown in Fig. 6.15-2(b), the pair correlation function for a liquid is zero for small values of r, has a maximum at the most probable nearest neighbor distance, and has diminishing maxima that correspond to second nearest neighbor distance, and so on. The first maximum is due to the 8 to 12 nearest neighbors that surround each molecule. The pair correlation function $g_2(r)$ is essentially zero at distances less than one molecular diameter because of the steep slope of the potential energy function at smaller intermolecular distances. The existence of this shell of nearest neighbors means that there will be a very low probability of finding any molecules between one molecular diameter and two. As the temperature of the liquid is raised, the maxima and the minima in the pair correlation function become less pronounced.

In general, a correlation function of order n is defined by the relation

$$\rho^{(n)}(r_1, \ldots, r_n) = \rho^n g^{(n)}(r_1, \ldots, r_n) \tag{6.15-3}$$

$g^{(n)}$ is called a correlation function because if the molecules were independent of one another, $\rho^{(n)}$ would equal simply ρ^n, and so the factor $g^{(n)}$ in Eq. (6.15-3) corrects for the nonindependence of the molecules. The simplest correlation function is $\rho^{(1)}(r_1)$. The quantity $\rho^{(1)}(r_1)\,dr_1$ is the probability that any one molecule will be found in dr_1. For a crystal this is a periodic function of r_1 with sharp maxima at the lattice sites, but in a fluid all points within V are equivalent and so $\rho^{(1)}(r_1)$ is independent of r_1. For a fluid, therefore,

$$\frac{1}{V}\int \rho^{(1)}(r_1)\,dr_1 = \rho^{(1)} = \frac{N}{V} = \rho \tag{6.15-4}$$

As for $g^{(2)}(r_1, r_2)$, it is particularly important since it can be determined experimentally. In a liquid of spherically symmetric molecules, $g^{(2)}(r_1, r_2)$ depends only upon the relative distance between molecules 1 and 2, that is, upon r_{12}. It is usual to denote r_{12} simply by r, and to write $g^{(2)}(r_{12}) = g(r)$. This is a standard notation.

Now $\rho g(r)\,dr$ is the "probability" of observing a second molecule in dr given that there is a molecule at the origin of r. Note that this "probability" is not normalized to unity, but

$$\int \rho g(r) 4\pi r^2\,dr = N - 1 \approx N \tag{6.15-5}$$

In fact, Eq. (6.15-5) shows that $\rho g(r)4\pi r^2$ is really the number of molecules between r and $r + dr$ about a central molecule. The function $g(r)$ can also be thought of as the factor that multiplies the bulk density ρ to give a local density $\rho(r) = \rho g(r)$ about some fixed molecule. Clearly $g \to 1$ as $r \to \infty$. The function $g(r)$ is called the radial distribution of the fluid. Figure 6.15-2(b) shows the radial distribution function for a Lennard–Jones 6–12 fluid as determined from molecular dynamics calculations.

■ EXAMPLE 6.15-1

Physical properties of a substance are primarily determined by intermolecular forces. Because liquids have a wide variation in their properties, it is desirable to attempt a qualitative classification of liquids in terms of the intermolecular forces. Give such a classification, based on type and symmetry of intermolecular energies.

Solution. The equilibrium properties of a substance, whether solid or fluid, are the result of a balance between the cohesive or potential energy on the one hand and the kinetic energy of the thermal motions on the other. The translational kinetic energy is the same for all molecules at a given temperature (principle of equipartion of energy), and so it is primarily the differences in strength and types of the intermolecular energies that cause the properties of one substance to differ from those of another at a fixed temperature. If liquids are compared at the same fraction of, say, their critical temperatures, it is the type and symmetry of intermolecular energies that determine their properties. The following classes of liquids may be distinguished, in increasing order of complexity:

1. *Noble gases.* The molecules are monatomic and so, in isolation, spherically symmetric. The forces between a pair of molecules is entirely central.
2. *Homonuclear diatomic molecules and heteronuclear diatomic molecules with negligible dipole moments.* Examples of the former are hydrogen, nitrogen, oxygen, and halogen. An example of the latter is carbon monoxide.
3. *Lower hydrocarbons and simple nonpolar substances.* Here, the forces depart greatly from spherical symmetry.
4. *Simple polar substances.* These molecules have direct electrostatic interaction between permanent dipole moments.
5. *Molecules of great polarity.* Water, hydrogen fluoride, and ammonia are in this class. The polarity is often localized in one part of the molecule.

There are important omissions from the above list of liquids, one being helium. The phase diagram of helium shows no triple point at which gas, liquid, and solid are in equilibrium, since the solid is stable only at pressures above 24 bar. There is, however, a second liquid phase, which is formed on cooling the normal liquid. This behavior is partly a consequence of the fact that ^{4}He obeys Bose–Einstein statistics, as seen in Chapter 4, but is also due to the light mass and weak intermolecular forces. ■

■ EXAMPLE 6.15-2

All theories of the liquid state basically treat liquids either as imperfect solids or as dense gases. The principal means of examining the structure of solids is X-ray diffraction. The scattered intensity can be expressed in terms of a radial distribution function $g(r)$ and a correlation function $h(r)$ defined by

$$g(r) = \frac{1}{\rho N}\langle \sum_{i \neq j} \delta(r + r_i - r_j) \rangle \tag{6.15-6}$$

$$h(r) = g(r) - 1 \tag{6.15-7}$$

where $\rho = N/V$ is the density, r is a radius vector, δ is a delta function, and the brackets denote thermodynamic average value. Sketch and discuss the typical $g(r)$ for (a) a solid crystal, (b) a perfect gas, and (c) a liquid.

Solution. Equation (6.15-6) says that in each configuration of the atoms of the sample, a delta function is recorded whenever any two atoms anywhere in the sample are separated by the vector r. This is done for each possible configuration, and a weighted average is calculated. If the sample is a crystal, the result will actually depend on both magnitude and direction. For a liquid, however, the result will be isotropic. In any case, the basic question is: If there is an atom at the origin, what is the probability of having an atom at r (in both magnitude and direction)?

(a) If $g(r)$ is computed for a solid, taking r in the direction of one of the edges of the unit cell, a narrow Gaussian distribution would be found about each lattice site. This is shown in Fig. 6.15-3(a).

(b) For a perfect gas, where an atom at the origin has no effect on the uniform distribution of the other atoms, the result is quite simply $g(r) = 1$, as shown in Fig. 6.15-3(a).

(c) Liquids are neither as structured as solids nor as structureless as perfect gases. This makes them extremely difficult to analyze. Nevertheless, a typical $g(r)$ for a liquid is sketched in Fig. 6.15-3(b). The main features are a region at small r where $g(r) = 0$ due to repulsion of other atoms by the one at the origin, a bump representing a shell of neighbors in the attractive part of the central atom's potential, a few further wiggles representing more neighbors and so on, and an asymptotic value of $g(r) = 1$ far away. ■

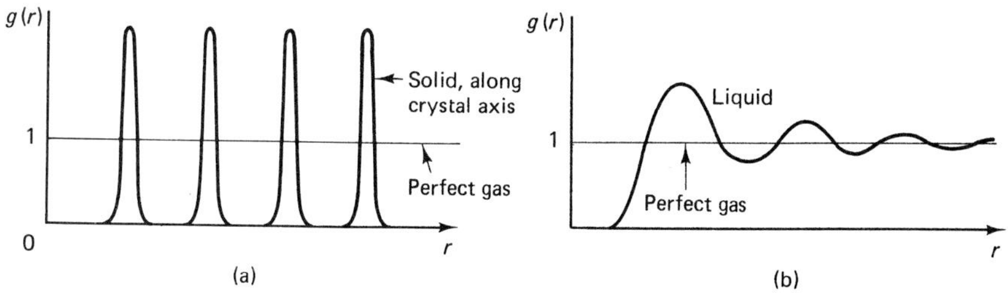

Figure 6.15-3 Radial distribution for solid, liquid, and perfect gas. (a) Solid and perfect gas. (b) Liquid and perfect gas.

6.16 CLAUSIUS VIRIAL THEOREM

The usefulness of the function $g^{(2)}(r)$ in the determination of thermodynamic functions may be shown by a simple example. If the mutual potential energy of a pair of molecules is $\epsilon_p(r)$, and if the potential energy of a given pair is independent of the presence of other molecules (i.e., if the potential is additive), the total potential energy of an atom in the field of its neighbors will clearly be

$$\int \epsilon_p(r)\rho g^{(2)}(r)\, d\tau$$

where $d\tau = dx\, dy\, dz$. The total potential energy of N atoms will be given by

$$E_p = \frac{1}{2} N \int \epsilon_p(r)\rho g^{(2)}(r)\, d\tau \tag{6.16-1}$$

The factor $\frac{1}{2}$ is included so as not to count mutual potentials of pairs twice, and $d\tau$ may of course be replaced by $4\pi r^2\, dr$, in which case r may be carried from 0 to ∞. The total energy of the system is then found by adding $\frac{3}{2}NkT$ to E_p;

$$E = \frac{3}{2} NkT + \frac{1}{2} N \int \epsilon_p(r)\rho g^{(2)}(r)\, dr\tau \tag{6.16-2}$$

By deriving an equation of state, the pressure may also be found if $g^{(2)}(r)$ is known. Consider a vessel of sides a, b, and c, in the x, y, and z directions containing N_A atoms of a monatomic fluid, each atom having a mass m [Fig. 6.16-1(a)]. As the atoms move about in the fluid, each will be subjected to a rapidly fluctuating force. At any one instant the force on atom i located at point (x_i, y_i, z_i) may be resolved into its Cartesian components X_i, Z_i, and Z_i. The response of the atom to this force is determined by Newton's second law:

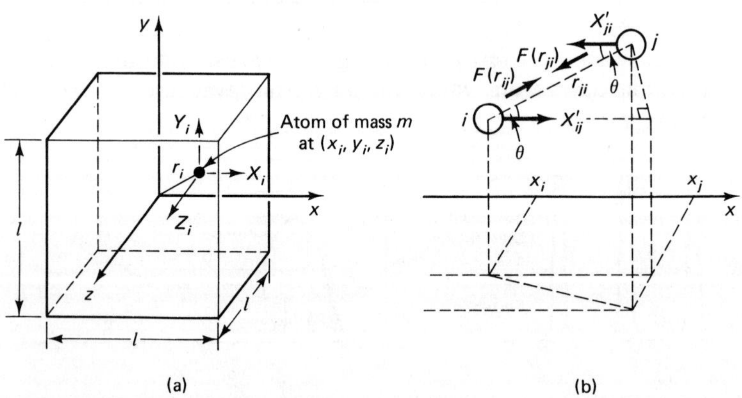

Figure 6.16-1 Atoms in a liquid. (a) Atom at (x_i, y_i, z_i) acted on by force X_i, Y_i, Z_i. (b) Interaction of two atoms i and j.

$$X_i = m\frac{d^2x_i}{dt^2} \tag{6.16-3}$$

or, multiplying both sides by x_i,

$$X_i x_i = m x_i \frac{d^2x_i}{dt^2} \tag{6.16-4}$$

Similar equations apply for the other two force components along the y and z directions. Making use of the rule for differentiating a product

$$\frac{d}{dt}\left(m x_i \frac{dx_i}{dt^2}\right) = m x_i \frac{d^2x_i}{dt^2} + m\left(\frac{dx_i}{dt}\right)^2 \tag{6.16-5}$$

substituting for mx_i, d^2x_i/dt^2, as given by (6.16-5), into (6.16-4) yields

$$X_i x_i = -m\left(\frac{dx_i}{dt}\right)^2 + m\frac{d}{dt}\left(x_i\frac{dx_i}{dt}\right) \tag{6.16-6}$$

In view of the identity

$$\frac{1}{2}\frac{d}{dt}\left[\frac{d(x_i^2)}{dt}\right] = \frac{d}{dt}\left(x_i\frac{dx_i}{dt}\right). \tag{6.16-7}$$

(6.16-6) can be rewritten as

$$X_i x_i + m\left(\frac{dx_i}{dt}\right)^2 = \frac{m}{2}\frac{d}{dt}\left[\frac{d(x_i^2)}{dt}\right] \tag{6.16-8}$$

Expressions similar to (6.16-8) apply to the y and z components. Adding together all three such expressions gives

$$\frac{m}{2}\frac{d}{dt}\left(\frac{d}{dt}r_i^2\right) = X_i x_i + Y_i y_i + Z_i z_i + m u_i^2 \tag{6.16-9}$$

where the distance r_i of atom i from the center of the box is given by $r_i^2 = x_i^2 + y_i^2 + z_i^2$, and where the speed u_i of atom i is related to the component speeds $u_{xi}(=dx_i/dt)$, u_{yi}, and u_{zi} by $u_i^2 = u_{xi}^2 + u_{yi}^2 + u_{zi}^2$.

Writing expressions similar to (6.16-9) for all N_A atoms present and adding all these expressions together, yields

$$\sum_i \frac{m}{2}\frac{d}{dt}\left(\frac{d}{dt}r_i^2\right) = \sum_i (X_i x_i + Y_i y_i + Z_i z_i) + \sum_i m u_i^2 \tag{6.16-10}$$

This represents what is happening at one instant of time in the fluid. As time goes by, the distances r_i will vary as an atom changes position in the box (it executes a random walk, in fact). However, when a time average is taken, the left-hand side of (6.16-10) vanishes, as can be seen by rewriting the summation (ignoring the contant term m) as

$$\frac{1}{2}\frac{d}{dt}\left(\frac{d}{dt}r_i^2\right) = \frac{d}{dt}\left(r_i\frac{dr_i}{dt}\right) \tag{6.16-11}$$

Now, the time average of the sum of such terms must be zero in a stationary fluid, since there is no net change of momentum. Thus, (6.16-10) results in

$$0=\left\langle\sum_i (X_i x_i + Y_i y_i + Z_i z_i)\right\rangle + \left\langle\sum_i m u_i^2\right\rangle \tag{6.16-12}$$

where the brackets indicate averaging.

So far, nothing has been said about the origin of the force on atom i. This force is made up of two contributions. First, there will be the forces acting on an atom due to all other atoms in the fluid (intermolecular forces). Second, there will be forces between the molecules of the assembly and the walls of the box. Therefore, the component force X_i is

$$X_i = X_i' + X_i'' \tag{6.16-13}$$

where X_i' is the x component of the resultant force acting on atom i attributable to all other atoms of the fluid, and X_i'' is the x component of the resultant force attributable to the walls. Substituting (6.16-13) and the two equivalent expressions for Y_i and Z_i into (6.16-13) gives

$$\left\langle\sum_i (x_i X_i' + x_i X_i'' + y_i Y_i' + y_i Y_i'' + z_i Z' + z_i Z_i')\right\rangle = -\left\langle\sum_i m u_i^2\right\rangle \tag{6.16-14}$$

where the summation is made over all N_A atoms in the box, and then this sum is time averaged.

Consider the wall forces first. The total force on all the molecules near any face of the container will be equal to the pressure p times the area of the face. This force on the molecules near the three faces of the container which pass through the origin (taken to be one of the corners of the box) will contribute nothing, since for every such molecule $x_i = y_i = z_i = 0$. The contribution from the other three faces will be

$$-ap(bc) - bp(ac) - cp(ab) = -3pV \tag{6.16-15}$$

where the products in the parentheses are the areas of the faces for which $x_i = a$, $y_i = b$, and $z_i = c$, respectively, and abc is the volume of the fluid. The signs are negative, since the forces on the atoms from these faces are in the negative directions. Thus, Eq. (6.16-14) becomes

$$\left\langle\sum_i (x_i X_i' + y_i Y_i' + z_i Z_i')\right\rangle - 3pV = -\left\langle\sum_i m u_i^2\right\rangle \tag{6.16-16}$$

As for the interaction between atoms, consider a pair of atoms i and j separated by distance r_{ji} as in Fig. 6.16-1(b). Let X_{ij}' denote the part of the net force X_i' acting on i (the first suffix in X_{ij}') which is due to atom j (the second suffix). Likewise, let X_{ji}' denote that part of the net force X_j' acting on j which is due to i. By Newton's third law, $X_{ji}' = -X_{ij}'$. The contribution which X_{ij}' and X_{ji}' make to the first pair term in (6.16-16) is thus

$$x_i X_{ij}' + x_j X_{ji}'(x_j - x_i) \tag{6.16-17}$$

the single X_{ji}' ensuring that each pair of atoms is counted only once. In the liquid inert gases the force between two atoms is directed along the line joining their centers. For such a central force,

$$X'_{ji} = F(r_{ji}) \cos\theta \tag{6.16-18}$$

where θ is the angle r_{ji} makes with the x axis, and $F(r_{ji})$ is the central force at interatomic separation r_{ji}. Now,

$$\cos\theta = \frac{x_j - x_i}{r_{ji}} \tag{6.16-19}$$

so that (6.16-18) and (6.16-19) enable (6.16-17) to be rewritten as

$$x_i X'_{ij} + x_i X'_{ji} = F(r_{ji}) \frac{(x_j - x_i)^2}{r_{ji}} \tag{6.16-20}$$

and likewise for the y and z contributions. With all three such relations, (6.16-16) becomes

$$\left\langle \sum_{\text{pairs}} \frac{F(r_{ji})}{r_{ji}} [(x_j - x_i)^2 + (y_j - y_i)^2 + (z_j - z_i)^2] \right\rangle - 3pV = -\left\langle \sum_i mu_i^2 \right\rangle \tag{6.16-21}$$

Since

$$(x_j - x_i)^2 + (y_j - y_i)^2 + (z_j - z_i)^2 = r_{ji}^2 \tag{6.16-22}$$

and

$$\frac{1}{2}\left\langle \Sigma\, mu_i^2 \right\rangle = \frac{3}{2} N_A kT \tag{6.16-23}$$

from the equipartition-of-energy theorem, (6.16-21) can be written as

$$\overline{\Sigma f(r_{ji}) r_{ji}} - 3pV = -3N_A kT$$

or

$$pV = RT + \frac{1}{3} \overline{\sum_{\text{pairs}} F(r_{ji}) r_{ji}} \tag{6.16-24}$$

where $RT = N_A kT$, and the bar is used for time average. Equation (6.16-24) is known as the *Clausius virial equation* for atoms with central forces. It says that to find the equation of state of a fluid all that is needed is to multiply the central force $F(r_{ji})$ between a pair of atoms by their distance r_{ji} apart between centers, sum these products over all pairs in the fluid, take the average of this result as measured at different times, and then substitute the result into (6.16-24). It makes no difference whether the fluid is a liquid or a gas for no specific atomic model is assumed. In a fluid composed of atoms of effectively zero size in which $F(r) = 0$ for all r, (6.16-24) reduces to the equation of state of a perfect gas, $pV = RT$, as it must if the analysis is correct.

Although (6.16-24) is valid as it stands, it is much more useful to have the radial density $\rho(r)$ in the result as this is the quantity that is determined experimentally from X-ray studies. If i is chosen as the reference atom (Fig.

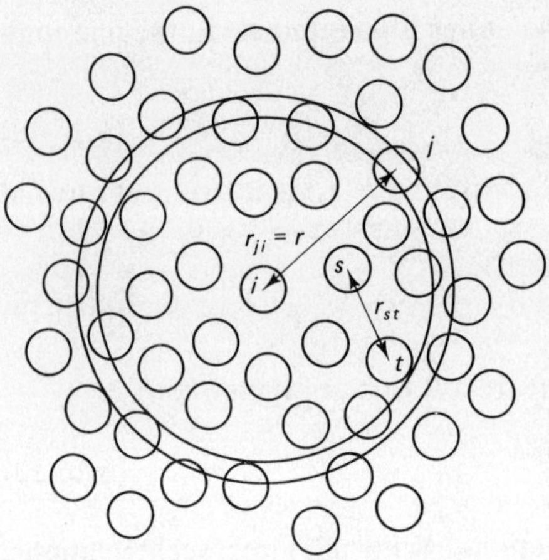

Figure 6.16-2 Distribution of atoms in a liquid about a particular atom i.

6.16-2), and j as one of the atoms in the spherical shell of radius $r = r_{ji}$, then $F(r_{ji}) = -dV(r)/dr$, where $V(r)$ is the interatomic pair potential. The atoms whose centers lie within the shell of radius r and thickness δr make a contribution of

$$\sum_{\text{shell}} r_{ji} F(r_{ji}) = r\left[-\frac{dV(r)}{dr}\right][4\pi r^2\, dr\, \rho(r)] \tag{6.16-25}$$

for the particular atom i. The term $4\pi r^2\, dr\, \rho(r)$ represents the number of atoms whose centers lie within the shell, each of which is at a distance r from i.

To find the contribution to atom i of all other atoms, (6.16-25) must be integrated from $r = 0$ to $r = \infty$:

$$\sum_{i} r_{ji} F(r_{ji}) = -\int_0^\infty \frac{dV(r)}{dr} \rho(r) 4\pi r^3\, dr \tag{6.16-26}$$

For a total of N_A atoms, (6.16-26) should be multiplied by $\frac{1}{2}N_A$, the factor of $\frac{1}{2}$ being required to ensure that each pair is only counted once. Substituting this modified version (6.16-26) into (6.16-24) leads to

$$pV_m = RT - \frac{N_A}{6}\int_0^\infty \frac{dV(r)}{dr} \rho(r) 4\pi r^3\, dr \tag{6.16-27}$$

as the more immediately useful form of the Clausius virial theorem. The calculation of the equation of state has thus been resolved by separating the effects of collisions with a wall (as in a dilute gas) from the effects of interactions between molecules.

It is worth mentioning that $\rho(r)$, which is related to the bulk density ρ and the $g_2(r)$ of Section 6.15 by $\rho(r) = \rho g_2(r)$, is dependent on the state of the liquid, that is, on, say, p and T. The properties of a liquid, as well as its equation of state, are embodied in $g_2(r)$ and $v(r)$. For an interatomic potential of the van der Waals type (a hard sphere repulsion plus a weak attraction), in which the attractive force in $v(r)$ falls off as $1/r^5$ or faster, (6.16-27) reduces to the van der Waals equation (the density of the fluid being such that there is a negligible chance of three or more atoms lying within a volume over which the attractive forces are significant).

Consider now the calculation of the internal energy, and suppose that the total potential energy of molecular interaction can be accurately represented as a sum of pair interactions. Also, let the pair interactions be independent of the molecular velocities. Given these conditions, the contribution to the internal energy from the kinetic energy of the molecules is $\frac{3}{2}NkT$, just the same as for a perfect gas of monatomic particles. However, in the case of the liquid (or the dense gas) there is an additional contribution to the internal energy resulting from the potential energy of interaction. Figure 6.16-2 shows the time-averaged distribution of atoms about a particulr atom i, where the potential energy (stored in the field) between atom i and each atom, such as j, lying within the shell of radius r and thickness dr is $V(r)$. Since there are $4\pi r^2\,dr\,\rho(r)$ atoms within the shell the potential energy of i due to all these atoms is $4\pi r^2\rho(r)V(r)\,dr$. Integrating from $r = 0$ to $r = \infty$ gives

$$\text{Potential energy of } i = \int_0^\infty V(r)\rho(r)4\pi r^2\,dr \tag{6.16-28}$$

To find the total potential energy of all N_A atoms present in 1 mol of liquid, (6.16-28) is multiplied by $\frac{1}{2}N_A$, the factor of $\frac{1}{2}$ being necessary to prevent each pair of atoms from being counted twice. Hence, the total molar internal energy of the liquid is

$$U_m = \frac{3}{2}RT + \frac{N_A}{2}\int_0^\infty V(r)\rho(r)4\pi r^2\,dr \tag{6.16-29}$$

where $\rho(r)$ is specified at different molar volumes and temperatures or at different pressures and temperatures, so that $U_m = U_m(V_m, \tau)$ or $U_m = U_m(\rho, \tau)$. Equations (6.16-27) and (6.16-29) enable the other thermodynamic properties to be obtained, at least in principle.

■ EXAMPLE 6.16-1

Liquid argon has a mass density of 0.98×10^3 kg/m^3 at a temperature of 143 K. Figure 6.16-3(a) shows the variation of $\rho(r)/\rho_0$. The average number density of the liquid as a whole is the value that $\rho(r)$ takes as $r \to \infty$. In other words, $\rho_0 = N_A/V_m$. Also shown is a plot of r^2 and the Lennard–Jones potential $v(r)$ for the interaction between two argon atoms, namely

$$v(r) = 4\epsilon\left[-\left(\frac{\sigma}{r}\right)^6 + \left(\frac{\sigma}{r}\right)^{12}\right] \tag{6.16-30}$$

where $\epsilon = 1.66 \times 10^{-21}$ J and $\sigma = 3.4 \times 10^{-10}$ m. The product $v(r) \times [\rho(r)/\rho_0]r^2$ is plotted in Fig. 6.16-3(b), and the area under the plot gives a value of -6.53×10^{-50} J m^3. (The area under the abscissa has a negative value.) Estimate the molar internal energy U_m of liquid argon.

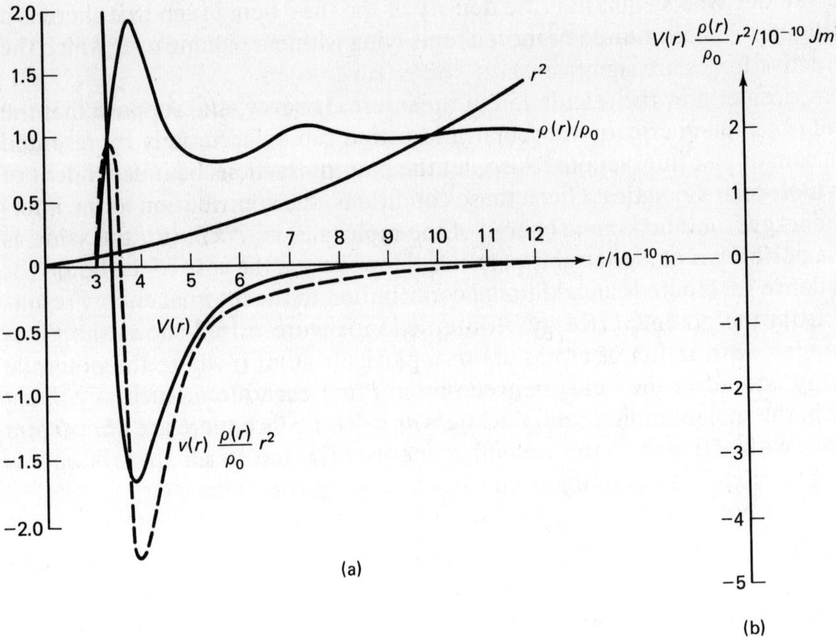

Figure 6.16-3 Radial density and potential energy functions for liquid argon.

Solution. Equation (6.16-29) may be written (upon multiplication and division by ρ_0) as

$$U_m = \frac{3}{2}RT + 2\pi\rho_0 N_A \int_0^\infty v(r)\frac{\rho(r)}{\rho_0} r^2\, dr \qquad (6.16\text{-}31)$$

where the integral has been evaluated graphically. Now, the mass of an atom of argon is 6.64×10^{-26} kg. Dividing the mass density of 0.98×10^3 kg/m^3 by the mass of the argon atom gives the number density $\rho_0 = 1.48 \times 10^{28}/\text{m}^3$. The value of $2\pi\rho_0 N_A$ is then $2\pi(1.48 \times 10^{28})(6.02 \times 10^{23}) = 5.6 \times 10^{52}$ m^{-3} mol^{-1}, so the second term on the right-hand side of (6.12-31) has a value of -3655 J mol^{-1}. At 143 K the value of $\frac{3}{2}RT$ is 1782 J mol^{-1}. Therefore, $U_m = -1.9 \times 10^3$ J mol^{-1}. This is in satisfactory agreement with the experimental value of -1.7×10^3 J mol^{-1}, in view of the uncertainty in the precise form of $\rho(r)/\rho_0$. ■

6.17 COMPUTER SIMULATION: MONTE CARLO STRATEGY

Basically, the aim is to find closed-form expressions for $g(r)$ [or $\rho(r) = \rho_0 g(r)$] that improve the dilute gas approximation, shown in Fig. 6.17-1(a), making it look more like Fig. 6.17-1(b), by including higher-order correlations in order to put more wiggles in $g(r)$ after the first big hump.

So far $\rho(r)$ and $V(r)$ have been treated as if they were independent of each other, yet a moment's thought will show that this is very unlikely to be the case. Consider the two hypothetical forms of $V(r)$ shown in Fig. 6.17-2. If $V(r)$ has the form shown in Fig. 6.17-2(a), there is a hard sphere repulsive force at $r \leq r_0$ and a very strong, but short-ranged, attractive force at $r = r_1$. It would be surprising if the radial densities arising out of these interactions were the same in both cases. A more closely knit group of atoms may be expected to exist in a liquid in which $V(r)$ has the form of Fig. 6.17-2(a) than in a liquid in which $V(r)$ has the form of Fig. 6.17-2(b). In Fig. 6.17-2(a) there is a constant attractive force over the range $r_0 < r < r_1$ and a hard sphere repulsive force at $r \leq r_0$. In Fig. 6.17-2(b), there is very strong, but short-ranged, attractive force at $r = r_1$ and a hard sphere repulsive force at $r \leq r_0$. The connection between $\rho(r)$ and $V(r)$ is, however, not as simple as these arguments might suggest. It will be remembered that $\rho(r)$ is the number density of atoms that are present on average at a radial distance r from a particular (reference) atom. To find $\rho(r)$ it is necessary to elucidate all possible atomic configurations consistent with the external constraints (p, T) and, from this, calculate the mean configuration. Now, the probability that a collection of N_A interacting atoms has a particular configuration (specified by the position and momentum of all the atoms) depends on the total energy of that configuration. This energy is the sum of the kinetic and potential energies of all N_A atoms. The mutual potential

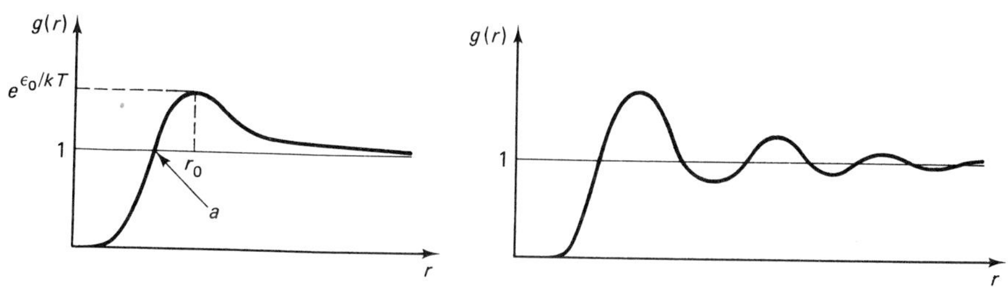

Figure 6.17-1 Radial distribution function for dilute gas and classical liquid.

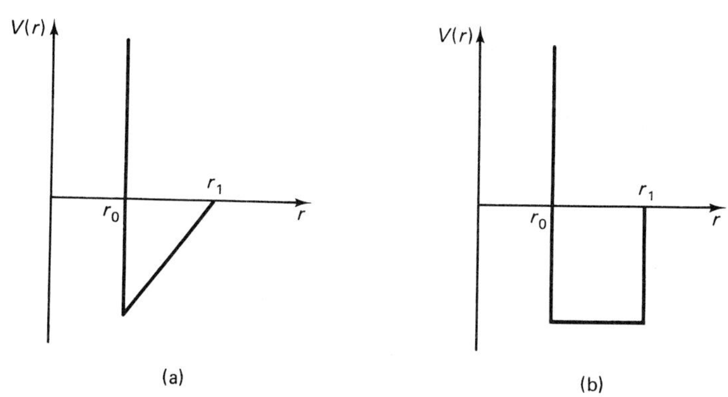

Figure 6.17-2 Hypothetical models for the potential energy $V(r)$ of a pair of atoms a distance r apart.

energy of all the atoms depends, of course, on how the atoms are arranged in that configuration. The radial density depends on the potential energy which in turn depends on how the atoms are arranged. One way to break this loop is to use simulation techniques.

There are two computer simulation approaches that are employed in statistical mechanics: the *molecular dynamics method* and the *Monte Carlo method*. In the molecular dynamics (MD) method (developed by Alder and coworkers), calculations are done on a system of 10^2 to 10^3 molecules (the larger the size of each molecule, the smaller the number of molecules that can be handled). The molecules are in a box whose volume is fixed to correspond to a typical liquid density. The molecules are assigned initial positions, orientations, and momenta; the momenta are chosen to be consistent with some desired temperature. Newton's equations for the several hundred molecules,

$$\frac{d\mathbf{r}_i}{dt} = \mathbf{v}_i \tag{6.17-1}$$

$$m\frac{d\mathbf{v}_i}{dt} = -\boldsymbol{\nabla}\phi \tag{6.17-2}$$

are solved by stepwise numerical integration on a modern computer, and the thermodynamic properties obtained by time averaging. An explicit form for the intermolecular potential v is assumed. Generally, v is taken as the sum of pairwise interactions v_{ij} as in

$$V \simeq \sum_{i=1}^{N} \sum_{j>1} v_{ij}(n_{ij}) \tag{6.17-3}$$

where v_{ij} is an empirical potential (such as the Lennard–Jones) or a quantum–mechanically calculated potential. The computer then solves the classical–mechanical equations of motion to find the new configurations of the molecules at successive small intervals of time. Typically, each time interval is about 10^{-15} s, and the system is followed for 10^{-11} s. The MD method provides a "movie" of molecular motions. Moreover, it is not restricted to equilibrium properties and can be used to calculate transport properties (for example, viscosity).

The Monte Carlo (MC) method (developed by Metropolis and coworkers and named after the European gambling resort) is quite similar to the MD method. The main difference is that, in the MC method, successive configurations of the system are not found by solving the equations of motion. Instead, the computer gives one molecule (picked at random) small changes in position and orientation, from (x, y, z) to $(x + u, y + v, z + w)$, where u, v, and w are randomly chosen in the interval $(-\delta, \delta)$. If the potential energy V_2 of the new configuration is less than the potential energy V_1 of the original configuration, the new configuration is rejected unless $\exp[-(V_2 V_1)/kT]$ is greater than a randomly chosen number r that lies between 0 and 1. This procedure generates a sequence of configurations such that the sequence is proportional to the Boltzmann factor $e^{-v/kT}$. The system's thermodynamic properties and radial distribution function are then found by averaging over the sequence of configurations (typically, 10^5 to 10^6). Unlike the MD method, the MC method is restricted to equilibrium properties. The basic Monte Carlo strategy is now outlined.

Monte Carlo methods are ways of efficiently evaluating integrals of high dimension such as the classical partition function for a gas of A atoms at a tempertaure $1/\beta$ interacting through a pairwise potential $v(r_{ij})$:

$$Z = \int d^3 r_1 \cdots d^3 r_A \, e^{-\beta \Sigma_{i<j} v(r_{ij})} \tag{6.17-4}$$

The straightforward evaluation of an integral like this by the standard quadrature formulas is out of the question except for the very smallest values of A. To see why, suppose that the quadrature allows each coordinate to take on 10 different values (not a very fine discretization), so that the integrand must be evaluated at 10^{3A} points. For a modest value of $A = 20$ and a very fast computer capable of some 10^7 evaluations per second, this would take some 10^{53} s, more than 10^{34} times the age of the universe! So, the essential idea is not to evaluate the integrand at every one of a large number of quadrature points, but rather at only a representative random sampling of abscissae. This is analogous to predicting the results of an election on the basis of a poll of a small number of voters. The Monte Carlo strategy turns out to be very appropriate for a broad class of problems in statistical and quantum mechanics.

It is easiest to illustrate the basic ideas in a one-dimensional situation. Suppose that the integral

$$I = \int_0^1 f(x)\, dx \tag{6.17-5}$$

is to be evaluated for some particular function f. The definite integral between two limits can be calculated by quadrature formulas that employ values of f at particular values of x (e.g., equally spaced). However, an alternative way of evaluating I is to think of it as the average of f over the interval $[0, 1]$. In this light, a plausible quadrature formula is

$$I \approx \frac{1}{N} \sum_{i=1}^{N} f(x_i) \tag{6.17-6}$$

Here, the average of f is evaluated by considering its values at N abscissae, x_i, chosen at random with equal probability anywhere within the interval $[0, 1]$. There are methods for generating such "random" numbers, but for now it is sufficient to suppose that there is a computer function (e.g., RND in BASIC) that provides as many of them as are required, one after the other.

Considering $f_i = f(x_i)$ as a random variable, the uncertainty associated with the quadrature (6.17-6) may be estimated from the usual statistical results in Chapter 1 for large N, namely,

$$\sigma_I^2 \approx \frac{1}{N}\sigma_f^2 = \frac{1}{N}\left[\frac{1}{N}\sum_{i=1}^{N} f_i^2 - \left(\frac{1}{N}\sum_{i=1}^{N} f_i\right)^2\right] \tag{6.17-7}$$

where σ_f^2 is the variance in f; that is, a measure of the extent to which f deviates from its average value over the region of integration. Equation (6.17-7) reveals two very important aspects of Monte Carlo quadrature. First, the uncertainty in the estimate of the integral, σ_I, decreases as $N^{-1/2}$. Hence, if more points are used, a more accurate answer will be obtained, although the error decreases rather slowly with the number of points. This is to be contrasted with a method like the trapezoidal rule, in which the error scales like N^{-2}, and which affords a much greater accuracy for a given amount of numerical work. (This advantage, however, vanishes in multidimensional cases.) A second important point to be made about (6.17-7) is that the precision is greater if σ_f is smaller, that is, if f is as smooth as possible. One limit of this is when f is a constant, in which case only one point is needed to define its average. The other limit is when f is zero except for a very narrow peak about some value of x. If the x_i have an equal probability to lie anywhere between 0 and 1, it is probable that all but a few of them will lie outside the peak of f, and that only these few of the f_i will be nonzero; this will lead to a poor estimate of I.

Since Eq. (6.17-7) shows that the uncertainty in a Monte Carlo quadrature is proportional to the variance of the integrand, it is easy to devise a general scheme for reducing the variance and improving the efficiency of the method. Let the integrand be multiplied and divided by a positive weight function $w(x)$, normalized so that

$$\int_0^1 dx\, w(x) = 1 \tag{6.17-8}$$

The integral can then be written as

$$\int_0^1 dx\, w(x) \frac{f(x)}{w(x)} \tag{6.17-9}$$

If now a change of variable is made from x to

$$y(x) = \int_0^x dx'\, w(x') \tag{6.17-10}$$

so that

$$\frac{dy}{dx} = w(x); \qquad y(x=0) = 0; \qquad y(x=1) = 1$$

then the integral becomes

$$I = \int_0^1 dy \frac{f[x(y)]}{w[x(y)]} \tag{6.17-11}$$

The Monte Carlo evaluation proceeds by averaging the values of f/w at a random sampling of points uniformly distributed in y over the interval $[0, 1]$:

$$I \approx \frac{1}{N} \sum_{i=1}^{N} \frac{f[x(y_i)]}{w[x(y_i)]} \tag{6.17-12}$$

The potential benefit of the change of variable is clear. If the chosen w behaves approximately as f does, that is, it is large where f is large and small where f is small, then the integrand in (6.17-11), f/w, can be made very smooth, with a consequent reduction in the variance of the Monte Carlo estimate (6.17-12). This benefit is, of course, contingent upon being able to find an appropriate w and upon being able to invert the relation (6.17-10) to obtain $x(y)$.

For an illustration of the Monte Carlo evaluation of an integral, consider

$$\int_0^1 \frac{dx}{1+x^2} = \frac{\pi}{4} = 0.78540 \tag{6.17-13}$$

A program in BASIC to calculate this integral for values of N, without using a weight function, is given in Koonin's *Computational Physics*. Results of running this program for various values of N are given in the first three columns of Table 6.17-1. The calculated result is equal to the exact value within a few standard deviations and the quadrature becomes more precise as N increases.

Now let the following weight function be chosen,

$$w(x) = \tfrac{1}{3}(4 - 2x)$$

which is positive definite, decreases monotonically over the range of integration (as does f), and is normalized correctly. Moreover, since $f/w = \frac{3}{4}$ at both $x = 0$ and $x = 1$, w will approximate the behavior of f. According to (6.13-10), the new integration variable is

$$y = \tfrac{1}{3}x(4 - x)$$

which can be inverted to give

$$x = 2 - (4 - 3y)^{1/2}$$

The BASIC code then evaluates I according to (6.17-12), and the results for various values of N are shown in the last two columns of Table 6.17-1. The improvement over the $w = 1$ case is evident.

TABLE 6.17-1 MONTE CARLO EVALUATION OF THE INTEGRAL (6.17-13)

	$w(x) = 1$		$w(x) = \frac{1}{3}(4 - 2x)$	
N	I	σ_I	I	σ_I
10	0.81491	0.04638	0.79982	0.00481
20	0.73535	0.03392	0.79071	0.00392
50	0.79606	0.02259	0.78472	0.00258
100	0.79513	0.01632	0.78838	0.00194
200	0.78677	0.01108	0.78529	0.00140
500	0.78242	0.00719	0.78428	0.00091
1000	0.78809	0.00508	0.78524	0.00064
2000	0.78790	0.00363	0.78648	0.00045
5000	0.78963	0.00227	0.78530	0.00028

In equilibrium statistical mechanics, properties are generally expressed as averages over well-defined, but often complicated probability distribution functions. These averages are of the generic form,

$$\langle f \rangle = \frac{\int dx\, \rho(x) f(x)}{\int dx\, \rho(x)} \tag{6.17-14}$$

where x represents a set of coordinates, $f(x)$ is the quantity to be averaged expressed as a function of those coordinates, and $\rho(x)$ describes the equilibrium distribution of possible coordinate values for the system under study. In the discussion of the Monte Carlo quadrature, it is seen that two basic operations are involved: generating abscissae randomly distributed over the integration volume with a specified distribution $w(x)$ (which may perhaps be unity) and then evaluating the function f/w at these abscissae. The second operation is straightforward, but it is not obvious how to generate "random" numbers on a deterministic computer. In the early 1950s Metropolis et al. (A. Rosenbluth, M. Rosenbluth, A. Teller, and E. Teller) devised an ingenious stochastic method for the evaluation of averages of the type shown in Eq. (6.17-14). For the integration, a random walk in x space is designed in such a way that the probability of visiting a particular point x is proportional to the statistical weight for that point, $\rho(x)$. So constructed, the set of points visited in an N-step random walk, (x_n), $n = 1, \ldots, N$, can be used to estimate $\langle f \rangle$ according to the expression

$$\langle f \rangle = \frac{\sum_{n=1}^{N} f(x_n)}{N} \tag{6.17-15}$$

For finite N, this estimate has a statistical uncertainty which is proportional to $\sigma_f/\sqrt{N}$, where σ_f is the standard deviation of f, and is defined by

$$\sigma_f = (\langle f^2 \rangle - \langle f \rangle^2)^{1/2} \tag{6.17-16}$$

The Metropolis method has a number of remarkable features. In essence, it represents the exchange of a well-defined problem, that of computing integrals, for one involving the study of an artificially constructed random process. This exchange, however, produces an algorithm with significant advantages. Chief among these advantages is that the resulting method applies with equal ease to small- or large-dimensional problems, producing the $1/\sqrt{N}$ convergence rate in either case. The method has the additional feature of variable resolution: short random walks rapidly produce order-of-magnitude estimates, whereas more extensive random walks permit further refinement. This variable resolution allows the precision of a calculation to be adjusted to match the particular problem under study.

To describe a simple realization of the Metropolis et al. algorithm, suppose that it is desired to generate a set of points in a (possibly multidimensional) space of variables X distributed with probability density $w(X)$. The

Metropolis algorithm generates a sequence of points, $X_0, X_1, \ldots$, as those visited successively by a random walker moving through X space; as the walk becomes longer and longer, the points it connects approximate more closely the desired distribution. The rules by which the random walk proceeds through configuration space are as follows. Suppose that the walker is at a point X_n in the sequence. To generate X_{n+1}, it makes a trial step to a new point X_t. This new point can be chosen in any convenient manner, for example, uniformly at random within a multidimensional cube of small side δ about X_n. This trial step is then "accepted" or "rejected" according to the ratio

$$r = \frac{w(X_t)}{w(X_n)} \tag{6.17-17}$$

If r is larger than 1, then the step is accepted (i.e., letting $X_{n+1} = X_t$), while if r is less than 1, the step is accepted with probability r. This latter is conveniently accomplished by comparing r with a random number η uniformly distributed in the interval $[0, 1]$ and accepting the step if $\eta < r$. If the trial step is not accepted, then it is rejected, letting $X_{n+1} = X_n$. This generates X_{n+1}, and then X_{n+2} can be generated by the same process, making a trial step from X_{n+1}. Any arbitrary point, X_0, can be used as the starting point for the random walk.

Figure 6.17-3 shows the results of a Monte Carlo calculation of a particular thermodynamic property, the radial distribution function, for a chosen thermodynamic state of a simple classical fluid. The radial distribution function denotes the ratio of the density observed in the fluid as a function of the distance from a specified atom compared with the overall bulk fluid density. In one set of classical mechanics calculations (CM), the atoms in the fluid are assumed to interact by means of pairwise additive Lennard–Jones potentials,

$$V_{LJ}\left(\frac{r}{\sigma}\right) = 4\epsilon\left[\left(\frac{\sigma}{r}\right)^{12} - \left(\frac{\sigma}{r}\right)^{6}\right] \tag{6.17-18}$$

where ϵ and σ are energy and distance parameters, respectively. The results

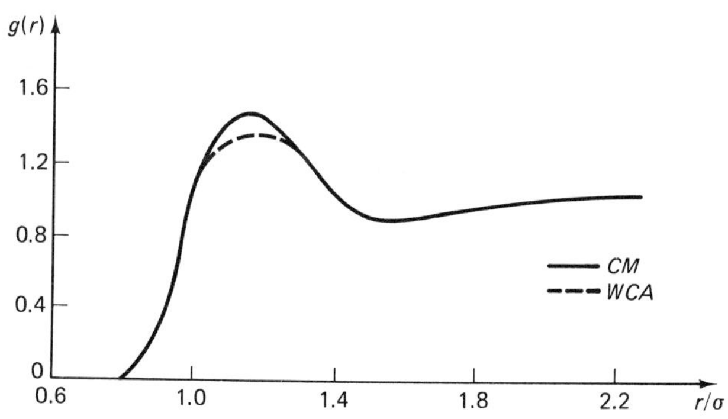

Figure 6.17-3 Radial distribution function $g(r)$ for Lennard–Jones fluid obtained from Monte Carlo. $kt/\epsilon = 5.0$; $\rho\sigma^3 = 0.365$. (D. Chandler, *Annu. Rev. Phys. Chem.* 29, 441, 1978).

obtained by using the full interaction are denoted by CM. Those obtained by using the repulsive portion of the Lennard–Jones interaction,

$$V_{\mathrm{WCA}}\left(\frac{r}{\sigma}\right) = 4\epsilon\left[\left(\frac{\sigma}{r}\right)^{12} - \left(\frac{\sigma}{r}\right)^{6}\right] + \epsilon \quad \text{for} \quad r < \sigma$$
$$= 0, \quad \text{otherwise} \tag{6.17-19}$$

are marked WCA (for Weeks–Chandler–Anderson). The agreement of the two results is a measure of the quality of the assumption of the dominance of short-ranged, repulsive forces for this example.

6.18 STATISTICAL MECHANICS OF FLUIDS; RELATION BETWEEN $g(r)$ AND $u(r)$

The structure and properties of a fluid at equilibrium may be described by canonical molecular distribution functions. These define the probability of occurrence of a particular arrangement of molecules in an assembly of fixed number, N, and fixed volume, V. This section outlines (derives would be too strong a word) some of the closed-form expressions that have been obtained in the literature for the radial distribution function to better fit the characteristics of a fluid.

First, consider some properties of the two-particle distribution function, $n^{(2)}$. A particularly useful result is

$$\int_V n^{(2)}(r_1, r_2)\, dr_1\, dr_2 = \langle N^2\rangle - \langle N\rangle \tag{6.18-1}$$

where the r's are vectors (having direction as well as magnitude). For a stationary liquid in thermal equilibrium, $n^{(2)}$ does not depend on the choice of r_1, but only on the difference $r = r_1 - r_2$. Thus, for an isotropic body, $n^{(2)}$ depends only on the scalar $|r|$. The radial distribution function $g(r)$, or probability of finding a molecule at r if there is a molecule at the origin, can be variously expressed as

$$g(r) = \left(\frac{V}{N}\right)^2 n^2(r) \tag{6.18-2}$$

$$g(r) = \exp\left[-\frac{\phi(r)}{kT}\right] \tag{6.18-3}$$

$$g(r) = \frac{1}{\rho N}\langle \sum_{i \neq j} \delta\,(r + r_i - r_j)\rangle \tag{6.18-4}$$

where $\phi(r)$ is known as a potential mean force, and δ is a delta function. A general shape of $g(r)$ for a monatomic liquid is shown in Fig. 6.18-1. At low r it is almost zero due to the high energies for the atoms to overlap. At a distance roughly equal to the atomic diameter, there is a pronounced peak, which denotes a sphere of nearest neighbors, and at higher values of r, there are oscil-

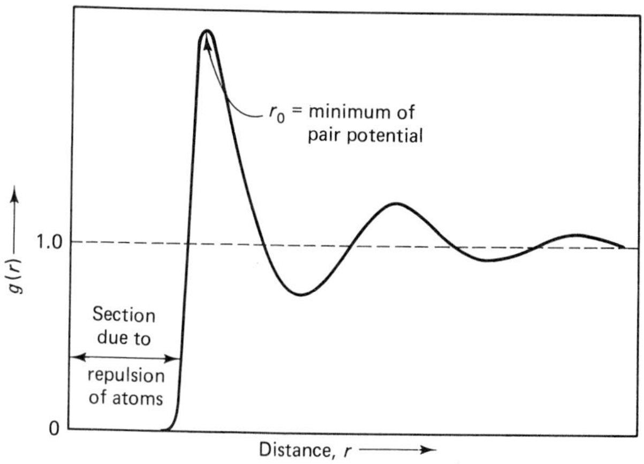

Figure 6.18-1 Typical form for pair distribution $g(r)$. [Principal peak of $g(r)$ occurs at a position close to the principal minimum of $u(r)$.]

lations representing more distant neighbors. These oscillations decrease in amplitude with increasing r, and eventually $g(r)$ approaches the (unit) mean density of the system.

Important properties may be calculated on the basis of pair theory if the radial distribution function $g(r)$ or the potential function $u(r)$ is known. Before discussing these results, experimental methods for studying $g(r)$ should be mentioned, and some useful potentials listed. The experimental methods include X-ray and neutron scattering, while two potentials often used in numerical calculations are

the Lennard–Jones potential:

$$u(r) = 4\epsilon\left[\left(\frac{\sigma}{r}\right)^{12} - \left(\frac{\sigma}{r}\right)^{6}\right] \qquad (6.18\text{-}5)$$

the Hard–Sphere potential:

$$\left.\begin{aligned} u(r) &= +\infty \quad \text{for } r < \sigma \\ &= 0 \quad\quad \text{for } r \geq \sigma \end{aligned}\right\} \qquad (6.18\text{-}6)$$

They are shown in Fig. 6.18-2.

The pair potential $u(r)$ is regarded as a basic quantity from which all other liquid properties can be calculated. This is because if $g(r)$ can be obtained from $u(r)$, then it would be possible, for example, to calculate the equation of state from first principles. The first step is to derive a relation between $n^{(2)}$, $n^{(3)}$ and $u(r)$ in a classical system. If a suitable approximation is made for $n^{(3)}$, this relation reduces to one between $u(r)$ and $g(r)$ only. Several such approximations are described below. Figure 6.18-3 shows the positions of any three atoms in a liquid and defines the vectors to be used in the discussion. These vectors connect any three atoms in the liquid. The origin is denoted by 0, and the full circles mark the three atoms (indexed 1, 2, and 3).

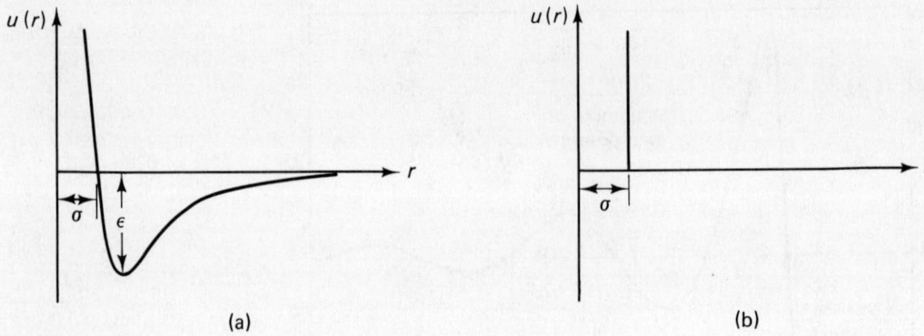

Figure 6.18-2 Lennard–Jones and Hard–Sphere potentials.

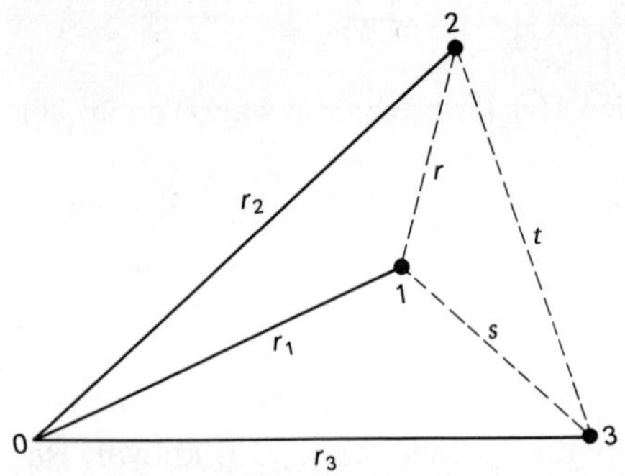

Figure 6.18-3 Vectors connecting any three atoms in a liquid. (Origin of vectors is at 0. Dark circles mark the atoms.)

The mean force acting on molecule 1 at position r_1 is, from the definition of the potential,

$$-\frac{\partial\phi(r)}{\partial r_1}=\frac{kT}{g(r)}\frac{\partial g(r)}{\partial r_1} \tag{6.18-7}$$

However, the mean force may be evaluated in another way: The probability of a third molecule being in dr_3 at r_3 when the first two molecules are at r_1 and r_2 is

$$P(r_1,r_2,r_3)\,dr_3=\frac{n^{(3)}(r_1,r_2,r_3)\,dr_3}{n^{(2)}(r_1,r_2)} \tag{6.18-8}$$

The force on molecule 1 due to molecule 3 is given by (using $s=r_3-r_1$):

$$\frac{-\partial u(s)}{\partial r_1}P(r_1,r_2,r_3)\,dr_3 \tag{6.18-9}$$

and the force on molecule 1 due to molecule 2 is

$$-\frac{\partial u(r)}{\partial r_1} \tag{6.18-10}$$

Thus, the mean force on molecule 1 due to all other molecules is the sum

$$-\frac{\partial u(r)}{\partial r_1}-\int\frac{\partial u(s)}{\partial r_1}P(r_1,r_2,r_3)\,dr_3 \tag{6.18-11}$$

Equating this to the right-hand side of Eq. (6.18-7) yields the general relation

$$\frac{\partial g(r)}{\partial r_1} \cdot \frac{kT}{g(r)} + \frac{\partial u(r)}{\partial r_1} = -\int \frac{\partial u(s)}{\partial r_1} P(r_1, r_2, r_3)\, dr_3 \tag{6.18-12}$$

which can be more conveniently rewritten as

$$\frac{\partial}{\partial r}\left[\ln g(r) + \frac{u(r)}{kT}\right] = -\frac{1}{kT}\int \frac{\partial u(s)}{\partial r_1} P(r_1, r_2, r_3)\, dr_3 \tag{6.18-13}$$

To proceed any further requires some approximation on $P(r_1, r_2, r_3)$, the probability pertaining to the third molecule. Several methods of attack are possible:

The Yvon–Born–Green Equation (YBG)

An approximation for $n^{(3)}$, due to Kirkwood, was employed by Yvon, Born, and Green to reduce Eq. (6.18-13). It states that the probability of finding molecule 3 at r_3, molecule 2 at r_2, and molecule 1 at r_1 is the product of the product of the separate pairs (1, 2), (2, 3), and (3, 1) at the appropriate spacings. Specifically, the approximation is

$$n^{(3)}(r_1, r_2, r_3) \simeq \frac{n^{(2)}(r)n^{(2)}(s)n^2(t)}{\rho^3} \tag{6.18-14}$$

Since $g(r)$ is a function of density, the approximation is not just the $\rho \to 0$ limit of $n^{(3)}$, but includes some terms to all orders in the density. After some reduction, Eq. (6.18-13) with Eq. (6.18-14) gives

$$\ln g(r) + \frac{u(r)}{kT} = -\rho \int_V E(|r_1 - r|)h(r_1)\, dr_1 = -\frac{W(r)}{kT} \tag{6.18-15}$$

where

$$E(r) = \frac{1}{kT}\int_\infty^r g(x)\frac{\partial u(x)}{\partial x}\, dx \tag{6.18-16}$$

and

$$h(r) = g(r) - 1 \tag{6.18-17}$$

Equation (6.18-15) is known as the YBG equation for $g(r)$ and $u(r)$.

The Hypernetted Chain Equation (HNC)

This approximation is obtained by carrying out a cluster expansion for $n^{(3)}$ and then neglecting certain terms that make the expansion difficult. The parts that are retained are written conveniently in terms of another function, $c(r)$, called the direct correlation function, defined by

$$h(r) = c(r) + \rho \int c(|r - r'|)h(r')\, dr' \tag{6.18-18}$$

Equation (6.18-18) was introduced by Arnstein and Zernike on the premise that the total correlation, $h(r)$, is the sum of a direct effect, $c(r)$, of molecule 1 on molecule 2, plus an indirect effect of all other molecules. The latter effect is expressed through the convolution of $h(r)$ and $c(r)$. The final result is

$$\ln g(r) + \frac{u(r)}{kT} = \rho \int c(|r - r'|)h(r')dr' = h(r) - c(r) = -\frac{W(r)}{kT} \qquad (6.18\text{-}19)$$

This relation between $g(r)$ and $u(r)$ is accurate at low densities. It can also be used at high densities because the expansion is summed to all orders in the density.

The Percus–Yevick Equation (PY)

This equation is obtained in a manner similar to the HNC equation through a cluster expansion, using a pictorial representation of an integral whose integrand is a product of f_{ij} functions. For example, a connection between two root points (open circles) denotes the function f_{12}, and a field point (black circle) forming a triangle with two root points denotes $f_{12} \int f_{13} f_{23}\, dn_3$ (see Fig. 6.18-4). Using the topological properties of cluster diagrams, the Percus–Yevick equation is obtained:

$$\ln g(r) + \frac{u(r)}{kT} = \ln[g(r) - c(r)] = -\frac{W(r)}{kT} \qquad (6.18\text{-}20)$$

This equation is accurate at low densities only, but since the expansion is summed to all orders in the density, it may also be useful at high densities. The radial distribution function, as obtained from the above approximations or computated by means of Monte Carlo or molecular dynamics, might be thought sufficient for determining intermolecular potential functions and other properties. Unfortunately, at liquid densities, $g(r)$ is rather insensitive to the detailed form of the interaction potential and is determined primarily by the short-range repulsive forces. No foolproof theory exists to date, and an assembly of hard spheres is quite capable of giving a good qualitative description of the main features of the radial distribution function.

Figure 6.18-4

■ **EXAMPLE 6.18-1**

(a) In a scattering experiment, the distribution of scattered intensity contains information on the distribution of atoms, and on $g(r)$. Sketch a typical structure factor for the experiment.

(b) Give examples from the literature for the calculated values of $g(r)$ compared to neutron diffraction data.

Solution

(a) Thermodynamic properties (p, V, T, C_p, etc.) of a liquid may be expressed in terms of two functions: the pair potential $u(r)$, assumed independent of p, V, T, etc., and the pair distribution function $g(r)$, which is a function of density and pressure. These two functions may be calculated from first principles or measured in a scattering experiment. In the latter case, the structure factor, $S(Q)$, has the typical shape shown in Fig. 6.18-5(a) for the liquid region. The value $g(r)$ may then be calculated from $u(r)$ by choosing a form for $u(r)$ to give a good fit to the function $S(Q)$.

(b) An example from the literature is liquid argon for which Khan (*Phys. Rev.*, 1964, 134, A367) has calculated $g(r)$ using a Lennard–Jones potential

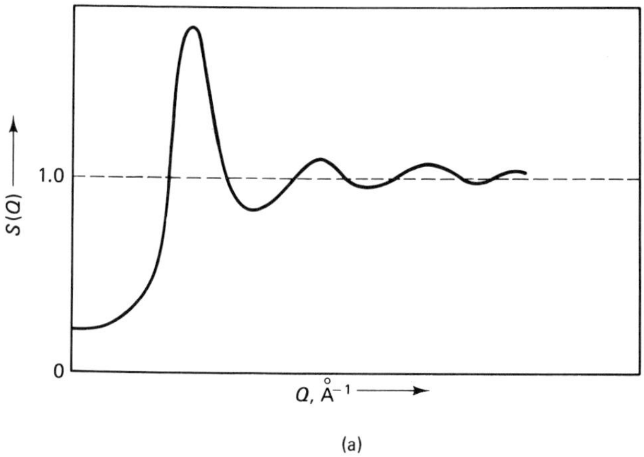

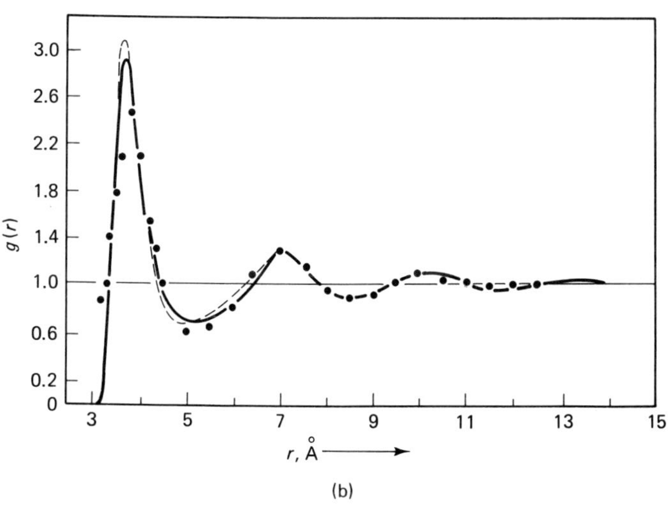

Figure 6.18-5 Structure factor curve and radial distribution function curve for liquid. (a) Typical shape of $S(Q)$ for a liquid. (b) Calculated values of $g(r)$ for argon at 84.4° compared to neutron-diffraction data.

and the PY and HNC equations. These values of $g(r)$ are compared to neutron diffraction data in Fig. 6.18-5(b). The full curve shows the HNC calculation, the dotted line shows the PY calculation, and the full circles indicate the experimental points (from Henshaw, *Phys. Rev.*, 1957, 105, 976). ■

6.19 TRANSPORT PROCESSES

So far statistical mechanics has been mainly concerned with equilibrium situations. Attention is now given to situations that lack equilibrium in character, such as transport processes. These phenomena result from the transport of mass, energy, momentum, and electric charge by molecular motion. They are not in the same category as thermodynamic properties. A property like internal energy, for example, indicates how energy is distributed among available molecular energy states, but it does not state the rate at which energy is transported from a higher temperature region to a lower temperature region. The study of transport processes is an important part of physical science and is usually known as *kinetic theory*. It has already been seen that the energies of simple monatomic molecules may be computed in terms of translational kinetic energies. The energy distribution thus gives the distribution of molecular velocities. With this basic information, it is possible, in principle, to calculate the various transport rates due to molecular motion.

For simplicity, consider first the self-diffusion of a gas into a radioactive isotope of the same gas. In this case both gases have atoms of effectively the same size and mass. One method of studying the interdiffusion is to start with the two gases separated by a dividing partition, as in Fig. 6.19-1(a). Once the partition is removed, the diffusion proceeds until at time $t = \infty$ the concentration of both species is constant throughout the chamber [Fig. 6.19-1(c)].

At some intermediate time t, depicted in Fig. 6.19-1(b), the radioactive gas will have partially penetrated the nonradioactive gas, and vice versa. The concentration of the radioactive gas can be found by moving a suitable detector along the axis of the cylinder. Such experiments yield the concentration gradient dn/dx and the flux j, at, say, $x = x_p$. Measurements show that it is proportional to $|dn/dx|$, with diffusion occurring down the concentration gradient. These observations are summed up in the relation

$$j = -D\frac{dn}{dx} \tag{6.19-1}$$

a result known as *Fick's law of diffusion*. The constant of proportionality D is called the diffusion coefficient (here the self-diffusion coefficient). Since dn/dx is positive at $x = x_p$ in Fig. 6.19-1(b) and since D is always supposed to have a positive value, it follows that the minus sign in Eq. (6.19-1) indicates diffusion in the $-x$ direction. Conversely, if dn/dx is negative (as it is for the nonradioactive atoms), Eq. (6.19-1) gives a positive value to j, indicating diffusion in the x direction. As a numerical example, if a concentration gradient of 1.6×10^{22} m^{-4} produces a flux of -2.5×10^{15} m^{-2} s^{-1}, then $D = (2.5 \times 10^{15}$

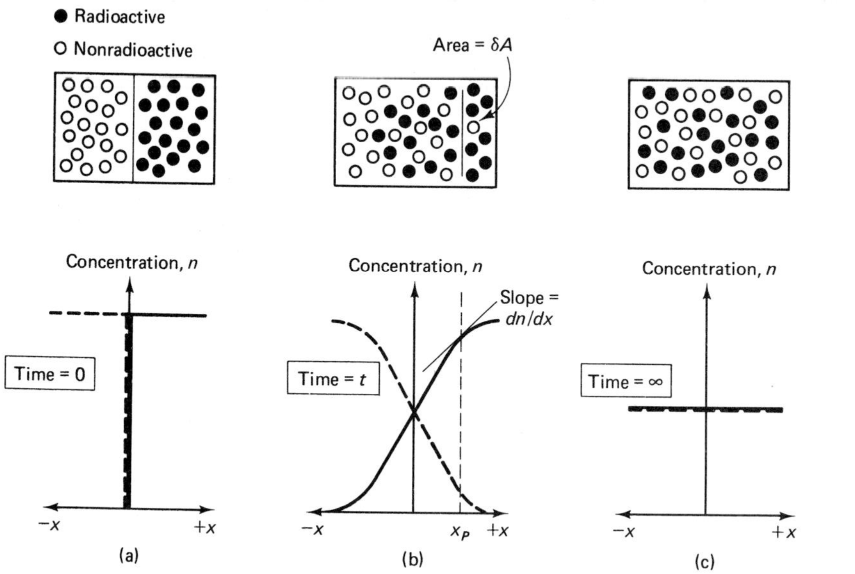

Figure 6.19-1 Interdiffusion of a gas and its radioactive equivalent. (a) Conditions at time $t = 0$. (b) Conditions upon removal of barrier. (c) Conditions at equilibrium.

$m^{-2}\,s^{-1})/1.6 \times 10^{22}\,m^{-4}) = 1.67 \times 10^{-7}\,m^2\,s^{-1}$. As another example of transport process, consider thermal conduction defined by the *Fourier relation*

$$j = \frac{q}{A} = -k\frac{\partial T}{\partial x} \tag{6.19-2}$$

where j is the heat flux per unit area and $\partial T/\partial x$ is the temperature gradient in the direction of heat flow. The positive constant k is called the thermal conductivity, and the minus sign is inserted to indicate that heat flows downhill on the temperature scale, that is, from a high-temperature region to a low-temperature region.

Consider next viscous action in a fluid as evidenced by a shear stress between fluid layers. The shear stress is assumed to be proportional to the normal velocity gradient, that is,

$$j = \tau = \mu\frac{\partial u}{\partial z} \tag{6.19-3}$$

where τ is the shear stress, and the proportionality constant μ is defined as the *dynamic viscosity,* a typical set of units for the dynamic viscosity being Newtons-second per square meter in the SI system. Equation (6.19-3) defines a *Newtonian fluid.*

Figure 6.19-2 shows the relation of dynamic viscosity to fluid shear. The physical mechanism of viscosity is one of momentum exchange. Molecules transport momentum from regions of high macroscopic velocity to regions of low velocity. This transport tends to pull the slower moving fluid layers along, and the "drag" or shear force is the quantity τ.

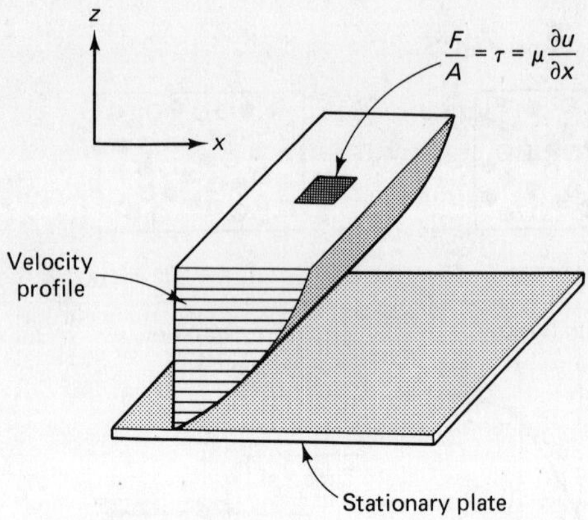

Figure 6.19-2 Shearing stress and viscous flow.

Finally, if the system is an electrical conductor, that is, if the conserved quantity is electrical charge, and the intensive property is electrical potential V_{elec}, then

$$j = \text{flux of charge} = -\sigma \frac{dV_{elec}}{dx} \tag{6.19-4}$$

where σ is the electrical conductivity of the system, more formally known as the coefficient of electrical conduction. However, flux of charge is current density j. Also the electrical field ε_F is $\varepsilon_F = -dV_{el}/dx$, so that

$$j = \sigma\varepsilon_F \tag{6.19-5}$$

which is *Ohm's law.*

It is noted that in all the transport processes considered, the molecular transport rates obey the general relation:

$$\text{flux of conserved quantity} = -\begin{pmatrix}\text{transport}\\ \text{coefficient}\end{pmatrix} \times \begin{pmatrix}\text{gradient of}\\ \text{intensive property}\end{pmatrix} \tag{6.19-6}$$

The conserved quantities, along with the gradients of intensive properties, are listed in Table 6.19-1. The task is to relate the transport coefficients to suitable molecular parameters.

TABLE 6.19-1 CHARACTERISTICS OF MAIN TRANSPORT PROCESSES

Process	Requirement at macroscopic level	Characteristic
Diffusion	Density gradient	Transport of mass
Viscous flow	Shearing stress	Transport of momentum
Thermal conduction	Temperature gradient	Transport of energy
Electrical conduction	Potential gradient	Transport of charge

6.20 MOLECULAR FLUX

The number of molecules striking an area, per unit area, and per unit time, is labeled the molecular flux. Consider an imaginary plane, as shown in Fig. 6.20-1(a), and let the flux of molecules through the element of area dA be calculated. It is assumed that the molecular speed distribution is Maxwellian, and, furthermore, that the direction of the incoming molecules that strike dA is random, so that there is no preferred direction of molecular motion. For the moment it is also assumed there is no collision between molecules.

In Fig. 6.20-1(b), let the directions of molecular velocities be distributed uniformly, and imagine that there is attached to each molecule a vector representing the magnitude and direction of its velocity. Let all these vectors be transferred to a common origin and construct a sphere of arbitrary radius r with center at the origin. The velocity vectors, prolonged if necessary, intersect the surface of the sphere in as many points as there are molecules and the assumption of uniform distribution in direction means that these points are distributed uniformly over the surface of the sphere. The fraction of the molecules traveling toward the area element in Fig. 6.20-1(a) in the specified direction (ϕ, θ) is

$$\frac{(r \sin\theta\, d\phi)(r\, d\theta)}{4\pi^2} = \frac{\sin\theta\, d\phi\theta}{4\pi} \tag{6.20-1}$$

These molecules have a speed v, so that in time dt, the number striking the area dA is

$$(dA)(\cos\theta)(dn_v)(dt) \tag{6.20-2}$$

where dn_v designates the number of molecules per unit volume and per unit velocity and direction. The quantity $v\,dt$ is the distance moved in time dt, and $dA \cos\theta$ is the projected area in the direction of the velocity v. Thus, the

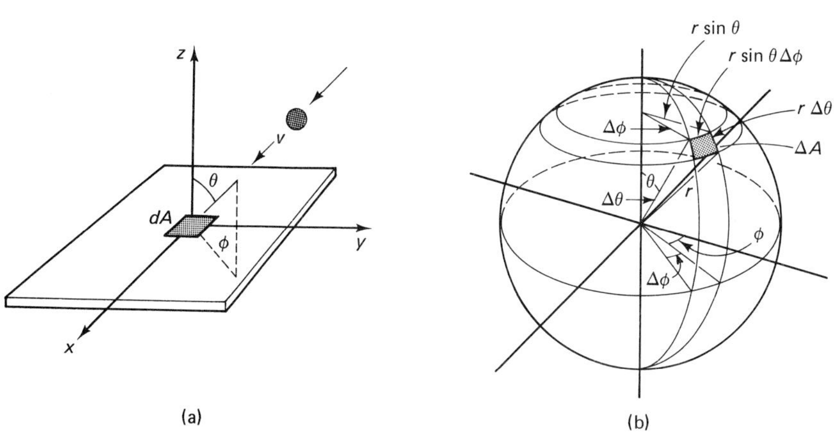

Figure 6.20-1 Molecular flux evaluation.

number of molecules contained in the volume (dA) $(\cos\theta)$ $(v\,dt)$ is the number that strikes dA.

Now, dn_v is the number density of molecules in this imaginary volume and is the product of (6.20-1) and the Maxwellian speed distribution $f(v)n\,dv$, so that the number of molecules striking the area dA in time dt as given by (6.20-2) is, finally,

$$(dA)(\cos\theta)\left(\frac{v\sin\theta\,d\phi\,d\theta}{4\pi}\right)f(v)n\,dv\,dt \tag{6.20-3}$$

The molecular flux per unit area and time is then obtained by integrating this expression over all directions and speeds, that is,

$$J = \frac{n}{4\pi}\int_0^{\pi/2}\sin\theta\cos\theta\,d\theta\int_0^{\infty} vf(v)\,dv \tag{6.20-4}$$

which gives

$$J = \frac{n}{4}\int_0^{\infty} vf(v)\,dv = \frac{n\bar{v}}{4} \tag{6.20-5}$$

where the last integral is, by definition, the average speed. Equation (6.20-5) is an important result. It should be noted, however, that the foregoing analysis does not consider the effects of collisions among molecules. (The same expression for the flux will be obtained later, taking into account the collisions between molecules.)

■ EXAMPLE 6.20-1

Calculate the mean speed, the molecular flux, and the impact rate on an 0.001 mm^2 area for oxygen at 300 K and 1 atm of pressure.

Solution. The molecular density may be computed from

$$p = nkT \tag{6.20-6}$$

where p = 1 atm = 1.013×10^5 N/m^{-2}, k = 1.381×10^{-23} J K^{-1}, T = 300 K, so that

$$n = \frac{1.013 \times 10^5}{(1.381 \times 10^{-23})(300)} = 2.45 \times 10^{25}\ \text{molecules/m}^3$$

The mean speed is given by

$$\bar{v} = \left(\frac{8kT}{\pi m}\right)^{1/2} \tag{6.20-7}$$

where $m = M/N_A = 32/6.022 \times 10^{26} = 5.31 \times 10^{-26}$ kg. Thus,

$$\bar{v} = \left[\frac{8 \times 1.381 \times 10^{-23} \times 300}{\pi(5.31 \times 10^{-26})}\right]^{1/2} = 446\ \text{m s}^{-1}$$

The molecular flux is

$$J = \frac{n\bar{v}}{4} = \frac{(2.45 \times 10^{25})(446)}{4} = 2.73 \times 10^{27}\ \text{molecules/m}^{-2}\ \text{s}^{-1}$$

Multiplying the flux by the area of 0.001 mm^2 gives

$$JA = (2.73 \times 10^{27})(0.001 \times 10^{-6}) = 2.73 \times 10^{18} \text{ molecules/s}^{-1}$$

This is the collision frequency. ■

■ EXAMPLE 6.20-2

The rate at which a gas escapes from a container through a small pinhole (of diameter less than the mean free path) is called *molecular effusion*. The molecules that effuse are mostly made up of those with greater speeds. As a result, the average escaping energy is greater than the translational energy $3kT/2$ of a molecule in the gas. (a) Show that the energy of the effused molecule is $2kT$. (b) For the effusion of a gas of two different types with molecular densities and masses n_1 and n_2 and m_1 and m_2, respectively, show that the ratio of effusion rates per unit area is $(n_1/n_2)(m_2/m_1)^{1/2}$.

Solution

(a) The mean energy $\bar{\varepsilon}$ of a molecule is found by calculating the total energy E transported from the container in a time t and dividing this energy by the number of molecules effusing from the container in this time. The molecules have speeds of between v and $v + dv$ and are moving at angles between θ and $\theta + d\theta$ as seen in Fig. 6.20-2. Each molecule of speed v has kinetic energy $mv^2/2$.

The energy transported from the container in time t is the molecular flux, Eq. (6.20-4), times the kinetic energy of each molecule, $\frac{1}{2}mv^2$,

$$\frac{mnt}{4}\int_0^{\theta} \sin\theta \cos\theta \, d\theta \int_0^{\infty} v^3 f(v)\, dv \qquad (6.20\text{-}8)$$

where ϕ has been integrated out, leaving the variables θ and v to work with. Now, from (6.20-4), the number of molecules effusing in time t per unit area is

$$\frac{nt}{2}\int_0^{\infty} \sin\theta \cos\theta \, d\theta \int_0^{\infty} v f(v)\, dv \qquad (6.20\text{-}9)$$

Dividing (6.20-8) by (6.20-9) gives

$$\bar{\varepsilon} = \frac{m}{2}\frac{\int_0^{\infty} v^3 f(v)\, dv}{\int_0^{\infty} v f(v)\, dv} \qquad (6.20\text{-}10)$$

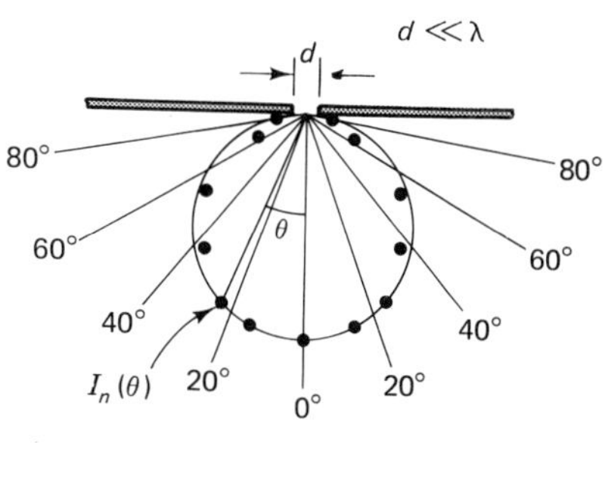

Figure 6.20-2 Effusion through a pinhole.

Using the Maxwell–Boltzmann speed distribution

$$f(v) = 4\pi\left(\frac{m}{2\pi kT}\right)^{3/2} v^2 e^{-mv^2/2kT} \qquad (6.20\text{-}11)$$

(6.20-10) becomes

$$\bar{\varepsilon} = \frac{m}{2}\frac{\int_0^\infty v^5 e^{-mv^2/2kT}\,dv}{\int_0^\infty v^3 e^{-mv^2/2kT}\,dv} \qquad (6.20\text{-}12)$$

Letting $\alpha = m/2kT$ and making use of the integrals listed in Chapter 1, Eq. (6.20-12) gives

$$\bar{\varepsilon} = \frac{m}{\alpha} = 2kT \qquad (6.20\text{-}13)$$

(b) Consider a mixture of two gases, with molecular masses and densities given by m_1 and m_2, and n_1 and n_2, respectively. If J_1 and J_2 are the rates of effusion per unit area of the two gases, then according to (6.20-5) and (6.20-7),

$$J_1 = n_1\left(\frac{kT}{2\pi m_1}\right)^{1/2} \qquad (6.20\text{-}14)$$

$$J_2 = n_2\left(\frac{kT}{2\pi m_2}\right)^{1/2} \qquad (6.20\text{-}15)$$

The ratio of the densities in the container is n_1/n_2. The ratio of the densities in the gas effusing from the container is

$$\frac{J_1}{J_2} = \left(\frac{n_1}{n_2}\right)\left(\frac{m_2}{m_1}\right)^{1/2} \qquad (6.20\text{-}16)$$

If $m_2 > m_1$, then $(J_1/J_2) > (n_1/n_2)$. That is, there is a greater concentration of the lighter gas in the gas that has escaped. The ratio $(J_1/J_2)/(n_1/n_2) = (m_2/m_1)^{1/2}$ is called the *separation factor f*. As the gas in a container effuses, the efficiency of the separation decreases, since the gas in the container is progressively becoming poorer in the lighter species. Ultimately a point will be reached at which the ratio J_1/J_2 is less than the original value of n_1/n_2 at the start of the process. This technique is applied to the separation of ^{235}U from ^{238}U, using more than 4000 stages. The uranium is combined with fluorine to form the gas uranium hexafluoride, UF_6. The separation factor f equals 1.0043, and (n_1/n_2) equals 0.0073. ■

6.21 COLLISIONS AND MEAN FREE PATH

So far discussion has centered on the number of molecules striking a surface per unit area and time. However, no transport can take place at the molecular level without collisions of the molecules, and to describe transport processes properly the influence of these collisions must be taken into account.

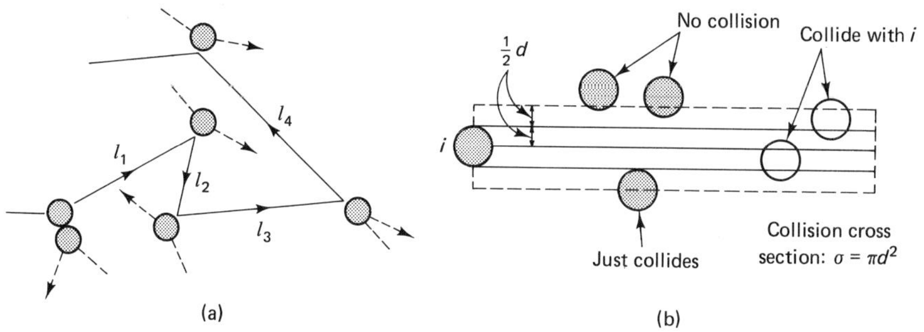

Figure 6.21-1 Molecular cross section and free paths.

Consider in imagination the motion of a molecule i in the gas over a time t. It can be expected that the molecule follows a zigzag path like that shown in Fig. 6.21-1(a). Between successive collisions with other particules the particle i will travel freely through distances l_1, l_2, l_3, and so on, distances known as free paths.

To simplify the analysis, consider the system to consist of many molecules at rest through which moves a single molecule of speed v and diameter d. The moving molecule will strike a stationary molecule when the distance between the centers of the molecules is less than or equal to the diameter, as depicted in Fig. 6.21-1(b). Thus, the number of collisions in time dt is equal to the number of molecules contained in the volume $\pi d^2 v\,dt$, or $n\pi d^2 v\,dt$. The quantity πd^2 is the collision cross section σ. The number of collisions per unit time is the collision frequency z, and is given by

$$z = n\pi d^2 v = n\sigma v = n\sigma\langle v\rangle \tag{6.21-1}$$

where a molecule is considered to have an average speed $\langle v\rangle$.

To find the mean distance λ traveled between collisions, the total distance distance $l_1 + l_2 + l_3 + \cdots = \langle v\rangle t$ must be divided by the number of collisions occurring in time t, that is

$$\lambda = \frac{\langle v\rangle t}{\text{number of collisions in time } t} \tag{6.21-2}$$

$$= \frac{\langle v\rangle t}{n\langle v\rangle t} = \frac{1}{n\sigma} \tag{6.21-3}$$

or, since $n = p/kT$,

$$\lambda = \frac{kT}{p\sigma} = \frac{kT}{p(\pi d^2)} \tag{6.21-4}$$

As an example, air at sea level is typically at $p = 1.0 \times 10^5$ Pa, $T = 300$ K. Assuming $d = 3 \times 10^{-10}$ m, (6.21-3) gives $\lambda = 1.46 \times 10^{-7}$ m. At a height of 200 km, $\lambda = 2.1 \times 10^2$ m (the mean intermolecular spacing is 5.0×10^{-6} m), while at a height of 600 km, $\lambda = 1 \times 10^5$ m (62 miles).

The foregoing analysis is oversimplified in that it does not take into account the fact that all of the molecules are in motion and that the collision frequency must be governed by the relative velocity between molecules. Also, the MB velocity distribution should be taken into account. To analyze the situation more properly, consider two molecules i and j to have velocities v_i and v_j as shown in Fig. 6.21-2(a). To bring molecule j to rest, its velocity must be canceled, and the relative velocity v_r obtained, as in (b) and (c). All the v_r's thus obtained are averaged, and the average value, $\langle v_r \rangle$, is then used in Eq. (6.21-2) to calculate the mean free path λ.

In finding $\langle v_r \rangle$, it will be assumed that all the molecules have a common speed $\langle v \rangle$. This means that when the tails of their velocity vectors are brought to a common point, as in Fig. 6.21-2(c), their tips will lie on a circle of radius $\langle v \rangle$ and that

$$v_r = 2\langle v \rangle \sin\frac{\theta}{2} \tag{6.21-5}$$

To find the average value of v_r, the average value of $2\langle v \rangle \sin \theta/2$ must be found. Now, the fraction of the atoms moving at angles between θ and $\theta + d\theta$ will be the area of a thin band shown in Fig. 6.21-2(c), $(2\pi u \sin\theta \times u\, d\theta)$, divided by the area of the sphere $(4\pi u^2)$. This gives $\frac{1}{2}\sin\theta\, d\theta$ for the distribution probability of v_r. The average v_r is then the product of (6.21-5) and $\frac{1}{2}\sin\theta\, d\theta$:

$$\begin{aligned}\langle v_r \rangle &= \int_0^\pi v_r\left(\frac{1}{2}\sin\theta\, d\theta\right) = \langle v \rangle \int_0^\pi \sin\left(\frac{\theta}{2}\right) \sin\theta\, d\theta \\ &= \langle v \rangle \int_0^\pi \sin\left(\frac{\theta}{2}\right) 2 \sin\left(\frac{\theta}{2}\right) \cos\left(\frac{\theta}{2}\right) d\theta \\ &= 4\langle v \rangle \int_0^{\pi/2} \sin^2\theta' \cos\theta'\, d\theta' \end{aligned} \tag{6.21-6}$$

or

$$\langle v_r \rangle = \frac{4}{3}\langle v \rangle \tag{6.21-7}$$

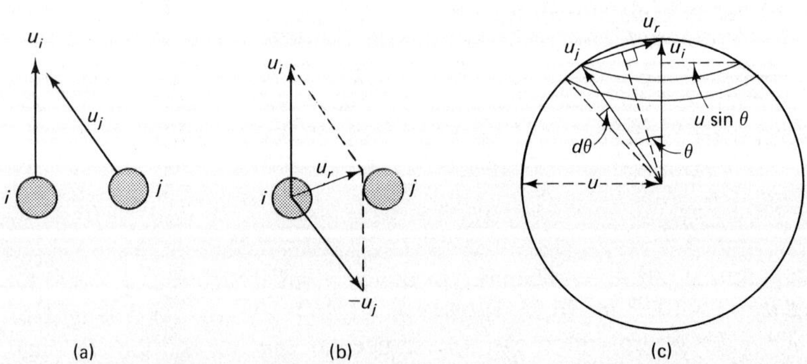

Figure 6.21-2 Collision of two moving molecules.

Hence, the denominator in (6.21-2) becomes $n\sigma(4\langle v\rangle t/3)$, while the numerator remains $\langle v\rangle t$. The mean free path is

$$\lambda = \frac{3}{4}\left(\frac{1}{n\sigma}\right) = \frac{3}{4}\left(\frac{1}{n\pi d^2}\right) \tag{6.21-8}$$

and the collision frequency is

$$z = \frac{4n\sigma}{3}\langle c\rangle = \frac{4n\pi d^2}{3}\langle v\rangle \tag{6.21-9}$$

The random relative motion of the molecules increases the collision rate by a factor of $\frac{4}{3}$. A yet more precise calculation is to stipulate that the collisions between molecules soon produces a Maxwell–Boltzmann speed distribution, in which case the collision frequency becomes

$$z = \sqrt{2}n\sigma\langle v\rangle = \sqrt{2}n\pi d^2\langle v\rangle \tag{6.21-10}$$

and the mean free path

$$\lambda = \frac{1}{\sqrt{2}}\left(\frac{1}{n\sigma}\right) = \frac{1}{\sqrt{2}}\left(\frac{1}{n\pi d^2}\right) \tag{6.21-11}$$

Interestingly, the value of λ calculated via Eq. (6.21-11) only differs by 6% from the value calculated via (6.21-8). Many of the properties of a gas are remarkably immune to the precise form of the speed distribution function. For most practical purposes it is quite sufficient to calculate λ by means of Eq. (6.21-3). Indeed, in view of the supposed billiard-ball nature of the collisions (hard sphere molecules which never approach closer than a distance d between centers) it is questionable whether there is much to be gained by introducing a factor of 3/4 or $1/\sqrt{2}$ into (6.21-3).

The foregoing has dealt with the number of collisions z made per unit time by a single atom i. Because there are n atoms present per unit volume and because each of them makes z collisions per unit time it follows from $z = n\sigma\langle v\rangle$ that

$$z_{AA} = \tfrac{1}{2}n^2\sigma\langle v\rangle = \tfrac{1}{2}n^2\pi d^2\langle v\rangle \tag{6.21-12}$$

or, from $n\sigma = 1/\lambda$ that

$$z_{AA} = \frac{n\langle v\rangle}{2\lambda} \tag{6.21-13}$$

where z_{AA} is the collision frequency of the same species per unit volume, and the factor of 1/2 has been introduced into Eqs. (6.21-12) and (6.21-13) to prevent each collision from being counted twice. Without it atom A_1 striking atom A_2 is regarded as a different event from atom A_2 striking atom A_1; in reality they are, of course, a common event.

Since it has been previously established that the average speed is

$$\langle v\rangle = \left(\frac{8kT}{\pi m}\right)^{1/2} \tag{6.21-14}$$

it follows from (6.21-12) that

$$z_{AA} = \pi d^2 \left(\frac{2kT}{\pi m}\right)^{1/2} n^2 \tag{6.21-15}$$

In the more exact analysis, $\langle v \rangle$ in $z = n\sigma\langle v \rangle = n\pi d^2 \langle v \rangle$ would be replaced by $\langle v_r \rangle = \sqrt{2}\langle v \rangle$, giving

$$z_{AA} = \pi d^2 \left(\frac{4kT}{\pi m}\right)^{1/2} n^2 \tag{6.21-16}$$

Finally, if there are two different species present with molecular diameters of d_1 and d_2 but with the same mass m, it is seen from Fig. 6.21-1(b) that d in $z = n\sigma\langle v \rangle = n\pi d^2 \langle v \rangle$ must be replaced by $\frac{1}{2}(d_A + d_B)$. In addition, n^2 in Eq. (6.21-12) must be replaced by $n_A n_B$. Thus, (6.21-16) becomes

$$z_{AB} = \frac{1}{4}\pi(d_A + d_B)^2 \left(\frac{4kT}{\pi m}\right)^{1/2} n_A n_B \tag{6.21-17}$$

This result is of value for the reaction rate in chemical reactions occurring in the gaseous phase.

6.22 DISTRIBUTION OF FREE PATHS

It is of further interest to discuss the distribution of free paths. Consider once again a large number of molecules, and assume that a stream of N molecules impinges upon the large group. As some of the molecules have collisions, they are removed from the stream, as depicted on Fig. 6.22-1(a). The target area σ of a single molecule is the area of a circle of radius d, the exclusion radius, that is,

$$\sigma = \pi d^2 \tag{6.22-1}$$

This area is now called the *microscopic collision cross section* (of one molecule). If there are n target molecules per unit volume, the product $n\sigma$ is called

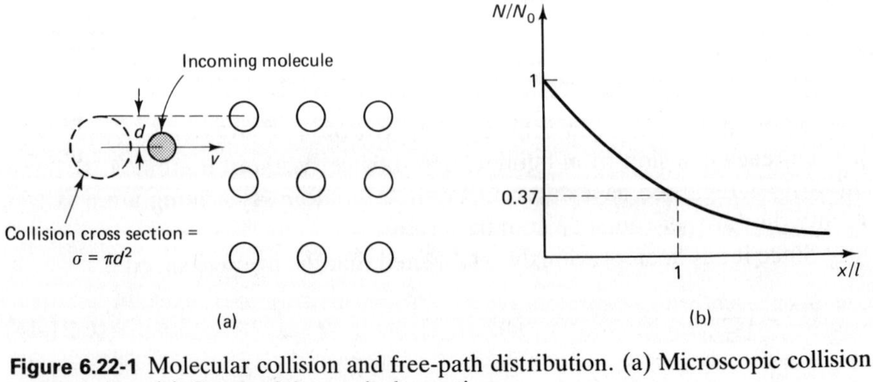

Figure 6.22-1 Molecular collision and free-path distribution. (a) Microscopic collision cross section. (b) Graph of the survival equation.

the *macroscopic collision cross section*. Since the number density n, in the SI system, is the number of molecules per cubic meter and the collision cross section σ is the number of square meters per molecule, the unit of the product $n\sigma$ is 1 square meter per cubic meter ($1\ m^2\ m^{-3} = 1\ m^{-1}$). More generally, in any system, the unit of macroscopic collision cross section is a *reciprocal length*, not an area.

Let ΔN be the number of collisions, each of the ΔN collisions diverting a molecule from its original path and scattering it out of the beam, thereby decreasing the number of molecules remaining in the beam. Thus, ΔN is negative and is proportional to the number in the stream, N, and the distance traveled, Δx:

$$\Delta N = \alpha N\, \Delta x = -n\sigma N\, \Delta x \qquad (6.22\text{-}1a)$$

where the constant α is a collision probability which is none other than the macroscopic cross section. Equation (6.22-1a) gives

$$\frac{\Delta N}{N} = -n\sigma\, \Delta x \qquad (6.22\text{-}2)$$

Now, N decreases in stepwise fashion as individual molecules make collisions, but if N is very large, it can be considered to be a continuous function of x and so

$$\frac{dN}{N} = -n\sigma\, dx \qquad (6.22\text{-}3)$$

Then

$$\ln N = -n\sigma x + \text{const} \qquad (6.22\text{-}4)$$

and if $N = N_0$, when $x = 0$,

$$N = N_0 \exp(-n\sigma x) \qquad (6.22\text{-}5)$$

This is known as the *survival equation*. It represents the number of molecules N, out of an initial number N_0, that has not yet made a collision after traveling a distance x. Inserting this expression for N in Eq. (6.22-2) gives

$$\Delta N = N_0 n\sigma \exp(-n\sigma x)\, \Delta x \qquad (6.22\text{-}6)$$

In this equation, ΔN is the number of molecules making their first collision after having traveled a distance between x and $x + \Delta x$.

To calculate the average distance traveled by the N_0 molecules before they make their first collision (the mean free path, λ), multiply x by the number of particles ΔN that travel the distance x before colliding, sum over all values of x, and divide by the total number N_0. In other words,

$$\lambda = \frac{\Sigma x\, \Delta N}{N_0} = n\sigma \int_0^{\infty} x \exp(-n\sigma x)\, dx \qquad (6.22\text{-}7)$$

The definite integral equals $1/n^2\sigma^2$, so

$$\lambda = \frac{1}{n\sigma} \qquad (6.22\text{-}8)$$

The mean free path is inversely proportional to the macroscopic collision cross section. Since the unit of macroscopic collision cross section is the reciprocal of the unit of length, the unit of mean free path is the unit of length.

Note that the mean free path does not depend on the molecular speed, but on n and σ. This may be understood by thinking of a man shooting bullets randomly into a thick forest. All of the bullets will eventually hit trees, but some will travel farther than others. Figure 6.22-2 is a heuristic derivation of Eq. (6.22-8). The deer readily understands that the average distance traveled will depend inversely on both the denseness of the woods (n) and on the size of the trees (σ).

Note also that a molecular diameter d equals 2×10^{-10} m. At standard conditions, there are about 3×10^{25} molecules/m^{-3} in a gas. The macroscopic collision cross section is then $n\sigma = n\pi d^2 \approx 3 \times 10^{25} \times 3.14 \times 4 \times 10^{-20} \approx 40 \times 10^5$ m^{-1}, and the mean free path is approximately 2.5×10^{-7} m. This is smaller than the wavelength of visible light. The average intermolecular separation at standard conditions is about 3×10^{-9} m, so the mean free path is much larger than the average intermolecular separation. In terms of the mean free path, the survival equation can be written

$$N = N_0 \exp(-n\sigma x) = N_0 \exp\left(\frac{-x}{\lambda}\right) \tag{6.22-9}$$

The graph of this equation is shown in Fig. 6.22-1(b) in which the dimensionless ratio N/N_0 is plotted as a function of x/λ. The ordinate of the curve is the fractional number of molecules with free paths longer than the mean free path. The fraction with free paths longer than the mean is exp(-1) or 37%, while the number with free paths shorter than the mean is 63%.

Figure 6.22-2 Experimental verification of $\lambda = 1/n\sigma$. Man shoots bullets in all directions, hitting nothing but trees. Deer observes that mean-free path of bullets is independent of velocity but inversely proportional to density and cross section of trees.

More complete theories of the mean free path take into account the relative motion of all the molecules of a gas, that is, they consider the "target" molecules, as well as the "bullet" molecules, to be in motion. The only change in the end result is to introduce a small correction factor in Eq. (6.22-8). The inverse dependence on the number of molecules per unit volume and on the collision cross section remains unaltered. On the assumption that all molecules have the same speed, Clausius obtained

$$\lambda = \frac{3}{4}\frac{1}{n\sigma} = \frac{0.75}{n\sigma} \tag{6.22-10}$$

If the molecules have a Maxwellian velocity distribution, then

$$\lambda = \frac{1}{\sqrt{2}}\frac{1}{n\sigma} = \frac{0.707}{n\sigma} \tag{6.22-11}$$

One final remark: In the preceding discussion, the target molecules and bullet molecules were considered identical hard spheres, each of diameter d. These dimensions are somewhat different for the mean free path of an electron, moving among the neutral or ionized molecules of a plasma, or among the fixed metallic ions in a metallic conductor. The diameter of an electron is so much smaller than that of a molecule that the electron can be considered a geometrical point, and the center-to-center distance in a collision (see Fig. 6.21-1) becomes $d/2$ rather than d, where d is the molecular diameter. Furthermore, the velocities of the electrons are so much greater than those of the molecules that the latter can be assumed at rest, and the correction for relative velocities need not be made. From these considerations, the electronic mean free path λ_e is

$$\lambda_e = 4\frac{1}{n\sigma} \tag{6.22-12}$$

where n is the number density of molecules and $n\sigma$ is the macroscopic collision cross section of electrons with molecules or ions.

■ EXAMPLE 6.22-1

The effective diameter of an oxygen molecule is 3.5×10^{-10} m. What is the mean free path for oxygen molecules at 300 K and 1-atm pressure, and what fraction of molecules have free paths greater than 2λ?

Solution. The collision cross section is

$$\sigma = \pi d^2 = \pi(3.5 \times 10^{-10})^2 = 3.85 \times 10^{-19}\ \text{m}^2$$

The molecular density is

$$n = \frac{p}{kT} = \frac{1.013 \times 10^5}{(1.35 \times 10^{23})(300)} = 2.45 \times 10^{26}\ \text{molecules/m}^3$$

The mean free path is calculated from (6.22-11):

$$\lambda = \frac{1}{\sqrt{2}\sigma n} = \frac{1}{\sqrt{2}(3.85 \times 10^{19})(2.45 \times 10^{25})}$$
$$= 7.50 \times 10^{-8} \text{ m}$$

The fraction of molecules having free paths of distance x or greater is given by Eq. (6.22-9).

$$\frac{N}{N_0} = e^{-x/\lambda}$$

For $x = 2\lambda$, this gives

$$\frac{N}{N_0} = e^{-2\lambda/\lambda} = 0.135$$

or 13.5% of the molecules have free paths greater than twice the mean free path. ■

6.23 ELECTRICAL CONDUCTIVITY

The classical theory of metallic conduction developed by Drude in 1900 assumes that the free electrons in a metallic conductor can be considered an ideal gas and that their average random speed $\langle v \rangle$ is the same as that of gas molecules of the same mass, at the same temperature. (This assumption is not very realistic and it will be amended by quantum mechanics.)

In the absence of an electric field, a gas ion will, of course, follow straight-line paths in between collisions with gas atoms. The effect of an electrical field of strength $\mathscr{E}$ will be to subject an ion of charge q to a force $F = q\mathscr{E}$, with the result that the positive ions tend to move in the field direction and the negative ions in the opposite direction, as shown in Fig. 6.23-1. Let $\mathscr{E}$ be the applied field, v_e the velocity acquired by a given electron, $\bar{v}_t$ the average thermal velocity (equivalent to that for a classical gas), and q/m the ratio of the charge to the mass of the ion or electron (more precisely, for an electron, m is the effective mass m^*). When the field is applied, an ion of charge q will be subject

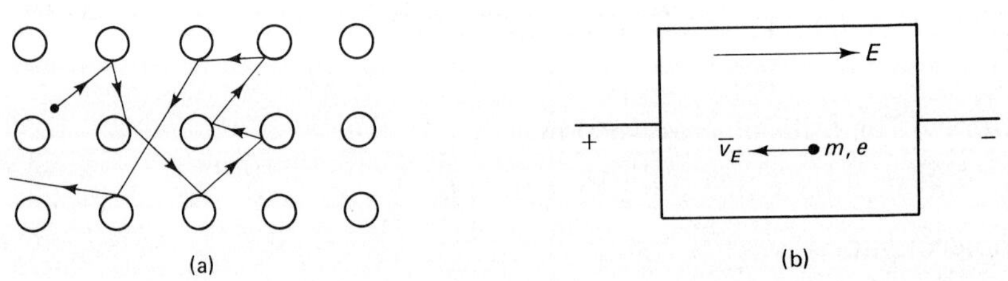

Figure 6.23-1 Electron motion. (a) In zero electric field. (b) In presence of an electric field.

to a force $F = q\mathscr{E}$. Newton's second law of motion gives

$$\frac{d^2x}{dt^2} = \frac{q}{m}\mathscr{E} \qquad (6.23\text{-}1)$$

If an ion of mass m travels freely for a time t_1 before it suffers a collision, it will acquire a final speed of $(F/m)t_1$ upon integration of Eq. (6.23-1). As seen from Fig. 6.23-2, after each collision, which occurs at times t_1, t_2, t_3, and so on, an ion stops drifting. Between collisions it accelerates at a constant rate. The effect of the collision at $t = t_1$ is to remove the extra speed acquired by q from the field. The ion does not stop moving; it merely stops drifting. The ion may travel freely for a time t_2, say, collide and stop drifting yet again, and so on. The distance traveled at the end of interval t is given by a second integration of (6.23-1), namely,

$$X_d = \frac{1}{2}\frac{q}{m}\mathscr{E}t^2 \qquad (6.23\text{-}2)$$

Now, the mean time interval τ between n collisions is

$$\tau = \frac{t_1 + t_2 + t_3 + \cdots + t_n}{n} \qquad (6.23\text{-}3)$$

while the mean distance $\bar{x}_d$ traveled between collisions is

$$\bar{x}_d = \frac{1}{2}q\frac{\mathscr{E}}{m}\overline{t^2} \qquad (6.23\text{-}4)$$

where (for a sequence of n collisions)

$$\overline{t^2} = \frac{t_1^2 + t_2^2 + t_3^2 + \cdots + t_n^2}{n} \qquad (6.23\text{-}5)$$

The drift speed v_d, defined as the mean distance $\bar{x}_d$ traveled between collisions divided by the mean time interval τ between collisions, is thus

$$v_d = \frac{\bar{x}_d}{\tau} = \frac{q\mathscr{E}\overline{t^2}}{2m\tau} \qquad (6.23\text{-}6)$$

To evaluate v_d, it is necessary to know how the free times $t_1, t_2, \ldots,$ are dis-

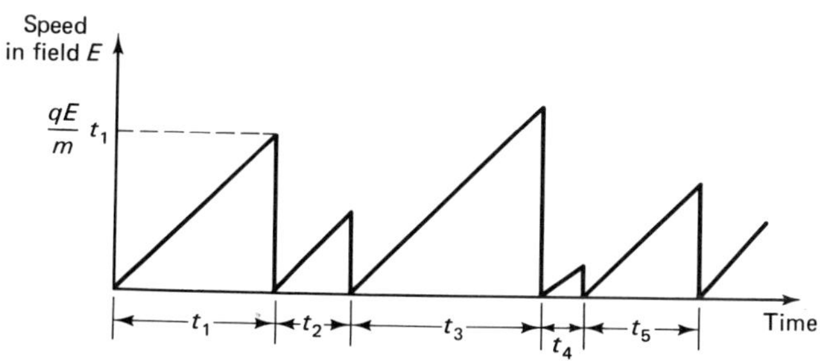

Figure 6.23-2 Ion speeds in an electron field.

tributed. Their distribution is identical to the distribution of free paths in Section 6.22 with the substitution of t for x. The probabiity that a molecule or an ion has a free time at least equal to τ is $e^{-t/\tau}$ and the fraction of molecules or ions with free times between t and $t + dt$ is

$$\frac{dN}{N} = \frac{1}{\tau} e^{-t/\tau}\, dt \tag{6.23-7}$$

Using this in

$$\overline{t^2} = \frac{1}{N} \int_0^\infty t^2\, dN \tag{6.23-8}$$

gives

$$\overline{t^2} = \frac{1}{\tau} \int_0^\infty t^2 e^{-t/\tau}\, dt \tag{6.23-9}$$

On evaluating the integral by parts (applied twice over),

$$\overline{t^2} = 2\tau^2 \tag{6.23-10}$$

Substituting this result into (6.23-6) gives

$$v_d = q\frac{Et}{m} \tag{6.23-11}$$

For an electron ($q = e$), this is

$$v_d = \frac{e}{m} E\tau \tag{6.23-12}$$

Note that the drift velocity is actually of the average velocity due to the random thermal motion (zero in any specific direction) and the average velocity due to the field.

For a total of n electrons per unit volume taking part in the conduction process, the total current density will be the total charge times the average drift velocity. Thus,

$$J = ne\bar{v}_D \tag{6.23-13}$$

Substituting the value of $\bar{v}_D$ from (6.23-11) into (6.23-13), gives

$$J = \frac{ne^2\tau}{m} E \tag{6.23-14}$$

This is recognized as *Ohm's law* where the resistivity ρ is given by

$$\rho = \frac{m}{ne^2\tau} \tag{6.23-15}$$

or in terms of the conductivity σ,

$$\sigma = \frac{ne^2\tau}{m} = ne\mu \tag{6.23-16}$$

with

$$\mu = \frac{e\tau}{m} \tag{6.23-17}$$

μ is a parameter, called the *mobility,* defined as the average drift velocity per unit electric field,

$$\mu = \frac{\overline{v}_d}{E} \tag{6.23-18}$$

If there is conduction by positive carriers as well as negative carriers, the conductivity is given by

$$\sigma = nq_n \mu_n + pq_p \mu_p \tag{6.23-19}$$

where μ_n and μ_p are the mobilities of negative and positive charge carriers, q_n and q_p are their charges, and n and p are the numbers of these carriers per unit volume.

It should be noted that the thermal velocity is much larger than the drift velocity. An estimate of $\overline{v}_t$ can be made easily from kinetic theory which states that

$$\tfrac{1}{2}m\overline{v}_t^2 = \tfrac{3}{2}kT \tag{6.23-20}$$

where k is Boltzmann's constant. Substituting values of m and k and $T = 300$ K gives $\overline{v}_t = 1.2 \times 10^5$ m s^{-1}. On the other hand, for a typical field value of $E = 10$ V m^{-1}, the drift velocity $\overline{v}_d = 17.6 \times 10^{-3}$ m s^{-1} for $\tau = 10^{-14}$ s. All the parameters in (6.23-16) can be determined experimentally, and substituting for n, e, m, and τ gives values of σ which agree very well with those obtained by direct measurement.

In terms of the band theory previously discussed, consider a solid subject to an electric field. The electrons would like to move up the field direction, thereby gaining energy. However, the opportunity for any particular electron to gain energy is governed by whether or not there is an available and empty energy level at the new energy. If there is a level that can be occupied by the electron in its excited state, the electron will move under the influence of the field. If there is no available level, the electron cannot obtain energy from the field and so will not respond to it. Conductivity will occur only in an unfilled band, as in sodium, or in a filled band only if it overlaps the next higher unoccupied band, as in magnesium. If a band is just filled, as, for example, in diamond, in general no response to a field will be possible. If, however, the energy gap between a filled band and the next higher unfilled band is not large, the possibility exists of excitation of electrons over the gap. (This excitation could take place because of either thermal or field energy. In such a case conductivity would be observed, but its value would be a function of the degree of excitation between the two bands, that is, of the temperature or field. This is the situation in a semiconductor.) The various possibilities are summarized in Fig. 6.23-3.

The response to an electric field can also be described in terms of the

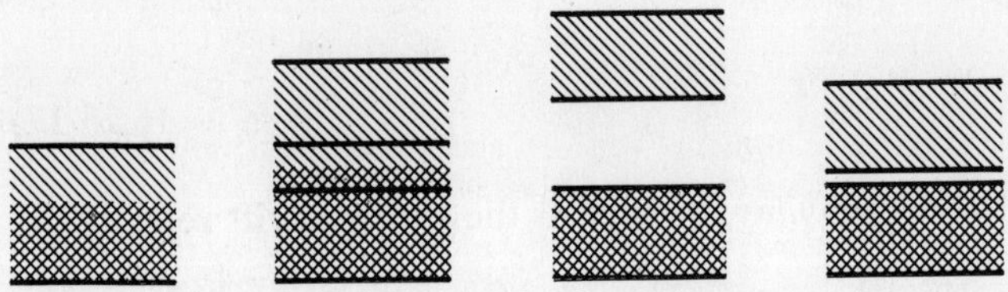

Figure 6.23-3 Electron occupation and band configuration. (a) and (b) Conductors. (c) Insulator. (d) Semiconductors.

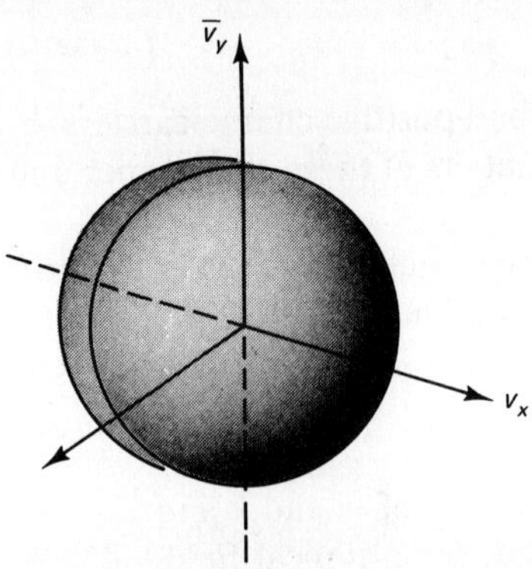

Figure 6.23-4 Displacement of Fermi surface in an electric field.

Fermi surface discussed in Chapter 4. In Fig. 6.23-4 the densely shaded sphere represents the Fermi surface in the absence of an external field. Electrons have velocities in all directions but no net current flows because the distribution is spherically symmetrical. Now consider the application of an electric field. The electrons all acquire an average velocity v_d in the $-X$ direction in the field, the altered distribution being shown as the displaced sphere in Fig. 6.23-4. The distribution is no longer symmetrical about the origin, and net current flow will occur.

■ EXAMPLE 6.23-1

The behavior of electrons bound in a solid is different from that of electrons in free space. Electrons in free space are accelerated indefinitely by an applied field, but in a metal, some process must be operative to restrict the current to the observed values. This is done by ascribing to the electron a time, τ, called the *relaxation time,* during which it can be accelerated. At the end of time τ the electron is "scattered" into a new state, thereby losing all the momentum gained in the field direction. In the standard Kronig–Penney model treatment, an electron remains in its particular momentum state indefinitely, and τ is infinite. To account for the nonclassical properties of the electron in a periodic lattice, a quantity m^*, the effective mass of the electron, is used. Show that

$$m^* = \frac{h^2}{4\pi^2}\left(\frac{d^2\varepsilon}{dk^2}\right)^{-1}$$

where ε is the particle energy, and $k = 2\pi/\lambda$ is the wave number of propagation constant.

Solution. Consider an electron in a conductor in an electric field $\mathscr{E}$. The electron will be accelerated by the field. The classical mechanics energy–momentum relation for a free particle is simply

$$\varepsilon = \frac{p^2}{2m} \tag{6.23-21}$$

and the acceleration a of a charge e in a field $\mathscr{E}$ is

$$a = \frac{e\mathscr{E}}{m} \tag{6.23-22}$$

In wave mechanics, the de Broglie wave constitutes a quantum so that v and ω are given in terms of the particle energy ε by

$$\varepsilon = hv = \frac{h\omega}{2\pi} = \frac{p^2}{2m} \tag{6.23-23}$$

Now the wave number k and the angular frequency ω of the de Broglie wave must be related, because the momentum and energy of the particle are related. For any wave, the group velocity (the velocity at which amplitude modulation of the wave and hence information moves) is given by

$$v_g = \frac{d\omega}{dk} \tag{6.23-24}$$

Now,

$$\omega = \frac{2\pi\varepsilon}{h} = \frac{\pi p^2}{mh} = \frac{hk^2}{4\pi m} \tag{6.23-25}$$

Therefore, the group velocity is

$$v_g = \frac{h}{4\pi m}\cdot 2k = \frac{hk}{2\pi m} = \frac{\varepsilon k}{\omega m} = \frac{mv}{m} = v \tag{6.23-26}$$

In other words, the group velocity of the de Broglie waves, the velocity with which they transmit information, is the same as the velocity of the particle. Thus, for the electron

$$v = \frac{2\pi}{h}\frac{d\varepsilon}{dk} \tag{6.23-27}$$

This relation now provides a means of calculating the velocity of an electron in the band model and consequently the acceleration. The acceleration is simply

$$a = \frac{2\pi}{h}\frac{d}{dt}\left(\frac{d\varepsilon}{dk}\right) = \frac{2\pi}{h}\frac{d^2\varepsilon}{dk^2}\frac{dk}{dt} \tag{6.23-28}$$

Consider the electron to be acted on by a field during a time dt. If the velocity of the electron is v, the distance traveled in dt is $v\,dt$ so that

$$d\varepsilon = e\mathscr{E}\cdot v\,dt = e\mathscr{E}\frac{2\pi}{h}\frac{d\varepsilon}{dK}\,dt$$

That is,

$$\frac{dk}{dt} = \frac{2\pi e\mathscr{E}}{h} \tag{6.23-29}$$

Combining Eqs. (6.23-28) and (6.23-29) gives the acceleration

$$a = \frac{4\pi^2 e\mathscr{E}}{h^2}\frac{d^2\varepsilon}{dk^2} \tag{6.23-30}$$

Now if this is compared with (6.23-22) for a free, classical particle, it is obvious that the two are identical in form if a new quantity is defined

$$m^* = \frac{h^2}{4\pi^2}\left(\frac{d^2\varepsilon}{dk^2}\right)^{-1} \tag{6.23-31}$$

so as to write in all cases

$$a = \frac{e\mathscr{E}}{m^*} \tag{6.23-32}$$

The quantity m^* is the *effective mass* of the electron. Some values of m^*/m are given in Table 6.23-1. ■

TABLE 6.23-1 VALUES OF THE ELECTRON EFFECTIVE MASS

Metal	m^*/m	Metal	m^*/m
Ni	28	Zr	2.24
Fe	12	Cr	2.93
Pd	43	Li	1.53
Pt	22	Na	0.94
Cu	1.47	K	0.58
Mb	1.33	Be	1.62
Ti	3.15		

■ **EXAMPLE 6.23-2**

(a) Show that the resistivity ρ of a material is given by

$$\rho = \frac{m_e \overline{v}}{ne^2\lambda} \tag{6.23-33}$$

where m_e is the mass of the electron, $\overline{v}$ is the average velocity due to thermal excitation, n is the number of electrons per unit volume, and λ is the mean free path of the electron.

(b) Calculate the resistivity of silver ($A = 108$), whose density is 10.5×10^3 kg m^{-3}, assuming each atom contributes one electron (the valence electron) for conduction. Take λ equal to 100 times the atomic spacing d, and the drift velocity v_d equal to the velocity corresponding to the Fermi energy, $\varepsilon_{F,0} =$ 5.5 eV.

(c) What is the resistivity of silicon ($A = 28$), which has a density of 2.42×10^3 kg m^{-3} and a valence of 2? Assume the Fermi energy for silicon is about 5 eV.

Solution

(a) The model for electrical current conduction in a metal regards the metal as a contained electron gas, the container being the metallic lattice. When an electrical potential is applied across the material, the electrons, which are moving in a random manner, are accelerated in the direction of the applied electric field, and after many collisions with the heavy lattice ions acquire a net average drift velocity, v_d, which results in a net electrical current. For simplicity, assume the volume V is to be a box of cross-sectional area A and length l. When a potential ϕ is applied across the material along the side of length l, the force on an electron is

$$F = e\mathscr{E} = e\frac{\phi}{l} \tag{6.23-34}$$

and the electron's acceleration is

$$a = \frac{F}{m_e} = \frac{e\phi}{m_e l} \tag{6.23-35}$$

The average time between the electron's collisions with the lattice is $t = \lambda/\overline{v}$. The velocity (over and above its random velocity) acquired during this time interval is the drift velocity

$$v_d = at = \frac{e\phi\lambda}{m_e l\overline{v}} \tag{6.23-36}$$

The current, I, is the rate at which charge is transported through the cross-sectional area A, so

$$I = (ne)Av_d = \frac{ne^2\lambda A}{m_e l\overline{v}}\,\phi \tag{6.23-37}$$

which is Ohm's law, $I = \phi/R$. Defining the resistivity ρ of the material by

$$R = \rho\frac{1}{A} \tag{6.23-38}$$

gives

$$\rho = \frac{m_e\overline{v}}{ne^2\lambda} \tag{6.23-39}$$

(b) From $\varepsilon = \frac{1}{2}mv^2$, the velocity corresponding to the Fermi energy is given by

$$v_F = \left(\frac{2\varepsilon_{F,0}}{m_e c^2}\right)^{1/2} c = \left(\frac{2 \times 5.5 \text{ eV}}{0.511 \times 10^6 \text{ eV}}\right)^{1/2} (3 \times 10^8 \text{ m s}^{-1}) = 1.39 \times 10^6 \text{ m s}^{-1}$$

where $m_e c^2 = 0.511 \times 10^6$ eV is the mass of the electron, and $c = 3 \times 10^8$ m s^{-1} is the speed of light.

Since each atom contributes one electron, the atomic density equals the electron density:

$$n = \frac{6.02 \times 10^{26} \text{ atoms/ kmol}^{-1})(10.5 \times 10^3 \text{ kg m}^{-3})}{108 \text{ kg/kmol}} (1 \text{ electron/atom})$$

$$= 5.85 \times 10^{28} \text{ electrons/m}^{-3}$$

If each atom has a volume of approximately d^3, then

$$d = \left(\frac{1}{5.85 \times 10^{28} \text{ m}^{-3}}\right)^{1/3} = 2.58 \times 10^{-10} \text{ m}$$

With λ equal to 100 times the atomic spacing, this gives $\lambda = 2.58 \times 10^{-8}$ m. The resistivity of silver is thus

$$\rho = \frac{m_e v_F}{e^2 n \lambda} = \frac{(9.11 \times 10^{-31} \text{ kg})(1.39 \times 10^6 \text{ m s}^{-1})}{(1.6 \times 10^{-19} \text{ C})^2 (5.85 \times 10^{28} \text{ m}^{-3})(2.58 \times 10^{-8} \text{ m})}$$

$$= 3.29 \times 10^{-8} \ \Omega \text{ m}$$

which is in fairly good agreement with what is observed (1.5×10^{-8} Ω m at 0°C and 6.87×10^{-8} Ω m at 800°C).

(c) Similar calculations for silicon give

$$v_F = \left(\frac{2\varepsilon_{F,0}}{m_e c^2}\right)^{1/2} c = \left(\frac{2 \times 5 \text{ eV}}{0.511 \times 10^6 \text{ eV}}\right)^{1/2} (3 \times 10^8 \text{ m s}^{-1}) = 13.3 \times 10^5 \text{ m s}^{-1}$$

$$n = \frac{(6.02 \times 10^{26} \text{ atoms/kmol}^{-1})(2.42 \times 10^3 \text{ kg/m}^3)}{28 \text{ kg kmol}^{-1}} (2 \text{ electrons/atom}^{-1})$$

$$= 1.04 \times 10^{29} \text{ electrons/m}^{-3}$$

The number of atoms per cubic meter is one-half the electron density, and if each atom takes up a volume of approximately d^3,

$$\lambda = 100d = 100\left[\frac{1}{\frac{1}{2}(1.04 \times 10^{29} \text{ m}^{-3}}\right]^{1/3} = 2.68 \times 10^{-8} \text{ m}$$

The resistivity is then

$$\rho = \frac{m_e v_F}{e^2 n \lambda} = \frac{(9.11 \times 10^{-31} \text{ kg})(13.3 \times 10^5 \text{ m s}^{-1})}{(1.6 \times 10^{-19} \text{ C})^2 (1.04 \times 10^{29} \text{ m}^{-3})(2.68 \times 10^{-8} \text{ m})}$$

$$= 1.7 \times 10^{-8} \ \Omega \text{ m}$$

This is considerably smaller than the resistance of silicon, which is found to be 10^3 Ω m at room temperature. The reason for this discrepancy is the assumption that valence electrons will produce conduction. However, silicon is a semiconductor, so there is a gap between the valence and conduction bands. Only those electrons having energies sufficient for them to be found in the conduction band will support electrical conduction. ■

6.24 COLLISION DISTANCE ABOVE A SURFACE

As previously remarked, transport processes involve collisions of materials as well as the flux of a conserved property. In fact, no transport can take place at the molecular level without collisions. When the expression $J = \frac{1}{4}n\overline{v}$ for the number of gas molecules striking a unit surface area per unit time was derived in Section 6.20, collisions were not taken into account in the $\theta\phi v$ cylinder of Fig. 6.24-1(a).

Previously, it was assumed that the molecules were geometrical points and that all $\theta\phi v$ molecules in the slant cylinder would arrive at the area ΔA without having made a collision. This cannot be correct, because on the average each molecule travels only a distance λ without colliding with another molecule. These molecular collisions will not, however, affect the total flux of $\theta\phi v$ molecules arriving at the surface, because for every collision that scatters a $\theta\phi v$ molecule out of the number originally in the cylinder, there will be another collision that results in an identical $\theta\phi v$ molecule at essentially the same point. This can be shown by redoing the calculation of Section 6.20, allowing for molecular collisions to take place. First, consider the number of collisions occurring within a volume $dV = r^2 \sin\theta \, dr \, d\phi$ at (r, θ, ϕ) in Fig. 6.24-1(b).

It was established in Section 6.21 that the number of collisions per unit volume per unit time in a gas is given by Eq. (6.21-13), namely,

$$z_{AA} = \frac{n\overline{v}}{2\lambda} \tag{6.24-1}$$

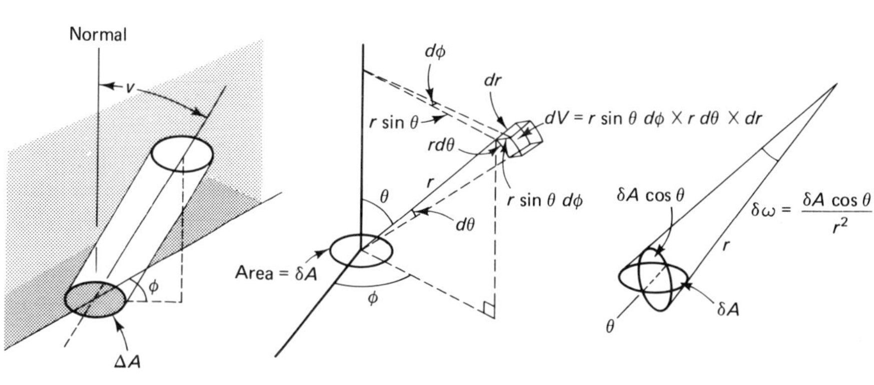

Figure 6.24-1 Collisions in transport theory. (a) $\theta\phi v$ molecules in a cylinder arriving at ΔA during Δt. (b) Last mean-free path before crossing a surface. (c) Solid angle $\delta\omega$.

Since the volume element at (r, θ, ϕ) is

$$dV = r^2 \sin\theta \, dr \, d\theta \, d\phi, \tag{6.24-2}$$

the number of collisions per unit time within the volume is given by the product of (6.24-1) and (6.24-2) times 2, or

$$\frac{n\bar{v}}{\lambda} r^2 \sin\theta \, dr \, d\theta \, d\phi \tag{6.24-3}$$

(The factor 2 is due to the fact that it takes two molecules to produce a collision.) The expression (6.24-3) represents the number of molecules per unit time coming straight from a collision within the volume dV given by (6.24-2). These molecules have their velocity vectors distributed uniformly in space. From a point of view of an observer located within the small volume dV, it is seen that the fraction of these $(n\bar{u}/\lambda)r^2 \sin\theta \, dr \, d\theta$ particles which move off toward ΔA per unit time is given by $(\delta A \cos\theta/r^2)/4\pi$, this being the fraction of 4π subtended at dV by ΔA in Fig. 6.24-1(b). Thus,

$$\left(\frac{\delta A \cos\theta}{4\pi r^2}\right)\left(\frac{nu}{\lambda}\right) r^2 \sin\theta \, dr \, d\theta \tag{6.24-4}$$

molecules are moving in the direction to strike δA per unit time. But only those with a free path in excess of r can do so. Since the fraction of atoms with a free path in excess of r is $e^{-r/\lambda}$, it follows that the number of atoms originating from dV which strike δA per unit time is

$$\left(\frac{\delta A \cos\theta}{4\pi r^2}\right)\left(\frac{nu}{\lambda}\right) r^2 \sin\theta \, dr \, d\theta \, d\phi \times e^{-r/\lambda}$$
$$= \frac{\delta A n\bar{u}}{4\pi\lambda} e^{-r/\lambda} \, dr \, \sin\theta \cos\theta \, d\theta \, d\phi \tag{6.24-5}$$

To obtain the total number of collisions I_n per unit area of surface per unit time, (6.24-5) is divided through by δA and integrated from $r = 0$ to $r = \infty$, from $\theta = 0$ to $\theta = \pi/2$, and from $\phi = 0$ to $\phi = 2\pi$. This gives

$$I_n = \frac{n\bar{u}}{4\pi} \int_0^{\pi/2} \sin\theta \cos\theta \, d\theta \tag{6.24-6}$$

Now,

$$\int_0^\infty \frac{1}{\lambda} e^{-r/\lambda} \, dr = \frac{\lambda}{\lambda} = 1 \tag{6.24-7}$$

$$\int_0^{\pi/2} \sin\theta \cos\theta \, d\theta = \frac{1}{2} \tag{6.24-8}$$

$$\int_0^{2\pi} d\phi = 2\pi \tag{6.24-9}$$

so that (6.24-6) becomes

$$J = \tfrac{1}{4} n\bar{v} \tag{6.24-10}$$

which is the same result as previously obtained. The reason why the earlier analysis worked is that it implicitly assumed that if an atom was lost by collision from the sloping cylinder of Fig. 6.24-1(a) its place was taken by another one possessing the same velocity. Thus, the simple arguments that ignore intermolecular collisions usually give the same results as do the arguments that explicitly consider these collisions.

There is, however, an observation to be made. Each point in the gas is characterized by a certain value of T, $\overline{v}$, and n. Now, imagine that the quantities T and $\overline{v}$ are properties of the space within which the molecules move, and assume that if a molecule undergoes a collision at a point x, it immediately assumes the properties characteristic of that point, that is, the characteristics of the molecules in a Maxwellian gas having temperature $T(x)$ and average velocity $\overline{v}(x)$. This is equivalent to assuming that in each collision the memory of a molecule is completely destroyed. It can trace its history back only to its last collision. Beyond that it has no idea from whence it came. It is as if the molecule were reborn in the collision.

In light of the foregoing, it is desirable to calculate an average collision distance, that is, the average distance above the plane shown in Fig. 6.24-1(b) at which the molecules have their last collision before crossing δA. The molecules arriving at the surface will, on the average, have started their last free paths a distance λ away from the surface. The perpendicular distance y from the surface, for any θ molecule of Fig. 6.24-1(b), however, is

$$y = r\cos\theta \tag{6.24-11}$$

The average value of y, or $\overline{y}$, is obtained by multiplying it by the flux J_θ of all the molecules arriving at the angle θ, dividing by the total flux, and integrating over θ. The flux J_θ is obtained from (6.24-5) by integrating over r and ϕ:

$$J_\theta = \left[\frac{n\overline{v}}{4\pi\lambda}\int_0^\infty e^{-r/\lambda}\,dr\int_0^{\pi/2} d\phi\right]\sin\theta\cos\theta\,d\theta \tag{6.24-12}$$

Multiplying (6.24-11) by (6.24-12), summing over all values of θ, and dividing by the total flux $J = \frac{1}{4}n\overline{v}$ thus gives

$$\overline{y} = \frac{1}{\pi\lambda}\int_0^\infty re^{-r/\lambda}\,dr\int_0^{\pi/2}\sin\theta\cos^2\theta\,d\theta \tag{6.24-13}$$

Again, making use of

$$\int_0^\infty re^{-r/\lambda}\,dr = \lambda^2 \tag{6.24-14}$$

$$\int_0^{2\pi} d\phi = 2\pi \tag{6.24-15}$$

$$\int_0^{\pi/2}\sin\theta\cos^2\theta\,d\theta = \frac{1}{3} \tag{6.24-16}$$

(6.24-13) becomes

$$\bar{y} = \frac{2}{3}\lambda \tag{6.24-17}$$

In other words, on the average, a molecule crossing a surface makes its last collision before crossing at a distance equal to two-thirds of a mean free path above (or below) the surface. This result will be of importance in the next section.

6.25 GENERALIZED CONCEPT OF TRANSPORT

The preceding results may be applied to the calculation of a generalized coefficient in molecular transport processes. Consider a field of molecules, as shown in Fig. 6.25-1, which is uniform in density (n = const), but in which there may be a gradient in some molecular property Γ in the z direction. Γ might be energy, momentum, concentration, electric charge, and so on. The transport of the quantity Γ across the area dA is governed by the variation of Γ in the z direction, that is,

$$\Gamma = \Gamma\big|_{z=0} + \frac{d\Gamma}{dz}\bigg|_{z=0} dz \tag{6.25-1}$$

in the immediate neighborhood of the plane dA. For pressure and temperature conditions where λ is small, (6.25-1) may be assumed to hold within a few mean free paths of $z = 0$.

The downward transport of Γ through dA is the molecular flux, $\frac{1}{4}n\bar{v}$, times the value of the property Γ at its last collision before hitting dA:

$$J_{\downarrow} = \frac{1}{4}n\bar{v}\left[\Gamma_0 + \left(\frac{d\Gamma}{dz}\right)_0 \frac{2}{3}\lambda\right] \tag{6.25-2}$$

Similarly, the upward transport of Γ is given by

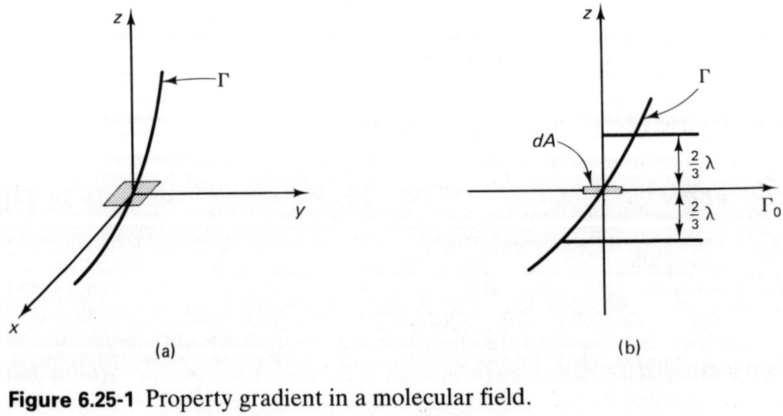

Figure 6.25-1 Property gradient in a molecular field.

$$J_{\uparrow} = \frac{1}{4} n\bar{v}\left[\Gamma_0 - \left(\frac{d\theta}{dz}\right)_0 \frac{2}{3}\lambda\right] \qquad (6.25\text{-}3)$$

The net transport rate in the positive z direction is the algebraic sum of these two relations, or

$$J = \frac{1}{3} n\bar{v}\lambda \frac{d\Gamma}{dz} \qquad (6.25\text{-}4)$$

The quantity $\frac{1}{3}n\bar{v}\lambda$ is a *generalized transport coefficient* whose value can be calculated from the proper molecular parameters once Γ is specified. Examples are given below.

Thermal Conductivity

Thermal conduction is defined by the Fourier relation

$$J_Q = \frac{qz}{A} = -K\frac{\partial T}{\partial z} \qquad (6.25\text{-}5)$$

where J_Q, the heat flux per unit area, is proportional to the temperature gradient in a direction z. The proportionality constant K is the *thermal conductivity*. In this process the quantity being transported is molecular energy. Energy is transported from a high-temperature region to a low-temperature region by the molecular motion, the average molecular energy being expressed in terms of the number of degrees of freedom f and the temperature as

$$\epsilon_i = \frac{f}{2}kT = \frac{3}{2}kT \qquad (6.25\text{-}6)$$

where $f = 3$ for translation.

The energy transport rate can be calculated from the generalized transport coefficient of (6.25-4):

$$J_Q = \frac{q}{A}J = -\frac{1}{3}n\bar{v}\lambda\frac{d}{dz}\left(\frac{f}{2}kT\right) \qquad (6.25\text{-}7)$$

Comparing (6.25-5) with (6.25-7) gives

$$K = \frac{1}{3}n\bar{v}\lambda\left(\frac{f}{2}k\right) \qquad (6.25\text{-}8)$$

This result can also be expressed in terms of the specific heat c_v of the gas by noting that the kinetic energy of an ideal gas is simply its internal energy $u = c_v T$, so that

$$K = \frac{1}{3}\frac{n\bar{v}\lambda_v}{N_0} = \frac{1}{3}\frac{\bar{v}c_v}{N_0\sigma} \qquad (6.25\text{-}9)$$

where N_0 is Avogadro's number and J is the collision cross section. From Equation (6.25-9) is seen that the thermal energy transport rate is directly proportional to the mean molecular speed. A typical set of units for thermal conductivity is kilojoules per second-meter-degree centigrade.

Dynamic Viscosity

Viscous action in a fluid is represented as a shear stress between fluid layers, the shear stress being proportional to a normal velocity gradient $\partial u/\partial z$ as stated in Section 6.19:

$$\tau = \mu \frac{\partial u}{\partial z} \tag{6.25-10}$$

where τ is the shear stress, and the proportionality constant μ is defined as the *dynamic viscosity*. (A typical set of units for the dynamic viscosity would be newtons-second per square meter in the SI system.) The physical mechanism of viscosity is one of momentum exchange. Molecules transport momentum from regions of high macroscopic velocity to regions of low velocity. This transport tends to pull the slower-moving fluid layers along, and the drag or shear force is the quantity τ. The momentum of a molecule due to the macroscopic velocity u is

$$\rho = mu \tag{6.25-11}$$

The net momentum transport through an area element may again be written with the aid of Eq. (6.25-4):

$$J_p = \tau = -\frac{1}{3} n\bar{v}\lambda \frac{\partial}{\partial z}(mu) = -\frac{1}{3} n\bar{v}\lambda m \frac{\partial u}{\partial z} \tag{6.25-12}$$

The net momentum transport is negative when $\partial u/\partial z$ is positive. This indicates a net transport downward through dA and a shear stress acting to the right. Comparing (6.25-10) with (6.25-12) gives

$$\mu = \frac{1}{3} nm\bar{v}\lambda = \frac{1}{3}\frac{m\bar{v}}{\sigma} \tag{6.25-13}$$

The viscosity, like thermal conductivity, is directly proportional to the mean molecular speed.

Diffusion of Like Molecules

Consider a container where there are tagged and untagged molecules, all of the same size and same mass as in the case of the self-diffusion of a gas into a radioactive isotope of the same gas. If the tagged molecules are placed at one end of the container and the untagged molecules at the other end, and the partition is removed, a molecular concentration gradient will have been impressed on the system. The total number of molecules in the system remains the same, but the molecules are no longer uniformly distributed throughout the system volume. There is a net flow of tagged molecules from the region of high concentration to the region of low concentration; this flow results from the tagged molecule density gradient and the random molecular motion. The physical situation described is mutual or self-diffusion, and the property being transported is molecular identity, in this instance, tagged (radioactive) molecules. If the concentration of tagged molecules is n_1/n, where n is the total number of molecules per unit volume, the transport per unit area per unit time is again given by (6.25-4):

$$J_{N1} = -\frac{1}{3} n\bar{v}\lambda \frac{\partial}{\partial z}\left(\frac{n_1}{n}\right) \tag{6.25-14}$$

The diffusion coefficient D is then defined by

$$J_{N1} = -D\frac{\partial n_1}{\partial z} \tag{6.25-15}$$

This gives

$$D = \frac{1}{3}\bar{v}\lambda \tag{6.25-16}$$

A remark concerning concentration and molecular identity is in order. The concentration parameter is the one chosen to measure the fraction of molecules which have a certain identity. Since the total molecular density is constant, the total net flux of molecules across some plane is zero while the flux of tagged molecules is not zero. The rate of change of the concentration in tagged molecules is the quantity that governs the transport process.

The theory of transport processes predicts that for all gases, certain combinations of properties should equal unity. For example, it follows from (6.25-13) and (6.25-16) that

$$\frac{\rho D}{\mu} = 1 \tag{6.25-17}$$

where $\rho(=mn)$ is the density of the gas. Also, from (6.25-9) and (6.25-13),

$$\frac{K}{c_v\,\mu} = 1 \tag{6.25-18}$$

where $c_v = C_{v,\mathrm{m}}/N_A m$ (which is the heat capacity at constant volume per unit mass of the gas). Experimental values of $\rho D/\mu$ lie between about 1.3 and 1.5 (e.g., it is 1.37 for hydrogen), whereas the values of $K/c_v\,\mu$ lie close to 2.5 for the inert gases but decrease with increasing molecular complexity (e.g., it is 1.55 for ethylene). The discrepancy is due partly to molecular repulsions which tend to reduce μ. A more important factor is that, in thermal conduction, the transport of the translational kinetic energy takes place more efficiently than the transfer of the rotational and vibrational forms of thermal energy.

■ EXAMPLE 6.25-1

Calculate the diffusion coefficient, dynamic viscosity, and thermal conductivity of nitrogen at STP, and compare them with experimental values. Assume that nitrogen has molecular specific heat $c_{v,m} = \frac{5}{2}R$ and a molecular diameter $d = 3.5 \times 10^{-10}$ m.

Solution. Since all three transport coefficients involve the mean free path and the mean speed, it is well to calculate λ and $\bar{u}$ straight away:

$$\lambda = \frac{kT}{\pi d^2 \rho} = 9.7 \times 10^{-8} \text{ m}$$

$$\bar{u} = \left(\frac{8kT}{\pi m}\right)^{1/2} = 453 \text{ m s}^{-1}$$

Next, the number density $n = N/V$ is calculated from

$$\rho = \tfrac{1}{3} mn\bar{u}^2 = nkT$$

or

$$n = \frac{\rho}{kT} = 2.68 \times 10^{25} \text{ m}^{-3}$$

Then (6.25-16), (6.25-13), and (6.25-9) give

$$D = \frac{1}{3}\lambda\bar{u} = 1.46 \times 10^{-5} \text{ m}^2 \text{ s}^{-1}$$

$$\mu = \frac{1}{3} nm\bar{u}\lambda = 1.83 \times 10^{-5} \text{ Pa s}$$

$$K = \frac{1}{3}\frac{n\bar{u}c_v\lambda}{N_A} = 1.35 \times 10^{-2} \text{ W m}^{-1} \text{ K}^{-1}$$

The experimentally determined values of the transport coefficients are $D = 1.7 \times 10^{-5}$ m^2 s^{-1}, $\mu = 1.75 \times 10^{-5}$ Pa s, and $K = 2.4 \times 10^{-2}$ W m^{-1} K^{-1}. The predicted and the measured values agree to within a factor of 2. The biggest discrepancy occurs in the thermal conductivity. ■

■ **EXAMPLE 6.25-2**

Some of the properties that characterize a metal are: (1) high electric conductivity σ which, above a certain temperature, is inversely proportional to the temperature T; (2) high thermal conductivity K which, at sufficiently high temperatures, is independent of T; (3) validity of the Wiedemann–Franz law, according to which the ratio of thermal and electrical conductivities is equal to a universal constant multiplied by the absolute temperature T:

$$\frac{K}{\sigma} = CT \qquad (6.25\text{-}19)$$

where K is the thermal conductivity, σ the electrical, and C a constant having a value of 2.31×10^{-8} W Ω K^{-2}; (4) temperature-independent carrier concentration; (5) high optical absorption and thus high reflexivity (metallic glance); and (6) ductility (metals can be rolled and forged).

Derive an expression for C by applying the kinetic theory of gases to the random thermal motion of the electrons in a metal.

Solution. Drude in 1900 applied the kinetic theory of a dilute gas to predict the behavior of a dense electron gas. Implicit in the Drude theory is the assumption that electrons are not only the agencies of electrical conduction but are also responsible for the transport of thermal energy in a solid.

Consider a region filled with an electron gas, with n electrons per unit volume. Let the region be divided by an imaginary plane as in Fig. 6.25-2. Let the average thermal velocity be $\bar{v}_t$, the mean free path be λ, and let a thermal gradient be along the x direction as indicated.

Suppose that the average energy of a given electron crossing the plane is E. The energy at planes a distance λ away will differ from E by an amount $\lambda\, dE/dx$, so that for each electron crossing, the energy flow is $2\lambda\, dE/dx$. From kinetic theory, the number of particles flowing in a given direction is $\frac{1}{6}n\bar{v}_t$. Therefore, the net energy flow across the central plane will be

$$\left(\frac{1}{6}n\bar{v}_t\right)\left(2\lambda\frac{dE}{dx}\right)=\frac{n\bar{v}_t}{3}\frac{dE}{dx}\lambda \tag{6.25-20}$$

and this can be equated to the product of the thermal conductivity K and the temperature gradient:

$$\frac{n\bar{v}_t}{3}\frac{dE}{dx}\lambda = K\frac{dT}{dx} \tag{6.25-21}$$

Now,

$$n\frac{dE}{dx}=n\frac{dE}{dT}\frac{dT}{dx}=C_V\frac{dT}{dx} \tag{6.25-22}$$

where C_V is the specific heat at constant volume. Equations (6.25-21) and (6.25-22) together give

$$K=\frac{1}{3}C_V\bar{v}_t\lambda \tag{6.25-23}$$

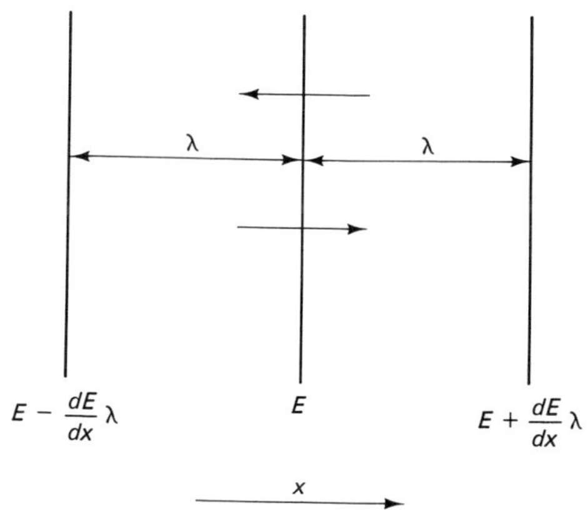

Figure 6.25-2 Electron gas in a region subject to a thermal gradient. (Electrons with energy E crossing an imaginary plane from a distance λ.)

The mean free path λ can be written in terms of the average time τ between collisions and the average thermal velocity $\bar{v}_t$ as $\lambda = \tau \bar{v}_t$. The electric conductivity σ is given by

$$\sigma = \frac{ne^2\tau}{m} = ne\mu \tag{6.25-24}$$

where

$$\mu = \frac{e\tau}{m} \tag{6.25-25}$$

is the mobility. From Eqs. (6.25-23) and (6.25-24), therefore,

$$\frac{K}{\sigma} = \frac{1}{3}\frac{C_V \bar{v}_t \lambda m}{ne^2\tau} = \frac{1}{3}\frac{C_V m \bar{v}_t^2}{ne^2} \tag{6.25-26}$$

Also, from kinetic theory,

$$C_V = \frac{3}{2} nk \tag{6.25-27}$$

$$m\bar{v}_t^2 = 2\left(\frac{3}{2} kT\right) \tag{6.25-28}$$

where k is Boltzmann's constant. Equations (6.25-26), (6.25-27), and (6.25-28) then give

$$\frac{K}{\sigma} = \frac{3}{2}\left(\frac{k}{e}\right)^2 T = CT \tag{6.25-29}$$

where C is a constant, called the *Lorenz number,* or *Wiedemann–Franz constant.* It has a theoretical value of 1.22×10^{-8} W Ω K^{-2}, which is about half the average experimental value of 2.31×10^{-8} W Ω K^{-2}. Nevertheless, it is of the same order and considered acceptable in view of the simplicity of the model. Lorenz numbers for silver, copper, and aluminum are 2.45, 2.37, and 2.02, respectively. ■

6.26 BOLTZMANN TRANSPORT EQUATION

So far, kinetic theory and transport processes were introduced to shed light on questions arising in statistical mechanics. The part of kinetic theory that is of general interest is the derivation of the transport coefficients of gases. These coefficients describe the reaction of a gas to mild but very common departures from thermal equilibrium, the most common being viscosity, heat conduction, and diffusion. The element that brings the kinetics into the simple approach is the concept of a mean free path, λ. For a particle moving among stationary obstacles of cross section σ and density n, the mean free path is given by

$$\lambda = \frac{1}{n\sigma} \tag{6.26-1}$$

It is of interest to observe that the results of the previous sections could be obtained with the help of the Maxwellian velocity distribution. If that distribution is valid at any position in space the mean velocity of the molecules in any direction is zero; the same is true for the various components of current. From this point of view, the essential feature of transport processes is the disturbance of the distribution due to external conditions. This disturbance was introduced in an intuitive way by the device of the mean free path. However, more reliable answers can be obtained by calculating the modification of the Maxwellian distribution first and then computing the transport results as an average over the disturbed distribution. This is the method followed here.

The Boltzmann transport equation has as its objective, the determination of the Boltzmann distribution function $f(\mathbf{v}, \mathbf{r})$ in six-dimensional space indicating how many molecules have velocities within $dv_x\,dv_y\,dv_z$ of $\mathbf{v}$ and positions within $dr_x\,dr_y\,dr_z$ of $\mathbf{r}$. In other words, the distribution function $f(\mathbf{v}, \mathbf{r})$ is defined by

$$f(\mathbf{v}, \mathbf{r})d\mathbf{v}\,d\mathbf{r} = \text{number of particles in } d\mathbf{v}\,d\mathbf{r} \tag{6.26-2}$$

The rate of change of this function is estimated through *binary collisions*, that is,

$$\frac{df}{dt} = \left(\frac{\partial f}{\partial t}\right)_{\text{coll}} \tag{6.26-3}$$

The left-hand side of Eq. (6.26-3) may be evaluated by recalling Liouville's theorem that in phase space, representative points of a system move in the manner of an incompressible fluid. The velocity $\mathbf{v}$ makes the representative points change their positions and an acceleration $\mathbf{a}$ makes them change their velocities. In hydrodynamics, this is taken care of by introducing the hydrodynamic total derivative, which indicates changes taking place at a point moving with the fluid. The left-hand side of Eq. (6.26-3) is thus

$$Df \equiv \frac{\partial f}{\partial t} + \mathbf{v}\cdot\frac{\partial f}{\partial \mathbf{r}} + \mathbf{a}\cdot\frac{\partial f}{\partial \mathbf{v}} \tag{6.26-4}$$

The right-hand side of Eq. (6.26-3) may be complicated. However, since this term tends to make the distribution relax toward thermal equilibrium, it is appropriate to introduce a relaxation time $\tau(\mathbf{r}, \mathbf{v})$, and take

$$\left(\frac{\partial f}{\partial t}\right)_{\text{coll}} = -\frac{f - f_0}{\tau} \tag{6.26-5}$$

where f_0 is the distribution function in thermal equilibrium. With the substitutions (6.26-4) and (6.26-5), (6.26-3) becomes a *simplified Boltzmann equation* which reads

$$\frac{\partial f}{\partial t} + \mathbf{v}\cdot\frac{\partial f}{\partial \mathbf{r}} + \mathbf{a}\cdot\frac{\partial f}{\partial \mathbf{v}} = \left(\frac{\partial f}{\partial t}\right)_{\text{coll}} = -\frac{f - f_0}{\tau} \tag{6.26-6}$$

It is recognized that $\partial f/\partial \mathbf{r}$ is the gradient with respect to $\mathbf{r}$, that is, the vector with components $\partial f/\partial x, \partial f/\partial y, \partial f/\partial z$. Similarly, $\partial f/\partial \mathbf{v}$ is the gradient with respect to $\mathbf{v}$, that is, the vector with components $\partial f/\partial v_x, \partial f/\partial v_y, \partial f/\partial v_z$. If the particles are under the influence of a force $\mathbf{F}$, then $\mathbf{a} = \mathbf{F}/m$, and an alternate form for (6.26-6) is

$$\frac{\partial f}{\partial t} + \mathbf{v} \cdot \frac{\partial f}{\partial \mathbf{r}} + \frac{\mathbf{F}}{m} \cdot \frac{\partial f}{\partial \mathbf{v}} = -\frac{f - f_0}{\tau} \tag{6.26-7}$$

It is assumed that the number of particles is conserved. If not, terms representing the generation and recombination of particles must be added to the right-hand side of Eq. (6.26-6). Such is the case, for example, in junction transistors and in nuclear reactor theory. In the absence of collisions, (6.26-6) becomes

$$\frac{\partial f}{\partial t} + \mathbf{v} \cdot \frac{\partial f}{\partial \mathbf{r}} + \mathbf{a} \cdot \frac{\partial f}{\partial \mathbf{v}} = 0 \tag{6.26-8}$$

Equation (6.26-8) asserts that f remains unchanged to an observer moving along with the molecules in phase space. (This is a special case of Liouville's theorem, proved in Chapter 3.) Equation (6.26-8) is a linear partial differential equation satisfied by f. It is the *Boltzmann equation without collisions.* (In plasma physics, it is sometimes called the *Vlasov equation.*)

For a first illustration, suppose that an external electric field perturbs the velocity distribution and by so doing creates a systematic drift of electrons in an electron gas. At any instant, and any location, the rate at which the distribution function changes is the sum of two kinds of contributions:

$$\left(\frac{\partial f}{\partial t}\right)_{\text{field}} + \left(\frac{\partial f}{\partial t}\right)_{\text{scattering}} = \left(\frac{\partial f}{\partial t}\right)_{\text{total}} \tag{6.26-9}$$

The first term is the influence of the field in changing the distribution, while the second is the effect of scattering in an attempt to restore the equilibrium distribution. Since force is rate of change of momentum, and an electron of speed $\mathbf{v}$ has momentum mv and experiences a force $-e\mathscr{E}$ in an electric field $\mathscr{E}$, then

$$\left(\frac{\partial f}{\partial t}\right)_{\text{field}} = \left(\frac{\partial \mathbf{v}}{\partial t}\right) \cdot \left(\frac{\partial f}{\partial \mathbf{v}}\right) = \left(\frac{-e\mathscr{E}}{m}\right)\left(\frac{\partial f}{\partial \mathbf{v}}\right) \tag{6.26-10}$$

Thus, the effect of an electric field in changing the distribution depends on the derivative of the distribution function with respect to velocity. Assuming

$$\left(\frac{\partial f}{\partial t}\right)_{\text{scattering}} = -\frac{f - f_0}{\tau}, \tag{6.26-11}$$

Eqs. (6.26-10) and (6.26-11) together with Eq. (6.26-6) give

$$\frac{\partial f}{\partial t} + \frac{e\mathscr{E}}{m} \cdot \left(\frac{\partial f}{\partial \mathbf{v}}\right) + \frac{f - f_0}{\tau} = 0 \tag{6.26-12}$$

for the relation between f_0 and the distribution function perturbed by an electric field. Figure 6.26-1(b) shows a distribution function which has been dis-

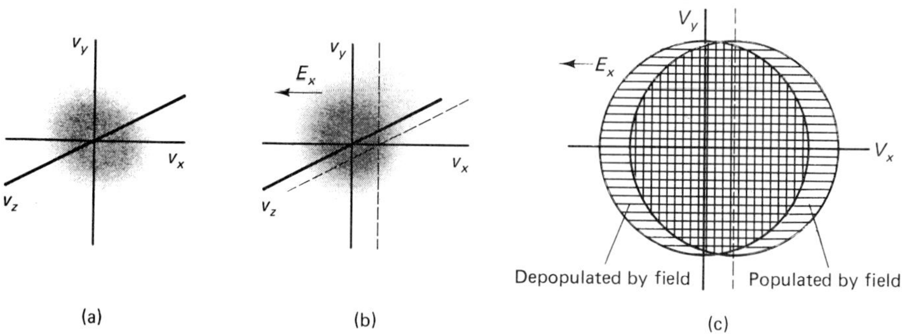

Figure 6.26-1 (a), (b) Velocity distribution change for a Boltzmann gas upon application of an electric field. (c) Displacement of Fermi sphere for a Fermi gas upon application of an electric field.

placed by an electric field in the negative x direction. The location of the center of the new distribution and its shape as compared to the original distribution [Fig. 6.26-1(a)] depend on the competition of the field and the scattering. When a constant electric field is applied for a duration long compared to τ (so that a steady state is reached), then the first term in Eq. (6.26-12) is zero, and the second and third terms are equal and opposite. A suitable integration over the displaced distribution then results in a nonzero drift velocity for all electrons and a finite electrical conductivity.

Figure 6.26-1(a) is for a diffuse Maxwellian distribution rather than a sharply defined Fermi sphere in velocity space or k space as in Fig. 6.26-1(c). In the latter case, application of an electric field totally depopulates a thin curved section of **v** space on one side of the Fermi sphere and totally populates the mirror image section on the opposite side of the occupied region. All of the states involved in this interchange have a velocity commensurate with the Fermi energy. The Boltzmann equation approach is not restricted to electrons; it can be used whenever a small systematic drift is superimposed upon a random motion of larger amplitude. Thus, it appears in the formal theory of thermal conduction by phonons and in the exact treatment of transport processes.

■ EXAMPLE 6.26-1

Equation (6.26-6) written as

$$\frac{\partial f}{\partial t} + \mathbf{v}\cdot\frac{\partial f}{\partial \mathbf{r}} + \mathbf{a}\cdot\frac{\partial f}{\partial \mathbf{v}} = -\frac{f - f_0}{\tau} \tag{6.26-13}$$

is the Boltzmann transport equation in the relaxation time approximation. In the steady state, $\partial f/\partial t = 0$. If the introduction of a relaxation time is not justified, the collision term would have to be treated in detail. In general, (6.26-13) leads to an integrodifferential equation.

To apply the general kinetic formulation to transport processes such as the electrical and thermal conductivity in an electron gas, consider a specimen with an electric field $\mathscr{E}$ in the x direction and a temperature gradient dT/dx. Assuming steady-state (dc) conditions, solve approximately the Boltzmann equation for the distribution function and find the flux of electric charge and energy.

Solution. For the steady state, $\partial f/\partial t = 0$, so the transport equation becomes, with $a = e\mathscr{E}/m$,

$$u\frac{\partial f}{\partial x} + \frac{e\mathscr{E}}{m}\frac{\partial f}{\partial u} = -\frac{f - f_0}{\tau} \tag{6.26-14}$$

where u is the x component of the velocity, and e is the charge on the electron. Equation (6.26-14) can be rewritten as

$$f = f_0 - \tau\left(\frac{e\mathscr{E}}{m}\frac{\partial f}{\partial u} + u\frac{\partial f}{\partial x}\right) \tag{6.26-15}$$

Assuming small fields and small temperature gradients so that the change in the distribution will be small and terms in f involving squares and cross products of the perturbations may be neglected, (6.26-15) becomes

$$f = f_0 - \tau\left(\frac{e\varepsilon}{m}\frac{\partial f_0}{\partial u}\right) \tag{6.26-16}$$

with $(f - f_0)/f_0 \ll 1$ in regard to x. What this means is that the calculated f should not be inconsistent with the distribution f_0 assumed immediately after a collision. For example, as a result of a collision, the velocity $\mathbf{v}$ of a molecule can change by arbitrarily large amounts in an infinitesimal time, but its position cannot change instantaneously. In other words, the new distribution $f(\mathbf{v}, \mathbf{r}, t)$ is independent of $\mathbf{r}$ and t.

The current density is the flux of charge through an element of area. It is given by

$$\begin{aligned} J_e &= e\int uf\,d\mathbf{v} = e\int u\left[f_0 - \tau\left(\frac{e\mathscr{E}}{m}\cdot\frac{\partial f_0}{\partial u}\right)\right]d\mathbf{v} \\ &= -\frac{\tau e^2\mathscr{E}}{m}\int u\left(\frac{\partial f_0}{\partial u}\right)d\mathbf{v} \end{aligned} \tag{6.26-17}$$

where $\int uf_0\,d\mathbf{v} = 0$ because f is an even function of the velocity component u. Also, in taking τ out of the integral, it has been assumed that the relaxation time is independent of the velocity. Now,

$$\frac{\partial f_0}{\partial u} = \frac{\partial f_0}{\partial \epsilon}\cdot\frac{d\epsilon}{du} = mu\frac{\partial f_0}{\partial \epsilon} \tag{6.26-18}$$

where $\epsilon = \frac{1}{2}mu^2$ is the energy. Thus, Eq. (6.26-17) becomes

$$J_e = -\tau e^2\varepsilon\int u^2\left(\frac{\partial f_0}{\partial \epsilon}\right)d\mathbf{v} \tag{6.26-19}$$

At this point, a Maxwellian distribution function is first assumed:

$$f_0 = n\left(\frac{m}{2\pi kT}\right)^{3/2} e^{-mv^2/2kT} \tag{6.26-20}$$

where v is the magnitude of the velocity: $v^2 = v_x^2 + v_y^2 + v_z^2$. For a Maxwellian distribution,

$$\frac{\partial f_0}{\partial \epsilon} = -\frac{f_0}{kT} \tag{6.26-21}$$

so that, from (6.26-19),

$$J_e = \frac{\tau e^2 \varepsilon}{kT}\int u^2 f_0\, d\mathbf{v} \tag{6.26-22}$$

Noting that $\int u^2 f_0\, d\mathbf{v}$ is the average $\langle v^2\rangle$ and that the kinetic energy in the x-direction of motion is $\frac{1}{2}m\int u^2 f_0\, d\mathbf{v} = \frac{1}{2}nkT$, Eq. (6.26-22) yields

$$J_e = \frac{ne^2\tau\varepsilon}{m} \tag{6.26-23}$$

and since the electrical conductivity is $\sigma = J_e/\varepsilon$, this gives

$$\sigma = \frac{ne^2\tau}{m} \tag{6.26-24}$$

The preceding is for ions (or electrons) in a gas where these ions are scattered by collisions with the neutral gas molecules. For the case of electrons in a metal, the Fermi-Dirac distribution function

$$f_0(\epsilon) = \frac{1}{e^{(\epsilon-\epsilon_F)/kT} + 1} \tag{6.26-25}$$

is as shown in Fig. 6.26-2, along with $\partial f_0/\partial\epsilon$. This derivative is different from zero only in the region the Fermi energy. Mathematically,

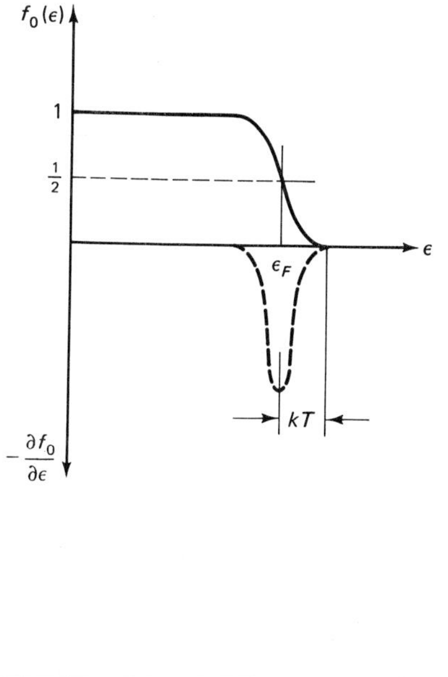

Figure 6.26-2 FD distribution function and its derivative with respect to energy.

$$\frac{\partial}{\partial \epsilon} \frac{1}{e^{(\epsilon - \epsilon_F)/kT} + 1} = \delta(\epsilon - \epsilon_F) \tag{6.26-26}$$

where $\delta(x)$ is the Dirac delta function defined by its property

$$\int_{-\infty}^{\infty} f(x)\delta(x - a)\, dx = f(a) \tag{6.26-27}$$

To proceed with (6.26-19) for the current, the simplest strategy is to look at Fig. 6.26-2 and note that only the region within kT of ϵ_F makes any contribution, meaning

$$\int_0^{\infty} \frac{\partial f_0}{\partial \epsilon}\, d\epsilon = 0 - 1 = -1 \tag{6.26-28}$$

Therefore only the electrons with energy close to the Fermi energy are active in the conductivity of the metal. Equation (6.26-19) contains the term $\int u^2(\partial f_0/\partial \epsilon)\, d\mathbf{v}$, which is really

$$\int_{-\infty}^{\infty} v_x^2 \frac{\partial f_0}{\partial \epsilon}\, dv_x = \int_{-\infty}^{\infty} \frac{v_x}{m} \frac{\partial f_0}{\partial v_x}\, dv_x = \frac{1}{m}\,[f_0 v_x]_{-\infty}^{\infty} - \frac{1}{m}\int_{-\infty}^{\infty} f_0\, dv_x$$

where integration has been performed by parts. The first term on the right vanishes since $f_0 = 0$ for $v_x = \pm\infty$, and the second term is $-n/m$ since the integral is simply equal to n. Thus, (6.29-19) gives for a Fermi gas,

$$J_e = -\tau_F e^2 \varepsilon\left(-\frac{n}{m}\right) = \frac{ne^2 \tau_F \varepsilon}{m} \tag{6.26-29}$$

and the conductivity is

$$\sigma = \frac{J_e}{\mathscr{E}} = \frac{ne^2 \tau_F}{m} \tag{6.26-30}$$

which is the same as that for a Boltzmann gas, except that no assumption has to be made that τ is independent of velocity; it is only the value of the relaxation time at the Fermi surface that governs. ■

6.27 DIFFUSION AS A RANDOM WALK PROBLEM

It is possible to look upon the diffusion problem as a random walk executed by a labeled molecule. This phenomenon, also called *Brownian motion,* reveals clearly the statistical fluctuations that occur in a system in thermal equilibrium. It serves as a protype problem where analysis provides considerable insight into the mechanisms that are responsible for the existence of fluctuations and dissipations of energy. Examples of this nature are the random motion of the mirror mounted on the suspension fiber of a galvanometer, or the fluctuating current existing in an electric resistor. Such fluctuations constitute the background of *noise* that imposes limitations on the accuracy of physical measurements.

In 1827 the English botanist Robert Brown observed the zigzag motion of pollen grains suspended in water. This motion of suspended particles is similar to the diffusion of molecules, except that the particles are much larger than the molecules. An insight into the process can be obtained by considering the statistical problem of *random walk*. In the one-dimensional version of this problem, a man flips a coin and takes one step to the right if the result is heads or one step to the left if the result is tails. It is of interest to determine the root-mean-square distance. Suppose the man takes N steps each of size unity. After N steps, the man is distance x_N from the origin. Since the probabilities of a right step and a left step are equal, the average distance $\bar{x}_N$ will be zero, but not the root-mean-square distance. From

$$x_{N+1} = x_N \pm 1 \tag{6.27-1}$$

$$x_{N+1}^2 = x_N^2 \pm 2x_N + 1 \tag{6.27-2}$$

the average is

$$(x_{N+1}^2)_{\text{av}} = (x_N^2)_{\text{av}} + 1 \tag{6.27-3}$$

where the fact that $(x_N)_{\text{av}}$ is zero has been taken into account. For $N = 1$, $x_1 = \pm 1$, and

$$(x_1^2)_{\text{av}} = 1$$

Equation (6.27-3) then gives

$$(x_2^2)_{\text{av}} = 1 + 1 = 2$$

and in general,

$$(x_N^2)_{\text{av}} = N \tag{6.27-4}$$

If the step size is l, the above argument gives

$$(x_N^2)_{\text{av}} = Nl^2 \tag{6.27-5}$$

and

$$x_{\text{RMS}} = N^{1/2}\, l \tag{6.27-6}$$

This result can be applied to error theory. The probable resulting error due to the combination of a large number N of small random errors of size l is given by Eq. (6.27-6). It can be applied to the problem of self-diffusion by taking N to be the number of collisions made and l to be the mean free path. If the mean speed of the molecules is $\bar{v}$, the number of collisions made in time t is $N = \bar{v}t/l$; so

$$(x^2)_{\text{av}} = \bar{v}lt \tag{6.27-7}$$

Thus, the mean square distance is proportional to the time. The first complete theory of Brownian motion was given by Einstein in 1905. In 1908, Jean Perrin made exhaustive quantitative observations of the paths of many suspended particles of different sizes. From these observations, which were in good agreement with Einstein's theory, Perrin calculated Avogadro's number, using the following derivation due to Langevin (1908).

Consider a particle moving under the influence of a fluctuating external force F_0 and a viscous drag force proportional to the velocity. Newton's law of motion is then

$$F_0 - b\frac{dx}{dt} = m\frac{d^2x}{dt^2} \tag{6.27-8}$$

where b is related to the viscosity. For a sphere of radius a in a fluid which has a coefficient of viscosity η, the viscous force is given by Stokes' law as $6\pi\eta a(dx/dt)$. Equation (6.27-8) can be solved by noting that

$$\frac{d(x^2)}{dt} = 2x\frac{dx}{dt} \tag{6.27-9}$$

$$\frac{d^2}{dt^2}x^2 = 2x\frac{d^2x}{dt^2} + 2\left(\frac{dx}{dt}\right)^2 \tag{6.27-10}$$

Multiplying (6.27-8) by x gives

$$xF_0 - bx\frac{dx}{dt} = mx\frac{d^2x}{dt^2}$$

and, using (6.27-9) and (6.27-10),

$$xF_0 - \frac{1}{2}b\frac{d}{dt}x^2 = \frac{1}{2}m\frac{d^2}{dt^2}x^2 - m\left(\frac{dx}{dt}\right)^2 \tag{6.27-11}$$

(6.27-11) is now averaged over a large number of particles. Since the position x is unrelated to the random force F_0, the average of xF_0 will be $m(dx/dt)^2 = kT$. Equation (6.27-11) thus becomes

$$\frac{1}{2}m\frac{d^2}{dt^2}(x^2)_{\text{av}} + \frac{1}{2}b\frac{d}{dt}(x^2)_{\text{av}} = kT \tag{6.27-12}$$

where the time differentiation and the averaging over particles have been interchanged since they are independent of each other. Letting $f = (d/dt)(x^2)_{\text{av}}$, Eq. (6.27-12) becomes

$$\frac{1}{2}m\frac{df}{dt} + \frac{1}{2}bf = kT \tag{6.27-13}$$

The solution of this equation is

$$f = \frac{2kT}{b} + Ce^{-bt/m} \tag{6.27-14}$$

as can be verified by direct substitution.

For the very small masses that must be used in order to observe Brownian motion, the quantity bt/m is very large except for a very short time, so the second term in (6.27-14) can be neglected. This gives

$$f = \frac{d}{dt}(x^2)_{av} = \frac{2kT}{b}$$

or

$$(x^2)_{av} = \frac{2kT}{b}t \tag{6.27-15}$$

where $(x^2)_{av}$ was chosen to be zero for $t = 0$. Using Stokes' law for $b = 6\pi\eta a$ then gives

$$(x^2)_{av} = \frac{kT}{3\pi\eta a}t \tag{6.27-16a}$$

or

$$(x^2)_{av} = \frac{RT}{3\pi\eta a N_A}t \tag{6.27-16b}$$

Equation (6.27-16) is the same as that obtained by Einstein in 1905. He pointed out that it could be used to obtain Avogadro's number. From Eq. (6.27-8), it is seen that for a macroscopic particle of radius a moving with velocity v through a liquid of viscosity η, Stokes' law yields a frictional force $-bv$, where $b = 6\pi\eta a$. Beyond a time t greater than b/m, the particle behaves like a diffusing particle executing a random walk, with $\langle v^2\rangle$ being proportional to t.

The connection with the diffusion process can be established by writing the partial differential equation

$$\frac{\partial n}{\partial t} = D\frac{\partial^2 n}{\partial x^2} \tag{6.27-17}$$

where D is the coefficient of self-diffusion. Suppose that at time $t = 0$ in an infinite medium,

$$n(x,0) = N\delta(x) \tag{6.27-18}$$

where $\delta(x)$ is the Dirac delta function. There are N particles concentrated at $x = 0$ where $t = 0$. Solving (6.27-17) gives

$$n(x,t) = N(4\pi Dt)^{-1/2}e^{-x^2/4Dt} \tag{6.27-19}$$

which easily establishes that

$$\langle x^2\rangle = 2Dt \tag{6.27-20}$$

This result is of the same form as Eq. (6.27-15), showing that the diffusion coefficient is

$$D = \frac{kT}{b} = \frac{kT}{6\pi\eta a} \tag{6.27-21}$$

Thus, calculations of the mean-square displacement $\langle x^2(t)\rangle$ either by macroscopic reasoning based on the diffusion equation, or by random walk evaluation, yield the same result.

■ EXAMPLE 6.27-1

What is the root-mean-square speed of a grain of gamboge having a radius of 2×10^{-7} m and a mass of 3×10^{-17} kg (some 10^9 times the mass of a water molecule), and how did Perrin perform his experiment for the determination of Avogadro's number?

Solution. Consider each grain as being a body of mass M and to move at a mean-square speed $\overline{U^2}$ such that

$$\frac{1}{2} M\overline{U^2} = \frac{1}{2} m\overline{u^2} = \frac{3}{2} kT \tag{6.27-22}$$

where m is the molecular mass of the medium in which the larger particle is suspended. In the present case, a grain of the resin gamboge with a radius of 2×10^{-7} m has a mass of 3×10^{-17} kg and hence a root-mean-square speed $(\overline{U^2})^{1/2} = (3kT/M)^{1/2} = 2.0 \times 10^{-2}$ m/s at 300 K. The combination of such slow speed with such large relative size makes it possible to observe the motion of a suspended particle with a microscope.

In 1908 Jean Perrin made a series of measurements of Avogadro's number. In order to use Eq. (6.27-16) he needed a large number of small but visible particles of equal radius a. He found that he could make emulsions of gamboge (prepared from a dried vegetable latex) and mastic which, after several months of separation by centrifuging, contained grains of nearly equal size. In one series of measurements, he watched individual particles as they moved about and recorded their positions at equal time intervals. Equation (6.27-16) deals with the component Δx of the displacement along an arbitrarily chosen x direction. It says that a "snapshot" of the suspension is to be taken at time $t = 0$ and again at time $t = \tau$, and then $\overline{\Delta x^2}$ deduced (by summing all the Δx^2 and dividing by the number of particles). With only three particles observed in Fig. (6.27-1), it may not seem possible to reasonably deduce $\overline{\Delta x^2}$, and

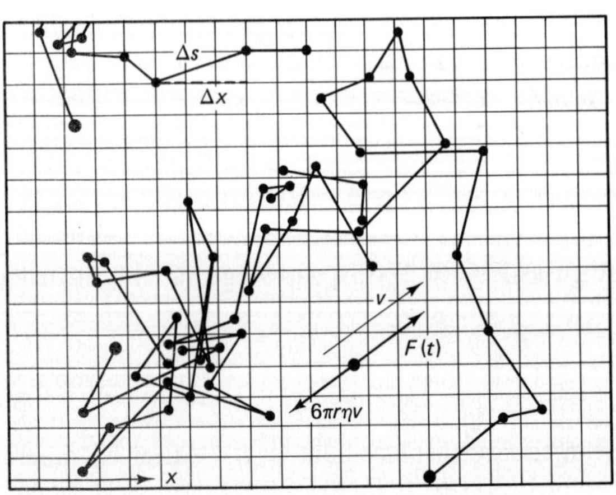

Figure 6.27-1 Brownian motion. (From J. Perrin, (1909). *Ann. Chim. Physique* 18, 1–114.)

hence check Eq. (6.27-16). However, it can be assumed that one particle observed for N successive intervals of time τ (where N is a large number) will have the same value of $\overline{\Delta x^2}$ as all N particles observed during a single interval τ. The assumption that the time-averaged value of some property of a single particle has the same value as that found by averaging the property over all the particles at a single time is essentially the ergodic hypothesis that was discussed in Chapter 1. It is frequently employed in physics and applies to particles that are in thermal equilibrium with one another.

For the Brownian motion depicted in Fig. 6.27-1, the dots indicate the position of a particle at 30-s intervals. The size of the grid's divisions are micrometers (1 micrometer = 10^{-6} m). Looking at the first 20 displacements (beginninng at the right-hand end of the track) it is seen that Δx has the following values (all in units of 10^{-6} m): 2.4, 1.2, −1.6, −0.9, −4.0, −1.5, 1.7, 1.0, 0.3, 1.3, −2.9, −3.1, −0.5, 1.5, 0.7, 1.9, −0.2, 0.1, −2.7. Squaring each of these 20 values of Δx, summing the Δx^2, and dividing by 20 gives $\overline{\Delta x^2} = 3.00 \times 10^{-12}$ m^2. Perrin actually performed this experiment with (100 different) particles of known size. With $T = 293$ K, $\tau = 30$ s, $\eta = 1.0 \times 10^{-3}$ kg m^{-1} s^{-1} (for water), and $r = 4.3 \times 10^{-6}$ m, he deduced the value of $k = 1.38 \times 10^{-23}$ J K^{-1}, or, since $N_A = R/k$, the value of N_A. (Perrin received the Nobel prize in 1926.) ■

6.28 SUMMARY

The starting point for the study of condensed matter is atomic bonding. The chemical nature of each atom is determined by the number of protons and neutrons within its nucleus. The nature of atomic bonding is determined by the behavior of the electrons orbiting the nucleus. There are three kinds of strong (primary) bonds responsible for the cohesion of solids. The *ionic* bond involves electron transfer and is nondirectional. The *covalent* bond involves electron sharing and is highly directional. The *metallic* bond involves sharing of delocalized electrons and is nondirectional. The resulting electron cloud or gas gives rise to a high electrical conductivity. Energy must be supplied to break bonds; it is released when atoms are joined. Engineering materials are classified on the basis of a particular bonding type or combinations of types. *Metals* involve metallic bonding. *Ceramics* and glasses involve ionic bonding, but usually in conjunction with a strong covalent character. *Polymers* have strong covalent bonds along polymeric chains and weaker secondary bonds between adjacent chains. *Semiconductors* are predominantly covalent in nature, with some semiconducting compounds having a significant ionic character. These four categories comprise the fundamental types of engineering materials. (*Composites* are simply combinations of the first three fundamental types.)

Crystals are solids with a long-range periodicity, that is, their atomic-scale structure is regular and repeating. This regularity allows the structure to be defined in terms of a fundamental structural unit, called the unit cell. There are seven crystal systems, corresponding to the possible unit cell shapes. Based on these systems, there are 14 Bravais lattices that represent the possible arrangement of points in space. Standard methods for describing the geometry of crystalline structures include notations for lattice positions, directions, and planes. The three primary crystal structures observed for common metals are the body-centered cubic (bcc), the face-centered cubic (fcc), and the hexagonal close-packed (hcp). X-ray diffraction, using Bragg's law, is the standard experimental tool for analyzing crystal structures in both single-crystal and polycrystalline (powdered) materials.

Another way of classifying engineering materials is based on electrical conduction. The magnitude of the electrical conductivity depends on both the number of charge carriers available and the relative mobility of these carriers. Various charged species can serve as carriers, but primary interest is in the electron. In a solid, energy bands exist corresponding to discrete energy levels in isolated atoms. Metals are classified as *conductors* because of their high values of electrical conductivity. The delocalized electrons are capable of transporting charge in a valence band. Metals have a partially filled valence band. Thermal energy, even at room temperature, is sufficient to promote large number of electrons above the Fermi level into the upper half of the valence band. Any lattice irregularity (such as thermal vibrations, solute atoms, or dislocations from plastic strain) shortens the mean free path of electrons and therefore reduces the drift velocity, decreases the mobility, and lowers the conductivity. Important examples of conductors include the thermocouples and the superconductors. Ceramics, glasses, and polymers are *insulators* because their electrical conductivity is typically 20 orders of magnitude lower than that of metallic conductors. This is because there is a large energy gap (greater than 2 eV) between their filled valence bands and their conduction bands. Thermal energy is insufficient to promote significant number of electrons above the Fermi level into a conduction band. Important examples of insulators are the ferroelectrics and piezoelectrics.

A *semiconductor* has a sufficiently small energy gap (usually less than 2 eV) that a fair number of electrons can be activated from the valence band to the conduction band by heat, by light, or by other electrons. Intrinsic semiconductors have equal numbers of negative carriers (electrons) and positive carriers (holes). Extrinsic semiconductors are produced by the addition of impurities (doping). Donors provide electrons to the conduction band, while acceptors receive electrons from the valence band. Thus, the donors and acceptors produce negative and positive charge carriers, respectively. The major electrical properties needed to specify an intrinsic semiconductor are band gap, electron mobility, hole mobility, and electron density (equal to hole density) at room temperature. For extrinsic semiconductors, either the donor level (for n-type material) or the acceptor level (for p-type material) must be specified. There are two main categories of semiconducting devices. The first includes thermistors, photoconductors, pressure gages, and photomultipliers, all of which depend on the conduction variations of the semiconducting material. The second category, which involves junctions between n- and p-type semiconductors, includes light-emitting diodes (LEDs), rectifiers, Zener diodes, and transistors.

Unlike the ideal crystal (emphasis on structural order modified slightly by thermal motion of the atoms) and the ideal gas (description of the thermal motion of the atoms on the basis of random atomic positions and motions), liquids have no ideal model. A liquid is specified on the p–V and p–T diagrams as a substance being in a region where the volume change on melting is small compared with the volume change upon evaporation. This implies that the latent heat of fusion is small compared to the latent heat of vaporization, so that the interatomic binding is similar to that in a solid. The temperatures at which most elements are liquid are high enough so that the system may be discussed from a classical point of view. Liquids do not fall into any simple classification and often must be treated as general *many body* systems. For this reason they are often discussed in terms of the general laws of statistical thermodynamics. Almost all the properties of a liquid may be calculated from the hypothesis that the interatomic forces and energies in a classical liquid are dominated by the sum of *pair* interactions. That is, if three or more atoms are close together in a dense liquid, the energy of the system is the sum of the energies obtained by taking the atoms in pairs. Thus, *three-body* or higher-order forces are neglected in this theory. The *pair theory of liquids* is an attempt to derive expressions for liquid properties (equilibrium, transport, macroscopic, and microscopic) on the basis of the pair approximation. A full test of the pair theory should indicate where it is able to predict experimental data adequately and where it fails to do so. In some cases (the most important of which are the diffusion and viscosity coefficients) the pair theory has not been developed to the point where rigorous expressions may be derived. The starting point in the study of liquids is the molecular pair distribution function, which is a particular member of a general class of distribution functions. These functions describe the probability of occurrence of a particular arrangement of molecules (in thermal equilibrium) in the liquid. A two-particle distribution function is often written as a *pair distribution function* $g(r)$ normalized to unity at large r and defined by

$$g(r) = \left(\frac{V}{N}\right)^2 n^{(2)}(r) \tag{6.28-1}$$

where $g(r)$ is the probability of finding a molecule at r if there is a molecule at the origin. The general shape of $g(r)$ for a monatomic liquid shows it to be almost zero at low r (due to high energies required to force atoms to overlap). At a distance roughly equal to the atomic diameter, there is a pronounced peak in $g(r)$, which denotes a sphere of nearest neighbors. At higher values of r there are oscillations representing more distant neighbors. The oscillations decrease in amplitude with increasing r, and eventually $g(r)$ approaches the (unit) mean density of the system.

A pair potential, $u(r)$, can be calculated, in principle, from the basic properties of the atoms composing the liquid. $g(r)$ may then be calculated from $u(r)$. When this is possible, the equation of state may be obtained from first principles. A first step is to derive a relation between $n^{(2)}$, $n^{(3)}$, and $u(r)$ in the classical system. Then an approximation is made for $n^{(3)}$ to reduce the problem to one between $u(r)$ and $g(r)$ only. Conversely, the relation between

$g(r)$ and $u(r)$ may be used to obtain $u(r)$ if $g(r)$ has been measured experimentally. An equation such as

$$g(r) = \exp\left[-\frac{u(r)}{kT}\right] \tag{6.28-2}$$

gives the radial distribution function for a dilute gas. However, it does not have the form (oscillations) that describes a liquid. An approach is to try to find closed-form expressions for $g(r)$ that make it look more like one for liquids. Basically, this involves including higher-order correlations between particles. The three equations arrived at are all of the form

$$g(r) + \frac{u(r)}{kT} = W(r) \tag{6.28-3}$$

The closed-form expressions for $W(r)$ all involve $g(r)$ itself in rather complicated ways, but they do include contributions from all orders in density. The first of these, the Yvon–Born–Green (YBG) equation, has the virtue that the approximation used to arrive at it has clear physical significance. However, it turns out to be the least accurate of the three in predicting the properties of reasonably dense fluids. The hyphernetted chain (HNC) equation and the Percus–Yevick (PY) equation, are obtained by summing certain classes of cluster integrals. The approximations consist of throwing away those clusters that cannot be summed. In all three cases, the results give a way of predicting $g(r)$ given $u(r)$ and vice versa.

Kinetic theory of gases is a complete molecular theory which is capable in principle of predicting experimental results under a wide variety of conditions. The part of kinetic theory that has retained general interest is the derivation of the transport coefficients of gases. These transport coefficients describe the reaction of a gas to mild but very common departures from thermal equilibrium. The most common ones are described below.

Viscosity

If a fluid is subject to a gradient in the z direction of its body velocity u which is perpendicular to the z direction, a shearing stress τ is set up which is given by

$$\tau = \mu \frac{\partial u}{\partial z} \tag{6.28-4}$$

where μ is the coefficient of viscosity.

Heat Conduction

If a body is subject to a temperature gradient $\partial T/\partial x$, heat will flow down the gradient according to Fourier's law

$$\frac{Q}{A} = -k\frac{\partial T}{\partial x} \tag{6.28-5}$$

where Q/A is an energy current density and k is the thermal conductivity. The minus sign indicates that heat flows from a high-temperature to a low-temperature region.

Diffusion

If a density gradient $\partial n/\partial x$ exists for species particles within a body, and if the density gradient is balanced in such a way that there is no gradient of the total pressure, the species will spread out according to Fick's law

$$j_n = -D\frac{\partial n}{\partial x} \tag{6.28-6}$$

where j_n is the particle current density, and D is the coefficient of diffusion. The minus sign indicates that diffusion occurs down the concentration gradient.

Electrical Conductivity

If a system is an electrical conductor and the conserved quantity is electrical charge under the influence of an electrical potential V, the flux of charge is

$$j_e = -\sigma\frac{\partial V}{\partial x} = \sigma\mathscr{E} \tag{6.28-7}$$

where σ is the electrical conductivity, V is the voltage, and $\mathscr{E} = -\partial V/\partial x$ is the electric field.

The empirical definitions (6.28-4) to (6.28-7) apply to a wide range of cases, not just gases. In particular, they apply to liquids and liquid solutions. For the case of gases, the determination of transport coefficients can be obtained from simple reasoning. The number of molecules crossing a unit surface of an imaginary plane in unit time from the negative to the positive side is

$$j = \frac{1}{4}n\overline{v} \tag{6.28-8}$$

where $\overline{v}$ is the mean speed. There is another current j_- from the positive to the negative side of the plane. In equilibrium both currents are exactly the same. The elementary theory of transport coefficients assumes that the result (6.28-8) is not seriously modified by the nonequilibrium aspects of an actual situation and proceeds to compute the transport of quantities across the plane. The element that brings the kinetics into the simple approach is the concept of a mean free λ. For a particle moving among stationary obstacles of cross section σ and density n, the mean free path is given by

$$\lambda = \frac{1}{n\sigma} \tag{6.28-9}$$

The mean free path may also be defined as the mean speed divided by the mean collision rate. This allows the collision rate to determine gaseous properties. The mean collision rate in the case of hard spheres having a constant collision cross section and a Maxwellian distribution is $\sqrt{2}n\sigma\bar{v}$. Using this result, the mean free path becomes

$$\lambda = \frac{1}{\sqrt{2}n\sigma} = \frac{0.707}{n\sigma} \tag{6.28-10}$$

Equations (6.28-9) and (6.28-10) are dimensionally the same. The difference between them is that, since the molecules as well as their obstacles (other molecules), are in motion, the mean free path is decreased by the factor 0.707. This factor is specific to the hard sphere model. The elementary way of estimating the transport of a quantity across an imaginary plane is to assume that molecules crossing the plane will have made their last collision within a fraction $\alpha(0<\alpha<1)$ of the mean free path λ. For example, in the case of viscosity, the quantity that is transported is momentum mu. Considering an imaginary horizontal plane, the downward transport of momentum as evaluated by means of Eq. (6.28-8) is

$$j_- = \frac{1}{4} n\bar{v}(mu)_{z=\alpha\lambda} = \frac{1}{4} n\bar{v}m\left(u + \alpha\lambda\frac{\partial u}{\partial z}\right) \tag{6.28-11}$$

The upward transport of momentum is

$$j = \frac{1}{4} n\bar{v}(mu)_{z=-\alpha\lambda} = \frac{1}{4} n\bar{v}m\left(u - \alpha\lambda\frac{\partial u}{\partial z}\right) \tag{6.28-12}$$

The net current density of momentum is thus

$$\frac{1}{4} n\bar{v}m\left(u + \alpha\lambda\frac{\partial u}{\partial z}\right) - \frac{1}{4} n\bar{v}m\left(u - \alpha\lambda\frac{\partial u}{\partial z}\right) = \frac{1}{2} n\bar{v}m\alpha\lambda\frac{\partial u}{\partial z} \tag{6.28-13}$$

Comparing this with the definition (6.28-4) for viscosity, it is seen that

$$\mu = \frac{1}{2} n\bar{v}m\alpha\lambda \tag{6.28-14}$$

Substituting for the mean free path from (6.28-10) gives

$$\mu = \frac{1}{\sqrt{8}}\frac{m\bar{v}\alpha}{\sigma} \tag{6.28-15}$$

Furthermore, for a Maxwell gas with mean speed,

$$\bar{v} = \left(\frac{8}{\pi}\frac{kT}{m}\right)^{1/2} \tag{6.28-16}$$

Equation (6.28-15), with $\alpha = 1$, yields

$$\mu = \left(\frac{mkT}{\pi}\right)^{1/2}\frac{1}{\sigma} \tag{6.28-17}$$

which is an important result, for it shows that μ does not depend on pressure, but only on temperature. Similar calculations can be made for the other transport coefficients, with minor alterations. For example, in the case of heat conduction, the quantity that is being transported across an imaginary plane is the energy instead of the momentum.

A more exact determination of transport coefficients may be obtained from the Boltzmann transport equation. Its object is the Boltzmann distribution function which is a density $f(\mathbf{r}, \mathbf{v}, t)$ indicating how many molecules at time t have position and velocity in a range $d^3\mathbf{r}\, d^3\mathbf{v}$ near $\mathbf{r}$ and $\mathbf{v}$. The rate of change of this function is estimated through binary collisions by setting

$$\frac{df}{dt} = \left(\frac{\partial f}{\partial t}\right)_{\text{coll}} \tag{6.28-18}$$

where df/dt is the hydrodynamic total derivative

$$\frac{df}{dt} = \frac{\partial f}{\partial t} + \mathbf{v} \cdot \frac{\partial f}{\partial \mathbf{r}} + \mathbf{a} \cdot \frac{\partial f}{\partial \mathbf{v}} \tag{6.28-19}$$

The compact form (6.28-18) hides a complicated equation. An explicit expression for df/dt may be written in terms of integrals involving f and describing the rate at which molecules enter and leave the range $d^3\mathbf{r}\, d^3\mathbf{v}$ as a result of collisions. The Boltzmann equation (6.28-18) is then an integrodifferential equation with both partial derivatives of f and integrals over f entering into the equation for the unknown function f. This equation is the basis of the more exact treatment of transport processes.

Excessive complications can be avoided by considering suitable approximations for $(\partial f/\partial t)_{\text{coll}}$. If the effect of collisions is always to restore a local equilibrium situation described by $f^{(0)}(\mathbf{r}, \mathbf{v}, t)$, then the effect of collisions is simply to restore $f(\mathbf{r}, \mathbf{v}, t)$ to the local equilibrium value $f^{(0)}$ exponentially with a relaxation time τ that is of the order of the time between molecular collisions. Symbolically, this means

$$\left(\frac{\partial f}{\partial t}\right)_{\text{coll}} = \frac{f - f^{(0)}}{\tau} \tag{6.28-20}$$

With the substitutions (6.28-19) and (6.28-20), (6.28-18) becomes a *simplified Boltzmann* which reads

$$\frac{\partial f}{\partial t} + \mathbf{v} \cdot \frac{\partial f}{\partial \mathbf{r}} + \mathbf{a} \cdot \frac{\partial f}{\partial \mathbf{v}} = -\frac{f - f^{(0)}}{\tau} \tag{6.28-21}$$

This is simply a linear partial differential equation for f and is the Boltzmann equation with the relaxation-time assumption.

According to (6.28-20), $(\partial f/\partial t)_{\text{coll}} = 0$ if $f = f^{(0)}$; also for fixed $\mathbf{r}$ and $\mathbf{v}$, f changes in the manner of

$$f(t) = f^{(0)}(t) + [f(0) - f^{(0)}(0)] \exp(-t/\tau) \tag{6.28-22}$$

For a situation slightly removed from equilibrium, $f = f^{(0)} + f^{(1)}$, where $f^{(1)} << f^{(0)}$. The collision term on the right side of Eq. (6.28-21) then becomes just $-f^{(1)}/\tau$. The left side is also small and can be evaluated approximately by neglecting terms in $f^{(1)}$. This perturbation procedure is illustrated in the derivation of some of the transport coefficients.

SYMBOLS IN CHAPTER 6

B	Second virial coefficient
C	Third virial coefficient
D	Diffusion coefficient
E, ε	Energy
f	Distribution function
$g(r)$	Radial distribution function
J	Current density
K, k	Wave number, thermal conductivity, Boltzmann constant
M, m	Molecular weight, mass
N, n	Number of particles, particle density
p	Pressure
q	Charge
t	Time
u	Velocity component (in x direction); internal energy
V, v	Potential; volume; velocity
Γ	General molecular property
Λ, λ	de Broglie wavelength; mean free path
μ	Dynamic viscosity; mobility, Frequency, Hz
ρ	Resistivity
σ	Conductivity
τ	Relaxation time
ϕ	Work function
ψ	Wave function
ω	Circular frequency $2\pi\nu$
$\mathscr{E}$	Electric field

PROBLEMS

6.1. The molar internal energy of a crystal lattice is given as

$$U_m = -\left(\frac{Le^2 M}{4\pi\epsilon_0 R_0}\right)\left(\frac{1-R^*}{R_0}\right)$$

where R^* is the range of the repulsive interaction, R_0 is the nearest neighbor separation, and M is the Madelung constant. Calculate the internal energy of sodium chloride for which R^* is 32.1 pm and the nearest neighbor separation is 282 pm. The Madelung constant for a rock-salt lattice is 1.7475.

6.2. A virial expansion has the form

$$\frac{pV_m}{RT} = 1 + B(T)V_m + \frac{C(T)}{V_m^2} + \cdots$$

where B is the second virial coefficient, and C the third. V_m is the molar volume. The ideal gas satisfies $pV_m/RT = 1$, and so the virial coefficients must arise from the effects of the intermolecular forces in a real gas. The coefficient B may be calculated from

$$B(T) = 2\pi L\int_0^\infty \{1 - e^{-V(R)/kT}\}R^2\,dR$$

where $V(R)$ is the intermolecular energy. Van der Waals' equation for n moles of a gas is

$$p = \frac{nRT}{V \cdot nb} - \frac{an^2}{V^2}$$

where a and b are corrections for the pressure and the volume. Which part of this equation corresponds to the second virial coefficient?

6.3. Van der Waals' equation, per mole volume V_m of gas, is

$$P = \frac{RT}{V_m - b} - \frac{a}{V_m^2}$$

where a and b are coefficients that take into account the attraction and repulsion between molecules. Discuss whether this equation bears any resemblance to a true equation of state by predicting isotherms that resemble the experimental isotherms.

6.4. The reduced variables of a gas are defined as the actual variables divided by the critical variable:

reduced pressure: $p_r = p/p_c$

reduced volume: $V_r = V_m/V_{m,c}$

reduced temperature: $T_r = T/T_c$

Show (from observation of experimental data) that real gases in the same state of reduced volume and temperature exert approximately the same reduced pressure. This is called the principle of corresponding states.

6.5. What is the consequence of the long-range order and the short-range order on the radial distribution function of solids and liquids? Why is the viscosity of a liquid expected to decrease with increasing temperature?

6.6. In ionic crystals, the force of attraction between ions is the coulomb force $-e^2/4\eta\epsilon_0 r_0^2$, and the energy of a molecule is given by

$$V = -\frac{e^2}{4\pi\epsilon_0 r_0} + \frac{Ce^2}{mr_0^m}$$

Find the minimum energy V_0 and the Madelung constant for solid sodium in which each Cl^{-1} ion is surrounded by 6 Na ions at distance r, 12 Cl^- ions at $r\sqrt{2}$, 8 Na ions at $r\sqrt{3}$, and so on.

6.7. The thermal expansion for a material is defined as the fractional change in either length or volume per unit temperature rise, for example, for the linear coefficient

$$\alpha_L = \frac{1}{L}\frac{dL}{dT}$$

Explain, on the basis of an interatomic potential energy diagram, why weakly bonded materials such as the alkali metals, have a high coefficient of expansion, whereas strongly bonded materials such as glass and quartz, have a low coefficient of expansion.

6.8. The ability of a material to conduct heat is given by its thermal conductivity K which is defined as the quantity of heat flow Q, per unit area A, per second t, per unit temperature gradient dT/dx, that is,

$$dQ/dt = -KA(dT/dx)$$

The negative sign indicates that the heat flow is in the direction of decreasing temperature. At a given temperature a material will contain a "gas" of energetic free electrons if the material is metallic or semiconducting, plus a concentration of lattice phonons. Explain how, for most materials, the properties of thermal conduction are determined by the electronic and crystalline state, that is, by the mean free paths of the electrons and phonons.

6.9. Ohm's law in a solid,

$$J = \sigma\varepsilon = ne\mu\varepsilon, \qquad \sigma = ne\mu$$

shows that the electrical conductivity σ depends on two factors, the number n of carriers per unit volume and their mobility μ. What is the relative contribution of n and μ to the whole spectrum of σ values for metal conductors and semiconductors?

6.10. Two atoms, separated by a distance r, exert attractive and repulsive forces on each other, the attractive potential energy being $-\alpha/r^m$, and the repulsive potential energy being β/r^m. The net potential energy $V(r)$ is, therefore,

$$V(r) = -\frac{\alpha}{r^m} + \frac{\beta}{r^n}$$

Show that the condition for a stable bond is when n is greater than m.

6.11. Explain the electrical properties of semiconductors as determined by (a) the existence of a gap in the energy states and (b) the presence of impurities.

6.12. Show that the occupation numbers for the three statistics are in the order of

$$\left(\frac{N_i}{g_i}\right)_{\text{FD}} < \left(\frac{N_i}{g_i}\right)_{\text{MB}} < \left(\frac{N_i}{g_i}\right)_{\text{BE}}$$

and that the degeneracy parameter

$$\Lambda = \frac{(2\pi mkT)^{3/2}}{h^3} \cdot \frac{V}{N}$$

is the criterion for obeying Maxwell–Boltzmann statistics. Apply this to (a) a laboratory gas, (b) a valence–electron gas, and (c) a plasma.

6.13. Show that the pressure of a classical gas is given by

$$pV = \frac{2}{3}E$$

Compare this to the pressure of a phonon or photon gas (of zero rest mass).

6.14. Find the partition function and thermodynamic properties of a phonon gas. In particular, show that for a phonon or photon gas,

$$pV = \frac{1}{3}E$$

6.15. The partition functions and internal energy of a phonon gas are given by

$$z = \frac{qN}{x_{\max}^3}\int_0^{x_{\max}} x^2 \ln(1 - e^{-x})\, dx$$

$$E = kT^2\left(\frac{\partial z}{\partial T}\right)_V = 3NkT\left(\frac{3}{x_{\max}^3}\int_0^{x_{\max}} \frac{x^3\, dx}{e^x - 1}\right)$$

where $x = h\nu/kT$ and $x_{\max} = h\nu_{\max}/kT$. Discuss the asymptotic behavior of the thermodynamic properties of a phonon gas at high and low temperatures.

6.16. For a solid whose valence electrons are completely independent of lattice actions, the constant-volume heat capacity is

$$C_V^{\text{total}} = C_V^{\text{electron}} + C_V^{\text{lattice}}$$

Show that for temperatures very much less than the Debye temperature,

$$C_V^{\text{total}} = \frac{\pi^2}{2} kN_e\left(\frac{kT}{\epsilon_F}\right) + \frac{12\pi^4}{5} kN\left(\frac{T}{\Theta_D}\right)^3$$

where N_e is the number of electrons, ϵ_F is the Fermi energy, and Θ_D the Debye temperature $h\nu_{\text{max}}/k$. Show that

$$\frac{C_V^{\text{total}}}{T} = \gamma + AT^2$$

where $\gamma = \pi^2 k^2 N_e/2\epsilon_F$ and $A = 12\pi^4 kN/5\Theta_D^3$. What is the valence–electron contribution to the heat capacity of a solid at room temperature and below? How can this contribution be detected at very low temperatures?

6.17. What is the number of vibrational states with frequencies in the interval between ν and $\nu + d\nu$ for (a) photons in a cavity and (b) phonons in a solid?

6.18. The values of the electrical and thermal conductivities of a solid are largely determined by the number of scattering collisions of the electrons with the lattice phonons. (a) Show that the density of states for a phonon gas may be written as

$$g(\nu) = \frac{9N\nu^2}{\nu_{\text{max}}^3}$$

where ν_{max} is the upper limit of the allowed frequencies. (b) Find the number of phonons in a solid at the high-temperature and low-temperature limits.

6.19. **(a)** Show that the energy and pressure of a valence–electron gas at absolute zero temperature is

$$E_0 = \frac{3}{5} N_e \epsilon_F, \qquad p_0 = \frac{2}{5}\frac{N_e}{V}\epsilon_F$$

(b) Determine the electron energy and pressure of silver at absolute zero temperature. Silver has a Fermi energy of 9.0×10^{-19} J or 5.6 eV and a volume electron density of 5.86×10^{28} electrons/m^3.

6.20. **(a)** Give expressions for the thermodynamic properties of a valence–electron gas at temperatures above absolute zero in terms of a series expansion in the variable kT/ϵ_F. What is the order of kT/ϵ_F and the correction for temperatures above absolute zero?

6.21. At a temperature above absolute zero, the thermodynamic properties of an electron gas are in terms of a series expansion in the variable kT/ϵ_F. What is the size of this variable for silver? (Silver is monovalent.) What is the constant-volume heat capacity of an electron gas compared to that of a thermally perfect gas of structureless *MB* particles?

6.22. The average energy per particle of a boson gas is

$$\langle\epsilon\rangle = \frac{E}{N} = \frac{3}{2}kT\left[1 - \frac{1}{2^{5/2}}\cdot\frac{Nh^3}{V(2\pi mkT)^{3/2}}\right]$$

while for a fermion gas, it is

$$\langle\epsilon\rangle = \frac{E}{N} = \frac{3}{2}kT\left[1 + \frac{1}{2^{5/2}}\cdot\frac{Nh^3}{V(2\pi mkT)^{3/2}}\right]$$

Explain why a Bose gas has a lower gas pressure (lower average momentum) at the same temperature than a classical gas, while a Fermi gas has a higher gas pressure (higher average momentum).

6.23. Thermionic emission is the process whereby electrons can be removed from a metal by the application of heat. In order for an electron to escape from a metal it must acquire at least sufficient energy to take it from the Fermi level to the vacuum level (the level corresponding to field-free space outside the metal). This minimum requirement (see figure) is the work function, ϕ, and is a property of any metal. In the treatment of the interactions of particles with potential steps, it is known that electrons may be reflected at the potential discontinuity. Thus, any expression for the number of electrons which emerge from the metal must contain a term $(1-\rho)$, where ρ is a reflection coefficient. The thermionic current is then

$$J = \frac{4\pi mek^2}{h^3} T^2(1-\rho) \exp\left(-\frac{\phi}{kT}\right) = A(1-\rho)T^2 \exp\left(-\frac{\phi}{kT}\right)$$

where A is called the Richardson or Richardson–Dushman constant. It has a value of 120 A K m^{-2}. It can be seen that the current has an exponential dependence on the work function. At 1500°C a metal filament emits a current of 0.1 A m^{-2}. If the work function is 3.55 eV, what is the percentage of electrons with $E \geq \phi$ that strike the barrier but fail to escape?

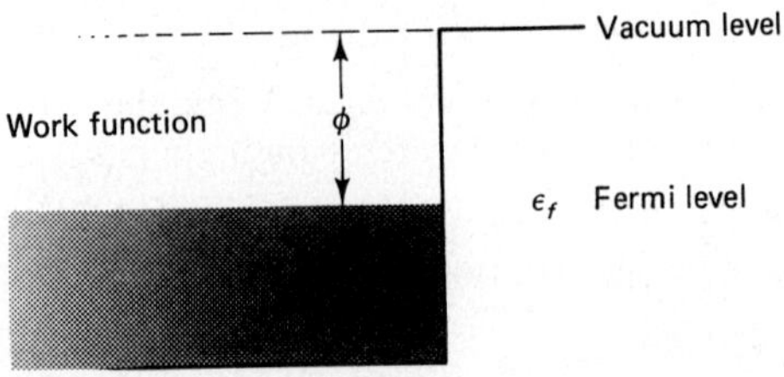

Problem 6.23 Work function for a free electron metal.

6.24. One of the reasons why the electrical properties of solids are so important can be seen by considering the electrical resistivity of various materials. The most striking feature is the enormous range of resistivity value that are covered, starting at 10^{-8} Ω m and going up to 10^{20} Ω m; indeed, no other physical parameter covers such a wide range. A successful theory must therefore explain why the resistance of a metal increases as the temperature rises, why the addition of small amounts of impurities to pure semiconductors improves their conductive properties, and why the illumination of certain crystals with light of a suitable wavelength causes an increase in conductivity. What were the salient features of the Drude theory and the points on which it broke down?

6.25. A particle suspended in a fluid is acted on by random forces. This is the basis of Brownian motion, or the random walk problem. (a) Show that the average of the square of the particle's displacement is proportional to the time interval, that is,

$$(x^2)_{\text{avg}} = \alpha t$$

where α is a proportionality constant. (b) Show that α is related to the viscosity μ of the fluid by

$$\alpha = \frac{6kT}{\mu}$$

(c) What use did J. Perrin have in mind when he measured $(x^2)_{\text{avg}}$ for a fixed time interval t and temperature T?

6.26. The mean free path of a molecule is the average distance it travels between collisions. Estimate the mean free path of a molecule in a gas at standard conditions if the molecular diameter is 4 Å.

6.27. A molecule has an average speed of 454 m/s and moves randomly a distance equal to its mean free path every time it undergoes a collision. What is its root-mean-square displacement after 10 s? (After N collisions, the average value of the square of the particle's displacement in a time interval t is proportional to t.)

6.28. The statement that at constant temperature the pressure p at a distance z above the surface of the earth is related to the pressure p_0 at sea level by

$$p = p_0 e^{-mgz/kT}$$

is known as the *law of atmospheres*. Show this result from (a) Boltzmann's distribution in statistical mechanics and (b) the ideal gas law in the kinetic theory of gases.

6.29. Kinetic theory shows that the pressure exerted on the walls of a rectangular box of volume $V = abc$, containing N molecules each of mass m, is

$$p = \frac{2}{3}\frac{N}{V}\left[\frac{1}{2}m(v^2)_{\text{avg}}\right] = \frac{2}{3}\frac{N}{V}\left[\frac{3}{2}kT\right]$$

$$= \frac{NkT}{V}$$

where the quantity $\frac{3}{2}kT$ represents the average kinetic energy of a molecule. (a) What is the average kinetic energy of a molecule at room temperature (300 K)? (b) If the molecule is O_2, what is the rms speed of the molecule?

6.30. The resistivity ρ of a material is given by

$$\rho = \frac{m_e \bar{v}}{ne^2\lambda}$$

where m_e is the mass of the electron (9.11×10^{-31} kg), $\bar{v}$ is the average velocity due to thermal excitation, λ is the electron mean free path, n is the electron density (number of electrons per unit volume), and e is the electron charge (1.602×10^{-19} C).
Determine the resistivity of silver ($A = 108$), whose density is 10.5×10^3 kg/m^3, assuming each atom contributes one electron (the valence electron) for conduction. For purposes of this estimation take λ equal to 100 times the atomic spacing d, and a velocity corresponding to the Fermi energy, $E_{F,0} = 5.5$ eV.

6.31. Calculate the resistivity of silicon ($A = 28$), which has a density of 2.42×10^3 kg/m^3, a valence of 2, and a Fermi energy of 5 eV. How does this compare with the actual resistivity at room temperature, and what is the reason for the discrepancy?

6.32. Cu, Ag, and Au have resistivities 1.8×10^{-8}, 1.6×10^{-8}, 2.4×10^{-8} Ω m and densities 8.93, 10.5, 19.35 gm/cm^3, respectively. Their Fermi energies are 7.04, 5.51, and 5.51 eV. (a) What is the maximum random velocity of electrons in Cu, Ag, and Au? (b) What is the mean free time between collisions in Cu, Ag, and Au? (c) What is the electron mean free path in the three metals, respectively?

6.33. Draw the density-of-states function $g(E)$ and the electron distribution $n(E) = g(E)F_{FD}$ for a semiconductor in the region immediately above the conduction band and below the valence band.

6.34. Show that the number n of electrons per unit volume of a crystal that are in the conduction band at a temperature T is given by

$$n_e = Ce^{-E_g/2kT}$$

where $C = 2(2\pi m_e kT)^{3/2}/h^3$ and $E_g = E_c - E_v$ is the size of the energy gap.

6.35. The electron density in the conduction band of a crystal is

$$n_e = 2\left(\frac{2\pi m_e kT}{h^2}\right)^{3/2} e^{-E_g/2kT}$$

What is the ratio of the electron densities in the conduction bands of the insulator carbon ($E_g = 5.33$ eV) and the semiconductor silicon ($E_g = 1.14$ eV) at room temperature (300 K)?

6.36. The conductivity σ has two contributions, one from the electrons and the other from the holes:

$$\sigma = n_n e \mu_n + n_p e \mu_p$$

where n_n and n_p are the densities of the negative and positive charge carriers, and μ is the mobility of the charge carriers. Find the conductivity of silicon ($E_g = 1.1$ eV) at room temperature. The mobility of electrons is $\mu_n = 0.13\ \mathrm{m^2/V\ s}$ and the mobility of holes is $\mu_p = 0.05\ \mathrm{m^2/V\ s}$.

6.37. An impurity atom that donates one extra electron to the conduction band is added to silicon in the ratio of one impurity atom to 10^{10} silicon atoms. Determine the conductivity of the doped silicon, assuming that the mobility of the donated electrons is the same as the mobility of the host electrons and that the donor density is much greater than the intrinsic density of the host electrons and holes. Silicon ($A = 28$), has a density of $2.42 \times 10^3\ \mathrm{kg/m^3}$ and a valence of 2.

6.38. Measurements on certain metals indicate that positive charge carriers (holes) seem to be responsible for conduction just as electrons are. The Hall effect furnishes this evidence. When a conductor is placed in a magnetic field perpendicular to the direction of current flow, a voltage is developed across the specimen in the direction perpendicular to both the current and the magnetic field, as shown in Fig. Prob. 6.38. The voltage is called the Hall voltage. Show that

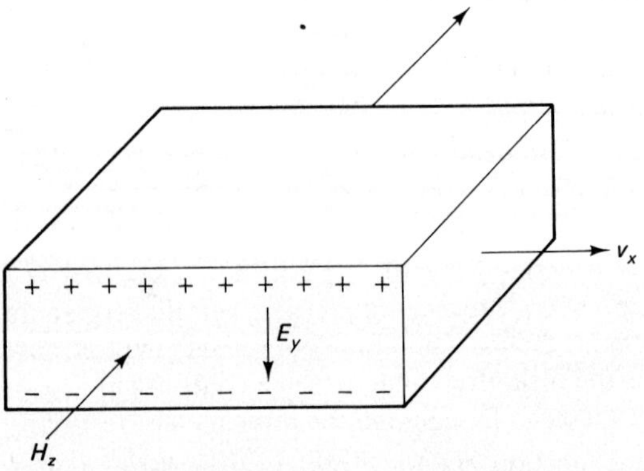

Problem 6.38 Hall effect. (Electrons flowing in the x direction in the presence of the magnetic field H_z are deflected toward the bottom, setting up an electric field that corrects the effect of the magnetic field.)

$$\mu = R_H \sigma$$

and therefore, in a given experiment, if the Hall constant R_H and the conductivity σ are measured, the mobility μ and the carrier concentration n can be found. Discuss the results of some Hall effect measurements at room temperature for various conductors.

6.39. Silver ($A = 108$), has a resistivity of $3.3 \times 10^{-8}\ \Omega$ m, a Fermi energy of 5.5 eV, and a density of 10.5×10^3 kg/m^3. Find the ratio of the mean free path λ to the interatomic spacing d for the conduction electrons if each silver atom contributes one free electron.

6.40. The separation between (111) planes in a simple cubic crystal is 2.5 Å. A particle (mass $= m_e$) is trapped in a rectangular box within the crystal. The potential box has the following dimensions: the length is defined by the separation between the (110) and (110) planes, the width is equivalent to d_{100}, and the height is equivalent to d_{001}. Calculate the value of the ground-state energy. The energy of a particle in a three-dimensional rectangular well is given by

$$\frac{h^2}{8m}\left[\frac{n_x^2}{a_x^2} + \frac{n_y^2}{a_y^2} + \frac{n_z^2}{a_z^2}\right]$$

6.41. Aluminum has a density of 2700 kg m^{-3} and an atomic weight of 26.9. (a) What is the number of atoms per cubic meter? (b) If d_{100} for this *fcc* (face-centered cubic) metal is 4 Å, calculate d_{100} and d_{111}.

6.42. A cube has several symmetry axes. An axis through the center of opposite faces is said to have fourfold symmetry. This is because a 90° rotation about this axis leaves the cube appearing exactly as it was before it was rotated. An axis through the center of diagonally opposite edges has twofold symmetry. Calculate the separation between planes which are normal to the threefold axis of symmetry in a *fcc* cubic structure whose unit cell length is 4 Å.

6.43. In X-ray crystallography, shown below, ray 2 travels a distance $2d \sin\theta$ further than ray 1. Constructive interference occurs if $2d \sin\theta = n\lambda$, where $n = 1, 2, \ldots,$ is the order of the diffraction, and λ is the wavelength. This is called the Bragg diffraction formula. Calculate the first-order diffraction angle that would be expected if a beam of X rays ($\lambda = 1$ Å) is directed toward the (100) planes of the cube in the previous problem.

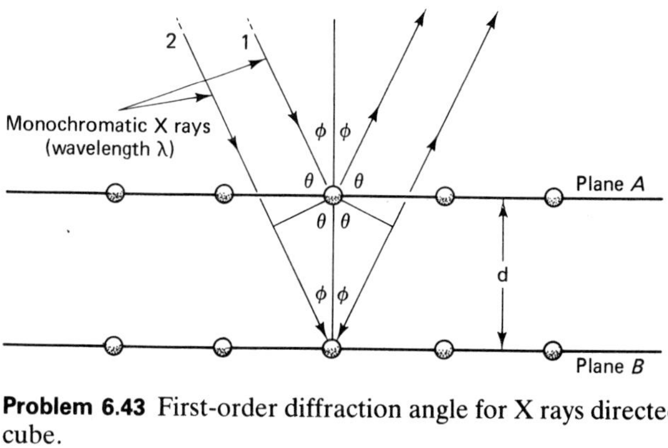

Problem 6.43 First-order diffraction angle for X rays directed at (100) planes of a cube.

6.44. Interference or diffraction experiments must be performed using apertures comparable to the de Broglie wavelength. (a) What is the de Broglie wavelength of a 0.01-kg pellet having a velocity of 10 m/s? (b) Determine the accelerating potential necessary to give an electron a de Broglie wavelength of 1 Å, which is the size of the interatomic spacing of atoms in a crystal.

6.45. An object of 2.5 Å in size is to be observed. (a) What is the minimum-energy photon that can be used? (b) If electrons are used, what is the energy required? How does the resolving power of electrons compare with that of photons?

6.46. Diamond has an energy gap E_g of approximately 5.5 eV. What is the relative probability of thermally exciting electrons into the conduction band at room temperature?

6.47. What are the respective probabilities of exciting electrons into the conduction band for insulators, conductors, and semiconductors? Why must semiconductor material used for devices be extremely pure?

6.48. Where is the Fermi level in an intrinsic semiconductor? Does it change as the temperature changes? By what factor is it displaced from the midpoint of the energy gap?

6.49. What is the position of the Fermi level in an extrinsic semiconductor? How does it vary with temperature and how is it affected by the presence of dopants?

6.50. In what manner does the position of the Fermi level affect the conductivity of a semiconductor? The position of the Fermi level in a given material can be calculated precisely if the impurity concentrations, energy gap, and temperature are known.

Appendix A

Constants and Conversion Factors

Atomic mass unit	u	$1u = 1.660540 \times 10^{-27}$ kg $= 931.494$ MeV/c^2
Avogadro constant	N_A	6.0231×10^{23} molecules/mole
Boltzmann constant	k	1.38066×10^{-23} J K^{-1} $= 8.6174 \times 10^{-5}$ eV K^{-1}
Electric field constant (permittivity of vacuum)	ε_o	$8.8541878 \times 10^{-12}$ C^2/N m^2
Coulomb constant	$\frac{1}{4}\pi\varepsilon_o$	8.9875×10^{9} N m^2 C^{-2}
Coulomb energy	ke^2	1.44 MeV fm = 14.4 eV Å
Faraday constant	F	9.648456×10^{4} C mol^{-1}
Gravitational constant	G	6.673×10^{-11} N m^2 kg^{-2} $= 6.673 \times 10^{-11}$ m^3 s^{-2} kg^{-1}
Magnetic fill constant (permeability of vacuum)	μ_o	1.256637×10^{-6} M m^{-1}
Planck constant	h	6.62608×10^{-34} J s $= 4.13567 \times 10^{-15}$ eV s
	$\hbar = h/2\pi$	1.054573×10^{-34} J s $= 6.58212 \times 10^{-16}$ eV s
	hc	1.239842×10^{-6} eV m
Speed of light (vacuum)	c	2.99792458×10^{8} m s^{-1}
Molar gas volume (ideal gas at 273.15 K and 101.325 Pa)	V_m	22.4138×10^{-3} m^3 mol^{-1}
Molar gas constant	R	8.31441 J mol^{-1} K^{-1} 0.082057 L atm mol^{-1} K^{-1}
Electron charge (quantum of charge)	e	1.602177×10^{-19} C $= 1.602177 \times 10^{-19}$ A s
Electron rest mass	m_e	9.10939×10^{-31} kg $= 0.510999$ MeV/C^2 $= 5.48580 \times 10^{-4}$ u
Electron radius (classical)	r_e	$2.8179380 \times 10^{-15}$ m
Electron charge-mass ratio	e/m_e	1.7588047×10^{11} C kg^{-1}
Electron magnetic moment	μ_e	9.284832×10^{-24} J T^{-1}
Electron compton wavelength	λ_c	$2.4263089 \times 10^{-12}$ m

Proton rest mass	m_p	1.672623×10^{-27} kg = 938.272 MeV/C^2 = 1.007276 u
Proton charge-mass ratio	e/m_p	9.578756×10^{7} C kg^{-1}
Proton magnetic moment	μ_p	1.410617×10^{-26} J T^{-1}
Proton compton wavelength	$\lambda_{c,p}$	1.321410×10^{-15} m
Neutron rest mass	m_n	1.674929×10^{-27} kg = 939.566 MeV/C^2 = 1.008665 u
Neutron compton wavelength	$\lambda_{c,n}$	1.319591×10^{-15} m
Bohr magneton	μ_B	9.27402×10^{-24} J T^{-1} = 5.788383×10^{-5} eV/T
Bohr radius	a_o	5.291770×10^{-11} m
Electron-proton mass ratio	m_e/m_p	5.446174×10^{-4}
Fine structure constant	α	7.297350×10^{-3}
Magnetic flux quantum	Φ_o	2.067850×10^{-15} Wb
Nuclear magneton	μ_N	5.05079×10^{-27} J T^{-1} = 3.152452×10^{-8} eV/T
Quantum of ciculation	h/m_e	7.273891×10^{-4} J s kg^{-1}
Rydberg constant	R_∞	1.097373×10^{7} m^{-1}
Thomson cross section	σ_e	0.665245×10^{-28} m^2
Zeeman splitting constant	μ_B/hc	4.668583×10 m^{-1} T^{-1}
Atomic specific heat constant	h/k	4.79928×10^{-11} s K
First radiation constant	c_1	3.741832×10^{-16} W m^2
Second radiation constant	c_2	1.438786×10^{-2} m K
Stefan–Boltzmann constant	σ	5.67032×10^{-8} W m^{-2} K^{-4}
Wein displacement constant	$\lambda_{max}T$	0.289779×10^{-2} m K

Conversion Factors

Length: 1 m = 10^{10} Å = 3.281 ft = 39.37 in.

Mass: 1 kg = 2.205 lbm

Force: 1 N = 0.2248 lbf

Pressure: 1 Pa = 1 N m^2 = 9.869×10^{-6} atm = 0.1450×10^{-3} lbf/in.^{2}; 1 bar = 10^5 N m^{-2}; 1 torr = 133.322 N m^{-2}

Viscosity: 1 Pa s = 10 poise

Energy: 1 J = 1 N m = 1 W s = 1 V C = 0.2389 cal = 6.242×10^{18} eV = 0.7377 ft lbf = 948×10^{-6} Btu

Temperature: °C = K − 273 = (°F − 32)/1.8; kT at room temperature = 0.0258 eV ≃ (1/40) eV

Current: 1 A = 1 C s = 1 V/Ω

Other useful conversions: 1 eV = 1.6021×10^{-19} J; 1 Å = 10^{-10} m;
1 barn (bn) = 10^{-28} m^2
1 u = 1.66054×10^{-27} kg = 931.494 MeV/C^2
1 eV/C = 5.34429×10^{-28} kg m/s
1 eV/C^2 = 1.782663×10^{-36} kg

SI Prefix	Power of Ten	Abbreviation
tera	10^{12}	T
giga	10^{9}	G
mega	10^{8}	M
kilo	10^{3}	k
centi	10^{-2}	c
milli	10^{-3}	m
micro	10^{-6}	μ
nano	10^{-9}	n
pico	10^{-12}	p

Appendix B

Periodic Table of the Elements

The atom is the basic unit of internal structure for the study of materials. It is sufficient to use a relatively simple planetary model of atomic structure, that is, *electrons* orbiting about a *nucleus,* as shown in Fig. B-1. It is not necessary to consider the detailed structure of the nucleus, aside from the number of *protons* and *neutrons* that serve as the basis of the chemical identification of a given atom.

Each proton or neutron has a mass of approximately 1.66×10^{-24} g. This value is referred to as an *atomic mass unit* (amu). It is convenient to express the mass of elemental materials in these units. For instance, the most common isotope of carbon is ^{12}C, which contains in its nucleus six protons and six neutrons for an *atomic mass* of 12 amu. (The amu is formally defined as one twelfth of the atomic mass of carbon − 12.) It is useful to note that there are 6.023×10^{23} amu per gram. This large value (known as *Avogadro's number*) represents the number of protons or neutrons necessary to produce a mass of 1 g. Avogadro's number of ^{12}C atoms would have a mass of 12.00 g.

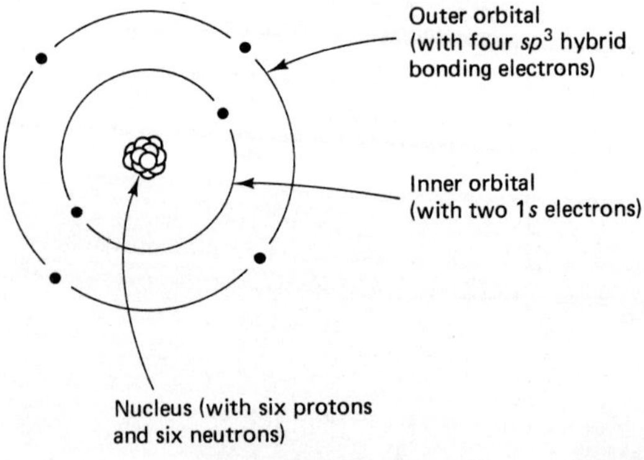

Figure B-1 Planetary model of ^{12}C atom.

METALS

NONMETALS

I A	II A	III B	IV B	V B	VI B	VII B	VIII	VIII	VIII	I B	II B	III A	IV A	V A	VI A	VII A	0
1 H 1.0079 $1s^1$																	2 He 4.002 $1s^2$
3 Li 6.939 $2s^2$	4 Be 9.012 $2s^2$											5 B 10.811 $2s^22p^1$	6 C 12.011 $2s^22p^2$	7 N 14.007 $2s^22p^3$	8 O 15.999 $2s^22p^4$	9 F 18.998 $2s^22p^5$	10 Ne 20.183 $2s^22p^6$
11 Na 22.990 $3s^1$	12 Mg 24.312 $3s^2$											13 Al 26.98 $3s^23p^1$	14 Si 28.086 $3s^23p^2$	15 P 30.97 $3s^23p^3$	16 S 32.064 $3s^23p^4$	17 Cl 35.453 $3s^23p^5$	18 Ar 39.95 $3s^23p^6$
19 K 39.102 $4s^1$	20 Ca 40.08 $4s^2$	21 Sc 44.96 $3d4s^2$	22 Ti 47.90 $3d^24s^2$	23 V 50.94 $3d^34s^2$	24 Cr 52.00 $3d^54s$	25 Mn 54.94 $3d^54s^2$	26 Fe 55.85 $3d^64s^2$	27 Co 58.93 $3d^74s^2$	28 Ni 58.71 $3d^84s^2$	29 Cu 63.54 $3d^{10}4s$	30 Zn 65.37 $3d^{10}4s^2$	31 Ga 69.72 $4s^24p$	32 Ge 72.59 $4s^24p^2$	33 As 74.92 $4s^24p^3$	34 Se 78.96 $4s^24p^4$	35 Br 79.91 $4s^24p^5$	36 Kr 83.80 $4s^24p^6$
37 Rb 85.47 $4p^65s$	38 Sr 87.62 $4p^65s^2$	39 Y 88.91 $4d5s^2$	40 Zr 91.22 $4d^25s^2$	41 Nb 92.91 $4d^45s$	42 Mo 95.94 $4d^55s$	43 Tc 99 $4d^55s^2$	44 Ru 101.07 $4d^75s$	45 Rh 102.91 $4d^85s$	46 Pd 106.4 $4d^{10}$	47 Ag 107.87 $4d^{10}5s$	48 Cd 112.40 $4d^{10}5s^2$	49 In 114.82 $5s^25p$	50 Sn 118.69 $5s^25p^2$	51 Sb 121.75 $5s^25p^3$	52 Te 127.60 $5s^25p^4$	53 I 126.90 $5s^25p^5$	54 Xe 131.30 $5s^25p^6$
55 Cs 132.91 $5p^66s$	56 Ba 137.34 $5p^66s^2$	57 La 138.91 $5d6s^2$	72 Hf 178.49 $5d^26s^2$	73 Ta 180.95 $5d^36s^2$	74 W 183.85 $5d^46s^2$	75 Re 186.2 $5d^56s^2$	76 Os 190.2 $5d^66s^2$	77 Ir 192.2 $5d^76s^2$	78 Pt 195.1 $5d^96s$	79 Au 196.97 $5d^{10}6s$	80 Hg 200.59 $5d^{10}6s^2$	81 Tl 204.37 $6s^26p$	82 Pb 207.19 $6s^26p^2$	83 Bi 208.98 $6s^26p^3$	84 Po 210 $6s^26p^4$	85 At 210 $6s^26p^5$	86 Rn 222 $6s^26p^6$
87 Fr 223 $6p^67s$	88 Ra 226 $7s^2$	89 Rc 227 $6d7s^2$															

58 Ce 140.12 $4f5d6s^2$	59 Pr 140.91 $4f^36s^2$	60 Nd 144.24 $4f^46s^2$	61 Pm 145 $4f^56s^2$	62 Sm 150.35 $4f^66s^2$	63 Eu 151.96 $4f^76s^2$	64 Gd 157.25 $4f^75d6s^2$	65 Tb 158.92 $4f^96s^2$	66 Dy 162.50 $4f^{10}6s^2$	67 Ho 164.93 $4f^{11}6s^2$	68 Er 167.26 $4f^{12}6s^2$	69 Tm 168.93 $4f^{13}6s^2$	70 Yb 173.04 $4f^{14}6s^2$	71 Lu 174.97 $4f^{14}5d6s^2$
90 Th 232.04 $6d^27s^2$	91 Pa 231 $5f^26d7s^2$	92 U 238.03 $5f^36d7s^2$	93 Np 237 $5f^46d7s^2$	94 Pu 242 $5f^67s^2$	95 Am 243 $5f^77s^2$	96 Cm 247 $5f^76d7s^2$	97 Bk 249 $5f^86d7s^2$	98 Cf 251 $5f^{10}7s^2$	99 Es 254 $5f^{11}7s^2$	100 Fm 253 $5f^{12}7s^2$	101 Md 256 $5f^{13}7s^2$	102 No 253 $5f^{14}7s^2$	103 Lw 257 $5f^{14}6d7s^2$

Naturally occurring carbon actually has an atomic mass of 12.011 amu because of the presence of 1.1% of the isotope ^{13}C, in which the nucleus contains six protons and seven neutrons. (The number of six protons still identifies the element as carbon.)

In general, the number of protons in the nucleus, and, hence, the number of electrons associated with each neutral atom, is known as the *atomic number* of the element (six for carbon). Different numbers of neutrons (six or seven) identify different *isotopes*. The periodicity of chemical elements is based on this system of elemental numbers and atomic masses arranged in chemically similar *groups* (vertical columns). While chemical identification is done relative to the nucleus, atomic bonding involves the *electron orbitals*. The electron, with a mass of 0.911×10^{-27} g, makes a negligible contribution to the atomic mass of an element. However, this particle has a negative charge of 0.16×10^{-18} C, equal in magnitude to the $+0.16 \times 10^{-18}$ C charge of each proton. (The neutron is electrically neutral.) The periodic table arranges the atoms by sequentially higher atomic numbers, such that the vertical columns (groups) possess atoms of similar chemical and electronic characteristics. The electrons, particularly the outermost ones, establish the nature of the interatomic bonding and determine the electrical conductivity and optical characteristics of the material.

Appendix C

Fourier Analysis

Quantum theory of matter states that a moving particle has an associated wave nature that can be described as a *wave group* or a *wave packet* as shown in Fig. C-1. A simple wave group may be mathematically described as a superposition of a very large number of sinusoidal waves each differing slightly in wavelength. The group velocity for a wave packet associated with a moving particle is identical to the particle's velocity through space and is the velocity with which energy is transported by the wave. The special theory of relativity predicts that the magnitude of the group velocity can never exceed the speed of light c.

Consider the superposition of two waves of the same amplitude and in phase but differing in angular frequency by an amount of $\Delta\omega$ and in propagation constant by Δk. The equations for the two waves may be taken as

$$\Delta y_1 = A \sin(\omega t - kx) \tag{C-1}$$

$$\Delta y_2 = A \sin[(\omega + \Delta\omega)t - (k + \Delta k)x] \tag{C-2}$$

The resultant displacement is

$$\Delta y = A \sin(\omega t - kx) + A \sin[(\omega + \Delta\omega)t - (k + \Delta k)x] \tag{C-3}$$

Using the trigonometric identity

$$\sin a + \sin b = 2 \cos \frac{b - a}{2} \sin \frac{b + a}{2} \tag{C-4}$$

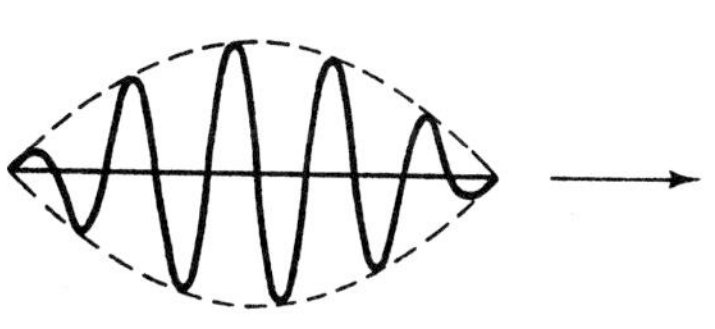

Figure C-1 Quantum theory of a moving particle.

(C-3) becomes

$$\Delta y = 2A \cos \frac{\omega t + \Delta\omega t - kx - \Delta kx - \omega t + kx}{2} \\ x \sin \frac{\omega t + \Delta\omega t - kx - \Delta kx + \omega t - kx}{2} \tag{C-5}$$

Assuming that

$$\Delta\omega t - \Delta kx \ll 2(\omega t - kx) \tag{C-6}$$

Eq. (C-5) gives

$$\Delta y = 2A \cos\left(\frac{\Delta\omega}{2} t - \frac{\Delta k}{2} x\right) \sin(\omega t - kx) \tag{C-7}$$

Equation (C-7) describes a sinusoidal wave of angular frequency ω, propagation constant k, with time and space dependent amplitude

$$2A \cos\left(\frac{\Delta\omega}{2} t - \frac{\Delta k}{2} x\right) \tag{C-8}$$

The successive amplitude maxima are known as *wave groups* and are pictured in Fig. C-2 at one instant of time. Since a wave travels through space with the phase velocity

$$v = \frac{\omega}{k} \tag{C-9}$$

the successive wave groups move through space, by analogy, with the group velocity

$$v_g = \frac{\Delta\omega/2}{\Delta k/2} = \frac{\Delta\omega}{\Delta k} \tag{C-10a}$$

and, in general, for waves differing infinitesimally in frequency and propagation constant,

$$v_g = \frac{d\omega}{dk} \tag{C-10b}$$

Periodic or single pulses of any shape other than sinusoidal may be represented by a series of sine and cosine waves of varying amplitude whose frequencies are simple ratios of a fundamental frequency. That is, any given function $f(x)$ that is repetitive between $-\pi$ and $+\pi$ can be expressed as

$$f(x) = \frac{a_0}{2} + \sum_{0}^{\infty} (a_n \cos nx + b_n \sin nx) \tag{C-11}$$

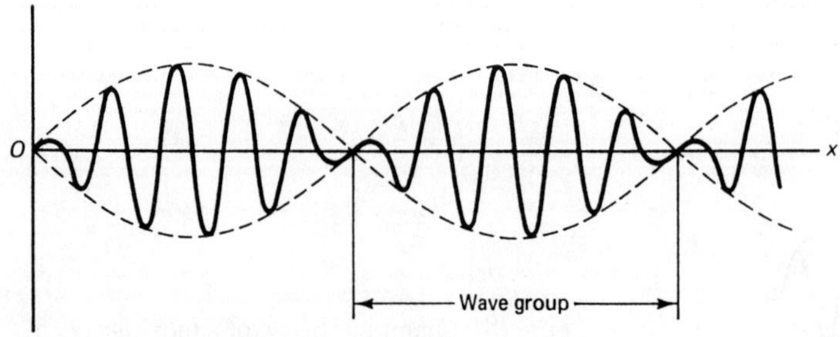

Figure C-2 A series of wave groups.

whcre

$$a_0 = \frac{1}{\pi}\int_{-\pi}^{\pi} f(x)\,dx \tag{C-12}$$

$$a_n = \frac{1}{\pi}\int_{-\pi}^{\pi} f(x)\,dx \tag{C-13}$$

$$b_n = \frac{1}{\pi}\int_{-\pi}^{\pi} f(x)\sin nx\,dx \tag{C-14}$$

For example, a sawtooth waveform where $f(x) = bx$ in the range $-\pi$ to $+\pi$ as in Fig. C-3 can be written

$$f(x) = 2b\left[\sin x - \frac{\sin 2x}{2} + \frac{\sin 3x}{3} - \cdots\right] \tag{C-15}$$

The Fourier components can be displayed as an amplitude-frequency spectrum in Fig. C-4.

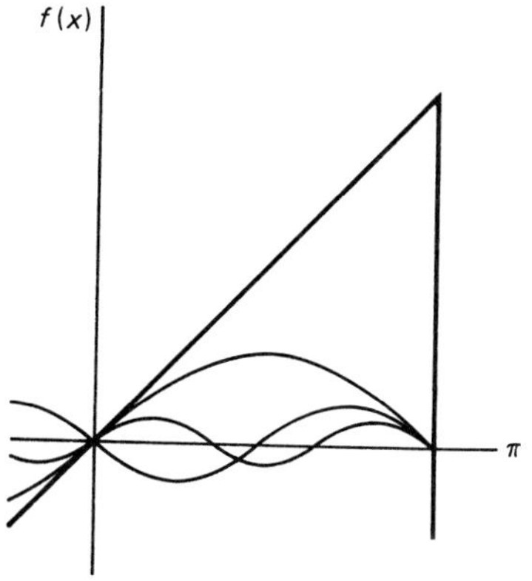

Figure C-3 Saw-tooth waveform with some of its sine components.

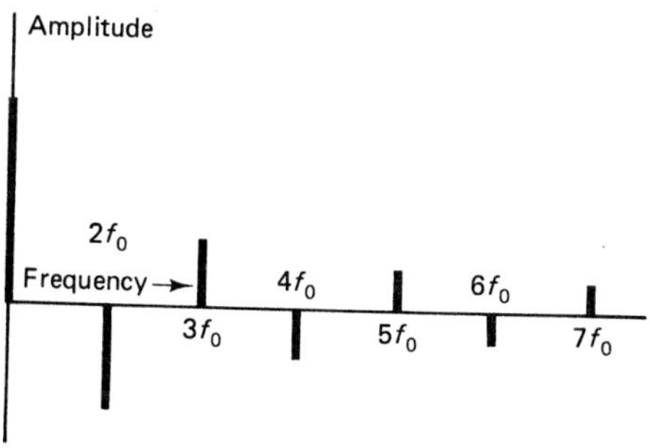

Figure C-4 Amplitude-frequency spectrum of the saw-tooth waveform.

The situation becomes more complex for localized waveforms. In this case, all components between $-\infty$ and $+\infty$ need to be specified. This requires the use of Fourier's integral theorem

$$f(x) = \frac{1}{2\pi}\int_{-\infty}^{\infty} f(k)\exp(ikx)\,dk \tag{C-16}$$

where k is the wave number ($= 2\pi/\lambda$), so that $\exp(ikx)$ represents the motion of a wave in space. As a consequence of Fourier's integral theorem, any spatial distribution of waves, $f(x)$, has a corresponding distribution of frequencies, $f(k)$, which can be represented in an amplitude-frequency spectrum. The distribution $f(k)$ in frequency space, can be calculated from the spatial distribution $f(x)$ using the inverse Fourier integral theorem.

$$f(k) = \frac{1}{2\pi}\int_{-\infty}^{\infty} f(x)\exp(-ikx)\,dx \tag{C-17}$$

The functions $f(x)$ and $f(k)$ in (C-16) and (C-17) are said to be *Fourier transforms* of each other.

To demonstrate these concepts, Fourier's inverse integral theorem will be used to calculate the amplitude-frequency spectrum corresponding to the Gaussian distribution shown in Fig. C-5. The Gaussian or normal distribution is well known in statistics as a means of expressing the spread of a continuous or near-continuous function, characterized by the standard deviation, σ, about the mean.

The expression of the Gaussian distribution, which could well be taken as the approximate representation of a localized distribution of matter (such as an electron particle) is

$$f(x) = A\exp(-x^2/2\sigma^2) \tag{C-18}$$

Its Fourier transform is

$$f(k) = \int_{-\infty}^{\infty} A'\exp(-x^2/2\sigma^2)\exp(-ikx)\,dx \tag{C-19}$$

$$= A'\int_{-\infty}^{\infty}\exp\left[-\left(\frac{x^2}{2\sigma^2} + ikx\right)\right]dx \tag{C-20}$$

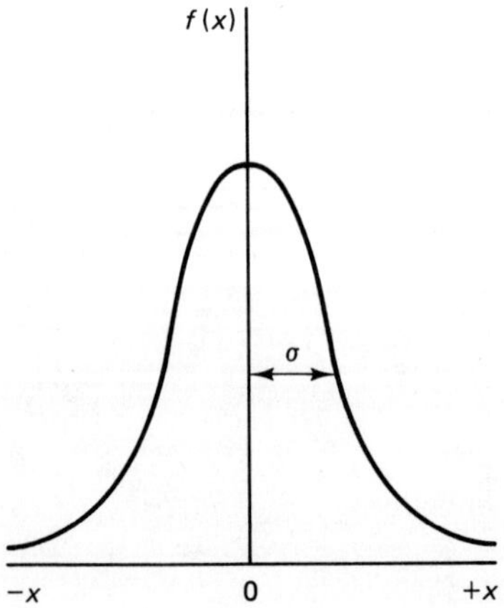

Figure C-5 Gaussian or normal distribution.

where the constant A' absorbs amplitude, $1/2\pi$, etc. To proceed with the integration, consider the expression

$$f(k) = A' \exp\left(-\frac{\sigma^2 k^2}{2}\right) \int_{-\infty}^{\infty} \exp\left[-\left(\frac{x^2}{2\sigma^2} + ikx - \frac{\sigma^2 k^2}{2}\right)\right] dx \tag{C-21}$$

$$= A' \exp\left(-\frac{\sigma^2 k^2}{2}\right) \int_{-\infty}^{\infty} \exp\left[-\left(\frac{x}{\sigma\sqrt{2}} + \frac{i\sigma k}{\sqrt{2}}\right)^2\right] dx \tag{C-22}$$

Letting

$$M = \left(\frac{x}{\sigma\sqrt{2}} + \frac{i\sigma k}{\sqrt{2}}\right); \qquad dM = \frac{dz}{\sigma\sqrt{2}} \tag{C-23}$$

Eq. (C-20) becomes

$$f(k) = A'' \exp\left(-\frac{\sigma^2 k^2}{2}\right) \int_{-\infty}^{\infty} \exp(-M^2)\, dM \tag{C-24}$$

where

$$A'' = \sigma\sqrt{2}\, A' \tag{C-25}$$

$$\int_{-\infty}^{\infty} \exp(-M^2) dM = \sqrt{\pi} \tag{C-26}$$

Thus,

$$f(k) = A''' \exp\left(-\frac{\sigma^2 k^2}{2}\right) \tag{C-27}$$

where $A''' = A''\sqrt{\pi}$. It can be seen from Fig. C-6 that the function (C-27) is similar in all respects to the original normal distribution (C-18) except that its standard deviation is $1/\sigma$.

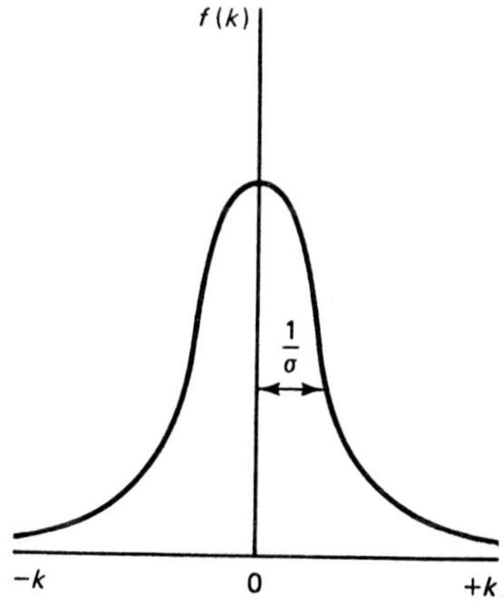

Figure C-6 Distribution of component frequencies of a Gaussian distribution in **k** space.

Uncertainty Principle

Heisenberg's uncertainty principle is an important consideration in any discipline that involves extremely small dimensions. In order that the wave nature of material particles be observable, it is necessary that the associated de Broglie wavelength be of the same order as the dimensions of the particle. For example, an electron traveling at a speed of 5×10^6 m s^{-1} has a wavelength $\lambda = h/mv = 1.5 \times 10^{-10}$ m, which is of the order of magnitude of atomic dimensions. In contrast, a macroscopic particle such as a golf ball of mass 0.10 kg and traveling with a speed of 50 m s^{-1} has a wavelength $\lambda = h/mv = 1.3 \times 10^{-34}$ m, which is negligible compared to its dimensions. It would not be expected to observe any wave aspects in the motion of a golf ball. This is an example of the *correspondence principle,* which asserts that the classical and quantum descriptions of nature must provide identical predictions in the limit of the macroscopic world of daily experience. If the amplitude of a de Broglie wave at a specific time at a particular position in space is interpretated as measure of the probability that the associated particle be found at that position, the associated wave corresponds to a wave packet or wave group shown in Fig. C-1. The envelope of the wave packet is proportional to the probability of finding the particle at position x. The wave group travels with the velocity of the particle; the shape of the wave group will in general be a function of time. A wave packet may be mathematically decomposed into a series of sinusoidal waves, each with constant amplitude and with a slightly different frequency. A fundamental property of this mathematical process is that the shorter the wave group is, the greater the frequency spread of the component sine waves that are required to reproduce the wave group.

The consequence of regarding a moving particle as a de Broglie wave group is that it places a fundamental limit to the accuracy with which the position of the particle can be specified. The shorter the wave packet associated with a particle, the more accurately its position in space can be defined. The assumption of a moving particle as a wave group also imposes a fundamental limit on the accuracy with which the momentum of the particle can be specified. From the de Broglie relation it can be seen that the momentum of a particle is proportional to the de Broglie frequency, so that if the frequency is not well-defined, neither is the momentum. Therefore, the longer the wave packet associated with a particle, the smaller is the range of frequencies required to describe it, and the better defined is the particle momentum. In other words, a long wave packet describes a particle whose momentum is well-defined but whose position is ill-defined; a short wave packet describes a particle whose momentum is ill-defined but whose position is well-defined. Mathematically, the wave number

$$k = \frac{2\pi}{\lambda} \tag{C-28}$$

may be expressed in terms of a momentum P through de Broglie's formula

$$P = \frac{h}{\lambda} \tag{C-29}$$

to give

$$k = \frac{2\pi P}{h} = \frac{P}{\hbar} \tag{C-30}$$

where $\hbar = h/2\pi$. From the example of the normal distribution, the standard deviation σ in $f(x)$ is the uncertainty Δx, whereas the standard deviation of the component wave distribution, $f(k)$, is $1/\sigma = \Delta k$. Thus,

$$\Delta k = \frac{\Delta P}{\hbar} = \frac{1}{\Delta x} \tag{C-31}$$

or

$$\Delta P = \Delta x = \hbar = \frac{h}{2\pi} \tag{C-32}$$

which is one form of the *uncertainty principle* put forward by Heisenberg. It states that it is conceptually impossible to simultaneously measure conjugate variables such as momentum P and position x accurately.

Another form of the uncertainty principle involves the energy E emitted in some processes and the time τ at which the energy is emitted. The uncertainty ΔE in E and the uncertainty $\Delta\tau$ in τ are related by

$$\Delta E = \Delta\tau = \frac{h}{4\pi} \tag{C-33}$$

This is rather easy to confirm for the case of the emission of electromagnetic radiation by an atomic system. The frequency ν is the number of waves in the time duration $\Delta\tau$ of emission. The uncertainty in the number of waves that can be created in a wave train is about one, so that

$$\Delta\nu \sim \frac{1}{\Delta\tau} \tag{C-34}$$

But

$$E = h\nu \tag{C-35}$$

or

$$\Delta E = h\Delta\nu \tag{C-36}$$

Equations (C-34) and (C-36) give

$$\Delta E \cdot \Delta\tau \sim h \tag{C-37}$$

A more rigorous derivation yields a factor of 4π in the uncertainty statement.

The uncertainty principle is an important consideration in the investigation of extremely small systems. An optical microscope is used to examine specimens greater than about 5 μm. Once the size of the sample is of the order of the wavelength of light used to examine it, diffraction becomes a problem, and radiation of smaller wavelengths must be used. This is achieved in an electron microscope, where the wavelength is a function of the accelerating voltage. However, as the wavelength is reduced, the energy increases, so a point is reached where any interaction between the specimen and the excitation beam will give rise to a large energy transfer. The resultant momentum change makes it impossible to obtain an accurate estimate of position. Similarly, in microcircuit manufacture, photolithography is the step where a circuit pattern is defined in a photoresist covering on a silicon slice. Visible light is used for patterns down to about 5 μm. Below that size, electron beam lithography must be used. But Heisenberg's uncertainty principle still holds. The electron beam spot has an approximate diameter but, as the energy is increased in order to go to smaller wavelengths, the uncertainty about spot size increases and it is this which ultimately limits the possibility to define ever smaller linewidths.

Appendix D

Density of States

A common problem occurring in statistical mechanics in several contexts is that of determining the number of states $f(x)dx$ in which a certain quantity x lies in the interval x to $x + dx$. For example, the number of eigenstates (microstates) per unit energy range in the vicinity of ε is equal to $D(\varepsilon)d\varepsilon$, where $D(\varepsilon)$ is called the *density of states*. The problem will be considered in general and the results related to particular cases.

Stationary Wave Solutions

The general case of wave motion is given by the equation

$$\nabla^2\psi - \frac{1}{c^2}\frac{\partial^2\psi}{\partial t^2} = 0 \tag{D-1}$$

where $\psi(\mathbf{r}, t)$ is the wave function and c is the velocity of the waves. To obtain standing or progressive waves, a solution of the form

$$\psi(\mathbf{r}, t) = \phi(\mathbf{r})T(t) \tag{D-2}$$

is assumed. Substituting in Eq. (D-1) gives

$$\nabla^2(\phi T) - \frac{1}{c^2}\frac{\partial^2(\phi T)}{\partial t^2} = T\nabla^2\phi - \frac{\phi}{c^2}\frac{\partial^2 T}{\partial t^2} = 0 \tag{D-3}$$

Dividing by ϕT and rearranging obtains

$$\frac{1}{\phi}\nabla^2\phi = \frac{1}{c^2 T}\frac{\partial^2 T}{\partial t^2} \tag{D-4}$$

The left-hand side of Eq. (D-4) depends only on $\mathbf{r}$, while the right-hand side depends only on T. The only way for this to hold is to have both sides equal to the same constant. Let this constant be equal to $-k^2$, so that

$$\frac{1}{\phi}\nabla^2\phi = -k^2 \tag{D-5}$$

$$\frac{1}{c^2 T}\frac{\partial^2 T}{\partial t^2} = -k^2 \tag{D-6}$$

Equation (D-5) can be rewritten in the form

$$\nabla^2\phi(x, y, z) + k^2\phi(x, y, z) = 0 \tag{D-7}$$

which describes standing or progressive waves. The value k is the magnitude of the three-dimensional wave vector $\mathbf{k}$. It is related to the wavelength λ by

$$k = \frac{2\pi}{\lambda} = \frac{2\pi\nu}{v} = \frac{\omega}{v} \tag{D-8}$$

where ν is the frequency (Hz), $\omega = 2\pi\nu$ is the circular frequency, and $v = \omega/k$ is the phase velocity. Equation (D-7) for ϕ is the same as the equation

$$\frac{\partial^2\psi}{\partial x^2} + \frac{\partial^2\psi}{\partial y^2} + \frac{\partial^2\psi}{\partial z^2} = -\frac{2m\varepsilon}{\hbar^2}\psi \tag{D-9}$$

from Schrödinger's time independent equation for a particle in a box, provided

$$k^2 = \frac{2m\varepsilon}{\hbar^2} \tag{D-10}$$

Consider waves described by the wave equation (D-7) inside a cubic enclosure of side L as shown in Fig. D-1a. The boundary conditions are

$$\phi(r) = 0 \qquad \text{on all faces of the cube} \tag{D-11}$$

so as to obtain *standing waves*. The standing wave solutions of Eq. (D-7) satisfying the boundary conditions (D-11) are given by

$$\phi = A\sin(k_x x)\sin(k_y y)\sin(k_z z) \tag{D-12}$$

where A is a constant,

$$k_x = \frac{n_x\pi}{Z}; \qquad k_y = \frac{n_y\pi}{Z}; \qquad k_z = \frac{n_z\pi}{Z} \tag{D-13}$$

and n_x, n_y, and n_z can have any positive integer value from 1 to infinity. They are called the *normal modes*. (The energy of each normal mode can vary independently.)

The values of k_x, k_y and k_z can be plotted in k space as shown in Fig. D-1(b). Each dot represents an independent standing wave solution of Eq. (D-1). For a fixed value of k_y and k_z, if n_x is increased by 1, then $\Delta k_x = \pi/L$, so that the distance between neighbor-

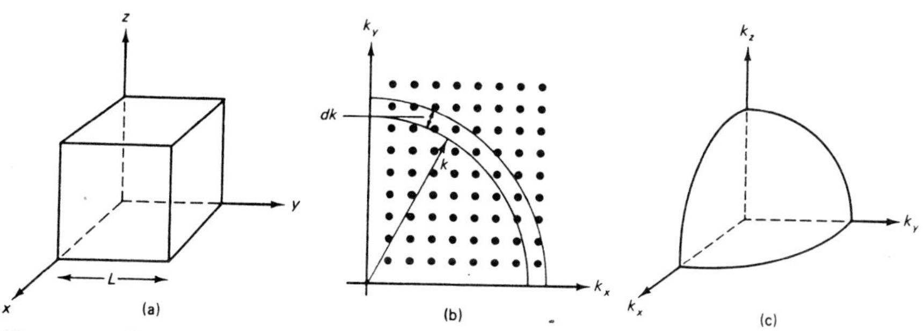

Figure D-1 Standing wave solutions in an enclosure.

ing dots in Fig. D-1(b) is π/L. In other words, there is one dot per n^3/L^3 volumes of k space, where

$$k^2 = k_x^2 + k_y^2 + k_z^2 \tag{D-13a}$$

It is desirable to have an expression for the number of normal modes of standing waves with wave vectors whose magnitude lies in the interval k to $k + dk$. This is the number of dots in **k** space lying between two spherical shells of radii k and $k + dk$ in the positive octant, as shown in Fig. D-1(c). Now, the volume of the region lying between the radii k and $k + dk$ in the positive octant is $(4\pi k^2 dk)/8$, as the number of standing wave solutions (normal modes) having a wave vector **k** whose magnitude lies in the range k to $k + dk$ is

$$D(k)dk = \frac{(\pi k^2 dk/2)}{(\pi^3/L^3)} = \frac{Vk^2/dk}{2\pi^2} \tag{D-14}$$

where $V = L^3$ is the volume of the cubic enclosure. Equation (D-14) is the basic result for the *density of states* upon which much of statistical mechanics results are based. An alternate way of deriving it is to consider the number of allowed **k** vectors of magnitude less than or equal to k, namely the volume $(1/8)(4\pi k^3/3)$, and divide this by π^3/L^3. The result, for the enclosure V, is $Vk^3/6\pi^2$. Differentiation of this with respect to k results in the same expression (D-14).

Equation (D-14) is valid whatever the shape of the enclosure and does not depend on the type of wave motion, but only on fitting standing waves in a box, subject to the boundary conditions that ϕ be zero everywhere on the inner surface of the enclosure. It can be rewritten in a number of different ways. For example, from $k = 2\pi/\lambda$,

$$dk = -\frac{2\pi d\lambda}{\lambda^2} \tag{D-15}$$

Hence, using this in (D-14), the number of standing waves (normal modes) $D(\lambda)d\lambda$ having wavelengths between λ and $\lambda + d\lambda$ is

$$D(\lambda)\, d\lambda = \frac{V(4\pi)d\lambda}{\lambda^4} \tag{D-16}$$

Alternately, the number of normal modes with circular frequencies in the range ω to $\omega + d\omega$ may be evaluated. From Eq. (D-14),

$$\frac{D(k)dk}{d\omega}d\omega = \frac{Vk^2}{2\pi^2}\frac{dk}{d\omega}d\omega \tag{D-17}$$

To eliminate k from (D-17), consider the phase velocity $c = \omega/k$ and the group velocity defined by

$$v_g = \frac{d\omega}{dk} \tag{D-18}$$

so that (D-17) becomes

$$D(\omega)d\omega = \frac{V\omega^2 d\omega}{2\pi^2 c^3} \tag{D-19}$$

$D(\omega)d\omega$ is the number of stationary wave solutions (normal modes) that have angular frequencies in the range ω to $\omega + d\omega$. The phase velocity c, equal to ω/k, is a constant for a nondispersive medium, so $dk = d\omega/c$. Equation (D-19) holds for a nondispersive medium, that is, for $v_g = c$. Since $\omega = 2\pi\nu$, the density of states may also be expressed in terms of frequency ν, as

$$D(\nu)d\nu = \frac{V(4\pi\nu^2)d\nu}{c^3} \tag{D-20}$$

where $D(\nu)d\nu$ is the member of stationary wave solutions having frequencies in the range ν to $\nu + d\nu$. Some special types of wave motion will now be considered.

Electromagnetic Waves

In the case of electromagnetic waves, $\psi(\mathbf{r}, t)$ stands for either the electric field intensity or the magnetising force. For example, the electric field of an electromagnetic wave in vacuum satisfies the wave equation

$$\nabla^2 \mathscr{E}(\mathbf{r}, t) = \frac{1}{c^2}\frac{\partial^2}{\partial t^2}\mathscr{E}(\mathbf{r}, t) \tag{D-21}$$

where c is the velocity of light in vacuum, which is constant and independent of frequency if there is no diffusion. With

$$\mathscr{E}(\mathbf{r}, t) = \mathscr{E}(\mathbf{r})e^{-i\omega t} \tag{D-22}$$

Eq. (D-21) reduces to

$$(\nabla^2 + k^2)\mathscr{E}(\mathbf{r}) = 0 \tag{D-23}$$

where, by comparison with (D-7) and (D-8),

$$k = \frac{\omega}{c} \tag{D-24}$$

Electromagnetic waves are transverse waves that can have two independent directions of polarization for each value of k. Hence, from Eq. (D-19), the number of normal modes of electromagnetic waves in an enclosure of volume V is

$$D(\omega)d\omega = 2\left(\frac{V\omega^2 d\omega}{2\pi^2 c^3}\right) = \frac{V\omega^2 d\omega}{\pi^2 c^3} \tag{D-25}$$

Elastic Waves in Solids (Sound Waves)

In a solid, elastic waves are propagated with a frequency-independent velocity, the velocity of sound, which is determined by the elastic stiffness of the medium. For each wave vector $\mathbf{k}$, there are three independent modes: longitudinal (compression) waves of velocity c_L and transverse (shear) waves of velocity c_T, in which the displacements are in two mutually orthogonal directions perpendicular to the direction of propagation. From Eq. (D-19), the number of stationary wave solutions (normal modes) in the angular frequency range ω to $\omega + d\omega$ is

$$D(\omega)d\omega = V\frac{\omega^2 d\omega}{2\pi^2}\left[\frac{1}{c_L^3} + \frac{2}{c_T^3}\right] \tag{D-26}$$

Schrödinger's Equation

For a particle of mass m and spin zero, not subject to any forces except that it is confined to a given volume V, Schrödinger's equation is just Eq. (D-7), together with the boundary conditions (D-1). From de Broglie's relation $\lambda = h/p$,

$$p = \frac{h}{\lambda} = \hbar k \tag{D-27}$$

where λ is the wavelength, p is the momentum, h is Planck's constant, and $k = 2\pi/\lambda = p/\hbar$. Using these in Eq. (D-14) gives the density of states

$$D(p)dp = V\frac{4\pi p^2 dp}{h^3} = V\frac{p^2 dp}{2\pi^2\hbar^3} \tag{D-28}$$

with momentum $\mathbf{p}$ of magnitude in the range p to $p + dp$. Equation (D-28) holds in both the relativistic and nonrelativistic limits. In the nonrelativistic case, when $p = mv$ and

$$\varepsilon = \tfrac{1}{2}mv^2 = \frac{p^2}{\Sigma m} \tag{D-29}$$

Eq. (D-28) becomes

$$D(\varepsilon)d\varepsilon = \frac{V}{4\pi^2}\left(\frac{2m}{\hbar^2}\right)^{3/2} \varepsilon^{1/2}\, d\varepsilon \tag{D-30}$$

This is the number of stationary wave solutions in which the particle in the box has kinetic energy between ε and $\varepsilon + d\varepsilon$. (For particles, the term *single-particle quantum state* is often used for stationary wave solution or normal mode. So is the term *orbital* used, particularly in chemistry.)

If a particle has spin s, there are $g = (2s + 1)$ states associated with each set of values of n_x, n_y, and n_z, and the right-hand side of Eqs. (D-28), and (D-30) must be multiplied by the spin degeneracy g to give

$$D(p)dp = g\frac{V(4\pi p^2)dp}{h^3} \tag{D-31}$$

$$D(\varepsilon)d\varepsilon = \frac{gV}{4\pi^2}\left(\frac{2m}{\hbar^2}\right)^{3/2} \varepsilon^{1/2}\, d\varepsilon \tag{D-32}$$

If the particle in the box is an electron or a proton of spin $s = \frac{1}{2}$, each set of quantum numbers n_x, n_y, and n_z can be associated with either $s = +\frac{1}{2}$ or $s = -\frac{1}{2}$. This doubles the number of single-particle states with a given value of n_x, n_y, and n_z. In other words, for fermions of spin $\frac{1}{2}$, such as electrons, g is equal to 2.

Photons

The possible single-particle (photon) states are the standing-wave solutions (normal modes) of the wave equation (D-7). This leads to Eq. (D-19) for the number of photon states in which the angular frequency of the photon is between ω and $\omega + d\omega$. Now, photons are bosons of spin $s = 1$. However, for photons, the number of spin states g is 2 and not $2s + 1 = 3$. The degeneracy of 2 comes from the fact that photons travel at the speed of light, and there are two possible independent polarization directions. Hence, Eq. (D-19) must be multiplied by 2, not 3, so that the number of photon states in which the photon has angular frequency between ω and $\omega + d\omega$ is

$$D(\omega)d\omega = \frac{V\omega^2\, d\omega}{\pi^2 c^3} \tag{D-33}$$

where c is the speed of light.

Appendix E

Heat Capacities of Solids; Debye Functions

For a majority of crystalline solids made up of atoms or monatomic ions, the temperature dependence of the atomic specific heat takes the form of a law of corresponding states:

$$\frac{c_V}{R} = f\left(\frac{T}{\theta_D}\right) \tag{E-1}$$

where $f(T/\theta_D)$ is a universal function for all monatomic crystals. The parameter θ_D, known as the *Debye characteristic temperature,* must be appropriately chosen for each temperature. According to (E-1), different crystals at the same multiple of their respective θ_D have the same heat capacity per gram-atom. For most solids at 1 atm, the heat capacities c_V and c_P are practically identical.

For calculations of the thermodynamic properties of a crystal made up of atoms, the crystal is regarded as one large molecule with $3N - 6$ internal degrees of freedom. For an atomic crystal these internal degrees of freedom all correspond to vibrations; for a molecular crystal some internal degrees of freedom will correspond to rotations or torsions. In principle, a normal mode analysis may be carried out on an atomic crystal. Since the normal modes are independent, the canonical partition function for the crystal has the form

$$\begin{aligned} \ln Z &= \ln \prod_{i=1}^{3N-6} \left[1 - \exp\left(-\frac{h\nu_i}{kT}\right)\right]^{-1} \\ &= \prod_{i=1}^{3N-6} \ln \left[1 - \exp\left(-\frac{h\nu_i}{kT}\right)\right]^{-1} \end{aligned} \tag{E-2}$$

Since $3N - 6$ is a very large number for a molecular crystal, the distribution of frequencies can be represented by a probability density $P(\nu)$, so that $P(\nu)\,d\nu$ is the fraction of frequencies in the range between ν and $\nu + d\nu$. The canonical partition function can be written in the form

$$\ln Z = -3N \int_0^\infty P(\nu) \ln \left[1 - \exp\left(-\frac{h\nu}{kT}\right)\right] d\nu$$

With the usual relations $E = kT^2(\partial \ln Z/\partial \tau)$ and $c_V = (\partial E/\partial \tau)$, the specific heat can then be obtained:

$$c_V = 3Nk \int_0^\infty \left(\frac{h\nu}{kT}\right)^2 P(\nu) \frac{\exp(h\nu/kT)}{[\exp(h\nu/kT) - 1]^2} d\nu \tag{E-3}$$

The calculation of the heat capacity of an idealized crystal done by Einstein (1907) assumes that all the vibrational frequencies of the solid have the single value ν_E, in which case the integrand in (E-3) is nonzero only for this frequency, and so

$$c_V = 3Nk \left(\frac{\theta_E}{T}\right)^2 \frac{e^{-\theta_E/T}}{(1 - e^{-\theta_E/T})^2} \tag{E-4}$$

where $\theta_E = h\nu_E/k$ is the characteristic vibrational temperature. This theoretical result explains why the heat capacity decreases from the classical value of $3R$ at high temperatures to zero as $T \rightarrow 0$. It also explains why c_V rises more slowly with increasing temperature for, say, diamond and graphite than for silver and copper. However, the theory predicts too rapid a decrease in c_V near absolute zero.

A more accurate theory proposed by Debye (1912) introduces a spectrum of frequencies for the atomic crystal, resulting in an expression for the heat capacity that more faithfully fits the experimental data at very low temperatures. For an elastic continuous medium, the probability density for vibrations is proportional to the square of the frequency:

$$P(\nu) = \alpha\nu^2 \tag{E-5}$$

In a truly continuous medium, the spectral density would increase without limit with increasing frequency. For a real crystal, it is impossible to achieve oscillations of wavelengths much shorter than the smallest interatomic spacings. There is some $\lambda_{\min}$ or $\nu_{\max}$ above which $P(\nu) = 0$. This is also consistent with the fact that the number of degrees of freedom is finite, equal to $3N$. Thus,

$$\int_0^\infty P(\nu) d\nu = \alpha \int_0^{\nu_{\max}} \nu^2 \, d\nu = \alpha \frac{\nu_{\max}^3}{3} = 3N \tag{E-6}$$

Equation (E-6) yields $\nu_{\max}$, which, along with ν_E, is shown in Fig. E-1.

In terms of $\nu_{\max}$, the heat capacity for the Debye crystal is

$$c_V = \frac{9Nkh^2}{\nu_{\max}^3 \, k^2T^2} \int_0^{\nu_{\max}} \frac{\nu^4 e^{h\nu/kT}}{(e^{h\nu/kT} - 1)^2} d\nu \tag{E-7}$$

Defining a Debye characteristic temperature

$$\theta_D = \frac{h\nu_{\max}}{k} \tag{E-8}$$

and changing the integration variable to

$$x = \frac{h\nu}{kT} \tag{E-9}$$

Eq. (E-7) can be expressed as

$$\frac{c_V}{Nk} = 9\left(\frac{T}{\theta_o}\right)^3 \int_0^{\theta_D/T} x^4 e^x (e^x - 1)^{-2} \, dx \tag{E-10}$$

The right side of (E-10) is of the form $f(T/\theta_D)$, which is in agreement with the law of

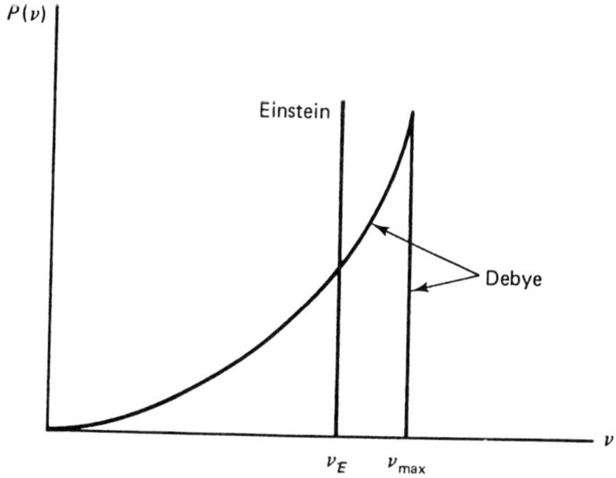

Figure E-1 Frequency spectrum for Debye and Einstein atomic crystal.

corresponding states for the capacity, (E-1). The exact form of the universal function predicted by the Debye model is determined by numerical integration.

In the high-temperature limit, as $T \to \infty$, $x \to 0$, the integrand in (E-10) approaches x^2 so that the definite integral approaches $(\theta_D/T)^3/3$. Thus, $(c_V/Nk) \to 3$, in agreement with the law of Dulong and Petit. At the low-temperature end, as $T \to 0$, $\theta_D/T \to \infty$, and the integrand in (E-10) falls off as $x^4 e^{-x}$ for large x, effectively losing its dependence on θ_D/T at low temperature. The integral is well approximated by

$$\int_0^\infty x^4 e^x (e^x - 1)^{-1}\, dx = \frac{4\pi^4}{15} \tag{E-11}$$

so that as $T \to 0$,

$$\frac{c_V}{Nk} \simeq \frac{12\pi^4}{5\theta_D^3} T^3 \tag{E-12}$$

which is Debyes' T^3 law.

Interestingly, noting that cosech $x = 1/\sinh x = 2/(e^x - e^{-x})$, the Debye heat capacity at any temperature can be written in the form

$$c_V = 72Nk \left(\frac{T}{\theta_D}\right)^3 \int_0^{\theta_D/2T} x^4 \operatorname{cosech}^2 x\, dx \tag{E-13}$$

In other words, the heat capacities of all dielectric solids is a universal function of T/θ_D. This function is tabulated in Table E-1 and shown in Fig. E-2 where it compares with the superimposed heat capacities of a range of solids.

Thermodynamic properties of a perfect crystal can also be expressed in terms of the following Debye function

$$D(u) = \frac{3}{u^3} \int_0^u \frac{x^3\, dx}{e^x - 1} \tag{E-14}$$

where the argument u is θ_D/T. For large values of u, the upper limit of the integral can be taken as infinity. Using the definite integral

$$\int_0^\infty \frac{x^3\, dx}{e^x - 1} = \frac{\pi^4}{15} \tag{E-15}$$

TABLE E-1 DEBYE FUNCTION (E-13)

T/θ_D	$\int_0^{\theta_D/2T} x^4 \operatorname{cosech}^2 x \, dx$	$\frac{C_V}{3Nk}$	T/θ_D	$\int_0^{\theta_D/2T} x^4 \operatorname{cosech}^2 x \, dx$	$\frac{C_V}{3Nk}$
0.01	3.2470	7.7928×10^{-5}	0.6	0.16856	0.87382
0.02	3.2470	6.2342×10^{-4}	0.7	0.10993	0.90494
0.03	3.2470	2.1041×10^{-3}	0.8	7.5361×10^{-2}	0.92604
0.04	3.2470	4.9874×10^{-3}	0.9	5.3778×10^{-2}	0.94090
0.05	3.2469	9.7407×10^{-3}	1.0	3.9656×10^{-2}	0.95174
0.06	3.2463	1.6829×10^{-2}	2.0	5.1438×10^{-3}	0.98761
0.07	3.2426	2.6693×10^{-2}	3.0	1.5347×10^{-3}	0.99449
0.08	3.2309	3.9701×10^{-2}	4.0	6.4901×10^{-4}	0.99688
0.09	3.2049	5.6073×10^{-2}	5.0	3.3267×10^{-4}	0.99801
0.1	3.1592	7.5821×10^{-2}	6.0	1.9263×10^{-4}	0.99859
0.2	1.9200	0.36864	7.0	1.2135×10^{-4}	0.99895
0.3	0.93781	0.60770	8.0	8.1317×10^{-5}	0.99922
0.4	0.48558	0.74585	9.0	5.7121×10^{-5}	0.99939
0.5	0.27514	0.82542	10.0	4.1646×10^{-5}	0.99950

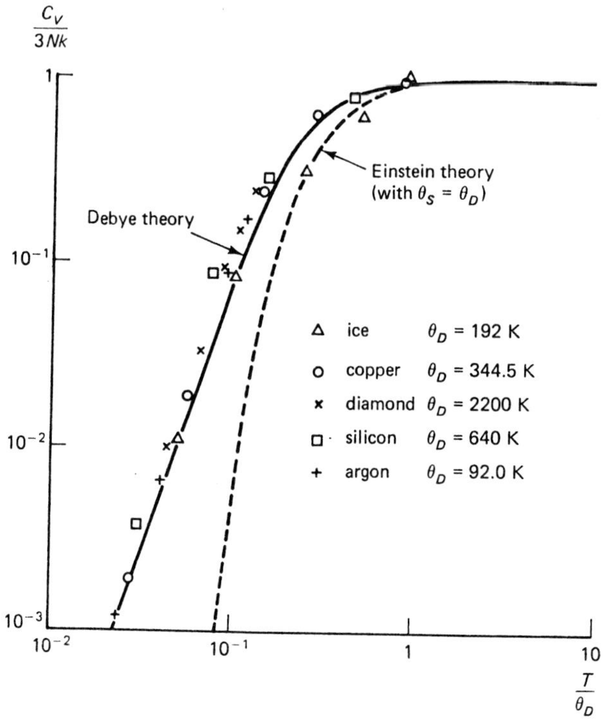

Figure E-2 Debye universal function (E-13) and experimental data.

the Debye function is then approximated by

$$D(u) = \frac{\pi^4}{5u^3} = \frac{19.482}{u^3} \tag{E-16}$$

The logarithm of the partition function is

$$\ln Z = -\frac{9N}{8}u - 3N\ln(1 - e^{-u}) + ND(u) \tag{E-17}$$

Using $E = kT^2(\partial \ln Z/\partial T)$, it is found that

$$E = \frac{9}{8}Nk\theta_D + 3NkTD\left(\frac{\theta_D}{T}\right) \tag{E-18}$$

whereupon

$$c_V = \left(\frac{\partial E}{\partial T}\right) = 3Nk\left[4D\left(\frac{\theta_D}{T}\right) - \frac{3\theta_D/T}{e^{\theta_D/T} - 1}\right] \tag{E-19}$$

Similarly, the entropy is shown to be

$$S = 3Nk\left[\frac{4}{3}D\left(\frac{\theta_D}{T}\right) - \ln(1 - e^{-\theta_D/T})\right] \tag{E-20}$$

The Debye function, as defined by Eq. (E-14), is tabulated in Table E-2.

TABLE E-2 DEBYE FUNCTION (E-14)

u	0.0	0.1	0.2	0.3	0.4	0.5	0.6	0.7	0.8	0.9	1.0
0.0	1.0000	0.9630	0.9270	0.8920	0.8580	0.8250	0.7929	0.7619	0.7318	0.7026	0.6744
1.0	0.6744	0.6471	0.6208	0.5954	0.5708	0.5471	0.5243	0.5023	0.4811	0.4607	0.4411
2.0	0.4411	0.4223	0.4042	0.3868	0.3701	0.3541	0.3388	0.3241	0.3100	0.2965	0.2836
3.0	0.2836	0.2712	0.2594	0.2481	0.2373	0.2269	0.2170	0.2076	0.1986	0.1900	0.1817
4.0	0.1817	0.1739	0.1664	0.1592	0.1524	0.1459	0.1397	0.1338	0.1281	0.1227	0.1176
5.0	0.1176	0.1127	0.1080	0.1036	0.09930	0.09524	0.09137	0.8768	0.08415	0.08079	0.7758
6.0	0.07758	0.07452	0.07160	0.06881	0.06615	0.06360	0.06118	0.05886	0.05664	0.05453	0.05251
7.0	0.05251	0.05057	0.04873	0.04696	0.04527	0.04366	0.04211	0.04063	0.03921	0.03786	0.03656
8.0	0.03656	0.03532	0.03413	0.03298	0.03189	0.03084	0.02983	0.02887	0.02794	0.02705	0.02620
9.0	0.02620	0.02538	0.02459	0.02384	0.02311	0.02241	0.02174	0.02109	0.02047	0.01987	0.01930
10.0	0.01930	0.01874	0.01821	0.01769	0.01720	0.01672	0.01626	0.01581	0.01538	0.01497	0.01457
11.0	0.01457	0.01418	0.01381	0.01345	0.01311	0.01277	0.01245	0.01213	0.01183	0.01153	0.01125
12.0	0.01125	0.01098	0.01071	0.01045	0.01020	0.00996	0.00973	0.00950	0.00928	0.00907	0.00886
13.0	0.00886	0.00866	0.00846	0.00827	0.00809	0.00791	0.00774	0.00757	0.00741	0.00725	0.00710
14.0	0.00710	0.00695	0.00680	0.00666	0.00652	0.00639	0.00626	0.00613	0.00601	0.00589	0.00577
15.0	0.00577	0.00566	0.00555	0.00544	0.00533	0.00523	0.00513	0.00503	0.00494	0.00485	0.00476

Appendix F

Chemical Potential for Fermi and Bose Gases; Fermi Integrals

Thermodynamic properties for the quantum gases contain μ, the chemical potential, as a parameter. For example, the internal energy of fermions or bosons is

$$\begin{aligned} U &= kT^2\left(\frac{\partial \ln \Xi}{\partial T}\right)_{V,\mu} + kT\mu\left(\frac{\partial \ln \Xi}{\partial \mu}\right)_{T,V} \\ &= \sum_s \varepsilon_s \left\{\exp\left[\frac{\varepsilon_s - \mu}{kT}\right] \pm 1\right\}^{-1} \end{aligned} \tag{F-1}$$

and the number of atoms is

$$N = \frac{2\pi V_g}{h^3}(2m)^{3/2}\int_0^\infty \varepsilon^{1/2}\left\{\exp\left[\frac{\varepsilon - \mu}{kT}\right] \pm 1\right\}^{-1} ds \tag{F-2}$$

where the upper sign is for fermions and the lower sign for bosons. Thermodynamic properties can be determined provided $\mu(T)$ is known. Equation (F-2) determines $\mu(T)$, since N (the number of atoms in the gas) is assumed to be known. The extrication of μ from this equation can be done by a variety of numerical methods, but some limiting cases should be considered first.

At high temperatures, it is expected that the distinction between Fermi and Bose gases disappears. Since the only algebraic difference is the ± 1 in Eq. (F-2), it will be neglected, and Eq. (F-2) is written in the simplified form for high T:

$$N = \frac{2\pi V_g}{h^3}(2m)^{3/2}\exp\left(\frac{\mu}{kT}\right)\int_0^\infty \varepsilon^{1/2}\exp\left(-\frac{\varepsilon}{kT}\right) d\varepsilon \tag{F-3}$$

This last integral can be performed explicitly and, after rearrangement, solved for $\mu(T)$:

$$\mu = -kT \ln\left[\left(\frac{2\pi mkT}{h^2}\right)^{3/2}\frac{gV}{N}\right] \tag{F-4}$$

As for the low-temperature limiting form of μ, it can readily be obtained by letting $T \to 0$, $\mu \to \mu_0$. Then for those states for which $\varepsilon > \mu_0$, $\exp[(\varepsilon - \mu_0)/kT] \to \exp[\infty]$ so that those states do not contribute to the integral in Eq. (F-2), and for states with $\varepsilon < \mu_0$, $\exp[(\varepsilon - \mu_0)/kT] \to \exp[-\infty] = 0$. The upper limit of integration may thus be taken as μ_0 in Eq. (F-2) which consequently reduces to

$$N = \frac{2\pi V_g}{h^3}(2m)^{3/2}\int_0^{\mu_0} \varepsilon^{1/2}\, d\varepsilon \tag{F-5}$$

and thus

$$\mu_0 = \frac{h^2}{8m}\left(\frac{6N}{\pi V_g}\right)^{2/3} \tag{F-6}$$

This value of μ is also called the Fermi energy ε_F, to which corresponds a temperature

$$T_F = \frac{\mu_0}{k} \tag{F-7}$$

called the Fermi degeneracy temperature. Similarly, for a Bose gas, a Bose degeneracy temperature can be obtained:

$$T_B = \frac{h^2}{2\pi mk}\left(\frac{N}{2.61\, V_g}\right)^{2/3} \tag{F-8}$$

For purposes of comparison of Bose and Fermi gases, it is convenient to introduce a common characteristic temperature T_0, given by

$$T_0 = \frac{h^2}{2mk}\left(\frac{N}{2\pi V_g}\right)^{2/3} \tag{F-9}$$

In terms of T_0, the Fermi degeneracy temperature T_F and the Bose degeneracy temperature T_B are

$$T_F = \left(\frac{3}{2}\right)^{2/3} T_0 = 1.3104 T_0 \tag{F-10}$$

$$T_B = \frac{1}{\pi}\left(\frac{2\pi}{2.612}\right)^{2/3} T_0 = 0.5714 T_0 \tag{F-11}$$

At any temperature above T_0, the chemical potential μ for fermions and for bosons can be written in the form

$$1 = \tau^{3/2}\int_0^{\infty} x^{1/2}\left\{\exp\left[x - \frac{\mu^*}{\tau}\right] \pm 1\right\}^{-1} dx \tag{F-12}$$

where $\tau = T/T_0$ and $\mu^* = \mu/kT_0$. Figures F-1 and F-2 show the chemical potentials and heat capacities of Fermi and Bose gases, plotted as functions of T/T_0. The chemical potential is negative for T above T_F and T_B, respectively. The c_V/Nk for a Bose gas shows a cusp at $T = T_B$, and at low T, c_V is proportional to $T^{3/2}$. The Fermi gas has a smooth c_V proportional to T at low T. For $T/T_0 \geq 2.5$, the c_V of both gases approach classical behavior.

The energies can also be expressed in terms of the parameters τ and μ^*. For fermions,

$$\mu_f^*(\tau) = \tau^{5/2}\int_0^{\infty} x^{3/2}\left\{\exp\left[x - \frac{\mu_f^*}{\tau}\right] + 1\right\}^{-1} dx \tag{F-13}$$

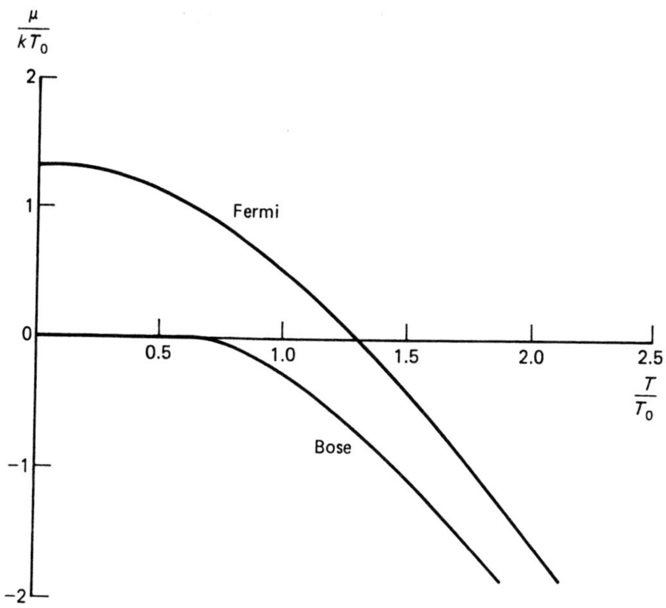

Figure F-1 Chemical potentials of Fermi and Bose gases.

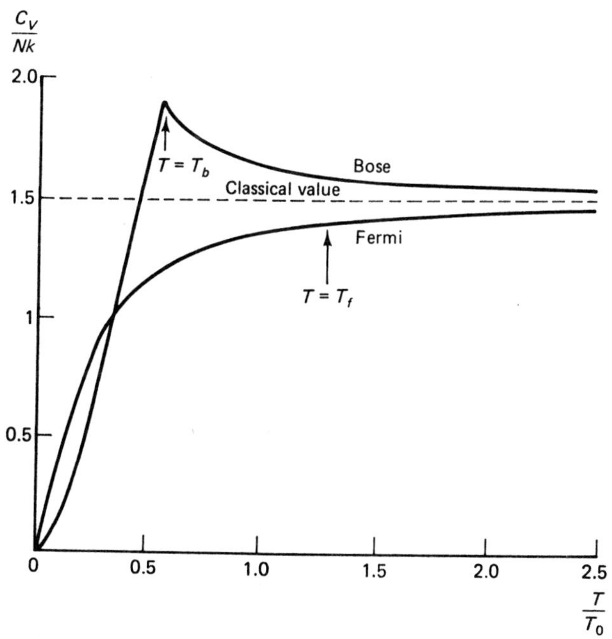

Figure F-2 Specific heats of Fermi and Bose gases.

For bosons, for $T > T_B$ and $\tau > 0.5714$,

$$\mu_b^*(\tau) = \tau^{5/2} \int_0^\infty x^{3/2} \left\{ \exp\left[x - \frac{\mu_b^*}{\tau} \right] - 1 \right\}^{-1} dx \tag{F-14}$$

and for $\tau < 0.5714$,

$$\mu_b^*(\tau) = \tau^{5/2} \int_0^\infty x^{3/2} \{\exp[x] - 1\}^{-1} \, dx = 1.7833 \, \tau^{5/2} \tag{F-15}$$

The computed values of μ_f^*, μ_b^*, u_f^*, and u_b^* are given in Table F-1.

In Fermi statistics, integrals of the form

$$F(\alpha) = \int_0^\infty \frac{f(x) \, dx}{[\exp(x - \alpha)] + 1} \tag{F-16}$$

are often encountered. They cannot be solved analytically, but by means of various

TABLE F-1 CHEMICAL POTENTIALS AND ENERGIES FOR FERMI AND BOSE GASES

τ	μ_f^*	μ_b^*	u_f^*	u_b^*
0.00	1.310 37	0.000 00	0.786 22	0.000 00
0.10	1.304 04	0.000 00	0.804 88	0.005 64
0.20	1.284 19	0.000 00	0.858 49	0.031 90
0.30	1.248 25	0.000 00	0.939 45	0.087 91
0.40	1.194 11	0.000 00	1.038 86	0.180 46
0.50	1.121 43	0.000 00	1.150 39	0.315 25
0.60	1.030 92	−0.001 70	1.270 09	0.494 67
0.70	0.923 72	−0.031 42	1.395 54	0.678 63
0.80	0.801 02	−0.091 97	1.525 17	0.855 08
0.90	0.663 98	−0.177 69	1.657 94	1.026 47
1.00	0.513 63	−0.284 68	1.793 13	1.194 24
1.10	0.350 91	−0.410 15	1.930 24	1.359 33
1.20	0.176 67	−0.551 92	2.068 88	1.522 35
1.30	−0.008 35	−0.708 31	2.208 78	1.683 75
1.40	−0.203 48	−0.877 95	2.349 73	1.843 83
1.50	−0.408 14	−1.059 71	2.491 56	2.002 84
1.60	−0.621 79	−1.252 65	2.634 13	2.160 95
1.70	−0.843 95	−1.455 96	2.777 34	2.318 30
1.80	−1.074 18	−1.688 94	2.921 10	2.475 01
1.90	−1.312 10	−1.890 99	3.065 34	2.631 16
2.00	−1.557 34	−2.121 57	3.210 01	2.786 83
3.00	−4.351 78	−4.812 44	4.672 05	4.326 55
4.00	−7.635 56	−8.034 51	6.149 21	5.850 00
5.00	−11.288 71	−11.645 54	7.633 56	7.365 94
6.00	−15.238 45	−15.564 19	9.121 97	8.877 67
7.00	−19.435 92	−19.737 50	10.612 96	10.386 78
8.00	−23.846 09	−24.128 19	12.105 69	11.894 11
9.00	−28.442 59	−28.708 55	13.599 66	13.400 18
10.00	−33.204 86	−33.457 18	15.094 55	14.905 32

series expansions they can be approximated for certain ranges of $\alpha = \mu/kT$. In particular, for $f(x) = x^n$,

$$F_n(\alpha) = \int_0^\infty \frac{x^n\,dx}{[\exp(x-\alpha)]+1} = \frac{\alpha^{n+1}}{n+1}\left[1 + \frac{\pi^2}{6}(n+1)n\alpha^{-2} + \cdots\right] \tag{F-17}$$

This series converges rapidly for $\alpha \gg 1$. For $n = 1/2$ and $\alpha \gg 1$, $F_{1/2}(\alpha)$ can be represented by

$$F_{1/2}(\alpha) = \frac{2}{3}\alpha^{3/2}\left(1 + \frac{\pi^2}{8\alpha^2} + \cdots\right) \tag{F-18}$$

and $\mu(T)$, the temperature dependence of the Fermi energy, is given by

$$\mu(T) = \mu_0\left[1 - \frac{\pi^2}{12}\left(\frac{kT}{\mu_0}\right)^2\right] \tag{F-19}$$

It is seen that μ decreases slightly with increasing temperature, but the relative drop of μ between 0 and 300 K amounts to little (assuming that $\mu_0 = 5$ eV). This is the case of metals, for which

$$\mu_0 = \frac{\hbar^2}{2m}(3\pi^2 n_i)^{2/3} \tag{F-20}$$

where $n_i = N/V$ is the electron concentration.

For $\mu \ll 0$, the Fermi–Dirac function goes over into the Maxwell–Boltzmann function

$$F(\varepsilon) = \left\{\exp\left[\frac{\varepsilon-\mu}{kt}\right]+1\right\}^{-1} \simeq \exp\left[-\frac{|\mu|}{kT}\right]\exp\left[-\frac{\varepsilon}{kT}\right] \tag{F-21}$$

where the value of the chemical potential is negative and less than $-kT$. The distribution of energy takes the form of a Maxwell–Boltzman distribution:

$$N(\varepsilon)d\varepsilon = \text{const.}\ \varepsilon^{1/2}\exp\left(-\frac{\varepsilon}{kT}\right)d\varepsilon \tag{F-22}$$

Under the assumption of $\alpha \ll 0$, the Fermi integral $F_{1/2}(\alpha)$ can be solved analytically:

$$\begin{aligned} F_{1/2}(\alpha) &\simeq \exp(-|\alpha|)\int_0^\infty x^{1/2}\exp(-x)\,dx \\ &= \frac{1}{2}\pi^{1/2}\exp(-|\alpha|) \end{aligned} \tag{F-23}$$

and an equation for the electron concentration obtained:

$$n_e = n_0 \exp\left(-\frac{|\mu|}{kT}\right) \tag{F-24}$$

showing that at low concentrations and high temperatures, the electron gas behaves like a classical gas obeying the Maxwell–Boltzmann distribution. Under these conditions the electron gas is said to be *nondegenerate*.

Deviations of the properties of the electron gas from those of an ideal gas are called *degeneracy*. The presence of a zero-point energy and thus a zero-point pressure are characteristic features of degeneracy. The criterion for degeneracy is defined by the condition

$$\mu(T) \simeq \mu_0 \gg kT$$

The electron gas is degenerate as long as its temperature T is appreciably below the degeneracy temperature T_F given by

$$T_F = \frac{2}{5}\frac{\mu_0}{k} \tag{F-25}$$

This in turn defines a critical electron concentration for which

$$n_e \gg n_{\text{crit}} = \frac{1}{3\pi^2}\left(\frac{2mk}{\hbar^2}\right)^{3/2} T^{3/2} \tag{F-26}$$

Degeneracy is favored by low temperatures and small masses.

The specific heat of the degenerate electron gas is much smaller than that of a classical gas. This is because only the fastest electrons and thus only a small fraction of electrons (those near μ_0) can take up thermal energy. The specific heat c_V is obtained from the temperature dependence of the interval energy $U(T)$ of the electron gas:

$$U(T) = \int_0^\infty \varepsilon N(\varepsilon)\, d\varepsilon = \frac{V}{2\pi^2}\left(\frac{2m}{\hbar^2}\right)^{3/2} \int_0^\infty \frac{\varepsilon^{3/2}\, d\varepsilon}{\exp\left[\dfrac{\varepsilon - \mu}{kT}\right] + 1} \tag{F-27}$$

With the substitution $x = \varepsilon/kT$ and $\alpha = \mu/kT$, this gives

$$\begin{aligned} U(T) &= \frac{V}{2\pi^2}\left(\frac{2m}{\hbar^2}\right)^{3/2} (kT)^{5/2} \int_0^\infty \frac{x^{3/2}\, dx}{\exp(x - \alpha) + 1} \\ &= \frac{V}{2\pi^2}\left(\frac{2m}{\hbar^2}\right)^{3/2} (kt)^{5/2} F_{3/2}(\alpha) \end{aligned} \tag{F-28}$$

where

$$F_{3/2}(\alpha) = \int_0^\infty \frac{x^{3/2}\, dx}{\exp(x - \alpha) + 1} \tag{F-29}$$

$$\simeq \frac{2}{5}\alpha^{5/2}\left[1 + \frac{5}{8}\left(\frac{\pi}{\alpha}\right)^2 + \cdots\right] \tag{F-30}$$

has been evaluated according to (F-17) for the case of $\alpha = \mu/kT \gg 1$. Additional values of $F_n(\alpha)$ are given in Table F-2. The internal energy or mean energy per electron is then

$$\langle u(T)\rangle = \frac{U(T)}{V} = \frac{3}{5}\mu_0\left[1 + \frac{5\pi^2}{12}\left(\frac{kT}{\mu_0}\right)^2 + \cdots\right] \tag{F-31}$$

whence

$$c_V = \frac{\pi^2}{2}\frac{Nk^2}{\mu_0} T = \gamma T \tag{F-32}$$

where

$$\gamma = \frac{\pi^2}{2}\frac{Nk^2}{\mu_0} \tag{F-33}$$

Unlike the classical, the electron gas has a temperature-dependent specific heat, which is directly proportional to the temperature.

TABLE F-2 FERMI INTEGRALS $F_n(\alpha) = \int_0^\infty \frac{x^n\,dx}{\exp(x-\mu)+1}$

α	$\frac{1}{1+e^{-\alpha}}$	$F_{-1/2}$	$\ln(1+e^{\alpha})$	$F_{1/2}$	F_1	$F_{3/2}$	F_2
−4	0.1799 −1	0.3204 −1	0.1815 −1	0.16128 −1	0.0182 0	2.42685 −2	0.0366 0
−3	0.4743 −1	0.8526 −1	0.4859 −1	0.43366 −1	0.0492 0	6.56115 −2	0.0990 0
−2	1.1920 −1	2.1918 −1	1.2693 −1	1.14588 −1	0.1310 0	1.75800 −1	0.2662 0
−1	2.6894 −1	5.2114 −1	3.1326 −1	2.90501 −1	0.3387 0	4.60848 −1	0.7050 0
0	5.000 −1	1.0722 0	6.9315 −1	6.78094 −1	0.8225 0	1.15280 0	1.8030 0
1	7.3106 −1	1.8204 0	1.3137 0	1.39638 0	1.8062 0	2.66168 0	4.3120 0
2	8.8079 −1	2.5954 0	2.1270 0	2.50246 0	3.5135 0	5.53725 0	9.4450 0
3	9.5257 −1	3.2852 0	3.0486 0	3.97699 0	6.0957 1	1.03537 1	1.8870 1
4	9.8201 −1	3.8743 0	4.0182 0	5.77073 0	9.6267 0	1.76277 1	3.4592 1
5	9.9331 −1	4.3832 0	5.0067 0	7.83797 9	1.4138 1	2.78024 1	5.8120 1
6	9.9753 −1	4.8338 0	6.0025 0	1.01443 1	1.9642 1	4.12610 1	9.1744 1
7	9.9909 −1	5.2416 0	7.0009 0	1.26646 1	2.6144 1	5.83422 1	1.3736 2
8	9.9967 −1	5.6170 0	8.0003 0	1.53805 1	3.3645 1	7.93526 1	1.9699 2
9	9.9988 −1	5.9674 0	9.0001 0	1.82776 1	4.2145 1	1.04574 2	2.7261 2
10	9.9995 −1	6.2972 0	1.0000 1	2.13445 1	5.1645 1	1.34270 2	3.6623 2
11	9.9998 −1	6.6096 0	1.1000 1	2.45718 1	6.2145 1	1.68688 2	4.7986 2
12	9.9999 −1	6.9076 0	1.2000 1	2.79518 1	7.3645 1	2.08062 2	6.1548 2
13	1.0000 0	7.1930 0	1.3000 1	3.14775 1	8.6145 1	2.52616 2	7.7510 2
14	1.0000 0	7.4672 0	1.4000 1	3.51430 1	9.9645 1	3.02564 2	9.6072 2
15	1.0000 0	7.7314 0	1.5000 1	3.89430 1	1.1415 2	3.58112 2	1.1743 3
16	1.0000 0	7.9868 0	1.6000 1	4.28730 1	1.2965 2	4.19458 2	1.4180 3
17	1.0000 0	8.2342 0	1.7000 1	4.69286 1	1.4615 2	4.86794 2	1.6936 3
18	1.0000 0	8.4744 0	1.8000 1	5.11061 1	1.6365 2	5.60305 2	2.0032 3
19	1.0000 0	8.7076 0	1.9000 1	5.50419 1	1.8215 2	6.40171 2	2.3488 3
20	1.0000 0	8.9350 0	2.0000 1	5.98128 1	2.0165 2	7.26568 2	2.7325 3

The last (separated) figure of each line gives the power of ten by which the corresponding value must be multiplied.

Appendix G

Semiconductors and Defective Solids

Almost all real crystals are inperfect. Although experimental identification and location of particular kinds of defect may be difficult, a simple classification is based on the dimensionality of the defect.

1. *Zero-dimensional or point defects.* These involve either a missing atom or *Schottky defect,* or a displaced atom or *Frenkel effect.* Both are prevalent in metals, covalent and ionic solids, but not in molecular solids.
2. *One-dimensional defects.* These are subdivided into *edge* or *screw dislocations.* An edge dislocation is a row of atoms marking the edges of an extra crystallographic plane, while a screw dislocation is a row of atoms about which a crystallographic plane spirals. Both types are common in crystals; they are important in crystal growth, as in the phenomenon of *twinning* and chiefly affect the strength and mechanical properties of the solid.
3. *Two-dimensional defects.* These include stacking faults in close-packed layers of atoms. They affect the mechanical properties of the solid and are less mobile than one-dimensional defects.
4. *Three-dimensional defects.* Crystals are really made up of many small microcrystals or domains, each of which may be nearly perfect or possess other types of defect. The edge of each domain may vary from a few hundred Å to a few centimeters. Domain boundries are lines of weakness along which crystals tend to break when subjected to great stress.

A semiconductor is a solid conductor with a resistivity that is usually in the range 10^{-2} to 10^{9} Ω cm at room temperature, whose electrical conductivity increases exponentially with temperature. Semiconduction is a property found in many solids, which may be giant covalent molecules such as Ge and Si, or more commonly in compounds of elements in the center of the periodic table. A substance is said to be an intrinsic semiconductor if its resistivity is in the semiconductor range when the substance is prepared in pure, stoichiometric crystals. The forbidden enegy gap between the highest filled band and the lowest unfilled band is between 0.2 and 2.2 eV for most intrinsic semiconductors. At absolute zero all semiconductors are insulators, but as the temperature is raised some electrons are excited to the lowest unfilled band by thermal energy, and the electrical conductivity rises. The electricity is conducted partly by the excited electrons and partly by the positive holes left in the highest filled band after excitation. The number of electrons excited is proportional to $\exp(-E_g/kT)$, where E_g is the width of the forbidden energy gap. The electrical properties of semiconducting materials at room temperature are given in Table G-1.

TABLE G-1 PROPERTIES OF SEMICONDUCTING MATERIALS AT 300 K

Semiconductor	Energy gap, E_g (eV)	Electron mobility, μ_c ($m^2 V^{-1} s^{-1}$)	Hole mobility, μ_h ($m^2 V^{-1} s^{-1}$)	Intrinsic carrier concentration, n_i (m^{-3})	Conductivity*, ($\Omega^{-1} m^{-1}$)	Electron effective mass m_c^* ($\times m$)	Hole effective mass, m_h^* ($\times m$)	Relative permittivity ε_r
Si	1.12	0.135	0.048	1.5×10^{16}	4.4×10^{-4}	0.26	0.38	11.8
Ge	0.67	0.39	0.19	2.4×10^{19}	2.2	0.12	0.23	16.1
GaAs	1.43	0.85	0.025	1.0×10^{13}	2.5×10^{-7}	0.06	0.44	10.9
GaP	2.25	0.03	0.015	1.4×10^{5}	9.9×10^{-1}	0.12	0.5	8.5
InP	1.28	0.4	0.01		1.25×10^{6}	0.07	0.4	12.4
InAs	0.36	3.3	0.045		3.3×10^{3}	0.027	0.018	14.6
InSb	0.18	8.0	0.045		1.67×10^{3}	0.013	0.18	16.7
CdS	2.5	0.03	0.002			0.21	0.8	7.1
PbS	0.37	0.06	0.02		2.0×10^{6}	0.25	0.25	161
PbSe	0.26	0.12	0.12		1.0×10^{5}	0.33	0.34	280
ZnS	3.6	0.01 (*n* type only)	—		1.0×10^{-8}	0.4	—	7.0
ZnTe	2.3	—	0.01 (*p* type only)		1.0	—		10.4

Appendix H

Thermal Expansion Coefficients

A considerable part of the thermal energy of a solid is in the form of atomic vibrations; the higher the temperature the greater the amplitude of the vibrations. Thermal expansion is the resulting expansion of the lattice structure, due to the injection of thermal energy. As shown in Fig. H-1, the lattice constant for a particular solid arises from the balance in forces between the repulsion of neighboring ion-cores due to their electron shells overlapping and the attraction of the particular bonding type (ionic, covalent, metallic). The equilibrium lattice constant corresponds to the minimum in the dependence of the potential energy upon interatomic distance.

The addition of energy to an atom is represented by the horizontal line A_1A_2. The extremes A_1 and A_2 correspond to the separations at which the kinetic energy has been transferred into potential energy. For this motion the mean interatomic distance A_0 is midway between A_1 and A_2. For higher temperatures and consequently higher vibrational energies (e.g., $A_1'A_2'$) the mean interatomic distance A_0' increases. The expan-

TABLE H-1 LINEAR EXPANSION COEFFICIENTS (K^{-1})

Potassium	K	83×10^{-6}
Aluminum	Al	23×10^{-6}
Iron	Fe	12×10^{-6}
Silicon	Si	2.3×10^{-6}
Germanium	Ge	5.5×10^{-6}
Quartz (fused)	SiO_2	0.5×10^{-6}
Crown glass		9×10^{-6}
Polyethylene		180×10^{-6}
Natural rubber		670×10^{-6}
Bakelite		40×10^{-6}

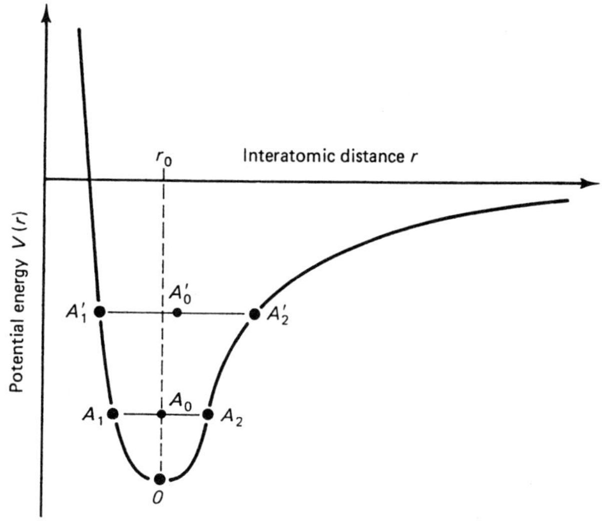

Figure H-1 Interatomic potential energy diagram.

sion of the lattice structure, due to injection of thermal energy occurs because the potential $V(r)$ is not symmetrical.

Thermal expansion is defined as the fractional change in either length or volume per unit temperature rise, for example, for the linear coefficient $(1/l)(dl/dT)$ listed in Table H-1. The expansion coefficient is largest for those materials with the shallowest potential curves. These are the relatively weakly bonded materials such as the alkali metals. Strongly bound insulators, on the other hand, such as ionically bound glass and quartz, or covalently bounded such as silicon, have low expansion coefficients. Molecular solids, such as polymers, have very weak attractive forces and consequently large thermal expansion. Interestingly, rubber contracts when heated, but this is due to the temperature dependence of its modulus of electricity.

Appendix I

Thermal Conductivity Coefficients

At a given temperature a material will contain a gas of energetic free electrons if the material is metallic or semiconducting, plus a concentration of lattice phonons. In equilibrium these electrons and phonons are moving in random directions and transporting energy, but with no net flow in any particular direction. Should the material be subjected to a gradient in temperature, the hot and cold regions will be characterized by different phonon concentrations and energies and different electron energies. As a result, there will be a natural diffusion of the more energetic electrons to the region where their mean energy is lower, and of phonons from the higher to the lower concentration. Phonons can be regarded as traveling down the sample with the velocity of sound in the material and with a characteristic mean free path. Unlike electrons, the number of phonons is not conserved, for phonons can be created at the hot end of the material and destroyed at the cold end. The energy transfer increases the mean electron kinetic energy and the phonon energy of the low-temperature region, thereby raising its temperature. It also decreases the electron and phonon energy of the high-temperature region, thereby lowering its temperature. This process will continue until equilibrium temperature is established. Should a temperature gradient be continually impressed upon the material, energy will eventually be transferred to the coolant at the low-temperature end, and heat will be continuously conducted along the material.

The ability of a material to conduct heat is indicated by its thermal conductivity K, which is defined as the quantity of heat flow per unit area A and time t, per unit temperature gradient dT/dx, that is,

$$\frac{dq}{dt} = -KA\left(\frac{dT}{dx}\right) \tag{I-1}$$

The minus sign is to indicate that heat flows in the direction of decreasing temperature. Table I-1 lists values of K for a range of materials. By simple kinetic theory, thermal conductivity K is related to the mean free path λ of the carrier by the relation

$$K = \frac{1}{3}cv\lambda \tag{I-2}$$

TABLE I-1 THERMAL CONDUCTIVITIES (W m^{-1} K^{-1})

Ag	420
Al	210
Fe	63
(Nichrome) Ni-Cr (70-30)	14
Ge	59
NaCl	7
Glass	0.8
Diamond	630
Polystyrene	0.1
Expanded polystyrene	0.03
Rubber	0.2
Wood	0.2
Concrete	0.9

where c is the specific heat and v is the velocity. For phonons, the velocity is

$$v = \left(\frac{Y}{\rho}\right)^{1/2} \tag{I-3}$$

where Y is Young's modulus for a longitudinal wave and the shear modulus for a transverse wave. Phonons have mean free paths that are about one-tenth of those of electrons. In a metal an electron may have a mean free path of 100 lattice spacings. In disordered alloys, the electron mean free path is considerably reduced. The best thermal conductors are metals in which the free electrons are dominant and very mobile. Nonmetallic materials conduct primarily by phonons and, in general, K is much smaller for these materials than for metals because of the reduced mean free path and velocity.

Appendix J

Magnetic Properties of Metals

At room temperature, different metals display various known types of magnetic properties. Thus, beryllium and metals of group I B–VI B are all diamagnetic, while nearly all the *A* metals are paramagnetic. Ferromagnetism and antiferromagnetism are only found in metals with unfilled *d* or *f* bands. Iron, cobalt, nickel, dysprosium, gadolinium, and a number of alloys and interstitial compounds have ferromagnetic forms, while α manganese, chromium, and a few other alloys are antiferromagnetic at low temperatures. Figure J-1 shows the dependence of the magnetic susceptibility on temperature for the four types of magnetic solids. The susceptibility is independent of temperature for diamagnetic solids and proportional to $1/T$ for paramagnetic solids. Both ferromagnetism and antiferromagnetism are lost above a transition temperature, the *Curie temperature* and *Néel temperature,* respectively.

The magnetic moments or spins of the electrons are all paired in diamagnetic metals, so that there is no net electron magnetism. In paramagnetic metals one or more electrons per atom have unpaired magnetic moments, but these moments are oriented

TABLE J-1 INTERATOMIC DISTANCES AND FERROMAGNETISM

Metal	Magnetic properties	No. of 3rd electrons	$r_{M\text{-}M}$ (Å)	$r_{3rd\ shell}$	$r_{4s\ shell}$
V	Paramagnetic	5	2.62	0.49	1.52
Cr	Antiferromagnetic	6	2.50	0.45	1.51
Mn	Antiferromagnetic	7	2.73	0.42	1.31
Fe	Ferromagnetic	8	2.48	0.39	1.22
Co	Ferromagnetic	9	2.51	0.36	1.14
Ni	Ferromagnetic	10	2.49	0.34	1.07
Cu	Diamagnetic	10	2.56	0.32	1.03

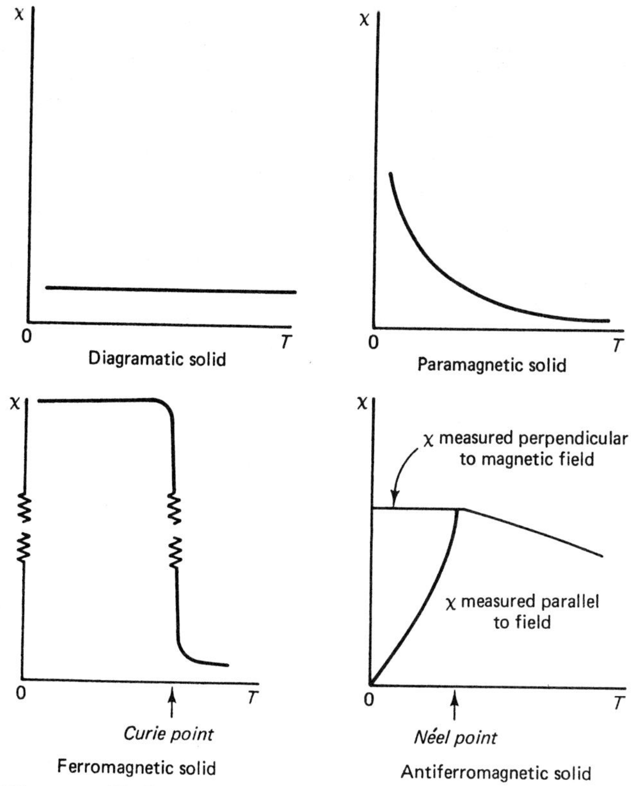

Figure J-1 Variation of magnetic susceptibility x with temperature.

at random in the solid lattice. Ferromagnetic metals contain unpaired electrons whose magnetic moments are oriented so as to reinforce one another. The orientation energy is opposed by the thermal energy of the electrons, and the two energies are equal at the Curie temperature. Antiferromagnetic metals contain unpaired electrons whose magnetic moments are oriented in such a way as to oppose one another in the lattice. Perfect crystals of ferromagnetic solids are spontaneously magnetized, but most real crystals are made up of a mosaic of randomly oriented microcrystals or domains. Energy is needed to orient these domains, which may be obtained by applying a magnetic field. The process is thermodynamically irreversible, as the magnetization of iron bars shows hysteresis. The relationship between interatomic distances and magnetic behavior is shown in Table J-1.

Appendix K

Types of Materials

Materials are classified into various groups that have common characteristics. Five categories are possible for the materials that are available to the practicing engineer.

Metals

Figure K-1 is the periodic table of the elements with the shaded areas indicating those elements that are inherently metallic. These elements are the bases of the various engineering alloys, including the irons and steels, aluminum alloys, magnesium alloys, titanium alloys, nickel alloys, zinc alloys, and copper alloys.

Metals are characterized by their high thermal and electrical conductivities. They are opaque and usually can be polished to a high luster. They owe their behavior to the fact that valence electrons are not bound, but instead can leave their parent atoms. Since these elecrons are independent, they can quickly transfer electric charge and thermal energy. The opacity and reflectivity of a metal arise from the response of these unbound electrons to electromagnetic vibrations at light frequencies.

Ceramics

The periodic table with ceramic compounds is indicated in Fig. K-2 by a combination of one or more metallic elements (lightly shaded) with one or more nonmetallic elements (in darker shade). Note that elements silicon (Si) and germanium (Ge) are included with the metals in this figure, but were not in Fig. K-1, because in elemental form, Si and Ge behave as semiconductors.

Ceramics are simply compounds that contain metallic and nonmetallic elements. There are many examples of ceramic materials, ranging from the cement of concrete, to glass, to electrical insulators, and to permanent magnets, to name but a few. Each of these materials is relatively hard and brittle, which is the general attribute of ceramics, as is higher electrical resistance by comparison with either metals or polymers. The rea-

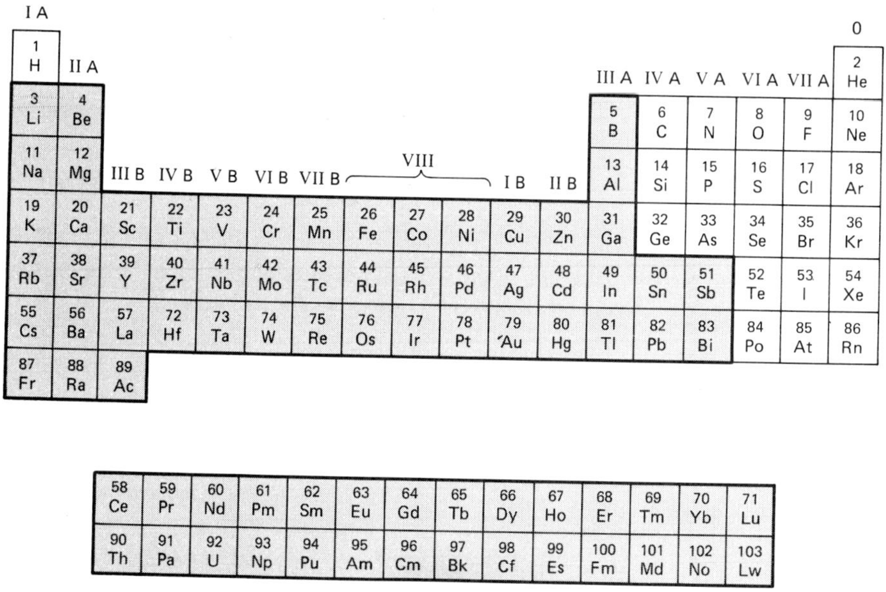

Figure K-1 Metallic elements.

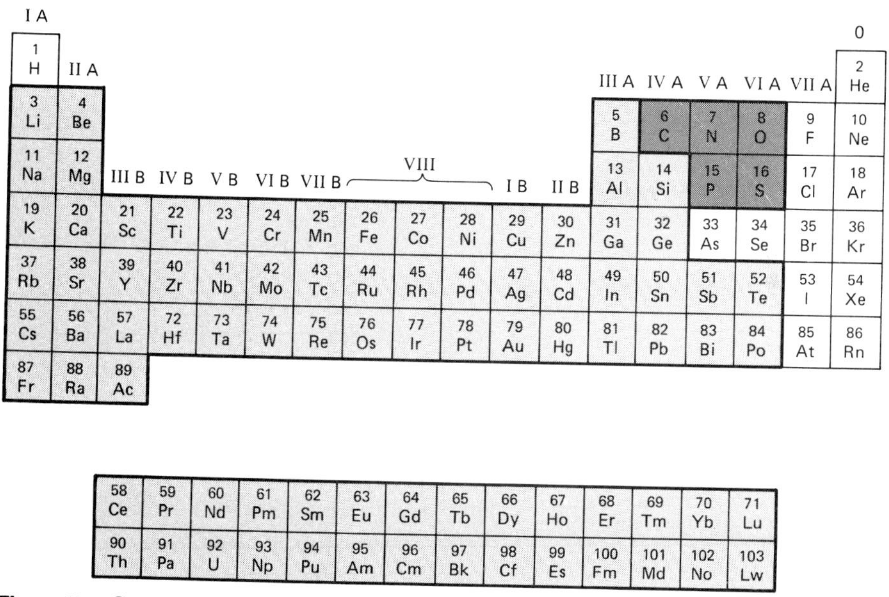

Figure K-2 Combinations for ceramic compounds.

son for these characteristics is again the electronic behavior of the constituent atoms: The metallic elements release their outermost electrons and give these electrons to the nonmetallic atoms, which retain them. The result is that the electrons are immobilized, as the typical ceramic material is a good insulator, both electrically and thermally.

Polymers

These man-made materials are a special branch of organic chemistry. An alternate name for this category is *plastics,* which describes the extensive formability of many polymers during fabrication. Polymers are long-chain molecules composed of many mers (hydrocarbon molecules such as ethylene) bonded together. Figure K-3 shows the relatively limited portion of the periodic table that is associated with commercial polymers. Many polymers (including polyethylene) are simply compounds of hydrogen and carbon. Others contain oxygen, nitrogen, fluorine, and silicone. Plastics commonly share with metals the desirable mechanical property of ductility. Unlike brittle ceramics, polymers are frequently lightweight, low-cost alternatives to metals in many applications.

Composites

These are materials that are a combination of individual materials from the previous categories. One of the best examples is fiberglass. This is a *composite* of glass fibers embedded in a polymer matrix. The high strength of the small-diameter glass fibers is combined with the ductility of the polymer matrix to produce a strong material capable of withstanding the bonding required of a structural material. Composites involve virtually the entire periodic table except for the noble gases. Fiberglass is typical of many man-made fiber-reinforced materials. Wood is another example of natural material with fiber-reinforced structure, so is concrete, which is an aggregate composite.

Semiconductors

A relatively small group of elements and compounds display an important electrical property, semiconductors, in which they are neither good electrical conductors nor

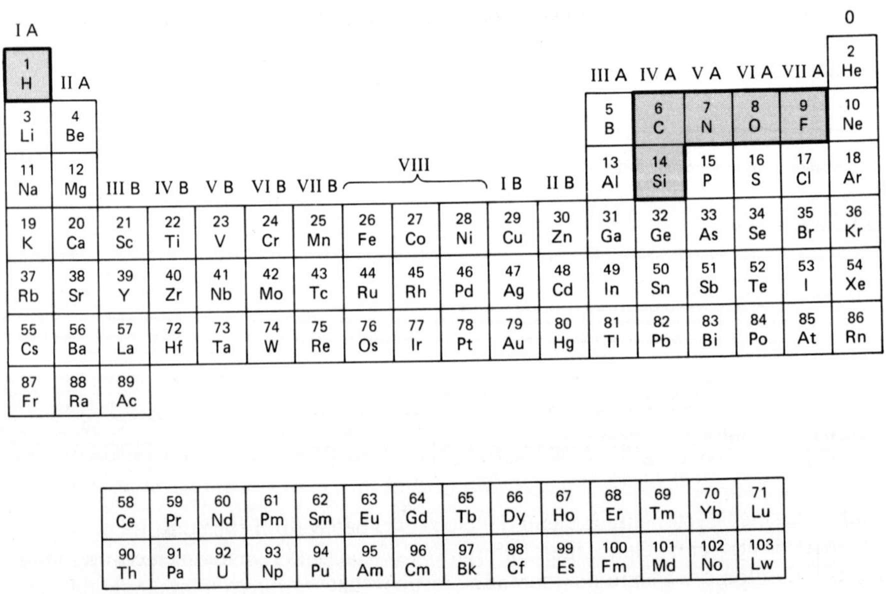

Figure K-3 Polymers.

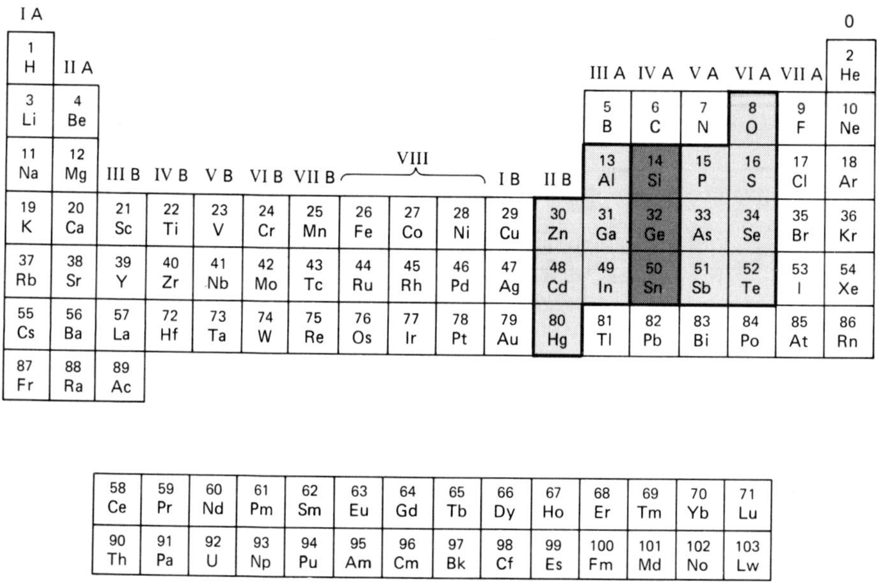

Figure K-4 Semiconductors and compounds.

good electrical insulators. Instead, their ability to conduct electricity is intermediate. Semiconductors, in general, do not fit into any of the previous four categories of materials based on atomic bonding. Metals are inherently good conductors; nonmetals (ceramics and polymers) are inherently poor conductors (but good insulators). Semiconductors are intermediate. Precise control of chemical purity allows precise control of electrical properties. Figure K-4 shows the cluster of elements of the periodic table immediately adjacent to column IV A, which forms compounds that are semiconducting. Examples include gallium arsenide (Ga As), which is used as a high-temperature rectifier and a laser material, and cadmium sulfide (CdS), which is used as a relatively low-cost solar cell in photovoltaics.

Appendix L

Monte Carlo Computer Simulation

This appendix describes atomistic model simulation programs that can be used to evaluate material properties and behavior. Computer simulation can be divided into two classes: (1) continuum models and (2) atomistic models. Continuum simulation involves the solution of partial differential equations using either the finite-difference method or the finite-element method. In atomistic simulation, the material is regarded as an aggregate of individual atoms, and each atom is treated on an individual basis.

The two atomistic simulations most often used are the *molecular dynamics* (MD) method, and the *Monte Carlo* (MC) method. In the MD method, the computational operation consists of solving $3N$ simultaneous Newtonian equations of motion for the N atoms in a computational cell. In the MC method, the operation consists of exchanging the positions of two atoms at a time in accordance with Boltzmann statistics. A sequence of configurations or states is generated by successive random displacements which derive, in roulette-wheel fashion, from a random-number generator. However, for an actual thermodynamic system, it is algorithmically arranged that the probability of configurations $\{\mathbf{r}_i\}$ appearing in the sequence is proportional to the Boltzmann factor $\exp[-V\{\mathbf{r}_i\}/kT]$, where $V\{\mathbf{r}_i\}$ is the potential energy of the configuration.

The name Monte Carlo was first applied to a class of mathematical methods used by scientists working on nuclear developments in Los Alamos in the 1940s. It is basically a means of solving probability problems in situations where the necessary integrals are too difficult to perform. An essential feature is the generation and use of a sequence of random numbers. In physics, the most obvious applications are to processes such as radioactive decay and Brownian motion, in which randomness is an inherent part of the phenomenon. The sequence of random numbers used in the Monte Carlo method could be generated by some random process, such as the rolling of a die or drawing numbers from a hat. In practice, however, the computer is used to generate the random numbers. Such computer-generated numbers are referred to as *pseudo-random,* since no sequence of numbers generated according to a well-defined arithmatic procedure can be regarded as truly random.

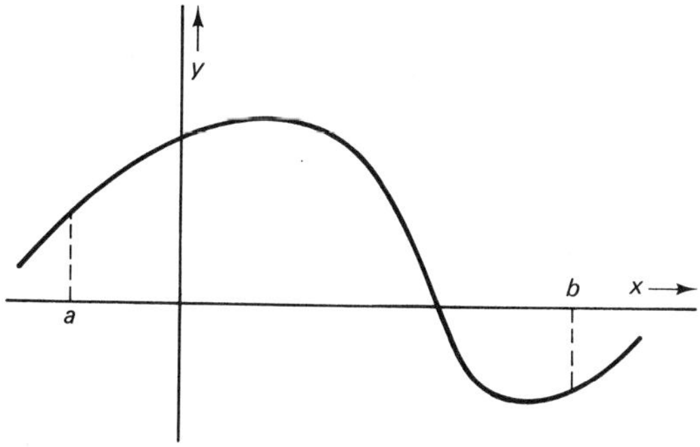

Figure L-1 Function $y(x)$ to be integrated between limits a and b by finding the average value of y at suitable x coordinates and then multiplying by $(b - a)$. The x coordinates can be uniformly spread, or they can be randomly distributed (Monte Carlo method).

To illustrate the technique, consider the expected value of a certain stochastic (chance) process, that is, the evaluation of the integral

$$I = \int_a^b y(x)\,dx \tag{L-1}$$

where $y(x)$ is any specified function, as shown in Fig. L-1.

A non-Monte Carlo method of evaluating (L-1) numerically is to divide the range from a to b into n equal steps, and then to use as an estimate of I the sum

$$\frac{b-a}{n}\sum_{i=1}^{n} y(x_i) \tag{L-2}$$

where

$$x_i = a + \left(i - \frac{1}{2}\right)\left(\frac{b-a}{n}\right) \tag{L-3}$$

The corresponding Monte Carlo method consists of evaluating the sum (L-2), but instead of calculating the $y(x_i)$ at equally spaced intervals, it is carried out at *randomly* chosen points

$$x_i = a + r_i(b - a) \tag{L-4}$$

where the r_i are members of a series of random numbers uniformly distributed in the range 0–1. Random numbers have a specified distribution, but the individual values are unpredictable. Sequences of random numbers can be obtained by drawing numbers out of a hat, from tables of random numbers (which can be found in most textbooks on statistics), from many pocket calculators, or from the number n of decays observed in a fixed time interval from a radioactive source. In the last example, if n is odd or even, a single binary digit is set as 0 or 1, respectively. By m successive applications of this procedure an m-digit binary number can be generated, which is then transformed into the range 0–1.

Most Monte Carlo calculations require a large number of random numbers, so it is sensible to obtain them from a computer. Such sequences are called pseudo-random, because they are generated by some algebraic algorithm. However, as far as Monte Carlo is concerned, they have many of the properties of true random number sequences. As an example, consider the integral

$$I = \int_0^1 x\,dx = \frac{1}{2} \tag{L-5}$$

by means of Monte Carlo. In other words, estimate the expected value of the process by sampling, and use this estimate as an approximation to I. In general terms,

$$\int_a^b f(x)\,dx \simeq \frac{b-a}{n}\,[f(x_1) + \cdots + f(x_n)] \tag{L-6}$$

where $x_1, x_2, \ldots, x_n$ are points selected at random in $[a, b]$. For the present, select $n = 50$, and choose 50 two-decimal numbers at random from a uniform distribution in the interval $[0, 1]$:

TABLE L-1 SELECTION FROM A TABLE OF 2500 RANDOM 5-DIGIT NUMBERS IN NBS HANDBOOK

.37	.68	.05	.63	.77
.40	.57	.86	.93	.41
.16	.54	.23	.74	.07
.26	.47	.38	.55	.11
.74	.97	.77	.53	.80
.09	.28	.52	.90	.20
.89	.16	.68	.63	.87
.74	.56	.88	.92	.42
.81	.02	.15	.64	.24
.39	.99	.53	.26	.71

This selection of numbers yields the estimate

$$\int_0^1 x\,dx = 0.539 \tag{L-7}$$

A more extensive calculation with $n = 100$ yields

$$\int_0^1 x\,dx = 0.527 \tag{L-8}$$

Behind those results is the *law of large numbers,* which says in effect that the probability of the sample average being close to the mean value I can be made arbitrarily close to unity if sufficiently large samples are taken. Furthermore, if

$$\left| \frac{1}{n} \sum_{i=1}^{n} f(x_i) - I \right| \tag{L-9}$$

is designated as the error, E, the "$n^{-1/2}$ law" says that it varies directly as σ and inversely as $n^{1/2}$, namely,

$$E = \frac{\lambda\sigma}{\sqrt{n}} \tag{L-10}$$

where σ is the standard deviation, n is the sample size, and λ is the confidence level, defined by the probability integral

$$PI = \frac{1}{\sqrt{2\pi}} \int_{-\lambda}^{\lambda} e^{-x^2/2}\,dx \tag{L-11}$$

Typical values of *PI* are given in Table L-2.

TABLE L-2 VALUES OF THE PROBABILITY INTEGRAL

λ	*PI*
.6745	.50
1.645	.90
1.960	.95
2.576	.99
3.291	.999
3.891	.9999

This means that, for instance, the error at 50% level of probability is $(0.6745)\sigma/\sqrt{n}$, the error at 90% level of probability is $(1.645)\sigma/\sqrt{n}$, and the error at 95% level of probability is $(1.960)\sigma/\sqrt{n}$. Referring to the case of $I = \int_0^1 x\,dx = 1/2$,

$$\sigma^2 = \int_0^1 x^2\,dx - \frac{1}{4} = \frac{1}{12} \tag{L-12}$$

$$\sigma = \frac{1}{2\sqrt{3}} \tag{L-13}$$

The error at 95% level is $(1.960)(1/2\sqrt{3})/\sqrt{50} = 0.08$ for $n = 50$. The error at 95% level is $(1.960)(1/2\sqrt{3})/(\sqrt{100}) = 0.057$ for $n = 100$.

Both the Monte Carlo and the classical integration methods yield approximate answers. The accuracy depends on the number of trials in the Monte Carlo method and on the number of intervals in the classical methods. If n random numbers are used to obtain an estimate of the integral I in Eq. (L-2), the accuracy of the answer in proportional to $1/\sqrt{n}$, whereas that for even the simplest numerical method is proportional to $1/n^2$. What then is the point of doing Monte Carlo calculations? For integrals over only one variable, the Monte Carlo technique would indeed be rarely used. Real life, however, often entails integration over a multidimensional region. In that case, the accuracy of the simplest numerical method is reduced to $n^{-2/d}$ (d being the number of dimensions), and more sophisticated numerical methods suffer a similar loss of accuracy; the error in the Monte Carlo method, however, remains at $n^{-1/2}$. Also, in many cases, the boundary of the multidimensional region over which the integration is to be performed may be so complicated that it is in practice impossible to set up the required network of points at which the integrand is to be evaluated for the numerical method. In contrast, for the Monte Carlo method, it is merely necessary to test whether the randomly chosen point is inside or outside the region of integration; then the point is rejected, or the integrand evaluated, respectively. Another feature of the Monte Carlo method is that, if calculations are repeated, slightly different answers will be obtained each time (provided the start is at a different place in the random sequence). Thus, Monte Carlo calculations simulate real situations, where the repetition of a particular experiment of limited accuracy or finite statistics is likely to produce a somewhat different result from the previous measurement.

As another example of performing integrals, consider a problem from mathematics: the evaluation of π. Let a circle be circumscribed by a square. The ratio of the area of the circle to the area of the square is $\pi/4$. It is plausible that if points were placed at random in the square, a fraction $\pi/4$ would lie inside the circle. Thus, one could measure $\pi/4$ by putting a round pan with diameter L inside a square pan with side L and collecting rain in both. It is also possible to program a computer to generate random pairs of cartesian coordinates to represent points in the square and count the fraction that lie in the circle. This fraction, as determined by many experiments, would be called the estimate for $\pi/4$. In Fig. L-2(a), the fraction of points that lie within the inscribed circle $(x - 1/2)^2 + (y - 1/2)^2 < 1/4$ provides the estimate for $\pi/4$.

In applying the Monte Carlo method, the integral

$$\int_0^1 x\sqrt{1-x^2}\,dx = \frac{\pi}{4} \tag{L-14}$$

is rewritten

$$\int_0^1 x\,f_u(x)\sqrt{1-x^2}\,dx = \frac{\pi}{4} \tag{L-15}$$

where

$$f_u(x) = 1 \qquad 0 \le x \le 1 \tag{L-16}$$

A uniform random variable x_i is drawn from $f_u(x)$, $\sqrt{1-x_i^2}$ is calculated, and this process is repeated N times to form

$$M = \frac{1}{N}\sum(1-x_i^2)^{1/2} \tag{L-17}$$

An algorithm for performing the Monte Carlo integration in FORTRAN is

```
      SUM = 0
      DO 10 I = 1, N
      SUM = SUM + SQRT(1. - RN(0)**2)
   10 CONTINUE
      MEAN = SUM/N
```

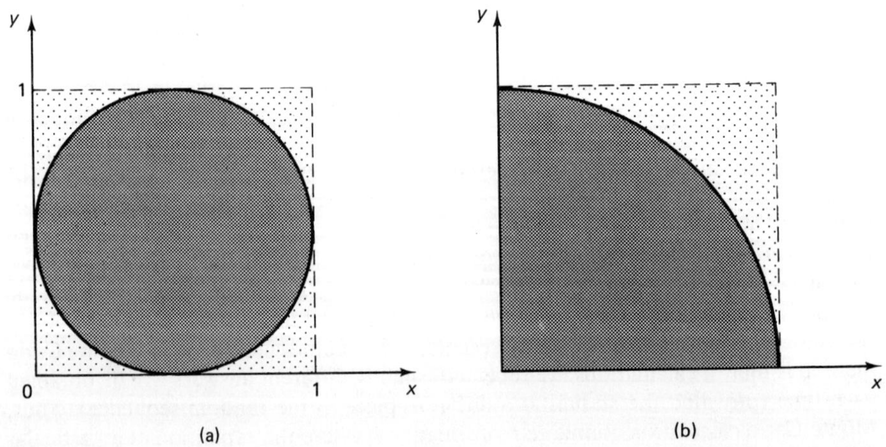

Figure L-2 Monte Carlo method for evaluating π. (Two random numbers are used to define the x- and y-coordinates of a point in the range 0–1. These points uniformly cover the circumscribing square. The fraction of points lying within the circle of radius 1/2 centered at (1/2, 1/2) provides an estimate of $\pi/4$. Alternately, the fraction can be found within the quadrant of the circle shown in (b).

where $RN(0)$ is a generic name for a routine to produce pseudorandom numbers uniform on (0,1). Table L-3 shows Monte Carlo measurements of the mean value of $f(x) = 4\sqrt{1-x^2}$ in the interval [0, 1]. A total of 10 measurements (runs) of $N = 10{,}000$ random numbers each were made. The mean value M and the standard deviation σ associated with each measurement are shown.

TABLE L-3 MEAN VALUE OF $f(x) = 4\sqrt{1-x^2}$ FROM A TOTAL OF 10 RUNS OF $n = 10{,}000$

Run α	M	σ
1	3.14892	0.88501
2	3.13255	0.89865
3	3.14042	0.88924
4	3.14600	0.88525
5	3.15257	0.88757
6	3.13972	0.89698
7	3.13107	0.89700
8	3.13585	0.89406
9	3.13442	0.89746
10	3.14047	0.89213

A convenient yardstick for the differences of these measurements is the *standard deviation of the means* σ_m, which is defined as

$$\sigma_m^2 = \langle M^2 \rangle - \langle M \rangle^2 \tag{L-18}$$

where

$$\langle M \rangle = \frac{1}{m} \sum_{\alpha=1}^{m} M_\alpha \tag{L-19}$$

and

$$\langle M^2 \rangle = \frac{1}{m} \sum_{\alpha=1}^{m} M_\alpha^2 \tag{L-20}$$

The value of σ_m is estimated by taking $m = 10$ measurements. Each measurement uses $N = 10{,}000$ random numbers and has its own M and σ. An inspection of Table L-3 shows that the measurements M vary in different runs. From the values of M in Table L-3 and the relation (L-18), the estimate is for $\sigma_m = 0.0068$. The standard deviation of the means, σ_m, is a measure of the error for a single measurement. The precise definition of σ_m is that $f_N(M_\alpha)$ has a 68% chance of being within σ_m of the "true" mean. The relationship between σ_m and σ is

$$\sigma_m \approx \frac{\sigma}{\sqrt{N}} \tag{L-21}$$

which becomes exact in the limit of a very large number of measurements. Equation (L-21) implies that the probable error decreases with the square root of the number of trials.

The evaluation of the integral in Eq. (L-14) by Monte Carlo can be approached in a somewhat different manner by considering the unit square in the $x - y$ plane and the circle with unit radius in Fig. L-2(b). Integrating over the unit square but counting only those pairs of x and y that lie within the quarter circle yields

$$\int_0^1 \int_0^1 f(x,y)\, g(x,y)\, dx\, dy = \frac{\pi}{4} \tag{L-22}$$

where $f(x,y) = 1$ inside unit square, $= 0$ otherwise, and $g(x,y) = 1$ for $x^2 + y^2 \leq 1$, $= 0$ for $x^2 + y^2 > 1$. Since x and y are independent,

$$f(x,y) = f_u(x) f_u(y) \tag{L-23}$$

so that $f(x,y)$ may be sampled by drawing two independent uniform random variables. Written in FORTRAN, the process is

```
   SUM = 0
   DO 10 I = 1, N
   X = RN(0)
   Y = RN(0)
   IF(X**2 + Y**2.LE.1.)SUM = SUM + 1
10 CONTINUE
   MEAN = SUM/N
```

It may seem odd that a deterministic computer can be used to generate sequences of "random" numbers. These pseudorandom sequences can be supplied in various languages. The following program illustrates the use of **rnd** and **randomize** functions in True BASIC to generate random numbers uniformly distributed in the interval [0, 1]:

```
PROGRAM random
RANDOMIZE                     ! generate different sequence each time
LET nrandom = 10              ! number of random numbers
CALL random1(nrandom)
END

SUB random1(nrandom)
    FOR i = 1 to nrandom
        LET x = rnd
        PRINT x
    NEXT i
END SUB
```

As with all statistical calculations, it is advisable to try for greater accuracy. Just as there are numerical methods of calculating the integral of (L-1) that are significantly better than the use of Eq. (L-2), while not involving much more computation, so there are various techniques for obtaining a more accurate answer if serious use of the Monte Carlo is contemplated. It is therefore important to incorporate variance-reducing techniques where possible, since Monte Carlo calculations often use large amounts of computer time, and the accuracy of the answer improves only slowly with n. Some refinements of the Monte Carlo method include the following.

(a) Stratification

This is a technique of dividing the range of integration into two equal x regions and generating half the Monte Carlo points in each. This ensures a more uniform distribution of x values than the straight Monte Carlo method and reduces the variance.

(b) Nonuniform sampling

Instead of generating points uniformly in x, it is advantageous to have a higher density where the integrand is varying more rapidly. The simplest implementation is to use a variant of method (a), in which the fractions of Monte Carlo points assigned to the two halves of the x range are unequal, but are so chosen to reduce the variance of the final answer.

(c) Importance sampling

Since the variance of the Monte Carlo answer is proportional to that of the integrand, it is advantageous to transform the integral so that the new integrand has less variation than the original one. Thus, the integral is written as

$$\int y(x)\,dx = \int \left[\frac{y(x)}{\omega(x)}\right] \omega(x)\,dx \tag{L-24}$$

$$= \int \left[\frac{y(x)}{\omega(x)}\right] dv(x) \tag{L-25}$$

where

$$v(x) = \int \omega(x)\,dx \tag{L-26}$$

The variance of the answer now depends on that of y/ω, rather than that on y itself. By choosing the shape of ω to resemble that of y, this variance can be made small.

(d) Antithetic variates

For each point x generated by Monte Carlo, another point x' is created according to some specified procedure, and the contribution $y(x) + y(x')$ is used for the integral. With x' suitably chosen, $y(x) + y(x')$ can be negatively correlated such that the variance on $y(x) + y(x')$ is reduced.

(e) Division and rejection

The division technique is used when the early stages of a sequence of processes are successful; then that configuration is copied a fixed number of times, and each of them is followed through. The rejection technique is employed in situations where the probability of each process in a chain is calculated separately, and a decision on whether to keep the whole configuration is made at the end of the sequence on the basis of the overall probability, which is the product of the individual ones.

As an example of *importance sampling*, consider the integral

$$I = \int_0^1 \frac{dx}{1+x^2} = \frac{\pi}{4} = 0.7854 \tag{L-27}$$

Let the integrand be multiplied and divided by a positive weight function $\omega(x)$ such that

$$\int_0^1 \omega(x)\,dx = 1 \tag{L-28}$$

The integral (L-27) can then be written as

$$I = \int_0^1 \left[\frac{f(x)}{\omega(x)}\right] \omega(x)\, dx \tag{L-29}$$

Now, make the change of variable from x to

$$y(x) = \int_0^x \omega(x)\, dx \tag{L-30}$$

so that

$$\frac{dy}{dx} = \omega(x); \qquad y(x=0) = 0; \qquad y(x=1) = 1 \tag{L-31}$$

then the integral becomes

$$I = \int_0^1 \left[\frac{f(x(y))}{\omega(x(y))}\right] dy \tag{L-32}$$

The Monte Carlo evaluation of this integral proceeds as before, namely, averaging the values of f/ω at a random sampling of points uniformly distributed in y over the interval [0, 1]:

$$I = \frac{1}{N} \sum_{i=1}^{N} \frac{f(x(y_i))}{\omega(x(y_i))} \tag{L-33}$$

The potential benefit of the change in variable is now clear: if ω is chosen so that it behaves approximately as f does, that is, it is large where f is large and small where f is small, then the integrand in (L-32) can be made very smooth, with a consequent reduction in the variance of the Monte Carlo estimate (L-33). Another way of expressing this is to realize that the uniform distribution of points in y means that the points are concentrated about the most "important" values of x where ω (and f) is large, and that little computing power is spent on calculating the integrand for "unimportant" values of x where ω and f are small. Table L-4 shows the results of the evaluation of the integral (L-27) using two different weight functions,

$$\omega(x) = 1 \tag{L-34}$$

and

$$\omega(x) = \frac{1}{3}(4 - 2x) \tag{L-35}$$

The latter weight function is positive definite, decreases monotonically over the range of integration (as does f), and is normalized correctly: $f/\omega = 3/4$ at both $x = 0$ and $x = 1$. According to (L-30), the new integration variable is

$$y = \int_0^x \frac{1}{3}(4 - 2x)\, dx = \frac{x}{3}(4 - x) \tag{L-36}$$

Results of running the BASIC code for various values of N are shown in Table L-4. The marked improvement over the $\omega = 1$ case is evident.

TABLE L-4 EVALUATION OF INTEGRAL (L-29) USING TWO DIFFERENT WEIGHT FUNCTIONS

	$\omega(x) = 1$		$\omega(x) = \frac{1}{3}(4 - 2x)$	
N	I	σ_I	I	σ_I
10	0.81491	0.04638	0.79982	0.00418
20	0.73535	0.03392	0.79071	0.00392
50	0.79606	0.02259	0.78472	0.00258
100	0.79513	0.01632	0.78838	0.00194
200	0.78677	0.01108	0.78529	0.00140
500	0.78242	0.00719	0.78428	0.00091
1000	0.78809	0.00508	0.78524	0.00064
2000	0.78790	0.00363	0.78648	0.00045
5000	0.78963	0.00227	0.78530	0.00028

Appendix M

Metropolis Algorithm and Ensemble Averaging

Assemblies of particles may be classified into few-body and many-body problems, where the approach in each case is quite different. Examples of the few-body problem include the atom, the nucleus, the molecule, and the solar system. Each of these systems is an assembly of a small number of particles whose structure and internal motion may be resolved exactly. Examples of many-body problems include liquids, gases, plasmas, chemical fluids, electrons in an ion lattice, ion lattices, and superfluids. Because of the complexity involved in solving the deterministic equations of motion of a many-particle assembly, statistical mechanics employs ensemble averaging to replace time averaging. Equivalently, in computational mechanics, an alternative approach to obtaining thermodynamic variables of interest is to compute the average over an ensemble of sytems by means of the Metropolis procedure.

For simplicity, consider the problem in two-dimensional space where the state of a system of N particles may be described by a point in $4N$-dimensional phase space. A volume element in such a position-velocity space is

$$dx^{2N}\,dv^{2N} = (dx_1\,dy_1\,dx_2\,dy_2\ldots dx_N\,dy_N)(dv_{x1}\,dv_{y1}\,dv_{x2}\,dv_{y2}\ldots dv_{xN}\,dv_{yN})$$

where the subscripts refer to each particle belonging to the system. The equilibrium thermodynamic value of a microscopic variable which is defined at every point in phase space is obtained from the ensemble average

$$\langle u\rangle = \int u\,\exp(-E/kT)dx^{2N}\,dv^{2N} \Big/ \int \exp(-E/kT)dx^{2N}\,dv^{2N} \tag{M-1}$$

where E is the energy in each phase space belonging to the ensemble. Since a many-thousand-dimensional integral cannot be evaluated by integro-difference techniques, an estimate will be obtained by *importance sampling* of the integrand at a set of representative points in phase space. Such an approach to the evaluation of integrals constitute the *Monte Carlo method.* Its application, however, involves *rejection* techniques.

Consider the simulation of a system of N particles confined in a volume V at a fixed temperature T. Since only a finite number m of the total number of M configurations can be generated, an estimate of the mean value $\langle u \rangle$ can be estimated from

$$\langle u \rangle = \sum_{s=1}^{M} u_s \exp(-E_s/kT) \Big/ \sum_{s=1}^{M} \exp(-E_s/kT) \tag{M-2}$$

$$\simeq \sum_{s=1}^{M} u_s \exp(-E_s/kT) \Big/ \sum_{s=1}^{M} \exp(-E_s/kT) \tag{M-3}$$

where E_s and u_s are the energy and value of the thermodynamic property as in configuration s. A crude Monte Carlo procedure is to generate a configuration at random, calculate E_s, u_s, and the product $u_s \exp(-E_s/kT)$, and evaluate the corresponding contribution of the configuration to the sums in (M − 3). However, such a sampling would mean generating random numbers according to a specified distribution and is difficult to implement in case of a complicated weight function in many dimensions. An alternative is to use the Metropolis, Rosenbluth, Rosenbluth, Teller, and Teller algorithm. To describe a simple realization, suppose that it is desired to generate a set of points in a (multidimensional) space of variables X distributed with a probability density $\omega(X)$. The Metropolis algorithm generates a response of points $X_0, X_1, \ldots$ as those visited successively by a *random walker* moving through X space; as the walk becomes longer and longer, the points it connects approximate more closely the desired distribution.

The sequence by which the random walk proceeds through configuration space is as follows. Let the walker be at point X_n to start with. To generate X_{n+1}, a trial step to a point X_t is made. This new point can be chosen in any convenient manner, for example, uniformly at random within a multidimensional cube of small side δ about X_n. This trial step is then *accepted* or *rejected* according to the ratio

$$r = \frac{\omega(X_t)}{\omega(X_n)} \tag{M-4}$$

If the probability r is larger than one, the step is accepted, meaning that $X_{n+1} = X_t$. If r is less than one, it is compared with a random number η uniformly distributed in the interval [0,1], and the step is accepted if $r > \eta$. If $r < \eta$, the trial step is not accepted, and so $X_{n+1} = X_n$. This generates X_{n+1}, and the procedure goes on to generate X_{n+2}, making a trial effect from X_{n+1}. Any arbitrary point X_0 can be used as the starting point for the random walk. The following subroutine (after Koonin: *Computational Physics*) illustrates the application of the Metropolis algorithm to sample a two-dimensional distribution in the variables X1 and X2. Each call to the subroutine executes another step of the random walk and returns the next values of X1 and X2; the main program must initialize these variables, as well as the value of DELTA, and define the distribution FNW(X1,X2).

```
1000 'subroutine to take a step in the Metropolis algorithm
1010 X1T=X1+DELTA*(2*RND-1)      'take a trial step in square about (X1,X2)
1020 X2T=X2+DELTA*(2*RND-1)
1030 R=FNW(X1T,X2T)/FNW(X1,X2)   'compute the ratio
1040 IF RND>R THEN GOTO 1060     'step rejected
1050   X1=X1T:  X2=X2T           'step accepted
1060 RETURN
```

After a sufficient number of steps, the Metropolis algorithm indeed generates a sequence of points distributed according to ω. In particular, if applied to (M-3), the procedure generates states with a probability proportional to the Boltzmann probablity distribution. For a choice of ω to be

$$\exp(-E_s/kT)\Big/\sum_s \exp(-E_s/kT) \tag{M-5}$$

implies that $\langle\mu\rangle$ can be written as

$$\langle\mu\rangle \simeq \frac{1}{m}\sum_s u_s \tag{M-6}$$

In statistical mechanics, the expressions *Monte Carlo Method* and *Metropolis sampling* are virtually synonymous. The common original objective is the numerical evluation of integrals. In the context of a system of spins or particles, the Metropolis algorithm most often takes on a form represented by the following steps:

1. Start with an initial configuration.
2. Make a random trial change. For example, choose a spin at random and attempt to flip it. Or choose a particle at random and attempt to displace it a random distance.
3. Compute ΔE, the change in the energy of the system due to the trial change.
4. If ΔE is less than or equal to zero, accept the new configuration and go to step 8.
5. If ΔE is positive, compute the transition probability $\omega = \exp(-\Delta E/kT)$.
6. Generate a random number η in the interval [0,1].
7. If $\omega \geq \eta$, accept the new configuration, otherwise retain the previous configuration.
8. Determine the value of the desired thermodynamic properties.
9. Repeat steps 2 through 8 to obtain a sufficient number of trials.
10. Compute averages over configurations that are statistically independent of each other.

As yet another example of generating a thermal equilibrium ensemble at a specified temperature, for an array of atoms, consider a binary alloy that is depicted schematically in Fig. M-1. The open circles represent one type of atom and the filled circles another type. It is traditional to call these two types A and B atoms. Assume that the atoms occupy sites on a triangular lattice and that the interaction energies $E(AA)$, $E(BB)$, and $E(AB)$ are known. The algorithm proceeds as a series of repetitions of the following steps:

1. Select two atoms I and J from the array at random.
2. Compute the interaction energy $EB(I)$ for atom I and interaction $EB(J)$ for atom J.
3. Interchange the positions of atoms I and J.
4. Compute the interaction energy $EA(I)$ for atom I and interaction energy $EA(J)$ in their new positions.
5. Determine $\Delta E = [EA(I) + EA(J)] - [EB(I) + EB(J)]$.
6. If ΔE is negative, retain the new configuration given by the interchange.
7. If ΔE is positive, compute $BF = \exp(-\Delta E/kT)$.
8. Select a random number R from the unit interval [0,1].
9. If R is less than BF, retain the new configuration obtained by the interchange; otherwise reject the configuration.

Figure M-1 Binary alloy, in which open circles represent type-A atoms and solid circles represent type-B atoms.

This procedure generates an ensemble of configuration states. The ensemble average energy provides an estimate of the averge energy for the system at temperature T.

Early Monte Carlo simulations were done between 1955 and 1970 on first- and second-generation mainframe computers. These early works are of interest to someone just starting to do Monte Carlo calculations because they were performed on small systems of up to 500 atoms. Because of this, most of the early experiments can now be duplicated on a 64-K personal computer, which is equivalent, in most respects, to a mainframe computer of the 1955–1970 era.

Bibliography

Alberty, R. A.: *Physical Chemistry,* 6th ed., Wiley, 1983.

Allis, W. P., and Herlin, M. A.: *Thermodynamics and Statistical Mechanics,* McGraw-Hill, 1952.

Andrews, F. C.: *Equilibrium Statistical Mechanics,* Wiley, 1963.

Ashcroft, N. W., and Mermin, N. D.: *Solid State Physics,* Holt, Rinehart and Winston, 1976.

Atkins, P.W.: *Physical Chemistry,* 3rd ed., Freeman, 1977.

Blakemore, J. S.: *Solid State Physics,* Saunders, 1974.

Blinder, S. M.: *Advanced Physical Chemistry,* Macmillan, 1969.

Bowen, H. J. M.: *Properties of Solids and Their Atomic Structure,* McGraw-Hill, 1967.

Brillouin, L.: *Science and Information Theory,* 2nd ed., Academic Press, 1962.

Bube, R. H.: *Electrons in Solids: An Introductory Survey,* Academic Press, 1981.

Busch, G., and Schade, H.: *Lectures on Solid State Physics,* Pergamon, 1976.

Colclaser, R. A., and Diehl-Nagle, S.: *Materials and Devices for Electrical Engineers and Physicists,* McGraw-Hill, 1985.

Crawford, F. H.: *Heat, Thermodynamics, and Statistical Physics,* Harcourt, Brace & World, 1963.

Davidson, N.: *Statistical Mechanics,* McGraw-Hill, 1962.

Desloge, E. A.: *Statistical Physics,* Holt, Rinehart and Winston, 1966.

Egelstaff, P. A.: *An Introduction to the Liquid State,* Academic Press, 1967.

Eisberg, R., and Resnick, R.: *Quantum Physics of Atoms, Molecules, Solids, Nuclei and Particles,* 2nd ed., Wiley, 1985.

Eyring, H., Henderson, D., Stover, B. J., and Eyring, E. M.: *Statistial Mechanics and Dynamics,* Wiley, 1963, 1982.

Fay, J. A.: *Molecular Thermodynamics,* Addison-Wesley, 1965.

Fowler, R. H., and Guggenheim, E. A.: *Statistical Thermodynamics,* Cambridge University Press, 1949.

Gautreau, R., and Savin, W.: *Modern Physics,* Schaum's Outline Series, McGraw-Hill, 1978.

Goodstein, D. L.: *State of Matter,* Dover, 1975, 1985.

Grossman, L. M.: *Thermodynamics and Statistical Mechanics,* McGraw-Hill, 1969.

Hill, T. L.: *An Introduction to Statistical Thermodynamics,* Addison-Wesley, 1960.

Huang, K.: *Statistical Mechanics,* Wiley, 1963.

Hutchison, T. S., and Baird, D. C.: *The Physics of Engineering Solids,* 2nd ed., Wiley, 1963.

Incropera, F. P.: *Introduction to Molecular Structure and Thermodynamics,* Wiley, 1974.

Jackson, E. A.: *Equilibrium Statistical Mechanics,* Prentice-Hall, 1968.

Katz, A.: *Principles of Statistical Mechanics,* Freeman, 1967.

Kauzmann, W.: *Thermodynamics and Statistics,* Benjamin, 1967.

King, A. L.: *Thermophysics,* Freeman, 1962.

Kittel, C.: *Elementary Statistical Physics,* Wiley, 1958.

Kittel, C.: *Introduction to Solid State Physics,* 6th ed., Wiley, 1986.

Knox, J. H.: *Molecular Thermodynamics,* Wiley-Interscience, 1971.

Knuth, E. L.: *Introduction to Statistical Thermodynamics,* McGraw-Hill, 1966.

Landau, L. D., and Lifshitz, E. M.: *Statistical Physics,* Addison-Wesley, 1959.

Landsberg, P. T.: *Thermodynamics and Statistical Mechanics,* Oxford University Press, 1978.

Lay, J. E.: *Thermodynamics: A Macroscopic-Microscopic Treatment,* Merrill, 1963, 1964.

Lee, J. F., Sears, F. W., and Turcotte, D. L.: *Statistical Thermodynamics,* Addison-Wesley, 1963.

Levine, I. N.: *Physical Chemistry,* 2nd ed., McGraw-Hill, 1983.

Lindsay, R. B.: *Introduction to Physical Statistics,* Wiley, 1941.

Lovell, M. C., Avery, A. J., and Vernon, M. W.: *Physical Properties of Materials,* Van Nortrand Reinhold, 1976.

McChesney, M.: *Thermodynamics of Electrical Processes,* Wiley-Interscience, 1971.

McClelland, B. J.: *Statistical Thermodynamics,* Chapman and Hall, 1973.

McQuarrie, D. A.: *Statistical Thermodynamics,* Harper & Row, 1973.

Mandl, F.: *Statistical Physics,* Wiley, 1971, 1981.

Mayer, J. E., and Mayer, M. G.: *Statistical Mechanics,* Wiley, 1940.

Morse, P. M.: *Thermal Physics,* Benjamin, 1964.

Nash, L. K.: *Elementary Statistical Thermodynamics,* Addison-Wesley, 1968.

Pippard, A. B.: *The Elements of Classical Thermodynamics,* Cambridge University Press, 1957.

Present, R. D.: *Introduction to the Kinetic Theory of Gases,* McGraw-Hill, 1958.

Pierce, F. J.: *Microscopic Thermodynamics,* International Textbook Co., 1966, 1968.

Reif, F.: *Statistical and Thermal Physics,* McGraw-Hill, 1965.

Rice, O. K.: *Statistical Mechanics, Thermodynamics, and Kinetics,* Freeman, 1967.

Robertson, B. C.: *Modern Physics for Applied Science,* Wiley, 1981.

Roser, W. G. V.: *An Introduction to Statistical Physics,* Ellis Horwood Ltd., 1982.

Rudden, M. N., and Wilson, J.: *Elements of Solid State Physics,* Wiley, 1980.

Rushbrooke, G. S.: *Introduction to Statistical Mechanics,* Oxford University Press, 1949.

Sandin, T. R.: *Modern Physics,* Addison-Wesley, 1989.

Schrödinger, E.: *Statistical Thermodynamics,* 2nd ed., Cambridge University Press, 1952.

Sears, F. W., and Salinger, G. L.: *Thermodynamics, Kinetic Theory, and Statistical Thermodynamics,* 3rd ed., Addison-Wesley, 1975.

Shackelford, J. F.: *Introduction to Materials Science for Engineers,* 2nd ed., Macmillan, 1988.

Solymar, L., and Walsh, D.: *Lectures on the Electrical Properties of Materials,* 4th ed., Oxford University Press, 1988.

Sommerfeld, A.: *Thermodynamics and Statistical Mechanics,* Academic Press, 1956.

Sonntag, R. E., and Van Wylen, G. J.: *Fundamentals of Statistical Thermodynamics,* Wiley, 1966.

Soo, S. L.: *Analytical Thermodynamics,* Prentice-Hall, 1962.

Stiddard, M. H. B.: *The Elementary Language of Solid State Physics,* Academic Press, 1975.

Tabor, D.: *Gases, Liquids, and Solids,* Penguin, 1969.

Ter Haar, D.: *Elements of Statistical Mechanics,* Holt, Rinehart and Winston, 1954.

Tien, C. L., and Lienhard, J. H.: *Statistical Thermodynamics,* Hemisphere, 1971, 1979.

Tipler, P. A.: *Foundations of Modern Physics,* Worth, 1969.

Tolman, R. C.: *The Principles of Statistical Mechanics,* Oxford University Press, 1938.

Tribus, M.: *Thermostatics and Thermodynamics,* Van Nostrand, 1961.

Turner, R. E., and Betts, D. S.: *Introductory Statistical Mechanics,* Sussex University Press, 1974.

Van Vlanck, L. H.: *Elements of Materials Science and Engineering,* 6th ed., Addison-Wesley, 1989.

Walton, . J.: *The Three Phases of Matter,* 2nd ed., Clarendon Press, 1983.

Wannier, G. H.: *Statistical Physics,* Dover, 1966, 1987.

Wilson, A. H.: *Thermodynamics and Statistical Mechanics,* Cambridge University Press, 1957.

Zemansky, M. W.: *Heat and Thermodynamics,* 4th ed., McGraw-Hill, 1957.

INDEX

J

K

L

M

N

O

P

ISBN 0-06-043884-3

90000

9 780060 438845